Hans Föllmer, Alexander Schied
Stochastic Finance

Also of Interest

Probability Theory. A First Course in Probability Theory and Statistics
Werner Linde, 2024
ISBN 978-3-11-132484-5, e-ISBN (PDF) 978-3-11-132506-4,
e-ISBN (EPUB) 978-3-11-132517-0

Non-Stationary Stochastic Processes Estimation. Vector Stationary Increments, Periodically Stationary Multi-Seasonal Increments
Maksym Luz, Mikhail Moklyachuk, 2024
ISBN 978-3-11-132533-0, e-ISBN (PDF) 978-3-11-132562-0,
e-ISBN (EPUB) 978-3-11-132625-2

Bitcoin: A Game-Theoretic Analysis
Micah Warren, 2023
ISBN 978-3-11-077283-8, e-ISBN (PDF) 978-3-11-077284-5,
e-ISBN (EPUB) 978-3-11-077305-7

Brownian Motion. A Guide to Random Processes and Stochastic Calculus
René L. Schilling, 2021
ISBN 978-3-11-074125-4, e-ISBN (PDF) 978-3-11-074127-8,
e-ISBN (EPUB) 978-3-11-074149-0

Probability Theory and Statistics with Real World Applications. Univariate and Multivariate Models Applications
Peter Zörnig, 2024
ISBN 978-3-11-133220-8, e-ISBN (PDF) 978-3-11-133227-7,
e-ISBN (EPUB) 978-3-11-133232-1

Continuous-Parameter Time Series
Peter J. Brockwell, Alexander M. Lindner, 2024
ISBN 978-3-11-132499-9, e-ISBN (PDF) 978-3-11-132503-3,
e-ISBN (EPUB) 978-3-11-132520-0

Hans Föllmer, Alexander Schied

Stochastic Finance

An Introduction in Discrete Time

5th revised and extended edition

DE GRUYTER

Mathematics Subject Classification 2020
Primary: 60-01, 91-01, 91-02; Secondary: 46N10, 60E15, 60G40, 60G42, 62P05, 91B05, 91B08, 91B16, 91B50, 91B70, 91G10, 91G15, 91G20, 91G30, 91G70

Authors
Prof. Dr. Hans Föllmer
Humboldt-Universität zu Berlin
Institut für Mathematik
Rudower Chaussee 25
12489 Berlin
Germany
foellmer@math.hu-berlin.de

Prof. Dr. Alexander Schied
University of Waterloo
Dept. of Statistics and Actuarial Science
200 University Avenue West
Walterloo, ON, N2L 3G1
Canada
aschied@uwaterloo.ca

ISBN 978-3-11-104481-1
e-ISBN (PDF) 978-3-11-104528-3
e-ISBN (EPUB) 978-3-11-104635-8

Library of Congress Control Number: 2025935556

Bibliographic information published by the Deutsche Nationalbibliothek
The Deutsche Nationalbibliothek lists this publication in the Deutsche Nationalbibliografie; detailed bibliographic data are available on the Internet at http://dnb.dnb.de.

Cover image: created by the authors
Typesetting: VTeX UAB, Lithuania

www.degruyter.com
Questions about General Product Safety Regulation:
productsafety@degruyterbrill.com

Preface to the fifth edition

This fifth edition marks a significant expansion of our book, increasing its length by about 15 %. We have thoroughly revised the entire text, adding numerous details to improve clarity and completeness. Moreover, this edition introduces a new chapter on the universal portfolio strategy, as well as new sections on stop-loss insurance contracts and the automatic Fatou property of law-invariant risk measures. The Appendix has been reorganized, with a new section on relative entropy added. Additionally, this edition includes 34 new exercises.

We have also incorporated numerous comments and suggestions from colleagues and students over the past 9 years. We are grateful to Erick Delage, Xiaoxue Deng, Stefan Gerhold, Mario Ghossoub, Peng Liu, Raphael Sayer, Ruodu Wang, and Xiaole Xue for their valuable input. Finally, we sincerely thank Steven Elliot and Nadja Schedensack of De Gruyter for their encouragement in preparing this fifth edition and for their helpful support.

March 2025

Hans Föllmer
Alexander Schied

https://doi.org/10.1515/9783111045283-202

Preface to the fourth edition

In this fourth edition, we have made a number of minor improvements, added some background material, and included more than two dozen new exercises. We have also addressed many comments and suggestions we received from colleagues and students during the past five years. It is a pleasure to thank Aurélien Alfonsi, Martin Barbie, Nike Blessing, Mengyu Du, Sabina Gadzhieva, Nina Gantert, Adrian Gorniak, Ria Grindel, Nils Hansen, Timo Hirscher, Marius Kaiser, Alexander Kalinin, Thomas Knispel, Mourad Lazgham, Tobias Lutz, Simon Maurer, Heiko Preuß, Max von Renesse, Uwe Schmock, Qun Wang, and Ilan Zlotin. Special thanks are due to Stefan Gerhold. Finally, we thank our publisher De Gruyter for encouraging us to prepare a fourth edition and for their helpful support.

March 2016

Hans Föllmer
Alexander Schied

https://doi.org/10.1515/9783111045283-203

Corollary 1.44. *For every complete market model, there exists a partition of* Ω *into at most* $d+1$ *atoms of* $(\Omega, \mathcal{F}, P)$.

Example 1.45. Consider the simple situation where the sample space Ω consists of two elements ω^+ and ω^-, and where the measure P is such that

$$p := P[\,\{\omega^+\}\,] \in (0,1).$$

We assume that there is one single risky asset, which takes at time $t = 1$ the two values b and a with the respective probabilities p and $1-p$, where a and b are such that $0 \le a < b$:

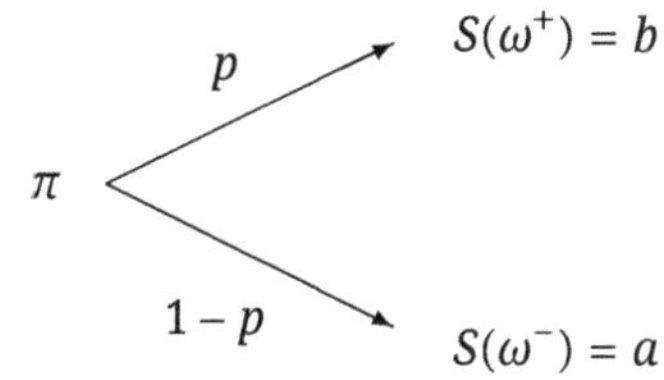

This model does not admit arbitrage if and only if

$$\pi(1+r) \in \{E_Q[\,S\,] \mid Q \approx P\} = \{q\,b + (1-q)a \mid q \in (0,1)\} = (a,b); \tag{1.24}$$

see also Example 1.10. In this case, the model is also complete: Any risk-neutral measure P^* must satisfy

$$\pi(1+r) = E^*[\,S\,] = p^*b + (1-p^*)a,$$

and this condition uniquely determines the parameter $p^* = P^*[\,\{\omega^+\}\,]$ as

$$p^* = \frac{\pi(1+r) - a}{b-a} \in (0,1).$$

Hence, $|\mathcal{P}| = 1$, and completeness follows from Theorem 1.42. Alternatively, we can directly verify completeness by showing that a given contingent claim C is attainable if (1.24) holds. Observe that the condition

$$C(\omega) = \xi^0\,S^0(\omega) + \xi\,S(\omega) = \xi^0(1+r) + \xi\,S(\omega) \quad \text{for all } \omega \in \Omega$$

is a system of two linear equations for the two real variables ξ^0 and ξ. The solution is given by

$$\xi = \frac{C(\omega^+) - C(\omega^-)}{b-a} \quad \text{and} \quad \xi^0 = \frac{C(\omega^-)b - C(\omega^+)a}{(b-a)(1+r)}.$$

Therefore, the unique arbitrage-free price of C is

$$\pi(C) = \overline{\pi} \cdot \overline{\xi} = \frac{C(\omega^+)}{1+r} \cdot \frac{\pi(1+r)-a}{b-a} + \frac{C(\omega^-)}{1+r} \cdot \frac{b-\pi(1+r)}{b-a}.$$

For a call option $C = (S - K)^+$ with strike $K \in [a, b]$, we have

$$\pi((S-K)^+) = \frac{b-K}{b-a} \cdot \pi - \frac{(b-K)a}{b-a} \cdot \frac{1}{1+r}. \tag{1.25}$$

Note that this price is *independent* of p and *increasing* in r, while the classical discounted expectation with respect to the "objective" measure P,

$$E\left[\frac{C}{1+r}\right] = \frac{p(b-K)}{1+r},$$

is *decreasing* in r and *increasing* in p.

In this example, one can illustrate how options can be used to modify the risk of a position. Consider the particular case in which the risky asset can be bought at time $t = 0$ for the price $\pi = 100$. At time $t = 1$, the price is either $S(\omega^+) = b = 120$ or $S(\omega^-) = a = 90$, both with positive probability. If we invest in the risky asset, the corresponding returns are given by

$$R(S)(\omega^+) = +20\,\% \quad \text{or} \quad R(S)(\omega^-) = -10\,\%.$$

Now consider a call option $C := (S - K)^+$ with strike $K = 100$. Choosing $r = 0$, the price of the call option is

$$\pi(C) = \frac{20}{3} \approx 6.67$$

from formula (1.25). Hence, the return

$$R(C) = \frac{(S-K)^+ - \pi(C)}{\pi(C)}$$

on the initial investment $\pi(C)$ equals

$$R(C)(\omega^+) = \frac{20 - \pi(C)}{\pi(C)} = +200\,\%$$

or

$$R(C)(\omega^-) = \frac{0 - \pi(C)}{\pi(C)} = -100\,\%,$$

according to the outcome of the market at time $t = 1$. Here, we see a dramatic increase of both profit opportunity and risk; this is sometimes referred to as the *leverage effect* of options.

On the other hand, we could reduce the risk of holding the asset by holding a married put, that is, a combination

$$\widetilde{C} := (K - S)^+ + S$$

of a put option and the asset itself. This "portfolio insurance" will of course involve an additional cost. If we choose our parameters as above, then the put-call parity in Remark 1.22 yields that the price of the put option $(K - S)^+$ is equal to 20/3. Thus, in order to hold both S and a put, we must invest the capital $100 + 20/3$ at time $t = 0$. At time $t = 1$, we have an outcome of either 120 or of 100 so that the return of $\widetilde{C}$ is given by

$$R(\widetilde{C})(\omega^+) = +12.5\,\% \quad \text{and} \quad R(\widetilde{C})(\omega^-) = -6.25\,\%.$$ ◇

Exercise 1.4.1. We consider the following three market models:
(A) $\Omega = \{\omega_1, \omega_2\}$ with $r = \frac{1}{9}$ and one risky asset with prices

$$\pi^1 = 5 \quad S^1(\omega_1) = \frac{20}{3}, \quad S^1(\omega_2) = \frac{49}{9}.$$

(B) $\Omega = \{\omega_1, \omega_2, \omega_3\}$ with $r = \frac{1}{9}$ and one risky asset with prices

$$\pi^1 = 5 \quad S^1(\omega_1) = \frac{20}{3}, \quad S^1(\omega_2) = \frac{49}{9}, \quad S^1(\omega_3) = \frac{10}{3}.$$

(C) $\Omega = \{\omega_1, \omega_2, \omega_3\}$ with $r = \frac{1}{9}$ and two risky assets with prices

$$\pi = \begin{pmatrix} 5 \\ 10 \end{pmatrix}, \quad S(\omega_1) = \begin{pmatrix} \frac{20}{3} \\ \frac{40}{3} \end{pmatrix}, \quad S(\omega_2) = \begin{pmatrix} \frac{20}{3} \\ \frac{80}{9} \end{pmatrix}, \quad S(\omega_3) = \begin{pmatrix} \frac{40}{9} \\ \frac{80}{9} \end{pmatrix}.$$

Each of these models is endowed with a probability measure that assigns strictly positive probability to each element of the corresponding sample space Ω.

(a) Which of these models are arbitrage-free? For those that are, describe the set $\mathscr{P}$ of equivalent risk-neutral measures. For those that are not, find an arbitrage opportunity.
(b) Discuss the completeness of those models that are arbitrage-free. For those that are not complete, find a nonattainable contingent claim. ◇

Exercise 1.4.2. Let $\Omega = \{\omega_1, \omega_2, \omega_3\}$ be endowed with a probability measure P such that $P[\{\omega_i\}] > 0$ for $i = 1, 2, 3$ and consider the market model with $r = 0$ and one risky asset with prices $\pi^1 = 1$ and $0 < S^1(\omega_1) < S^1(\omega_2) < S^1(\omega_3)$. We suppose that the model is arbitrage-free.

(a) Describe the following objects as subsets of three-dimensional Euclidean space $\mathbb{R}^3$:
   (i) the set $\mathscr{P}$ of equivalent risk neutral measures;
   (ii) the set $\widetilde{\mathscr{P}}$ of absolutely continuous risk neutral measures;
   (iii) the set of attainable contingent claims.
(b) Find an example for a nonattainable contingent claim.

(c) Show that the supremum

$$\sup_{\widetilde{P}\in\widetilde{\mathcal{P}}} \widetilde{E}[\,C\,] \tag{1.26}$$

is attained for every contingent claim C.

(d) Let C be a contingent claim. Give a direct and elementary proof of the fact that the map that assigns to each $\widetilde{P} \in \widetilde{\mathcal{P}}$ the expectation $\widetilde{E}[\,C\,]$ is constant if and only if the supremum (1.26) is attained in some element of $\mathcal{P}$. ◇

Exercise 1.4.3. Let $\Omega = \{\omega_1, \dots, \omega_N\}$ be endowed with a probability measure P such that $P[\{\omega_i\}] > 0$ for $i = 1, \dots, N$. On this probability space, we consider a market model with interest rate $r = 0$ and with one risky asset whose prices satisfy $\pi^1 = 1$ and

$$0 < S^1(\omega_1) < S^1(\omega_2) < \cdots < S^1(\omega_N).$$

Show that there are strikes $K_1, \dots, K_{N-2} > 0$ and prices π_{K_i} such that the corresponding call options $(S^1 - K_i)^+$ complete the market in the following sense: the market model extended by the risky assets with prices

$$\pi^i := \pi_{K_{i-1}} \quad \text{and} \quad S^i := (S^1 - K_{i-1})^+, \quad i = 2, \dots, N-2,$$

is arbitrage-free and complete. ◇

Exercise 1.4.4. Let $\Omega = \{\omega_1, \dots, \omega_{N+1}\}$ be endowed with a probability measure P such that $P[\{\omega_i\}] > 0$ for $i = 1, \dots, N+1$.

(a) On this probability space, we consider a nonredundant and arbitrage-free market model with d risky assets and prices $\overline{\pi} \in \mathbb{R}^{d+1}$ and $\overline{S}$, where $d < N$. Show that this market model can be extended by additional assets with prices $\pi^{d+1}, \dots, \pi^N$ and $S^{d+1}, \dots, S^N$ in such a way that the extended market model is arbitrage-free and complete.

(b) Let specifically $N = 2$, $d = 1$, $\pi^1 = 2$, and

$$S^1(\omega_i) = \begin{cases} 1 & \text{for } i = 1, \\ 2 & \text{for } i = 2, \\ 3 & \text{for } i = 3. \end{cases}$$

We suppose furthermore that the risk-free interest rate r is chosen such that the model is arbitrage-free. Find a nonattainable contingent claim. Then find an extended model that is arbitrage-free and complete. Finally, determine the unique equivalent risk-neutral measure P^* in the extended model. ◇

1.5 Geometric characterization of arbitrage-free models

The "fundamental theorem of asset pricing" in the form of Theorem 1.7 states that a market model is arbitrage-free if and only if the origin is contained in the set

$$M_b(Y,P) := \left\{ E_Q[\,Y\,] \;\middle|\; Q \approx P,\ \frac{dQ}{dP} \text{ is bounded},\ E_Q[\,|Y|\,] < \infty \right\} \subseteq \mathbb{R}^d,$$

where $Y = (Y^1,\dots,Y^d)$ is the random vector of discounted net gains defined in (1.2). The aim of this section is to give a geometric description of the set $M_b(Y,P)$ as well as of the larger set

$$M(Y,P) := \{E_Q[\,Y\,] \mid Q \approx P,\ E_Q[\,|Y|\,] < \infty\}.$$

To this end, it will be convenient to work with the distribution

$$\mu := P \circ Y^{-1}$$

of Y with respect to P. That is, μ is a Borel probability measure on $\mathbb{R}^d$ such that

$$\mu(A) = P[\,Y \in A\,] \quad \text{for each Borel set } A \subseteq \mathbb{R}^d.$$

If ν is a Borel probability measure on $\mathbb{R}^d$ such that $\int |y|\,\nu(dy) < \infty$, we will call $\int y\,\nu(dy)$ its *barycenter*.

Lemma 1.46. *We have*

$$M_b(Y,P) = M_b(\mu) := \left\{ \int y\,\nu(dy) \;\middle|\; \nu \approx \mu,\ \frac{d\nu}{d\mu} \textit{ is bounded},\ \int |y|\,\nu(dy) < \infty \right\},$$

and

$$M(Y,P) = M(\mu) := \left\{ \int y\,\nu(dy) \;\middle|\; \nu \approx \mu,\ \int |y|\,\nu(dy) < \infty \right\}.$$

Proof. If $\nu \approx \mu$ is a Borel probability measure on $\mathbb{R}^d$, then the Radon–Nikodym derivative of ν with respect to μ evaluated at the random variable Y defines a probability measure $Q \approx P$ on $(\Omega, \mathscr{F})$:

$$\frac{dQ}{dP}(\omega) := \frac{d\nu}{d\mu}(Y(\omega)).$$

Clearly, $E_Q[\,Y\,] = \int y\,\nu(dy)$. This shows that $M(\mu) \subseteq M(Y,P)$ and $M_b(\mu) \subseteq M_b(Y,P)$.

Conversely, if $\widetilde{Q}$ is a given probability measure on $(\Omega, \mathscr{F})$ that is equivalent to P, then the Radon–Nikodym theorem in Appendix B shows that the distribution $\widetilde{\nu} := \widetilde{Q} \circ Y^{-1}$

must be equivalent to μ, whence $\mathcal{M}(Y,P) \subseteq \mathcal{M}(\mu)$. Moreover, it follows from Proposition B.7 that the density $d\widetilde{\nu}/d\mu$ is bounded if $d\widetilde{Q}/dP$ is bounded, and so $\mathcal{M}_b(Y,P) \subseteq \mathcal{M}_b(\mu)$ also follows. □

By the preceding lemma, the characterization of the two sets $\mathcal{M}_b(Y,P)$ and $\mathcal{M}(Y,P)$ is reduced to a problem for Borel probability measures on $\mathbb{R}^d$. Here and in the sequel, we do not need the fact that μ is the distribution of the lower bounded random vector Y of discounted net gains; our results are true for arbitrary μ such that $\int |y|\,\mu(dy) < \infty$; see also Remark 1.9.

Proposition 1.47. *For every Borel probability measure ν on $\mathbb{R}^d$, there exists a smallest closed set $S \subseteq \mathbb{R}^d$ such that $\nu(S^c) = 0$. This set is called the* support *of ν, it will be denoted by* $\operatorname{supp}\nu$, *and it can be characterized as follows:* $\operatorname{supp}\nu$ *is the unique closed set S that satisfies $\nu(S^c) = 0$ and $\nu(G \cap S) > 0$ for any open set $G \subseteq \mathbb{R}^d$ with $G \cap S \neq \emptyset$.*

Proof. For $x \in \mathbb{R}^d$ and $r > 0$, let $B_r(x) := \{y \in \mathbb{R}^d \mid |x - y| < r\}$ denote the open ball with radius r and center x. We define

$$U := \bigcup_{\substack{r>0,\, r\in\mathbb{Q},\, x\in\mathbb{Q}^d \\ \nu(B_r(x))=0}} B_r(x).$$

Then U is open and satisfies $\nu(U) = 0$, due to the σ-additivity of ν and the fact that the definition of U involves a union of countably many balls with ν-measure zero. Now we claim that U^c is equal to the smallest closed set $S \subseteq \mathbb{R}^d$ such that $\nu(S^c) = 0$. Indeed, the complement of any such set S is open and can be written as follows:

$$S^c = \bigcup_{\substack{r>0,\, r\in\mathbb{Q},\, x\in\mathbb{Q}^d \\ B_r(x)\subseteq S^c}} B_r(x).$$

Every ball $B_r(x)$ on the right-hand side satisfies $\nu(B_r(x)) = 0$. Therefore, $S^c \subseteq U$, and hence $U^c \subseteq S$.

Now we turn to the proof of the alternative characterization of $\operatorname{supp}\nu$. Let G be an open set with $G \cap \operatorname{supp}\nu \neq \emptyset$ and assume by way of contradiction that $\nu(G \cap \operatorname{supp}\nu) = 0$. Then $0 = \nu(G \cap \operatorname{supp}\nu) + \nu(G \cap (\operatorname{supp}\nu)^c) = \nu(G)$, and so it follows that $\operatorname{supp}\nu \subseteq G^c$, which is a contradiction to our assumption $G \cap \operatorname{supp}\nu \neq \emptyset$. Finally, we show that if S_1, S_2 are closed sets with $\nu(S_i^c) = 0$ and the property that $\nu(G \cap S_i) > 0$ for all open sets G with $G \cap S_i \neq \emptyset$, then $S_1 = S_2$. Indeed, taking $G = S_2^c$ yields that $S_2^c \cap S_1 = \emptyset$, and in turn that $S_1 \subseteq S_2$, for otherwise we would have the contradiction $\nu(S_2^c) \geq \nu(S_2^c \cap S_1) > 0$. Reversing the roles of S_1 and S_2 yields $S_2 \subseteq S_1$ and hence $S_1 = S_2$. □

We denote by

$$\Gamma(\mu) := \operatorname{conv}(\operatorname{supp}\mu)$$

the *convex hull of the support of* μ. That is, $\Gamma(\mu)$ is the smallest convex set containing $\operatorname{supp}\mu$ (see Appendix A.1 for a definition and properties of the convex hull).

Example 1.48. Take $d = 1$, and consider the measure

$$\mu = \frac{1}{2}(\delta_{-1} + \delta_{+1}).$$

Clearly, the support of μ is equal to $\{-1, +1\}$, and so $\Gamma(\mu) = [-1, +1]$. A measure ν is equivalent to μ if and only if

$$\nu = \alpha\delta_{-1} + (1 - \alpha)\delta_{+1}$$

for some $\alpha \in (0, 1)$. Hence, $M_b(\mu) = M(\mu) = (-1, +1)$. ◇

The previous example gives the correct intuition, namely that one always has the inclusions

$$M_b(\mu) \subseteq M(\mu) \subseteq \Gamma(\mu).$$

But while the first inclusion will turn out to be an identity, the second inclusion is usually strict. Characterizing $M(\mu)$ in terms of $\Gamma(\mu)$ will involve the following concept.

Definition 1.49. The *relative interior* of a convex set $C \subseteq \mathbb{R}^d$ is the set of all points $x \in C$ such that for all $y \in C$ there exists some $\varepsilon > 0$ with

$$x - \varepsilon(y - x) \in C.$$

The relative interior of C is denoted $\operatorname{ri} C$. △

The preceding definition can be rephrased as follows: A point x belongs to the relative interior of a convex set $C \subseteq \mathbb{R}^d$ if the straight line joining x and any other point $y \in C$ can be extended within C beyond the point x. That is, there exists $\lambda > 1$ such that $\lambda x + (1-\lambda)y \in C$. By letting $\varepsilon = \lambda - 1$, one sees the equivalence of the preceding formulation and the one in Definition 1.49.

Remark 1.50. Let C be a nonempty convex subset of $\mathbb{R}^d$, and consider the *affine hull*, $\operatorname{aff} C$, spanned by C, that is, the smallest affine set containing C. If we identify $\operatorname{aff} C$ with some $\mathbb{R}^n$, then the relative interior of C is equal to the topological interior of C, considered as a subset of $\operatorname{aff} C \cong \mathbb{R}^n$. In particular, each nonempty convex set has nonempty relative interior. See, for instance, § 6 of [273] for proofs. ◇

Exercise 1.5.1. Let C be a nonempty convex subset of $\mathbb{R}^d$ and denote by $\overline{C}$ its closure. Use Remark 1.50 to prove the following statements:

(a) The relative interior $\operatorname{ri} C$ is equal to the topological interior of C if the latter is nonempty.

(b) For $x \in \operatorname{ri} C$,

$$\alpha x + (1-\alpha)y \in \operatorname{ri} C \quad \text{for all } y \in \overline{C} \text{ and all } \alpha \in (0,1]. \tag{1.27}$$

In particular, ri C is convex. Moreover, show that the operations of taking the closure or the relative interior of a convex set C are consistent with each other:

$$\operatorname{ri} \overline{C} = \operatorname{ri} C \quad \text{and} \quad \overline{\operatorname{ri} C} = \overline{C}. \tag{1.28}$$

Deduce in particular that ri C is nonempty. ◇

Exercise 1.5.2. Show that $\Gamma(\mu)$ has nonempty interior if the nonredundance condition (1.9) is satisfied. ◇

The following exercise shows that in the preceding exercise the nonredundance condition (1.9) cannot be replaced by the corresponding condition (1.10) for the random vector Y.

Exercise 1.5.3. On $\Omega = \{\omega_1, \omega_2\}$, we consider a financial market model with $r = 0$ and two risky assets with prices

$$\pi = \begin{pmatrix} 2 \\ 3/2 \end{pmatrix}, \qquad S(\omega_1) = \begin{pmatrix} 3 \\ 1/2 \end{pmatrix}, \qquad S(\omega_2) = \begin{pmatrix} 1 \\ 2 \end{pmatrix}.$$

(a) Show that this model satisfies (1.10) but not (1.9).
(b) Find $\Gamma(\mu)$ and show that it has empty interior. ◇

After these preparations, we can now state the announced geometric characterization of the set $M_b(\mu)$. Note that the proof of this characterization relies on the "fundamental theorem of asset pricing" in the form of Theorem 1.7.

Theorem 1.51. *The set of all barycenters of probability measures $\nu \approx \mu$ coincides with the relative interior of the convex hull of the support of μ. More precisely,*

$$M_b(\mu) = M(\mu) = \operatorname{ri} \Gamma(\mu).$$

Proof. In a first step, we show the inclusion $\operatorname{ri} \Gamma(\mu) \subseteq M_b(\mu)$. Suppose we are given $m \in \operatorname{ri} \Gamma(\mu)$. Let $\widetilde{\mu}$ denote the translated measure

$$\widetilde{\mu}(A) := \mu(A + m) \quad \text{for Borel sets } A \subseteq \mathbb{R}^d$$

where $A + m := \{x + m \mid x \in A\}$. Then $M_b(\widetilde{\mu}) = M_b(\mu) - m$, and analogous identities hold for $M(\widetilde{\mu})$ and $\Gamma(\widetilde{\mu})$. It follows that there is no loss of generality in assuming that $m = 0$, that is, we must show that $0 \in M_b(\mu)$ if $0 \in \operatorname{ri} \Gamma(\mu)$.

We claim that $0 \in \operatorname{ri}\Gamma(\mu)$ implies the following "no-arbitrage" condition:

$$\text{If } \xi \in \mathbb{R}^d \text{ is such that } \xi \cdot y \geq 0 \text{ for } \mu\text{-a. e. } y, \text{ then } \xi \cdot y = 0 \text{ for } \mu\text{-a. e. } y. \tag{1.29}$$

If (1.29) is false, then we can find some $\xi \in \mathbb{R}^d$ such that $\xi \cdot y \geq 0$ for μ-a. e. y but $\mu(\{y \mid \xi \cdot y > \delta\}) > 0$ for some $\delta > 0$. In this case, the support of μ is contained in the closed set $\{y \mid \xi \cdot y \geq 0\}$ but not in the hyperplane $\{y \mid \xi \cdot y = 0\}$. We conclude that $\xi \cdot y \geq 0$ for all $y \in \operatorname{supp}\mu$ and that there exists at least one $y^* \in \operatorname{supp}\mu$ such that $\xi \cdot y^* > 0$. In particular, $y^* \in \Gamma(\mu)$ so that our assumption $m = 0 \in \operatorname{ri}\Gamma(\mu)$ implies the existence of some $\varepsilon > 0$ such that $-\varepsilon y^* \in \Gamma(\mu)$. By Proposition A.4, $-\varepsilon y^*$ can be represented as a convex combination

$$-\varepsilon y^* = \alpha_1 y_1 + \cdots + \alpha_n y_n$$

of certain $y_1, \dots, y_n \in \operatorname{supp}\mu$ with $\alpha_1, \dots \alpha_n \geq 0$ and $\alpha_1 + \cdots + \alpha_n = 1$. It follows that

$$0 > -\varepsilon\xi \cdot y^* = \alpha_1 \xi \cdot y_1 + \cdots + \alpha_n \xi \cdot y_n,$$

in contradiction to our assumption that $\xi \cdot y \geq 0$ for all $y \in \operatorname{supp}\mu$. Hence, (1.29) must be true.

Due to (1.29), we can now apply the "fundamental theorem of asset pricing" in the form of Theorem 1.7 to $\Omega := \mathbb{R}^d$, $P := \mu$, and to the random variable $Y(y) := y$. It yields a probability measure $\mu^* \approx \mu$ whose density $d\mu^*/d\mu$ is bounded and which satisfies $\int |y| \, \mu^*(dy) < \infty$ and $\int y \, \mu^*(dy) = 0$. This proves the inclusion $\operatorname{ri}\Gamma(\mu) \subseteq M_b(\mu)$.

Clearly, $M_b(\mu) \subseteq M(\mu)$. So, the theorem will be proved if we can show the inclusion $M(\mu) \subseteq \operatorname{ri}\Gamma(\mu)$. To this end, suppose by way of contradiction that $\nu \approx \mu$ is such that

$$\int |y| \, \nu(dy) < \infty \quad \text{and} \quad m := \int y \, \nu(dy) \notin \operatorname{ri}\Gamma(\mu).$$

Again, we may assume without loss of generality that $m = 0$. We now apply the separating hyperplane theorem in the form of Proposition A.5 to $\mathscr{C} := \operatorname{ri}\Gamma(\mu)$, which is a nonempty convex set by part (b) of Exercise 1.5.1. We thus obtain some $\xi \in \mathbb{R}^d$ such that $\xi \cdot y \geq 0$ for all $y \in \operatorname{ri}\Gamma(\mu)$ and $\xi \cdot y^* > 0$ for at least one $y^* \in \operatorname{ri}\Gamma(\mu)$. We deduce from (1.27) that $\xi \cdot y \geq 0$ holds also for all $y \in \Gamma(\mu)$. Moreover, we conclude as above that $\xi \cdot y_0$ must be strictly positive for at least one $y_0 \in \operatorname{supp}\mu$. Hence, the open set $\{y \mid \xi \cdot y > 0\}$ has a nonempty intersection with $\operatorname{supp}\mu$, and so Proposition 1.47 yields that

$$\xi \cdot y \geq 0 \text{ for } \mu\text{-a. e. } y \in \mathbb{R}^d \quad \text{and} \quad \mu(\{y \mid \xi \cdot y > 0\}) > 0. \tag{1.30}$$

By the equivalence of μ and ν, (1.30) is also true for ν instead of μ, and so

$$\xi \cdot m = \xi \cdot \int y \, \nu(dy) = \int \xi \cdot y \, \nu(dy) > 0, \tag{1.31}$$

in contradiction to our assumption that $m = 0$. We conclude that $M(\mu) \subseteq \operatorname{ri}\Gamma(\mu)$. □

Theorem 1.51 does *not* extend to the set

$$\widetilde{M}(\mu) := \left\{ \int y\, \nu(dy) \,\middle|\, \nu \ll \mu \text{ and } \int |y|\, \nu(dy) < \infty \right\}.$$

Already the simple case $\mu := \frac{1}{2}(\delta_{-1} + \delta_{+1})$ serves as a counterexample, because here $\widetilde{M}(\mu) = [-1, +1]$ while $\operatorname{ri} \Gamma(\mu) = (-1, +1)$. In this case, we have an identity between $\widetilde{M}(\mu)$ and $\Gamma(\mu)$. But also this identity fails in general as can be seen by considering the normalized Lebesgue measure λ on $[-1, +1]$. For this choice, one finds $\widetilde{M}(\lambda) = (-1, +1)$ but $\Gamma(\lambda) = \operatorname{supp} \mu = [-1, +1]$. The following result states that $\widetilde{M}(\mu)$ is always contained in the closure of $\Gamma(\mu)$.

Proposition 1.52. *The barycenter of each probability measure $\nu \ll \mu$ is contained in the closure of $\Gamma(\mu)$. That is,*

$$\widetilde{M}(\mu) \subseteq \overline{\Gamma}(\mu).$$

Proof. We suppose by way of contradiction that there exists $\nu \ll \mu$ whose barycenter $m := \int y\, \nu(dy)$ does not belong to $\overline{\Gamma}(\mu)$. As the proof of Theorem 1.51, we may assume without loss of generality that $m = 0$. Then the closed convex set $\overline{\Gamma}(\mu)$ satisfies

$$\inf_{x \in \overline{\Gamma}(\mu)} |x| > 0.$$

Proposition A.5 hence yields the existence of some $\xi \in \mathbb{R}^d$ such that $\xi \cdot y > 0$ for all $y \in \overline{\Gamma}(\mu)$ and in particular for all y in the support of μ. It follows from Proposition 1.47 that $\mu(\{y \mid \xi \cdot y \leq 0\}) = 0$. Since $\nu \ll \mu$, we must have $\nu(\{y \mid \xi \cdot y \leq 0\}) = 0$ and hence $\xi \cdot y > 0$ for ν-a. e. $y \in \mathbb{R}^d$. As in (1.31), we now obtain $\xi \cdot m > 0$, in contradiction to our assumption $m = 0$. □

From Theorem 1.51, we obtain the following geometric characterization of the absence of arbitrage.

Corollary 1.53. *Let μ be the distribution of the discounted price vector $S/(1+r)$ of the risky assets. Then the market model is arbitrage-free if and only if the price system π belongs to the relative interior $\operatorname{ri} \Gamma(\mu)$ of the convex hull of the support of μ.*

Exercise 1.5.4. Prove Corollary 1.53. ◇

1.6 Contingent initial data

The idea of hedging contingent claims develops its full power only in a dynamic setting in which trading may occur at several times. The corresponding discrete-time theory is presented in Chapter 5. The introduction of additional trading periods requires more sophisticated techniques than those we have used so far. In this section, we will introduce some of these techniques in an extended version of our previous market model in which initial prices, and hence strategies, are contingent on scenarios. In this context, we are going to characterize the absence of arbitrage strategies. The results will be used as building blocks in the multi-period setting of Part II; their study can be postponed until Chapter 5.

Suppose that we are given a σ-algebra $\mathscr{F}_0 \subset \mathscr{F}$ that specifies the information that is available to an investor at time $t = 0$. The prices for our $d + 1$ assets at time 0 will be modeled as nonnegative $\mathscr{F}_0$-measurable random variables $S_0^0, S_0^1, \dots, S_0^d$. Thus, the price system $\overline{\pi} = (\pi^0, \pi^1, \dots, \pi^d)$ of our previous discussion is replaced by the vector

$$\overline{S}_0 = (S_0^0, \dots, S_0^d).$$

The portfolio $\overline{\xi}$ chosen by an investor at time $t = 0$ will also depend on the information available at time 0. Thus, we assume that

$$\overline{\xi} = (\xi^0, \xi^1, \dots, \xi^d)$$

is an $\mathscr{F}_0$-measurable random vector. The asset prices observed at time $t = 1$ will be denoted by

$$\overline{S}_1 = (S_1^0, S_1^1, \dots, S_1^d).$$

They are modeled as nonnegative random variables that are measurable with respect to a σ-algebra $\mathscr{F}_1$ such that $\mathscr{F}_0 \subset \mathscr{F}_1 \subseteq \mathscr{F}$. The σ-algebra $\mathscr{F}_1$ describes the information available at time 1, and in this section we can assume that $\mathscr{F} = \mathscr{F}_1$.

A riskless bond could be included by taking $S_0^0 \equiv 1$ and by assuming S_1^0 to be $\mathscr{F}_0$-measurable and P-a. s. strictly positive. However, in the sequel it will be sufficient to assume that S_0^0 is $\mathscr{F}_0$-measurable, S_1^0 is $\mathscr{F}_1$-measurable, and that

$$P[\, S_0^0 > 0 \text{ and } S_1^0 > 0 \,] = 1. \tag{1.32}$$

Under this condition, we can take the 0th asset as *numéraire*, and we denote by

$$X_t^i := \frac{S_t^i}{S_t^0}, \quad i = 1, \dots, d, \; t = 0, 1,$$

the discounted asset prices and by

$$Y = X_1 - X_0$$

the vector of the discounted net gains.

Definition 1.54. An *arbitrage opportunity* is a portfolio $\overline{\xi}$ such that $\overline{\xi} \cdot \overline{S}_0 \le 0$, $\overline{\xi} \cdot \overline{S}_1 \ge 0$ P-a. s., and $P[\, \overline{\xi} \cdot \overline{S}_1 > 0 \,] > 0$. △

By our assumption (1.32), any arbitrage opportunity $\bar{\xi} = (\xi^0, \xi)$ satisfies

$$\xi \cdot Y \geq 0 \; P\text{-a. s.} \quad \text{and} \quad P[\, \xi \cdot Y > 0\,] > 0. \tag{1.33}$$

In fact, the existence of a d-dimensional $\mathscr{F}_0$-measurable random vector ξ with (1.33) is equivalent to the existence of an arbitrage opportunity. This can be seen as in Lemma 1.4.

The space of discounted net gains that can be generated by some portfolio is given by

$$\mathscr{K} := \{\xi \cdot Y \mid \xi \in L^0(\Omega, \mathscr{F}_0, P; \mathbb{R}^d)\}.$$

Here, $L^0(\Omega, \mathscr{F}_0, P; \mathbb{R}^d)$ denotes the space of $\mathbb{R}^d$-valued random variables that are P-a. s. finite and $\mathscr{F}_0$-measurable modulo the equivalence relation (H.2) of coincidence up to P-null sets. The spaces $L^p(\Omega, \mathscr{F}_0, P; \mathbb{R}^d)$ for $p > 0$ are defined in the same manner. We denote by $L^p_+ := L^p_+(\Omega, \mathscr{F}_1, P)$ the cone of all nonnegative elements in the space $L^p := L^p(\Omega, \mathscr{F}_1, P)$. With this notation, the absence of arbitrage opportunities is equivalent to the condition

$$\mathscr{K} \cap L^0_+ = \{0\}.$$

We will denote by

$$\mathscr{K} - L^0_+$$

the convex cone of all $Z \in L^0$ that can be written as the difference of some $\xi \cdot Y \in \mathscr{K}$ and some $U \in L^0_+$.

The following definition involves the notion of the *conditional expectation*:

$$E_Q[\, Z \mid \mathscr{F}_0\,]$$

of a random variable Z with respect to a probability measure Q, given the σ-algebra $\mathscr{F}_0 \subseteq \mathscr{F}$; see Appendix B. If $Z = (Z^1, \dots, Z^n)$ is a random vector, then $E_Q[\, Z \mid \mathscr{F}_0\,]$ is shorthand for the random vector with components $E_Q[\, Z^i \mid \mathscr{F}_0\,]$, $i = 1, \dots, n$.

Definition 1.55. A probability measure Q satisfying

$$E_Q[\, X^i_t\,] < \infty \quad \text{for } i = 1, \dots, d \text{ and } t = 0, 1$$

and

$$X_0 = E_Q[\, X_1 \mid \mathscr{F}_0\,] \quad Q\text{-a. s.}$$

is called a *risk-neutral measure* or *martingale measure*. We denote by $\mathscr{P}$ the set of all risk-neutral measures P^* that are equivalent to P. △

Remark 1.56. The definition of a martingale measure Q means that for each asset $i = 0, \dots, d$, the discounted price process $(X_t^i)_{t=0,1}$ is a *martingale* under Q with respect to the σ-fields $(\mathscr{F}_t)_{t=0,1}$. The systematic discussion of martingales in a multi-period setting will begin in Section 5.2. The martingale aspect will be crucial for the theory of dynamic hedging in Part II. ◇

As the main result of this section, we can now state an extension of the "fundamental theorem of asset pricing" in Theorem 1.7 to our present setting. In the context of Section 1.2, where $\mathscr{F}_0 = \{\emptyset, \Omega\}$, the following arguments simplify considerably, and they yield an alternative proof of Theorem 1.7, in which the separation argument in $\mathbb{R}^d$ is replaced by a separation argument in L^1.

Theorem 1.57. *The following conditions are equivalent:*
(a) $\mathscr{K} \cap L_+^0 = \{0\}$.
(b) $(\mathscr{K} - L_+^0) \cap L_+^0 = \{0\}$.
(c) *There exists a measure $P^* \in \mathscr{P}$ with a bounded density dP^*/dP.*
(d) $\mathscr{P} \neq \emptyset$.

Proof. (d)⇒(a): Suppose by way of contradiction that there exist both a $P^* \in \mathscr{P}$ and some $\xi \in L^0(\Omega, \mathscr{F}_0, P; \mathbb{R}^d)$ with nonzero payoff $\xi \cdot Y \in \mathscr{K} \cap L_+^0$. For large enough $c > 0$, $\xi^{(c)} := \mathbb{1}_{\{|\xi| \le c\}} \xi$ will be bounded, and the payoff $\xi^{(c)} \cdot Y$ will still be nonzero and in $\mathscr{K} \cap L_+^0$. However,

$$E^*[\,\xi^{(c)} \cdot Y\,] = E^*[\,\xi^{(c)} \cdot E^*[\,Y \mid \mathscr{F}_0\,]\,] = 0,$$

which is the desired contradiction. Here, we have used the boundedness of $\xi^{(c)}$ to guarantee the integrability of $\xi^{(c)} \cdot Y$ and hence the validity of the formula $E^*[\,\xi^{(c)} \cdot Y \mid \mathscr{F}_0\,] = \xi^{(c)} \cdot E^*[\,Y \mid \mathscr{F}_0\,]$.

(a)⇔(b): It is obvious that (a) is necessary for (b). In order to prove sufficiency, suppose that we are given some $Z \in (\mathscr{K} - L_+^0) \cap L_+^0$. Then there exists a random variable $U \ge 0$ and a random vector $\xi \in L^0(\Omega, \mathscr{F}_0, P; \mathbb{R}^d)$ such that

$$0 \le Z = \xi \cdot Y - U.$$

This implies that $\xi \cdot Y \ge U \ge 0$, which according to condition (a), can only happen if $\xi \cdot Y = 0$. Hence, also $U = 0$ and in turn $Z = 0$.

(b)⇒(c): This is the difficult part of the proof. The assertion will follow by combining Lemmas 1.59, 1.60, 1.62, and 1.70. □

Remark 1.58. If Ω is discrete, or if there exists a decomposition of Ω in countable many atoms of $(\Omega, \mathscr{F}_0, P)$, then the martingale measure P^* can be constructed by applying the result of Theorem 1.7 separately on each atom. In the general case, the idea of patching together conditional martingale measures would involve subtle arguments of measurable selection; see [78]. Here, we present a different approach, which is based on separation arguments in $L^1(P)$. It is essentially due to W. Schachermayer [285]; our version uses in addition arguments by Y. Kabanov and C. Stricker [196]. ◇

We start with the following simple lemma, which helps taking care of the integrability condition in Definition 1.55.

Lemma 1.59. *For the proof of the implication* (b)⇒(c) *in Theorem* 1.57, *we may assume without loss of generality that*

$$E[\,|X_t|\,] < \infty \quad \textit{for } t = 0, 1. \tag{1.34}$$

Proof. Define a probability measure $\widetilde{P}$ by

$$\frac{d\widetilde{P}}{dP} := c\big(1 + |X_0| + |X_1|\big)^{-1}$$

where c is chosen such that the right-hand side integrates to 1. Clearly, (1.34) holds for $\widetilde{P}$. Moreover, condition (b) of Theorem 1.57 is satisfied by P if and only if it is satisfied by the equivalent measure $\widetilde{P}$. If $P^* \in \mathcal{P}$ is such that the density $dP^*/d\widetilde{P}$ is bounded, then so is the density

$$\frac{dP^*}{dP} = \frac{dP^*}{d\widetilde{P}} \cdot \frac{d\widetilde{P}}{dP}.$$

Therefore, the implication (b)⇒(c) holds for P if and only if it holds for $\widetilde{P}$. □

From now on, we will always assume (1.34). Our goal is to construct a suitable $Z \in L^\infty$ such that

$$\frac{dP^*}{dP} := \frac{Z}{E[\,Z\,]}$$

defines an equivalent risk-neutral measure P^*. The following simple lemma gives a criterion for this purpose, involving the convex cone:

$$\mathcal{C} := (\mathcal{K} - L^0_+) \cap L^1.$$

Lemma 1.60. *Suppose* $Z \in L^\infty$ *is such that*

$$E[\,Z\,W\,] \le c \quad \textit{for all } W \in \mathcal{C} \tag{1.35}$$

for some constant $c \ge 0$. *Then:*
(a) $E[\,Z\,W\,] \le 0$ *for all* $W \in \mathcal{C}$, *that is, we can take* $c = 0$.
(b) $Z \ge 0$ *P-a. s.*

(c) *If Z does not vanish P-a. s., then*

$$\frac{dQ}{dP} := \frac{Z}{E[\,Z\,]}$$

defines a risk-neutral measure $Q \ll P$.

Proof. (a): Note that $\mathscr{C}$ is a cone, that is, $W \in \mathscr{C}$ implies that $\alpha W \in \mathscr{C}$ for all $\alpha \geq 0$. This property excludes the possibility that $E[\,ZW\,] > 0$ for some $W \in \mathscr{C}$ if (1.35) holds.

(b): $\mathscr{C}$ contains the function $W := -\mathbb{1}_{\{Z<0\}}$. Hence, by part (a),

$$E[\,Z^-\,] = E[\,Z\,W\,] \leq 0,$$

where $Z^- := \max\{-Z, 0\} = -(0 \wedge Z)$ denotes thenegative part of the random variable Z.

(c): For all $\xi \in L^\infty(\Omega, \mathscr{F}_0, P; \mathbb{R}^d)$ and $\alpha \in \mathbb{R}$, we have $\alpha\xi \cdot Y \in \mathscr{C}$ by our integrability assumption (1.34). Thus, a similar argument as in the proof of (a) yields $E[\,Z\,\xi \cdot Y\,] = 0$. Since ξ is bounded, we may conclude that

$$0 = E[\,Z\,\xi \cdot Y\,] = E[\,\xi \cdot E[\,ZY \mid \mathscr{F}_0\,]\,].$$

As ξ is arbitrary, this yields $E[\,ZY \mid \mathscr{F}_0\,] = 0$ P-almost surely. Proposition B.8 now implies

$$E_Q[\,Y \mid \mathscr{F}_0\,] = \frac{1}{E[\,Z \mid \mathscr{F}_0\,]} E[\,ZY \mid \mathscr{F}_0\,] = 0 \quad Q\text{-a. s.},$$

which concludes the proof. □

In view of the preceding lemma, the construction of risk-neutral measures $Q \ll P$ with bounded density is reduced to the construction of elements of the set

$$\mathscr{Z} := \{Z \in L^\infty \mid 0 \leq Z \leq 1,\ P[\,Z > 0\,] > 0, \text{and } E[\,Z\,W\,] \leq 0 \text{ for all } W \in \mathscr{C}\,\}.$$

In the following lemma, we will construct such elements by applying a separation argument suggested by the condition

$$\mathscr{C} \cap L^1_+ = \{0\},$$

which follows from condition (b) of Theorem 1.57. This separation argument needs the additional assumption that $\mathscr{C}$ is closed in L^1. Showing that this assumption is indeed satisfied in our situation will be one of the key steps in our proof; see Lemma 1.70 below.

Lemma 1.61. *Assume that $\mathscr{C}$ is closed in L^1 and satisfies $\mathscr{C} \cap L^1_+ = \{0\}$. Then for each nonzero $F \in L^1_+$ there exists some $Z \in \mathscr{Z}$ such that $E[\,FZ\,] > 0$.*

Proof. Let $\mathscr{B} := \{F\}$ so that $\mathscr{B} \cap \mathscr{C} = \emptyset$. Since the set $\mathscr{C}$ is nonempty, convex, and closed in L^1, we may apply the Hahn–Banach separation theorem in the form of Theorem H.4 to obtain a continuous linear functional ℓ on L^1 such that

$$\sup_{W \in \mathscr{C}} \ell(W) < \ell(F).$$

Since the dual space of L^1 can be identified with L^∞, there exists some $Z \in L^\infty$ such that $\ell(F) = E[\, FZ\,]$ for all $F \in L^1$; see (H.5). By scaling ℓ, we may assume without loss of generality that $\|Z\|_\infty \leq 1$. By construction, Z satisfies the assumptions of Lemma 1.60, and so $Z \in \mathscr{Z}$. Moreover, $E[\, FZ\,] = \ell(F) > 0$ since the constant function $W \equiv 0$ is contained in $\mathscr{C}$. □

We will now use an exhaustion argument to conclude that $\mathscr{Z}$ contains a strictly positive element Z^* under the assumptions of Lemma 1.61. After normalization, Z^* will then serve as the density of our desired risk-neutral measure $P^* \in \mathscr{P}$.

Lemma 1.62. *Under the assumptions of Lemma* 1.61*, there exists* $Z^* \in \mathscr{Z}$ *with* $Z^* > 0$ *P-a. s.*

Proof. As a first step, we claim that $\mathscr{Z}$ is *countably convex*: If $(\alpha_k)_{k\in\mathbb{N}}$ is a sequence of nonnegative real numbers summing up to 1, and if $Z^{(k)} \in \mathscr{Z}$ for all k, then

$$Z := \sum_{k=1}^{\infty} \alpha_k Z^{(k)} \in \mathscr{Z}.$$

Indeed, we have $0 \leq Z \leq 1$, $P[\, Z > 0\,] > 0$, and for $W \in \mathscr{C}$,

$$\sum_{k=1}^{\infty} |\alpha_k Z^{(k)} W| \leq |W| \in L^1,$$

so that Lebesgue's dominated convergence theorem implies that

$$E[\, ZW\,] = \sum_{k=1}^{\infty} \alpha_k E[\, Z^{(k)} W\,] \leq 0.$$

For the second step, let

$$c := \sup\{P[\, Z > 0\,] \mid Z \in \mathscr{Z}\}.$$

We choose $Z^{(n)} \in \mathscr{Z}$ such that $P[\, Z^{(n)} > 0\,] \to c$. Then

$$Z^* := \sum_{n=1}^{\infty} 2^{-n} Z^{(n)} \in \mathscr{Z}$$

by step one, and

$$\{Z^* > 0\} = \bigcup_{n=1}^{\infty} \{Z^{(n)} > 0\}.$$

Hence, $P[Z^* > 0] = c$.

In the final step, we show that $c = 1$. Then Z^* will be as desired. Suppose by way of contradiction that $P[Z^* = 0] > 0$, so that $W := \mathbb{1}_{\{Z^*=0\}}$ is a nonzero element of L^1_+. Lemma 1.61 yields some $Z \in \mathscr{Z}$ with $E[\,WZ\,] > 0$. Hence,

$$P[\,\{Z > 0\} \cap \{Z^* = 0\}\,] > 0,$$

and so

$$P\Big[\,\frac{1}{2}(Z + Z^*) > 0\,\Big] > P[\,Z^* > 0\,] = c,$$

in contradiction to the maximality of $P[Z^* > 0]$. □

Thus, we have completed the proof of the implication (b)⇒(c) of Theorem 1.57 up to the requirement that $\mathscr{C}$ is closed in L^1. Let us pause here in order to state general versions of two of the arguments we have used so far. The first is known as the *Halmos–Savage theorem.*

Theorem 1.63. *Let $\mathcal{Q}$ be a set of probability measures that are all absolutely continuous with respect to a given measure P. Suppose moreover that $\mathcal{Q} \approx P$ in the sense that $Q[\,A\,] = 0$ for all $Q \in \mathcal{Q}$ implies that $P[\,A\,] = 0$. Then there exists a countable subfamily $\widetilde{\mathcal{Q}} \subseteq \mathcal{Q}$ that satisfies $\widetilde{\mathcal{Q}} \approx P$. In particular, there exists an equivalent measure in $\mathcal{Q}$ as soon as $\mathcal{Q}$ is countably convex in the following sense: If $(\alpha_k)_{k\in\mathbb{N}}$ is a sequence of nonnegative real numbers summing up to 1, and if $Q_k \in \mathcal{Q}$ for all k, then $\sum_{k=1}^{\infty} \alpha_k Q_k \in \mathcal{Q}$.*

Exercise 1.6.1. Prove Theorem 1.63 by modifying the exhaustion argument used in the proof of Lemma 1.62. ◇

An inspection of Lemmas 1.60, 1.61, and 1.62 shows that the particular structure of $\mathscr{C} = (\mathscr{K} - L^0_+) \cap L^1$ was only used for part (c) of Lemma 1.60. All other arguments relied only on the fact that $\mathscr{C}$ is a closed convex cone in L^1 that contains all bounded negative functions and no nontrivial positive function. Thus, we have in fact proved the following *Kreps–Yan theorem*, which was obtained independently in [323] and [223].

Theorem 1.64. *Suppose $\mathscr{C}$ is a closed convex cone in L^1 satisfying*

$$\mathscr{C} \supseteq -L^\infty_+ \quad \textit{and} \quad \mathscr{C} \cap L^1_+ = \{0\}.$$

Then there exists $Z \in L^\infty$ such that $Z > 0$ P-a. s. and $E[\,W\,Z\,] \le 0$ for all $W \in \mathscr{C}$.

Let us now turn to the closedness of our set $\mathscr{C} = (\mathscr{K} - L^0_+) \cap L^1$. The following example illustrates that we cannot expect $\mathscr{C}$ to be closed without assuming the absence of arbitrage opportunities.

Example 1.65. Let P be the Lebesgue measure on the Borel field $\mathcal{F}_1$ of $\Omega = [0,1]$, and take $\mathcal{F}_0 = \{\emptyset, \Omega\}$ and $Y(\omega) = \omega$. This choice clearly violates the no-arbitrage condition, that is, we have $\mathcal{K} \cap L^0_+ \neq \{0\}$. The convex set $\mathcal{C} = (\mathcal{K} - L^0_+) \cap L^1$ is a proper subset of L^1. More precisely, $\mathcal{C}$ does not contain any function $F \in L^1$ with $F \geq 1$: If we could represent F as $\xi \cdot Y - U$ for a nonnegative function U, then it would follow that

$$\xi \cdot Y = F + U \geq 1,$$

which is impossible for any ξ. However, as we show next, the closure of $\mathcal{C}$ in L^1 coincides with the *full* space L^1. In particular, $\mathcal{C}$ cannot be closed. Let $F \in L^1$ be arbitrary, and observe that

$$F_n := (F^+ \wedge n)\, \mathbb{1}_{[\frac{1}{n},1]} - F^-$$

converges to F in L^1 as $n \uparrow \infty$. Moreover, each F_n belongs to $\mathcal{C}$ as

$$(F^+ \wedge n)\, \mathbb{1}_{[\frac{1}{n},1]} \leq n^2 \cdot Y.$$

Consequently, F is contained in the L^1-closure of $\mathcal{C}$. ◇

In the special case $\mathcal{F}_0 = \{\emptyset, \Omega\}$, we can directly go on to the proof that $\mathcal{C}$ is closed, using a simplified version of Lemma 1.70 below. In this way, we obtain an alternative proof of Theorem 1.7. In the general case, we need some preparation. Let us first prove a "randomized" version of the Bolzano–Weierstraß theorem. It yields a simple construction of a *measurable* selection of a convergent subsequence of a given sequence in $L^0(\Omega, \mathcal{F}_0, P; \mathbb{R}^d)$.

Lemma 1.66. *Let (ξ_n) be a sequence in $L^0(\Omega, \mathcal{F}_0, P; \mathbb{R}^d)$ with $\liminf_n |\xi_n| < \infty$. Then there exists $\xi \in L^0(\Omega, \mathcal{F}_0, P; \mathbb{R}^d)$ and a strictly increasing sequence (σ_m) of $\mathcal{F}_0$-measurable integer-valued random variables such that*

$$\xi_{\sigma_m(\omega)}(\omega) \to \xi(\omega) \quad \textit{for P-a. e. } \omega \in \Omega.$$

Proof. Let $\Lambda(\omega) := \liminf_n |\xi_n(\omega)|$, and define $\sigma_m := m$ on the P-null set $\{\Lambda = \infty\}$. On $\{\Lambda < \infty\}$ we let $\sigma^0_1 := 1$, and we define $\mathcal{F}_0$-measurable random indices σ^0_m by

$$\sigma^0_m := \inf\left\{ n > \sigma^0_{m-1} \;\middle|\; \big||\xi_n| - \Lambda\big| \leq \frac{1}{m} \right\}, \qquad m = 2, 3, \dots.$$

We use recursion on $i = 1, \dots, d$ to define the ith component ξ^i of the limit ξ and to extract a new subsequence σ^i_m of random indices. Let

$$\xi^i = \liminf_{m \uparrow \infty} \xi^i_{\sigma^{i-1}_m},$$

which is already defined if $i = 1$. This ξ^i can be used in the construction of σ^i_m: Let $\sigma^i_1 := 1$ and, for $m = 2, 3, \dots$,

$$\sigma^i_m(\omega) := \inf\Big\{\sigma^{i-1}_n(\omega) \;\Big|\; \sigma^{i-1}_n(\omega) > \sigma^i_{m-1}(\omega) \text{ and } |\xi^i_{\sigma^{i-1}_n}(\omega) - \xi^i(\omega)| \le \frac{1}{m}\Big\}.$$

Then $\sigma_m := \sigma^d_m$ yields the desired sequence of random indices. □

It may happen that

$$\xi \cdot Y = \widetilde{\xi} \cdot Y \quad P\text{-a. s.},$$

although ξ and $\widetilde{\xi}$ are two different portfolios in $L^0(\Omega, \mathscr{F}_0, P; \mathbb{R}^d)$.

Remark 1.67. We could exclude this possibility by the following assumption of *nonredundance*:

$$\xi \cdot Y = \widetilde{\xi} \cdot Y \ P\text{-a. s.} \quad \Longrightarrow \quad \xi = \widetilde{\xi} \ P\text{-a. s.} \tag{1.36}$$

Under this assumption, we can immediately move on to the final step in Lemma 1.70. ◇

Without assumption (1.36), it will be convenient to have a suitable linear space $N^\perp$ of "reference portfolios" that are uniquely determined by their payoff. The construction of $N^\perp$ is the purpose of the following lemma. We will assume that the spaces L^0 and $L^0(\Omega, \mathscr{F}_0, P; \mathbb{R}^d)$ are endowed with the topology of convergence in P-measure, which is generated by the metric d of (H.3).

Lemma 1.68. *Define two linear subspaces N and $N^\perp$ of $L^0(\Omega, \mathscr{F}_0, P; \mathbb{R}^d)$ by*

$$\begin{aligned} N &:= \{\eta \in L^0(\Omega, \mathscr{F}_0, P; \mathbb{R}^d) \mid \eta \cdot Y = 0 \ P\text{-a. s.}\}, \\ N^\perp &:= \{\xi \in L^0(\Omega, \mathscr{F}_0, P; \mathbb{R}^d) \mid \xi \cdot \eta = 0 \ P\text{-a. s. for all } \eta \in N\}. \end{aligned}$$

(a) *Both N and $N^\perp$ are closed in $L^0(\Omega, \mathscr{F}_0, P; \mathbb{R}^d)$ and, in the following sense, invariant under the multiplication with scalar functions $g \in L^0(\Omega, \mathscr{F}_0, P)$: If $\eta \in N$ and $\xi \in N^\perp$, then $g\eta \in N$ and $g\xi \in N^\perp$.*
(b) *If $\xi \in N^\perp$ and $\xi \cdot Y = 0$ P-a. s., then $\xi = 0$, that is, $N \cap N^\perp = \{0\}$.*
(c) *Every $\xi \in L^0(\Omega, \mathscr{F}_0, P; \mathbb{R}^d)$ has a unique decomposition $\xi = \eta + \xi^\perp$, where $\eta \in N$ and $\xi^\perp \in N^\perp$.*

Remark 1.69. For the proof of this lemma, we will use a projection argument in Hilbert space. Let us sketch a more probabilistic construction of the decomposition $\xi = \eta + \xi^\perp$. Take a regular conditional distribution of Y given $\mathscr{F}_0$, that is, a stochastic kernel K from $(\Omega, \mathscr{F}_0)$ to $\mathbb{R}^d$ such that $K(\omega, A) = P[\,Y \in A \mid \mathscr{F}_0\,](\omega)$ for all Borel sets $A \subseteq \mathbb{R}^d$ and P-a. e. ω (see, e. g., § 44 of [21]). If one defines $\xi^\perp(\omega)$ as the orthogonal projection of $\xi(\omega)$ onto the linear hull $L(\omega)$ of the support of the measure $K(\omega, \cdot)$, then $\eta := \xi - \xi^\perp$ satisfies

$\eta \cdot Y = 0$ P-a. s., and any $\tilde{\eta}$ with the same property must be P-a. s. perpendicular to $L(\omega)$. However, carrying out the details of this construction involves certain measurability problems; this is why we use the projection argument below. ◇

Proof of Lemma 1.68. (a): The closedness of N and $N^{\perp}$ follows immediately from the metrizability of $L^0(\Omega, \mathscr{F}_0, P; \mathbb{R}^d)$ (see Appendix H) and the fact that every sequence that converges in measure has an almost surely converging subsequence. The invariance under the multiplication with $\mathscr{F}_0$-measurable scalar functions is obvious.

(b): Suppose that $\xi \in N \cap N^{\perp}$. Then taking $\eta := \xi$ in the definition of $N^{\perp}$ yields $\xi \cdot \xi = |\xi|^2 = 0$ P-a. s.

(c): Any given $\xi \in L^0(\Omega, \mathscr{F}_0, P; \mathbb{R}^d)$ can be written as

$$\xi(\omega) = \xi^1(\omega)\, e_1 + \cdots + \xi^d(\omega)\, e_d,$$

where e_i denotes the ith Euclidean unit vector, and where $\xi^i(\omega)$ is the ith component of $\xi(\omega)$. Consider e_i as a constant element of $L^0(\Omega, \mathscr{F}_0, P; \mathbb{R}^d)$, and suppose that we can decompose e_i as

$$e_i = n_i + e_i^{\perp} \quad \text{where } n_i \in N \text{ and } e_i^{\perp} \in N^{\perp}. \tag{1.37}$$

Since by part (a) both N and $N^{\perp}$ are invariant under the multiplication with $\mathscr{F}_0$-measurable functions, we can then obtain the desired decomposition of ξ by letting

$$\eta(\omega) := \sum_{i=1}^{d} \xi^i(\omega)\, n_i(\omega) \quad \text{and} \quad \xi^{\perp}(\omega) := \sum_{i=1}^{d} \xi^i(\omega)\, e_i^{\perp}(\omega).$$

Uniqueness of the decomposition follows from $N \cap N^{\perp} = \{0\}$.

It remains to construct the decomposition (1.37) of e_i. The constant e_i is an element of the space $H := L^2(\Omega, \mathscr{F}_0, P; \mathbb{R}^d)$, which becomes a Hilbert space if endowed with the natural inner product

$$(\eta, \xi)_H := E[\, \eta \cdot \xi\,], \quad \eta, \xi \in L^2(\Omega, \mathscr{F}_0, P; \mathbb{R}^d).$$

Observe that both $N \cap H$ and $N^{\perp} \cap H$ are closed subspaces of H, because convergence in H implies convergence in $L^0(\Omega, \mathscr{F}_0, P; \mathbb{R}^d)$. Therefore, we can define the corresponding orthogonal projections

$$\pi^0 : H \to N \cap H \quad \text{and} \quad \pi^{\perp} : H \to N^{\perp} \cap H.$$

Thus, letting $n_i := \pi^0(e_i)$ and $e_i^{\perp} := \pi^{\perp}(e_i)$ will be the desired decomposition (1.37) once we know that $e_i = \pi^0(e_i) + \pi^{\perp}(e_i)$. To prove this, we need only show that $\zeta := e_i - \pi^0(e_i)$ is contained in $N^{\perp}$. We assume by way of contradiction that ζ is not contained in $N^{\perp} \cap H$. Then there exists some $\eta \in N$ such that $P[\, \zeta \cdot \eta > 0\,] > 0$. Clearly,

$$\tilde{\eta} := \eta \, \mathbb{1}_{\{\zeta\cdot\eta>0,\, |\eta|\le c\}}$$

is contained in $N \cap H$ for each $c > 0$. But if c is large enough, then $0 < E[\,\tilde{\eta}\cdot\zeta\,] = (\tilde{\eta}, \zeta)_H$, which contradicts the fact that ζ is by construction orthogonal to $N \cap H$. □

After these preparations, we can now complete the proof of Theorem 1.57 by showing the closedness of $\mathscr{C} = (\mathscr{K} - L^0_+) \cap L^1$ in L^1. This is an immediate consequence of the following lemma, since convergence in L^1 implies convergence in L^0, that is, convergence in P-measure. Recall that we have already proved the equivalence of the conditions (a) and (b) in Theorem 1.57.

Lemma 1.70. *If $\mathscr{K} \cap L^0_+ = \{0\}$, then $\mathscr{K} - L^0_+$ is closed in L^0.*

Proof. Suppose $W_n \in (\mathscr{K} - L^0_+)$ converges in L^0 to some W as $n \uparrow \infty$. By passing to a suitable subsequence, we may assume without loss of generality that $W_n \to W$ P-almost surely. We can write $W_n = \xi_n \cdot Y - U_n$ for $\xi_n \in N^\perp$ and $U_n \in L^0_+$.

In a first step, we will prove the assertion given the fact that

$$\liminf_{n\uparrow\infty} |\xi_n| < \infty \quad P\text{-a. s.,} \tag{1.38}$$

which will be established afterwards. Assuming (1.38), Lemma 1.66 yields $\mathscr{F}_0$-measurable integer-valued random variables $\sigma_1 < \sigma_2 < \cdots$ and some $\xi \in L^0(\Omega, \mathscr{F}_0, P; \mathbb{R}^d)$ such that P-a. s. $\xi_{\sigma_n} \to \xi$. It follows that

$$U_{\sigma_n} = \xi_{\sigma_n} \cdot Y - W_{\sigma_n} \longrightarrow \xi \cdot Y - W =: U \quad P\text{-a. s.,} \tag{1.39}$$

so that $U \in L^0_+$ and $W = \xi \cdot Y - U \in \mathscr{K} - L^0_+$. Thus, $\mathscr{K} - L^0_+$ is closed in L^0 if (1.38) is satisfied.

Let us now show that $A := \{\liminf_n |\xi_n| = +\infty\}$ satisfies $P[\,A\,] = 0$ as claimed in (1.38). Let

$$\zeta_n := \begin{cases} \frac{\xi_n}{|\xi_n|} & \text{if } |\xi_n| > 0, \\ e_1 & \text{otherwise,} \end{cases}$$

where e_1 is the unit vector $(1, 0, \dots, 0)$. Using Lemma 1.66 on the sequence (ζ_n) yields $\mathscr{F}_0$-measurable integer-valued random variables $\tau_1 < \tau_2 < \cdots$ and some $\zeta \in L^0(\Omega, \mathscr{F}_0, P; \mathbb{R}^d)$ such that P-a. s. $\zeta_{\tau_n} \to \zeta$. The convergence of (W_n) implies that

$$0 \le \mathbb{1}_A \frac{U_{\tau_n}}{|\xi_{\tau_n}|} = \mathbb{1}_A\left(\zeta_{\tau_n} \cdot Y - \frac{W_{\tau_n}}{|\xi_{\tau_n}|}\right) \longrightarrow \mathbb{1}_A\, \zeta \cdot Y \quad P\text{-a. s.}$$

Hence, our assumption $\mathscr{K} \cap L^0_+ = \{0\}$ yields $(\mathbb{1}_A \zeta) \cdot Y = 0$. Below we will show that $\mathbb{1}_A \zeta \in N^\perp$, so that

$$\zeta = 0 \quad P\text{-a. s. on } A. \tag{1.40}$$

On the other hand, the fact that $|\zeta_n| = 1$ P-a. s. implies that $|\zeta| = 1$ P-a. s., which can only be consistent with (1.40) if $P[\,A\,] = 0$.

It remains to show that $\mathbb{1}_A\zeta \in N^\perp$. To this end, we first observe that each ζ_{τ_n} belongs to $N^\perp$ since, for each $\eta \in N$,

$$\zeta_{\tau_n} \cdot \eta = \sum_{k=1}^{\infty} \mathbb{1}_{\{\tau_n = k\}} \frac{1}{|\xi_k|} \xi_k \cdot \eta = 0 \quad P\text{-a. s.}$$

The closedness of $N^\perp$ implies $\zeta \in N^\perp$, and $A \in \mathscr{F}_0$ yields $\mathbb{1}_A\zeta \in N^\perp$. □

If in the proof of Lemma 1.70 $W_n = \xi_n \cdot Y$ for all n, then $U = 0$ in (1.39), and $W = \lim_n W_n$ is itself contained in $\mathscr{K}$. We thus get the following lemma, which will be useful in Chapter 5.

Lemma 1.71. *Suppose that $\mathscr{K} \cap L^0_+ = \{0\}$. Then $\mathscr{K}$ is closed in L^0.*

In fact, it is possible to show that $\mathscr{K}$ is *always* closed in L^0; see [285, 311]. But this stronger result will not be needed here.

As an alternative to the randomized Bolzano–Weierstraß theorem in Lemma 1.66, we can use the following variant of Komlós' *principle of subsequences*. It yields a convergent sequence of convex combinations of a sequence in $L^0(\Omega, \mathscr{F}_0, P; \mathbb{R}^d)$, and this will be needed later on. Recall from Appendix A.1 the notion of the *convex hull*

$$\operatorname{conv} A = \left\{ \sum_{i=1}^{n} \alpha_i x_i \;\middle|\; x_i \in A,\ \alpha_i \geq 0,\ \sum_{i=1}^{n} \alpha_i = 1,\ n \in \mathbb{N} \right\}$$

of a subset A of a linear space, which in our case will be $L^0(\Omega, \mathscr{F}_0, P; \mathbb{R}^d)$.

Lemma 1.72. *Let (ξ_n) be a sequence in $L^0(\Omega, \mathscr{F}_0, P; \mathbb{R}^d)$ such that $\sup_n |\xi_n| < \infty$ P-almost surely. Then there exists a sequence of convex combinations*

$$\eta_n \in \operatorname{conv}\{\xi_n, \xi_{n+1} \dots\}$$

that converges P-almost surely to some $\eta \in L^0(\Omega, \mathscr{F}_0, P; \mathbb{R}^d)$.

Proof. We can assume without loss of generality that $\sup_n |\xi_n| \leq 1$ P-a. s.; otherwise we consider the sequence $\widetilde{\xi}_n := \xi_n / \sup_n |\xi_n|$. Then (ξ_n) is a bounded sequence in the Hilbert space $H := L^2(\Omega, \mathscr{F}_0, P; \mathbb{R}^d)$. Since the closed unit ball in H is weakly compact, the sequence (ξ_n) has an accumulation point $\eta \in H$; here, weak sequential compactness follows from the Banach–Alaoglu theorem in the form of Theorem H.10 and the fact that the dual H' of the Hilbert space H is isomorphic to H itself. For each n, the accumulation point η belongs to the L^2-closure $\mathscr{C}_n$ of $\operatorname{conv}\{\xi_n, \xi_{n+1}, \dots\}$, due to the fact that a closed convex set in H is also weakly closed; see Theorem H.7. Thus, we can find $\eta_n \in \operatorname{conv}\{\xi_n, \xi_{n+1}, \dots\}$ such that

$$E[\,|\eta_n - \eta|^2\,] \le \frac{1}{n^2}.$$

This sequence (η_n) converges P-a. s. to η. □

Remark 1.73. The original result by Komlós [213] is more precise. It states that for any bounded sequence (ξ_n) in $L^1(\Omega, \mathscr{F}, P; \mathbb{R}^d)$ there is a subsequence (ξ_{n_k}) that satisfies a strong law of large numbers, that is,

$$\lim_{N\uparrow\infty} \frac{1}{N} \sum_{k=1}^{N} \xi_{n_k}$$

exists P-almost surely; see also [321]. ◇

2 Preferences

In a complete financial market model, the price of a contingent claim is determined by arbitrage arguments, without involving the preferences of economic agents. In an incomplete model, such claims may carry an intrinsic risk that cannot be hedged away. In order to determine desirable strategies in view of such risks, the preferences of an investor should be made explicit, and this is often done in terms of an expected utility criterion.

The paradigm of expected utility is the theme of this chapter. We begin with a general discussion of preference relations on a set $\mathscr{X}$ of alternative choices and their numerical representation by some functional U on $\mathscr{X}$. In the financial context, such choices can usually be described as payoff profiles. These are defined as functions X on an underlying set of scenarios with values in some set of payoffs. Thus, we are facing risk or even uncertainty. In the case of *risk*, a probability measure is given on the set of scenarios. In this case, we can focus on the resulting payoff distributions. We are then dealing with preferences on "lotteries," that is, on probability measures on the set of payoffs.

In Sections 2.2 and 2.3, we discuss the conditions—or *axioms*—under which such a preference relation on lotteries μ can be represented by a functional of the form

$$\int u(x)\,\mu(dx),$$

where u is a utility function on the set of payoffs. This formulation of preferences on lotteries in terms of expected utility goes back to D. Bernoulli [30]; the axiomatic theory was initiated by J. von Neumann and O. Morgenstern [256]. Section 2.4 characterizes uniform preference relations that are shared by a given class of functions u. This involves the general theory of probability measures on product spaces with given marginals, which will be discussed in Section 2.6.

In Section 2.5, we return to the more fundamental level where preferences are defined on payoff profiles, and where we are facing *uncertainty* in the sense that no probability measure is given a priori. Savage [284] clarified the conditions under which such preferences on a space of functions X admit a representation of the form

$$U(X) = E_Q[\,u(X)\,],$$

where Q is a "subjective" probability measure on the set of scenarios. We are going to concentrate on a robust extension of the Savage representation, which was introduced by Gilboa and Schmeidler [166] and later extended by Maccheroni, Marinacci, and Rustichini [241]. Here, the utility functional is of the form

$$U(X) = \inf_{Q\in\mathcal{Q}}\big(E_Q[\,u(X)\,] + \alpha(Q)\big).$$

https://doi.org/10.1515/9783111045283-003

It thus involves a whole class $\mathcal{Q}$ of probability measures Q, which are taken more or less seriously according to their penalization $\alpha(Q)$. The axiomatic approach to the robust Savage representation is closely related to the construction of coherent and convex risk measures, which will be the topic of Chapter 4.

2.1 Preference relations and their numerical representation

Let $\mathscr{X}$ be some nonempty set. An element $x \in \mathscr{X}$ will be interpreted as a possible choice of an economic agent. If presented with two choices $x, y \in \mathscr{X}$, the agent might prefer one over the other. This will be formalized as follows.

Definition 2.1. A *preference order* (or *preference relation*) on $\mathscr{X}$ is a binary relation $\succ$ with the following two properties:
- *Asymmetry:* If $x \succ y$, then $y \not\succ x$.
- *Negative transitivity*: If $x \succ y$ and $z \in \mathscr{X}$, then either $x \succ z$ or $z \succ y$ or both must hold. △

Negative transitivity states that if a clear preference exists between two choices x and y, and if a third choice z is added, then there is still a choice that is either least preferable (y if $z \succ y$) or most preferable (x if $x \succ z$).

Definition 2.2. A preference order $\succ$ on $\mathscr{X}$ induces a corresponding *weak preference order* $\succeq$ defined by

$$x \succeq y \;:\Longleftrightarrow\; y \not\succ x,$$

and an *indifference relation* $\sim$ given by

$$x \sim y \;:\Longleftrightarrow\; x \succeq y \text{ and } y \succeq x.$$ △

Thus, $x \succeq y$ means that either x is preferred to y or there is no clear preference between the two.

Remark 2.3. It is easy to check that the asymmetry and the negative transitivity of $\succ$ are equivalent to the following two respective properties of $\succeq$:
(a) *Completeness:* For all $x, y \in \mathscr{X}$, either $y \succeq x$ or $x \succeq y$ or both are true.
(b) *Transitivity:* If $x \succeq y$ and $y \succeq z$, then also $x \succeq z$.
Conversely, any complete and transitive relation $\succeq$ induces a preference order $\succ$ via the negation of $\succeq$, that is,

$$y \succ x \quad :\Longleftrightarrow \quad x \not\succeq y.$$

The indifference relation $\sim$ is an equivalence relation, that is, it is reflexive, symmetric, and transitive. ◇

Exercise 2.1.1. Prove the assertions in the preceding remark. ◇

Definition 2.4. A *numerical representation* of a preference order $\succ$ is a function $U : \mathscr{X} \to \mathbb{R}$ such that

$$y \succ x \iff U(y) > U(x) \tag{2.1}$$

holds for all $x, y \in \mathscr{X}$. △

Clearly, (2.1) is equivalent to

$$y \succeq x \iff U(y) \geq U(x).$$

Note that such a numerical representation U is not unique: If f is any strictly increasing function, then $\widetilde{U}(x) := f(U(x))$ is again a numerical representation.

Definition 2.5. Let $\succ$ be a preference relation on $\mathscr{X}$. A subset $\mathscr{Z}$ of $\mathscr{X}$ is called *order dense* if for any pair $x, y \in \mathscr{X}$ such that $x \succ y$ there exists some $z \in \mathscr{Z}$ with $x \succeq z \succeq y$. △

The following theorem characterizes those preference relations for which there exists a numerical representation.

Theorem 2.6. *A preference relation $\succ$ admits a numerical representation if and only if $\mathscr{X}$ contains a countable, order dense subset $\mathscr{Z}$. In particular, any preference order admits a numerical representation if $\mathscr{X}$ is countable.*

Proof. Suppose first that we are given a countable order dense subset $\mathscr{Z}$ of $\mathscr{X}$. For $x \in \mathscr{X}$, let

$$\mathscr{Z}_{\succ}(x) := \{z \in \mathscr{Z} \mid z \succ x\} \quad \text{and} \quad \mathscr{Z}_{\prec}(x) := \{z \in \mathscr{Z} \mid x \succ z\}.$$

The relation $x \succeq y$ implies that $\mathscr{Z}_{\succ}(x) \subseteq \mathscr{Z}_{\succ}(y)$ and $\mathscr{Z}_{\prec}(x) \supseteq \mathscr{Z}_{\prec}(y)$. If the strict relation $x \succ y$ holds, then at least one of these inclusions is also strict. To see this, pick $z \in \mathscr{Z}$ with $x \succeq z \succeq y$, so that either $x \succ z \succeq y$ or $x \succeq z \succ y$. In the first case, $z \in \mathscr{Z}_{\prec}(x) \setminus \mathscr{Z}_{\prec}(y)$, while $z \in \mathscr{Z}_{\succ}(y) \setminus \mathscr{Z}_{\succ}(x)$ in the second case.

Next, take any strictly positive probability distribution μ on $\mathscr{Z}$, and let

$$U(x) := \sum_{z \in \mathscr{Z}_{\prec}(x)} \mu(z) - \sum_{z \in \mathscr{Z}_{\succ}(x)} \mu(z).$$

By the above, $U(x) > U(y)$ if and only if $x \succ y$ so that U is the desired numerical representation.

For the proof of the converse assertion, take a numerical representation U and let $\mathscr{J}$ denote the countable set

$$\mathscr{J} := \{[a, b] \mid a, b \in \mathbb{Q},\ a < b,\ U^{-1}([a, b]) \neq \emptyset\}.$$

For every interval $I \in \mathcal{J}$, we can choose some $z_I \in \mathcal{X}$ with $U(z_I) \in I$, and thus define the countable set

$$A := \{z_I \mid I \in \mathcal{J}\}.$$

At first glance, it may seem that A is a good candidate for an order dense set. However, it may happen that there are $x, y \in \mathcal{X}$ such that $U(x) < U(y)$ and for which there is no $z \in \mathcal{X}$ with $U(x) < U(z) < U(y)$. In this case, an order dense set must contain at least one z with $U(z) = U(x)$ or $U(z) = U(y)$, a condition that cannot be guaranteed by A.

To address this issue, let us define the set C of all pairs (x, y) that do not admit any $z \in A$ with $y \succ z \succ x$:

$$C := \{(x,y) \mid x, y \in \mathcal{X} \setminus A,\ y \succ x \text{ and } \nexists z \in A \text{ with } y \succ z \succ x\}.$$

Then $(x, y) \in C$ implies the apparently stronger fact that we cannot find any $z \in \mathcal{X}$ such that $y \succ z \succ x$: Otherwise, we could find $a, b \in \mathbb{Q}$ such that

$$U(x) < a < U(z) < b < U(y),$$

so $I := [a, b]$ would belong to $\mathcal{J}$, and the corresponding z_I would be an element of A with $y \succ z_I \succ x$, contradicting the assumption that $(x, y) \in C$.

It follows that all intervals $(U(x), U(y))$ with $(x, y) \in C$ are disjoint and nonempty. Hence, there can be only countably many of them. For each such interval J, we pick now exactly one pair $(x^J, y^J) \in C$ such that $U(x^J)$ and $U(y^J)$ are the endpoints of J, and we denote by B the countable set containing all x^J and all y^J.

Finally, we claim that $\mathcal{Z} := A \cup B$ is an order dense subset of $\mathcal{X}$. Indeed, if $x, y \in \mathcal{X} \setminus \mathcal{Z}$ with $y \succ x$, then either there is some $z \in A$ such that $y \succ z \succ x$, or $(x, y) \in C$. In the latter case, there will be some $z \in B$ with $U(y) = U(z) > U(x)$ and, consequently, $y \succeq z \succ x$. □

The following example shows that even in a seemingly straightforward situation, a given preference order may *not* admit a numerical representation.

Example 2.7. Let $\succ$ be the usual lexicographical order on $\mathcal{X} := [0, 1] \times [0, 1]$, that is, $(x_1, x_2) \succ (y_1, y_2)$ if and only if either $x_1 > y_1$, or if $x_1 = y_1$ and simultaneously $x_2 > y_2$. One easily checks that $\succ$ is asymmetric and negative transitive, and hence a preference order. We show now that $\succ$ does *not* admit a numerical representation. To this end, let $\mathcal{Z}$ be any order-dense subset of $\mathcal{X}$. Then, for each $x \in [0, 1]$ there must be some $(z_1, z_2) \in \mathcal{Z}$ such that $(x, 1) \succeq (z_1, z_2) \succeq (x, 0)$. It follows that $z_1 = x$ and that $\mathcal{Z}$ is uncountable. Theorem 2.6 thus implies that there cannot be a numerical representation of the lexicographical order $\succ$. ◇

Definition 2.8. Let $\mathcal{X}$ be a topological space. A preference relation $\succ$ is called *continuous* if for all $x \in \mathcal{X}$,

$$\mathscr{B}_{\succ}(x) := \{y \in \mathscr{X} \mid y \succ x\} \quad \text{and} \quad \mathscr{B}_{\prec}(x) := \{y \in \mathscr{X} \mid x \succ y\} \tag{2.2}$$

are open subsets of $\mathscr{X}$. △

Remark 2.9. Clearly, every preference order that admits a continuous numerical representation is itself continuous. Under some mild conditions on the underlying space $\mathscr{X}$, the converse statement is also true; see Theorem 2.15 below. ◇

Example 2.10. The lexicographical order of Example 2.7 is not continuous for the Euclidean topology on $[0,1] \times [0,1]$: If $(x_1, x_2) \in [0,1] \times [0,1]$ is given, then

$$\{(y_1, y_2) \mid (y_1, y_2) \succ (x_1, x_2)\} = (x_1, 1] \times [0,1] \cup \{x_1\} \times (x_2, 1],$$

which is typically not an open subset of $[0,1] \times [0,1]$. ◇

Recall that a topological space $\mathscr{X}$ is called a *topological Hausdorff space* if any two distinct points in $\mathscr{X}$ have disjoint open neighborhoods. In this case, all singletons $\{x\}$ are closed. Clearly, every metric space is a topological Hausdorff space.

Proposition 2.11. *Let $\succ$ be a preference order on a topological Hausdorff space $\mathscr{X}$. Then the following properties are equivalent:*
(a) *$\succ$ is continuous.*
(b) *The set $\{(x,y) \mid y \succ x\}$ is open in $\mathscr{X} \times \mathscr{X}$.*
(c) *The set $\{(x,y) \mid y \succeq x\}$ is closed in $\mathscr{X} \times \mathscr{X}$.*

Proof. (a)⇒(b): We have to show that for any pair

$$(x_0, y_0) \in M := \{(x,y) \mid y \succ x\}$$

there exist open sets $U, V \subseteq \mathscr{X}$ such that $x_0 \in U$, $y_0 \in V$, and $U \times V \subseteq M$. Consider first the case in which there exists some $z \in \mathscr{B}_{\succ}(x_0) \cap \mathscr{B}_{\prec}(y_0)$ for the notation $\mathscr{B}_{\succ}(x_0)$ and $\mathscr{B}_{\prec}(y_0)$ introduced in (2.2). Then $y_0 \succ z \succ x_0$, so that $U := \mathscr{B}_{\prec}(z)$ and $V := \mathscr{B}_{\succ}(z)$ are open neighborhoods of x_0 and y_0, respectively. Moreover, if $x \in U$ and $y \in V$, then $y \succ z \succ x$, and thus $U \times V \subseteq M$.

If $\mathscr{B}_{\succ}(x_0) \cap \mathscr{B}_{\prec}(y_0) = \emptyset$, we let $U := \mathscr{B}_{\prec}(y_0)$ and $V := \mathscr{B}_{\succ}(x_0)$. If $(x,y) \in U \times V$, then $y_0 \succ x$ and $y \succ x_0$ by definition. We want to show that $y \succ x$ in order to conclude that $U \times V \subseteq M$. To this end, suppose that $x \succeq y$. Then $y_0 \succ y$ by negative transitivity, hence $y_0 \succ y \succ x_0$. But then $y \in \mathscr{B}_{\succ}(x_0) \cap \mathscr{B}_{\prec}(y_0) \neq \emptyset$, and we have a contradiction.

(b)⇒(c): First, note that the mapping $\phi(x,y) := (y,x)$ is a homeomorphism of $\mathscr{X} \times \mathscr{X}$. Then observe that the set $\{(x,y) \mid y \succeq x\}$ is just the complement of the open set $\phi(\{(x,y) \mid y \succ x\})$.

(c)⇒(a): Since $\mathscr{X}$ is a topological Hausdorff space, $\{x\} \times \mathscr{X}$ is closed in $\mathscr{X} \times \mathscr{X}$, and so is the set

$$\{x\} \times \mathscr{X} \cap \{(x,y) \mid y \succeq x\} = \{x\} \times \{y \mid y \succeq x\}.$$

Hence, $\{y \mid y \succeq x\}$ is closed in $\mathscr{X}$, and its complement $\{y \mid x \succ y\}$ is open. The same argument applies to $\{y \mid y \succ x\}$. □

Example 2.12. For $x_0, y_0 \in \mathbb{R}$ with $x_0 < y_0$, consider the set $\mathscr{X} := (-\infty, x_0] \cup [y_0, \infty)$ endowed with the usual order $>$ on $\mathbb{R}$. Then, with the notation introduced in (2.2), $\mathscr{B}_\prec(y_0) = (-\infty, x_0]$ and $\mathscr{B}_\succ(x_0) = [y_0, \infty)$. Hence,

$$\mathscr{B}_\succ(x_0) \cap \mathscr{B}_\prec(y_0) = \emptyset$$

despite $y_0 \succ x_0$, a situation we had to consider in the preceding proof. ◇

Recall that the topological space $\mathscr{X}$ is called *connected* if $\mathscr{X}$ cannot be written as the union of two disjoint and nonempty open sets. Assuming that $\mathscr{X}$ is connected will rule out the situation occurring in Example 2.12.

Proposition 2.13. *Let $\mathscr{X}$ be a connected topological space with a continuous preference order $\succ$. Then every dense subset $\mathscr{Z}$ of $\mathscr{X}$ is also order dense in $\mathscr{X}$. In particular, there exists a numerical representation of $\succ$ if $\mathscr{X}$ is separable.*

Proof. Take $x, y \in \mathscr{X}$ with $y \succ x$, and consider $\mathscr{B}_\succ(x)$ and $\mathscr{B}_\prec(y)$ as defined in (2.2). Since $y \in \mathscr{B}_\succ(x)$ and $x \in \mathscr{B}_\prec(y)$, neither $\mathscr{B}_\succ(x)$ nor $\mathscr{B}_\prec(y)$ are empty sets. Moreover, negative transitivity implies that $\mathscr{X} = \mathscr{B}_\succ(x) \cup \mathscr{B}_\prec(y)$. Hence, the open sets $\mathscr{B}_\succ(x)$ and $\mathscr{B}_\prec(y)$ cannot be disjoint, as $\mathscr{X}$ is connected. Thus, the open set $\mathscr{B}_\succ(x) \cap \mathscr{B}_\prec(y)$ must contain some element z of the dense subset $\mathscr{Z}$, which then satisfies $y \succ z \succ x$. Therefore, $\mathscr{Z}$ is an order dense subset of $\mathscr{X}$.

Separability of $\mathscr{X}$ means that there exists a countable dense subset $\mathscr{Z}$ of $\mathscr{X}$, which then is order dense. Hence, the existence of a numerical representation follows from Theorem 2.6. □

Remark 2.14. Consider the situation of Example 2.12, where $\mathscr{X} := (-\infty, x_0] \cup [y_0, \infty)$, and suppose that x_0 and y_0 are both irrational. Then $\mathscr{Z} := \mathbb{Q} \cap \mathscr{X}$ is dense in $\mathscr{X}$, but there exists no $z \in \mathscr{Z}$ such that $y_0 \succeq z \succeq x_0$. Hence, $\mathscr{Z}$ is not order dense in $\mathscr{X}$. This example shows that the assumption of topological connectedness is essential for Proposition 2.13. ◇

Let us quote without proof the following general result on the existence of a continuous numerical representation; see [88, Propositions 3 and 4].

Theorem 2.15. *Let $\mathscr{X}$ be a topological space that satisfies at least one of the following two properties:*

- *$\mathscr{X}$ has a countable base of open sets.*
- *$\mathscr{X}$ is separable and connected.*

Then every continuous preference order on $\mathscr{X}$ admits a continuous numerical representation.

The preceding result will not be needed in the sequel. For our purposes, namely for the proof of the von Neumann–Morgenstern representation in the next section and for the proof of the robust Savage representation in Section 2.5, the following lemma will be sufficient.

Lemma 2.16. *Let $\mathscr{X}$ be a connected metric space with a continuous preference order $\succ$. If $U : \mathscr{X} \to \mathbb{R}$ is a continuous function, and if its restriction to some dense subset $\mathscr{Z}$ is a numerical representation for the restriction of $\succ$ to $\mathscr{Z}$, then U is also a numerical representation for $\succ$ on $\mathscr{X}$.*

Proof. We have to show that $y \succ x$ if and only if $U(y) > U(x)$. In order to verify the "only if" part, take $x, y \in \mathscr{X}$ with $y \succ x$. As in the proof of Proposition 2.13, we obtain the existence of some $z_0 \in \mathscr{Z}$ with $y \succ z_0 \succ x$. Repeating this argument yields $z_0' \in \mathscr{Z}$ such that $z_0 \succ z_0' \succ x$. Now we take two sequences (z_n) and (z_n') in $\mathscr{Z}$ with $z_n \to y$ and $z_n' \to x$. By continuity of $\succ$, eventually

$$z_n \succ z_0 \succ z_0' \succ z_n',$$

and thus

$$U(z_n) > U(z_0) > U(z_0') > U(z_n').$$

The continuity of U implies that $U(z_n) \to U(y)$ and $U(z_n') \to U(x)$, whence

$$U(y) \geq U(z_0) > U(z_0') \geq U(x).$$

For the proof of the converse implication, suppose that $x, y \in \mathscr{X}$ are such that $U(y) > U(x)$. Since U is continuous,

$$\mathscr{U}_{>}(x) := \{z \in \mathscr{X} \mid U(z) > U(x)\} \quad \text{and} \quad \mathscr{U}_{<}(y) := \{z \in \mathscr{X} \mid U(z) < U(y)\}$$

are both nonempty open subsets of $\mathscr{X}$. Moreover, $\mathscr{U}_{<}(y) \cup \mathscr{U}_{>}(x) = \mathscr{X}$. Connectedness of $\mathscr{X}$ implies that $\mathscr{U}_{<}(y) \cap \mathscr{U}_{>}(x) \neq \emptyset$. As above, a repeated application of the preceding argument yields $z_0, z_0' \in \mathscr{Z}$ such that

$$U(y) > U(z_0) > U(z_0') > U(x).$$

Since $\mathscr{Z}$ is a dense subset of $\mathscr{X}$, we can find sequences (z_n) and (z_n') in $\mathscr{Z}$ with $z_n \to y$ and $z_n' \to x$ as well as with $U(z_n) > U(z_0)$ and $U(z_n') < U(z_0')$. Since U is a numerical representation of $\succ$ on $\mathscr{Z}$, we have

$$z_n \succ z_0 \succ z_0' \succ z_n'.$$

Hence, by the continuity of $\succ$, neither $z_0 \succ y$ nor $x \succ z_0'$ can be true, and negative transitivity yields $y \succ x$. □

2.2 Von Neumann–Morgenstern representation

Suppose that each possible choice for our economic agent corresponds to a probability distribution on a given set of scenarios. Thus, the set $\mathscr{X}$ can be identified with a subset $\mathscr{M}$ of the set $\mathscr{M}_1(S, \mathscr{S})$ of all probability distributions on a measurable space $(S, \mathscr{S})$. In the context of the theory of choice, the elements of $\mathscr{M}$ are sometimes called *lotteries*. We will assume in the sequel that $\mathscr{M}$ is convex. The aim of this section is to characterize those preference orders $\succ$ on $\mathscr{M}$ that admit a numerical representation U of the form

$$U(\mu) = \int u(x)\,\mu(dx) \quad \text{for all } \mu \in \mathscr{M}, \tag{2.3}$$

where u is a real function on S.

Definition 2.17. A numerical representation U of a preference order $\succ$ on $\mathscr{M}$ is called a *von Neumann–Morgenstern representation* if it is of the form (2.3). △

Any von Neumann–Morgenstern representation U is *affine* on $\mathscr{M}$ in the sense that

$$U(\alpha\mu + (1-\alpha)\nu) = \alpha U(\mu) + (1-\alpha)U(\nu)$$

for all μ, $\nu \in \mathscr{M}$ and $\alpha \in [0,1]$. It is easy to check that affinity of U implies the following two properties, or axioms, for a preference order $\succ$ on $\mathscr{M}$. The first property says that a preference $\mu \succ \nu$ is preserved in any convex combination, independent of the context described by another lottery λ.

Definition 2.18. A preference relation $\succ$ on $\mathscr{M}$ satisfies the *independence axiom* if, for all μ, $\nu \in \mathscr{M}$, the relation $\mu \succ \nu$ implies

$$\alpha\mu + (1-\alpha)\lambda \succ \alpha\nu + (1-\alpha)\lambda$$

for all $\lambda \in \mathscr{M}$ and all $\alpha \in (0,1]$. △

The independence axiom is also called the *substitution axiom*. It can be illustrated by introducing a *compound lottery* that represents the distribution $\alpha\mu + (1-\alpha)\lambda$ as a two-step procedure. First, we sample either lottery μ or λ with probability α and $1-\alpha$, respectively. Then the lottery drawn in this first step is realized. Clearly, this is equivalent to playing directly the lottery $\alpha\mu + (1-\alpha)\lambda$. With probability $1-\alpha$, the distribution λ is drawn and in this case there is no difference to the compound lottery where ν is replaced by μ. The only difference occurs when μ is drawn, and this happens with probability α. Thus, if $\mu \succ \nu$ then it seems reasonable to prefer the compound lottery with μ over the one with ν.

Definition 2.19. A preference relation $\succ$ on $\mathscr{M}$ satisfies the *Archimedean axiom* if for any triple $\mu \succ \lambda \succ \nu$ there are $\alpha, \beta \in (0,1)$ such that

$$\alpha\mu + (1-\alpha)\nu \succ \lambda \succ \beta\mu + (1-\beta)\nu.$$

△

The Archimedean axiom derives its name from its similarity to the Archimedean principle in real analysis: For every small $\varepsilon > 0$ and each large x, there is some $n \in \mathbb{N}$ such that $n\varepsilon > x$. Sometimes it is also called the *continuity axiom*, because it can act as a substitute for the continuity of $\succ$ in a suitable topology on $\mathcal{M}$. More precisely, suppose that $\mathcal{M}$ is endowed with a topology for which convex combinations are continuous curves, that is, $\alpha\mu + (1-\alpha)\nu$ converges to ν or μ as $\alpha \downarrow 0$ or $\alpha \uparrow 1$, respectively. Then continuity of our preference order $\succ$ in this topology automatically implies the Archimedean axiom.

Remark 2.20. As an axiom for consistent behavior in the face of risk, the Archimedean axiom is less intuitive than the independence axiom. Consider the following three deterministic distributions: μ yields 1000 €, λ yields 10 €, and ν is the lottery where one dies for sure. Even for small $\alpha \in (0,1)$ it is not clear that someone would prefer the gamble $\alpha\nu + (1-\alpha)\mu$, which involves the probability α of dying, over the conservative 10 € yielded by λ. Note, however, that most people would not hesitate to drive a car for a distance of 50 km in order to receive a premium of 1000 €. ◇

Our first goal is to show that the Archimedean axiom and the independence axiom imply the existence of an affine numerical representation.

Theorem 2.21. *Suppose that $\succ$ is a preference relation on $\mathcal{M}$ satisfying both the Archimedean and the independence axiom. Then there exists an affine numerical representation U of $\succ$. Moreover, U is unique up to positive affine transformations, that is, any other affine numerical representation $\widetilde{U}$ with these properties is of the form $\widetilde{U} = aU + b$ for some $a > 0$ and $b \in \mathbb{R}$.*

The affinity of a numerical representation does not always imply that it is also of von Neumann–Morgenstern form; see Exercise 2.2.1 and Example 2.26 below. In two important cases, however, such an affine numerical representation will already be of von Neumann–Morgenstern form. This is the content of the following two corollaries, which we state before proving Theorem 2.21. For the first corollary, we need the notion of a *simple probability distribution*. This is a probability measure μ on S that can be written as a finite convex combination of Dirac masses, that is, there exist $x_1, \dots, x_N \in S$ and $\alpha_1, \dots, \alpha_N \in (0,1]$ with $\sum_{i=1}^N \alpha_i = 1$ such that

$$\mu = \sum_{i=1}^{N} \alpha_i \delta_{x_i}.$$

Corollary 2.22. *Suppose that $\mathcal{M}$ is the set of all simple probability distributions on S and that $\succ$ is a preference order on $\mathcal{M}$ that satisfies both the Archimedean and the independence axiom. Then there exists a von Neumann–Morgenstern representation U. Moreover, both U and u are unique up to positive affine transformations.*

Proof. Let U be an affine numerical representation, which exists by Theorem 2.21. We define $u(x) := U(\delta_x)$, for $x \in S$. If $\mu \in \mathcal{M}$ is of the form $\mu = \alpha_1\delta_{x_1} + \cdots + \alpha_N\delta_{x_N}$, then affinity of U implies

$$U(\mu) = \sum_{i=1}^{N} \alpha_i U(\delta_{x_i}) = \int u(x)\,\mu(dx).$$

This is the desired von Neumann–Morgenstern representation. The uniqueness claim is obvious. □

On a finite set S, every probability measure is simple. Thus, we obtain the following result as a special case.

Corollary 2.23. *Suppose that $\mathcal{M}$ is the set of all probability distributions on a finite set S and that $\succ$ is a preference order on $\mathcal{M}$ that satisfies both the Archimedean and the independence axiom. Then there exists a von Neumann–Morgenstern representation, and it is unique up to positive affine transformations.*

For the proof of Theorem 2.21, we need the following auxiliary lemma. Its first assertion states that taking convex combination is monotone with respect to a preference order $\succ$ satisfying our two axioms. Its second part can be regarded as an "intermediate value theorem" for straight lines in $\mathcal{M}$, and (c) is the analogue of the independence axiom for the indifference relation $\sim$.

Lemma 2.24. *Under the assumptions of Theorem 2.21, the following assertions are true:*
(a) *If $\mu \succ \nu$, then $\alpha \mapsto \alpha\mu + (1-\alpha)\nu$ is strictly increasing with respect to $\succ$. More precisely, $\beta\mu + (1-\beta)\nu \succ \alpha\mu + (1-\alpha)\nu$ for $0 \le \alpha < \beta \le 1$.*
(b) *If $\mu \succ \nu$ and $\mu \succeq \lambda \succeq \nu$, then there exists a unique $\alpha \in [0,1]$ with $\lambda \sim \alpha\mu + (1-\alpha)\nu$.*
(c) *If $\mu \sim \nu$, then $\alpha\mu + (1-\alpha)\lambda \sim \alpha\nu + (1-\alpha)\lambda$ for all $\alpha \in [0,1]$ and all $\lambda \in \mathcal{M}$.*

Proof. (a): Let $\lambda := \beta\mu + (1-\beta)\nu$. The independence axiom implies that $\lambda \succ \beta\nu + (1-\beta)\nu = \nu$. Hence, for $\gamma := \alpha/\beta$,

$$\beta\mu + (1-\beta)\nu = (1-\gamma)\lambda + \gamma\lambda \succ (1-\gamma)\nu + \gamma\lambda = \alpha\mu + (1-\alpha)\nu.$$

(b): Part (a) guarantees that α is unique if it exists. To show existence, we need only to consider the case $\mu \succ \lambda \succ \nu$, for otherwise we can take either $\alpha = 0$ or $\alpha = 1$.

The natural candidate is

$$\alpha := \sup\{\gamma \in [0,1] \mid \lambda \succeq \gamma\mu + (1-\gamma)\nu\}.$$

If $\lambda \sim \alpha\mu + (1-\alpha)\nu$ is not true, then one of the following two possibilities must occur:

$$\lambda \succ \alpha\mu + (1-\alpha)\nu, \quad \text{or} \quad \lambda \prec \alpha\mu + (1-\alpha)\nu. \tag{2.4}$$

In the first case, we apply the Archimedean axiom to obtain some $\beta \in (0,1)$ such that

$$\lambda \succ \beta[\, \alpha\mu + (1-\alpha)\nu \,] + (1-\beta)\mu = \gamma\mu + (1-\gamma)\nu \tag{2.5}$$

for $\gamma = 1 - \beta(1-\alpha)$. Since $\gamma > \alpha$, it follows from the definition of α that $\gamma\mu + (1-\gamma)\nu \succ \lambda$, which contradicts (2.5). If the second case in (2.4) occurs, the Archimedean axiom yields some $\beta \in (0,1)$ such that

$$\beta(\alpha\mu + (1-\alpha)\nu) + (1-\beta)\nu = \beta\alpha\mu + (1-\beta\alpha)\nu \succ \lambda. \tag{2.6}$$

Clearly, $\beta\alpha < \alpha$, so that the definition of α yields some $\gamma \in (\beta\alpha, \alpha]$ with $\lambda \succeq \gamma\mu + (1-\gamma)\nu$. Part (a) and the fact that $\beta\alpha < \gamma$ imply that

$$\lambda \succeq \gamma\mu + (1-\gamma)\nu \succ \beta\alpha\mu + (1-\beta\alpha)\nu,$$

which contradicts (2.6).

(c): We must exclude both of the following two possibilities:

$$\alpha\mu + (1-\alpha)\lambda \succ \alpha\nu + (1-\alpha)\lambda \quad \text{and} \quad \alpha\nu + (1-\alpha)\lambda \succ \alpha\mu + (1-\alpha)\lambda. \tag{2.7}$$

To this end, we may assume that there exists some $\rho \in \mathcal{M}$ with $\rho \not\sim \mu \sim \nu$; otherwise the result is trivial. Let us assume that $\rho \succ \mu \sim \nu$; the case in which $\mu \sim \nu \succ \rho$ is similar. Suppose that the first possibility in (2.7) would occur. The independence axiom yields

$$\beta\rho + (1-\beta)\nu \succ \beta\nu + (1-\beta)\nu = \nu \sim \mu$$

for all $\beta \in (0,1)$. Therefore,

$$\alpha[\, \beta\rho + (1-\beta)\nu \,] + (1-\alpha)\lambda \succ \alpha\mu + (1-\alpha)\lambda \quad \text{for all } \beta \in (0,1). \tag{2.8}$$

Using our assumption that the first possibilities in (2.7) is occurring, we obtain from part (b) a unique $\gamma \in (0,1)$ such that, for any fixed β,

$$\begin{aligned} \alpha\mu + (1-\alpha)\lambda &\sim \gamma(\alpha[\, \beta\rho + (1-\beta)\nu \,] + (1-\alpha)\lambda) + (1-\gamma)[\, \alpha\nu + (1-\alpha)\lambda \,] \\ &= \alpha[\, \beta\gamma\rho + (1-\beta\gamma)\nu \,] + (1-\alpha)\lambda \\ &\succ \alpha\mu + (1-\alpha)\lambda, \end{aligned}$$

where we have used (2.8) for β replaced by $\beta\gamma$ in the last step. This is a contradiction. The second possibility in (2.7) is excluded by an analogous argument. □

Proof of Theorem 2.21. For the construction of U, we first fix two lotteries λ and ρ with $\lambda \succ \rho$ and define

$$\mathcal{M}(\lambda, \rho) := \{\mu \in \mathcal{M} \mid \lambda \succeq \mu \succeq \rho\};$$

the assertion is trivial if no such pair $\lambda \succ \rho$ exists. If $\mu \in \mathcal{M}(\lambda, \rho)$, part (b) of Lemma 2.24 yields a unique $\alpha \in [0,1]$ such that $\mu \sim \alpha\lambda + (1-\alpha)\rho$, and we put $U(\mu) := \alpha$. To prove that

U is a numerical representation of $\succ$ on $\mathscr{M}(\lambda,\rho)$, we must show that for $\nu,\mu \in \mathscr{M}(\lambda,\rho)$ we have $U(\mu) > U(\nu)$ if and only if $\mu \succ \nu$. To prove sufficiency, we apply part (a) of Lemma 2.24 to conclude that

$$\mu \sim U(\mu)\lambda + (1-U(\mu))\rho \succ U(\nu)\lambda + (1-U(\nu))\rho \sim \nu.$$

Hence, $\mu \succ \nu$. Conversely, if $\mu \succ \nu$ then the preceding arguments already imply that we cannot have $U(\nu) > U(\mu)$. Thus, it suffices to rule out the case $U(\mu) = U(\nu)$. But if $U(\mu) = U(\nu)$, then the definition of U yields $\mu \sim \nu$, which contradicts $\mu \succ \nu$. We conclude that U is indeed a numerical representation of $\succ$ restricted to $\mathscr{M}(\lambda,\rho)$.

Let us now show that $\mathscr{M}(\lambda,\rho)$ is a convex set. Take $\mu,\nu \in \mathscr{M}(\lambda,\rho)$ and $\alpha \in [0,1]$. Then

$$\lambda \succeq \alpha\lambda + (1-\alpha)\nu \succeq \alpha\mu + (1-\alpha)\nu,$$

using the independence axiom to handle the cases $\lambda \succ \nu$ and $\lambda \succ \mu$, and part (c) of Lemma 2.24 for $\lambda \sim \nu$ and for $\lambda \sim \mu$. By an analogous argument, it follows that $\alpha\mu + (1-\alpha)\nu \succeq \rho$, which implies the convexity of the set $\mathscr{M}(\lambda,\rho)$.

Therefore, $U(\alpha\mu + (1-\alpha)\nu)$ is well-defined; we proceed to show that it equals $\alpha U(\mu) + (1-\alpha)U(\nu)$. To this end, we apply part (c) of Lemma 2.24 twice:

$$\begin{aligned}\alpha\mu + (1-\alpha)\nu &\sim \alpha\big(U(\mu)\lambda + (1-U(\mu))\rho\big) + (1-\alpha)\big(U(\nu)\lambda + (1-U(\nu))\rho\big)\\ &= \big[\alpha U(\mu) + (1-\alpha)U(\nu)\big]\lambda + \big[1-\alpha U(\mu) - (1-\alpha)U(\nu)\big]\rho.\end{aligned}$$

The definition of U and the uniqueness in part (b) of Lemma 2.24 imply that

$$U\big(\alpha\mu + (1-\alpha)\nu\big) = \alpha U(\mu) + (1-\alpha)U(\nu).$$

So, U is indeed an affine numerical representation of $\succ$ on $\mathscr{M}(\lambda,\rho)$.

In a further step, we now show that the affine numerical representation U on $\mathscr{M}(\lambda,\rho)$ is unique up to positive affine transformations. So, let $\widetilde{U}$ be another affine numerical representation of $\succ$ on $\mathscr{M}(\lambda,\rho)$, and define

$$\hat{U}(\mu) := \frac{\widetilde{U}(\mu) - \widetilde{U}(\rho)}{\widetilde{U}(\lambda) - \widetilde{U}(\rho)}, \qquad \mu \in \mathscr{M}(\lambda,\rho).$$

Then $\hat{U}$ is a positive affine transformation of $\widetilde{U}$, and $\hat{U}(\rho) = 0 = U(\rho)$ as well as $\hat{U}(\lambda) = 1 = U(\lambda)$. Hence, affinity of $\hat{U}$ and the definition of U imply

$$\hat{U}(\mu) = \hat{U}\big(U(\mu)\lambda + (1-U(\mu))\rho\big) = U(\mu)\hat{U}(\lambda) + (1-U(\mu))\hat{U}(\rho) = U(\mu)$$

for all $\mu \in \mathscr{M}(\lambda,\rho)$. Thus, $\hat{U} = U$.

Finally, we have to show that U can be extended as a numerical representation to the full space $\mathscr{M}$. To this end, we first take $\widetilde{\lambda},\widetilde{\rho} \in \mathscr{M}$ such that $\mathscr{M}(\widetilde{\lambda},\widetilde{\rho}) \supseteq \mathscr{M}(\lambda,\rho)$. By the arguments in the first part of this proof, there exists an affine numerical representation $\widetilde{U}$ of $\succ$ on $\mathscr{M}(\widetilde{\lambda},\widetilde{\rho})$, and we may assume that $\widetilde{U}(\lambda) = 1$ and $\widetilde{U}(\rho) = 0$; otherwise we apply a positive affine transformation to $\widetilde{U}$. By the previous step of the proof, $\widetilde{U}$ coincides with U on $\mathscr{M}(\lambda,\rho)$, and so $\widetilde{U}$ is the unique consistent extension of U. Since each lottery belongs to some set $\mathscr{M}(\widetilde{\lambda},\widetilde{\rho})$, the affine numerical representation U can be uniquely extended to all of $\mathscr{M}$. □

Remark 2.25. In the proof of the preceding theorem, we did not use the fact that the elements of $\mathscr{M}$ are probability measures. All that was needed was convexity of the set $\mathscr{M}$, the Archimedean, and the independence axiom. Yet, even the concept of convexity can be generalized by introducing the notion of a *mixture space*; see, for example, [135, 224], or [180]. ◇

Let us now return to the problem of constructing a von Neumann–Morgenstern representation for preference relations on distributions. If $\mathscr{M}$ is the set of all probability measures on a finite set S, any affine numerical representation is already of this form, as we saw in the proof of Corollary 2.23. However, the situation becomes more involved if we take an infinite set S. In fact, the following examples show that in this case a von Neumann–Morgenstern representation may not exist.

Exercise 2.2.1. Let $\mathscr{M}$ be the set of probability measures μ on $S := \{1, 2, \dots\}$ for which $U(\mu) := \lim_{k\uparrow\infty} k^2\,\mu(k)$ exists as a finite real number. Show that U is affine and induces a preference order on $\mathscr{M}$ that satisfies both the Archimedean and the independence axiom. Show next that U does not admit a von Neumann–Morgenstern representation. ◇

Example 2.26. Let $\mathscr{M}$ be set the of all Borel probability measures on $S = [0, 1]$, and denote by λ the Lebesgue measure on S. According to the Lebesgue decomposition theorem, which is recalled in Theorem B.9, every $\mu \in \mathscr{M}$ can be decomposed as

$$\mu = \mu_s + \mu_a,$$

where μ_s is singular with respect to λ, and μ_a is absolutely continuous. We define a function $U : \mathscr{M} \to [0, 1]$ by

$$U(\mu) := \int x\mu_a(dx).$$

It is easily seen that U is an affine function on $\mathscr{M}$. Hence, U induces a preference order $\succ$ on $\mathscr{M}$, which satisfies both the Archimedean and the independence axioms. But $\succ$ cannot have a von Neumann–Morgenstern representation: Since $U(\delta_x) = 0$ for all x, the only possible choice for u in (2.3) would be $u \equiv 0$. So, the preference relation would be trivial in the sense that $\mu \sim \lambda$ for all $\mu \in \mathscr{M}$, in contradiction for instance to $U(\lambda) = \frac{1}{2}$ and $U(\delta_{\frac{1}{2}}) = 0$. ◇

One way to obtain a von Neumann–Morgenstern representation is to assume additional continuity properties of $\succ$, where continuity is understood in the sense of Definition 2.8. As we have already remarked, the Archimedean axiom holds automatically if taking convex combinations is continuous for the topology on $\mathscr{M}$. This is indeed the case for the weak topology on the set $\mathscr{M}_1(S, \mathscr{S})$ of all probability measures on a separable metric space S, endowed with the σ-field $\mathscr{S}$ of Borel sets. The space S will be fixed for the remainder of this section, and we will simply write $\mathscr{M}_1(S) = \mathscr{M}_1(S, \mathscr{S})$.

Theorem 2.27. *Let $\mathscr{M} := \mathscr{M}_1(S)$ be the space of all probability measures on S endowed with the weak topology, and let $\succ$ be a continuous preference order on $\mathscr{M}$ satisfying the independence axiom. Then there exists a von Neumann–Morgenstern representation*

$$U(\mu) = \int u(x)\,\mu(dx)$$

for which the function $u : S \to \mathbb{R}$ is bounded and continuous. Moreover, U and u are unique up to positive affine transformations.

Proof. Let $\mathscr{M}_s$ denote the set of all simple probability distributions on S. Since continuity of $\succ$ implies the Archimedean axiom, we deduce from Corollary 2.22 that $\succ$ restricted to $\mathscr{M}_s$ has a von Neumann–Morgenstern representation.

Let us show that the function u in this representation is bounded. For instance, if u is not bounded from above, then there are $x_0, x_1, \ldots \in S$ such that $u(x_0) < u(x_1)$ and $u(x_n) > n$. Now let

$$\mu_n := \Big(1 - \frac{1}{\sqrt{n}}\Big)\delta_{x_0} + \frac{1}{\sqrt{n}}\delta_{x_n}.$$

Clearly, $\mu_n \to \delta_{x_0}$ weakly as $n \uparrow \infty$. The continuity of $\succ$ together with the assumption that $\delta_{x_1} \succ \delta_{x_0}$ imply that $\delta_{x_1} \succ \mu_n$ for all large n. However, $U(\mu_n) > (1-n^{-1/2})u(x_0) + n^{1/2} \to \infty$, in contradiction to the fact that $U(\delta_{x_1}) > U(\mu_n)$ for sufficiently large n.

Suppose that the function u is not continuous. Then there exists some $x \in S$ and a sequence $(x_n)_{n\in\mathbb{N}}$ in S such that $x_n \to x$ but $u(x_n) \nrightarrow u(x)$. By taking a subsequence if necessary, we can assume that $u(x_n)$ converges to some number $a \neq u(x)$. Suppose that $u(x) - a =: \varepsilon > 0$. Then there exists some m such that $|u(x_n) - a| < \varepsilon/3$ for all $n \geq m$. Let $\mu := \frac{1}{2}(\delta_x + \delta_{x_m})$. For all $n \geq m$,

$$\begin{aligned} U(\delta_x) = a + \varepsilon &> \frac{1}{2}(a+\varepsilon) + \frac{1}{2}\Big(a + \frac{1}{3}\varepsilon\Big) > \frac{1}{2}u(x) + \frac{1}{2}u(x_m) \\ = U(\mu) &> \frac{1}{2}(a+\varepsilon) + \frac{1}{2}\Big(a - \frac{1}{3}\varepsilon\Big) = a + \frac{\varepsilon}{3} > U(\delta_{x_n}). \end{aligned}$$

Therefore, $\delta_x \succ \mu \succ \delta_{x_n}$, although δ_{x_n} converges weakly to δ_x, in contradiction to the continuity of $\succ$. The case $u(x) < a$ is excluded in the same manner.

Let us finally show that

$$U(\mu) := \int u(x)\,\mu(dx) \quad \text{for } \mu \in \mathcal{M}$$

defines a numerical representation of $\succ$ on all of $\mathcal{M}$. Since u is bounded and continuous, U is continuous with respect to the weak topology on $\mathcal{M}$. Moreover, Theorem G.3 states that $\mathcal{M}_s$ is a dense subset of the connected metrizable space $\mathcal{M}$. So, the proof is completed by an application of Lemma 2.16. □

The scope of the preceding theorem is limited insofar as it involves only *bounded* functions u. This will not be flexible enough for our purposes. In the next section, for instance, we will consider risk-averse preferences that are defined in terms of *concave* functions u on the space $S = \mathbb{R}$. Such a function cannot be bounded unless it is constant. Thus, we must relax the conditions of the previous theorem. We will present two approaches. In our first approach, we fix some point $x_0 \in S$ and denote by $\overline{B}_r(x_0)$ the closed metric ball of radius r around x_0. The space of *boundedly supported measures* on S is given by

$$\begin{aligned}\mathcal{M}_b(S) &:= \bigcup_{r>0} \mathcal{M}_1(\overline{B}_r(x_0)) \\ &= \{\mu \in \mathcal{M}_1(S) \mid \mu(\overline{B}_r(x_0)) = 1 \text{ for some } r \geq 0\}.\end{aligned}$$

Clearly, this definition does not depend on the particular choice of x_0.

Corollary 2.28. *Let $\succ$ be a preference order on $\mathcal{M}_b(S)$ whose restriction to each space $\mathcal{M}_1(\overline{B}_r(x_0))$ is continuous with respect to the weak topology. If $\succ$ satisfies the independence axiom, then there exists a von Neumann–Morgenstern representation*

$$U(\mu) = \int u(x)\,\mu(dx)$$

with a continuous function $u : S \to \mathbb{R}$. Moreover, U and u are unique up to positive affine transformations.

Proof. Theorem 2.27 yields a von Neumann–Morgenstern representation of the restriction of $\succ$ to $\mathcal{M}_1(\overline{B}_r(x_0))$ in terms of some continuous function $u_r : \overline{B}_r(x_0) \to \mathbb{R}$. The uniqueness part of the theorem implies that the restriction of u_r to some smaller ball $\overline{B}_{r'}(x_0)$ must be a equal to $u_{r'}$ up to a positive affine transformation. Thus, it is possible to find a unique continuous extension $u : S \to \mathbb{R}$ of $u_{r'}$ that defines a von Neumann–Morgenstern representation of $\succ$ on each set $\mathcal{M}_1(\overline{B}_r(x_0))$. □

Our second variant of Theorem 2.27 includes measures with *unbounded* support, but we need stronger continuity assumptions. Let ψ be a continuous function with values in $[1, \infty)$ on the separable metric space S. We use ψ as a *gauge function* and define

$$\mathscr{M}_1^{\psi}(S) := \left\{ \mu \in \mathscr{M}_1(S) \;\middle|\; \int \psi(x)\,\mu(dx) < \infty \right\}.$$

A suitable space of continuous test functions for measures in $\mathscr{M}_1^{\psi}(S)$ is provided by

$$C_\psi(S) := \{f \in C(S) \mid \exists c : |f(x)| \le c \cdot \psi(x) \text{ for all } x \in S\}.$$

These test functions can now be used to define a topology on $\mathscr{M}_1^{\psi}(S)$ in precisely the same way one uses the set of bounded continuous function to define the weak topology: A sequence (μ_n) in $\mathscr{M}_1^{\psi}(S)$ converges to some $\mu \in \mathscr{M}_1^{\psi}(S)$ if and only if

$$\int f\,d\mu_n \longrightarrow \int f\,d\mu \quad \text{for all } f \in C_\psi(S).$$

To be rigorous, one should first define a neighborhood base for the topology and then check that this topology is metrizable, so that it suffices indeed to consider the convergence of sequences; the reader will find all necessary details in Appendix G. We will call this topology the *ψ-weak topology* on $\mathscr{M}_1^{\psi}(S)$. If we take the trivial case $\psi \equiv 1$, $C_\psi(S)$ consists of all bounded continuous functions, and we recover the standard weak topology on $\mathscr{M}_1^{1}(S) = \mathscr{M}_1(S)$. However, by taking ψ as some unbounded function, we can also include von Neumann–Morgenstern representations in terms of unbounded functions u. The following theorem is a version of Theorem 2.27 for the ψ-weak topology. Its proof is analogous to that of Theorem 2.27, and we leave it to the reader to fill in the details.

Theorem 2.29. *Let $\succ$ be a preference order on $\mathscr{M}_1^{\psi}(S)$ that is continuous in the ψ-weak topology and satisfies the independence axiom. Then there exists a numerical representation U of von Neumann–Morgenstern form*

$$U(\mu) = \int u(x)\,\mu(dx)$$

with a function $u \in C_\psi(S)$. Moreover, U and u are unique up to positive affine transformations.

So far, we have presented the classical theory of expected utility, starting with the independence axiom and the Archimedean axiom. However, it is well known that in reality people may not behave according to this paradigm.

Example 2.30 (Allais paradox). The so-called *Allais paradox* questions the descriptive aspect of expected utility by considering the following lotteries. Lottery

$$\nu_1 = 0.33\,\delta_{2500} + 0.66\,\delta_{2400} + 0.01\,\delta_0$$

yields 2500 € with a probability of 0.33, 2400 € with probability 0.66, and draws a blank with the remaining probability of 0.01. Lottery

$$\mu_1 := \delta_{2400}$$

yields 2400 € for sure. When asked, most people prefer the sure amount—even though lottery ν_1 has the larger expected value, namely 2409 €.

Next, consider the following two lotteries μ_2 and ν_2:

$$\mu_2 := 0.34\,\delta_{2400} + 0.66\,\delta_0 \quad \text{and} \quad \nu_2 := 0.33\,\delta_{2500} + 0.67\,\delta_0.$$

Here, people tend to prefer the slightly riskier lottery ν_2 over μ_2, in accordance with the expectations of ν_2 and μ_2, which are 825 € and 816 €, respectively.

This observation is due to M. Allais [5]. It was confirmed by D. Kahnemann and A. Tversky [198] in empirical tests where 82 % of interviewees preferred μ_1 over ν_1 while 83 % chose ν_2 rather than μ_2. This means that at least 65 % chose both $\mu_1 \succ \nu_1$ and $\nu_2 \succ \mu_2$. As pointed out by M. Allais, this simultaneous choice leads to a "paradox" in the sense that it is inconsistent with the von Neumann–Morgenstern paradigm. More precisely, any preference relation $\succ$ for which $\mu_1 \succ \nu_1$ and $\nu_2 \succ \mu_2$ are both valid violates the independence axiom, as we will show now. If the independence axiom was satisfied, then necessarily

$$\alpha\mu_1 + (1-\alpha)\nu_2 \succ \alpha\nu_1 + (1-\alpha)\nu_2 \succ \alpha\nu_1 + (1-\alpha)\mu_2$$

for all $\alpha \in (0,1)$. By taking $\alpha = 1/2$, we would arrive at

$$\frac{1}{2}(\mu_1 + \nu_2) \succ \frac{1}{2}(\nu_1 + \mu_2),$$

which is a contradiction to the fact that

$$\frac{1}{2}(\mu_1 + \nu_2) = \frac{1}{2}(\nu_1 + \mu_2).$$

Therefore, the independence axiom was violated by at least 65 % of the people who were interviewed. This effect is empirical evidence against the von Neumann–Morgenstern theory as a *descriptive* theory. Even from a *normative* point of view, there are good reasons to go beyond our present setting, and this will be done in Section 2.5. In particular, we will take a second look at the Allais paradox in Remark 2.72. ◇

2.3 Expected utility

In this section, we focus on individual financial assets under the assumption that their payoff distributions at a fixed time are known, and without any regard to hedging opportunities in the context of a financial market model. Such asset distributions may be viewed as lotteries with monetary outcomes in some interval on the real line. Thus, we take $\mathscr{M}$ as a fixed set of Borel probability measures on a fixed interval $S \subseteq \mathbb{R}$. In this setting, we discuss the paradigm of expected utility in its standard form, where the function u appearing in the von Neumann–Morgenstern representation has additional properties suggested by the monetary interpretation. We introduce risk aversion and certainty equivalents, and illustrate these notions with a number of examples.

Throughout this section, we assume that $\mathscr{M}$ is convex and contains all point masses δ_x for $x \in S$. We assume also that each $\mu \in \mathscr{M}$ has a well-defined expectation:

$$m(\mu) := \int x \, \mu(dx) \in \mathbb{R}.$$

Remark 2.31. For an asset whose (discounted) random payoff has a known distribution μ, the expected value $m(\mu)$ is often called the *fair price* of the asset. For an insurance contract where μ is the distribution of payments to be received by the insured party in dependence of some random damage within a given period, the expected value $m(\mu)$ is also called the *fair premium*. Typically, actual asset prices and actual insurance premiums will be different from these values. In many situations, such differences can be explained within the conceptual framework of expected utility, and in particular in terms of *risk aversion*. ◇

Definition 2.32. A preference relation $\succ$ on $\mathscr{M}$ is called *monotone* if

$$x > y \quad \text{implies} \quad \delta_x \succ \delta_y.$$

The preference relation is called *risk averse* if for $\mu \in \mathscr{M}$,

$$\delta_{m(\mu)} \succ \mu \qquad \text{unless } \mu = \delta_{m(\mu)}.$$

△

It is easy to characterize these properties within the class of preference relations that admit a von Neumann–Morgenstern representation.

Proposition 2.33. *Suppose the preference relation $\succ$ has a von Neumann–Morgenstern representation*

$$U(\mu) = \int u \, d\mu.$$

Then:

(a) *$\succ$ is monotone if and only if u is strictly increasing.*
(b) *$\succ$ is risk averse if and only if u is strictly concave.*

Proof. (a): Monotonicity is equivalent to

$$u(x) = U(\delta_x) > U(\delta_y) = u(y) \quad \text{for } x > y.$$

(b): If $\succ$ is risk-averse, then

$$\delta_{\alpha x+(1-\alpha)y} \succ \alpha\,\delta_x + (1-\alpha)\delta_y$$

holds for all distinct $x, y \in S$ and $\alpha \in (0,1)$. Hence,

$$u(\alpha x + (1-\alpha)y) > \alpha\,u(x) + (1-\alpha)u(y),$$

that is, u is strictly concave. Conversely, if u is strictly concave, then Jensen's inequality implies risk aversion:

$$U(\delta_{m(\mu)}) = u\Big(\int x\,\mu(dx)\Big) \geq \int u(x)\,\mu(dx) = U(\mu)$$

with equality if and only if $\mu = \delta_{m(\mu)}$. □

Remark 2.34. In view of the monetary interpretation of the state space S, it is natural to assume that the preference relation $\succ$ is *monotone*. The assumption of *risk aversion* is more debatable, at least from a *descriptive* point of view. In fact, there is considerable empirical evidence that agents tend to switch between risk aversion and risk seeking behavior, depending on the context. In particular, they may be risk averse after prior gains, and they may become risk seeking if they see an opportunity to compensate prior losses. Tversky and Kahneman [316] propose to describe such a behavioral pattern by a function u of the form

$$u(x) = \begin{cases} (x-c)^{\gamma} & \text{for } x \geq c, \\ -\lambda(c-x)^{\gamma} & \text{for } x < c, \end{cases} \tag{2.9}$$

where c is a given benchmark level, and their experiments suggest parameter values λ around 2 and γ slightly less than 1. Nevertheless, one can insist on risk aversion from a *normative* point of view, and in the sequel we explore some of its consequences. ◇

Definition 2.35. A function $u : S \to \mathbb{R}$ is called a *utility function* if it is strictly concave, strictly increasing, and continuous on S. A von Neumann–Morgenstern representation

$$U(\mu) = \int u\,d\mu \tag{2.10}$$

in terms of a utility function u is called an *expected utility representation.* △

Any increasing concave function $u : S \to \mathbb{R}$ is necessarily continuous on every interval $(a, b] \subseteq S$; see Proposition A.7. Hence, the condition of continuity in the preceding definition is only relevant if S contains its lower boundary point. Note that any utility function $u(x)$ decreases at least linearly as $x \downarrow \inf S$. Therefore, u cannot be bounded from below unless $\inf S > -\infty$.

From now on, we will consider a fixed preference relation $\succ$ on $\mathscr{M}$ that admits a von Neumann–Morgenstern representation

$$U(\mu) = \int u\, d\mu$$

in terms of a strictly increasing continuous function $u : S \to \mathbb{R}$. The intermediate value theorem applied to the continuous function u yields for any $\mu \in \mathscr{M}$ a unique real number $c(\mu)$ for which

$$u(c(\mu)) = U(\mu) = \int u\, d\mu. \tag{2.11}$$

It follows that

$$\delta_{c(\mu)} \sim \mu,$$

that is, there is indifference between the lottery μ and the sure amount of money $c(\mu)$.

Definition 2.36. The *certainty equivalent* of the lottery $\mu \in \mathscr{M}$ with respect to u is defined as the number $c(\mu)$ of (2.11), and

$$\varrho(\mu) := m(\mu) - c(\mu)$$

is called the *risk premium* of μ. △

If u is a utility function, risk aversion implies that

$$u(c(\mu)) = U(\mu) < U(\delta_{m(\mu)}) = u(m(\mu))$$

for every lottery μ with $\mu \neq \delta_{m(\mu)}$. Hence, monotonicity yields that

$$c(\mu) < m(\mu) \quad \text{for } \mu \neq \delta_{m(\mu)}.$$

In particular, the risk premium $\varrho(\mu)$ associated with a utility function is always nonnegative, and it is strictly positive as soon as the distribution μ carries any risk.

Remark 2.37. The certainty equivalent $c(\mu)$ can be viewed as an upper bound on any price of μ that would be acceptable to an economic agent with utility function u. Thus, the fair price $m(\mu)$ must be reduced at least by the risk premium $\varrho(\mu)$ if one wants the agent to buy the asset distribution μ. Alternatively, suppose that the agent holds an asset with distribution μ. Then the risk premium may be viewed as the amount that the agent would be ready to pay for replacing the asset by its expected value $m(\mu)$. ◇

Example 2.38 (St. Petersburg paradox). Consider the lottery

$$\mu = \sum_{n=1}^{\infty} 2^{-n} \delta_{2^{n-1}},$$

which may be viewed as the payoff distribution of the following game. A fair coin is tossed until a head appears. If the head appears on the nth toss, the payoff will be 2^{n-1} €. Up to the early 18th century, it was commonly accepted that the price of a lottery should be computed as the *fair price*, that is, as the expected value $m(\mu)$. In the present example, the fair price is given by $m(\mu) = \infty$, but it is hard to find someone who is ready to pay even 20 €. In view of this "paradox," posed by Nicholas Bernoulli in 1713, Gabriel Cramer and Daniel Bernoulli [30] independently introduced the idea of determining an acceptable price as the certainty equivalent with respect to some utility function. For the two utility functions,

$$u_1(x) = \sqrt{x} \quad \text{and} \quad u_2(x) = \log x$$

proposed, respectively, by G. Cramer and by D. Bernoulli, these certainty equivalents are given by

$$c_1(\mu) = \frac{3}{2} + \sqrt{2} \approx 2.91 \quad \text{and} \quad c_2(\mu) = 2, \tag{2.12}$$

and this is within the range of prices people are usually ready to pay. Note, however, that for any utility function that is unbounded from above we could modify the payoff in such a way that the paradox reappears. For example, we could replace the payoff 2^n by $u^{-1}(2^n)$ for $n \geq 1000$, so that $\int u\,d\mu = +\infty$. The choice of a utility function that is bounded from above would remove this difficulty, but would create others; see the discussion on pages 80–84. ◇

Exercise 2.3.1. Prove (2.12). ◇

Given the preference order $\succ$ on $\mathscr{M}$, we can now try to determine those distributions in $\mathscr{M}$ that are maximal with respect to $\succ$. As a first illustration, consider the following simple *optimization problem*. Let X be an integrable random variable on some probability space $(\Omega, \mathscr{F}, P)$ with nondegenerate distribution $\mu \in \mathscr{M}$. We assume that X is bounded from below by some number a in the interior of S. What is the best mix

$$X_\lambda := (1-\lambda)X + \lambda c$$

of the risky payoff X and the certain amount c that also belongs to the interior of S? If we evaluate X_λ by its expected utility $E[\,u(X_\lambda)\,]$ and denote by μ_λ the distribution of X_λ under P, then we are looking for a maximum of the function f on $[0,1]$ defined by

$$f(\lambda) := U(\mu_\lambda) = \int u\,d\mu_\lambda = E[\,u((1-\lambda)X + \lambda c)\,].$$

If u is a utility function, f is strictly concave and attains its maximum in a unique point $\lambda^* \in [0,1]$.

Proposition 2.39. *Let u be a utility function.*

(a) *We have* $\lambda^* = 1$ *if* $E[\,X\,] \le c$, *and* $\lambda^* > 0$ *if* $c \ge c(\mu)$.
(b) *If u is differentiable, then*

$$\lambda^* = 1 \quad \Longleftrightarrow \quad E[\,X\,] \le c$$

and

$$\lambda^* = 0 \quad \Longleftrightarrow \quad c \le \frac{E[\,Xu'(X)\,]}{E[\,u'(X)\,]}.$$

Proof. (a): Jensen's inequality yields that

$$f(\lambda) \le u(E[\,X_\lambda\,]) = u((1-\lambda)E[\,X\,] + \lambda c),$$

with equality if and only if $\lambda = 1$. It follows that $\lambda^* = 1$ if the right-hand side is increasing in λ, that is, if $E[\,X\,] \le c$.

Strict concavity of u implies

$$\begin{aligned} f(\lambda) &\ge E[\,(1-\lambda)u(X) + \lambda u(c)\,] \\ &= (1-\lambda)u(c(\mu)) + \lambda u(c), \end{aligned}$$

with equality if and only if $\lambda \in \{0,1\}$. The right-hand side is increasing in λ if $c \ge c(\mu)$, and this implies $\lambda^* > 0$.

(b): Clearly, we have $\lambda^* = 0$ if and only if the right-hand derivative f'_+ of f satisfies $f'_+(0) \le 0$; see Appendix A.1 for the definition of f'_+ and f'_-. Note that the difference quotients

$$\frac{u(X_\lambda) - u(X)}{\lambda} = \frac{u(X_\lambda) - u(X)}{X_\lambda - X} \cdot (c - X)$$

satisfy P-a. s.

$$\left|\frac{u(X_\lambda) - u(X)}{\lambda}\right| \le u'_+(a \wedge c)|c - X| \in L^1(P)$$

and that they converge to

$$u'_+(X)(c-X)^+ - u'_-(X)(c-X)^-$$

as $\lambda \downarrow 0$. By Lebesgue's theorem, this implies

$$f'_+(0) = E[\,u'_+(X)(c-X)^+\,] - E[\,u'_-(X)(c-X)^-\,].$$

If u is differentiable, or if the countable set $\{x \mid u'_+(x) \neq u'_-(x)\}$ has μ-measure 0, then we can conclude

$$f'_+(0) = E[\,u'(X)(c-X)\,],$$

that is, $f'_+(0) \leq 0$ if and only if

$$c \leq \frac{E[\,Xu'(X)\,]}{E[\,u'(X)\,]}.$$

In the same way, we obtain

$$f'_-(1) = u'_-(c)E[\,(X-c)^-\,] - u'_+(c)E[\,(X-c)^+\,].$$

If u is differentiable at c, then we can conclude

$$f'_-(1) = u'(c)(c - E[\,X\,]).$$

This implies $f'_-(1) < 0$, and hence $\lambda^* < 1$, if and only if $E[\,X\,] > c$. □

Exercise 2.3.2. As above, let X be a random variable with a nondegenerate distribution $\mu \in \mathscr{M}$. Show that for a differentiable utility function u we have

$$m(\mu) > c(\mu) > \frac{E[\,u'(X)X\,]}{E[\,u'(X)\,]}.$$ ◇

Example 2.40 (Demand for a risky asset). Let $S = S^1$ be a risky asset with price $\pi = \pi^1$. Given an initial wealth w, an agent with utility function $u \in C^1$ can invest a fraction $(1-\lambda)w$ into the asset and the remaining part λw into a risk-free bond with interest rate r. The resulting payoff is

$$X_\lambda = \frac{(1-\lambda)w}{\pi}(S-\pi) + \lambda w \cdot r.$$

The preceding proposition implies that there will be no investment into the risky asset if and only if

$$E\left[\frac{S}{1+r}\right] \leq \pi.$$

In other words, the price of the risky asset must be below its expected discounted payoff in order to attract any risk averse investor, and in that case it will indeed be optimal for the investor to invest at least some amount. Instead of the simple linear profiles X_λ, the investor may wish to consider alternative forms of investment. For example, this may involve derivatives such as $\max\{S, K\} = K + (S-K)^+$ for some threshold K. In order to discuss such nonlinear payoff profiles, we need an extended formulation of the optimization problem; see Section 3.3 below. ◇

Example 2.41 (Demand for insurance). Suppose an agent with utility function $u \in C^1$ considers taking at least some partial insurance against a random loss Y, with $0 \leq Y \leq w$

and $P[\,Y \neq E[\,Y\,]\,] > 0$, where w is a given initial wealth. If insurance of λY is available at the insurance premium $\lambda\pi$, the resulting final payoff is given by

$$X_\lambda := w - Y + \lambda(Y - \pi) = (1 - \lambda)(w - Y) + \lambda(w - \pi).$$

By Proposition 2.39, full insurance is optimal if and only if $\pi \leq E[\,Y\,]$. In reality, however, the insurance premium π will exceed the "fair premium" $E[\,Y\,]$. In this case, it will be optimal to insure only a fraction $\lambda^* Y$ of the loss, with $\lambda^* \in [0,1)$. This fraction will be strictly positive as long as

$$\pi < \frac{E[\,Yu'(w - Y)\,]}{E[\,u'(w - Y)\,]} = w - \frac{E[\,(w - Y)u'(w - Y)\,]}{E[\,u'(w - Y)\,]}.$$

Since the right-hand side is strictly larger than $E[\,Y\,]$ due to Exercise 2.3.2, risk aversion may create a demand for insurance even if the insurance premium π lies above the "fair" price $E[\,Y\,]$. As in the previous example, the agent may wish to consider alternative forms of insurance such as a stop-loss contract, whose payoff has the nonlinear structure $(Y-K)^+$ of a call option. See Theorem 3.42 for a situation in which such stop-loss contracts appear in a natural manner as solutions to an actuarial optimization problem. ◇

Let us take another look at the risk premium $\varrho(\mu)$ of a lottery μ. For an approximate calculation, we consider the Taylor expansion of a sufficiently smooth and strictly increasing function $u(x)$ at $x = c(\mu)$ around $m := m(\mu)$, and we assume that μ has finite variance $\operatorname{var}(\mu)$. On the one hand,

$$u(c(\mu)) \approx u(m) + u'(m)(c(\mu) - m) = u(m) - u'(m)\varrho(\mu).$$

On the other hand,

$$\begin{aligned} u(c(\mu)) &= \int u(x)\,\mu(dx) \\ &= \int \Big[u(m) + u'(m)(x - m) + \frac{1}{2}u''(m)(x - m)^2 + r(x) \Big]\,\mu(dx) \\ &\approx u(m) + \frac{1}{2}u''(m)\operatorname{var}(\mu), \end{aligned}$$

where $r(x)$ denotes the remainder term in the Taylor expansion of u. It follows that

$$\varrho(\mu) \approx -\frac{u''(m(\mu))}{2 \cdot u'(m(\mu))}\operatorname{var}(\mu) =: \frac{1}{2}\,\alpha(m(\mu))\operatorname{var}(\mu). \tag{2.13}$$

Thus, $\alpha(m(\mu))$ is the factor by which an economic agent with von Neumann–Morgenstern preferences described by u weighs the risk, measured by $\frac{1}{2}\operatorname{var}(\mu)$, in order to determine the risk premium he or she is ready to pay.

Definition 2.42. Suppose that u is a twice continuously differentiable and strictly increasing function on S. Then

$$\alpha(x) := -\frac{u''(x)}{u'(x)}$$

is called the *Arrow–Pratt coefficient of absolute risk aversion* of u at level x. △

Example 2.43. The following classes of utility functions u and their corresponding coefficients of risk aversion are standard examples.

(a) **Constant absolute risk aversion (CARA):** $\alpha(x)$ equals some constant $\alpha > 0$. Since $\alpha(x) = -(\log u')'(x)$, it follows that $u(x) = a - b \cdot e^{-\alpha x}$. Using an affine transformation, u can be normalized to

$$u(x) = 1 - e^{-\alpha x}.$$

(b) **Hyperbolic absolute risk aversion (HARA):** $\alpha(x) = (1-\gamma)/x$ on $S = (0, \infty)$ for some $\gamma < 1$. Up to affine transformations, we have

$$u(x) = \log x \quad \text{for } \gamma = 0,$$
$$u(x) = \frac{1}{\gamma} x^{\gamma} \quad \text{for } \gamma \neq 0.$$

Sometimes, these functions are also called CRRA utility functions, because their "relative risk aversion" $x\alpha(x)$ is constant. Of course, these utility functions can be shifted to any interval $S = (a, \infty)$. The "risk-neutral" limiting case $\gamma = 1$ would correspond to an affine function u. ◇

Exercise 2.3.3. Compute the coefficient of risk aversion for the S-shaped utility function in (2.9). Sketch the graphs of u and its risk aversion for $\lambda = 2$ and $\gamma = 0.6$. ◇

Proposition 2.44. *Suppose that u and $\tilde{u}$ are two strictly increasing functions on S that are twice continuously differentiable, and that α and $\tilde{\alpha}$ are the corresponding Arrow–Pratt coefficients of absolute risk aversion. Then the following conditions are equivalent:*

(a) $\alpha(x) \geq \tilde{\alpha}(x)$ *for all* $x \in S$.

(b) $u = F \circ \tilde{u}$ *for a strictly increasing concave function* F.

(c) *The respective risk premiums ϱ and $\tilde{\varrho}$ associated with u and $\tilde{u}$ satisfy $\varrho(\mu) \geq \tilde{\varrho}(\mu)$ for all $\mu \in \mathcal{M}$.*

Proof. (a)⇒(b): Since $\tilde{u}$ is strictly increasing, we may define its inverse function, w. Then $F(t) := u(w(t))$ is strictly increasing, twice differentiable, and satisfies $u = F \circ \tilde{u}$. For showing that F is concave, we calculate the first two derivatives of w:

$$w' = \frac{1}{\tilde{u}'(w)}, \quad w'' = \tilde{\alpha}(w) \cdot \frac{1}{\tilde{u}'(w)^2}.$$

Now we can calculate the first two derivatives of F:

$$F' = u'(w) \cdot w' = \frac{u'(w)}{\widetilde{u}'(w)} > 0$$

and

$$\begin{aligned} F'' &= u''(w)(w')^2 + u'(w)w'' \\ &= \frac{u'(w)}{\widetilde{u}'(w)^2}\left[\,\widetilde{\alpha}(w) - \alpha(w)\,\right] \\ &\le 0. \end{aligned} \tag{2.14}$$

This proves that F is concave.

(b)⇒(c): Jensen's inequality implies that the respective certainty equivalents $c(\mu)$ and $\widetilde{c}(\mu)$ satisfy

$$\begin{aligned} u(c(\mu)) &= \int u\,d\mu = \int F \circ \widetilde{u}\,d\mu \\ &\le F\Big(\int \widetilde{u}\,d\mu\Big) = F(\,\widetilde{u}(\,\widetilde{c}(\mu))) = u(\,\widetilde{c}(\mu)). \end{aligned} \tag{2.15}$$

Hence, $\varrho(\mu) = m(\mu) - c(\mu) \ge m(\mu) - \widetilde{c}(\mu) = \widetilde{\varrho}(\mu)$.

(c)⇒(a): If condition (a) is false, there exists an open interval $O \subseteq S$ such that $\widetilde{\alpha}(x) > \alpha(x)$ for all $x \in O$. Let $\widetilde{O} := \widetilde{u}(O)$, and denote again by w the inverse of $\widetilde{u}$. Then the function $F(t) = u(w(t))$ will be strictly convex in the open interval $\widetilde{O}$ by (2.14). Thus, if μ is a measure with support in O, the inequality in (2.15) is reversed and is even strict—unless μ is concentrated at a single point. It follows that $\varrho(\mu) < \widetilde{\varrho}(\mu)$, which contradicts condition (c). □

As an application of the preceding proposition, we will now investigate the structure of those continuous and strictly increasing functions u on $\mathbb{R}$ whose associated certainty equivalents have the following *translation property*:

$$c(\mu_t) = c(\mu) + t \quad \text{for all } \mu \in \mathscr{M} \text{ and all } t \in \mathbb{R}, \tag{2.16}$$

where the translation μ_t of $\mu \in \mathscr{M}$ by $t \in \mathbb{R}$ is defined by

$$\int g(x)\,\mu_t(dx) = \int g(x+t)\,\mu(dx) \quad \text{for all bounded measurable } g.$$

Here, we also assume that $\mathscr{M}$ is closed under translation, that is, $\mu_t \in \mathscr{M}$ for all $\mu \in \mathscr{M}$ and $t \in \mathbb{R}$.

Lemma 2.45. *Suppose the certainty equivalent associated with a continuous and strictly increasing function $u : \mathbb{R} \to \mathbb{R}$ satisfies the translation property* (2.16). *Then u belongs to $C^\infty(\mathbb{R})$.*

Proof. Let λ denote the Lebesgue measure on $[0,1]$. Then

$$f(t) := u(c(\lambda) + t) = u(c(\lambda_t)) = \int u\, d\lambda_t = \int_t^{1+t} u(y)\, dy, \tag{2.17}$$

and this implies $f \in C^1(\mathbb{R})$ with

$$f'(t) = u(1+t) - u(t). \tag{2.18}$$

Thus, $u(x) = f(x - c(\lambda))$ is in $C^1(\mathbb{R})$, which implies that $f' \in C^1(\mathbb{R})$ by (2.18), hence $f \in C^2(\mathbb{R})$. Iterating the argument, we get $u \in C^\infty(\mathbb{R})$. □

The following proposition implies in particular that a utility function u that satisfies the translation property (2.16) is necessarily a CARA utility function of exponential type as in part (a) of Example 2.43.

Proposition 2.46. *Suppose the certainty equivalent associated with a continuous and strictly increasing function $u : \mathbb{R} \to \mathbb{R}$ satisfies the translation property* (2.16). *Then u has constant absolute risk aversion and is hence either affine or an exponential function. More precisely, there are constants $a \in \mathbb{R}$ and $b, \alpha > 0$ such that $u(x)$ equals one of the following three functions:*

$$\begin{cases} a - be^{-\alpha x}, \\ a + bx, \\ a + be^{\alpha x}. \end{cases}$$

Proof. For $t \in \mathbb{R}$, let $u_t(x) := u(x+t)$, and denote by $c_t(\mu)$ the corresponding certainty equivalent. For $\mu \in \mathcal{M}$, we have

$$u_t((c_t(\mu)) = \int u_t\, d\mu = \int u(x+t)\, \mu(dx) = \int u\, d\mu_t = u(c(\mu_t)) = u(c(\mu) + t) = u_t(c(\mu)).$$

It follows that $c_t(\mu) = c(\mu)$ for all $t \in \mathbb{R}$ and $\mu \in \mathcal{M}$. Therefore,

$$\varrho_t(\mu) := m(\mu) - c_t(\mu) = m(\mu) - c(\mu) = \varrho(\mu)$$

for all $t \in \mathbb{R}$. Since u is smooth by Lemma 2.45, we may apply Proposition 2.44 to conclude that the respective Arrow–Pratt coefficients $\alpha_t(x) = -u_t''(x)/u_t'(x)$ and $\alpha(x) = -u''(x)/u'(x)$ are equal for all t and x. But $\alpha_t(x) = \alpha(x+t)$, and so $\alpha(x)$ does not depend on x. If $\alpha > 0$, we see as in Example 2.43 that u is of the form $u(x) = a - be^{-\alpha x}$. If $\alpha = 0$, u must be linear. And if $\alpha < 0$, we must have $u(x) = a + be^{\alpha x}$. □

Now we focus on the case in which u is a utility function and preferences have the expected utility representation (2.10). In view of the underlying axioms, the paradigm of expected utility has a certain plausibility on a normative level, that is, as a guideline of rational behavior in the face of risk. But this guideline should be applied with care: If pushed too far, it may lead to unplausible conclusions. In the remaining part of this section, we discuss some of these issues. From now on, we assume that S is unbounded from above, so that $w+x \in S$ for any $x \in S$ and $w \geq 0$. So far, we have implicitly assumed that the preference relation $\succ$ on lotteries reflects the views of an economic agent in a given set of conditions, including a fixed level $w \geq 0$ of the agent's initial wealth. In particular, the utility function may vary as the level of wealth changes, and so it should really be indexed by w. Usually one assumes that u_w is obtained by simply shifting a fixed utility function u to the level w, that is, $u_w(x) := u(w+x)$. Thus, a lottery μ is declined at a given level of wealth w if and only if

$$\int u(w+x)\,\mu(dx) < u(w).$$

Let us now return to the situation of Proposition 2.39 where μ is the distribution of an integrable random variable X on $(\Omega, \mathscr{F}, P)$ that is bounded from below by some number a in the interior of S. We view X as the net payoff of some financial bet, and we assume that the bet is *favorable* in the sense that

$$m(\mu) = E[\,X\,] > 0.$$

Remark 2.47. Even though the favorable bet X might be declined at a given level w due to risk aversion, it follows from Proposition 2.39 that it would be optimal to accept the bet at some smaller scale, that is, there is some $\gamma^* > 0$ such that

$$E[\,u(w+\gamma^* X)\,] > u(w).$$

On the other hand, it follows from Proposition 2.49 below that the given bet X becomes acceptable at a sufficiently high level of wealth whenever the utility function is unbounded from above. ◇

Sometimes it is assumed that some favorable bet is declined at *every* level of wealth. The assumption that such a bet exists is not as innocent as it may look. In fact, it has rather drastic consequences. In particular, we are going to see that it rules out all utility functions in Example 2.43 except for the class of exponential utilities.

Example 2.48. For any exponential utility function $u(x) = 1 - e^{-\alpha x}$ with constant risk aversion $\alpha > 0$, the induced preference order on lotteries does not depend at all on the initial wealth w. To see this, note that

$$\int u(w+x)\,\mu(dx) < \int u(w+x)\,\nu(dx)$$

is equivalent to

$$\int e^{-\alpha x}\,\mu(dx) > \int e^{-\alpha x}\,\nu(dx). \qquad \diamond$$

Let us now show that the rejection of some favorable bet μ at every wealth level w leads to a not quite plausible conclusion: At high levels of wealth, the agent would reject a bet ν with huge potential gain even though the potential loss is just a negligible fraction of the initial wealth.

Proposition 2.49. *If the favorable bet μ is rejected at any level of wealth, then the utility function u is bounded from above, and there exists $A > 0$ such that the bet*

$$\nu := \frac{1}{2}(\delta_{-A} + \delta_{\infty})$$

is rejected at any level of wealth.

Proof. We have assumed that X is bounded from below, that is, μ is concentrated on $[a, \infty)$ for some $a < 0$, where a is in the interior of S. Moreover, we can choose $b > 0$ such that

$$\widetilde{\mu}(B) := \mu(B \cap [a, b]) + \delta_b(B) \cdot \mu((b, \infty))$$

is still favorable. Since u is increasing, we have

$$\int u(w + x)\,\widetilde{\mu}(dx) \le \int u(w + x)\,\mu(dx) < u(w)$$

for any $w \ge 0$, that is, also the lottery $\widetilde{\mu}$ is rejected at any level of wealth. It follows that

$$\int_{[0,b]} [\,u(w + x) - u(w)\,]\,\widetilde{\mu}(dx) < \int_{[a,0)} [\,u(w) - u(w + x)\,]\,\widetilde{\mu}(dx).$$

Let us assume for simplicity that u is differentiable; the general case requires only minor modifications. Then the previous inequality implies

$$u'(w + b)\,m^{+}(\widetilde{\mu}) < u'(w + a)\,m^{-}(\widetilde{\mu}),$$

where

$$m^{+}(\widetilde{\mu}) := \int_{[0,b]} x\,\widetilde{\mu}(dx) > \int_{[a,0]} (-x)\,\widetilde{\mu}(dx) =: m^{-}(\widetilde{\mu}),$$

due to the fact that $\widetilde{\mu}$ is favorable. Thus,

$$\frac{u'(w + b)}{u'(w - |a|)} < \frac{m^{-}(\widetilde{\mu})}{m^{+}(\widetilde{\mu})} =: \gamma < 1$$

for any w, hence

$$u'(x + n(|a| + b)) < \gamma^n u'(x)$$

for any x in the interior of S. This exponential decay of the derivative implies $u(\infty) := \lim_{x\uparrow\infty} u(x) < \infty$. More precisely, if $A := n(|a| + b)$ for some n, then

$$\begin{aligned} u(\infty) - u(x) &= \sum_{k=0}^{\infty} \int_{x+kA}^{x+(k+1)A} u'(y)\, dy \\ &= \sum_{k=0}^{\infty} \int_{x-A}^{x} u'(z + (k+1)A)\, dz \\ &< \sum_{k=0}^{\infty} \gamma^{(k+1)n} \int_{x-A}^{x} u'(z)\, dz \\ &= \frac{\gamma^n}{1-\gamma^n}(u(x) - u(x-A)). \end{aligned}$$

Take n such that $\gamma^n \leq 1/2$. Then we obtain

$$u(\infty) - u(x) < u(x) - u(x - A),$$

that is,

$$\frac{1}{2}(u(\infty) + u(x - A)) < u(x)$$

for all x such that $x - A \in S$. □

Example 2.50. For an exponential utility function $u(x) = 1 - e^{-\alpha x}$, the bet ν defined in the preceding lemma is rejected at any level of wealth as soon as $A > \frac{1}{\alpha}\log 2$. ◇

Suppose now that the lottery $\mu \in \mathcal{M}$ is played not only once but n times in a row. For instance, one can think of an insurance company selling identical policies to a large number of individual customers. More precisely, let $(\Omega, \mathcal{F}, P)$ be a probability space supporting a sequence $X_1, X_2, \ldots$ of independent random variables with common distribution μ. The value of X_i will be interpreted as the outcome of the ith drawing of the lottery μ. The accumulated payoff of n successive independent repetitions of the financial bet X_1 is given by

$$Z_n := \sum_{i=1}^{n} X_i,$$

and we assume that this accumulated payoff takes values in S; this is the case if, for example, $S = [0, \infty)$.

Remark 2.51. It may happen that an agent refuses the single favorable bet X at any level of wealth but feels tempted by a sufficiently large series $X_1, \ldots, X_n$ of independent repetitions of the same bet. It is true that, by the weak law of large numbers, the probability

$$P[\,Z_n < 0\,] = P\left[\,\frac{1}{n}\sum_{i=1}^{n} X_i < m(\mu) - \varepsilon\,\right]$$

(for $\varepsilon := m(\mu)$) of incurring a cumulative loss at the end of the series converges to 0 as $n \uparrow \infty$. Nevertheless, the decision of accepting n repetitions is not consistent with the decision to reject the single bet at any wealth level w. In fact, for $W_k := w + Z_k$ we obtain

$$\begin{aligned} E[\,u(W_n)\,] &= E[\,E[\,u(W_{n-1} + X_n) \mid X_1, \ldots, X_{n-1}\,]\,] \\ &= E\left[\,\int u(W_{n-1} + x)\,\mu(dx)\,\right] \\ &< E[\,u(W_{n-1})\,] < \cdots < u(w), \end{aligned}$$

that is, the bet described by Z_n should be rejected as well. ◇

Let us denote by μ_n the distribution of the accumulated payoff Z_n. The lottery μ_n has the mean $m(\mu_n) = n \cdot m(\mu)$, the certainty equivalent $c(\mu_n)$, and the associated risk premium $\varrho(\mu_n) = n \cdot m(\mu) - c(\mu_n)$. We are interested in the asymptotic behavior of these quantities for large n. Kolmogorov's law of large numbers states that the average outcome $\frac{1}{n}Z_n$ converges P-a. s. to the constant $m(\mu)$. Therefore, one might guess that a similar averaging effect occurs on the level of the relative certainty equivalents

$$c_n := \frac{c(\mu_n)}{n} \tag{2.19}$$

and of the relative risk premiums

$$\varrho_n := \frac{\varrho(\mu_n)}{n} = m(\mu) - c_n.$$

Does c_n converge to $m(\mu)$, and is there a successive reduction of the relative risk premiums ϱ_n as n grows to infinity? Applying our heuristic (2.13) to the present situation yields

$$\varrho_n \approx \frac{1}{2n}\,\alpha(m(\mu_n))\,\mathrm{var}(\mu_n) = \frac{1}{2}\,\alpha(n \cdot m(\mu))\,\mathrm{var}(\mu).$$

Thus, one should expect that ϱ_n tends to zero only if the Arrow–Pratt coefficient $\alpha(x)$ becomes arbitrarily small as x becomes large, that is, if the utility function is *decreasingly risk averse*. This guess is confirmed by the following two examples.

Example 2.52. Suppose that $u(x) = 1 - e^{-\alpha x}$ is a CARA utility function with constant risk aversion $\alpha > 0$ and assume that μ is such that $\int e^{-\alpha x}\,\mu(dx) < \infty$. Then, with the notation introduced above,

$$\int e^{-\alpha x}\,\mu_n(dx) = E\left[\,\prod_{i=1}^{n} e^{-\alpha X_i}\,\right] = \left(\int e^{-\alpha x}\,\mu(dx)\right)^n.$$

Hence, the certainty equivalent of μ_n is given by

$$c(\mu_n) = -\frac{n}{\alpha}\log\int e^{-\alpha x}\,\mu(dx) = n\cdot c(\mu).$$

It follows that c_n and ϱ_n are independent of n. In particular, the relative risk premiums are not reduced if the lottery is drawn more than once. ◇

The second example displays a different behavior. It shows that for HARA utility functions the relative risk premiums will indeed decrease to 0. In particular, the lottery μ_n will become attractive for large enough n as soon as the price of the single lottery μ is less than $m(\mu)$.

Example 2.53. Suppose that μ is a nondegenerate lottery concentrated on $(0,\infty)$, and that u is a HARA utility function of index $\gamma \in [0,1)$. If $\gamma > 0$, then $u(x) = \frac{1}{\gamma}x^\gamma$ and $c(\mu_n) = E[\,(Z_n)^\gamma\,]^{1/\gamma}$, hence

$$c_n = \frac{c(\mu_n)}{n} = E\Big[\Big(\frac{1}{n}Z_n\Big)^\gamma\Big]^{1/\gamma} < m(\mu).$$

If $\gamma = 0$, then $u(x) = \log x$, and the relative certainty equivalent satisfies

$$\log c_n = \log c(\mu_n) - \log n = E\Big[\log\Big(\frac{1}{n}Z_n\Big)\Big].$$

Thus, we have

$$u(c_n) = E\Big[u\Big(\frac{1}{n}Z_n\Big)\Big]$$

for any $\gamma \in [0,1)$. By symmetry,

$$\frac{1}{n+1}Z_{n+1} = E[\,X_k \mid Z_{n+1}\,] \quad \text{for } k = 1,\dots,n+1;$$

see, for example, part II of § 20 in [21]. It follows that

$$\frac{1}{n+1}Z_{n+1} = E\Big[\,\frac{1}{n}Z_n \,\Big|\, Z_{n+1}\Big]. \tag{2.20}$$

Since u is strictly concave and since μ is nondegenerate, we get

$$\begin{aligned} u(c_{n+1}) &= E\Big[\,u\Big(E\Big[\,\frac{1}{n}Z_n \,\Big|\, Z_{n+1}\Big]\Big)\Big] \\ &> E\Big[\,E\Big[\,u\Big(\frac{1}{n}Z_n\Big) \,\Big|\, Z_{n+1}\Big]\Big] \\ &= u(c_n), \end{aligned}$$

that is, the relative certainty equivalents are *strictly increasing* and the relative risk premiums ϱ_n are *strictly decreasing*. By Kolmogorov's law of large numbers,

$$\frac{1}{n}Z_n \longrightarrow m(\mu) \quad P\text{-a. s.} \tag{2.21}$$

Thus, by Fatou's lemma (we assume for simplicity that μ is concentrated on $[\varepsilon, \infty)$ for some $\varepsilon > 0$ if $\gamma = 0$),

$$\liminf_{n\uparrow\infty} u(c_n) \geq E\Big[\liminf_{n\uparrow\infty} u\Big(\frac{1}{n}Z_n\Big)\Big] = u(m(\mu)),$$

hence

$$\lim_{n\uparrow\infty} c_n = m(\mu) \quad \text{and} \quad \lim_{n\uparrow\infty} \varrho_n = 0.$$

Suppose that the price of μ is given by $\pi \in (c(\mu), m(\mu))$. At initial wealth $w = 0$, the agent would decline a single bet. But, in contrast to the situation in Remark 2.51, a series of n repetitions of the same bet would now become attractive for large enough n, since $c(\mu_n) = nc_n > n\pi$ for

$$n \geq n_0 := \min\{k \in \mathbb{N} \mid c_k > \pi\} < \infty.$$

◇

Remark 2.54. The identity (2.20) can also be written as

$$\frac{1}{n+1}Z_{n+1} = E\Big[\frac{1}{n}Z_n \,\Big|\, \mathscr{A}_{n+1}\Big] = E[\,X_1 \mid \mathscr{A}_{n+1}\,]$$

where $\mathscr{A}_{n+1} = \sigma(Z_{n+1}, Z_{n+2}, \dots)$. This means that the stochastic process $\frac{1}{n}Z_n$, $n = 1, 2\dots$, is a *backwards martingale*, sometimes also called *reversed martingale*. In particular, Kolmogorov's law of large numbers (2.21) can be regarded as a special case of the convergence theorem for backwards martingales; see part II of § 20 in [21]. ◇

Exercise 2.3.4. Investigate the asymptotics of c_n in (2.19) for a HARA utility function $u(x) = \frac{1}{\gamma}x^\gamma$ with $\gamma < 0$ and a suitable lottery μ concentrated on $(0, \infty)$. ◇

2.4 Stochastic orders

So far, we have considered preference relations on distributions defined in terms of a fixed utility function u. In this section, we focus on the question whether one distribution is preferred over another, regardless of the choice of a particular utility function.

For simplicity, we take $S = \mathbb{R}$ as the set of possible payoffs. Let $\mathscr{M}$ be the set of all $\mu \in \mathscr{M}_1(\mathbb{R})$ with well-defined and finite expectation

$$m(\mu) = \int x\,\mu(dx).$$

Recall from Definition 2.35 that a *utility function* on $\mathbb{R}$ is a strictly concave and strictly increasing function $u : \mathbb{R} \to \mathbb{R}$. Since each concave function u is dominated by an affine function, the existence of $m(\mu)$ implies the existence of the integral $\int u\,d\mu$ as an extended real number in $[-\infty, \infty)$.

Definition 2.55. Let ν and μ be lotteries in $\mathcal{M}$. We say that the lottery μ dominates ν in the *increasing concave order* and we write

$$\mu \succcurlyeq_{\text{icv}} \nu$$

if

$$\int u\,d\mu \geq \int u\,d\nu \quad \text{for all utility functions } u. \qquad \triangle$$

Thus, $\mu \succcurlyeq_{\text{icv}} \nu$ holds if and only if every risk-averse agent will prefer μ over ν, regardless of which utility function the agent is actually using. In this sense, $\mu \succcurlyeq_{\text{icv}} \nu$ expresses a uniform preference for μ over ν. Sometimes, $\succcurlyeq_{\text{icv}}$ is also called *second-order stochastic dominance* or *stochastic dominance of the second kind*; the notion of *first-order stochastic dominance* will be introduced in Definition 2.67.

Remark 2.56. The binary relation $\succcurlyeq_{\text{icv}}$ is a *partial order* on $\mathcal{M}$, that is, $\succcurlyeq_{\text{icv}}$ satisfies the following three properties:
- *Reflexivity:* $\mu \succcurlyeq_{\text{icv}} \mu$ for all $\mu \in \mathcal{M}$.
- *Transitivity:* $\mu \succcurlyeq_{\text{icv}} \nu$ and $\nu \succcurlyeq_{\text{icv}} \lambda$ imply $\mu \succcurlyeq_{\text{icv}} \lambda$.
- *Antisymmetry:* $\mu \succcurlyeq_{\text{icv}} \nu$ and $\nu \succcurlyeq_{\text{icv}} \mu$ imply $\mu = \nu$.

The first two properties are obvious, the third is derived in part (b) of Remark 2.58. Moreover, $\succcurlyeq_{\text{icv}}$ is *monotone* and *risk-averse* in the sense that

$$\delta_y \succcurlyeq_{\text{icv}} \delta_x \text{ for } y \geq x, \quad \text{and} \quad \delta_{m(\mu)} \succcurlyeq_{\text{icv}} \mu \text{ for all } \mu \in \mathcal{M}.$$

Note, however, that $\succcurlyeq_{\text{icv}}$ is *not* a weak preference relation in the sense of Definition 2.2, since it is not complete in the sense of Remark 2.3. ◇

In the following theorem, we will give a number of equivalent formulations of the statement $\mu \succcurlyeq_{\text{icv}} \nu$. One of them needs the notion of a *stochastic kernel* on $\mathbb{R}$. This is a mapping

$$Q : \mathbb{R} \longrightarrow \mathcal{M}_1(\mathbb{R})$$

such that $x \mapsto Q(x, A)$ is measurable for each fixed Borel set $A \subseteq \mathbb{R}$. A stochastic kernel is sometimes also called a *Markov kernel* or a *probability kernel*. See Appendix D for the notion of a quantile function, which will be used in condition (e).

Theorem 2.57. *For any pair $\mu, \nu \in \mathcal{M}$, the following conditions are equivalent:*

(a) $\mu \succcurlyeq_{\text{icv}} \nu$.

(b) $\int f\,d\mu \geq \int f\,d\nu$ *for all increasing concave functions* f.

(c) *For all* $c \in \mathbb{R}$,

$$\int (c-x)^+\,\mu(dx) \leq \int (c-x)^+\,\nu(dx).$$

(d) *If* F_μ *and* F_ν *denote the cumulative distribution functions of* μ *and* ν, *then*

$$\int_{-\infty}^{c} F_\mu(x)\,dx \leq \int_{-\infty}^{c} F_\nu(x)\,dx \quad \textit{for all } c \in \mathbb{R}.$$

(e) *If* q_μ *and* q_ν *are quantile functions for* μ *and* ν, *then*

$$\int_0^t q_\mu(s)\,ds \geq \int_0^t q_\nu(s)\,ds \quad \textit{for } 0 < t \leq 1.$$

(f) *There exists a probability space* $(\Omega, \mathcal{F}, P)$ *with random variables* X_μ *and* X_ν *having respective distributions* μ *and* ν *such that*

$$E[\,X_\nu \mid X_\mu\,] \leq X_\mu \quad P\text{-a. s.}$$

(g) *There exists a stochastic kernel* $Q(x, dy)$ *on* $\mathbb{R}$ *such that* $Q(x,\cdot) \in \mathcal{M}$ *and* $m(Q(x,\cdot)) \leq x$ *for all* x *and such that* $\nu = \mu Q$, *where* μQ *denotes the measure*

$$\mu Q(A) := \int Q(x,A)\,\mu(dx) \quad \textit{for Borel sets } A \subseteq \mathbb{R}.$$

Below we will show the following implications between the conditions of the theorem:

$$\text{(e)} \iff \text{(d)} \iff \text{(c)} \iff \text{(b)} \iff \text{(a)} \Longleftarrow \text{(g)} \Longleftarrow \text{(f)}. \tag{2.22}$$

The difficult part is the proof that (b) implies (f). It will be deferred to Section 2.6, where we will prove a multidimensional variant of this result; cf. Theorem 2.94.

Proof of (2.22). (e)⇔(d): Lemma D.11 states that for every $\lambda \in \mathcal{M}$ and $t \in (0,1)$,

$$\int_0^t q_\lambda(s)\,ds = \sup_{c\in\mathbb{R}}\left(ct - \int_{-\infty}^{c} F_\lambda(x)\,dx \right).$$

Hence, (d) implies (e). Moreover, Proposition A.9(b) applied to the preceding identity gives that for $c \in \mathbb{R}$,

$$\int_{-\infty}^{c} F_\lambda(x)\,dx = \sup_{t>0}\left(ct - \int_{0}^{t} q_\lambda(s)\,ds \right).$$

Thus, (e) also implies (d).

(d)⇔(c): By Fubini's theorem,

$$\begin{aligned}\int_{-\infty}^{c} F_\mu(y)\,dy &= \int_{-\infty}^{c} \int_{(-\infty,y]} \mu(dz)\,dy \\ &= \int\int \mathbb{1}_{\{z\le y\le c\}}\,dy\,\mu(dz) \\ &= \int (c-z)^+\,\mu(dz).\end{aligned}$$

(c)⇔(b): Condition (b) implies (c) because $f(x) := -(c-x)^+$ is concave and increasing. In order to prove the converse assertion, we take an increasing concave function f and let $h := -f$. Then h is convex and decreasing, and its increasing right-hand derivative $h' := h'_+$ can be regarded as a "cumulative distribution function" of a nonnegative Radon measure γ on $\mathbb{R}$,

$$h'(b) = h'(a) + \gamma((a,b]) \quad \text{for } a \le b;$$

see Appendix A.1 and Remark D.8. By Exercise 1.3.3(a),

$$h(x) = h(b) - h'(b)\,(b-x) + \int_{(-\infty,b]} (z-x)^+\,\gamma(dz) \quad \text{for } x \le b.$$

Using that $\int_{(-\infty,b]}(b-x)\,\mu(dx) = \int (b-x)^+\,\mu(dx)$, Fubini's theorem, $h'(b)\le 0$, and condition (c), we obtain that

$$\begin{aligned}-\int_{(-\infty,b]} f\,d\mu &= \int_{(-\infty,b]} h\,d\mu \\ &= h(b)\mu((-\infty,b]) - h'(b)\int (b-x)^+\,\mu(dx) + \int_{(-\infty,b]}\int (z-x)^+\,\mu(dx)\,\gamma(dz) \\ &\le h(b)\mu((-\infty,b]) - h'(b)\int (b-x)^+\,\nu(dx) + \int_{(-\infty,b]}\int (z-x)^+\,\nu(dx)\,\gamma(dz) \\ &= \int_{(-\infty,b]} h\,d\nu + h(b)[\mu((-\infty,b]) - \nu((-\infty,b])] \\ &= -\int_{(-\infty,b]} f\,d\nu + h(b)[\nu((b,\infty)) - \mu((b,\infty))].\end{aligned}$$

Taking $b \uparrow \infty$ yields $\int f\,d\mu \geq \int f\,d\nu$. Indeed, the convex decreasing function h decays at most linearly, and the existence of first moments for μ and ν implies that $b\mu((b,\infty)) \to 0$ and $b\nu((b,\infty)) \to 0$ for $b \uparrow \infty$.

(a)⇔(b): That (b) implies (a) is obvious. For the proof of the converse implication, choose any utility function u_0 for which both $\int u_0\,d\mu$ and $\int u_0\,d\nu$ are finite. For instance, one can take

$$u_0(x) := \begin{cases} x - e^{x/2} + 1 & \text{if } x \leq 0, \\ \sqrt{x+1} - 1 & \text{if } x \geq 0. \end{cases}$$

Then, for f concave and increasing and for $\alpha \in [0,1)$,

$$u_\alpha(x) := \alpha f(x) + (1-\alpha)u_0(x)$$

is a utility function. Hence,

$$\int f\,d\mu = \lim_{\alpha \uparrow 1} \int u_\alpha\,d\mu \geq \lim_{\alpha \uparrow 1} \int u_\alpha\,d\nu = \int f\,d\nu.$$

(f)⇒(g): By considering the joint distribution of X_μ and X_ν, we may reduce our setting to the situation in which $\Omega = \mathbb{R}^2$ and where X_μ and X_ν are the respective projections on the first and second coordinates, that is, for $\omega = (x,y) \in \Omega = \mathbb{R}^2$ we have $X_\mu(\omega) = x$ and $X_\nu(\omega) = y$. Let $Q(x,dy)$ be a *regular conditional distribution* of X_ν given X_μ, that is, a stochastic kernel on $\mathbb{R}$ such that

$$P[X_\nu \in A \mid X_\mu](\omega) = Q(X_\mu(\omega), A) \tag{2.23}$$

for all Borel sets $A \subseteq \mathbb{R}$ and for P-a. e. $\omega \in \Omega$ (see, e. g., Theorem 44.3 of [21] for an existence proof). Equation (2.23) implies that for any Borel sets $A \subseteq \mathbb{R}$,

$$\nu(A) = P[X_\nu \in A] = E[P[X_\nu \in A \mid X_\mu]] = E[Q(X_\mu, A)] = \int Q(x,A)\,\mu(dx) = \mu Q(A), \tag{2.24}$$

so that $\nu = \mu Q$. Next, (2.23) clearly extends as follows from indicator functions to all measurable and suitably integrable functions $f : \mathbb{R} \to \mathbb{R}$,

$$E[f(X_\nu) \mid X_\mu](\omega) = \int f(y)\,Q(X_\mu(\omega), dy) \quad \text{for } P\text{-a. e. } \omega \in \Omega.$$

Therefore, condition (f) implies that

$$X_\mu(\omega) \geq E[X_\nu \mid X_\mu](\omega) = \int y\,Q(X_\mu(\omega), dy) \quad \text{for } P\text{-a. e. } \omega \in \Omega.$$

Thus, Q satisfies

$$\int y\, Q(x, dy) \leq x \quad \text{for } \mu\text{-a. e. } x.$$

By modifying Q on a μ-null set (e. g., by putting $Q(x,\cdot) := \delta_x$ there), this inequality can be achieved for all $x \in \mathbb{R}$.

(g)⇒(a): Let u be a utility function. Jensen's inequality applied to the measure $Q(x, dy)$ implies

$$\int u(y)\, Q(x, dy) \leq u(m(Q(x,\cdot))) \leq u(x).$$

Hence,

$$\int u\, d\nu = \int \int u(y)\, Q(x, dy)\, \mu(dx) \leq \int u\, d\mu,$$

completing the proof of the set of implications (2.22). □

Remark 2.58. Let us note some consequences of the preceding theorem.

(a) Taking in condition (b) the increasing concave function $f(x) = x$ yields

$$m(\mu) \geq m(\nu) \quad \text{if } \mu \succcurlyeq_{\text{icv}} \nu,$$

that is, the expectation $m(\cdot)$ is increasing with respect to $\succcurlyeq_{\text{icv}}$.

(b) Theorem 2.57 implies that the following three conditions are equivalent for $\nu, \mu \in \mathcal{M}$:

(i) $\mu \succcurlyeq_{\text{icv}} \nu$ and $\nu \succcurlyeq_{\text{icv}} \mu$.

(ii) For all $c \in \mathbb{R}$,

$$\int (c - x)^+\, \mu(dx) = \int (c - x)^+\, \nu(dx).$$

(iii) For all $c \in \mathbb{R}$,

$$\int_{-\infty}^{c} F_\mu(x)\, dx = \int_{-\infty}^{c} F_\nu(x)\, dx. \tag{2.25}$$

Differentiating both sides of (2.25) with respect to c gives the identity $F_\mu = F_\nu$ and in turn $\mu = \nu$. It follows that $\succcurlyeq_{\text{icv}}$ is antisymmetric and that a measure $\mu \in \mathcal{M}$ is uniquely determined by the integrals $\int (c - x)^+\, \mu(dx)$ for all $c \in \mathbb{R}$. ◇

The following proposition characterizes the partial order $\succcurlyeq_{\text{icv}}$ considered on the set of all normal distributions $N(m, \sigma^2)$. Recall that the standard normal distribution $N(0, 1)$ is defined by its density function

$$\varphi(x) = \frac{1}{\sqrt{2\pi}} e^{-x^2/2}, \quad x \in \mathbb{R}.$$

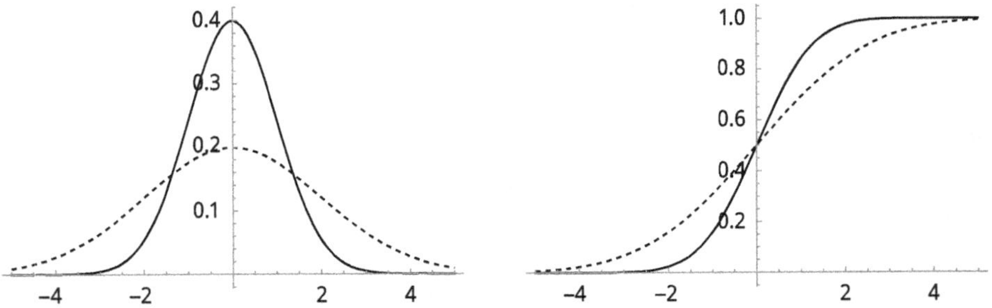

Figure 2.1: Densities and cumulative distribution functions of $N(0,1)$ (solid) and $N(0,4)$ (dashed).

The corresponding cumulative distribution function is usually denoted

$$\Phi(x) = \int_{-\infty}^{x} \varphi(y)\,dy, \quad x \in \mathbb{R}. \tag{2.26}$$

More generally, the normal distribution $N(m,\sigma^2)$ with mean $m \in \mathbb{R}$ and variance $\sigma^2 > 0$ is given by the density function

$$\frac{1}{\sqrt{2\pi\sigma^2}} \cdot \exp\left(-\frac{(x-m)^2}{2\sigma^2}\right), \quad x \in \mathbb{R},$$

and it has the cumulative distribution function $x \mapsto \Phi((x-m)/\sigma)$; see Figure 2.1 for an illustration.

Proposition 2.59. *For two normal distributions, we have $N(m,\sigma^2) \succcurlyeq_{\mathrm{icv}} N(\widetilde{m},\widetilde{\sigma}^2)$ if and only if both $m \geq \widetilde{m}$ and $\sigma^2 \leq \widetilde{\sigma}^2$ hold.*

Proof. In order to prove necessity, note that $N(m,\sigma^2) \succcurlyeq_{\mathrm{icv}} N(\widetilde{m},\widetilde{\sigma}^2)$ implies that

$$e^{-\alpha m+\alpha^2\sigma^2/2} = \int e^{-\alpha x}\,N(m,\sigma^2)(dx) \leq \int e^{-\alpha x}\,N(\widetilde{m},\widetilde{\sigma}^2)(dx) = e^{-\alpha\widetilde{m}+\alpha^2\widetilde{\sigma}^2/2}.$$

Hence, for $\alpha > 0$,

$$m - \frac{1}{2}\alpha\sigma^2 \geq \widetilde{m} - \frac{1}{2}\alpha\widetilde{\sigma}^2,$$

which gives $m \geq \widetilde{m}$ by letting $\alpha \downarrow 0$ and $\sigma^2 \leq \widetilde{\sigma}^2$ for $\alpha \uparrow \infty$.

We show sufficiency first in the case $m = \widetilde{m} = 0$. The cumulative distribution function of $N(0,\sigma^2)$ is given by $\Phi(x/\sigma)$. Since $\varphi'(x) = -x\varphi(x)$,

$$\frac{d}{d\sigma}\int_{-\infty}^{c} \Phi\left(\frac{x}{\sigma}\right)dx = \int_{-\infty}^{c} \varphi\left(\frac{x}{\sigma}\right)\cdot\frac{-x}{\sigma^2}\,dx = \varphi\left(\frac{c}{\sigma}\right) > 0.$$

Note that interchanging differentiation and integration is justified by dominated convergence. Thus, we have shown that $\sigma \mapsto \int_{-\infty}^{c} \Phi(x/\sigma)\,dx$ is strictly increasing for all c, and $N(0,\sigma^2) \succcurlyeq_{\text{icv}} N(0,\tilde{\sigma}^2)$ follows from part (d) of Theorem 2.57.

Now we turn to the case of arbitrary expectations m and $\widetilde{m}$. Let u be a utility function. Then

$$\int u\,dN(m,\sigma^2) = \int u(m+x)\,N(0,\sigma^2)(dx) \geq \int u(\widetilde{m}+x)\,N(0,\sigma^2)(dx),$$

because $m \geq \widetilde{m}$. Since $x \mapsto u(\widetilde{m}+x)$ is again a utility function, we obtain from the preceding step of the proof that

$$\int u(\widetilde{m}+x)\,N(0,\sigma^2)(dx) \geq \int u(\widetilde{m}+x)\,N(0,\tilde{\sigma}^2)(dx) = \int u\,dN(\widetilde{m},\tilde{\sigma}^2),$$

and $N(m,\sigma^2) \succcurlyeq_{\text{icv}} N(\widetilde{m},\tilde{\sigma}^2)$ follows. □

Remark 2.60. Let us indicate an alternative proof for the sufficiency part of Proposition 2.59 that uses condition (g) instead of (d) in Theorem 2.57. To this end, we define a stochastic kernel by $Q(x,\cdot) := N(x+\widetilde{m}-m,\hat{\sigma}^2)$, where $\hat{\sigma}^2 := \tilde{\sigma}^2 - \sigma^2 > 0$. Then $m(Q(x,\cdot)) = x + \widetilde{m} - m \leq x$ and

$$N(m,\sigma^2)\,Q = N(m,\sigma^2) * N(\widetilde{m}-m,\hat{\sigma}^2) = N(m+\widetilde{m}-m,\sigma^2+\hat{\sigma}^2) = N(\widetilde{m},\tilde{\sigma}^2),$$

where $*$ denotes convolution. Hence, $N(m,\sigma^2) \succcurlyeq_{\text{icv}} N(\widetilde{m},\tilde{\sigma}^2)$ follows. ◇

Exercise 2.4.1. Let $\nu, \mu \in \mathcal{M}$ have the cumulative distribution functions F_ν and F_μ.

(a) Suppose that $m(\mu) \geq m(\nu)$. Show that the following property ensures that $\mu \succcurlyeq_{\text{icv}} \nu$: there exists $x_0 \in \mathbb{R}$ such that

$$F_\mu(x) \leq F_\nu(x) \text{ for } x \leq x_0 \quad \text{and} \quad F_\mu(y) \geq F_\nu(y) \text{ for } y \geq x_0. \tag{2.27}$$

The property (2.27) states that the two cumulative distribution functions F_μ may cross each other only once. It is therefore sometimes called the *single-crossing property*. See Figure 2.1 for a case in which it applies.

(b) Construct an example showing that $\mu \succcurlyeq_{\text{icv}} \nu$ does not imply the single-crossing property. ◇

The following corollary investigates the relation $\mu \succcurlyeq_{\text{icv}} \nu$ for lotteries with the same expectation. A multidimensional version of this result will be given in Corollary 2.95 below.

Corollary 2.61. *For all $\mu, \nu \in \mathcal{M}$ the following conditions are equivalent:*

(a) $\mu \succcurlyeq_{\text{icv}} \nu$ *and* $m(\mu) = m(\nu)$.

(b) $\int f\,d\mu \geq \int f\,d\nu$ *for all (not necessarily increasing) concave functions* f.

(c) $m(\mu) \geq m(\nu)$ *and* $\int (x-c)^+\,\mu(dx) \leq \int (x-c)^+\,\nu(dx)$ *for all* $c \in \mathbb{R}$.

(d) *There exists a probability space $(\Omega, \mathscr{F}, P)$ with random variables X_μ and X_ν having respective distributions μ and ν such that*

$$E[X_\nu \mid X_\mu] = X_\mu \quad P\text{-a. s.}$$

(e) *There exists a "mean-preserving spread" Q, that is, a stochastic kernel on $\mathbb{R}$ such that $m(Q(x,\cdot)) = x$ for all $x \in S$, such that $\nu = \mu Q$.*

Proof. (a)⇒(e): Condition (g) of Theorem 2.57 yields a stochastic kernel Q such that $\nu = \mu Q$ and $m(Q(x,\cdot)) \le x$. Due to the assumption $m(\mu) = m(\nu)$, Q must satisfy $m(Q(x,\cdot)) = x$ at least for μ-a. e. x. By modifying Q on the μ-null set where $m(Q(x,\cdot)) < x$ (e. g., by putting $Q(x,\cdot) := \delta_x$ there), we obtain a kernel as needed for condition (e).

(e)⇒(b): Since

$$\int f(y)\, Q(x,dy) \le f(m(Q(x,\cdot))) = f(x)$$

by Jensen's inequality, we obtain

$$\int f\, d\nu = \int \int f(y)\, Q(x,dy)\, \mu(dx) \le \int f\, d\mu.$$

(b)⇒(c): Just take the concave functions $f(x) = -(x-c)^+$, and $f(x) = x$.

(c)⇒(a): Note that

$$\int (x-c)^+\, \mu(dx) = \int_{(c,\infty)} x\, \mu(dx) - c + c\,\mu((-\infty,c]).$$

The finiteness of $m(\mu)$ implies that $c\,\mu((-\infty,c]) \to 0$ as $c \downarrow -\infty$. Hence, we deduce from the second condition in (c) that $m(\mu) \le m(\nu)$, that is, the two expectations are in fact identical. Now we can apply the following identity, which is in fact the put-call parity from Remark 1.22:

$$\int (c-x)^+\, \mu(dx) = c - m(\mu) + \int (x-c)^+\, \mu(dx).$$

It yields that our condition (c) implies the third condition of Theorem 2.57, and thus $\mu \succcurlyeq_{\text{icv}} \nu$.

(d)⇔(a): Condition (d) implies both $m(\mu) = m(\nu)$ and condition (f) of Theorem 2.57, and this implies our condition (a). Conversely, assume that (a) holds. Then Theorem 2.57 provides random variables X_μ and X_ν having the respective distributions μ and ν such that $E[X_\nu \mid X_\mu] \le X_\mu$. Since X_μ and X_ν have the same mean, this inequality must in fact be an almost-sure equality, and we obtain condition (d). □

Let us denote by

$$\mathrm{var}(\mu) := \int (x - m(\mu))^2 \mu(dx) = \int x^2 \mu(dx) - m(\mu)^2 \in [0, \infty]$$

the *variance* of a lottery $\mu \in \mathcal{M}$.

Exercise 2.4.2. Let μ and ν be two lotteries in $\mathcal{M}$ such that $m(\mu) = m(\nu)$ and $\mu \succcurlyeq_{\mathrm{icv}} \nu$. Show that $\mathrm{var}(\mu) \le \mathrm{var}(\nu)$. ◇

In the financial context, comparisons of portfolios with known payoff distributions often use a *mean-variance* approach based on the relation

$$\mu \succcurlyeq_{\mathrm{mv}} \nu \quad :\Longleftrightarrow \quad m(\mu) \ge m(\nu) \text{ and } \mathrm{var}(\mu) \le \mathrm{var}(\nu).$$

For normal distributions μ and ν, we have seen that the relation $\mu \succcurlyeq_{\mathrm{mv}} \nu$ is equivalent to $\mu \succcurlyeq_{\mathrm{icv}} \nu$. Beyond this special case, the equivalence typically fails as illustrated by the following example and also by Proposition 2.65 below.

Example 2.62. Let μ be the uniform distribution on the interval $[-1, 1]$. Then $m(\mu) = 0$ and $\mathrm{var}(\mu) = 1/3$. We furthermore let $\nu = p\delta_{-1/2} + (1-p)\delta_2$ for $p = 4/5$. Then $m(\nu) = 0$ and $1 = \mathrm{var}(\nu) > \mathrm{var}(\mu)$. However,

$$\frac{1}{16} = \int \left(-\frac{1}{2} - x\right)^+ \mu(dx) > \int \left(-\frac{1}{2} - x\right)^+ \nu(dx) = 0,$$

so $\mu \succcurlyeq_{\mathrm{icv}} \nu$ does not hold. ◇

Remark 2.63. Let μ and ν be two lotteries in $\mathcal{M}$. We will say that μ dominates ν in the *convcave order*, and we will write $\mu \succcurlyeq_{\mathrm{cv}} \nu$ if

$$\int f\,d\mu \ge \int f\,d\nu \quad \text{for all concave functions } f \text{ on } \mathbb{R}. \tag{2.28}$$

Note that $\mu \succcurlyeq_{\mathrm{cv}} \nu$ implies that $m(\mu) = m(\nu)$, because both $f(x) = x$ and $\tilde{f}(x) = -x$ are concave. Corollary 2.61 shows that $\succcurlyeq_{\mathrm{cv}}$ coincides with our uniform partial order $\succcurlyeq_{\mathrm{icv}}$ if we compare two measures that have the same mean. The stochastic orders $\succcurlyeq_{\mathrm{cv}}$ and $\succcurlyeq_{\mathrm{icv}}$ were used in [278, 279] to express the view that μ is *less risky* than ν. The inverse relation $\mu \succcurlyeq_{\mathrm{cx}} \nu$, defined by

$$\int f\,d\mu \ge \int f\,d\nu \quad \text{for all convex functions } f \text{ on } \mathbb{R}, \tag{2.29}$$

is called the *convex order*. It is sometimes also called the *balayage order*. ◇

The following class of asset distributions is widely used in finance.

Definition 2.64. A real-valued random variable Y on some probability space $(\Omega, \mathscr{F}, P)$ is called *log-normally distributed* with parameters $\alpha \in \mathbb{R}$ and $\sigma \geq 0$ if it can be written as

$$Y = \exp(\alpha + \sigma X), \tag{2.30}$$

where X has a standard normal law $N(0,1)$. △

Clearly, any log-normally distributed random variable Y on $(\Omega, \mathscr{F}, P)$ takes P-a. s. strictly positive values. Recall from above the standard notations φ and Φ for the density and the cumulative distribution function of the standard normal law $N(0,1)$. We obtain from (2.30) the cumulative distribution function

$$P[\,Y \leq y\,] = \Phi\left(\frac{\log y - \alpha}{\sigma}\right), \quad 0 < y < \infty,$$

and the density

$$\psi(y) = \frac{1}{\sigma \cdot y}\,\varphi\left(\frac{\log y - \alpha}{\sigma}\right) \cdot \mathbb{1}_{(0,\infty)}(y) \tag{2.31}$$

of the log-normally distributed random variable Y. Its pth moment is given by the formula

$$E[\,Y^p\,] = \exp\left(p\alpha + \frac{1}{2}p^2\sigma^2\right).$$

In particular, the law μ of Y has the expectation

$$m(\mu) = E[\,Y\,] = \exp\left(\alpha + \frac{1}{2}\sigma^2\right) \tag{2.32}$$

and the variance

$$\operatorname{var}(\mu) = \exp(2\alpha + \sigma^2)(\exp(\sigma^2) - 1).$$

See Figure 2.2 for an illustration.

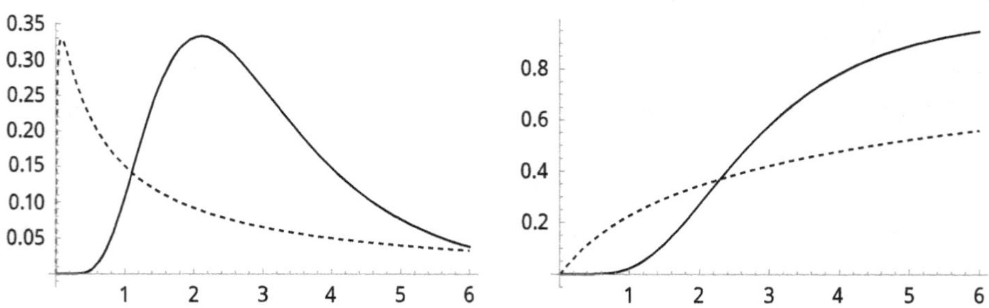

Figure 2.2: Densities (left) and cumulative distribution functions (right) for the log-normal distributions with parameters $\sigma = 2$ and $\alpha = 1.5$ (dashed) and $\sigma = 0.5$ and $\alpha = 1$ (solid).

Proposition 2.65. *Let μ and $\widetilde{\mu}$ be two log-normal distributions with parameters (α, σ) and $(\widetilde{\alpha}, \widetilde{\sigma})$, respectively. Then $\mu \succcurlyeq_{\text{icv}} \widetilde{\mu}$ holds if and only if $\sigma^2 \le \widetilde{\sigma}^2$ and $\alpha + \frac{1}{2}\sigma^2 \ge \widetilde{\alpha} + \frac{1}{2}\widetilde{\sigma}^2$.*

Proof. According to (2.32), the condition $\alpha + \frac{1}{2}\sigma^2 \ge \widetilde{\alpha} + \frac{1}{2}\widetilde{\sigma}^2$ is equivalent to $m(\mu) \ge m(\widetilde{\mu})$. First, suppose that $\sigma^2 \le \widetilde{\sigma}^2$ and $m(\mu) \ge m(\widetilde{\mu})$. We define a kernel $Q(x, \cdot)$ as the law of $x \cdot \exp(\lambda + \beta Z)$ where Z is a standard normal random variable and the parameters $\lambda \in \mathbb{R}$ and $\beta > 0$ will be specified later. Now suppose that μ is represented by (2.30) with X independent of Z, and let f denote a bounded measurable function. It follows that

$$\int f \, d(\mu Q) = E[f(e^{\alpha + \sigma X} \cdot e^{\lambda + \beta Z})] = E[f(e^{\alpha + \lambda + (\sigma^2 + \beta^2)^{1/2} U})],$$

where

$$U = \frac{\sigma X + \beta Z}{\sqrt{\sigma^2 + \beta^2}}$$

is also $N(0, 1)$-distributed. Thus, μQ is a log-normal distribution with parameters $(\alpha + \lambda, \sqrt{\sigma^2 + \beta^2})$. By taking $\beta := \sqrt{\widetilde{\sigma}^2 - \sigma^2}$ and $\lambda := \widetilde{\alpha} - \alpha$, we can represent $\widetilde{\mu}$ as $\widetilde{\mu} = \mu Q$. With this parameter choice,

$$\lambda = \widetilde{\alpha} - \alpha = \log m(\widetilde{\mu}) - \log m(\mu) - \frac{1}{2}(\widetilde{\sigma}^2 - \sigma^2) \le -\frac{\beta^2}{2}.$$

We have thus $m(Q(x, \cdot)) \le x$ for all x, and so $\mu \succcurlyeq_{\text{icv}} \widetilde{\mu}$ follows from condition (g) of Theorem 2.57.

As to the converse implication, the inequality $m(\mu) \ge m(\widetilde{\mu})$ is already clear from (2.32). To prove $\sigma^2 \le \widetilde{\sigma}^2$, let $\nu := \mu \circ \log^{-1}$ and $\widetilde{\nu} := \widetilde{\mu} \circ \log^{-1}$ so that $\nu = N(\alpha, \sigma^2)$ and $\widetilde{\nu} = N(\widetilde{\alpha}, \widetilde{\sigma}^2)$. For $\varepsilon > 0$, we define the concave increasing function $f_\varepsilon(x) := \log(\varepsilon + x)$. If u is a concave increasing function on $\mathbb{R}$, the function $u \circ f_\varepsilon$ is a concave and increasing function on $[0, \infty)$. It can be extended linearly to $(-\infty, 0)$ so as to obtain a concave increasing function v_ε on the full real line, which is needed to apply the definition of $\succcurlyeq_{\text{icv}}$. Therefore, using monotone convergence,

$$\int u \, d\nu = \lim_{\varepsilon \downarrow 0} \int v_\varepsilon \, d\mu \ge \lim_{\varepsilon \downarrow 0} \int v_\varepsilon \, d\widetilde{\mu} = \int u \, d\widetilde{\nu}. \tag{2.33}$$

Consequently, $\nu \succcurlyeq_{\text{icv}} \widetilde{\nu}$ and Proposition 2.59 yields $\sigma^2 \le \widetilde{\sigma}^2$. □

Remark 2.66. Let $\nu = N(\alpha, \sigma^2)$, $\widetilde{\nu} = N(\widetilde{\alpha}, \widetilde{\sigma}^2)$, and denote by μ and $\widetilde{\mu}$ the respective images of ν and $\widetilde{\nu}$ under the map $x \mapsto e^x$. Then $\mu \succcurlyeq_{\text{icv}} \widetilde{\mu}$ implies that $\nu \succcurlyeq_{\text{icv}} \widetilde{\nu}$. This can either be seen from (2.33) or alternatively as follows. Since $\mu \succcurlyeq_{\text{icv}} \widetilde{\mu}$, Proposition 2.65 states that $\sigma^2 \le \widetilde{\sigma}^2$ and

$$\widetilde{\alpha} + \frac{1}{2}\widetilde{\sigma}^2 \le \alpha + \frac{1}{2}\sigma^2 \le \alpha + \frac{1}{2}\widetilde{\sigma}^2,$$

which yields that $\widetilde{\alpha} \le \alpha$. Thus, Proposition 2.59 implies $\nu \succcurlyeq_{\text{icv}} \widetilde{\nu}$.

However, $\nu \succcurlyeq_{\text{icv}} \widetilde{\nu}$ does not imply $\mu \succcurlyeq_{\text{icv}} \widetilde{\mu}$. To see why, suppose that $\alpha \geq \widetilde{\alpha}$ and choose $\widetilde{\sigma}^2 > 2(\alpha - \widetilde{\alpha}) + \sigma^2$. Then $\widetilde{\sigma}^2 > \sigma^2$, and so $\nu \succcurlyeq_{\text{icv}} \widetilde{\nu}$ by Proposition 2.59. But Proposition 2.65 implies that $\mu \succcurlyeq_{\text{icv}} \widetilde{\mu}$ cannot hold because we have $\alpha + \frac{1}{2}\sigma^2 < \widetilde{\alpha} + \frac{1}{2}\widetilde{\sigma}^2$. ◇

Proposition 2.65 can also be proved from the Black–Scholes formula for option prices, which will be derived in Section 5.7. Below we will sketch the corresponding alternative proof of Proposition 2.65.

Second proof of Proposition 2.65. Let

$$Y_{m,\sigma} := m \cdot \exp\left(\sigma X - \frac{\sigma^2}{2}\right)$$

for a standard normally distributed random variable X. Then

$$E[\,(Y_{m,\sigma} - c)^+\,] = m\,\Phi(d_+) - c\,\Phi(d_-) \quad \text{with } d_\pm = \frac{\log\frac{x}{c} \pm \frac{1}{2}\sigma^2}{\sigma};$$

see Example 5.58 in Section 5.7. Calculating the derivative of this expectation with respect to $\sigma > 0$, one finds that

$$\frac{d}{d\sigma}E[\,(Y_{m,\sigma} - c)^+\,] = \frac{d}{d\sigma}(m\,\Phi(d_+) - c\,\Phi(d_-)) = x\,\varphi(d_+) > 0;$$

see (5.50) in Chapter 5. The law $\mu_{m,\sigma}$ of $Y_{m,\sigma}$ satisfies $m(\mu_{m,\sigma}) = m$ for all $\sigma > 0$. Condition (c) of Corollary 2.61 implies that $\mu_{m,\sigma}$ is decreasing with respect to $\succcurlyeq_{\text{icv}}$ as a function of $\sigma > 0$, and hence also with respect to $\succcurlyeq_{\text{cv}}$. That is, $\mu_{m,\sigma} \succcurlyeq_{\text{cv}} \mu_{m,\widetilde{\sigma}}$ if and only if $\sigma \leq \widetilde{\sigma}$. For two different expectations m and $\widetilde{m}$, we simply use the monotonicity of the function $u(y) := (y - c)^+$ to conclude that

$$\begin{aligned}\int u\,d\mu_{m,\sigma} &= E[u(m \cdot \exp(\sigma X - \sigma^2/2))\,] \\ &\geq E[u(\widetilde{m} \cdot \exp(\sigma X - \sigma^2/2))\,] \\ &\geq \int u\,d\mu_{\widetilde{m},\widetilde{\sigma}},\end{aligned}$$

provided that $m \geq \widetilde{m}$ and $0 < \sigma \leq \widetilde{\sigma}$. □

The partial order $\succcurlyeq_{\text{icv}}$ was defined in terms of integrals against increasing concave functions. By taking the larger class of all concave functions as integrands, we arrived at the partial order $\succcurlyeq_{\text{cv}}$ defined by (2.28) and characterized in Corollary 2.61. In the remainder of this section, we will briefly discuss the *increasing order*, which is induced by comparing the integrals with respect to all increasing integrands.

Definition 2.67. Let μ and ν be two arbitrary probability measures on $\mathbb{R}$. We say that μ dominates ν in the increasing order and we write $\mu \succcurlyeq_{\text{i}} \nu$ if

$$\int f\,d\mu \geq \int f\,d\nu \quad \text{for all bounded increasing functions } f \in C(\mathbb{R}). \qquad \triangle$$

The increasing order is sometimes also called *first-order stochastic dominance*, *stochastic dominance of the first kind*, or simply *stochastic dominance*. It is indeed a partial order on $\mathcal{M}_1(\mathbb{R})$: Reflexivity and transitivity are obvious, and antisymmetry follows, for example, from the equivalence (a)⇔(b) below. As will be shown by the following theorem, the relation $\mu \succcurlyeq_{\mathrm{i}} \nu$ means that the distribution μ is "higher" than the distribution ν. In our one-dimensional situation, we can provide a complete proof of this fact by using elementary properties of cumulative distribution functions. For instance, part (b) of the following theorem implies that the cumulative distribution functions of μ and ν never cross if $\mu \succcurlyeq_{\mathrm{i}} \nu$; this may be compared with the single-crossing property explored in Exercise 2.4.1.

Theorem 2.68. *For $\mu, \nu \in \mathcal{M}_1(\mathbb{R})$ the following conditions are equivalent:*
(a) $\mu \succcurlyeq_{\mathrm{i}} \nu$.
(b) *The cumulative distribution functions of μ and ν satisfy $F_\mu(x) \leq F_\nu(x)$ for all x.*
(c) *Any pair of quantile functions for μ and ν satisfies $q_\mu(t) \geq q_\nu(t)$ for a. e. $t \in (0,1)$.*
(d) *There exists a probability space $(\Omega, \mathcal{F}, P)$ with random variables X_μ and X_ν with distributions μ and ν such that $X_\mu \geq X_\nu$ P-a. s.*
(e) *There exists a stochastic kernel $Q(x, dy)$ on $\mathbb{R}$ such that $Q(x, (x, \infty)) = 0$ and such that $\nu = \mu Q$.*
(f) *$\int f\,d\mu \geq \int f\,d\nu$ for all increasing, bounded, and measurable functions f.*

In particular, $\mu \succcurlyeq_{\mathrm{i}} \nu$ implies $\mu \succcurlyeq_{\mathrm{icv}} \nu$.

Proof. (a)⇒(b): Note that $F_\mu(x) = \mu((-\infty, x])$ can be written as

$$F_\mu(x) = 1 - \int \mathbb{1}_{(x,\infty)}(y)\,\mu(dy).$$

For each x, we can construct a sequence of increasing continuous functions with values in $[0,1]$ that increase to $\mathbb{1}_{(x,\infty)}$. Hence,

$$\int \mathbb{1}_{(x,\infty)}(y)\,\mu(dy) \geq \int \mathbb{1}_{(x,\infty)}(y)\,\nu(dy) = 1 - F_\nu(x).$$

(b)⇔(c): This follows from the definition of a quantile function and from Lemma D.5.

(c)⇒(d): Let $(\Omega, \mathcal{F}, P)$ be a probability space supporting a random variable U with a uniform distribution on $(0,1)$. Then $X_\mu := q_\mu(U)$ and $X_\nu := q_\nu(U)$ satisfy $X_\mu \geq X_\nu$ P-almost surely. Moreover, it follows from Lemma D.7 that they have the distributions μ and ν.

(d)⇒(e): As in the proof of the implication (f)⇒(g) in Theorem 2.57, we let $Q(x, dy)$ be a regular conditional distribution of X_ν given X_μ. By (2.24), we then have $\nu = \mu Q$. Our assumption $X_\mu \geq X_\nu$ P-a. s. implies that for $r \in \mathbb{Q}$,

$$0 = P[X_\mu \leq r < X_\nu] = E\big[P[X_\nu > r \mid X_\mu]\,\mathbb{1}_{\{X_\mu \leq r\}}\big] = E\big[Q(X_\mu, (r,\infty))\,\mathbb{1}_{\{X_\mu \leq r\}}\big],$$

and hence P-a. s.

$$Q(X_\mu, (r,\infty))\,\mathbb{1}_{\{X_\mu \le r\}} = 0. \tag{2.34}$$

Using that $\mathbb{Q}$ is countable, we conclude that (2.34) holds with probability one simultaneously for all $r \in \mathbb{Q}$. Hence, $Q(X_\mu, (X_\mu, \infty)) = 0$ P-almost surely. Thus, the condition $Q(x,(x,\infty)) = 0$ is satisfied for μ-a. e. x. By modifying Q on a μ-null set (e. g., by putting $Q(x,\cdot) := \delta_x$ there), this condition can be achieved for all $x \in \mathbb{R}$.

(e)⇒(f): Condition (e) implies that $x \ge y$ for $Q(x,\cdot)$-a. e. y. Hence, if f is bounded, measurable, and increasing, then

$$\int f(y)\,Q(x,dy) \le \int f(x)\,Q(x,dy) = f(x).$$

Therefore,

$$\int f\,d\nu = \int \int f(y)\,Q(x,dy)\,\mu(dx) \le \int f\,d\mu.$$

(f)⇒(a): Obvious.

Finally, due to the equivalence (a) ⇔ (b) above and the equivalence (a) ⇔ (d) in Theorem 2.57, $\mu \succcurlyeq_{\mathrm{i}} \nu$ implies $\mu \succcurlyeq_{\mathrm{icv}} \nu$. □

Remark 2.69. It is clear from conditions (d) or (e) of Theorem 2.68 that the set of bounded, increasing, and continuous functions in Definition 2.67 can be replaced by the set of all increasing functions for which the two integrals make sense. Thus, $\mu \succcurlyeq_{\mathrm{i}} \nu$ for $\mu, \nu \in \mathcal{M}$ implies $\mu \succcurlyeq_{\mathrm{icv}} \nu$, and in particular $m(\mu) \ge m(\nu)$. Moreover, condition (d) shows that $\mu \succcurlyeq_{\mathrm{i}} \nu$ together with $m(\mu) = m(\nu)$ implies $\mu = \nu$. ◇

The study of stochastic orders can be extended into several directions. For instance, in Section 2.6 we prove versions of Theorem 2.57 and Corollary 2.61 for the concave order on lotteries on $\mathbb{R}^d$; see Theorem 2.94 and Corollary 2.95. These will in particular establish the remaining implication (b)⇒(f) in the proof of Theorem 2.57. In the same section, we also study an increasing order for lotteries on a general Polish space endowed with a continuous preference relation and we provide a corresponding characterization along the lines of Theorem 2.68; see Theorem 2.96. Another possible direction consists in extending the notions of first and second-order stochastic dominance to the concept of nth-*order stochastic dominance*, where $n \ge 2$ is an integer; for details, we refer to [137, 251, 301, 319].

2.5 Robust preferences in the face of model uncertainty

In this section, we discuss the structure of preferences for assets on a more fundamental level. Instead of assuming that the distributions of assets are known and that preferences are defined on a set of probability measures, we will take as our basic objects the assets themselves. An asset will be viewed as a function that associates real-valued payoffs to possible scenarios. More precisely, $\mathscr{X}$ will denote a set of bounded measurable functions X on some measurable set $(\Omega, \mathscr{F})$. We emphasize that no a priori probability measure is given on $(\Omega, \mathscr{F})$. In other words, we are facing uncertainty instead of risk.

We assume that $\mathscr{X}$ is endowed with a preference relation $\succ$. In view of the financial interpretation, it is natural to assume that $\succ$ is *monotone* in the sense that

$$Y \succeq X \quad \text{if } Y(\omega) \geq X(\omega) \text{ for all } \omega \in \Omega.$$

Under a suitable condition of *continuity*, we could apply the results of Section 2.1 to obtain a numerical representation of $\succ$. L. J. Savage introduced a set of additional axioms that guarantee that there is a numerical representation of the special form

$$U(X) = E_Q[\, u(X)\,] = \int u(X(\omega))\, Q(d\omega) \quad \text{for all } X \in \mathscr{X} \tag{2.35}$$

where Q is a probability measure on $(\Omega, \mathscr{F})$ and u is a function on $\mathbb{R}$. The measure Q specifies the subjective view of the probabilities of events that is implicit in the preference relation $\succ$. Note that the function $u : \mathbb{R} \to \mathbb{R}$ is determined by restricting U to the class of constant functions on $(\Omega, \mathscr{F})$. Clearly, the monotonicity of $\succ$ is equivalent to the condition that u is an increasing function.

Definition 2.70. A numerical representation of the form (2.35) will be called a *Savage representation* of the preference relation $\succ$. △

Remark 2.71. Let $\mu_{Q,X}$ denote the distribution of X under the subjective measure Q. Clearly, the preference order $\succ$ on $\mathscr{X}$ given by (2.35) induces a preference order on

$$\mathscr{M}_Q := \{\mu_{Q,X} \mid X \in \mathscr{X}\}$$

with von Neumann–Morgenstern representation

$$U_Q(\mu_{Q,X}) := U(X) = E_Q[\, u(X)\,] = \int u\, d\mu_{Q,X},$$

that is,

$$U_Q(\mu) = \int u(x)\, \mu(dx) \quad \text{for } \mu \in \mathscr{M}_Q.$$

On this level, Section 2.3 specifies the conditions on U_Q that guarantee that u is a (strictly concave and strictly increasing) utility function. ◇

Remark 2.72. Even if an economic agent with preferences $\succ$ accepts the view that scenarios $\omega \in \Omega$ are generated in accordance with a given objective probability measure P on $(\Omega, \mathscr{F})$, the preference order $\succ$ on $\mathscr{X}$ may be such that the subjective measure Q appearing in the Savage representation (2.35) is different from the objective measure P.

Suppose, for example, that P is the Lebesgue measure restricted to $\Omega = [0,1]$, and that $\mathscr{X}$ is the space of bounded right-continuous increasing functions on $[0,1]$. Let $\mu_{P,X}$ denote the distribution of X under P. By Lemma D.7, every probability measure on $\mathbb{R}$ with bounded support is of the form $\mu_{P,X}$ for some $X \in \mathscr{X}$, that is,

$$\mathscr{M}_b(\mathbb{R}) = \{\mu_{P,X} \mid X \in \mathscr{X}\}.$$

Suppose the agent agrees that, objectively, $X \in \mathscr{X}$ can be identified with the lottery $\mu_{P,X}$, so that the preference relation on $\mathscr{X}$ could be viewed as a preference relation on $\mathscr{M}_b(\mathbb{R})$ with numerical representation

$$U^*(\mu_{P,X}) := U(X).$$

This does not imply that U^* satisfies the assumptions of Section 2.2; in particular, the preference relation on $\mathscr{M}_b(\mathbb{R})$ may violate the independence axiom. In fact, the agent might take a pessimistic view and distort P by putting more emphasis on unfavorable scenarios. For example, the agent could replace P by the subjective measure

$$Q := \alpha\delta_0 + (1-\alpha)P$$

for some $\alpha \in (0,1)$ and specify preferences by a Savage representation in terms of u and Q. In this case,

$$\begin{aligned} U^*(\mu_{P,X}) &= E_Q[\, u(X)\,] = \int u \, d\mu_{Q,X} \\ &= \alpha u(X(0)) + (1-\alpha)E_P[\, u(X)\,] \\ &= \alpha u(X(0)) + (1-\alpha)\int u \, d\mu_{P,X}. \end{aligned}$$

Note that $X(0) = \ell(\mu_{P,X})$ for

$$\ell(\mu) := \inf(\operatorname{supp}\mu) = \sup\{a \in \mathbb{R} \mid \mu((-\infty, a)) = 0\},$$

where $\operatorname{supp}\mu$ is the support of μ. Hence, replacing P by Q corresponds to a nonlinear distortion on the level of lotteries: $\mu = \mu_{P,X}$ is distorted to the lottery $\mu^* = \mu_{Q,X}$ given by

$$\mu^* = \alpha\,\delta_{\ell(\mu)} + (1-\alpha)\mu,$$

and the preference relation on lotteries has the numerical representation

$$U^*(\mu) = \int u(x)\,\mu^*(dx) \quad \text{for } \mu \in \mathscr{M}_b(\mathbb{R}).$$

Let us now show that such a subjective distortion of objective lotteries would be consistent with the Allais paradox. Consider the lotteries μ_i and ν_i, $i = 1, 2$, described in Example 2.30. Clearly,

$$\mu_1^* = \mu_1 \quad \text{and} \quad \nu_1^* = \alpha\,\delta_0 + (1-\alpha)\nu_1,$$

while

$$\mu_2^* = \alpha\,\delta_0 + (1-\alpha)\mu_2 \quad \text{and} \quad \nu_1^* = \alpha\,\delta_0 + (1-\alpha)\nu_1.$$

For the particular choice $u(x) = x$, we have $U^*(\nu_2) > U^*(\mu_2)$, and for $\alpha > 9/2409$ we obtain $U^*(\mu_1) > U^*(\nu_1)$, in accordance with the observed preferences $\nu_2 \succ \mu_2$ and $\mu_1 \succ \nu_1$ described in Example 2.30.

For a systematic discussion of preferences described in terms of a subjective distortion of lotteries, we refer to [207]. In Section 4.8, we will discuss the role of distortions in the context of risk measures, and in particular the connection to Yaari's "dual theory of choice under risk" [322]. ◇

Even in its general form (2.35), however, the paradigm of expected utility has a limited scope as illustrated by the following example.

Example 2.73 (Ellsberg paradox). You are faced with a choice between two urns, each containing 100 balls that are either red or black. In the first urn, the proportion p of red balls is know; assume, for example, $p = 0.49$. In the second urn, the proportion $\tilde{p}$ is unknown. Suppose that you get 1000 € if you draw a red ball and 0 € otherwise. In this case, most people would choose the first urn. Naturally, they make the same choice if you get 1000 € for drawing a black ball and 0 € for a red one. But this behavior is not compatible with Savage's paradigm of subjective expected utility: For any subjective probability $\tilde{p}$ of drawing a red ball in the second urn, the first choice would imply $p > \tilde{p}$, the second would yield $1 - p > 1 - \tilde{p}$, and this is a contradiction. ◇

For this reason, we are going to make one further conceptual step beyond the Savage representation before we start to prove a representation theorem for preferences on $\mathscr{X}$. Instead of a single measure Q, let us consider a whole class $\mathcal{Q}$ of measures on $(\Omega, \mathscr{F})$. Our aim is to characterize those preference relations on $\mathscr{X}$ that admit a representation of the form

$$U(X) = \inf_{Q \in \mathcal{Q}} E_Q[\, u(X)\,]. \tag{2.36}$$

This may be viewed as a *robust* version of the paradigm of expected utility: The agent has in mind a whole collection of possible probabilistic views of the given set of scenarios and takes a *worst-case approach* in evaluating the expected utility of a given payoff.

It will be convenient to extend the discussion to the following framework where payoffs can be lotteries. Let $\mathscr{X}$ denote the space of all bounded measurable functions on $(\Omega, \mathscr{F})$. We are going to embed $\mathscr{X}$ into a certain space $\widetilde{\mathscr{X}}$ of functions $\widetilde{X}$ on $(\Omega, \mathscr{F})$ with values in the convex set

$$\mathscr{M}_b(\mathbb{R}) = \{\mu \in \mathscr{M}_1(\mathbb{R}) \mid \mu([-c, c]) = 1 \text{ for some } c \geq 0\}$$

of boundedly supported Borel probability measures on $\mathbb{R}$. More precisely, $\widetilde{\mathscr{X}}$ is defined as the convex set of all those stochastic kernels $\widetilde{X}(\omega, dy)$ from $(\Omega, \mathscr{F})$ to $\mathbb{R}$ for which there exists a constant $c \geq 0$ such that

$$\widetilde{X}(\omega, [-c, c]) = 1 \quad \text{for all } \omega \in \Omega.$$

In economics, the elements of $\widetilde{\mathscr{X}}$ are sometimes called *acts*; see, for example, [224]. The space $\mathscr{X}$ can be embedded into $\widetilde{\mathscr{X}}$ by virtue of the mapping

$$\mathscr{X} \ni X \longmapsto \delta_X \in \widetilde{\mathscr{X}}. \tag{2.37}$$

In this way, $\mathscr{X}$ can be identified with the set of all $\widetilde{X} \in \widetilde{\mathscr{X}}$ for which the measure $\widetilde{X}(\omega, \cdot)$ is a Dirac measure. A preference order on $\mathscr{X}$ defined by (2.36) clearly extends to $\widetilde{\mathscr{X}}$ by

$$\widetilde{U}(\widetilde{X}) = \inf_{Q \in \mathcal{Q}} \int \int u(y)\, \widetilde{X}(\omega, dy)\, Q(d\omega) = \inf_{Q \in \mathcal{Q}} E_Q[\, \widetilde{u}(\widetilde{X})\,] \tag{2.38}$$

where $\widetilde{u}$ is the affine function on $\mathscr{M}_b(\mathbb{R})$ defined by

$$\widetilde{u}(\mu) = \int u \, d\mu, \qquad \mu \in \mathscr{M}_b(\mathbb{R}).$$

Remark 2.74. Restricting the preference order $\succ$ on $\widetilde{\mathscr{X}}$ obtained from (2.38) to the constant maps $\widetilde{X}(\omega) = \mu$ for $\mu \in \mathscr{M}_b(\mathbb{R})$, we obtain a preference order on $\mathscr{M}_b(\mathbb{R})$, and on this level we know how to characterize risk aversion by the property that u is strictly concave. ◇

Example 2.75. Let us show how the Ellsberg paradox fits into our extended setting, and how it can be resolved by a suitable choice of the set $\mathcal{Q}$. For $\Omega = \{0, 1\}$ define

$$\widetilde{X}_0(\omega) := p\,\delta_{1000} + (1-p)\delta_0, \quad \widetilde{X}_1(\omega) := (1-p)\delta_{1000} + p\,\delta_0,$$

and

$$\widetilde{Z}_i(\omega) := \delta_{1000} \cdot \mathbb{1}_{\{1-i\}}(\omega) + \delta_0 \cdot \mathbb{1}_{\{i\}}(\omega), \quad i = 0, 1.$$

Take

$$\mathcal{Q} := \{q\,\delta_1 + (1-q)\delta_0 \mid a \leq q \leq b\}$$

with $[a, b] \subseteq [0, 1]$. For any increasing function u, the functional

$$\widetilde{U}(\widetilde{X}) := \inf_{Q \in \mathcal{Q}} E_Q[\, \widetilde{u}(\widetilde{X})\,]$$

satisfies

$$\widetilde{U}(\widetilde{X}_i) > \widetilde{U}(\widetilde{Z}_i), \quad i = 0, 1,$$

as soon as $a < p < b$, in accordance with the preferences described in Example 2.73. ◇

Let us now formulate those properties of a preference order $\succ$ on the convex set $\widetilde{\mathscr{X}}$ that are crucial for a representation of the form (2.38). For $\widetilde{X}, \widetilde{Y} \in \widetilde{\mathscr{X}}$, and $\alpha \in (0,1)$, (2.38) implies

$$\begin{aligned}\widetilde{U}(\alpha\widetilde{X} + (1-\alpha)\widetilde{Y}) &= \inf_{Q\in\mathcal{Q}} (\alpha E_Q[\,\tilde{u}(\widetilde{X})\,] + (1-\alpha) E_Q[\,\tilde{u}(\widetilde{Y})\,]) \\ &\geq \alpha \widetilde{U}(\widetilde{X}) + (1-\alpha)\widetilde{U}(\widetilde{Y}).\end{aligned}$$

In contrast to the Savage case $\mathcal{Q} = \{Q\}$, we can no longer expect equality, except for the case of *certainty* $\widetilde{Y}(\omega) \equiv \mu$. If $\widetilde{X} \sim \widetilde{Y}$, then $\widetilde{U}(\widetilde{X}) = \widetilde{U}(\widetilde{Y})$, and the lower bound reduces to $\widetilde{U}(\widetilde{X}) = \widetilde{U}(\widetilde{Y})$. Thus, $\succ$ satisfies the following two properties:

Uncertainty aversion: If $\widetilde{X}, \widetilde{Y} \in \widetilde{\mathscr{X}}$ are such that $\widetilde{X} \sim \widetilde{Y}$, then

$$\alpha\widetilde{X} + (1-\alpha)\widetilde{Y} \succeq \widetilde{X} \quad \text{for all } \alpha \in [0,1].$$

Certainty independence: For $\widetilde{X}, \widetilde{Y} \in \widetilde{\mathscr{X}}, \widetilde{Z} \equiv \mu \in \mathscr{M}_b(\mathbb{R})$, and $\alpha \in (0,1]$, we have

$$\widetilde{X} \succ \widetilde{Y} \iff \alpha\widetilde{X} + (1-\alpha)\widetilde{Z} \succ \alpha\widetilde{Y} + (1-\alpha)\widetilde{Z}.$$

Remark 2.76. In order to motivate the term "uncertainty aversion," consider the situation of Example 2.75. Suppose that an agent is indifferent between the choices $\widetilde{Z}_0$ and $\widetilde{Z}_1$, which both involve the same kind of Knightian uncertainty. For $\alpha \in (0,1)$, the convex combination $\widetilde{Y} := \alpha\widetilde{Z}_0 + (1-\alpha)\widetilde{Z}_1$, which is weakly preferred to both $\widetilde{Z}_0$ and $\widetilde{Z}_1$ in the case of uncertainty aversion, takes the form

$$\widetilde{Y}(\omega) = \begin{cases} \alpha\,\delta_{1000} + (1-\alpha)\delta_0 & \text{for } \omega = 1, \\ \alpha\,\delta_0 + (1-\alpha)\delta_{1000} & \text{for } \omega = 0, \end{cases}$$

that is, uncertainty is reduced in favor of risk. For $\alpha = 1/2$, the resulting lottery $\widetilde{Y}(\omega) \equiv \frac{1}{2}(\delta_{1000} + \delta_0)$ is independent of the scenario ω, that is, Knightian uncertainty is completely replaced by the risk of a coin toss. ◇

Remark 2.77. The axiom of "certainty independence" extends the independence axiom for preferences on lotteries to our present setting, but only under the restriction that one of the two contingent lotteries $\widetilde{X}$ and $\widetilde{Y}$ is *certain*, that is, does not depend on the scenario $\omega \in \Omega$. Without this restriction, the extended independence axiom would lead to the Savage representation in its original form (2.35); see Exercise 2.5.3 below. There are good reasons for *not* requiring full independence for all $\widetilde{Z} \in \widetilde{\mathscr{X}}$. As an example, take $\Omega = \{0,1\}$ and define $\widetilde{X}(\omega) = \delta_\omega$, $\widetilde{Y}(\omega) = \delta_{1-\omega}$, and $\widetilde{Z} = \widetilde{X}$. An agent may prefer

$\widetilde{X}$ over $\widetilde{Y}$, thus expressing the implicit view that scenario 1 is somewhat more likely than scenario 0. At the same time, the agent may like the idea of *hedging* against the occurrence of scenario 0, and this could mean that the certain lottery

$$\frac{1}{2}(\widetilde{Y} + \widetilde{Z})(\cdot) \equiv \frac{1}{2}(\delta_0 + \delta_1)$$

is preferred over the contingent lottery

$$\frac{1}{2}(\widetilde{X} + \widetilde{Z})(\cdot) \equiv \widetilde{X}(\cdot),$$

thus violating the independence assumption in its unrestricted form. In general, the role of $\widetilde{Z}$ as a hedge against scenarios unfavorable for $\widetilde{Y}$ requires that $\widetilde{Y}$ and $\widetilde{Z}$ are *not comonotone*, that is,

$$\exists\, \omega, \eta \in \Omega: \quad \widetilde{Y}(\omega) \succ \widetilde{Y}(\eta),\ \widetilde{Z}(\omega) \prec \widetilde{Z}(\eta). \tag{2.39}$$

Thus, the wish to hedge would still be compatible with the following enforcement of certainty independence, called

Comonotonic independence: For $\widetilde{X}, \widetilde{Y}, \widetilde{Z} \in \widetilde{\mathscr{X}}$, and $\alpha \in (0,1]$,

$$\widetilde{X} \succ \widetilde{Y} \quad \Longleftrightarrow \quad \alpha\widetilde{X} + (1-\alpha)\widetilde{Z} \succ \alpha\widetilde{Y} + (1-\alpha)\widetilde{Z}, \tag{2.40}$$

whenever $\widetilde{Y}$ and $\widetilde{Z}$ are *comonotone* in the sense that (2.39) does *not* occur.

The consequences of requiring comonotonic independence will be analyzed in Exercise 2.5.4 below. It is also relevant to weaken the axiom of certainty independence and to require instead

Weak certainty independence: if for $\widetilde{X}, \widetilde{Y} \in \widetilde{\mathscr{X}}$, and for some $\nu \in \mathscr{M}_b(S)$ and $\alpha \in (0,1]$, we have $\alpha\widetilde{X} + (1-\alpha)\nu \succ \alpha\widetilde{Y} + (1-\alpha)\nu$, then

$$\alpha\widetilde{X} + (1-\alpha)\mu \succ \alpha\widetilde{Y} + (1-\alpha)\mu \quad \text{for all } \mu \in \mathscr{M}_b(S).$$

The consequences of requiring weak certainty independence will be analyzed in Theorem 2.88 below. ◇

From now on, we assume that $\succ$ is a given preference order on $\widetilde{\mathscr{X}}$. The set $\mathscr{M}_b(\mathbb{R})$ will be regarded as a subset of $\widetilde{\mathscr{X}}$ by identifying a constant function $\widetilde{Z} \equiv \mu$ with its value $\mu \in \mathscr{M}_b(\mathbb{R})$. We assume that $\succ$ possesses the following properties:
- *Uncertainty aversion.*
- *Certainty independence.*
- *Monotonicity:* If $\widetilde{Y}(\omega) \succeq \widetilde{X}(\omega)$ for all $\omega \in \Omega$, then $\widetilde{Y} \succeq \widetilde{X}$. Moreover, $\succ$ is compatible with the usual order on $\mathbb{R}$, that is, $\delta_y \succ \delta_x$ if and only if $y > x$.

- *Continuity:* The following analogue of the *Archimedean axiom* holds on $\widetilde{\mathscr{X}}$: If $\widetilde{X}, \widetilde{Y}, \widetilde{Z} \in \widetilde{\mathscr{X}}$ are such that $\widetilde{Z} \succ \widetilde{Y} \succ \widetilde{X}$, then there are $\alpha, \beta \in (0,1)$ with

$$\alpha\widetilde{Z} + (1-\alpha)\widetilde{X} \succ \widetilde{Y} \succ \beta\widetilde{Z} + (1-\beta)\widetilde{X}.$$

Moreover, for all $c > 0$ the restriction of $\succ$ to $\mathscr{M}_1([-c,c])$ is continuous with respect to the weak topology.

Let us denote by

$$\mathscr{M}_{1,f} := \mathscr{M}_{1,f}(\Omega, \mathscr{F})$$

the class of all set functions $Q : \mathscr{F} \to [0,1]$ that are normalized to $Q[\,\Omega\,] = 1$ and that are *finitely additive*, that is, $Q[\,A \cup B\,] = Q[\,A\,] + Q[\,B\,]$ for all disjoint $A, B \in \mathscr{F}$; see Definition G.20. By $E_Q[\,X\,]$, we denote the integral of X with respect to $Q \in \mathscr{M}_{1,f}$; see Appendix G.4. With $\mathscr{M}_1 = \mathscr{M}_1(\Omega, \mathscr{F})$ we denote the σ-additive members of $\mathscr{M}_{1,f}$, that is, the class of all probability measures on $(\Omega, \mathscr{F})$. Note that the inclusion $\mathscr{M}_1 \subseteq \mathscr{M}_{1,f}$ is typically strict as is illustrated by Example G.23.

Theorem 2.78. *Consider a preference order $\succ$ on $\widetilde{\mathscr{X}}$ satisfying the four properties of uncertainty aversion, certainty independence, monotonicity, and continuity.*

(a) *There exists a strictly increasing function $u \in C(\mathbb{R})$ and a convex set $\mathscr{Q} \subseteq \mathscr{M}_{1,f}(\Omega, \mathscr{F})$ such that*

$$\widetilde{U}(\widetilde{X}) = \min_{Q \in \mathscr{Q}} E_Q\Big[\int u(x)\,\widetilde{X}(\cdot, dx)\Big]$$

is a numerical representation of $\succ$. Moreover, u is unique up to positive affine transformations.

(b) *If the induced preference order $\succ$ on $\mathscr{X}$, viewed as a subset of $\widetilde{\mathscr{X}}$ as in (2.37), satisfies the following additional continuity property:*

$$X \succ Y \text{ and } X_n \nearrow X \quad \Longrightarrow \quad X_n \succ Y \quad \text{for all large } n, \tag{2.41}$$

then the set functions in $\mathscr{Q}$ are in fact probability measures, that is, each $Q \in \mathscr{Q}$ is σ-additive. In this case, the induced preference order on $\mathscr{X}$ has the robust Savage representation

$$U(X) = \min_{Q \in \mathscr{Q}} E_Q[\,u(X)\,] \quad \text{for } X \in \mathscr{X}$$

with $\mathscr{Q} \subseteq \mathscr{M}_1(\Omega, \mathscr{F})$.

Remark 2.79. Even without its axiomatic foundation, the robust Savage representation is highly plausible as it stands, since it may be viewed as a worst-case approach to the problem of model uncertainty. This aspect will be of particular relevance in our discussion of risk measures in Chapter 4. ◇

The proof of Theorem 2.78 needs some preparation.

When restricted to $\mathcal{M}_b(\mathbb{R})$, viewed as a subset of $\widetilde{\mathcal{X}}$, the axiom of certainty independence is just the *independence axiom* of the von Neumann–Morgenstern theory. Thus, the preference relation $\succ$ on $\mathcal{M}_b(\mathbb{R})$ satisfies the assumptions of Corollary 2.28, and we obtain the existence of a continuous function $u : \mathbb{R} \to \mathbb{R}$ such that

$$\widetilde{u}(\mu) := \int u(x)\,\mu(dx) \tag{2.42}$$

is a numerical representation of $\succ$ on the set $\mathcal{M}_b(\mathbb{R})$. Moreover, u is unique up to positive affine transformations. The second part of our monotonicity assumption implies that u is strictly increasing. Without loss of generality, we assume $u(0) = 0$ and $u(1) = 1$.

Remark 2.80. In view of the representation (2.42), it follows as in (2.11) that any $\mu \in \mathcal{M}_b(\mathbb{R})$ admits a unique *certainty equivalent* $c(\mu) \in \mathbb{R}$ for which

$$\mu \sim \delta_{c(\mu)}.$$

Thus, if $X \in \mathcal{X}$ is defined for $\widetilde{X} \in \widetilde{\mathcal{X}}$ as $X(\omega) := c(\widetilde{X}(\omega))$, then the first part of our monotonicity assumption yields

$$\widetilde{X} \sim \delta_X, \tag{2.43}$$

and so the preference relation $\succ$ on $\widetilde{\mathcal{X}}$ is uniquely determined by its restriction to $\mathcal{X}$. ◇

Lemma 2.81. *There exists a unique extension $\widetilde{U}$ of the functional $\widetilde{u}$ in (2.42) as a numerical representation of $\succ$ on $\widetilde{\mathcal{X}}$.*

Proof. For $\widetilde{X} \in \widetilde{\mathcal{X}}$, let $c > 0$ be such that $\widetilde{X}(\omega, [-c, c]) = 1$ for all $\omega \in \Omega$. Then

$$\widetilde{u}(\delta_{-c}) \le \widetilde{u}(\widetilde{X}(\omega)) \le \widetilde{u}(\delta_c) \quad \text{for all } \omega \in \Omega,$$

and our monotonicity assumption implies that

$$\delta_c \succeq \widetilde{X} \succeq \delta_{-c}.$$

We will show below that there exists a unique $\alpha \in [0, 1]$ such that

$$\widetilde{X} \sim (1 - \alpha)\delta_{-c} + \alpha\delta_c. \tag{2.44}$$

Once this has been achieved, the only possible choice for $\widetilde{U}(\widetilde{X})$ is

$$\widetilde{U}(\widetilde{X}) := \widetilde{u}((1 - \alpha)\delta_{-c} + \alpha\delta_c) = (1 - \alpha)\widetilde{u}(\delta_{-c}) + \alpha\widetilde{u}(\delta_c).$$

This definition of $\widetilde{U}$ provides a numerical representation of $\succ$ on $\widetilde{\mathcal{X}}$.

The proof of the existence of a unique $\alpha \in [0,1]$ with (2.44) is similar to the proof of Lemma 2.24. Uniqueness follows from the monotonicity

$$\beta > \alpha \quad \Longrightarrow \quad (1-\beta)\delta_{-c} + \beta\delta_c \succ (1-\alpha)\delta_{-c} + \alpha\delta_c, \tag{2.45}$$

which is an immediate consequence of the von Neumann–Morgenstern representation (2.42). Now we let

$$\alpha := \sup\{\gamma \in [0,1] \mid \widetilde{X} \succeq (1-\gamma)\delta_{-c} + \gamma\delta_c\}.$$

We have to exclude the two following cases:

$$\widetilde{X} \succ (1-\alpha)\delta_{-c} + \alpha\delta_c, \tag{2.46}$$

$$(1-\alpha)\delta_{-c} + \alpha\delta_c \succ \widetilde{X}. \tag{2.47}$$

In the case (2.46), our continuity axiom yields some $\beta \in (0,1)$ for which

$$\widetilde{X} \succ \beta((1-\alpha)\delta_{-c} + \alpha\delta_c) + (1-\beta)\delta_c = (1-\gamma)\delta_{-c} + \gamma\delta_c$$

where $\gamma = \beta\alpha + (1-\beta) > \alpha$, in contradiction to the definition of α.

If (2.47) holds, then the same argument as above yields $\beta \in (0,1)$ with

$$\beta\alpha\delta_c + (1-\beta\alpha)\delta_{-c} \succ \widetilde{X}.$$

By our definition of α, there must be some $\gamma \in (\beta\alpha, \alpha)$ with

$$\widetilde{X} \succeq (1-\gamma)\delta_{-c} + \gamma\delta_c \succ \beta\alpha\delta_c + (1-\beta\alpha)\delta_{-c},$$

where the second relation follows from (2.45). This, however, is a contradiction. □

Remark 2.82. Note that the proof of the preceding lemma relies only on the assumptions of continuity and monotonicity of $\succ$. Certainty independence and uncertainty aversion are not needed. ◇

Via the embedding (2.37), Lemma 2.81 induces a numerical representation U of $\succ$ on $\mathscr{X}$ given by

$$U(X) := \widetilde{U}(\delta_X). \tag{2.48}$$

The following proposition clarifies the properties of the functional U and provides the key to a robust Savage representation of the preference order $\succ$ on $\mathscr{X}$.

Proposition 2.83. *Given u of* (2.42) *and the numerical representation U on $\mathscr{X}$ constructed via Lemma* 2.81 *and* (2.48), *there exists a unique functional $\phi : \mathscr{X} \to \mathbb{R}$ such that*

$$U(X) = \phi(u(X)) \quad \textit{for all } X \in \mathscr{X}, \tag{2.49}$$

and such that the following four properties are satisfied:
- Monotonicity: *If* $Y(\omega) \geq X(\omega)$ *for all* ω, *then* $\phi(Y) \geq \phi(X)$.
- Concavity: *If* $\lambda \in [0,1]$, *then* $\phi(\lambda X + (1-\lambda)Y) \geq \lambda\phi(X) + (1-\lambda)\phi(Y)$.
- Positive homogeneity: $\phi(\lambda X) = \lambda\phi(X)$ *for* $\lambda \geq 0$.
- Cash invariance: $\phi(X + z) = \phi(X) + z$ *for all* $z \in \mathbb{R}$.

Proof. Denote by $\mathscr{X}_u$ the space of all $X \in \mathscr{X}$ that take values in a compact subset of the range $u(\mathbb{R})$ of u. Clearly, $\mathscr{X}_u$ coincides with the range of the nonlinear transformation $\mathscr{X} \ni X \mapsto u(X)$. Note that this transformation is bijective since u is continuous and strictly increasing due to our assumption of monotonicity. Thus, ϕ is well-defined on $\mathscr{X}_u$ via (2.49). We show next that this ϕ has the four properties of the assertion.

The monotonicity of ϕ on $\mathscr{X}_u$ is obvious. For the next step, we note first that the range $u(\mathbb{R})$ of u is an interval containing 0 as interior point, because the strictly increasing and continuous function $u : \mathbb{R} \to \mathbb{R}$ satisfies $u(0) = 0$ by assumption. It follows that $\lambda X \in \mathscr{X}_u$ for all $X \in \mathscr{X}_u$ and $\lambda \in (0,1]$. We will show next that $\phi(\lambda X) = \lambda\phi(X)$ for $X \in \mathscr{X}_u$ and $\lambda \in (0,1]$. Let $X_0 \in \mathscr{X}$ be such that $u(X_0) = X$. We define $\widetilde{Z} \in \widetilde{\mathscr{X}}$ by

$$\widetilde{Z} := \lambda\delta_{X_0} + (1-\lambda)\delta_0.$$

By (2.43), $\widetilde{Z} \sim \delta_Z$ where Z is given by

$$\begin{aligned} Z(\omega) &= c(\lambda\delta_{X_0(\omega)} + (1-\lambda)\delta_0) \\ &= u^{-1}(\lambda u(X_0(\omega)) + (1-\lambda)u(0)) \\ &= u^{-1}(\lambda u(X_0(\omega))), \end{aligned}$$

where we have used our convention $u(0) = 0$. It follows that $u(Z) = \lambda u(X_0) = \lambda X$, and so

$$\phi(\lambda X) = U(Z) = \widetilde{U}(\widetilde{Z}). \tag{2.50}$$

As in (2.44), one can find $\nu \in \mathscr{M}_b(\mathbb{R})$ such that $\nu \sim \delta_{X_0}$. Certainty independence implies that

$$\widetilde{Z} = \lambda\delta_{X_0} + (1-\lambda)\delta_0 \sim \lambda\nu + (1-\lambda)\delta_0.$$

Indeed, $\nu \sim \delta_{X_0}$ is equivalent to $\nu \not\succ \delta_{X_0}$ and $\nu \not\prec \delta_{X_0}$, and so certainty independence implies that we can have neither $\widetilde{Z} \succ \lambda\nu + (1-\lambda)\delta_0$ nor $\widetilde{Z} \prec \lambda\nu + (1-\lambda)\delta_0$. Hence,

$$\widetilde{U}(\widetilde{Z}) = \widetilde{u}(\lambda\nu + (1-\lambda)\delta_0) = \lambda\widetilde{u}(\nu) = \lambda U(X_0) = \lambda\phi(X).$$

This shows that $\phi(\lambda X) = \lambda\phi(X)$ for $X \in \mathscr{X}_u$ and $\lambda \in (0,1]$.

Since the range of u is an interval containing 0 as interior point, we can extend ϕ from $\mathscr{X}_u$ to all of $\mathscr{X}$ by positive homogeneity. More precisely, for arbitrary $X \in \mathscr{X}$, there is $\lambda \in (0,1]$ such that $\lambda X \in \mathscr{X}_u$, and we can define $\phi(X) := \frac{1}{\lambda}\phi(\lambda X)$. The preceding step of the proof guarantees that this definition of $\phi(X)$ is independent of the particular choice of λ. Clearly, the functional $\phi : \mathscr{X} \to \mathbb{R}$ obtained via this extension is also monotone and positively homogeneous.

Let us now show that the monotone and positively homogeneous functional ϕ constructed via extension in the preceding step is cash invariant. By taking any x such that $u(x) \neq 0$, we obtain from the positive homogeneity of ϕ that

$$\phi(1) = \frac{\phi(u(x))}{u(x)} = \frac{\widetilde{u}(\delta_x)}{u(x)} = 1.$$

Now take $X \in \mathscr{X}$ and $z \in \mathbb{R}$. By positive homogeneity, we may assume without loss of generality that $2X \in \mathscr{X}_u$ and $2z \in u(\mathbb{R})$. Then there are $X_0 \in \mathscr{X}$ such that $2X = u(X_0)$ as well as $z_0, x_0 \in \mathbb{R}$ with $2z = u(z_0)$ and $2\phi(X) = u(x_0)$. Note that $\delta_{X_0} \sim \delta_{x_0}$. Thus, as above, certainty independence yields

$$\widetilde{Z} := \frac{1}{2}(\delta_{X_0} + \delta_{z_0}) \sim \frac{1}{2}(\delta_{x_0} + \delta_{z_0}) =: \mu.$$

On the one hand, it follows that

$$\widetilde{U}(\widetilde{Z}) = U(\mu) = \frac{1}{2}u(x_0) + \frac{1}{2}u(z_0) = \phi(X) + z.$$

On the other hand, the same reasoning that lead to (2.50), shows that

$$\widetilde{U}(\widetilde{Z}) = \phi(X + z).$$

Putting both identities together gives $\phi(X + z) = \phi(X) + z$.

As to concavity, we need only show that $\phi(\frac{1}{2}X+\frac{1}{2}Y) \geq \frac{1}{2}\phi(X)+\frac{1}{2}\phi(Y)$ for $X, Y \in \mathscr{X}_u$, by Exercise 2.5.2 below. Let $X_0, Y_0 \in \mathscr{X}$ be such that $X = u(X_0)$ and $Y = u(Y_0)$. If $\phi(X) = \phi(Y)$, then $\delta_{X_0} \sim \delta_{Y_0}$, and uncertainty aversion gives

$$\widetilde{Z} := \frac{1}{2}(\delta_{X_0} + \delta_{Y_0}) \succeq \delta_{X_0},$$

which by the same arguments as above yields

$$\widetilde{U}(\widetilde{Z}) = \phi\left(\frac{1}{2}X + \frac{1}{2}Y\right) \geq \phi(X) = \frac{1}{2}(\phi(X) + \phi(Y)).$$

The case in which $\phi(X) > \phi(Y)$ can be reduced to the previous one by letting $z := \phi(X) - \phi(Y)$, and by replacing Y by $Y_z := Y + z$. Cash invariance then implies that

$$\begin{aligned}
\phi\left(\frac{1}{2}X + \frac{1}{2}Y\right) + \frac{1}{2}z &= \phi\left(\frac{1}{2}X + \frac{1}{2}Y_z\right) \\
&\geq \frac{1}{2}(\phi(X) + \phi(Y_z)) \\
&= \frac{1}{2}(\phi(X) + \phi(Y)) + \frac{1}{2}z.
\end{aligned}$$

□

A functional $\phi : \mathcal{X} \to \mathbb{R}$ satisfying the properties of monotonicity, concavity, positive homogeneity, and cash invariance is sometimes called a *coherent monetary utility functional.* The functional $\rho(X) := -\phi(X)$ is called a *coherent risk measure*. Functionals of this type will be studied in detail in Chapter 4.

Exercise 2.5.1. Let $\phi : \mathcal{X} \to \mathbb{R}$ be a functional that satisfies the properties of monotonicity and cash invariance stated in Proposition 2.83. Show that is Lipschitz continuous on $\mathcal{X}$ with respect to the supremum norm $\|\cdot\|$, that is,

$$|\phi(X) - \phi(Y)| \le \|X - Y\| \quad \text{for all } X, Y \in \mathcal{X}. \qquad \diamond$$

Exercise 2.5.2. Suppose that $\phi : \mathcal{X} \to \mathbb{R}$ is monotone, cash invariant, and satisfies $\phi(\frac{1}{2}(X+Y)) \ge \frac{1}{2}\phi(X)+\frac{1}{2}\phi(Y)$ for all $X, Y \in \mathcal{X}$. Use Exercise 2.5.1 to show that ϕ is concave. ◇

Let us now show that a function with the four properties established in Proposition 2.83 can be represented in terms of the class $\mathcal{M}_{1,f}$ of finitely additive and normalized set functions $Q : \mathcal{F} \to [0,1]$.

Proposition 2.84. *A functional $\phi : \mathcal{X} \to \mathbb{R}$ is monotone, concave, positively homogeneous, and cash invariant if and only if there exists a set $\mathcal{Q} \subseteq \mathcal{M}_{1,f}$ such that*

$$\phi(X) = \inf_{Q \in \mathcal{Q}} E_Q[X], \quad X \in \mathcal{X}.$$

Moreover, the set $\mathcal{Q}$ can always be chosen to be convex and such that the infimum above is attained, that is,

$$\phi(X) = \min_{Q \in \mathcal{Q}} E_Q[X], \quad X \in \mathcal{X}.$$

Proof. The necessity of the four properties is obvious. Conversely, we will construct for any $X \in \mathcal{X}$ a finitely additive set function Q_X such that $\phi(X) = E_{Q_X}[X]$ and $\phi(Y) \le E_{Q_X}[Y]$ for all $Y \in \mathcal{X}$. Then

$$\phi(X) = \min_{Q \in \mathcal{Q}_0} E_Q[X] \quad \text{for all } X \in \mathcal{X}, \tag{2.51}$$

where $\mathcal{Q}_0 := \{Q_X \mid X \in \mathcal{X}\}$. Clearly, (2.51) remains true if we replace $\mathcal{Q}_0$ by its convex hull $\mathcal{Q} := \operatorname{conv} \mathcal{Q}_0$.

To construct Q_X for a given $X \in \mathcal{X}$, we define three convex sets in $\mathcal{X}$ by

$$\mathcal{B} := \{Y \in \mathcal{X} \mid \phi(Y) > 1\},$$

$$\mathcal{C}_1 := \{Y \in \mathcal{X} \mid Y \le 1\}, \quad \text{and} \quad \mathcal{C}_2 := \left\{Y \in \mathcal{X} \;\middle|\; Y \le \frac{X}{\phi(X)}\right\}.$$

The convexity of $\mathcal{C}_1$ and $\mathcal{C}_2$ implies that the convex hull of their union is given by

$$\mathscr{C} := \operatorname{conv}(\mathscr{C}_1 \cup \mathscr{C}_2) = \{\alpha Y_1 + (1-\alpha)Y_2 \mid Y_i \in \mathscr{C}_i \text{ and } \alpha \in [0,1]\}.$$

If $Y \in \mathscr{C}$ is of the form $Y = \alpha Y_1 + (1-\alpha)Y_2$ for some $Y_i \in \mathscr{C}_i$ and $\alpha \in [0,1]$, then

$$\phi(Y) \le \phi(\alpha + (1-\alpha)Y_2) = \alpha + (1-\alpha)\phi(Y_2) \le 1,$$

and so $\mathscr{B}$ and $\mathscr{C}$ are disjoint. Let $\mathscr{X}$ be endowed with the supremum norm $\|Y\| = \sup_{\omega\in\Omega} |Y(\omega)|$. Then $\mathscr{C}_1$, and hence $\mathscr{C}$, contains the unit ball in $\mathscr{X}$. In particular, $\mathscr{C}$ has nonempty interior. Thus, we may apply the separation argument in the form of Theorem H.2, which yields a nonzero continuous linear functional ℓ on $\mathscr{X}$ such that

$$c := \sup_{Y\in\mathscr{C}} \ell(Y) \le \inf_{Z\in\mathscr{B}} \ell(Z).$$

Since $\mathscr{C}$ contains the unit ball, c must be strictly positive, and there is no loss of generality in assuming $c = 1$. In particular, $\ell(1) \le 1$ as $1 \in \mathscr{C}$. On the other hand, any constant $b > 1$ is contained in $\mathscr{B}$, and so

$$\ell(1) = \lim_{b\downarrow 1} \ell(b) \ge c = 1.$$

Hence, $\ell(1) = 1$.

If $A \in \mathscr{F}$, then $\mathbb{1}_{A^c} \in \mathscr{C}_1 \subseteq \mathscr{C}$, which implies that

$$\ell(\mathbb{1}_A) = \ell(1) - \ell(\mathbb{1}_{A^c}) \ge 1 - 1 = 0.$$

By Theorem G.21, there exists a finitely additive set function $Q_X \in \mathscr{M}_{1,f}(\Omega, \mathscr{F})$ such that $\ell(Y) = E_{Q_X}[\,Y\,]$ for any $Y \in \mathscr{X}$.

It remains to show that $E_{Q_X}[\,Y\,] \ge \phi(Y)$ for all $Y \in \mathscr{X}$, with equality for $Y = X$. By the cash invariance of ϕ, we need only consider the case in which $\phi(Y) > 0$. Then

$$Y_n := \frac{Y}{\phi(Y)} + \frac{1}{n} \in \mathscr{B},$$

and $Y_n \to Y/\phi(Y)$ uniformly, whence

$$\frac{E_{Q_X}[\,Y\,]}{\phi(Y)} = \lim_{n\uparrow\infty} E_{Q_X}[\,Y_n\,] \ge 1.$$

On the other hand, $X/\phi(X) \in \mathscr{C}_2 \subseteq \mathscr{C}$ yields the inequality

$$\frac{E_{Q_X}[\,X\,]}{\phi(X)} \le c = 1.$$ □

We are now ready to complete the proof of the first main result in this section.

Proof of Theorem 2.78. (a): By Remark 2.80, it suffices to consider the induced preference relation $\succ$ on $\mathscr{X}$ once the function u has been determined. According to Lemma 2.81 and the two Propositions 2.83 and 2.84, there exists a convex set $\mathcal{Q} \subseteq \mathcal{M}_{1,f}$ such that

$$U(X) = \min_{Q \in \mathcal{Q}} E_Q[\, u(X)\,]$$

is a numerical representation of $\succ$ on $\mathscr{X}$. This proves the first part of the assertion.

(b): The assumption (2.41) applied to $X \equiv 1$ and $Y \equiv b < 1$ gives that any sequence with $X_n \nearrow 1$ is such that $X_n \succ b$ for large enough n. We claim that this implies that $U(X_n) \nearrow u(1) = 1$. Otherwise, $U(X_n)$ would increase to some number $a < 1$. Since u is continuous and strictly increasing, we may take b such that $a < u(b) < 1$. But then $U(X_n) > U(b) = u(b) > a$ for large enough n, which is a contradiction.

In particular, we obtain that for any increasing sequence of events $A_n \in \mathscr{F}$ with $\bigcup_n A_n = \Omega$,

$$\lim_{n\uparrow\infty} \min_{Q\in\mathcal{Q}} Q[\,A_n\,] = \lim_{n\uparrow\infty} U(\mathbb{1}_{A_n}) = 1.$$

But this means that each $Q \in \mathcal{Q}$ satisfies $\lim_n Q[\,A_n\,] = 1$, which is equivalent to the σ-additivity of Q. □

The continuity assumption (2.41), required for *all* $X_n \in \mathscr{X}$, is actually quite strong. In a topological setting, our discussion of risk measures in Chapter 4 will imply the following version of the representation theorem.

Proposition 2.85. *Consider a preference order $\succ$ as in Theorem* 2.78. *Suppose that Ω is a Polish space with Borel σ-field $\mathscr{F}$ and that* (2.41) *holds if X_n and X are continuous. Then there exists a class of probability measures $\mathcal{Q} \subseteq \mathcal{M}_1(\Omega, \mathscr{F})$ such that the induced preference order on $\mathscr{X}$ has the* robust Savage representation

$$U(X) = \min_{Q\in\mathcal{Q}} E_Q[\, u(X)\,] \quad \textit{for continuous } X \in \mathscr{X}.$$

Proof. As in the proof of Theorem 2.78, the continuity property of $\succ$ implies the corresponding continuity property of U, and hence of the functional ϕ in (2.49). The result follows by combining Proposition 2.83, which reduces the representation of U to a representation of ϕ, with Proposition 4.27 applied to the coherent risk measure $\rho := -\phi$. □

Now we consider an alternative setting where we fix in advance a reference measure P on $(\Omega, \mathscr{F})$. In this context, $\mathscr{X}$ will be identified with the space $L^\infty(\Omega, \mathscr{F}, P)$, and the representation of preferences will involve measures that are absolutely continuous with respect to P. Note, however, that this passage from measurable functions to equivalence classes of random variables in $L^\infty(\Omega, \mathscr{F}, P)$, and from arbitrary probability measures to absolutely continuous measures, involves a certain loss of robustness in the face of model uncertainty.

Theorem 2.86. *Let $\succ$ be a preference relation as in Theorem 2.78, and assume that*

$$X \sim Y \quad \text{whenever } X = Y \text{ } P\text{-a. s.}$$

(a) *The following conditions are equivalent:*

  (i) *There exists a robust Savage representation of the form*

$$U(X) = \inf_{Q \in \mathcal{Q}} E_Q[\, u(X)\,], \quad X \in \mathscr{X},$$

  where $\mathcal{Q}$ consists of probability measures on $(\Omega, \mathscr{F})$ that are absolutely continuous with respect to P.

  (ii) *The preference relation $\succ$ satisfies the following condition of continuity from above:*

$$Y \succ X \text{ and } X_n \searrow X \text{ } P\text{-a. s.} \quad \Longrightarrow \quad Y \succ X_n \quad P\text{-a. s. for all large } n.$$

(b) *The following conditions are equivalent:*

  (i) *There exists a representation of the form*

$$U(X) = \min_{Q \in \mathcal{Q}} E_Q[\, u(X)\,], \quad X \in \mathscr{X},$$

  where $\mathcal{Q}$ consists of probability measures on $(\Omega, \mathscr{F})$ that are absolutely continuous with respect to P.

  (ii) *The preference relation $\succ$ satisfies the following condition of continuity from below:*

$$X \succ Y \text{ and } X_n \nearrow X \text{ } P\text{-a. s.} \quad \Longrightarrow \quad X_n \succ Y \quad P\text{-a. s. for all large } n.$$

Proof. As in the proof of Theorem 2.78, the continuity property of $\succ$ implies the corresponding continuity property of U, and hence of the functional ϕ in (2.49). The results follow by combining Proposition 2.83, which reduces the representation of U to a representation of ϕ, with Corollary 4.37 and Corollary 4.38 applied to the coherent risk measure $\rho := -\phi$. □

In the following two exercises, we explore the impact of replacing the axiom of certainty independence by stronger requirements as discussed in Remark 2.77.

Exercise 2.5.3. Show that the following conditions are equivalent:

(a) The preference relation $\succ$ satisfies the following unrestricted independence axiom on $\widetilde{X}$:

  Independence: For $\widetilde{X}, \widetilde{Y}, \widetilde{Z} \in \widetilde{\mathscr{X}}$, and $\alpha \in (0, 1]$, we have

$$\widetilde{X} \succ \widetilde{Y} \quad \Longleftrightarrow \quad \alpha\widetilde{X} + (1-\alpha)\widetilde{Z} \succ \alpha\widetilde{Y} + (1-\alpha)\widetilde{Z}.$$

(b) The functional ϕ is additive: $\phi(X+Y) = \phi(X) + \phi(Y)$ for $X \in \mathscr{X}$.
(c) The set $\mathcal{Q}$ has exactly one element $Q \in \mathscr{M}_{1,f}$, and so $U(X) = \phi(u(X))$ admits the Savage representation

$$U(X) = E_Q[\,u(X)\,], \quad X \in \mathscr{X}.$$

◇

Exercise 2.5.4. Two random variables $X, Y \in \mathscr{X}$ will be called *comonotone* if

$$[X(\omega) - X(\omega')][Y(\omega) - Y(\omega')] \geq 0 \quad \text{for all pairs } (\omega, \omega') \in \Omega \times \Omega.$$

Show that the following conditions are equivalent:
(a) The preference relation $\succ$ satisfies the axiom of comonotonic independence (2.40).
(b) The functional ϕ is *comonotonic* in the sense that $\phi(X+Y) = \phi(X) + \phi(Y)$ whenever $X, Y \in \mathscr{X}$ are comonotone. ◇

Now we discuss what happens if we replace the assumption of certainty independence by weak certainty independence as introduced in Remark 2.77. We thus assume from now on that $\succ$ is a preference relation on $\widetilde{\mathscr{X}}$ satisfying the following conditions:
- *Weak certainty independence.*
- *Uncertainty aversion.*
- *Monotonicity.*
- *Continuity.*

Exercise 2.5.5. Show that the restriction of $\succ$ to $\mathscr{M}_b(\mathbb{R})$ satisfies the independence axiom of von Neumann–Morgenstern theory, and hence admits a von Neumann–Morgenstern representation

$$\tilde{u}(\mu) := \int u(x)\,\mu(dx), \quad \mu \in \mathscr{M}_b(\mathbb{R}), \tag{2.52}$$

with a continuous and strictly increasing function $u : \mathbb{R} \to \mathbb{R}$. ◇

For simplicity, we will assume for the remainder of this section that the function u in (2.52) has an unbounded range $u(\mathbb{R})$ containing zero. The assumption of an unbounded range is satisfied automatically if the restriction of $\succ$ to $\mathscr{M}_b(\mathbb{R})$ is risk averse and u hence concave. Lemma 2.81 and Remark 2.82 imply that the numerical representation $\tilde{u}$ in (2.52) admits a unique extension $\widetilde{U} : \widetilde{\mathscr{X}} \to \mathbb{R}$ that is a numerical representation of $\succ$ on all of $\widetilde{\mathscr{X}}$. In analogy to (2.48), we define

$$U(X) := \widetilde{U}(\delta_X) \quad \text{for } X \in \mathscr{X}.$$

As in Remark 2.80, we see that $\widetilde{X} \sim \delta_X$ if $X \in \mathscr{X}$ is defined as $X(\omega) := c(\widetilde{X}(\omega))$ with

$$c(\mu) = u^{-1}\Big(\int u\,d\mu\Big)$$

denoting the certainty equivalent of a lottery $\mu \in \mathcal{M}_b(\mathbb{R})$. In particular, the preference relation $\succ$ on $\widetilde{\mathcal{X}}$ is uniquely determined by its restriction to $\mathcal{X}$. We now analyze the structure of U in analogy to Proposition 2.83. Our next result shows that U is again of the form $U(X) = \phi(u(X))$ for a functional $\phi : \mathcal{X} \to \mathbb{R}$. However, ϕ may no longer be positively homogeneous, since we have replaced certainty independence with weak certainty independence.

Proposition 2.87. *Under the above assumptions, there exists a unique functional* $\phi : \mathcal{X} \to \mathbb{R}$ *such that*

$$\widetilde{U}(X) = \phi(\widetilde{u}(\widetilde{X})) \quad \text{for all } \widetilde{X} \in \widetilde{\mathcal{X}}, \tag{2.53}$$

and such that the following three properties are satisfied:

- Monotonicity: *If* $Y(\omega) \geq X(\omega)$ *for all* ω, *then* $\phi(Y) \geq \phi(X)$.
- Concavity: *If* $\lambda \in [0,1]$, *then* $\phi(\lambda X + (1-\lambda)Y) \geq \lambda\phi(X) + (1-\lambda)\phi(Y)$.
- Cash invariance: $\phi(X+z) = \phi(X) + z$ *for all* $z \in \mathbb{R}$.

Proof. Let $\mathcal{X}_u$ denote the set of all $X \in \mathcal{X}$ whose range is relatively compact in $u(\mathbb{R})$. Since u is strictly increasing, we can define ϕ on $\mathcal{X}_u$ via

$$\phi(X) := \widetilde{U}(\delta_{u^{-1}(X)}), \quad X \in \mathcal{X}_u.$$

Then we have

$$\widetilde{U}(\widetilde{X}) = \widetilde{U}(\delta_{c(\widetilde{X})}) = \phi(u(c(\widetilde{X}))) = \phi(\widetilde{u}(\widetilde{X})), \tag{2.54}$$

and so (2.53) follows. Moreover, ϕ is monotone on $\mathcal{X}_u$ due to our monotonicity assumption.

We now prove that ϕ is cash invariant on $\mathcal{X}_u$. To this end, we assume first that $u(\mathbb{R}) = \mathbb{R}$. For $X \in \mathcal{X}$ and $z \in \mathbb{R}$, we let $X_0 := u^{-1}(2X)$, $z_0 := u^{-1}(2z)$, and $y := u^{-1}(0)$. Taking $a > 0$ such that $a \geq X_0(\omega) \geq -a$ for each ω, we see as in (2.44) that there exists $\beta \in [0,1]$ such that

$$\frac{1}{2}\delta_{X_0} + \frac{1}{2}\delta_y \sim \frac{\beta}{2}(\delta_a + \delta_y) + \frac{1-\beta}{2}(\delta_{-a} + \delta_y) = \frac{1}{2}\mu + \frac{1}{2}\delta_y,$$

where $\mu = \beta\delta_a + (1-\beta)\delta_{-a}$. Using weak certainty independence, we may replace δ_y by δ_{z_0} and obtain $\frac{1}{2}(\delta_{X_0} + \delta_{z_0}) \sim \frac{1}{2}(\mu + \delta_{z_0})$. Indeed, if we assume for instance that $\frac{1}{2}(\delta_{X_0} + \delta_{z_0}) \succ \frac{1}{2}(\mu + \delta_{z_0})$, then weak certainty aversion would yield the contradiction $\frac{1}{2}(\delta_{X_0} + \delta_y) \succ \frac{1}{2}(\mu + \delta_y)$. Hence, from (2.54),

$$\begin{aligned}\phi(X+z) &= \phi\Big(\frac{1}{2}u(X_0)+\frac{1}{2}u(z_0)\Big) = \phi\Big(\widetilde{u}\Big(\frac{1}{2}\delta_{X_0}+\frac{1}{2}\delta_{z_0}\Big)\Big)\\ &= \widetilde{U}\Big(\frac{1}{2}\delta_{X_0}+\frac{1}{2}\delta_{z_0}\Big) = \widetilde{U}\Big(\frac{1}{2}\mu+\frac{1}{2}\delta_{z_0}\Big)\\ &= \widetilde{u}\Big(\frac{1}{2}\mu+\frac{1}{2}\delta_{z_0}\Big) = \frac{1}{2}\widetilde{u}(\mu)+\frac{1}{2}u(z_0).\end{aligned}$$

Cash invariance now follows from $u(z_0) = 2z$ and the fact that

$$\begin{aligned}\frac{1}{2}\widetilde{u}(\mu) = \frac{1}{2}(\widetilde{u}(\mu)+u(\delta_y)) &= \widetilde{u}\Big(\frac{1}{2}\mu+\frac{1}{2}\delta_y\Big) = \widetilde{U}\Big(\frac{1}{2}\mu+\frac{1}{2}\delta_y\Big)\\ &= \widetilde{U}\Big(\frac{1}{2}\delta_{X_0}+\frac{1}{2}\delta_y\Big) = \phi\Big(\frac{1}{2}u(X_0)+\frac{1}{2}u(y)\Big) = \phi(X).\end{aligned}$$

Here, we have again applied (2.54). If $u(\mathbb{R})$ is not equal to $\mathbb{R}$, it is sufficient to consider the cases in which $u(\mathbb{R})$ contains $[0,\infty)$ or $(-\infty,0]$ and to work with positive or negative quantities X and z, respectively. Then the preceding argument establishes the cash invariance of ϕ on the spaces of positive or negative bounded measurable functions, and ϕ can be extended by translation to the entire space of bounded measurable functions.

Now we prove the concavity of ϕ by showing $\phi(\frac{1}{2}(X+Y)) \geq \frac{1}{2}\phi(X)+\frac{1}{2}\phi(Y)$. This is sufficient due to Exercise 2.5.2. Let $X_0 := u^{-1}(X)$ and $Y_0 := u^{-1}(Y)$ and suppose first that $\phi(X) = \phi(Y)$. Then $\delta_{X_0} \sim \delta_{Y_0}$ and uncertainty aversion implies that $\widetilde{Z} := \frac{1}{2}\delta_{X_0}+\frac{1}{2}\delta_{Y_0} \succeq \delta_{X_0}$. Hence, by using (2.54),

$$\phi\Big(\frac{1}{2}(X+Y)\Big) = \widetilde{U}(\widetilde{Z}) \geq \widetilde{U}(\delta_{X_0}) = \phi(X) = \frac{1}{2}\phi(X)+\frac{1}{2}\phi(Y).$$

If $\phi(X) \neq \phi(Y)$, we let $z := \phi(X)-\phi(Y)$ so that $Y_z := Y+z$ satisfies $\phi(Y_z) = \phi(X)$. Hence,

$$\phi\Big(\frac{1}{2}(X+Y)\Big)+\frac{1}{2}z = \phi\Big(\frac{1}{2}(X+Y_z)\Big) \geq \frac{1}{2}\phi(X)+\frac{1}{2}\phi(Y_z) = \frac{1}{2}\phi(X)+\frac{1}{2}\phi(Y)+\frac{1}{2}z.$$

The preceding argument was done under the implicit assumption that Y_z belongs again to $\mathscr{X}_u$. If this is not the case, we switch the roles of X and Y. □

A functional $\phi : \mathscr{X} \to \mathbb{R}$ satisfying the properties of monotonicity, concavity, and cash invariance is sometimes called a *concave monetary utility functional*, and $\rho := -\phi$ is called a *convex risk measure*. In Chapter 4, we will derive various representation results for convex risk measures. In particular, it will follow from Theorem 4.16 that every concave monetary utility functional $\phi : \mathscr{X} \to \mathbb{R}$ is of the form

$$\phi(X) = \min_{Q\in\mathscr{M}_{1,f}} (E_Q[X]+\alpha(Q)), \quad X \in \mathscr{X}, \tag{2.55}$$

where the *penalty function* $\alpha : \mathscr{M}_{1,f} \to \mathbb{R}\cup\{+\infty\}$ is bounded from below.

Exercise 2.5.6. Prove the representation (2.55) for the case in which $\mathscr{X}$ is the set of all functions $X : \Omega \to \mathbb{R}$ on a finite set Ω. To this end, one can either use biduality (Theorem H.9) or, more directly, a separation argument as given in Proposition A.5. ◇

Combining Proposition 2.87 with the representation (2.55) yields the final result of this section.

Theorem 2.88. *Consider a preference order $\succ$ on $\widetilde{\mathscr{X}}$ satisfying the four properties of uncertainty aversion, weak certainty independence, monotonicity, and continuity. Assume moreover that the function u in* (2.52) *has an unbounded range $u(\mathbb{R})$. Then there exists a penalty function $\alpha : \mathscr{M}_{1,f} \to \mathbb{R} \cup \{+\infty\}$ that is bounded from below such that*

$$\widetilde{U}(\widetilde{X}) = \min_{Q \in \mathscr{M}_{1,f}} \left(E_Q\left[\int u(x)\widetilde{X}(\cdot, dx) \right] + \alpha(Q) \right), \quad \widetilde{X} \in \widetilde{\mathscr{X}},$$

is a numerical representation of $\succ$.

2.6 Probability measures with given marginals

In this section, we study the construction of probability measures with given marginals. In particular, this will yield the missing implication in the characterization of uniform preferences in Theorem 2.57, but the results in this section are of independent interest. We focus on the following basic question: Suppose μ_1 and μ_2 are two probability measures on S, and Λ is a convex set of probability measures on $S \times S$; when does Λ contain some $\overline{\mu}$ that has μ_1 and μ_2 as marginals?

The answer to this question will be given in a general topological setting. Let S be a Polish space, and let us fix a continuous function ψ on S with values in $[1, \infty)$. As in Section 2.2 and in Appendix G, we use ψ as a *gauge function* in order to define the space of measures

$$\mathscr{M}_1^{\psi}(S) := \left\{ \mu \in \mathscr{M}_1(S) \,\middle|\, \int \psi(x)\,\mu(dx) < \infty \right\}$$

and the space of continuous test functions

$$C_\psi(S) := \{ f \in C(S) \mid \text{there exists } c \text{ such that } |f(x)| \le c \cdot \psi(x) \text{ for all } x \in S \}.$$

The ψ-weak topology on $\mathscr{M}_1^{\psi}(S)$ is the coarsest topology such that

$$\mathscr{M}_1^{\psi}(S) \ni \mu \mapsto \int f\,d\mu$$

is a continuous mapping for all $f \in C_\psi(S)$; see Appendix G.2. On the product space $S \times S$, we take the gauge function

$$\overline{\psi}(x,y) := \psi(x) + \psi(y),$$

and define the corresponding set $\mathcal{M}_1^{\overline{\psi}}(S \times S)$, which will be endowed with the $\overline{\psi}$-weak topology.

Theorem 2.89. *Suppose that $\Lambda \subseteq \mathcal{M}_1^{\overline{\psi}}(S \times S)$ is convex and closed in the $\overline{\psi}$-weak topology, and that μ_1, μ_2 are probability measures in $\mathcal{M}_1^{\psi}(S)$. Then there exists some $\overline{\mu} \in \Lambda$ with marginal distributions μ_1 and μ_2 if and only if*

$$\int f_1 \, d\mu_1 + \int f_2 \, d\mu_2 \leq \sup_{\lambda \in \Lambda} \int (f_1(x) + f_2(y)) \, \lambda(dx, dy) \quad \textit{for all } f_1, f_2 \in C_\psi(S).$$

Theorem 2.89 is due to V. Strassen [309]. Its proof boils down to an application of the Hahn–Banach theorem; the difficult part consists in specifying the right topological setting. First, let us investigate the relations between $\mathcal{M}_1^{\overline{\psi}}(S \times S)$ and $\mathcal{M}_1^{\psi}(S)$. To this end, we define mappings

$$\pi_i : \mathcal{M}_1^{\overline{\psi}}(S \times S) \to \mathcal{M}_1^{\psi}(S), \qquad i = 1, 2,$$

that yield the ith marginal distribution of a measure $\lambda \in \mathcal{M}_1^{\overline{\psi}}(S \times S)$:

$$\int f \, d(\pi_1 \lambda) = \int f(x) \, \lambda(dx, dy) \quad \text{and} \quad \int f \, d(\pi_2 \lambda) = \int f(y) \, \lambda(dx, dy),$$

for all $f \in C_\psi(S)$.

Lemma 2.90. *π_1 and π_2 are continuous and affine mappings from $\mathcal{M}_1^{\overline{\psi}}(S \times S)$ to $\mathcal{M}_1^{\psi}(S)$.*

Proof. Suppose that λ_n converges to λ in $\mathcal{M}_1^{\overline{\psi}}(S \times S)$. For $f \in C_\psi(S)$, let $\overline{f}(x,y) := f(x)$. Clearly, $\overline{f} \in C_{\overline{\psi}}(S \times S)$, and thus

$$\int f \, d(\pi_1 \lambda_n) = \int \overline{f} \, d\lambda_n \longrightarrow \int \overline{f} \, d\lambda = \int f \, d(\pi_1 \lambda).$$

Therefore, π_1 is continuous, and the same is true of π_2. Affinity is obvious. □

Now, let us consider the linear space

$$E := \{\alpha\mu - \beta\nu \mid \mu, \nu \in \mathcal{M}_1^{\psi}(S), \ \alpha, \beta \in \mathbb{R}\}$$

spanned by $\mathcal{M}_1^{\psi}(S)$. For $\rho = \alpha\mu - \beta\nu \in E$, the integral $\int f \, d\rho$ against a function $f \in C_\psi(S)$ is well-defined and given by

$$\int f \, d\rho = \alpha \int f \, d\mu - \beta \int f \, d\nu.$$

In particular, $\rho \mapsto \int f\,d\rho$ is a linear functional on E, so we can regard $C_\psi(S)$ as a subset of the algebraic dual E^* of E. Note that $\int f\,d\rho = \int f\,d\tilde{\rho}$ for all $f \in C_\psi(S)$ implies $\rho = \tilde{\rho}$, that is, $C_\psi(S)$ separates the points of E. We endow E with the coarsest topology $\sigma(E, C_\psi(S))$ for which all maps

$$E \ni \rho \mapsto \int f\,d\rho, \quad f \in C_\psi(S),$$

are continuous; see Definition H.5. With this topology, E becomes a locally convex topological vector space.

Lemma 2.91. *Under the above assumptions, $\mathcal{M}_1^\psi(S)$ is a closed convex subset of E, and the relative topology of the embedding coincides with the ψ-weak topology.*

Proof. The sets of the form

$$U_\varepsilon(\rho; f_1, \dots, f_n) := \bigcap_{i=1}^{n} \left\{ \tilde{\rho} \in E \,\middle|\, \left| \int f_i\,d\rho - \int f_i\,d\tilde{\rho} \right| < \varepsilon \right\}$$

with $\rho \in E$, $n \in \mathbb{N}$, $f_i \in C_\psi(S)$, and $\varepsilon > 0$ form a base of the topology $\sigma(E, C_\psi(S))$. Thus, if $U \subseteq E$ is open, then every point $\mu \in U \cap \mathcal{M}_1^\psi(S)$ has some neighborhood $U_\varepsilon(\mu; f_1, \dots, f_n) \subseteq U$. But $U_\varepsilon(\mu; f_1, \dots, f_n) \cap \mathcal{M}_1^\psi(S)$ is an open neighborhood of μ in the ψ-weak topology. Hence, $U \cap \mathcal{M}_1^\psi(S)$ is open in the ψ-weak topology. Similarly, one shows that every open set $V \subseteq \mathcal{M}_1^\psi(S)$ is of the form $V = U \cap \mathcal{M}_1^\psi(S)$ for some open subset U of E. This shows that the relative topology $\mathcal{M}_1^\psi(S) \cap \sigma(E, C_\psi(S))$ coincides with the ψ-weak topology.

Moreover, $\mathcal{M}_1^\psi(S)$ is an intersection of closed subsets of E:

$$\mathcal{M}_1^\psi(S) = \left\{ \rho \in E \,\middle|\, \int 1\,d\rho = 1 \right\} \cap \bigcap_{\substack{f \in C_\psi(S) \\ f \geq 0}} \left\{ \rho \in E \,\middle|\, \int f\,d\rho \geq 0 \right\}.$$

Therefore, $\mathcal{M}_1^\psi(S)$ is closed in E. □

Next, let E^2 denote the product space $E \times E$. We endow E^2 with the product topology for which the sets $U \times V$ with $U, V \in \sigma(E, C_\psi(S))$ form a neighborhood base. Clearly, E^2 is a locally convex topological vector space.

Lemma 2.92. *Every continuous linear functional ℓ on E^2 is of the form*

$$\ell(\rho_1, \rho_2) = \int f_1\,d\rho_1 + \int f_2\,d\rho_2$$

for some $f_1, f_2 \in C_\psi(S)$.

Proof. By linearity, ℓ is of the form $\ell(\rho_1, \rho_2) = \ell_1(\rho_1) + \ell_2(\rho_2)$, where $\ell_1(\rho_1) := \ell(\rho_1, 0)$ and $\ell_2(\rho_2) := \ell(0, \rho_2)$. By continuity of ℓ, the set

$$V := \ell^{-1}((-1,1))$$

is open in E^2 and contains the point $(0,0)$. Hence, there are two open neighborhoods $U_1, U_2 \subseteq E$ such that $(0,0) \in U_1 \times U_2 \subseteq V$. Therefore,

$$0 \in U_i \subseteq \ell_i^{-1}((-1,1)) \quad \text{for } i = 1,2,$$

that is, 0 is an interior point of $\ell_i^{-1}((-1,1))$. It follows that the ℓ_i are continuous at 0, which in view of their linearity implies continuity everywhere on E. Finally, we may conclude from Proposition H.6 that each ℓ_i is of the form $\ell_i(\rho) = \int f_i \, d\rho_i$ for some $f_i \in C_\psi(S)$. □

The proof of the following lemma uses the characterization of compact sets for the ψ-weak topology that is stated in Corollary G.14. It is here that we need our assumption that S is Polish.

Lemma 2.93. *If Λ is a closed convex subset of $\mathcal{M}_1^{\overline{\psi}}(S \times S)$, then*

$$H_\Lambda := \{(\pi_1\lambda, \pi_2\lambda) \mid \lambda \in \Lambda\}$$

is a closed convex subset of E^2.

Proof. Convexity follows from the affinity of the projections π_i, as established in Lemma 2.90. To prove closedness, it is enough to show that H_Λ is closed in $\mathcal{M}_1^\psi(S)^2 := \mathcal{M}_1^\psi(S) \times \mathcal{M}_1^\psi(S)$, because Lemma 2.91 implies that the relative topology induced by E^2 on $\mathcal{M}_1^\psi(S)^2$ coincides with the product topology for the ψ-weak topology. The ψ-weak topology is metrizable by Corollary G.12. So, let $(\mu_n, \nu_n) \in H_\Lambda$, $n \in \mathbb{N}$ be a sequence converging to some $(\mu, \nu) \in \mathcal{M}_1^\psi(S)^2$ in the product topology. Since both sequences $(\mu_n)_{n\in\mathbb{N}}$ and $(\nu_n)_{n\in\mathbb{N}}$ are relatively compact for the ψ-weak topology, Corollary G.14 yields functions $\phi_i : S \to [1,\infty]$, $i = 1,2$, such that sets of the form $K_k^i := \{\phi_i \le k\psi\}$, $k \in \mathbb{N}$, are relatively compact in S and such that

$$\sup_{n\in\mathbb{N}} \int \phi_1 \, d\mu_n + \sup_{n\in\mathbb{N}} \int \phi_2 \, d\nu_n < \infty.$$

For each n, there exists $\lambda_n \in \Lambda$ such that $\pi_1\lambda_n = \mu_n$ and $\pi_2\lambda_n = \nu_n$. Hence, if we let $\overline{\phi}(x,y) := \phi_1(x) + \phi_2(y)$, then

$$\sup_{n\in\mathbb{N}} \int \overline{\phi} \, d\lambda_n = \sup_{n\in\mathbb{N}} \Big(\int \phi_1 \, d\mu_n + \int \phi_2 \, d\nu_n \Big) < \infty.$$

Moreover, we claim that each set $\{\overline{\phi} \le k\overline{\psi}\}$ is relatively compact in $S \times S$. To prove this claim, let $l_i \in \mathbb{N}$ be such that

$$l_i \ge \sup_{x\in K_k^i} \psi(x).$$

Then, since $\psi \geq 1$,

$$\{\overline{\phi} \leq k\overline{\psi}\} \subseteq K^1_{k(1+l_2)} \times K^2_{k(1+l_1)},$$

and the right-hand side is a relatively compact set in $S \times S$.

It follows from Corollary G.14 that the sequence $(\lambda_n)_{n\in\mathbb{N}}$ is relatively compact for the $\overline{\psi}$-weak topology. Any accumulation point λ of this sequence belongs to the closed set Λ. Moreover, λ has marginal distributions μ and ν, since the projections π_i are continuous according to Lemma 2.90. Hence, $(\mu, \nu) \in H_\Lambda$. □

Proof of Theorem 2.89. Let $\mu_1, \mu_2 \in \mathcal{M}_1^\psi(S)$ be given. Since H_Λ is closed and convex in E^2 by Lemma 2.93, we may apply the Hahn–Banach separation theorem in the form of Theorem H.4 with $\mathcal{B} := \{(\mu_1, \mu_2)\}$ and $\mathcal{C} := H_\Lambda$: We conclude that $(\mu_1, \mu_2) \notin H_\Lambda$ if and only if there exists a linear functional ℓ on E^2 such that

$$\ell(\mu_1, \mu_2) > \sup_{(\nu_1,\nu_2)\in H_\Lambda} \ell(\nu_1, \nu_2) = \sup_{\lambda\in\Lambda} \ell(\pi_1\lambda, \pi_2\lambda).$$

Applying Lemma 2.92 to ℓ completes the assertion. □

We will now use Theorem 2.89 to establish the remaining implication (b)⇒(f) in the proof of Theorem 2.57. We consider here a more general, d-dimensional setting. To this end, let $x = (x^1, \dots, x^d)$ and $y = (y^1, \dots, y^d)$ be two d-dimensional vectors. We will say that $x \leq y$ if $x^i \leq y^i$ for all i. A function on $\mathbb{R}^d$ is called increasing, if it is increasing with respect to the partial order $\leq$. Let moreover $\mathcal{M}$ denote the set of Borel probability measures μ on $\mathbb{R}^d$ with $\int |x|\mu(dx) < \infty$. In analogy to Definition 2.55, we say that $\mu_1 \in \mathcal{M}$ dominates $\mu_2 \in \mathcal{M}$ in the *increasing concave order*, and we write $\mu_1 \succcurlyeq_{\text{icv}} \mu_2$, if

$$\int f\,d\mu_1 \geq \int f\,d\mu_2 \quad \text{for all increasing concave functions } f \text{ on } \mathbb{R}^d.$$

As in the one-dimensional case, the stochastic order $\succcurlyeq_{\text{icv}}$ is sometimes also called *second-order stochastic dominance* or *stochastic dominance of the second kind*.

Theorem 2.94. *For $\mu_1, \mu_2 \in \mathcal{M}$ the following assertions are equivalent:*

(a) $\mu_1 \succcurlyeq_{\text{icv}} \mu_2$.

(b) *There exists a probability space $(\Omega, \mathcal{F}, P)$ with random variables X_1 and X_2 having distributions μ_1 and μ_2, respectively, such that*

$$E[X_2 \mid X_1] \leq X_1 \quad P\text{-a. s.}$$

(c) *There exists a kernel $Q(x, dy)$ on $\mathbb{R}^d$ such that*

$$\int_{\mathbb{R}^d} y\,Q(x, dy) \leq x \quad \text{for all } x \in \mathbb{R}^d$$

and such that $\mu_2 = \mu_1 Q$.

Proof. (a)⇒(b): We will apply Theorem 2.89 with $S := \mathbb{R}^d$ and with the gauge functions $\psi(x) := 1 + |x|$ and $\overline{\psi}(x,y) := \psi(x) + \psi(y)$. We denote by $C_b(\mathbb{R}^d)$ the set of bounded and continuous functions on $\mathbb{R}^d$. Let

$$\Lambda := \bigcap_{f \in C_b(\mathbb{R}^d)} \left\{ \lambda \in \mathcal{M}_1^{\overline{\psi}}(\mathbb{R}^d \times \mathbb{R}^d) \,\Big|\, \int y f(x)\,\lambda(dx,dy) \le \int x f(x)\,\lambda(dx,dy) \right\}.$$

Each single set of the intersection is convex and closed in $\mathcal{M}_1^{\overline{\psi}}(\mathbb{R}^d \times \mathbb{R}^d)$, because the functions $g(x,y) := yf(x)$ and $\widetilde{g}(x,y) := xf(x)$ belong to $C_{\overline{\psi}}(\mathbb{R}^d \times \mathbb{R}^d)$ for $f \in C_b(S)$. Therefore, Λ itself is convex and closed.

Suppose we can show that Λ contains an element P that has μ_1 and μ_2 as marginal distributions. Then we can take $\Omega := \mathbb{R}^d \times \mathbb{R}^d$ with its Borel σ-algebra $\mathscr{F}$, and let X_1 and X_2 denote the canonical projections on the first and the second components, respectively. By definition, X_i will have the distribution μ_i, and

$$E[\,E[\,X_2 \mid X_1\,]f(X_1)\,] = E[\,X_2 f(X_1)\,] \le E[\,X_1 f(X_1)\,] \quad \text{for all } f \in C_b(\mathbb{R}^d).$$

By monotone class arguments (see, e. g., Theorem I.1.8 in [267]), we may thus conclude that

$$E[\,X_2 \mid X_1\,] \le X_1 \qquad P\text{-a. s.}$$

so that the assertion will follow.

It remains to prove the existence of P. To this end, we will apply Theorem 2.89 with the set Λ defined above. Take a pair $f_1, f_2 \in C_\psi(\mathbb{R}^d)$, and let

$$\widetilde{f}_2(x) := \inf\{g(x) \mid g \text{ is concave, increasing, and dominates } f_2\} \in \mathbb{R} \cup \{+\infty\}.$$

Then $\widetilde{f}_2$ is concave, increasing, and dominates f_2. In fact, $\widetilde{f}_2$ is the smallest function with these properties. From (a), we have

$$\begin{aligned}
\int f_1\,d\mu_1 + \int f_2\,d\mu_2 &\le \int f_1\,d\mu_1 + \int \widetilde{f}_2\,d\mu_2 \\
&\le \int (f_1 + \widetilde{f}_2)\,d\mu_1 \\
&\le \sup_{x \in \mathbb{R}^d} (f_1(x) + \widetilde{f}_2(x)) =: r_0.
\end{aligned}$$

We will establish the condition in Theorem 2.89 for our set Λ by showing that for $r < r_0$ we have

$$r < \sup_{\lambda \in \Lambda} \int (f_1(x) + f_2(y))\,\lambda(dx,dy).$$

To this end, let for $z \in \mathbb{R}^d$,

$$\Lambda_z := \left\{ \nu \in \mathcal{M}_1^{\psi}(\mathbb{R}^d) \,\middle|\, \int x\, \nu(dx) \le z \right\}$$

and

$$g_2(z) := \sup\left\{ \int f_2 \, d\nu \,\middle|\, \nu \in \Lambda_z \right\}.$$

Then g_2 is increasing and $g_2(z) \ge f_2(z)$, because $\delta_z \in \Lambda_z$. Moreover, if $\nu_1 \in \Lambda_{z_1}$ and $\nu_2 \in \Lambda_{z_2}$, then

$$\alpha\nu_1 + (1-\alpha)\nu_2 \in \Lambda_{\alpha z_1 + (1-\alpha) z_2}$$

for $\alpha \in [0,1]$. Therefore, g_2 is concave, and we conclude that $g_2 \ge \widetilde{f}_2$ (recall that $\widetilde{f}_2$ is the smallest increasing and concave function dominating f_2). Hence, $r < f_1(z) + g_2(z)$ for some $z \in \mathbb{R}^d$, that is, there exists some $\nu \in \Lambda_z$ such that the product measure $\lambda := \delta_z \otimes \nu$ satisfies

$$r < f_1(z) + \int f_2 \, d\nu = \int (f_1(x) + f_2(y))\, \lambda(dx, dy).$$

But $\lambda = \delta_z \otimes \nu \in \Lambda$.

(b)⇒(c): This follows as in the proof of the implication (f)⇒(g) of Theorem 2.57 by using regular conditional distributions.

(c)⇒(a): As in the proof of (g)⇒(a) of Theorem 2.57, this follows by an application of Jensen's inequality. □

As in Remark 2.63, we can also define the concave order order for $\mu_1, \mu_2 \in \mathcal{M}$. We will say that μ_1 dominates μ_2 in the *concave order*, and we will write $\mu_1 \succcurlyeq_{cv} \mu_2$, if

$$\int f \, d\mu_1 \ge \int f \, d\mu_2 \quad \text{for all concave functions } f \text{ on } \mathbb{R}^d.$$

By the same arguments as for Corollary 2.61, we obtain the following result from Theorem 2.94.

Corollary 2.95. *For $\mu_1, \mu_2 \in \mathcal{M}$, the following assertions are equivalent:*

(a) $\mu_1 \succcurlyeq_{cv} \mu_2$.

(b) *There exists a probability space $(\Omega, \mathcal{F}, P)$ with random variables X_1 and X_2 having distributions μ_1 and μ_2, respectively, such that*

$$E[X_2 \mid X_1] = X_1 \quad P\text{-a.s.}$$

(c) *There exists a kernel $Q(x, dy)$ on $\mathbb{R}^d$ such that*

$$\int y \, Q(x, dy) = x \quad \text{for all } x \in \mathbb{R}^d$$

(*i. e., Q is a* mean-preserving spread) *and such that $\mu_2 = \mu_1 Q$.*

We conclude this section with a generalization of Theorem 2.68. Let S be a Polish space that is endowed with a preference order $\succ$. We will assume that $\succ$ is continuous in the sense of Definition 2.8. A function on S will be called *increasing* if it is increasing with respect to $\succ$. Let μ_1 and μ_2 be two Borel probability measures on S. We say that μ_1 dominates μ_2 in the *increasing order*, and we write $\mu_1 \succcurlyeq_{\mathrm{i}} \mu_2$, if

$$\int f\,d\mu_1 \geq \int f\,d\mu_2 \quad \text{for all bounded, increasing, and measurable functions } f \text{ on } S.$$

Theorem 2.96. *For two Borel probability measures μ_1 and μ_2 on S, the following conditions are equivalent:*

(a) $\mu_1 \succcurlyeq_{\mathrm{i}} \mu_2$.

(b) *There exists a probability space $(\Omega, \mathscr{F}, P)$ with random variables X_1 and X_2 having distributions μ_1 and μ_2, respectively, such that $X_1 \succeq X_2$ P-a. s.*

(c) *There exists a kernel Q on S such that $\mu_2 = \mu_1 Q$ and*

$$Q(x, \{y \mid x \succeq y\}) = 1 \quad \textit{for all } x \in S.$$

Proof. (a)⇒(b): We will apply Theorem 2.89 with the gauge function $\psi \equiv 1$, so that $\mathscr{M}_1^{\psi}(S)$ is just the space $\mathscr{M}_1(S)$ of all Borel probability measures on S with the usual weak topology. Then $\overline{\psi} \equiv 2$, which is equivalent to taking $\overline{\psi} := 1$. Let

$$M := \{(x,y) \in S \times S \mid x \succeq y\}.$$

This set M is closed in $S \times S$. Hence, the portmanteau theorem in the form of Theorem G.4 implies that the convex set

$$\Lambda := \{\lambda \in \mathscr{M}_1(S \times S) \mid \lambda(M) = 1\}$$

is closed in $\mathscr{M}_1(S \times S)$. For $f_2 \in C_b(S)$, let

$$\widetilde{f}_2(x) := \sup\{f_2(y) \mid x \succeq y\}.$$

Then $\widetilde{f}_2$ is bounded, increasing, and dominates f_2. Therefore, if $f_1 \in C_b(S)$,

$$\begin{aligned}
\int f_1\,d\mu_1 + \int f_2\,d\mu_2 &\leq \int f_1\,d\mu_1 + \int \widetilde{f}_2\,d\mu_2 \\
&\leq \int (f_1 + \widetilde{f}_2)\,d\mu_1 \\
&\leq \sup_{x \in S}(f_1(x) + \widetilde{f}_2(x)) \\
&= \sup_{x \succeq y}(f_1(x) + f_2(y)).
\end{aligned}$$

If $x \succeq y$, then the product measure $\lambda := \delta_x \otimes \delta_y$ is contained in Λ, and so

$$\sup_{x \succeq y}(f_1(x) + f_2(y)) = \sup_{\lambda \in \Lambda} \int (f_1(x) + f_2(y))\, \lambda(dx, dy).$$

Hence, all assumptions of Theorem 2.89 are satisfied, and we conclude that there exists a probability measure $P \in \Lambda$ with marginals μ_1 and μ_2. Taking $\Omega := S \times S$ and X_i as the projection on the ith coordinate finishes the proof of (a)⇒(b).

(b)⇒(c) follows as in the proof of Theorem 2.57 by using regular conditional distributions.

(c)⇒(a) is proved as the corresponding implication of Theorem 2.68. □

3 Optimality and equilibrium

Consider an investor whose preferences can be expressed in terms of expected utility. In Section 3.1, we discuss the problem of constructing a portfolio that maximizes the expected utility of the resulting payoff. The existence of an optimal solution is equivalent to the absence of arbitrage opportunities. This leads to an alternative proof of the "fundamental theorem of asset pricing," and to a specific choice of an equivalent martingale measure defined in terms of marginal utility. Section 3.2 contains a detailed case study describing the interplay between exponential utility and relative entropy. In Section 3.3, the optimization problem is formulated for general contingent claims. Typically, optimal profiles will be nonlinear functions of a given market portfolio, and this is one source of the demand for financial derivatives. In Section 3.4, we solve the problem of identifying the most cost-efficient payoff satisfying a stochastic dominance constraint. In Section 3.5, we formulate and prove Arrow's theorem of the deductible, which identifies an optimal indemnity schedule as a stop-loss contract. In Section 3.6, we study the optimal investment problem for an economic agent who is averse against both risk and Knightian uncertainty. Section 3.7 introduces the idea of market equilibrium. Prices of risky assets will no longer be given in advance; they will be derived as equilibrium prices in a microeconomic setting, where different agents demand contingent claims in accordance with their preferences and with their budget constraints.

3.1 Portfolio optimization and the absence of arbitrage

Let us consider the one-period market model of Section 1.1 in which $d+1$ assets are priced at time 0 and at time 1. Prices at time 0 are given by the price system

$$\overline{\pi} = (\pi^0, \pi) = (\pi^0, \pi^1, \dots, \pi^d) \in \mathbb{R}_+^{d+1},$$

prices at time 1 are modeled by the price vector

$$\overline{S} = (S^0, S) = (S^0, S^1, \dots, S^d)$$

consisting of nonnegative random variables S^i defined on some probability space $(\Omega, \mathscr{F}, P)$. The 0th asset models a riskless bond, and so we assume that

$$\pi^0 = 1 \quad \text{and} \quad S^0 \equiv 1 + r$$

for some constant $r > -1$. At time $t = 0$, an investor chooses a portfolio

$$\overline{\xi} = (\xi^0, \xi) = (\xi^0, \xi^1, \dots, \xi^d) \in \mathbb{R}^{d+1}$$

where ξ^i represents the amount of shares of the ith asset. Such a portfolio $\overline{\xi}$ requires an initial investment $\overline{\pi} \cdot \overline{\xi}$ and yields at time 1 the random payoff $\overline{\xi} \cdot \overline{S}$.

https://doi.org/10.1515/9783111045283-004

Consider a risk-averse economic agent whose preferences are described in terms of a utility function $\tilde{u}$, and who wishes to invest a given amount w into the financial market. Recall from Definition 2.35 that a real-valued function $\tilde{u}$ is called a utility function if it is continuous, strictly increasing, and strictly concave. A rational choice of the investor's portfolio $\overline{\xi} = (\xi^0, \xi)$ will be based on the expected utility

$$E[\, \tilde{u}(\overline{\xi} \cdot \overline{S})\,] \tag{3.1}$$

of the payoff $\overline{\xi} \cdot \overline{S}$ at time 1, where the portfolio $\overline{\xi}$ satisfies the *budget constraint*

$$\overline{\pi} \cdot \overline{\xi} \le w. \tag{3.2}$$

Thus, the problem is to maximize the expected utility (3.1) among all portfolios $\overline{\xi} \in \mathbb{R}^{d+1}$ that satisfy the budget constraint (3.2). Here, we make the implicit assumption that the payoff $\overline{\xi} \cdot \overline{S}$ is P-a. s. contained in the domain of definition of the utility function $\tilde{u}$.

In a first step, we remove the constraint (3.2) by considering instead of (3.1) the expected utility of the *discounted net gain*

$$\frac{\overline{\xi} \cdot \overline{S}}{1+r} - \overline{\pi} \cdot \overline{\xi} = \xi \cdot Y$$

earned by a portfolio $\overline{\xi} = (\xi^0, \xi)$. Here, Y is the d-dimensional random vector with components

$$Y^i = \frac{S^i}{1+r} - \pi^i, \quad i = 1, \dots, d.$$

For any portfolio $\overline{\xi}$ with $\overline{\pi} \cdot \overline{\xi} < w$, adding the risk-free investment $w - \overline{\pi} \cdot \overline{\xi}$ would lead to the strictly better portfolio $(\xi^0 + w - \overline{\pi} \cdot \overline{\xi}, \xi)$. Thus, we can focus on portfolios $\overline{\xi}$ that satisfy $\overline{\pi} \cdot \overline{\xi} = w$, and then the payoff is an affine function of the discounted net gain:

$$\overline{\xi} \cdot \overline{S} = (1+r)(\xi \cdot Y + w).$$

Moreover, for any $\xi \in \mathbb{R}^d$ there exists a unique numéraire component $\xi^0 \in \mathbb{R}$ such that the portfolio $\overline{\xi} := (\xi^0, \xi)$ satisfies $\overline{\pi} \cdot \overline{\xi} = w$.

Let u denote the following transformation of our original utility function $\tilde{u}$:

$$u(y) := \tilde{u}((1+r)(y+w)).$$

Note that u is again a utility function, and that CARA and (shifted) HARA utility functions are transformed into utility functions in the same class.

Clearly, the original utility maximization problem with budget constraint (3.2) is equivalent to the problem of maximizing the expected utility $E[\, u(\xi \cdot Y)\,]$ over all $\xi \in \mathbb{R}^d$ such that $\xi \cdot Y$ is contained in a certain domain D.

Assumption 3.1. We assume one of the following two cases:

(a) $D = \mathbb{R}$. In this case, we will admit all portfolios $\xi \in \mathbb{R}^d$, but we assume that u is bounded from above.

(b) $D = [a, \infty)$ for some $a < 0$. In this case, we only consider portfolios that satisfy the constraint

$$\xi \cdot Y \geq a \quad P\text{-a. s.},$$

and we assume that the expected utility generated by such portfolios is finite, that is,

$$E[\, u(\xi \cdot Y)\,] < \infty \quad \text{for all } \xi \in \mathbb{R}^d \text{ with } \xi \cdot Y \geq a \; P\text{-a. s.}$$

Remark 3.2. Part (a) of this assumption is clearly satisfied in the case of an exponential utility function $u(x) = 1 - e^{-\alpha x}$. Domains of the form $D = [a, \infty)$ appear, for example, in the case of (shifted) HARA utility functions $u(x) = \log(x - b)$ for $b < a$ and $u(x) = \frac{1}{\gamma}(x - c)^\gamma$ for $c \leq a$ and $0 < \gamma < 1$. The integrability assumption in (b) holds if $E[\,|Y|\,] < \infty$, because any concave function is bounded from above by an affine function. ◇

In order to simplify the notation, let us denote by

$$\mathscr{S}(D) := \{\xi \in \mathbb{R}^d \mid \xi \cdot Y \in D \; P\text{-a. s.}\}$$

the set of admissible portfolios for D. Clearly, $\mathscr{S}(D) = \mathbb{R}^d$ if $D = \mathbb{R}$. Our aim is to find some $\xi^* \in \mathscr{S}(D)$ that is optimal in the sense that it maximizes the expected utility $E[\, u(\xi \cdot Y)\,]$ over all $\xi \in \mathscr{S}(D)$. In this case, ξ^* will be an optimal strategy for investments into the risky assets. Complementing ξ^* with a suitable numéraire component ξ^0 yields a portfolio $\overline{\xi}^* = (\xi^0, \xi^*)$ that maximizes the expected utility $E[\, \tilde{u}(\overline{\xi} \cdot \overline{S})\,]$ under the budget constraint $\overline{\pi} \cdot \overline{\xi} = w$. Our first result in this section will relate the existence of such an optimal portfolio to the absence of arbitrage opportunities.

Theorem 3.3. *Suppose that the utility function $u : D \to \mathbb{R}$ satisfies Assumption* 3.1. *Then there exists a maximizer of the expected utility*

$$E[\, u(\xi \cdot Y)\,], \quad \xi \in \mathscr{S}(D),$$

if and only if the market model is arbitrage-free. Moreover, there exists at most one maximizer if the market model is nonredundant in the sense of Definition 1.15.

Proof. The uniqueness part of the assertion follows immediately from the strict concavity of the function $\xi \mapsto E[\, u(\xi \cdot Y)\,]$ for nonredundant market models. As to existence, we may assume without loss of generality that our model is nonredundant. If the nonredundance condition (1.10) does not hold, then we define a linear space $N \subset \mathbb{R}^d$ by

$$N := \{\eta \in \mathbb{R}^d \mid \eta \cdot Y = 0 \; P\text{-a. s.}\}.$$

Clearly, Y takes P-a. s. values in the orthogonal complement $N^\perp$ of N. Moreover, the no-arbitrage condition (1.3) holds for all $\xi \in \mathbb{R}^d$ if and only if it is satisfied for all $\xi \in N^\perp$. By identifying $N^\perp$ with some $\mathbb{R}^n$, we arrive at a situation in which the nonredundance condition (1.10) is satisfied and where we may apply our result for nonredundant market models.

If the model admits arbitrage opportunities, then a maximizer ξ^* of the expected utility $E[\, u(\xi \cdot Y)\,]$ cannot exist: Adding to ξ^* some nonzero $\eta \in \mathbb{R}^d$ for which $\eta \cdot Y \geq 0$ P-a. s., which exists by Lemma 1.4, would yield a contradiction to the optimality of ξ^*, because then

$$E[\, u(\xi^* \cdot Y)\,] < E[\, u((\xi^* + \eta) \cdot Y)\,].$$

From now on, we assume that the market model is arbitrage-free. Let us first consider the case in which $D = [a, \infty)$ for some $a \in (-\infty, 0)$. Then $\mathscr{S}(D)$ is compact. In order to prove this claim, note first that $\mathscr{S}(D)$ is closed in $\mathbb{R}^d$. Thus, it suffices to show that it is bounded. To this end, suppose by way of contradiction that (ξ_n) is a sequence in $\mathscr{S}(D)$ such that $|\xi_n| \to \infty$. By choosing a subsequence if necessary, we may assume that $\eta_n := \xi_n / |\xi_n|$ converges to some unit vector $\eta \in \mathbb{R}^d$. Clearly,

$$\eta \cdot Y = \lim_{n \uparrow \infty} \frac{\xi_n \cdot Y}{|\xi_n|} \geq \lim_{n \uparrow \infty} \frac{a}{|\xi_n|} = 0 \quad P\text{-a. s.},$$

and so nonredundance implies that $\overline{\eta} := (-\pi \cdot \eta, \eta)$ is an arbitrage-opportunity.

In the next step, we show that our assumptions guarantee the continuity of the function

$$\mathscr{S}(D) \ni \xi \mapsto E[\, u(\xi \cdot Y)\,],$$

which, in view of the compactness of $\mathscr{S}(D)$, will imply the existence of a maximizer of the expected utility. By dominated convergence, it suffices to construct an integrable random variable that dominates $u(\xi \cdot Y)$ for all $\xi \in \mathscr{S}(D)$. Define $\eta \in \mathbb{R}^d$ by

$$\eta^i := \Big(\max_{\xi \in \mathscr{S}(D)} \xi^i \Big)^+ < \infty.$$

Then $\eta \cdot S \geq \xi \cdot S$ for $\xi \in \mathscr{S}(D)$, and hence

$$\xi \cdot Y = \frac{\xi \cdot S}{1+r} - \pi \cdot \xi \leq \frac{\eta \cdot S}{1+r} + \Big(\min_{\xi' \in \mathscr{S}(D)} \pi \cdot \xi' \Big)^-.$$

Note that $\eta \cdot Y$ is bounded from below by $-\pi \cdot \eta$ and that there exists some $\alpha \in (0, 1]$ such that $\alpha\pi \cdot \eta < |a|$. Hence, $\alpha\eta \in \mathscr{S}(D)$, and so $E[\, u(\alpha\eta \cdot Y)\,] < \infty$. Now we apply Lemma 3.4 below. Taking first $b := \alpha\pi \cdot \eta$ and then $b := (\min_{\xi' \in \mathscr{S}(D)} \pi \cdot \xi')^-$, we obtain that

$$E\left[u\left(\frac{\eta\cdot S}{1+r}+\left(\min_{\xi'\in\mathscr{S}(D)}\pi\cdot\xi'\right)^{-}\right)\right]<\infty.$$

This concludes the proof of the theorem in the case $D=[a,\infty)$.

Let us now turn to the case of a utility function on $D=\mathbb{R}$ that is bounded from above. We will reduce the assertion to a general existence criterion for minimizers of lower semicontinuous convex functions on $\mathbb{R}^d$, given in Lemma 3.5 below. It will be applied to the convex function $h(\xi):=-E[\,u(\xi\cdot Y)\,]$. Note that $h(0)<\infty$ by assumption. We must show that h is lower semicontinuous. Take a sequence $(\xi_n)_{n\in\mathbb{N}}$ in $\mathbb{R}^d$ converging to some ξ. By part (a) of Assumption 3.1, the random variables $-u(\xi_n\cdot Y)$ are uniformly bounded from below, and so we may apply Fatou's lemma,

$$\liminf_{n\uparrow\infty}h(\xi_n)=\liminf_{n\uparrow\infty}E[\,-u(\xi_n\cdot Y)\,]\ \geq E[\,-u(\xi\cdot Y)\,]=h(\xi).$$

Thus, h is lower semicontinuous.

By our nonredundance assumption, h is strictly convex and admits at most one minimizer. We claim that the absence of arbitrage opportunities is equivalent to the following condition:

$$\lim_{\alpha\uparrow\infty}h(\alpha\,\xi)=+\infty\quad\text{for all nonzero }\xi\in\mathbb{R}^d. \tag{3.3}$$

This is just the condition (3.4) required in Lemma 3.5. It follows from (1.3) and (1.10) that a nonredundant market model is arbitrage-free if and only if each nonzero $\xi\in\mathbb{R}^d$ satisfies $P[\xi\cdot Y<0]>0$. Since the utility function u is strictly increasing and concave, the set $\{\xi\cdot Y<0\}$ can be described as

$$\{\xi\cdot Y<0\}=\left\{\lim_{\alpha\uparrow\infty}u(\alpha\,\xi\cdot Y)=-\infty\right\}\quad\text{for }\xi\in\mathbb{R}^d.$$

The probability of the right-hand set is strictly positive if and only if

$$\lim_{\alpha\uparrow\infty}E[\,u(\alpha\xi\cdot Y)\,]=-\infty,$$

because u is bounded from above. This observation proves that the absence of arbitrage opportunities is equivalent to the condition (3.3) and completes the proof. □

Lemma 3.4. *If $D=[a,\infty)$, $b<|a|$, $0<\alpha\leq 1$, and X is a nonnegative random variable, then*

$$E[\,u(\alpha X-b)\,]<\infty\quad\Longrightarrow\quad E[\,u(X)\,]<\infty.$$

Proof. As in (A.5) in the proof of Proposition A.7 we obtain that, on $\{X>0\}$,

$$\frac{u(X)-u(0)}{X-0}\leq\frac{u(\alpha X)-u(0)}{\alpha X-0}\leq\frac{u(\alpha X-b)-u(-b)}{\alpha X-b-(-b)}.$$

Multiplying by X shows that $u(X)$ can be dominated by a multiple of $u(\alpha X - b)$ plus some constant. □

Lemma 3.5. *Suppose $h : \mathbb{R}^d \to \mathbb{R} \cup \{+\infty\}$ is a convex and lower semicontinuous function with $h(0) < \infty$. Then h attains its infimum provided that*

$$\lim_{\alpha \uparrow \infty} h(\alpha \xi) = +\infty \quad \textit{for all nonzero } \xi \in \mathbb{R}^d. \tag{3.4}$$

Moreover, if h is strictly convex on $\{h < \infty\}$, then also the converse implication holds; the existence of a minimizer implies (3.4).

Proof. First, suppose that (3.4) holds. We will show below that for $c > \inf h$ the level sets $\{x \mid h(x) \leq c\}$ of h are bounded, and hence compact, since they are closed due to lower semicontinuity. Once the compactness of the level sets is established, it follows that the set

$$\{x \in \mathbb{R}^d \mid h(x) = \inf h\} = \bigcap_{c > \inf h} \{x \in \mathbb{R}^d \mid h(x) \leq c\}$$

of minimizers of h is nonempty as an intersection of decreasing and nonempty compact sets.

Suppose $c > \inf h$ is such that the level set $\{h \leq c\}$ is not bounded, and take a sequence (x_n) in $\{h \leq c\}$ such that $|x_n| \to \infty$. By passing to a subsequence if necessary, we may assume that $x_n/|x_n|$ converges to some nonzero ξ. For any $\alpha > 0$,

$$\begin{aligned} h(\alpha\xi) &\leq \liminf_{n \uparrow \infty} h\left(\alpha \frac{x_n}{|x_n|}\right) = \liminf_{n \uparrow \infty} h\left(\frac{\alpha}{|x_n|} x_n + \left(1 - \frac{\alpha}{|x_n|}\right) 0\right) \\ &\leq \liminf_{n \uparrow \infty} \left(\frac{\alpha}{|x_n|} c + \left(1 - \frac{\alpha}{|x_n|}\right) h(0)\right) \\ &= h(0). \end{aligned}$$

Thus, we arrive at a contradiction to condition (3.4). This completes the proof of the existence of a minimizer under assumption (3.4).

In order to prove the converse implication, suppose that the strictly convex function h has a minimizer x^* but that there exists a nonzero $\xi \in \mathbb{R}^d$ violating (3.4), that is, there exists a sequence $(\alpha_n)_{n \in \mathbb{N}}$ and some $c < \infty$ such that $\alpha_n \uparrow \infty$ but $h(\alpha_n \xi) \leq c$ for all n. Let

$$x_n := \lambda_n x^* + (1 - \lambda_n)\alpha_n \xi$$

where λ_n is such that $|x^* - x_n| = 1$, which is possible for all large enough n. By the compactness of the Euclidean unit sphere centered in x^*, we may assume that x_n converges to some x. Then necessarily $|x - x^*| = 1$. As $\alpha_n \xi$ diverges, we must have that $\lambda_n \to 1$. By using our assumption that $h(\alpha_n \xi)$ is bounded, we obtain

$$h(x) \le \liminf_{n\uparrow\infty} h(x_n) \le \lim_{n\uparrow\infty} (\lambda_n h(x^*) + (1-\lambda_n)h(\alpha_n\xi)) = h(x^*).$$

Hence, x is another minimizer of h besides x^*, contradicting the strict convexity of h. Thus, (3.4) must hold if the strictly convex function h attains its infimum. □

Remark 3.6. Note that the proof of Theorem 3.3 under Assumption 3.1(a) did not use the fact that the components of Y are bounded from below. The result remains true for arbitrary Y. ◇

We turn now to a characterization of the solution ξ^* of our utility maximization problem for continuously differentiable utility functions.

Proposition 3.7. *Let u be a continuously differentiable utility function on D such that $E[\,u(\xi \cdot Y)\,]$ is finite for all $\xi \in \mathscr{S}(D)$. Suppose that ξ^* is a solution to the utility maximization problem, and that one of the following two sets of conditions is satisfied:*
- *u is defined on $D = \mathbb{R}$ and is bounded from above.*
- *u is defined on $D = [a,\infty)$, and ξ^* is an interior point of $\mathscr{S}(D)$.*

Then

$$u'(\xi^* \cdot Y)\,|Y| \in L^1(P), \tag{3.5}$$

and the following "first-order condition" holds:

$$E[\,u'(\xi^* \cdot Y)\,Y\,] = 0. \tag{3.6}$$

Proof. For $\xi \in \mathscr{S}(D)$ and $\varepsilon \in (0,1]$ let $\xi_\varepsilon := \varepsilon\xi + (1-\varepsilon)\xi^*$, and define

$$\Delta_\varepsilon := \frac{u(\xi_\varepsilon \cdot Y) - u(\xi^* \cdot Y)}{\varepsilon}.$$

The concavity of u implies that $\Delta_\varepsilon \ge \Delta_\delta$ for $\varepsilon \le \delta$, and so

$$\Delta_\varepsilon \nearrow u'(\xi^* \cdot Y)\,(\xi - \xi^*) \cdot Y \quad \text{as } \varepsilon \downarrow 0.$$

Our assumptions imply that $u(\xi \cdot Y) \in L^1(P)$ for all $\xi \in \mathscr{S}(D)$. In particular, we have $\Delta_1 \in L^1(P)$, so that monotone convergence and the optimality of ξ^* yield that

$$0 \ge E[\,\Delta_\varepsilon\,] \nearrow E[\,u'(\xi^* \cdot Y)\,(\xi - \xi^*) \cdot Y\,] \quad \text{as } \varepsilon \downarrow 0. \tag{3.7}$$

In particular, the expectation on the right-hand side of (3.7) is finite and nonpositive.

Both sets of assumptions imply that ξ^* is an interior point of $\mathscr{S}(D)$. Hence, for $\eta \in \mathbb{R}^d$ with $|\eta|$ sufficiently small, we still have $\xi := \eta + \xi^* \in D$. Then we deduce from (3.7) that

$$E[\,u'(\xi^* \cdot Y)\,\eta \cdot Y\,] \le 0.$$

Replacing η by $-\eta$ shows that the expectation must vanish. □

Remark 3.8. Let us comment on the assumption that the optimal ξ^* is an interior point of $\mathscr{S}(D)$:

(a) If the nonredundance condition (1.10) is not satisfied, then either each or none of the solutions to the utility maximization problem is contained in the interior of $\mathscr{S}(D)$. This can be seen by using the reduction argument given at the beginning of the proof of Theorem 3.3.

(b) Note that $\xi \cdot Y$ is bounded from below by $-\pi \cdot \xi$ in case ξ has only nonnegative components. Thus, the interior of $\mathscr{S}(D)$ is always nonempty.

(c) As shown by the following example, the optimal ξ^* need not be contained in the interior of $\mathscr{S}(D)$, and in this case, the first-order condition (3.6) will generally fail.

◇

Example 3.9. Take $r = 0$, $d = 1$, and let S^1 be integrable but unbounded. We choose $D = [a, \infty)$ with $a := -\pi^1$, and we assume that $P[\,S^1 \leq \varepsilon\,] > 0$ for all $\varepsilon > 0$. Then $\mathscr{S}(D) = [0, 1]$. If $0 < E[\,S^1\,] < \pi^1$, then Example 2.40 shows that the optimal investment is given by $\xi^* = 0$, and so ξ^* lies at the boundary of $\mathscr{S}(D)$. Thus, if u is sufficiently smooth,

$$E[\,u'(\xi^* \cdot Y)\,Y\,] = u'(0)(E[\,S^1\,] - \pi^1) < 0.$$

The intuitive reason for this failure of the first-order condition is that taking a short position in the asset would be optimal as soon as $E[\,S^1\,] < \pi^1$. This choice, however, is ruled out by the constraint $\xi \in \mathscr{S}(D)$. ◇

Proposition 3.7 yields a formula for the density of a particular equivalent risk-neutral measure. Recall that P^* is risk-neutral if and only if $E^*[\,Y\,] = 0$.

Corollary 3.10. *Suppose that the market model is arbitrage-free and that the assumptions of Proposition 3.7 are satisfied for a utility function $u : D \to \mathbb{R}$ and an associated maximizer ξ^* of the expected utility $E[u(\xi \cdot Y)]$. Then*

$$\frac{dP^*}{dP} = \frac{u'(\xi^* \cdot Y)}{E[u'(\xi^* \cdot Y)]} \tag{3.8}$$

defines an equivalent risk-neutral measure.

Proof. Proposition 3.7 states that $u'(\xi^* \cdot Y)Y$ is integrable with respect to P and that its expectation vanishes. Hence, we may conclude that P^* is an equivalent risk-neutral measure if we can show that P^* is well-defined by (3.8), that is, if $u'(\xi^* \cdot Y) \in L^1(P)$. Let

$$c := \sup\{u'(x) \mid x \in D \text{ and } |x| \leq |\xi^*|\} \leq \begin{cases} u'(a) & \text{for } D = [a, \infty), \\ u'(-|\xi^*|) & \text{for } D = \mathbb{R}, \end{cases}$$

which is finite by our assumption that u is continuously differentiable on all of D. Thus,

$$0 \le u'(\xi^* \cdot Y) \le c + u'(\xi^* \cdot Y)|Y| \cdot \mathbb{1}_{\{|Y|\ge 1\}} \le c + u'(\xi^* \cdot Y)|Y|,$$

and the right-hand side has a finite expectation by (3.5). □

Remark 3.11. Corollary 3.10 yields an independent and constructive proof of the "fundamental theorem of asset pricing" in the form of Theorem 1.7: Suppose that the model is arbitrage-free. If Y is P-a. s. bounded, then so is $u(\xi^* \cdot Y)$, and the measure P^* of (3.8) is an equivalent risk-neutral measure with a bounded density dP^*/dP. If Y is unbounded, then we may consider the bounded random vector

$$\widetilde{Y} := \frac{Y}{1 + |Y|},$$

which also satisfies the no-arbitrage condition (1.3). Let $\widetilde{\xi}^*$ be a maximizer of the expected utility $E[\, u(\xi \cdot \widetilde{Y})\,]$. Then an equivalent risk-neutral measure P^* is defined through the bounded density

$$\frac{dP^*}{dP} := c \cdot \frac{u'(\widetilde{\xi}^* \cdot \widetilde{Y})}{1 + |Y|},$$

where c is an appropriate normalizing constant. ◇

Example 3.12. Consider the exponential utility function

$$u(x) = 1 - e^{-\alpha x}$$

with constant absolute risk aversion $\alpha > 0$. The requirement that $E[\, u(\xi \cdot Y)\,]$ is finite for all $\xi \in \mathscr{S}(D)$ is equivalent to the condition

$$E[\, e^{\xi \cdot Y}\,] < \infty \quad \text{for all } \xi \in \mathbb{R}^d.$$

If ξ^* is a maximizer of the expected utility, then the density of the equivalent risk-neutral measure P^* in (3.8) takes the particular form

$$\frac{dP^*}{dP} = \frac{e^{-\alpha \xi^* \cdot Y}}{E[\, e^{-\alpha \xi^* \cdot Y}\,]}.$$

In fact, P^* is independent of α since ξ^* maximizes the expected utility $1 - E[\, e^{-\alpha \xi \cdot Y}\,]$ if and only if $\lambda^* := -\alpha \xi^*$ is a minimizer of the *moment generating function*

$$Z(\lambda) := E[\, e^{\lambda \cdot Y}\,], \quad \lambda \in \mathbb{R}^d,$$

of Y. In Corollary 3.21 below, the measure P^* will be characterized by the fact that it minimizes the *relative entropy* with respect to P among the risk-neutral measures in $\mathscr{P}$; see Definition C.1 for the notion of relative entropy. ◇

3.2 Exponential utility and relative entropy

In this section, we give a more detailed study of the problem of portfolio optimization with respect to a CARA utility function

$$u(x) = 1 - e^{-\alpha x}$$

for $\alpha > 0$. As in the preceding Section 3.1, the problem is to maximize the expected utility

$$E[\, u(\xi \cdot Y)\,]$$

of the discounted net gain $\xi \cdot Y$ earned by an investment into risky assets. The key assumption for this problem is that

$$E[\, u(\xi \cdot Y)\,] > -\infty \quad \text{for all } \xi \in \mathbb{R}^d. \tag{3.9}$$

Recall from Example 3.12 that the maximization of $E[\, u(\xi \cdot Y)\,]$ can be reduced to the minimization of the moment generating function

$$Z(\lambda) := E[\, e^{\lambda \cdot Y}\,], \quad \lambda \in \mathbb{R}^d,$$

which does not depend on the risk aversion α. The key assumption (3.9) is equivalent to the condition that

$$Z(\lambda) < \infty \quad \text{for all } \lambda \in \mathbb{R}^d. \tag{3.10}$$

Throughout this section, we will always assume that (3.10) holds. But we will not need the assumption that Y is bounded from below (which in our financial market model follows from assuming that asset prices are nonnegative); all results remain true for general random vectors Y; see also Remarks 1.9 and 3.6.

Lemma 3.13. *The condition* (3.10) *is equivalent to*

$$E[\, e^{\alpha |Y|}\,] < \infty \quad \textit{for all } \alpha > 0.$$

Proof. Clearly, the condition in the statement of the lemma implies (3.10). To prove the converse assertion, take a constant $c > 0$ such that $|x| \le c \sum_{i=1}^d |x^i|$ for $x \in \mathbb{R}^d$. By Hölder's inequality,

$$E[\, e^{\alpha|Y|}\,] \le E\left[\, \exp\left(\alpha c \sum_{i=1}^d |Y^i| \right) \right] \le \prod_{i=1}^d E[\, e^{\alpha c d |Y^i|}\,]^{1/d}.$$

In order to show that the ith factor on the right is finite, take $\lambda \in \mathbb{R}^d$ such that $\lambda^i = \alpha c d$ and $\lambda^j = 0$ for $j \neq i$. With this choice,

$$E[\,e^{\alpha c d|Y^i|}\,] \le E[\,e^{\lambda\cdot Y}\,] + E[\,e^{-\lambda\cdot Y}\,],$$

which is finite by (3.10). □

Definition 3.14. The *exponential family* of P with respect to Y is the set of measures

$$\{P_\lambda \mid \lambda \in \mathbb{R}^d\}$$

defined via

$$\frac{dP_\lambda}{dP} = \frac{e^{\lambda\cdot Y}}{Z(\lambda)}.$$

△

Example 3.15. Suppose that the risky asset S^1 has under P a Poisson distribution with parameter $\alpha > 0$, that is, S^1 takes values in $\{0, 1, \dots\}$ and satisfies

$$P[\,S^1 = k\,] = e^{-\alpha}\frac{\alpha^k}{k!}, \qquad k = 0, 1, \dots.$$

Then (3.10) is satisfied for $Y := S^1 - \pi^1$, and S^1 has under P_λ a Poisson distribution with parameter $e^\lambda \alpha$. Hence, the exponential family of P generates the family of all Poisson distributions. ◇

Example 3.16. Let Y have a standard normal distribution $N(0, 1)$. Then (3.10) is satisfied, and the distribution of Y under P_λ is equal to the normal distribution $N(\lambda, 1)$ with mean λ and variance 1. ◇

Remark 3.17. Two parameters λ and λ' in $\mathbb{R}^d$ determine the same element in the exponential family of P if and only if $(\lambda - \lambda')\cdot Y = 0$ P-almost surely. It follows that the mapping

$$\lambda \longmapsto P_\lambda$$

is injective provided that the nonredundance condition holds in the form

$$\xi \cdot Y = 0 \ P\text{-a.s.} \quad \Longrightarrow \quad \xi = 0. \tag{3.11}$$

◇

In the sequel, we will be interested in the barycenters of the members of the exponential family of P with respect to Y. These barycenters will be denoted by

$$m(\lambda) := E_\lambda[\,Y\,] = \frac{1}{Z(\lambda)}E[\,Ye^{\lambda\cdot Y}\,], \quad \lambda \in \mathbb{R}^d.$$

The next lemma shows that $m(\lambda)$ can be obtained as the gradient of the logarithmic moment generating function.

Lemma 3.18. *Z is a smooth function on $\mathbb{R}^d$, and the gradient of $\log Z$ at λ is the expectation of Y under P_λ:*

$$(\nabla \log Z)(\lambda) = E_\lambda[\, Y\,] = m(\lambda).$$

Moreover, the Hessian of $\log Z$ at λ equals the covariance matrix $(\mathrm{cov}_{P_\lambda}(Y^i, Y^j))_{i,j}$ of Y under the measure P_λ:

$$\frac{\partial^2}{\partial\lambda^i\,\partial\lambda^j} \log Z(\lambda) = \mathrm{cov}_{P_\lambda}(Y^i, Y^j) = E_\lambda[\, Y^i Y^j\,] - E_\lambda[\, Y^i\,] E_\lambda[\, Y^j\,].$$

In particular, $\log Z$ is convex.

Proof. Observe that

$$\left|\frac{\partial e^{\lambda\cdot x}}{\partial\lambda^i}\right| = |x^i|\, e^{\lambda\cdot x} \le \exp[(1+|\lambda|)\cdot|x|].$$

Hence, Lemma 3.13 and Lebesgue's dominated convergence theorem justify the interchanging of differentiation and integration (see the "differentiation lemma" in § 16 of [22] for details). The result thus follows from a short computation. □

Recall from Section 1.5 the notion of the convex hull $\Gamma(\nu)$ of the support of a measure ν on $\mathbb{R}^d$ and the definition of the relative interior $\mathrm{ri}\, C$ of a convex set C.

Corollary 3.19. *Denote by $\mu := P \circ Y^{-1}$ the distribution of Y under P. Then the function*

$$\lambda \longmapsto \lambda \cdot m_0 - \log Z(\lambda)$$

attains its maximum if and only if m_0 is contained in the relative interior of the convex hull of the support of μ, that is, if and only if

$$m_0 \in \mathrm{ri}\,\Gamma(\mu).$$

In this case, any maximizer λ^* satisfies

$$m_0 = m(\lambda^*) = E_{\lambda^*}[\, Y\,].$$

In particular, we have

$$\{m(\lambda) \mid \lambda \in \mathbb{R}^d\} = \mathrm{ri}\,\Gamma(\mu). \tag{3.12}$$

Moreover, if the nonredundance condition (3.11) *holds, then there exists at most one maximizer λ^*.*

Proof. Taking $\widetilde{Y} := Y - m_0$ reduces the problem to the situation where $m_0 = 0$. Applying Theorem 3.3 with the utility function $u(z) = 1 - e^{-z}$ shows that the existence of a maximizer λ^* of $-\log Z$ is equivalent to the absence of arbitrage opportunities. Corollary 3.10 states that $m(\lambda^*) = 0$ and that 0 belongs to $M(\mu)$, where $M(\mu)$ was defined in Lemma 1.46. An application of Theorem 1.51 completes the proof. □

It will turn out that the maximization problem of the preceding corollary is closely related to the concept of relative entropy as discussed in Appendix C. Recall from Definition C.1 that the *relative entropy* of a probability measure Q with respect to P is defined as

$$H(Q|P) := \begin{cases} E\left[\frac{dQ}{dP} \log \frac{dQ}{dP} \right] & \text{if } Q \ll P, \\ +\infty & \text{otherwise,} \end{cases}$$

where we use the convention $0 \log 0 = 0$. The following result shows in particular that P_λ is the unique minimizer of the relative entropy $H(Q|P)$ among all probability measures Q with $E_Q[\,Y\,] = E_\lambda[\,Y\,]$.

Proposition 3.20. *Let $m_0 \in \mathbb{R}^d$ be given. Then, for any probability measure Q on $(\Omega, \mathscr{F})$ with expectation $E_Q[\,Y\,] = m_0$,*

$$H(Q|P) \geq \sup_{\lambda \in \mathbb{R}^d} (\lambda \cdot m_0 - \log Z(\lambda)). \tag{3.13}$$

If, moreover, $\lambda_0 \in \mathbb{R}^d$ is such that $m(\lambda_0) = m_0$, then λ_0 realizes the supremum on the right-hand side of (3.13) *and*

$$\sup_{\lambda \in \mathbb{R}^d} (\lambda \cdot m_0 - \log Z(\lambda)) = \lambda_0 \cdot m_0 - \log Z(\lambda_0) = H(P_{\lambda_0}|P).$$

In particular, P_{λ_0} is the unique minimizer of the relative entropy $H(Q|P)$ within the class of all probability measures Q on $(\Omega, \mathscr{F})$ for which $E_Q[\,Y\,] = m_0$.

Proof. Let Q be a probability measure on $(\Omega, \mathscr{F})$ for which $E_Q[\,Y\,] = m_0$. We show first that

$$H(Q|P) = H(Q|P_\lambda) + \lambda \cdot m_0 - \log Z(\lambda) \quad \text{for all } \lambda \in \mathbb{R}^d. \tag{3.14}$$

To this end, note that both sides of (3.14) are infinite if $Q \not\ll P$. Otherwise, by (B.2),

$$\frac{dQ}{dP} = \frac{dQ}{dP_\lambda} \cdot \frac{dP_\lambda}{dP} = \frac{dQ}{dP_\lambda} \cdot \frac{e^{\lambda \cdot Y}}{Z(\lambda)}.$$

Taking logarithms and integrating with respect to Q yields (3.14).

Since $H(Q|P_\lambda) \geq 0$ according to (C.1), we get from (3.14) that

$$H(Q|P) \geq \lambda \cdot m_0 - \log Z(\lambda) \tag{3.15}$$

for all $\lambda \in \mathbb{R}^d$ and all measures Q such that $E_Q[\,Y\,] = m_0$. This proves (3.13).

Next, suppose that $\lambda_0 \in \mathbb{R}^d$ is such that $m(\lambda_0) = m_0$. Then, taking $Q := P_{\lambda_0}$ and applying (3.14) with $\lambda := \lambda_0$ yields equality in (3.15). It follows that P_{λ_0} minimizes the relative entropy on the set

$$M_0 := \{Q \mid E_Q[\,Y\,] = m_0\},$$

and that λ_0 maximizes the function $\lambda \mapsto \lambda \cdot m_0 - \log Z(\lambda)$. Finally, the strict convexity of the function $x \mapsto x \log x$ implies the strict convexity of the relative entropy on its effective domain $\{Q \mid H(Q|P) < \infty\}$. Hence, P_{λ_0} is the unique minimizer in M_0. □

Taking $m_0 = 0$ in Proposition 3.20 yields a special equivalent risk-neutral measure in our financial market model, namely the *entropy-minimizing risk-neutral measure* P_{λ_0}. Sometimes it is also called the *Esscher transform* of P. Recall our assumption (3.10).

Corollary 3.21. *Suppose the market model is arbitrage-free. Then there exists a unique equivalent risk-neutral measure $P^* \in \mathcal{P}$ that minimizes the relative entropy $H(\hat{P}|P)$ over all $\hat{P} \in \mathcal{P}$. The density of P^* is of the form*

$$\frac{dP^*}{dP} = \frac{e^{\lambda^* \cdot Y}}{E[\,e^{\lambda^* \cdot Y}\,]},$$

where λ^* denotes a minimizer of the moment generating function $E[\,e^{\lambda \cdot Y}\,]$ of Y.

Proof. This follows immediately from Corollary 3.19 and Proposition 3.20. □

By combining Proposition 3.20 with Remark 3.17 and Theorem 1.51, we obtain the following corollary. It clarifies the question of uniqueness in the representation of points in the relative interior of $\Gamma(P \circ Y^{-1})$ as barycenters of the exponential family.

Corollary 3.22. *If the nonredundance condition* (3.11) *holds, then*

$$\lambda \longmapsto m(\lambda)$$

is a bijective mapping from $\mathbb{R}^d$ to $\operatorname{ri}\Gamma(P \circ Y^{-1})$.

Remark 3.23. It follows from Corollary 3.19 and Proposition 3.20 that for all $m \in \operatorname{ri}\Gamma(P \circ Y^{-1})$,

$$\min_{E_Q[\,Y\,]=m} H(Q|P) = \max_{\lambda \in \mathbb{R}^d}(\lambda \cdot m - \log Z(\lambda)). \tag{3.16}$$

Here, the right-hand side is the *Fenchel–Legendre transform* of the convex function $\log Z$ evaluated at $m \in \mathbb{R}^d$. ◇

The following theorem shows that the variational principle (3.16) extends to all $m \in \mathbb{R}^d$, if we replace "min" and "max" by "inf" and "sup," respectively.

Theorem 3.24. *For all* $m \in \mathbb{R}^d$,

$$\inf_{E_Q[Y]=m} H(Q|P) = \sup_{\lambda \in \mathbb{R}^d} (\lambda \cdot m - \log Z(\lambda)).$$

The proof of Theorem 3.24 relies on the following lemma, which in turn uses results from the Appendices C and H. Let us introduce the following set of all densities of probability measures $Q \ll P$:

$$\mathscr{D} := \left\{ \frac{dQ}{dP} \;\middle|\; Q \ll P \right\} = \{ \varphi \in L^1 \mid \varphi \geq 0, \, E[\varphi] = 1 \} \subset L^1.$$

Then the relative entropy, restricted to the class of probability measures absolutely continuous with respect to P, can be identified with the following map:

$$\begin{aligned} F : \mathscr{D} &\longrightarrow [0, \infty] \\ \varphi &\longmapsto F(\varphi) = E[\varphi \log \varphi]. \end{aligned}$$

Lemma 3.25. *The map* $F : \mathscr{D} \to [0, \infty]$ *is lower semicontinuous with respect to the weak topology* $\sigma(L^1, L^\infty)$ *on* $\mathscr{D}$ *and for each* $\alpha \geq 0$, *the set*

$$\{ \varphi \in \mathscr{D} \mid F(\varphi) \leq \alpha \}$$

is weakly sequentially compact in L^1.

Proof. The set $\mathscr{D}$ is clearly convex and closed in L^1. Hence, $\mathscr{D}$ is also weakly closed in L^1 by Theorem H.7.

We show next that F is weakly lower semicontinuous on $\mathscr{D}$. To this end, note first that for $Z \in L^\infty$ the map $L^1 \ni X \mapsto E[ZX]$ is a (weakly) continuous linear functional on L^1 by (H.5). Moreover, Theorem C.5 states that for $\varphi \in \mathscr{D}$,

$$F(\varphi) = E[\varphi \log \varphi] = \sup_{Z \in L^\infty} (E[Z\varphi] - \log E[e^Z]).$$

Therefore, F is weakly lower semicontinuous as the supremum of weakly continuous functionals.

The weak lower semicontinuity of F implies that the level set $\{ \varphi \in \mathscr{D} \mid F(\varphi) \leq \alpha \}$ is weakly closed in L^1 for each $\alpha \geq 0$. Moreover, the criterion of de la Vallée Poussin stated in Proposition H.16 yields that the level set is also uniformly integrable. Applying the Eberlein–Šmulian and Dunford–Pettis theorems, as stated in Theorem H.13 and Theorem H.19, concludes the proof. □

Proof of Theorem 3.24. Due to Remark 3.23, the assertion holds for all $m \in \operatorname{ri}\Gamma(\mu)$, where $\mu = P \circ Y^{-1}$. In view of (3.13) and (3.12), it thus remains to prove that

$$\inf_{E_Q[\,Y\,]=m} H(Q|P) \le \sup_{\lambda\in\mathbb{R}^d}(\lambda \cdot m - \log Z(\lambda)) \tag{3.17}$$

for those m that do not belong to $\operatorname{ri}\Gamma(\mu)$. Since the right-hand side of (3.17) is just the Fenchel–Legendre transform of the convex function $\log Z$ evaluated at m, we denote it by $(\log Z)^*(m)$.

First, we consider the case in which m does not belong to the closure $\overline{\Gamma}(\mu)$ of the convex hull of the support of μ. In this case, Proposition 1.52 implies via a similar argument as in the proof of Lemma 1.46 that there is no $Q \ll P$ such that $E_Q[\,Y\,] = m$, and so the left-hand side of (3.17) is infinite. To show that the right-hand side is also infinite, we apply the separating hyperplane theorem in the form of Proposition A.5. It yields some $\xi \in \mathbb{R}^d$ such that

$$\xi \cdot m > \sup\{\xi \cdot x \mid x \in \overline{\Gamma}(\mu)\} \ge \sup\{\xi \cdot x \mid x \in \operatorname{supp}\mu\}.$$

By taking $\lambda_n := n\xi$, it follows that

$$\lambda_n \cdot m - \log Z(\lambda_n) \ge n\Big(\xi \cdot m - \sup_{y\in\operatorname{supp}\mu} \xi \cdot y\Big) \longrightarrow +\infty \quad \text{as } n \uparrow \infty.$$

Hence, the right-hand side of (3.17) is infinite if $m \notin \overline{\Gamma}(\mu)$.

It remains to prove (3.17) for $m \in \overline{\Gamma}(\mu) \setminus \operatorname{ri}\Gamma(\mu)$ with $(\log Z)^*(m) < \infty$. Recall from (1.28) that $\operatorname{ri}\Gamma(\mu) = \operatorname{ri}\overline{\Gamma}(\mu)$. Pick some $m_1 \in \operatorname{ri}\Gamma(\mu)$ and let

$$m_n := \frac{1}{n}\, m_1 + \left(1 - \frac{1}{n}\right) m.$$

Then $m_n \in \operatorname{ri}\Gamma(\mu)$ by (1.27). By the convexity of $(\log Z)^*$ and the fact that $(\log Z)^*(m_1) < \infty$ by Corollary 3.19, we have

$$\begin{aligned}\limsup_{n\uparrow\infty}(\log Z)^*(m_n) &\le \limsup_{n\uparrow\infty}\left(\frac{1}{n}(\log Z)^*(m_1) + \frac{n-1}{n}(\log Z)^*(m)\right)\\ &= (\log Z)^*(m).\end{aligned} \tag{3.18}$$

Due to Corollary 3.19 and Proposition 3.20, we also know that to each m_n there corresponds some $\lambda_n \in \mathbb{R}^d$ such that

$$m_n = E_{\lambda_n}[\,Y\,] \quad \text{and} \quad H(P_{\lambda_n}|P) = (\log Z)^*(m_n). \tag{3.19}$$

From (3.18) and (3.19), we conclude that

$$\limsup_{n\uparrow\infty} H(P_{\lambda_n}|P) = \limsup_{n\uparrow\infty}(\log Z)^*(m_n) \le (\log Z)^*(m) < \infty.$$

In particular, $H(P_{\lambda_n}|P)$ is uniformly bounded in n, and Lemma 3.25 implies that—after passing to a suitable subsequence if necessary—the densities dP_{λ_n}/dP converge weakly in $L^1(\Omega, \mathscr{F}, P)$ to a probability density φ. Let $dP_\infty = \varphi\, dP$. By the weak lower semicontinuity of

$$\frac{dQ}{dP} \longmapsto H(Q|P),$$

which was established in Lemma 3.25, we conclude that $H(P_\infty|P) \leq (\log Z)^*(m)$.

The theorem will be proved once we can show that $E_\infty[\,Y\,] = m$. To this end, let $\gamma := \sup_n (\log Z)^*(m_n)$, which is a finite nonnegative number by (3.18). Note that for numbers α, $c > 0$ that will be determined later,

$$Z := \alpha\, \mathbb{1}_{\{|Y|\geq c\}}|Y|$$

is such that $e^Z \in L^1(P)$, due to condition (3.10) and Lemma 3.13. Taking this Z in the first identity of (C.2) yields

$$\gamma \geq H(P_{\lambda_n}|P) \geq \alpha E_{\lambda_n}[\,|Y|\mathbb{1}_{\{|Y|\geq c\}}\,] - \log E[\exp(\alpha|Y|\mathbb{1}_{\{|Y|\geq c\}})\,] \quad \text{for all } n \in \mathbb{N}.$$

By taking α large so that $\gamma/\alpha < \varepsilon/2$ for some given $\varepsilon > 0$, and by choosing c such that

$$\frac{1}{\alpha}\log E[\,\exp(\alpha|Y|\mathbb{1}_{\{|Y|\geq c\}})\,] < \frac{\varepsilon}{2},$$

we obtain that

$$\sup_{n\in\mathbb{N}} E_{\lambda_n}[\,|Y|\mathbb{1}_{\{|Y|\geq c\}}\,] \leq \varepsilon. \tag{3.20}$$

Next, the map

$$L^1 \ni \psi \longmapsto E[\,\psi|Y|\mathbb{1}_{\{|Y|\geq c\}}\,]$$

is weakly lower semicontinuous as the supremum of the weakly continuous maps

$$\psi \longmapsto E[\,\psi|Y|\mathbb{1}_{\{k>|Y|\geq c\}}\,], \quad k > c.$$

Thus, the weak convergence $dP_{\lambda_n}/dP \to dP_\infty/dP$ implies that

$$E_\infty[\,|Y|\mathbb{1}_{\{|Y|\geq c\}}\,] \leq \liminf_{n\uparrow\infty} E_{\lambda_n}[\,|Y|\mathbb{1}_{\{|Y|\geq c\}}\,] \leq \varepsilon. \tag{3.21}$$

It implies moreover that

$$E_{\lambda_n}[\,Y\mathbb{1}_{\{|Y|<c\}}\,] \longrightarrow E_\infty[\,Y\mathbb{1}_{\{|Y|<c\}}\,].$$

We conclude that

$$\begin{aligned}|E_\infty[Y] - E_{\lambda_n}[Y]| \le |E_{\lambda_n}[\,Y\mathbb{1}_{\{|Y|<c\}}\,] - E_\infty[\,Y\mathbb{1}_{\{|Y|<c\}}\,]| \\ + E_\infty[\,|Y|\mathbb{1}_{\{|Y|\ge c\}}\,] + E_{\lambda_n}[\,|Y|\mathbb{1}_{\{|Y|\ge c\}}\,] \le 3\varepsilon\end{aligned}$$

if $n \in \mathbb{N}$ is sufficiently large. Thus,

$$m = \lim_{n\uparrow\infty} E_{\lambda_n}[\,Y\,] = E_\infty[\,Y\,],$$

as desired. □

3.3 Optimal contingent claims

In this section, we study the problem of maximizing the expected utility

$$E[\,u(X)\,]$$

under a given budget constraint in a broader context. The random variables X will vary in a general convex class $\mathscr{X} \subseteq L^0(\Omega, \mathscr{F}, P)$ of *admissible* payoff profiles. In the setting of our financial market model, this will allow us to explain the demand for *nonlinear* payoff profiles provided by financial derivatives.

In order to formulate the budget constraint in this general context, we introduce a linear pricing rule of the form

$$\Phi(X) = E^*[\,X\,] = E[\,\varphi X\,]$$

where P^* is a probability measure on $(\Omega, \mathscr{F})$ that is equivalent to P with density φ. For a given initial wealth $w \in \mathbb{R}$, the corresponding budget set is defined as

$$\mathscr{B} := \{X \in \mathscr{X} \cap L^1(P^*) \mid E^*[\,X\,] \le w\}. \tag{3.22}$$

Our optimization problem can now be stated as follows:

$$\textit{Maximize } E[\,u(X)\,] \textit{ over all } X \in \mathscr{B}. \tag{3.23}$$

Note, however, that we will need some extra conditions that guarantee that the expectations $E[\,u(X)\,]$ make sense and are bounded from above.

Remark 3.26. In general, our optimization problem would not be well posed without the assumption $P^* \approx P$. Note first that it should then be rephrased in terms of a class $\mathscr{X}$ of measurable functions on $(\Omega, \mathscr{F})$ since we can no longer pass to equivalence classes with respect to P. If P is not absolutely continuous with respect to P^*, then there exists $A \in \mathscr{F}$ such that $P[\,A\,] > 0$ and $P^*[\,A\,] = 0$. For any $X \in \mathscr{L}^1(P^*)$ and $c > 0$, the random variable $\widetilde{X} := X + c\,\mathbb{1}_A$ would satisfy $E^*[\,\widetilde{X}\,] = E^*[\,X\,]$ and $E[\,u(\widetilde{X})\,] > E[\,u(X)\,]$. Similarly, if $P^*[\,A\,] > 0$ and $P[\,A\,] = 0$ then

$$\hat{X} := X + c - \frac{c}{P^*[\,A\,]}\,\mathbb{1}_A$$

would have the same price as X but higher expected utility. In particular, the expectations in (3.23) would be unbounded in both cases if $\mathscr{X}$ is the class of *all* measurable functions on $(\Omega, \mathscr{F})$ and if the function u is not bounded from above. ◇

Remark 3.27. If a solution X^* with $-\infty < E[\,u(X^*)\,] < \infty$ exists then it is *unique*, since $\mathscr{B}$ is convex and u is strictly concave. Moreover, if $\mathscr{X} = L^0(\Omega, \mathscr{F}, P)$ or $\mathscr{X} = L^0_+(\Omega, \mathscr{F}, P)$ then X^* satisfies

$$E^*[\,X^*\,] = w$$

since $E^*[\,X^*\,] < w$ would imply that $X := X^* + w - E^*[\,X^*]$ is a strictly better choice, due to the strict monotonicity of u. ◇

Let us first consider the unrestricted case $\mathscr{X} = L^0(\Omega, \mathscr{F}, P)$ where any finite random variable on $(\Omega, \mathscr{F}, P)$ is admissible. The following heuristic argument identifies a candidate X^* for the maximization of the expected utility. Suppose that a solution X^* exists. For any $X \in L^\infty(P)$ and any $\lambda \in \mathbb{R}$,

$$X_\lambda := X^* + \lambda(X - E^*[\,X\,])$$

satisfies the budget constraint $E^*[\,X_\lambda\,] \le w$. A heuristic computation yields

$$\begin{aligned} 0 &= \frac{d}{d\lambda}\bigg|_{\lambda=0} E[\,u(X_\lambda)\,] \\ &= E[\,u'(X^*)(X - E^*[\,X\,])] \\ &= E[\,u'(X^*)X\,] - E[\,XE[\,u'(X^*)\,]\,\varphi] \\ &= E[\,X(u'(X^*) - c\,\varphi)\,] \end{aligned}$$

where $c := E[\,u'(X^*)\,]$. The identity

$$E[\,X\,u'(X^*)\,] = c\,E[\,X\,\varphi\,]$$

for all bounded measurable X implies $u'(X^*) = c\,\varphi$ P-almost surely. Thus, if we denote by

$$I := (u')^{-1}$$

the inverse function of the strictly decreasing function u', then X^* should be of the form

$$X^* = I(c\,\varphi).$$

We will now formulate a set of assumptions on our utility function u that guarantee that $X^* := I(c\,\varphi)$ is indeed a maximizer of the expected utility, as suggested by the preceding argument.

Theorem 3.28. *Suppose $u : \mathbb{R} \to \mathbb{R}$ is a continuously differentiable utility function that is bounded from above and whose derivative satisfies*

$$\lim_{x \downarrow -\infty} u'(x) = +\infty. \tag{3.24}$$

Assume moreover that $c > 0$ is a constant such that

$$X^* := I(c\,\varphi) \in L^1(P^*).$$

Then X^* is the unique maximizer of the expected utility $E[\,u(X)\,]$ among all those $X \in L^1(P^*)$ for which $E^*[\,X\,] \le E^*[\,X^*\,]$. In particular, X^* solves our optimization problem (3.23) for $\mathscr{X} = L^0(\Omega, \mathscr{F}, P)$ if c can be chosen such that $E^*[\,X^*\,] = w$.

Proof. Since u is bounded from above, its derivative satisfies

$$\lim_{x \uparrow \infty} u'(x) = 0,$$

in addition to (3.24). Hence, $(0, \infty)$ is contained in the range of u', and it follows that $I(c\,\varphi)$ is P-a. s. well-defined for all $c > 0$.

To show the optimality of $X^* = I(c\,\varphi)$, note that the concavity of u implies that for any $X \in L^1(P^*)$,

$$u(X) \le u(X^*) + u'(X^*)(X - X^*) = u(X^*) + c\,\varphi(X - X^*).$$

Taking expectations with respect to P yields

$$E[\,u(X)\,] \le E[\,u(X^*)\,] + c\,E^*[\,X - X^*\,]. \tag{3.25}$$

Hence, X^* is indeed a maximizer in the class $\{X \in L^1(P^*) \mid E^*[\,X\,] \le w\}$, where $w := E^*[\,X^*\,]$.

To show that X^* is the unique maximizer, we will apply Remark 3.27. To this end, we need to show that $E[\,u(X^*)\,]$ is finite. Since u is bounded from above, we immediately get $E[\,u(X^*)\,] < \infty$. On the other hand, applying (3.25) with $X := w$ gives

$$-\infty < u(w) = E[\,u(X)\,] \le E[\,u(X^*)\,],$$

and this concludes the proof. □

Example 3.29. Let $u(x) = 1 - e^{-\alpha x}$ be an exponential utility function with constant absolute risk aversion $\alpha > 0$. In this case,

$$I(y) = -\frac{1}{\alpha}\log\frac{y}{\alpha}.$$

It follows that

$$\begin{aligned} E^*[\,I(c\,\varphi)\,] &= -\frac{1}{\alpha}\log\frac{c}{\alpha} - \frac{1}{\alpha}\cdot E[\,\varphi\log\varphi\,] \\ &= -\frac{1}{\alpha}\log\frac{c}{\alpha} - \frac{1}{\alpha}\cdot H(P^*|P), \end{aligned}$$

where $H(P^*|P)$ denotes the relative entropy of P^* with respect to P; see Definition C.1. Hence, the utility maximization problem can be solved for any $w \in \mathbb{R}$ if and only if the relative entropy $H(P^*|P)$ is finite. In this case, the optimal profile is given by

$$X^* = -\frac{1}{\alpha}\log\varphi + w + \frac{1}{\alpha}H(P^*|P),$$

and the maximal value of expected utility is

$$E[\,u(X^*)\,] = 1 - \exp(-\alpha w - H(P^*|P)),$$

corresponding to the certainty equivalent

$$w + \frac{1}{\alpha}H(P^*|P).$$

Let us now return to the financial market model considered in Section 3.1, and let P^* be the entropy-minimizing risk-neutral measure constructed in Corollary 3.21. The density of P^* is of the form

$$\varphi = \frac{e^{-\alpha\xi^*\cdot Y}}{E[\,e^{-\alpha\xi^*\cdot Y}\,]},$$

where $\xi^* \in \mathbb{R}^d$ denotes a maximizer of the expected utility $E[\,u(\xi\cdot Y)\,]$; see Example 3.12. In this case, the optimal profile takes the form

$$X^* = \xi^*\cdot Y + w = \frac{\overline{\xi}^*\cdot\overline{S}}{1+r},$$

that is, X^* is the discounted payoff of the portfolio $\overline{\xi}^* = (\xi^0, \xi^*)$, where $\xi^0 = w - \xi^*\cdot\pi$ is determined by the budget constraint $\overline{\xi}\cdot\overline{\pi} = w$. Thus, the optimal profile is given by a *linear* profile in the given primary assets $S^0, \dots, S^d$: No derivatives are needed at this point. ◇

In many situations, it will be natural to restrict the discussion to payoff profiles that are nonnegative. For the rest of this section, we will make this restriction, and so the utility function u may be defined only on $[0,\infty)$. In several applications, we will also use an upper bound on payoff profiles given by an $\mathcal{F}$-measurable random variable W :

$\Omega \longrightarrow [0,\infty]$. We include the case $W \equiv +\infty$, where there is no real constraint, and define the convex class of admissible payoff profiles as

$$\mathscr{X} := \{X \in L^0(P) \mid 0 \le X \le W \ P\text{-a. s.}\}.$$

Thus, our goal is to maximize the expected utility $E[\,u(X)\,]$ over all $X \in \mathscr{B}$ where the budget set $\mathscr{B}$ is defined in terms of $\mathscr{X}$ and P^* as in (3.22), that is,

$$\mathscr{B} = \{X \in L^1(P^*) \mid 0 \le X \le W \ P\text{-a. s. and } E^*[\,X\,] \le w\}.$$

We first formulate a general existence result.

Proposition 3.30. *Let u be any utility function on $(0,\infty)$, and suppose that W satisfies $W < \infty$ P-a. s. and $E[\,(u(W))^+\,] < \infty$. Then there exists a unique $X^* \in \mathscr{B}$, which maximizes the expected utility $E[\,u(X)\,]$ over all $X \in \mathscr{B}$.*

Proof. Take a sequence (X_n) in $\mathscr{B}$ such that $E[\,u(X_n)\,]$ converges to the supremum of the expected utilities. Since $\sup_n |X_n| = \sup_n X_n \le W < \infty$ P-almost surely, we obtain from Lemma 1.72 a sequence

$$\widetilde{X}_n \in \operatorname{conv}\{X_n, X_{n+1}, \dots\}$$

of convex combinations that converge almost surely to some $\widetilde{X}$. Clearly, every $\widetilde{X}_n$ belongs to $\mathscr{B}$. Fatou's lemma implies that

$$E^*[\,\widetilde{X}\,] \le \liminf_{n\uparrow\infty} E^*[\,\widetilde{X}_n\,] \le w,$$

and so $\widetilde{X} \in \mathscr{B}$. Each $\widetilde{X}_n$ can be written as $\sum_{i=1}^m \alpha_i^n X_{n_i}$ for indices $n_i \ge n$ and coefficients $\alpha_i^n \ge 0$ summing up to 1. Hence,

$$u(\widetilde{X}_n) \ge \sum_{i=1}^m \alpha_i^n u(X_{n_i}),$$

and it follows that

$$E[\,u(\widetilde{X}_n)\,] \ge \inf_{m\ge n} E[\,u(X_m)\,].$$

Since $u(\widetilde{X}_n) \le u(W) \le (u(W))^+ \in L^1(P)$, Fatou's lemma and the preceding inequality yield

$$E[\,u(\widetilde{X})\,] \ge \limsup_{n\uparrow\infty} E[\,u(\widetilde{X}_n)\,] \ge \liminf_{n\uparrow\infty} E[\,u(X_n)\,] = \sup_{X\in\mathscr{B}} E[\,u(X)\,],$$

and so $\widetilde{X}$ is optimal. □

Remark 3.31. The argument used to prove the preceding proposition works just as well in the following general setting. Let $U : \mathscr{B} \to \mathbb{R} \cup \{-\infty\}$ be a concave functional on a set $\mathscr{B}$ of random variables defined on a probability space $(\Omega, \mathscr{F}, P)$ and with values in $\mathbb{R}^n$. Assume that:

- $\mathscr{B}$ is convex and closed under P-a. s. convergence.
- There exists a random variable $W \in L^0_+(\Omega, \mathscr{F}, P)$ with $|X| \le W < \infty$ P-a. s. for each $X = (X^1, \dots, X^n) \in \mathscr{B}$.
- $-\infty < \sup_{X \in \mathscr{B}} U(X) < \infty$.
- U is upper semicontinuous with respect to P-a. s. convergence.

Then there exists some $X^* \in \mathscr{B}$ that maximizes U on $\mathscr{B}$, and X^* is unique if U is strictly concave. As a special case, this includes the utility functionals

$$U(X) = \inf_{Q \in \mathcal{Q}} E_Q[\, u(X)\,],$$

appearing in a robust Savage representation of preferences on n-dimensional asset profiles, where u is a suitable function on $\mathbb{R}^n$ and $\mathcal{Q}$ is a set of probability measures equivalent to P; see Section 2.5. ◇

We turn now to a characterization of the optimal profile X^* in terms of the inverse of the derivative u' of u in case where u is continuously differentiable on $(0, \infty)$. Let

$$a := \lim_{x \uparrow \infty} u'(x) \ge 0 \quad \text{and} \quad b := u'(0+) = \lim_{x \downarrow 0} u'(x) \le +\infty.$$

We define

$$I^+ : (a, b) \longrightarrow (0, \infty),$$

as the continuous, bijective, and strictly decreasing inverse function of u' on (a, b), and we extend I^+ to the full half axis $[0, \infty]$ by setting

$$I^+(y) := \begin{cases} 0 & \text{for } y \ge b, \\ +\infty & \text{for } y \le a. \end{cases} \tag{3.26}$$

With this convention, $I^+ : [0, \infty] \to [0, \infty]$ is continuous. If $a = 0$ and $b = +\infty$, the function $I^+ : [0, \infty] \to [0, \infty]$ is bijective, and so the situation becomes particularly transparent. The conditions $a = 0$ and $b = +\infty$ are often referred to as the *Inada conditions*. They hold, for example, if u is a HARA utility function.

Remark 3.32. If u is a utility function defined on all of $\mathbb{R}$, the function I^+ is the inverse of the restriction of u' to $[0, \infty)$. Thus, I^+ is simply the positive part of the function $I = (u')^{-1}$. For instance, in the case of an exponential utility function $u(x) = 1 - e^{-\alpha x}$, we have $a = 0$, $b = \alpha$, and

$$I^+(y) = \left(\frac{1}{\alpha}\log\frac{y}{\alpha}\right)^- = (I(y))^+, \quad y \geq 0. \qquad \diamond$$

Theorem 3.33. *Assume that $X^* \in \mathscr{B}$ is of the form*

$$X^* = I^+(c\,\varphi) \wedge W$$

for some constant $c > 0$ such that $E^*[\,X^*\,] = w$. If $-\infty < E[\,u(X^*)\,] < \infty$ then X^* is the unique maximizer of the expected utility $E[\,u(X)\,]$ among all $X \in \mathscr{B}$.

Proof. In a first step, we consider the function

$$v(y,\omega) := \sup_{0\leq x\leq W(\omega)} (u(x) - xy) \tag{3.27}$$

defined for $y \in \mathbb{R}$ and $\omega \in \Omega$. Clearly, for each ω with $W(\omega) < \infty$ the supremum above is attained in a unique point $x^*(y) \in [0, W(\omega)]$, which satisfies

$$\begin{aligned} x^*(y) = 0 \quad &\Longleftrightarrow \quad u'(x) < y \quad \text{for all } x \in (0, W(\omega)),\\ x^*(y) = W(\omega) \quad &\Longleftrightarrow \quad u'(x) > y \quad \text{for all } x \in (0, W(\omega)). \end{aligned}$$

For $x^*(y) = 0$, the convention (3.26) hence implies $I^+(y) = 0$ and in turn $x^*(y) = I^+(y) \wedge W(\omega)$. Moreover, $y = u'(x^*(y))$ if $x^*(y)$ is an interior point of the interval $[0, W(\omega)]$. It follows that

$$x^*(y) = I^+(y) \wedge W(\omega),$$

or

$$X^* = x^*(c\,\varphi) \quad \text{on } \{W < \infty\}. \tag{3.28}$$

If $W(\omega) = +\infty$, then the supremum in (3.27) is not attained if and only if $u'(x) > y$ for all $x \in (0,\infty)$. By our convention (3.26), this holds if and only if $y \leq a$, and hence $I^+(y) = +\infty$. But our assumptions on X^* imply that $I^+(c\,\varphi) < \infty$ P-a. s. on $\{W = \infty\}$, and hence that

$$X^* = x^*(c\,\varphi) \quad P\text{-a. s. on } \{W = \infty\}. \tag{3.29}$$

Putting (3.27), (3.28), and (3.29) together yields

$$u(X^*) - X^* c\,\varphi = v(c\varphi,\cdot) \quad P\text{-a. s.}$$

Applied to an arbitrary $X \in \mathscr{B}$, this shows that

$$u(X^*) - c\,\varphi X^* \geq u(X) - c\,\varphi X \quad P\text{-a. s.}$$

Taking expectations gives

$$E[\,u(X^*)\,] \geq E[\,u(X)\,] + c \cdot E^*[\,X^* - X\,] \geq E[\,u(X)\,].$$

Hence, X^* maximizes the expected utility on $\mathscr{B}$. Uniqueness follows from Remark 3.27. □

In the following examples, we study the application of the preceding theorem to CARA and HARA utility functions. For simplicity, we consider only the case $W \equiv \infty$. The extension to a nontrivial bound W is straightforward.

Example 3.34. For an exponential utility function $u(x) = 1 - e^{-\alpha x}$, we have by Remark 3.32,

$$\varphi I^+(y\varphi) = \frac{1}{\alpha}\varphi\Big(\log\Big(\frac{y\varphi}{\alpha}\Big)\Big)^- = \frac{1}{y}h\Big(\frac{y\varphi}{\alpha}\Big),$$

where $h(x) = (x\log x)^-$. Since h is bounded by e^{-1}, it follows that $\varphi I^+(y\varphi)$ belongs to $L^1(P)$ for all $y > 0$. Thus, dominated convergence yields that

$$g(y) := E^*[\,I^+(y\varphi)\,] = \frac{1}{y}E\Big[\,h\Big(\frac{y\varphi}{\alpha}\Big)\Big]$$

decreases continuously from $+\infty$ to 0 as y increases from 0 to ∞, and there exists a unique c with $g(c) = w$. The corresponding profile

$$X^* := I^+(c\varphi)$$

maximizes the expected utility $E[\,u(X)\,]$ over all $X \geq 0$. Let us now return to the special situation of the financial market model of Section 3.1, and take P^* as the entropy-minimizing risk-neutral measure of Corollary 3.21. Then the optimal profile X^* takes the form

$$X^* = (\xi^* \cdot Y - K)^+,$$

where ξ^* is the maximizer of the expected utility $E[\,u(\xi \cdot Y)\,]$, and where K is given by

$$K = \frac{1}{\alpha}\log\frac{c}{\alpha} - \frac{1}{\alpha}\log E[\,e^{-\alpha\xi^*\cdot Y}\,] = \frac{1}{\alpha}\log\frac{c}{\alpha} + \frac{1}{\alpha}H(P^*|P).$$

Note that X^* is a *linear* combination of the primary assets only in the case where $\xi^* \cdot Y \geq K$ holds P-almost surely. In general, X^* is a *basket call option* on the attainable asset $w + (1+r)\xi^* \cdot Y \in \mathscr{V}$ with *strike price* $w + (1+r)K$. This illustrates how the optimization of expected utility may create a demand for *derivatives*. ◇

Example 3.35. If u is a HARA utility function of index $\gamma \in [0,1)$ then $u'(x) = x^{\gamma-1}$. Hence,

$$I^+(y) = y^{-\frac{1}{1-\gamma}}$$

and

$$I^+(c\,\varphi) = c^{-\frac{1}{1-\gamma}} \cdot \varphi^{-\frac{1}{1-\gamma}}.$$

In the *logarithmic* case $\gamma = 0$, we assume that the relative entropy $H(P|P^*)$ of P with respect to P^* is finite. Then

$$X^* = \frac{w}{\varphi} = w\,\frac{dP}{dP^*}$$

is the unique maximizer, and the maximal value of expected utility is

$$E[\,\log X^*\,] = \log w + H(P|P^*).$$

If $\gamma \in (0,1)$ and

$$E[\,\varphi^{-\frac{\gamma}{1-\gamma}}\,] = E^*[\,\varphi^{-\frac{1}{1-\gamma}}\,] < \infty,$$

then the unique optimal profile is given by

$$X^* = \frac{w\varphi^{-\frac{1}{1-\gamma}}}{E[\,\varphi^{-\frac{\gamma}{1-\gamma}}\,]},$$

and the maximal value of expected utility is equal to

$$E[\,u(X^*)\,] = \frac{1}{\gamma} w^\gamma E[\,\varphi^{-\frac{\gamma}{1-\gamma}}\,]^{1-\gamma}.$$ ◇

Exercise 3.3.1. In the context of Example 3.35, compute the maximal value of expected utility if φ has a log-normal distribution. ◇

The following corollary gives a simple condition on W that guarantees the existence of the maximizer X^* in Theorem 3.33.

Corollary 3.36. *If* $-\infty < E[\,u(W)\,] < \infty$ *and if* $0 < w < E^*[\,W\,] < \infty$, *then there exists a constant* $c > 0$ *such that*

$$X^* = I^+(c\,\varphi) \wedge W$$

satisfies $E^*[\,X^*\,] = w$. *In particular,* X^* *is the unique maximizer of the expected utility* $E[\,u(X)\,]$ *among all* $X \in \mathscr{B}$.

Proof. For any $\beta \in (0, \infty)$,

$$y \mapsto I^+(y) \wedge \beta$$

is a continuous decreasing function with $\lim_{y \uparrow b} I^+(y) \wedge \beta = 0$ and $I^+(y) \wedge \beta = \beta$ for all $y \le u'(\beta)$. Hence, dominated convergence implies that the function

$$g(y) := E^*[\, I^+(y\varphi) \wedge W\,],$$

is continuous and decreasing with

$$\lim_{y \uparrow \infty} g(y) = 0 < w < E^*[\, W\,] = \lim_{y \downarrow 0} g(y).$$

Hence, there exists some c with $g(c) = w$, and Theorem 3.33 yields that X^* is the unique maximizer. □

Let us now extend the discussion to the case where preferences themselves are uncertain. This additional uncertainty can be modeled by incorporating the choice of a utility function into the description of possible scenarios; for an axiomatic discussion see [206]. More precisely, we assume that preferences are described by a measurable function u on $[0, \infty) \times \Omega$ such that $u(\cdot, \omega)$ is a utility function on $[0, \infty)$. Such a function is sometimes called a *state-dependent, contingent*, or *random utility function.* We assume that $u(\cdot, \omega)$ is continuously differentiable on $(0, \infty)$. For each $\omega \in \Omega$, the inverse of $u'(\cdot, \omega)$ is extended as above to a function

$$I^+(\cdot, \omega) : [0, \infty] \longrightarrow [0, \infty].$$

Using exactly the same arguments as above, we obtain the following extension of Corollary 3.36 to the case of *random* or *state-dependent preferences*:

Corollary 3.37. *If* $E[\, u(W, \cdot)\,] < \infty$ *and if* $0 < w < E^*[\, W\,] < \infty$, *then there exists a constant* $c > 0$ *such that*

$$X^*(\omega) : = I^+(c\,\varphi(\omega), \omega) \wedge W(\omega)$$

is the unique maximizer of the expected utility

$$E[\, u(X, \cdot)\,] = \int u(X(\omega), \omega)\, P(d\omega)$$

among all $X \in \mathscr{B}$.

3.4 Optimal contingent claims under stochastic dominance constraints

So far, we have discussed the structure of asset profiles that are optimal with respect to a fixed utility function u. Let us now introduce an optimization problem with respect to the increasing concave order $\succcurlyeq_{\text{icv}}$ as discussed in Section 2.4. Recall that $\succcurlyeq_{\text{icv}}$ is also called second-order stochastic dominance. The partial order $\succcurlyeq_{\text{icv}}$ can be viewed as a reflexive and transitive relation on $L^1(\Omega, \mathscr{F}, P)$ by letting

$$\begin{aligned} X \succcurlyeq_{\text{icv}} Y :&\Longleftrightarrow \mu_X \succcurlyeq_{\text{icv}} \mu_Y \\ &\Longleftrightarrow E[\, u(X)\,] \ge E[\, u(Y)\,] \quad \text{for all utility functions } u, \end{aligned} \tag{3.30}$$

where μ_X and μ_Y denote the laws of X and Y under P. Note that $X \succcurlyeq_{\text{icv}} Y \succcurlyeq_{\text{icv}} X$ if and only if X and Y have the same distribution; see Remark 2.58(b). Thus, the relation $\succcurlyeq_{\text{icv}}$ is antisymmetric on the level of distributions but not on the level of financial positions.

Exercise 3.4.1. Suppose $X, Y \in L^1(\Omega, \mathscr{F}, P)$ are such that $E[X] = E[Y]$. Show that $X \succcurlyeq_{\text{icv}} Y$ if and only if $-X \succcurlyeq_{\text{icv}} -Y$. ◇

As in Section 3.3, we let $P^* \approx P$ be a probability measure with density $\varphi = dP^*/dP$ and the expectation with respect to P^* will serve as a pricing rule. Let us now fix a position

$$X_0 \in \mathscr{X} := L^1_+(\Omega, \mathscr{F}, P)$$

and let us try to minimize the cost over all positions $X \in \mathscr{X}$ that, with respect to $\succcurlyeq_{\text{icv}}$, are at least as attractive as X_0:

$$\text{Minimize } E^*[\, X\,] \text{ over all } X \succcurlyeq_{\text{icv}} X_0.$$

In order to describe the minimal cost and the minimizing profile, let us denote by F_φ and F_{X_0} the cumulative distribution functions and by q_φ and q_{X_0} quantile functions of φ and X_0; see Appendix D.

Theorem 3.38. *For any $X \in \mathscr{X}$ such that $X \succcurlyeq_{\text{icv}} X_0$,*

$$E^*[\, X\,] \ge \int_0^1 q_\varphi(1-s)\, q_{X_0}(s)\, ds. \tag{3.31}$$

The lower bound is attained by $X^* = f(\varphi)$, where f is a decreasing function on $(0, \infty)$ satisfying

$$f(x) = \begin{cases} q_{X_0}(1 - F_\varphi(x)) & \text{if } 0 < F_\varphi(x) < 1 \text{ and } F_\varphi(x-) = F_\varphi(x), \\ \dfrac{1}{F_\varphi(x) - F_\varphi(x-)} \displaystyle\int_{F_\varphi(x-)}^{F_\varphi(x)} q_{X_0}(1 - t)\, dt & \text{if } F_\varphi(x-) < F_\varphi(x). \end{cases}$$

The proof will use the following lemma, which yields another characterization of the increasing concave order $\succcurlyeq_{\text{icv}}$.

Lemma 3.39. *For two probability measures μ and ν on $\mathbb{R}$ with finite means, the following conditions are equivalent:*

(a) $\mu \succcurlyeq_{\text{icv}} \nu$.

(b) *For all decreasing functions $h : (0,1) \to [0,\infty)$,*

$$\int_0^1 h(t) q_\mu(t)\, dt \geq \int_0^1 h(t) q_\nu(t)\, dt, \tag{3.32}$$

where q_μ and q_ν are quantile functions of μ and ν.

(c) *The relation* (3.32) *holds for all bounded decreasing functions $h : (0,1) \to [0,\infty)$.*

Proof. Let us start by noting that the integrals in (3.32) are always well-defined but may take the value $-\infty$. To see why, consider the case of q_μ. Our claim is obvious if either $q_\mu(t) \leq 0$ or $q_\mu(t) \geq 0$ for all t. If neither is the case, then there exists $t_0 \in (0,1)$ such that $q_\mu(t) \leq 0$ for $t \leq t_0$ and $q_0(t) \geq 0$ for $t \geq t_0$. Since μ has a finite mean and h is decreasing and nonnegative,

$$\int_0^1 h(t) q_\mu(t)\, dt \leq \int_{t_0}^1 h(t) q_\mu(t)\, dt \leq h(t_0) \int_{t_0}^1 q_\mu(t)\, dt = h(t_0) \int_0^\infty x\, \mu(dx) < \infty.$$

The relation $\mu \succcurlyeq_{\text{icv}} \nu$ is equivalent to

$$\int_0^y q_\mu(t)\, dt \geq \int_0^y q_\nu(t)\, dt \quad \text{for all } y \in [0,1];$$

see Theorem 2.57. The implication (c)⇒(a) thus follows by taking $h = \mathbb{1}_{(0,t]}$. For the proof of (a)⇒(b), we may assume without loss of generality that h is left-continuous; otherwise we replace $h(t)$ by $h(t-)$ for the at most countably many $t \in (0,1)$ at which those two values are not the same. Then there exists a positive Radon measure η on $(0,1]$ such that $h(t) = \eta([t,1])$. Fubini's theorem yields

$$\int_0^1 h(t)\, q_\mu(t)\, dt = \int_0^1 \int_{[t,1]} q_\mu(t)\, \eta(ds)\, dt = \int_{(0,1]} \int_0^s q_\mu(t)\, dt\, \eta(ds)$$
$$\geq \int_{(0,1]} \int_0^s q_\nu(t)\, dt\, \eta(ds) = \int_0^1 \int_{[t,1]} q_\nu(t)\, \eta(ds)\, dt$$
$$= \int_0^1 h(t)\, q_\nu(t)\, dt. \qquad \square$$

Proof of Theorem 3.38. Using the first Hardy–Littlewood inequality in Theorem D.13, we see that

$$E^*[\,X\,] = E[\,X\varphi\,] \geq \int_0^1 q_\varphi(1-t)\, q_X(t)\, dt,$$

where q_X is a quantile function for X. Taking $h(t) := q_\varphi(1-t)$ and using Lemma 3.39 thus yields (3.31).

Let us now turn to the identification of the optimal profile. Note that the function f defined in the assertion satisfies

$$f(q_\varphi) = E_\lambda[\,g \mid q_\varphi\,] \tag{3.33}$$

where g is defined by $g(t) = q_{X_0}(1-t)$, and where $E_\lambda[\,\cdot \mid q_\varphi\,]$ denotes the conditional expectation with respect to q_φ under the Lebesgue measure λ on $(0,1)$; see Exercise 3.4.2 below. Let us show that $X^* = f(\varphi)$ satisfies $X^* \succcurlyeq_{\text{icv}} X_0$. Indeed, Lemma D.7 and Jensen's inequality for conditional expectations yield that for any utility function u,

$$E[\,u(X^*)\,] = E[\,u(f(\varphi))\,] = \int_0^1 u(f(q_\varphi(t)))\, dt = E_\lambda[\,u(E_\lambda[\,g \mid q_\varphi\,])\,]$$
$$\geq E_\lambda[\,E_\lambda[\,u(g) \mid q_\varphi\,]\,] = E_\lambda[\,u(g)\,] = \int_0^1 u(q_{X_0}(t))\, dt = E[\,u(X_0)\,].$$

Moreover, X^* attains the lower bound in (3.31):

$$E^*[\,X^*\,] = E[f(\varphi)\,\varphi\,] = \int_0^1 f(q_\varphi(t)) q_\varphi(t)\, dt = E_\lambda[\,E_\lambda[\,g \mid q_\varphi\,]\, q_\varphi\,]$$
$$= E_\lambda[\,g\, q_\varphi\,] = \int_0^1 q_{X_0}(1-t)\, q_\varphi(t)\, dt = \int_0^1 q_{X_0}(t)\, q_\varphi(1-t)\, dt.$$

Finally, we need to show that $X^* \in \mathscr{X}$. We clearly have $X^* \geq 0$. Moreover,

$$E[X^*] = E_\lambda[f(q_\varphi)] = E_\lambda[E_\lambda[g \mid q_\varphi]] = E_\lambda[g] = \int_0^1 q_{X_0}(1-t)\,dt = E[X_0], \tag{3.34}$$

and so X^* has the same expectation as X_0. □

Exercise 3.4.2. In this exercise, we prove formula (3.33). Let $(0,1)$ be equipped with its Borel σ-algebra and the Lebesgue measure λ. Now let $F : \mathbb{R} \to [0,1]$ be a cumulative distribution function, that is, F is right-continuous, increasing, and satisfies $\lim_{x\downarrow-\infty} F(x) = 0$ and $\lim_{x\uparrow\infty} F(x) = 1$. Let $q : (0,1) \to \mathbb{R}$ be a quantile function for F. Show that for any measurable function $\varphi : (0,1) \to [0,\infty)$ the conditional expectation $E_\lambda[\varphi \mid q]$ is λ-a. s. given by

$$E_\lambda[\varphi \mid q] = f(q),$$

where the function $f : \mathbb{R} \to [0,\infty]$ satisfies

$$f(x) = \begin{cases} \varphi(F(x)) & \text{if } 0 < F(x) < 1 \text{ and } F(x-) = F(x), \\ \dfrac{1}{F(x)-F(x-)} \displaystyle\int_{F(x-)}^{F(x)} \varphi(t)\,dt & \text{if } F(x-) < F(x). \end{cases}$$

To this end, argue in particular that $f(q)$ is λ-a. s. independent of the values $f(x)$ that are not specified by the above formula. ◇

Equation (3.34) in the proof of Theorem 3.38 shows that X^* and X_0 have the same expectation. Even more can be said if the distribution of φ is continuous.

Corollary 3.40. *Suppose that φ has a continuous distribution. Then the optimal profile X^* constructed in Theorem* 3.38 *has the same law under P as X_0 and also solves the following cost minimization problem:*

$$\text{minimize } E^*[X] \text{ over all } X \in \mathscr{X} \text{ that have the same law under } P \text{ as } X_0. \tag{3.35}$$

Proof. If φ has a continuous distribution, then $F_\varphi(\varphi)$, and hence also $1 - F_\varphi(\varphi)$, are uniformly distributed on $(0,1)$ by Lemma D.10. Therefore, Lemma D.7 implies that $X^* = q_{X_0}(1 - F_\varphi(\varphi))$ has the same law as X_0. Since any X with the same law as X_0 satisfies $X \succcurlyeq_{\text{icv}} X_0$ and $q_{X_0}(1 - F_\varphi(\varphi))$ is a decreasing function of φ, the result now follows from Theorem 3.38. □

A solution X^* of the minimization problem (3.35) is sometimes called a *cost-efficient payoff profile*, because it represents the least expensive random variable with a desired distribution.

Remark 3.41. The lower bound in (3.31) may be viewed as a "reservation price" for X_0 in the following sense. Let X_0 be a financial position, and let $\mathscr{X}$ be any class of financial positions such that $X \in \mathscr{X}$ is available at price $\pi(X)$. For a given relation $\succeq$ on $\mathscr{X} \cup \{X_0\}$,

$$\pi_R(X_0) := \inf\{\pi(X) \mid X \in \mathscr{X},\, X \succeq X_0\}$$

is called the *reservation price* of X_0 with respect to $\mathscr{X}$, π, and $\succeq$.

If $\mathscr{X}$ is the space of constants with $\pi(c) = c$, and if the relation $\succeq$ is of von Neumann–Morgenstern type with some utility function u, then $\pi_R(X_0)$ reduces to the certainty equivalent of X_0 with respect to u; see (2.11).

In the context of the optimization problem (3.23), where

$$X \succeq X_0 \;:\Longleftrightarrow\; E[\,u(X)\,] \geq E[\,u(X_0)\,],$$

the reservation price is given by $E^*[\,X^*\,]$, where X^* is the utility maximizer in the budget set defined by $w := E^*[\,X_0\,]$.

In the context of the financial market model of Chapter 1, we can take $\mathscr{X}$ as the space $\mathscr{V}$ of attainable claims with

$$V \succeq X_0 \;:\Longleftrightarrow\; V \geq X_0 \text{ } P\text{-a.s.}$$

and $\pi(V) = \overline{\xi} \cdot \overline{\pi}$ for $V = \overline{\xi} \cdot \overline{S}$. In this case, the reservation price $\pi_R(X_0)$ coincides with the upper bound $\pi_{\sup}(X_0)$ of the arbitrage-free prices for X_0; see Theorem 1.34. ◇

Exercise 3.4.3. Consider our one-period market model with interest rate $r = 0$, so that $\pi^0 = S^0 = 1$, and one risky asset with prices $\pi := \pi^1$ and $S := S^1 = \pi e^{\sigma Z + \alpha}$, where Z has a standard normal distribution, $\sigma > 0$, and $\alpha \in \mathbb{R}$. Then S is log-normally distributed.

(a) Determine $b \in \mathbb{R}$ such that $\varphi := e^{-bZ - \frac{b^2}{2}}$ is the density of a risk-neutral probability measure $P^* \in \mathscr{P}$, and keep this b fixed for the remainder of this exercise.

(b) Write φ in the form $\varphi = h(S)$, where h is a function on $(0, \infty)$. Then determine $\alpha_0 \in \mathbb{R}$ such that h is strictly decreasing for $\alpha > \alpha_0$ and strictly increasing for $\alpha < \alpha_0$.

(c) Suppose that $X_0 = (K - S)^+$ is the payoff of a put option with strike $K > 0$. Sow that the solution X^* of (3.35) is given as follows:

$$X^* = \begin{cases} \frac{K}{S}(S - \frac{\pi^2 e^{2\alpha}}{K})^+ & \text{if } \alpha > \alpha_0, \\ X_0 & \text{if } \alpha < \alpha_0. \end{cases}$$

(d) In analogy to (c), find the solution X^* of (3.35) if $X_0 = (S - K)^+$ is the payoff of a call option with strike $K > 0$. ◇

3.5 Stop-loss insurance contracts

In this section, we look at insurance contracts that are optimal with respect to the increasing concave order. The main result is Arrow's theorem of the deductible, which states that the optimal indemnity schedule takes the form of a stop-loss contract. We present two independent proofs of this result and provide a third one in the form of an exercise.

Similar to Example 2.41, we consider an agent with initial wealth $w > 0$ who is exposed to some random loss $Y \geq 0$ with $E[\,Y\,] < \infty$ and $P[\,Y \neq E[\,Y\,]\,] > 0$. For a premium $\pi > 0$, the agent can purchase insurance against this loss in the form of an insurance contract that pays the *indemnity* $I(Y(\omega))$ if the loss $Y(\omega)$ occurs. Here, the indemnity schedule I belongs to the class $\mathscr{I}$ of all measurable functions $I : [0,\infty) \to [0,\infty)$ satisfying the following two conditions:

$$0 \leq I(y) \leq y, \quad \text{for all } y \geq 0, \tag{3.36}$$

$$\pi \geq (1+\rho)E[\,I(Y)\,], \tag{3.37}$$

where $\rho \geq 0$ is a given constant. The first condition states that the indemnity can never be larger than the actual loss, which prevents moral hazard. The second condition provides a lower bound on the insurance premium in terms of the expectation $E[\,I(Y)\,]$, which corresponds to the fair actuarial value of the insurance contract. The factor ρ takes care of costs and margin for the insurance company underwriting the contract. We exclude the trivial case $\pi \geq \|Y\|_\infty$, in which the agent would not receive any benefit from insuring the loss.

After having insured the loss, the position of the agent is given by

$$X_I := w - \pi - Y + I(Y).$$

If the agent is risk averse with utility function u, a natural question is to maximize the expected utility $E[\,u(X_I)\,]$ over all functions $I \in \mathscr{I}$. It turns out that this problem can be solved, and that its solution does not depend on the specific choice of the utility function u. That is, the solution is optimal with respect to the increasing concave order. This is the content of *Arrow's theorem of the deductible*, which we state next. The name of this theorem derives from the fact that the optimal indemnity schedule I^* takes the form of a stop-loss contract with a certain deductible d, that is, $I^*(y) = (y-d)^+$. Thus, the loss Y is reduced to

$$Y - I^*(Y) = Y \wedge d.$$

Theorem 3.42. *There exists $d \geq 0$ such that the indemnity schedule $I^*(y) := (y-d)^+$ belongs to $\mathscr{I}$ and satisfies $X_{I^*} \succcurlyeq_{\text{icv}} X_I$ for all $I \in \mathscr{I}$. Moreover, the optimal indemnity schedule I^* is unique in the sense that $I^* = \tilde{I}$ holds $P \circ Y^{-1}$-a. s. whenever $\tilde{I} \in \mathscr{I}$ is such that $X_{\tilde{I}} \succcurlyeq_{\text{icv}} X_I$ for all $I \in \mathscr{I}$.*

Proof. Since $X_{\tilde{I}} \succcurlyeq_{\text{icv}} X_I$ holds if and only if $c + X_{\tilde{I}} \succcurlyeq_{\text{icv}} c + X_I$ for any given constant c, we may assume without loss of generality that $w = \pi$. If $\pi \geq (1+\rho)E[\,Y\,]$, then it is clearly optimal to take $d = 0$ so that $I^*(Y) = Y$. Otherwise, we choose $d > 0$ such that $(1+\rho)E[\,(Y-d)^+\,] = \pi$. Such a value d exists, because by dominated convergence the function $x \mapsto E[\,(Y-x)^+\,]$ decreases continuously from $E[\,Y\,]$ to zero as x increases from zero to infinity. It is then clear that $I^*(y) = (y-d)^+$ belongs to $\mathcal{I}$.

Now let $I \in \mathcal{I}$ be arbitrary. Our goal is to verify condition (e) in Theorem 2.57, namely

$$\int_0^t q_{X_{I^*}}(s)\,ds \geq \int_0^t q_{X_I}(s)\,ds \quad \text{for } 0 < t \leq 1, \tag{3.38}$$

where q_Z denotes the right-continuous quantile function of a random variable Z. Our assumption $w = \pi$ yields that

$$0 \geq X_{I^*} = -Y + I^*(Y) = -(d \wedge Y) = f(-Y),$$

where the function $f(y) = (-d) \vee y = \max\{-d, y\}$ is continuous and increasing. Hence, Lemma D.12 shows that $q_{X_{I^*}} = f(q_{-Y})$. We define

$$t_0 := \inf\{s \in (0,1) \mid q_{X_I}(s) > -d\} \wedge 1,$$

where $\inf \emptyset = +\infty$ as usual. For any $s \in (0, t_0)$, we have $q_{X_I}(s) \leq -d$. Since $X_I \geq -Y$, and hence $q_{X_I}(s) \geq q_{-Y}(s)$, we get $q_{-Y}(s) \leq -d$ and in turn $q_{X_{I^*}}(s) = f(q_{-Y}(s)) = -d \geq q_{X_I}(s)$. Hence, (3.38) holds for all $t \leq t_0$. In particular, we are done if $t_0 = 1$. Otherwise, we have $q_{X_I}(s) > -d$ for $s \in (t_0, 1)$, and hence $q_{X_{I^*}}(s) = f(q_{-Y}(s)) \leq f(q_{X_I}(s)) = q_{X_I}(s)$. It follows that

$$\int_0^t (q_{X_{I^*}}(s) - q_{X_I}(s))\,ds$$

decreases to the limit

$$\int_0^1 (q_{X_{I^*}}(s) - q_{X_I}(s))\,ds = E[X_{I^*}] - E[X_I] \geq 0$$

as t increases from t_0 to 1. This shows that (3.38) holds for all $t \in (0,1]$, and so $I^*(y) = (y-d)^+$ is indeed optimal. Finally, uniqueness follows from part (b) of Remark 2.58. □

In the following remark, we show that Theorem 3.42 can also be viewed as a corollary to Theorem 3.33.

Remark 3.43. The preceding proof of Theorem 3.42 consists in showing that

$$R^* := Y \wedge d \quad \text{and} \quad R_I := Y - I(Y)$$

satisfy

$$-R^* \succcurlyeq_{\mathrm{icv}} -R_I \tag{3.39}$$

for any $I \in \mathscr{I}$. For an alternative proof of (3.39), we assume $E[R^*] = E[R_I]$. Then (3.39) is equivalent to

$$R^* \succcurlyeq_{\mathrm{icv}} R_I; \tag{3.40}$$

see Exercise 3.4.1. But (3.40) can be seen as a special case of Theorem 3.33, taking $P^* = P$ and $W = Y$, since $\varphi \equiv 1$ implies that R^* maximizes the expected utility for any choice of the unitility function u. ◇

The following exercise provides yet another proof of Theorem 3.42.

Exercise 3.5.1. Let $I^*(y) = (y - d)^+$ be a stop-loss contract whose deductible d is such $I^* \in \mathscr{I}$, and let $I \in \mathscr{I}$ be arbitrary. Show that the cumulative distribution functions of X_{I^*} and X_I satisfy the following single-crossing property:

$$\begin{aligned} F_{X_{I^*}}(x) &\le F_{X_I}(x) \quad \text{for } x < w - \pi - d, \\ F_{X_{I^*}}(x) &\ge F_{X_I}(x) \quad \text{for } x \ge w - \pi - d. \end{aligned}$$

Then use Exercise 2.4.1 to conclude that I^* is an optimal indemnity schedule if d is chosen such that $(1 + \rho)E[\,(Y - d)^+\,] = \pi$. ◇

Combining Theorem 3.42 with Exercise 2.4.2 yields the following corollary.

Corollary 3.44. *If* $I^*(y) = (y-d)^+$ *is as in Theorem* 3.42 *and* $I \in \mathscr{I}$ *satisfies* $E[X_I] = E[X_{I^*}]$, *then* $\mathrm{var}(X_{I^*}) \le \mathrm{var}(X_I)$.

In this weaker form, the optimality of the stop-loss contract appears in the reinsurance literature at least since 1961; cf. Kahn [197], where the following simple and direct proof of Corollary 3.44 was given.

Proof of Corollary 3.44. With the notation introduced in Remark 3.43, it is sufficient to show $E[R_I^2] \ge E[(R^*)^2]$ for $I \in \mathscr{I}$. Since $R_I \le Y \le d$ on $\{Y \le d\}$, we have

$$\begin{aligned} E[R_I^2] &= E[(R_I - d)^2] - d^2 + 2dE[R_I] \\ &\ge E[(R_I - d)^2 \mathbb{1}_{\{Y \le d\}}] - d^2 + 2dE[R_I] \\ &\ge E[(Y - d)^2 \mathbb{1}_{\{Y \le d\}}] - d^2 + 2dE[R_I] \\ &= E[(d - Y \wedge d)^2 \mathbb{1}_{\{Y \le d\}}] - d^2 + 2dE[R^*] \\ &= E[(R^*)^2]. \end{aligned}$$

This completes the proof. □

3.6 Robust utility maximization

In this section, we consider the optimal investment problem for an economic agent who, as in Section 2.5, is averse against both risk and Knightian uncertainty. Under the assumptions of Theorem 2.78(b), the preferences of such an agent can be described by the following *robust utility functional:*

$$\inf_{Q \in \mathcal{Q}} E_Q[\, u(X) \,], \tag{3.41}$$

where $\mathcal{Q}$ is a set of probability measures on $(\Omega, \mathscr{F})$, u is a utility function, and X is a payoff profile. We assume that such a robust utility functional is given and that all measures in $\mathcal{Q}$ are absolutely continuous with respect to a fixed pricing measure P^* on $(\Omega, \mathscr{F})$. We will also assume that all payoff profiles X are nonnegative and, for simplicity, that the utility function u is finite on $[0, \infty)$. As in Section 3.3, the budget set for a given initial capital $w > 0$ is defined as

$$\mathscr{B} := \{X \in L^1_+(P^*) \mid E^*[\, X \,] \le w\}.$$

The problem of *robust utility maximization* can thus be stated as follows:

$$\text{maximize } \inf_{Q \in \mathcal{Q}} E_Q[\, u(X) \,] \text{ over all } X \in \mathscr{B}. \tag{3.42}$$

In the sequel, we will assume that $\mathcal{Q}$ is a convex set, which can be done without loss of generality. Moreover, we will assume throughout this section that $\mathcal{Q}$ is equivalent to P^* in the following sense: For all $A \in \mathscr{F}$,

$$P^*[\, A \,] = 0 \quad \iff \quad Q[\, A \,] = 0 \text{ for all } Q \in \mathcal{Q}. \tag{3.43}$$

Clearly, our problem (3.42) would not be well-posed without the implication $\Rightarrow$. Note that (3.43) implies that every measure $Q \in \mathcal{Q}$ is absolutely continuous with respect to P^* but not necessarily equivalent to P^*.

If $\mathcal{Q}$ consists of the single element $Q \approx P^*$, we have seen in Section 3.3 that the solution involves the Radon–Nikodym derivative dP^*/dQ. In our present situation, however, not every $Q \in \mathcal{Q}$ needs to be equivalent to P^*. The Radon–Nikodym derivative dP^*/dQ will therefore be understood in the sense of the Lebesgue decomposition; see Theorem B.9. With the convention $\frac{x}{0} := +\infty$ for $x > 0$, it can be expressed as

$$\frac{dP^*}{dQ} = \left(\frac{dQ}{dP^*}\right)^{-1} \quad P^*\text{-a.s.}$$

It will turn out in Theorem 3.53 below that the following concept can be a key for solving problem (3.42).

Definition 3.45. A measure $Q_0 \in \mathcal{Q}$ is called a *least-favorable measure* with respect to P^* if the density $\varphi_0 = dP^*/dQ_0$ satisfies

$$Q_0[\,\varphi_0 \le t\,] = \inf_{Q \in \mathcal{Q}} Q[\,\varphi_0 \le t\,] \quad \text{for all } t > 0. \qquad \triangle$$

Not every set $\mathcal{Q}$ is such that it admits a least-favorable measure with respect to a given measure P^*. But here are two examples in which least-favorable measures can be determined explicitly.

Example 3.46. Let Y be a measurable function on $(\Omega, \mathscr{F})$, and denote by μ its law under P^*. For $\nu \approx \mu$ given, let

$$\mathcal{Q} := \{Q \ll P^* \mid Q \circ Y^{-1} = \nu\}.$$

The interpretation behind the set $\mathcal{Q}$ is that an investor has full knowledge about the pricing measure P^* but is uncertain about the true distribution of market prices and has only the "weak information" that a certain functional Y of the stock price has the distribution ν under that true distribution. Define Q_0 as follows via its Radon–Nikodym derivative:

$$\frac{dQ_0}{dP^*} = \frac{d\nu}{d\mu}(Y).$$

Then $Q_0 \in \mathcal{Q}$, and the law of $\varphi_0 := dP^*/dQ_0 = d\mu/d\nu(Y)$ is the same for all $Q \in \mathcal{Q}$. Hence, Q_0 is a least-favorable measure. ◇

Example 3.47. For $\lambda \in (0,1)$ and a given probability measure $P \approx P^*$, let

$$\mathcal{Q}_\lambda := \left\{Q \ll P \;\middle|\; \frac{dQ}{dP} \le \frac{1}{\lambda}\right\}.$$

In Chapter 4, this set will play an important role as the maximal representing set for the risk measure $\mathrm{AV@R}_\lambda$. If $\varphi := dP^*/dP$ satisfies $P[\,\varphi > \lambda^{-1}\,] > 0$ and admits a continuous and strictly increasing cumulative distribution function $F_\varphi(x) := P[\,\varphi \le x\,]$, then $\mathcal{Q}_\lambda$ admits a least-favorable measure with respect to P^*, which is given by

$$\frac{dQ_0}{dP} = \frac{1}{\lambda} \cdot \frac{\varphi}{\varphi \vee q_\varphi(t_\lambda)},$$

where $q_\varphi = F_\varphi^{-1}$ is the (unique) quantile function of φ and t_λ is the unique solution to the equation

$$q_\varphi(t_\lambda)(t_\lambda - 1 + \lambda) = \int_0^{t_\lambda} q_\varphi(t)\,dt.$$

This will be proved in Corollary 8.28. In this case, one finds that

$$\varphi_0 = \frac{dP^*}{dQ_0} = \lambda(\varphi \vee q_\varphi(t_\lambda)). \qquad \diamond$$

Further examples for least-favorable measures can be found in Huber and Ronchetti [184].

Remark 3.48. The existence of a least-favorable measure of a set $\mathcal{Q}$ with respect to P^* is closely related to the concept of *submodularity* or *strong additivity* of the set function

$$c(A) := \sup_{Q \in \mathcal{Q}} Q[\,A\,], \quad A \in \mathscr{F},$$

which will be discussed in Sections 4.8 and 4.9. The Huber–Strassen theorem implies that a set $\mathcal{Q}$ that is weakly compact with respect to a Polish topology on Ω admits a least-favorable measure with respect to any other probability measure $P^* \approx \mathcal{Q}$ if and only if the set function c is *submodular* or *strongly subadditive*:

$$c(A \cup B) + c(A \cap B) \le c(A) + c(B), \quad A, B \in \mathscr{F}.$$

We refer to Huber and Strassen [185] and Lembcke [236]. Example 3.47 is a special case of this situation, since it will follow from Proposition 4.81 that the associated set function is strongly subadditive. ◇

Let us now show that every least-favorable measure Q_0 is equivalent to P^* if $\mathcal{Q}$ satisfies a mild closure condition.

Lemma 3.49. *Suppose that $\{dQ/dP^* \mid Q \in \mathcal{Q}\}$ is closed in $L^1(P^*)$. Then every least-favorable measure Q_0 is equivalent to P^*.*

Proof. Due to our assumption (3.43) and the closedness of $\{dQ/dP^* \mid Q \in \mathcal{Q}\}$, we may apply the Halmos–Savage theorem in the form of Theorem 1.63 and obtain a measure $Q_1 \approx P^*$ in $\mathcal{Q}$. We get

$$\begin{aligned} 1 = Q_0[\,\varphi_0 < \infty\,] &= \lim_{t\uparrow\infty} Q_0[\,\varphi_0 \le t\,] = \lim_{t\uparrow\infty} \inf_{Q\in\mathcal{Q}} Q[\,\varphi_0 \le t\,] \le \lim_{t\uparrow\infty} Q_1[\,\varphi_0 \le t\,] \\ &= Q_1[\,\varphi_0 < \infty\,]. \end{aligned}$$

Hence, also $P^*[\,\varphi_0 < \infty\,] = 1$. That is, $1/\varphi_0 = dQ_0/dP^* > 0$ P^*-almost surely. By Corollary B.5, Q_0 and P^* are hence equivalent. □

We have the following characterization of least-favorable measures.

Proposition 3.50. *For a measure $Q_0 \in \mathcal{Q}$ with $Q_0 \approx P^*$ and $\varphi_0 := dP^*/dQ_0$, the following conditions are equivalent:*

(a) *Q_0 is a least-favorable measure for P^*.*

(b) *If* $f : (0, \infty] \to \mathbb{R}$ *is decreasing and* $\inf_{Q \in \mathcal{Q}} E_Q[f(\varphi_0) \wedge 0] > -\infty$, *then*

$$\inf_{Q \in \mathcal{Q}} E_Q[f(\varphi_0)] = E_{Q_0}[f(\varphi_0)].$$

(c) *If* $g : (0, \infty] \to \mathbb{R}$ *is increasing and* $\sup_{Q \in \mathcal{Q}} E_Q[g(\varphi_0) \vee 0] < \infty$, *then*

$$\sup_{Q \in \mathcal{Q}} E_Q[g(\varphi_0)] = E_{Q_0}[g(\varphi_0)].$$

(d) Q_0 *minimizes the* g-divergence

$$I_g(Q|P^*) := \int g\left(\frac{dQ}{dP^*}\right) dP^*$$

among all $Q \in \mathcal{Q}$ *whenever* $g : [0, \infty) \to \mathbb{R}$ *is a continuous convex function such that* $I_g(Q|P^*)$ *is finite for some* $Q \in \mathcal{Q}$.

Proof. (a)⇔(b): According to the definition, Q_0 is a least-favorable measure if and only if $Q_0[\varphi_0 \le t] \le Q[\varphi_0 \le t]$ for all $t \ge 0$ and each $Q \in \mathcal{Q}$. By Theorem 2.68, this is equivalent to the fact that, for all $Q \in \mathcal{Q}$, $Q_0 \circ \varphi_0^{-1}$ stochastically dominates $Q \circ \varphi_0^{-1}$ in the sense of Definition 2.67. Hence, (a) is equivalent to the following condition:
(b′) If $f : (0, \infty] \to \mathbb{R}$ is decreasing and bounded, then

$$\inf_{Q \in \mathcal{Q}} E_Q[f(\varphi_0)] = E_{Q_0}[f(\varphi_0)].$$

If f is unbounded and $\inf_{Q \in \mathcal{Q}} E_Q[f(\varphi_0) \wedge 0] > -\infty$, then the conclusion of (b′) holds for $f_N := (-N) \vee f \wedge 0$ where $N \in \mathbb{N}$. Thus, for all $Q \in \mathcal{Q}$,

$$E_Q[f_N(\varphi_0)] \ge E_{Q_0}[f_N(\varphi_0)] \ge E_{Q_0}[f(\varphi_0) \wedge 0] > -\infty.$$

By sending N to infinity, monotone convergence yields that $E_Q[f(\varphi_0) \wedge 0] \ge E_{Q_0}[f(\varphi_0) \wedge 0]$ for every $Q \in \mathcal{Q}$. After using a similar argument for $0 \vee f(\varphi_0)$, we get

$$E_Q[f(\varphi_0)] = E_Q[f(\varphi_0) \vee 0] + E_Q[f(\varphi_0) \wedge 0] \ge E_{Q_0}[f(\varphi_0)] \quad \text{for all } Q \in \mathcal{Q}.$$

(b)⇔(c) follows by taking $f = -g$.

(b)⇒(d): Due to Jensen's inequality, $I_g(Q|P^*)$ is well-defined and larger than or equal to $g(1)$ for each $Q \ll P^*$. Now take any $Q_1 \in \mathcal{Q}$ with $I_g(Q_1|P^*) < \infty$, and denote by $g'_+(x)$ the right-hand derivative of g at $x \ge 0$. Suppose first that g'_+ is bounded. Since $g(y) - g(x) \ge g'_+(x)(y - x)$, we have

$$I_g(Q_1|P^*) - I_g(Q_0|P^*) \ge \int g'_+(\varphi_0^{-1})\left(\frac{dQ_1}{dP^*} - \frac{dQ_0}{dP^*}\right) dP^* = \int f(\varphi_0)\, dQ_1 - \int f(\varphi_0)\, dQ_0,$$

where $f(x) := g'_+(1/x)$ is a bounded decreasing function. Therefore, (b) implies that $\int f(\varphi_0)\,dQ_1 \geq \int f(\varphi_0)\,dQ_0$, and so Q_0 minimizes $I_g(\,\cdot\,|P^*)$ on $\mathcal{Q}$.

To prove the assertion for convex functions g with unbounded g'_+, let μ_Q denote the law of $\frac{dQ}{dP^*}$ under P^*. Then the preceding step of the proof implies in particular that $\int (x-c)^+\,\mu_Q(dx) \geq \int (x-c)^+\,\mu_{Q_0}(dx)$ for all $c \in \mathbb{R}$ and each $Q \in \mathcal{Q}$. Since in addition $\int x\,\mu_Q(dx) = 1$ for all $Q \in \mathcal{Q}$, the result now follows from the equivalence (b)$\Leftrightarrow$(c) of Corollary 2.61.

(d)$\Rightarrow$(b): It is enough to prove (b) for continuous bounded decreasing functions f. For such a function f let $g(x) := \int_1^x f(1/t)\,dt$. Then g is convex. For $Q_1 \in \mathcal{Q}$, we let $Q_t := tQ_1 + (1-t)Q_0$ and $h(t) := I_g(Q_t|P^*)$. The right-hand derivative of h satisfies $0 \leq h'_+(0) = \int f(\varphi_0)\,dQ_1 - \int f(\varphi_0)\,dQ_0$, and the proof is complete. □

Remark 3.51. Let us discuss the connection between least-favorable measures and the statistical test theory for *composite* hypotheses, extending the standard Neyman–Pearson theory as outlined in Appendix E. In a composite hypothesis testing problem, one tests a hypothesis $\mathcal{P}$ against a null hypothesis $\mathcal{Q}$ and allows both $\mathcal{P}$ and $\mathcal{Q}$ to be *sets* of probability measures. Our situation here corresponds to the special case in which $\mathcal{P}$ consists of the single element P^* and only the null hypothesis $\mathcal{Q}$ is composite. As in Remark E.4, one looks for a randomized statistical test $\psi^0 : \Omega \to [0,1]$ that maximizes the *power* $E^*[\,\psi\,]$ over all randomized tests ψ with a given *significance level*. For a composite null hypothesis $\mathcal{Q}$, the significance of a randomized test ψ can be defined as $\sup_{Q\in\mathcal{Q}} E_Q[\,\psi\,]$. Thus, the problem can be stated as follows:

$$\text{maximize } E^*[\,\psi\,] \text{ over all } \psi \in \mathcal{R} \text{ with } \sup_{Q\in\mathcal{Q}} E_Q[\,\psi\,] \leq \alpha, \tag{3.44}$$

where $\alpha \in (0,1)$ is a given significance level and $\mathcal{R}$ denotes the class of all randomized tests (see Appendix E). This problem can be solved in a straightforward manner as soon as $\mathcal{Q}$ admits a least-favorable measure Q_0 with respect to P^*. Indeed, choose $\kappa \in [0,1]$ and $c > 0$ such that

$$\psi^0 := \kappa \mathbb{1}_{\{\varphi_0 = c\}} + \mathbb{1}_{\{\varphi_0 > c\}}$$

satisfies $E_{Q_0}[\,\psi^0\,] = \alpha$. Since ψ^0 is an increasing function of φ_0, Proposition 3.50 yields that

$$\sup_{Q\in\mathcal{Q}} E_Q[\,\psi^0\,] = E_{Q_0}[\,\psi^0\,] = \alpha.$$

Moreover, if ψ is any other randomized test with $\sup_{Q\in\mathcal{Q}} E_Q[\,\psi\,] \leq \alpha$, then in particular $E_{Q_0}[\,\psi\,] \leq \alpha$, and so Theorem E.3 implies $E^*[\,\psi\,] \leq E^*[\,\psi^0\,]$, because ψ^0 is an optimal randomized test for the standard problem of testing the hypothesis P^* against the null hypothesis Q_0. Therefore, ψ^0 is a solution to (3.44). Note also that the likelihood of a type 1 error for the test ψ^0 is maximized by Q_0. The fact that Q_0 is independent of the significance level α is remarkable, and it explains why Q_0 is called a "least-favorable" measure. ◇

The next proposition prepares for the solution to the robust utility maximization problem (3.42).

Proposition 3.52. *Let $Q_0 \approx P^*$ be a least-favorable measure and $\varphi_0 = dP^*/dQ_0$.*
(a) *For any $X \in \mathscr{B}$, there exists a decreasing function $f \geq 0$ such that $f(\varphi_0) \in \mathscr{B}$ and*

$$\inf_{Q\in\mathcal{Q}} E_Q[\, u(f(\varphi_0))\,] \geq \inf_{Q\in\mathcal{Q}} E_Q[\, u(X)\,]. \tag{3.45}$$

(b) *Every solution X^* of (3.42) is of the form $X^* = f^*(\varphi_0)$ for a deterministic decreasing function $f^* : (0,\infty) \to [0,\infty)$.*

Proof. (a) Throughout the proof, we denote by $q_Y(t)$ a quantile function of a random variable Y with respect to the probability measure Q_0. Then we let F denote the cumulative distribution function of φ_0 under Q_0. As in Theorem 3.38, we consider a function f such that

$$f(t) = \begin{cases} q_X(1-F(t)) & \text{if } F(t) \in (0,1) \text{ and } F(t-) = F(t), \\[2ex] \dfrac{1}{F(t)-F(t-)} \displaystyle\int_{F(t-)}^{F(t)} q_X(1-s)\,ds & \text{if } F(t-) < F(t). \end{cases} \tag{3.46}$$

Then f can be chosen as a decreasing function and satisfies $f(q_{\varphi_0}) = E_\lambda[\, h \mid q_{\varphi_0}\,]$, where λ is the Lebesgue measure and $h(t) := q_X(1-t)$; see Exercise 3.4.2. Hence, Jensen's inequality for conditional expectations and Lemma D.12 show that

$$\begin{aligned} \inf_{Q\in\mathcal{Q}} E_Q[\, u(X)\,] &\leq E_{Q_0}[\, u(X)\,] = \int_0^1 u(q_X(t))\,dt = \int_0^1 u(h(t))\,dt \\ &\leq \int_0^1 u(E_\lambda[\, h \mid q_{\varphi_0}\,](t))\,dt = \int_0^1 u(f(q_{\varphi_0}(t)))\,dt \\ &= E_{Q_0}[\, u(f(\varphi_0))\,] = \inf_{Q\in\mathcal{Q}} E_Q[\, u(f(\varphi_0))\,], \end{aligned} \tag{3.47}$$

where in the final step we have used Proposition 3.50 and the fact that $u \circ f$ is decreasing. Thus, f satisfies (3.45).

It remains to show that $f(\varphi_0)$ satisfies the capital constraint. To this end, we first use the lower Hardy–Littlewood inequality from Theorem D.13:

$$w \geq E^*[\, X\,] = E_{Q_0}[\, \varphi_0 X\,] \geq \int_0^1 q_{\varphi_0}(t)\, q_X(1-t)\,dt\,. \tag{3.48}$$

Here, we may replace $q_X(1-t) = h(t)$ by $E_\lambda[\, h \mid q_{\varphi_0}\,](t) = f(q_{\varphi_0}(t))$. We then get

$$\begin{aligned}\int_0^1 q_{\varphi_0}(t)\, q_X(1-t)\, dt &= \int_0^1 q_{\varphi_0}(t) f(q_{\varphi_0}(t))\, dt \\ &= E_{Q_0}[\,\varphi_0 f(\varphi_0)\,] = E^*[\,f(\varphi_0)\,]\,.\end{aligned} \tag{3.49}$$

Thus, f is as desired.

(b) Now suppose X^* solves (3.42). In the first step, we show that X^* must be Q_0-a. s. equal to a $\sigma(\varphi_0)$-measurable random variable. To this end, we assume by way of contradiction that $Y := E_{Q_0}[\, X^* \mid \varphi_0\,]$ is not Q_0-a. s. equal to X^*. Then the strict concavity of u and Jensen's inequality imply that

$$E_{Q_0}[\,u(Y)\,] > E_{Q_0}[\,u(X^*)\,]\,. \tag{3.50}$$

If we define $\widetilde{f}$ as in (3.46) with q_Y replacing q_X and $\widetilde{h}(t) := q_Y(1-t)$, then we get as in (3.48) and (3.49) that

$$\begin{aligned}E^*[\,\widetilde{f}(\varphi_0)\,] = E_{Q_0}[\,\varphi_0\widetilde{f}(\varphi_0)\,] &= \int_0^1 q_{\varphi_0}(t) E_\lambda[\,\widetilde{h} \mid q_{\varphi_0}\,](t)\, dt = \int_0^1 q_{\varphi_0}(t) q_Y(1-t)\, dt \\ &\le E_{Q_0}[\,\varphi_0 Y\,] = E_{Q_0}[\,\varphi_0 E_{Q_0}[\,X^* \mid \varphi_0\,]\,] = E_{Q_0}[\,\varphi_0 X^*\,] \le w.\end{aligned}$$

Thus, $\widetilde{f}(\varphi)$ belongs to $\mathscr{B}$. Moreover, by (3.47) and (3.50),

$$\inf_{Q\in\mathcal{Q}} E_Q[\,u(\widetilde{f}(\varphi_0))\,] \ge E_{Q_0}[\,u(Y)\,] > E_{Q_0}[\,u(X^*)\,] \ge \inf_{Q\in\mathcal{Q}} E_Q[\,u(X^*)\,]\,,$$

in contradiction to the optimality of X^*. Thus, X^* is necessarily $\sigma(\varphi_0)$-measurable and can hence be written as a (not yet necessarily decreasing) function of φ_0.

If we define f^* as in (3.46) with X^* replacing X, then $f^*(\varphi_0)$ is the terminal wealth of yet another solution in $\mathscr{B}$. Clearly, we must have $E^*[\,X^*\,] = w = E^*[\,f^*(\varphi_0)\,]$. Thus, (3.48) and (3.49) yield that $E_{Q_0}[\,\varphi_0 X^*\,] = \int_0^1 q_{\varphi_0}(t)\, q_{X^*}(1-t)\, dt$. But then the "only if" part of the lower Hardy–Littlewood inequality, together with the $\sigma(\varphi_0)$-measurability of X^*, imply that X^* is a decreasing function of φ_0. □

We can now state and prove the main result of this section.

Theorem 3.53. *Suppose that $\mathcal{Q}$ admits a least-favorable measure $Q_0 \approx P^*$. Then the robust utility maximization problem* (3.42) *is equivalent to the standard utility maximization problem with respect to Q_0, that is, to* (3.42) *with $\mathcal{Q}$ replaced by $\{Q_0\}$. More precisely, $X^* \in \mathscr{B}$ solves the robust problem* (3.42) *if and only if it solves the standard problem for Q_0, and the values of the corresponding optimization problems are equal, whether there exists a solution or not:*

$$\sup_{X \in \mathscr{B}} \inf_{Q \in \mathcal{Q}} E_Q[\,u(X)\,] = \sup_{X \in \mathscr{B}} E_{Q_0}[\,u(X)\,],$$

for any initial capital $w > 0$.

Proof. Proposition 3.52 implies that in solving the robust utility maximization problem (3.42) we may restrict ourselves to payoff profiles that are decreasing functions of φ_0. By Propositions 3.50, the robust utility of a such a payoff profile is the same as the expected utility with respect to Q_0. On the other hand, taking $\mathcal{Q}_0 := \{Q_0\}$ in Proposition 3.52 implies that the standard utility maximization problem for Q_0 also requires only payoff profile that are decreasing functions of φ_0. Therefore, the two problems are equivalent, and Theorem 3.53 is proved. □

Theorem 3.53 states in particular that the robust utility maximization problem (3.42) has a solution if and only if the standard problem for Q_0 has a solution, and we refer to Theorem 3.33 (with $W = +\infty$) for a sufficient condition.

Example 3.54. Consider the situation of Example 3.47, where for $\lambda \in (0,1)$ a least-favorable measure of $\mathcal{Q}_\lambda$ with respect to P^* is given by

$$\frac{dQ_0}{dP} = \frac{1}{\lambda} \cdot \frac{\varphi}{\varphi \vee q_\varphi(t_\lambda)}.$$

Here, $\varphi = dP^*/dP$ satisfies $P[\,\varphi > \lambda^{-1}\,] > 0$ and admits a continuous and strictly increasing quantile function q_φ with respect to P, and t_λ is the solution to a certain nonlinear equation. Let us also assume for the sake of concreteness that $u(x) = \frac{1}{\gamma}x^\gamma$ is a HARA utility function with risk aversion $\gamma \in (0,1)$ and that $E[\,\varphi^{-\gamma/(1-\gamma)}\,] < \infty$.

It was shown in Example 3.35 that the standard utility maximization problem under P has the solution

$$X^* = \frac{w\varphi^{-\frac{1}{1-\gamma}}}{E[\,\varphi^{-\frac{\gamma}{1-\gamma}}\,]},$$

where $\varphi = dP^*/dP$. Note that X^* is large for small values of φ, that is, for low-price scenarios.

Now let us replace the single probability measure P by the entire set $\mathcal{Q}_\lambda$. By Theorem 3.53, the corresponding robust utility maximization problem is solved by

$$X_0^* = \frac{w\varphi_0^{-\frac{1}{1-\gamma}}}{E_{Q_0}[\,\varphi_0^{-\frac{\gamma}{1-\gamma}}\,]},$$

where

$$\varphi_0 = \frac{dP^*}{dQ_0} = \lambda(\varphi \vee q_\varphi(t_\lambda)).$$

It follows that

$$X_0^* = c(X^* \wedge r)$$

where $r > 0$ and $c > 1$ are certain constants. Thus, the effect of robustness is here that the standard optimal payoff profile X^* is cut off at a certain threshold. That is, one gives up the opportunity for high profits in low-price scenarios in favor of enhanced returns in all other scenarios. ◇

3.7 Microeconomic equilibrium

The aim of this section is to provide a brief introduction to the theory of market equilibrium. Prices of assets will no longer be given in advance. Instead, they will be derived from "first principles" in a microeconomic setting where different agents demand asset profiles in accordance with their preferences and with their budget constraints. These budget constraints are determined by a given price system. The role of equilibrium prices consists in adjusting the constraints in such a way that the resulting overall demand is matched by the overall supply of assets.

Consider a finite set $\mathscr{A}$ of *economic agents* and a convex set $\mathscr{X} \subseteq L^0(\Omega, \mathscr{F}, P)$ of *admissible claims*. At time $t = 0$, each agent $a \in \mathscr{A}$ has an initial endowment whose discounted payoff at time $t = 1$ is described by an admissible claim

$$W_a \in \mathscr{X}, \quad a \in \mathscr{A}.$$

The aggregated claim

$$W := \sum_{a \in \mathscr{A}} W_a$$

is also called the *market portfolio*. Agents may want to exchange their initial endowment W_a against some other admissible claim $X_a \in \mathscr{X}$. This could lead to a new allocation $(X_a)_{a \in \mathscr{A}}$ if the resulting total demand matches the overall supply.

Definition 3.55. A collection $(X_a)_{a \in \mathscr{A}}$ in $\mathscr{X}$ is called a *feasible allocation* if

$$\sum_{a \in \mathscr{A}} X_a = W \quad P\text{-a. s.} \tag{3.51}$$

Condition (3.51) is called the *market clearing condition*. △

The budget constraints will be determined by a linear pricing rule of the form

$$\Phi(X) := E[\,\varphi X\,], \quad X \in \mathscr{X},$$

where φ is a *price density*, that is, an integrable function on $(\Omega, \mathscr{F}, P)$ such that $\varphi > 0$ P-a. s. and $E[\,|W_a|\,\varphi\,] < \infty$ for all $a \in \mathscr{A}$. To any such φ, we can associate a normalized *price measure* $P^\varphi \approx P$ with density $\varphi E[\,\varphi\,]^{-1}$.

Remark 3.56. In the context of our one-period model of a financial market with d risky assets $S^1, \dots, S^d$ and a risk-free asset $S^0 \equiv 1+r$, P^φ is a risk-neutral measure if the pricing rule Φ is consistent with the given price vector $\overline{\pi} = (\pi^0, \pi)$, where $\pi^0 = 1$. In this section, the pricing rule will be derived as an equilibrium price measure, given the agents' preferences and endowments. In particular, this will amount to an endogenous derivation of the price vector π. In a situation where the structure of the equilibrium is already partially known in the sense that it is consistent with the given price vector π, the construction of a microeconomic equilibrium yields a specific choice of a martingale measure P^*, that is, of a specific extension of π from the space $\mathscr{V}$ of attainable payoffs to a larger space of admissible claims. ◇

The preferences of agent $a \in \mathscr{A}$ are described by a utility function u_a. Given the price density φ, an agent $a \in \mathscr{A}$ may want to exchange the endowment W_a for an admissible claim X_a^φ that maximizes the expected utility

$$E[\,u_a(X)\,]$$

over all X in the agent's *budget set*

$$\begin{aligned}\mathscr{B}_a(\varphi) &:= \{X \in \mathscr{X} \mid E[\,\varphi|X|\,] < \infty \text{ and } E[\,\varphi X\,] \le E[\,\varphi\, W_a\,]\} \\ &= \{X \in \mathscr{X} \mid X \in L^1(P^\varphi) \text{ and } E^\varphi[\,X\,] \le E^\varphi[\,W_a\,]\}.\end{aligned}$$

In this case, we will say that X_a^φ solves the *utility maximization problem* of agent $a \in \mathscr{A}$ with respect to the price density φ. The key problem is whether φ can be chosen in such a way that the requested profiles X_a^φ, $a \in \mathscr{A}$, form a feasible allocation.

Definition 3.57. A price density φ^* together with a feasible allocation $(X_a^*)_{a\in\mathscr{A}}$ is called an *Arrow–Debreu equilibrium* if each X_a^* solves the utility maximization problem of agent $a \in \mathscr{A}$ with respect to φ^*. △

Thus, the price density φ^* appearing in an Arrow–Debreu equilibrium decentralizes the crucial problem of implementing the global feasibility constraint (3.51). This is achieved by adjusting the budget sets in such a way that the resulting demands respect the market clearing condition, even though the individual demand is determined without any regard to this global constraint.

Example 3.58. Assume that each agent $a \in \mathscr{A}$ has an exponential utility function with parameter $\alpha_a > 0$, and let us consider the unconstrained case

$$\mathscr{X} = L^0(\Omega, \mathscr{F}, P).$$

In this case, there is a unique equilibrium, and it is easy to describe it explicitly. For a given pricing measure $P^* \approx P$ such that $W_a \in L^1(P^*)$ for all $a \in \mathscr{A}$, the utility maximization problem for agent $a \in \mathscr{A}$ can be solved if and only if the relative entropy $H(P^*|P)$ is finite, and in this case the optimal demand is given by

$$X_a^* = -\frac{1}{\alpha_a}\log\varphi^* + w_a^* + \frac{1}{\alpha_a}H(P^*|P)$$

where

$$w_a^* := E^*[\,W_a\,];$$

see Example 3.29. The market clearing condition (3.51) takes the form

$$W = -\frac{1}{\alpha}\log\varphi^* + \sum_{a\in\mathscr{A}} w_a^* + \frac{1}{\alpha}H(P^*|P)$$

where α is defined via

$$\frac{1}{\alpha} = \sum_{a\in\mathscr{A}}\frac{1}{\alpha_a}. \tag{3.52}$$

Thus, a normalized equilibrium price density must have the form

$$\varphi^* = \frac{e^{-\alpha W}}{E[\,e^{-\alpha W}\,]}, \tag{3.53}$$

and this shows *uniqueness*. As to *existence*, let us assume that

$$E[\,|W_a|e^{-\alpha W}\,] < \infty, \qquad a \in \mathscr{A};$$

this condition is satisfied if, for example, the random variables W_a are bounded from below. Define $P^* \approx P$ via (3.53). Then

$$H(P^*|P) = -\alpha E^*[\,W\,] - \log E[\,e^{-\alpha W}\,] < \infty,$$

and the optimal profile for agent $a \in \mathscr{A}$ with respect to the pricing measure P^* takes the form

$$X_a^* = w_a^* + \frac{\alpha}{\alpha_a}(W - E^*[\,W\,]). \tag{3.54}$$

Since

$$\sum_{a\in\mathscr{A}} w_a^* = E^*[\,W\,],$$

the allocation $(X_a^*)_{a\in\mathscr{A}}$ is feasible, and so we have constructed an Arrow–Debreu equilibrium. Thus, the agents share the market portfolio in a *linear* way, and in inverse proportion to their risk aversion.

Let us now return to our financial market model of Section 3.1. We assume that the initial endowment of agent $a \in \mathscr{A}$ is given by a portfolio $\overline{\eta}_a \in \mathbb{R}^{d+1}$ so that the discounted payoff at time $t = 1$ is

$$W_a = \frac{\overline{\eta}_a \cdot \overline{S}}{1+r}, \qquad a \in \mathscr{A}.$$

In this case, the market portfolio is given by $W = \overline{\eta} \cdot \overline{S}/(1+r)$ with $\overline{\eta} := \sum_a \overline{\eta}_a = (\eta^0, \eta)$. The optimal claim for agent $a \in \mathscr{A}$ in (3.54) takes the form

$$X_a^* = \overline{\eta}_a \cdot \overline{\pi} + \frac{\alpha}{\alpha_a}\eta \cdot \left(\frac{S}{1+r} - \pi\right),$$

where $\overline{\pi} = (1, \pi)$ and

$$\pi^i = E^*\left[\frac{S^i}{1+r}\right] \quad \text{for } i = 1, \dots, d.$$

Thus, we could have formulated the equilibrium problem within the smaller space $\mathscr{X} = \mathscr{V}$ of attainable payoffs, and the resulting equilibrium allocation would have been the same. In particular, the extension of $\mathscr{X}$ from $\mathscr{V}$ to the general space $L^0(\Omega, \mathscr{F}, P)$ of admissible claims does not create a demand for derivatives in our present example. ◇

From now on, we assume that the set of admissible claims is given by

$$\mathscr{X} = L^0_+(\Omega, \mathscr{F}, P),$$

and that the preferences of agent $a \in \mathscr{A}$ are described by a continuously differentiable utility function $u_a : (0, \infty) \to \mathbb{R}$. We define

$$u_a(0) := \lim_{x\downarrow 0} u_a(x) \in \mathbb{R} \cup \{-\infty\} \quad \text{and} \quad u_a'(0) := \lim_{x\downarrow 0} u_a'(x) \in (0, \infty].$$

The initial endowments W_a are assumed to be nonnegative. Moreover, we assume

$$P[\, W_a > 0\,] \neq 0 \quad \text{for all } a \in \mathscr{A}.$$

and

$$E[\, W\,] < \infty \quad \text{and} \quad E[\, u_a(W)\,] > -\infty \quad \text{for all } a \in \mathscr{A}. \tag{3.55}$$

A function $\varphi \in L^1(\Omega, \mathscr{F}, P)$ such that $\varphi > 0$ P-a. s. is a price density if

$$E[\,\varphi W\,] < \infty;$$

note that this condition is satisfied as soon as φ is bounded, due to our assumption (3.55). Given a price density φ, each agent faces exactly the optimization problem discussed in Section 3.3 in terms of the price measure $P^\varphi \approx P$. Thus, if $(X_a^*)_{a\in\mathscr{A}}$ is an equilibrium allocation with respect to the price density φ^*, feasibility implies $0 \le X_a^* \le W$, and so it follows as in the proof of Corollary 3.36 that

$$X_a^* = I_a^+(c_a\,\varphi^*), \quad a \in \mathscr{A}, \tag{3.56}$$

with positive constants $c_a > 0$. Note that the market clearing condition

$$W = \sum_{a\in\mathscr{A}} X_a^* = \sum_{a\in\mathscr{A}} I_a^+(c_a\,\varphi^*)$$

will determine φ^* as a continuous and *decreasing* function of W, and thus the optimal profiles X_a^* will be continuous and *increasing* functions of W.

Before we discuss the *existence* of an Arrow–Debreu equilibrium, let us first illustrate the structure of such equilibria by the following simple examples. In particular, they show that an equilibrium allocation will typically involve nonlinear derivatives of the market portfolio W.

Example 3.59. Let us consider the constrained version of Example 3.58, where agents $a \in \mathscr{A}$ have exponential utility functions with parameters $\alpha_a > 0$. Define

$$\underline{w} := \sup\{c \mid W \ge c\ P\text{-a.s.}\} \ge 0,$$

and let P^* be the measure defined via (3.53). For any agent $a \in \mathscr{A}$ such that

$$w_a^* := E^*[\,W_a\,] \ge \frac{\alpha}{\alpha_a}(E^*[\,W\,] - \underline{w}), \tag{3.57}$$

the unrestricted optimal profile

$$X_a^* = w_a^* + \frac{\alpha}{\alpha_a}(W - E^*[\,W\,])$$

satisfies $X_a^* \ge 0$ P-almost surely. Thus, if *all* agents satisfy the requirement (3.57) then the unrestricted equilibrium computed in Example 3.58 is a forteriori an Arrow–Debreu equilibrium in our present context. In this case, there is no need for nonlinear derivatives of the market portfolio.

If some agents do *not* satisfy the requirement (3.57), then the situation becomes more involved, and there is an equilibrium allocation that will need derivatives such as call options. Let us illustrate this effect in the simple setting where there are only two agents $a \in \mathscr{A} = \{1, 2\}$. Suppose that agent 1 satisfies condition (3.57), while agent 2 does not. For $c \ge 0$, we define the measure $P^c \approx P$ in terms of the density

$$\varphi^c := \begin{cases} \dfrac{1}{Z_1} e^{-\alpha_1 W} & \text{on } \{W \le c\}, \\ \dfrac{1}{Z_2} e^{-\alpha W} & \text{on } \{W \ge c\}, \end{cases}$$

where α is given by (3.52), and where the constants Z_1 and Z_2 are determined by the requirement of normalization $E[\,\varphi^c\,] = 1$ and the continuity condition

$$\log Z_2 - \log Z_1 = c(\alpha_1 - \alpha),$$

which is suggested by the fact that an equilibrium price density should be a continuous decreasing function of W. Note that $P^0 = P^*$ with P^* as in (3.53). Consider the following equation for the parameter c:

$$\frac{\alpha}{\alpha_2} E^c[\,(W-c)^+\,] = w_2^c, \tag{3.58}$$

where $w_2^c := E^c[\,W_2\,]$. Both sides are continuous in c. As c increases from 0 to $+\infty$, the left-hand side goes from $\frac{\alpha}{\alpha_2} E^*[\,W\,]$ to 0, while w_2^c goes from $w_2^0 < \frac{\alpha}{\alpha_2} E^*[\,W\,]$ to $E^\infty[\,W_2\,] > 0$, where the probability measure P^∞ has the density $\varphi^\infty = e^{-\alpha_1 W}/E[e^{-\alpha_1 W}]$. Thus, there exists a solution c of (3.58). Let us now check that for this solution c,

$$X_2^c := \frac{\alpha}{\alpha_2}(W-c)^+, \quad X_1^c := W - X_2^c$$

defines an equilibrium allocation with respect to the pricing measure P^c. Clearly, X_1^c and X_2^c are nonnegative and satisfy $X_1^c + X_2^c = W$. The budget condition for agent 2 is satisfied due to (3.58), and this implies the budget condition

$$E^c[\,X_1^c\,] = E^c[\,W\,] - w_2^c = w_1^c$$

for agent 1. Both are optimal since

$$X_a^c = I_a^+(\lambda_a\,\varphi^c) \quad \text{with} \quad \lambda_1 := \alpha_1 Z_1 \quad \text{and} \quad \lambda_2 := \alpha_2 Z_2 e^{\alpha c}; \tag{3.59}$$

see Exercise 3.7.1 for details. Thus, agent 2 demands $\frac{\alpha}{\alpha_2}$ shares of a call option on the market portfolio W with strike c, agent 1 demands the remaining part of W, and so the market is cleared.

In the general case of a finite set $\mathscr{A}$ of agents, there is an equilibrium price measure $\hat{P}$ with the following structure. There are levels $0 =: c_0 < \cdots < c_N = \infty$ with $1 \le N \le |\mathscr{A}|$ such that the price density $\hat{\varphi}$ is given by

$$\hat{\varphi} = \frac{1}{Z_i} e^{-\beta_i W} \quad \text{on } \{W \in [c_{i-1}, c_i]\}$$

for $i = 1, \ldots, N$, where

$$\beta_i := \left(\sum_{\alpha \in \mathscr{A}_i} \frac{1}{\alpha_a} \right)^{-1},$$

and where $\mathscr{A}_i$ $(i = 1, \ldots, N)$ are the increasing sets of agents that are active at the ith layer in the sense that $X_a > 0$ on $\{W \in (c_{i-1}, c_i]\}$. At each layer $(c_{i-1}, c_i]$, the active agents are sharing the market portfolio in inverse proportions to their risk aversion. Thus, the optimal profile $\hat{X}_a$ of any agent $a \in \mathscr{A}$ is given by an increasing piecewise linear function in W, and thus it can be implemented by a linear combination of call options with strikes c_i. More precisely, an agent $a \in \mathscr{A}_i$ takes β_i/α_a shares of the *call spread*

$$(W - c_{i-1})^+ - (W - c_i)^+,$$

that is, the agent goes long on a call option with strike c_{i-1} and short on a call option with strike c_i. ◇

Exercise 3.7.1. Prove (3.59). ◇

Example 3.60. Assume that all agents $a \in \mathscr{A}$ have preferences described by HARA utility functions so that

$$I_a^+(y) = y^{-\frac{1}{1-\gamma_a}}, \quad a \in \mathscr{A}$$

with $0 \leq \gamma_a < 1$; see Example 3.35. For a given price density φ, the optimal claims take the form

$$X_a = I_a^+(c_a \varphi) = b_a \, \varphi^{-\frac{1}{1-\gamma_a}} \tag{3.60}$$

with constants $b_a > 0$. If $\gamma_a = \gamma$ for all $a \in \mathscr{A}$, then the market clearing condition (3.51) implies

$$W = \sum_{a \in \mathscr{A}} X_a = \left(\sum_{a \in \mathscr{A}} b_a \right) \varphi^{-\frac{1}{1-\gamma}},$$

that is, the equilibrium price density φ^* must take the form

$$\varphi^* = \frac{1}{Z} W^{\gamma - 1},$$

where Z is the normalizing constant, and so the agents demand *linear* shares of the market portfolio W. If risk aversion varies among the agents then the structure of the equilibrium becomes more complex, and it will involve nonlinear *derivatives* of the market portfolio. Let us number the agents so that $\mathscr{A} = \{1, \ldots, n\}$ and $\gamma_1 \geq \cdots \geq \gamma_n$. Condition (3.60) implies

$$X_i = d_i X_n^{\beta_i}$$

with some constants d_i, and where

$$\beta_i := \frac{1-\gamma_n}{1-\gamma_i}$$

satisfies $\beta_1 \geq \cdots \geq \beta_n = 1$ with at least one strict inequality. Thus, each X_i is a convex increasing function of X_n. In equilibrium, X_n is a concave function of W determined by the condition

$$\sum_{i=1}^{n} d_i X_n^{\beta_i} = W, \tag{3.61}$$

and the price density φ^* takes the form

$$\varphi^* = \frac{1}{Z} X_n^{\gamma_n - 1}. \tag{3.62}$$

As an illustration, we consider the special case “Bernoulli vs. Cramer,” where $\mathscr{A} = \{1,2\}$ with $u_1(x) = \sqrt{x}$ and $u_2(x) = \log x$, that is, $\gamma_1 = \frac{1}{2}$ and $\gamma_2 = 0$; see Example 2.38. In this case, (3.61) is a quadratic equation in X_2, and hence

$$X_2 = \frac{1}{2d_1}\left(\sqrt{d_2^2 + 4d_1 W} - d_2\right) = \frac{1}{\sqrt{d_1}}(\sqrt{c+W} - \sqrt{c}),$$

where $c = d_2^2/4d_1$. We must have that $0 \leq X_2 \leq W$ and $X_1 = W - X_2$. Thus, agent 2 demands a payoff profile that is concave in W, whereas the one of agent 1 is convex. Moreover, by (3.62), the price density must be of the form

$$\varphi^* = \frac{1}{Z} \frac{1}{\sqrt{c+W} - \sqrt{c}}.$$

The preceding formulas state necessary conditions on the structure of an Arrow–Debreu equilibrium, provided that it exists. General conditions for the existence of an equilibrium will be given in Theorem 3.62. ◇

Let us now return to our general setting, and let us prove the existence of an Arrow–Debreu equilibrium. Consider the following condition:

$$\limsup_{x \downarrow 0} x\, u_a'(x) < \infty \quad \text{and} \quad E\left[u_a'\left(\frac{W}{|\mathscr{A}|} \right) \right] < \infty, \quad a \in \mathscr{A}. \tag{3.63}$$

Remark 3.61. Condition (3.63) is clearly satisfied if

$$u_a'(0) := \lim_{x \downarrow 0} u_a'(x) < \infty, \quad a \in \mathscr{A}. \tag{3.64}$$

But it also includes HARA utility functions u_a with parameter $\gamma_a \in [0,1)$ if we assume

$$E[\, W^{\gamma_a - 1} \,] < \infty, \qquad a \in \mathscr{A},$$

in addition to our assumption (3.55). ◇

Theorem 3.62. *Under assumptions* (3.55) *and* (3.63), *there exists an Arrow–Debreu equilibrium.*

In a first step, we are going to show that an equilibrium allocation maximizes a suitable weighted average

$$U^\lambda(X) := \sum_{a \in \mathscr{A}} \lambda_a E[\, u_a(X_a) \,]$$

of the individual utility functionals over all feasible allocations $X = (X_a)_{a \in \mathscr{A}}$. The weights λ_a are nonnegative, and without loss of generality we can assume that they are normalized so that the vector $\lambda := (\lambda_a)_{a \in \mathscr{A}}$ belongs to the convex compact set

$$\Lambda = \Big\{ \lambda \in [0,1]^{\mathscr{A}} \;\Big|\; \sum_{a \in \mathscr{A}} \lambda_a = 1 \Big\}.$$

In a second step, we will use a fixed-point argument to obtain a weight vector and a corresponding price density such that the maximizing allocation satisfies the individual budget constraints.

Definition 3.63. A feasible allocation $(X_a)_{a \in \mathscr{A}}$ is called *λ-efficient* for $\lambda \in \Lambda$ if it maximizes U^λ over all feasible allocations. △

Lemma 3.64. *For any $\lambda \in \Lambda$, there exists a unique λ-efficient allocation $(X_a^\lambda)_{a \in \mathscr{A}}$.*

Proof. Existence follows from the general argument in Remark 3.31 applied to the set $\mathscr{B}$ of all feasible allocations and to the concave functional U^λ. Note that

$$U^\lambda(X) \le \max_{a \in \mathscr{A}} E[\, u_a(W) \,]$$

for any feasible allocation, and that the right-hand side is finite due to our assumption (3.55). Moreover, by dominated convergence, U^λ is indeed continuous on $\mathscr{B}$ with respect to P-a. s. convergence.

Now we prove uniqueness. If $\lambda_a > 0$ for all $a \in \mathscr{A}$, then U^λ is strictly concave and uniqueness follows. Now suppose that there exists $a_0 \in \mathscr{A}$ for which $\lambda_{a_0} = 0$. Then any feasible allocation X that maximizes U^λ must satisfy $X_{a_0} = 0$ P-a. s., because otherwise we would have $U^\lambda(X) < U^\lambda(\widetilde{X})$ for the feasible allocation $\widetilde{X}$ defined by $\widetilde{X}_{a_0} := 0$, $\widetilde{X}_{a_1} := X_{a_1} + X_{a_0}$ for some $a_1 \in \mathscr{A}$ with $\lambda_{a_1} > 0$, and $\widetilde{X}_a := X_a$ for all remaining $a \in \mathscr{A}$. Thus, by eliminating successively all $a_0 \in \mathscr{A}$ with $\lambda_{a_0} = 0$, the situation is reduced to the case in which $\lambda_a > 0$ for all a and where uniqueness has already been established. □

The preceding lemma and part (a) of the following proposition show that the existence proof for an Arrow–Debreu equilibrium is reduced to the construction of a suitable vector $\lambda^* \in \Lambda$. Part (b) of the proposition is sometimes called the *first fundamental welfare theorem.*

Proposition 3.65. *Let $X^* = (X_a^*)_{a\in\mathscr{A}}$ be the equilibrium allocation in an Arrow–Debreu equilibrium.*

(a) *X^* is λ-efficient for the vector $\lambda = (c \cdot c_a^{-1})_{a\in\mathscr{A}}$, where $c^{-1} := \sum_a c_a^{-1}$.*

(b) *X^* is* Pareto optimal *in the class $\mathscr{B}$ of all feasible allocations. That is, there exists no $X = (X_a)_{a\in\mathscr{A}} \in \mathscr{B}$ such that $E[\, u_a(X_a)\,] \geq E[\, u_a(X_a^*)\,]$ for all $a \in \mathscr{A}$ and strict inequality holds for at least one a.*

The proof is based on the following lemma.

Lemma 3.66.

(a) *A feasible allocation $(X_a)_{a\in\mathscr{A}}$ is λ-efficient if and only if it P-a. s. satisfies the first-order conditions*

$$\lambda_a\, u_a'(X_a) \leq \varphi \quad \text{with equality on } \{X_a > 0\} \tag{3.65}$$

with respect to some price density φ. In this case, $(X_a)_{a\in\mathscr{A}}$ coincides with the allocation $(X_a^\lambda)_{a\in\mathscr{A}}$ constructed in Lemma 3.64, and the price density can be chosen as

$$\varphi^\lambda := \max_{a\in\mathscr{A}} \lambda_a\, u_a'(X_a^\lambda). \tag{3.66}$$

(b) *For each $a \in \mathscr{A}$, X_a^λ maximizes $E[\, u_a(X)\,]$ over all $X \in \mathscr{X}$ such that*

$$E[\, \varphi^\lambda X\,] \leq E[\, \varphi^\lambda X_a^\lambda\,].$$

Proof. (a) Let us first show sufficiency. If $X = (X_a)_{a\in\mathscr{A}}$ is a feasible allocation satisfying the first-order conditions, and $Y = (Y_a)_{a\in\mathscr{A}}$ is another feasible allocation then

$$\begin{aligned} U^\lambda(X) - U^\lambda(Y) &= \sum_{a\in\mathscr{A}} \lambda_a E[\, u_a(X_a) - u_a(Y_a)\,] \\ &\geq \sum_{a\in\mathscr{A}} \lambda_a E[\, u_a'(X_a)(X_a - Y_a)\,] \\ &\geq E\Big[\varphi\Big(\sum_{a\in\mathscr{A}} X_a - \sum_{a\in\mathscr{A}} Y_a\Big)\Big] = 0, \end{aligned}$$

using concavity of u_a in the second step and the first-order conditions in the third. This shows that X is λ-efficient.

Turning to necessity, consider the λ-efficient allocation $(X_a^\lambda)_{a\in\mathscr{A}}$ for $\lambda \in \Lambda$. We may assume without loss of generality that $\lambda_a > 0$ for all $a \in \mathscr{A}$, since it was shown in the proof of Lemma 3.64 that $X_a^\lambda = 0$ P-a. s. whenever $\lambda_a = 0$. Let $(X_a)_{a\in\mathscr{A}}$ be another feasible allocation. For $\varepsilon \in (0,1]$, let $Y_a^\varepsilon := \varepsilon X_a + (1-\varepsilon)X_a^\lambda$. Since $(Y_a^\varepsilon)_{a\in\mathscr{A}}$ is feasible, λ-efficiency of $(X_a^\lambda)_{a\in\mathscr{A}}$ yields

$$\begin{aligned} 0 &\geq \frac{1}{\varepsilon}\sum_{a\in\mathscr{A}} \lambda_a E[\, u_a(Y_a^\varepsilon) - u_a(X_a^\lambda)\,] \\ &\geq \frac{1}{\varepsilon}\sum_{a\in\mathscr{A}} \lambda_a E[\, u_a'(Y_a^\varepsilon)(Y_a^\varepsilon - X_a^\lambda)\,] \\ &= \sum_{a\in\mathscr{A}} \lambda_a E[\, u_a'(Y_a^\varepsilon)(X_a - X_a^\lambda)\,]. \end{aligned} \tag{3.67}$$

Let us first assume (3.64); in part (c) of the proof we show how to modify the argument under the less restrictive condition (3.63). Using dominated convergence and (3.64), we may let $\varepsilon \downarrow 0$ in the above inequality to conclude that for

$$\varphi_a^\lambda := \lambda_a u_a'(X_a^\lambda)$$

we have

$$\sum_{a\in\mathscr{A}} E[\, \varphi_a^\lambda X_a\,] \leq \sum_{a\in\mathscr{A}} E[\, \varphi_a^\lambda X_a^\lambda\,] \leq E[\, \varphi^\lambda W\,]. \tag{3.68}$$

Note that φ^λ is a price density, since by (3.64)

$$0 < \varphi^\lambda \leq \max\{\lambda_a u_a'(0) \mid a \in \mathscr{A}\} < \infty.$$

Take a feasible allocation $(X_a)_{a\in\mathscr{A}}$ such that

$$\sum_{a\in\mathscr{A}} \varphi_a^\lambda X_a = \varphi^\lambda\, W\,; \tag{3.69}$$

for example, we can enumerate $\mathscr{A} := \{1,\dots,|\mathscr{A}|\}$ and take $X_a := W\mathbb{1}_{\{T=a\}}$ where

$$T(\omega) := \min\{a \mid \varphi_a^\lambda(\omega) = \varphi^\lambda(\omega)\}.$$

In view of (3.68), we see that

$$\sum_{a\in\mathscr{A}} E[\, \varphi_a^\lambda X_a^\lambda\,] = E[\, \varphi^\lambda\, W\,]. \tag{3.70}$$

This implies $\varphi_a^\lambda = \varphi^\lambda$ on $\{X_a^\lambda > 0\}$, which is equivalent to the first-order condition (3.65) with respect to φ^λ.

(b) In order to show optimality of X_a^λ, we may assume without loss of generality that $P[\, X_a^\lambda > 0\,] > 0$, and hence $\lambda_a > 0$. Thus, the first-order condition with respect to φ^λ takes the form

$$X_a^\lambda = I_a^+(\lambda_a^{-1}\varphi^\lambda), \tag{3.71}$$

due to our convention (3.26) and the fact that $I_a^+ \geq 0$, which guarantees the validity of (3.71) also on $\{X_a^\lambda = 0\}$. By Corollary 3.36, X_a^λ solves the optimization problem for agent $a \in \mathscr{A}$ under the constraint

$$E[\,\varphi^\lambda X\,] \leq E[\,\varphi^\lambda X_a^\lambda\,].$$

(c) If (3.64) is replaced by (3.63), then we first need an additional argument in order to pass from (3.67) to (3.68). Note first that by Fatou's lemma,

$$\begin{aligned}\liminf_{\varepsilon\downarrow 0} \sum_{a\in\mathscr{A}} \lambda_a E[\,u_a'(Y_a^\varepsilon)X_a\,] &\geq \sum_{a\in\mathscr{A}} \lambda_a \liminf_{\varepsilon\downarrow 0} E[\,u_a'(Y_a^\varepsilon)X_a\,]\\ &\geq \sum_{a\in\mathscr{A}} \lambda_a E[\,u_a'(X_a^\lambda)X_a\,].\end{aligned}$$

On the other hand, since

$$\kappa := \max_{a\in\mathscr{A}} \sup_{0<x\leq 1} x\,u_a'(x) < \infty$$

by (3.63), we have $xu_a'(x) \leq \kappa + xu_a'(1) \leq \kappa(1+x)$ for all $x \geq 0$. This implies

$$0 \leq u_a'(X_a)X_a \leq V := \kappa(1+W) \in L^1(P), \tag{3.72}$$

and also

$$0 \leq u_a'(Y_a^\varepsilon)\,X_a^\lambda \leq u_a'((1-\varepsilon)\,X_a^\lambda)X_a^\lambda \leq (1-\varepsilon)^{-1}\,V,$$

since $Y_a^\varepsilon \geq (1-\varepsilon)X_a^\lambda$. Thus, dominated convergence implies

$$E[\,u_a'(Y_a^\varepsilon)\,X_a^\lambda\,] \longrightarrow E[\,u_a'(X_a^\lambda)\,X_a^\lambda\,], \quad \varepsilon \downarrow 0,$$

and this concludes the proof of (3.68).

By (3.72), we have

$$\varphi_a^\lambda X_a^\lambda := \lambda_a\,u_a'(X_a^\lambda)\,X_a^\lambda \in L^1(P).$$

Hence, $E[\,\varphi^\lambda W\,] < \infty$ follows by taking in (3.68) a feasible allocation $(X_a)_{a\in\mathscr{A}}$ that is as in (3.69). We furthermore get (3.70), which yields as in part (b) the first-order conditions (3.65).

It remains to show that φ^λ is integrable in order to conclude that φ^λ is a price density. Our assumption (3.63) implies

$$F := \max_{a\in\mathscr{A}} u_a'\left(\frac{W}{|\mathscr{A}|}\right) \in L^1(P), \tag{3.73}$$

and so it is enough to show that $F \geq \varphi^\lambda$. Since $X_a^\lambda = I_a^+(\varphi^\lambda/\lambda_a)$, feasibility and $\lambda_a \leq 1$ imply

$$W \leq \sum_{a\in\mathscr{A}} I_a^+(\varphi^\lambda) \leq |\mathscr{A}| \max_{a\in\mathscr{A}} I_a^+(\varphi^\lambda),$$

hence

$$\begin{aligned} F &\geq \max_{a\in\mathscr{A}} u_a'\Big(\max_{b\in\mathscr{A}} I_b^+(\varphi^\lambda)\Big) \\ &\geq u_{a_0}'(I_{a_0}^+(\varphi^\lambda)) = \varphi^\lambda \quad \text{on } \Big\{\max_{a\in\mathscr{A}} I_a^+(\varphi^\lambda) = I_{a_0}^+(\varphi^\lambda)\Big\}. \end{aligned}$$

□

Proof of Proposition 3.65. (a) As noted in (3.56), an Arrow–Debreu equilibrium $((X_a^*)_{a\in\mathscr{A}}, \varphi^*)$ satisfies $X_a^* = I_a^+(c_a\varphi^*)$ for $a \in \mathscr{A}$. Applying the function u_a' to both sides of this identity yields the first-order conditions (3.65) in the form $cc_a^{-1}u_a'(X_a^*) \leq c\varphi^*$ P-a. s. and with equality on $\{X_a^* > 0\}$. Therefore, Lemma 3.66(a) yields the assertion.

To prove part (b), note first that $\lambda_a = cc_a^{-1} > 0$ for all $a \in \mathscr{A}$. Thus, if $X = (X_a)_{a\in\mathscr{A}}$ is a feasible allocation such that $E[\,u_a(X_a)\,] \geq E[\,u_a(X_a^*)\,]$ for all $a \in \mathscr{A}$ and strict inequality holds for at least one a, then also $U^\lambda(X) > U^\lambda(X^*)$, in contradiction to the fact that X^* is λ-efficient. □

After these preliminaries, we are now in a position to prove the existence of an Arrow–Debreu equilibrium. Note that for each $\lambda \in \Lambda$ the λ-efficient allocation $(X_a^\lambda)_{a\in\mathscr{A}}$ and the price density φ^λ would form an Arrow–Debreu equilibrium if

$$E[\,\varphi^\lambda W_a\,] = E[\,\varphi^\lambda X_a^\lambda\,] \quad \text{for all } a \in \mathscr{A}. \tag{3.74}$$

If this is not the case, then we can replace λ by the vector $g(\lambda) = (g_a(\lambda))_{a\in\mathscr{A}}$ defined by

$$g_a(\lambda) := \lambda_a + \frac{1}{E[\,V\,]} \cdot E[\,\varphi^\lambda(W_a - X_a^\lambda)\,],$$

where V is given by (3.72). Note that $g(\lambda) \in \Lambda$: Since the first-order conditions (3.65) together with (3.72) imply

$$E[\,\varphi^\lambda X_a^\lambda\,] = \lambda_a E[\,u_a'(X_a^\lambda) X_a^\lambda\,] \leq \lambda_a E[\,V\,],$$

we have $g_a(\lambda) \geq 0$, and $\sum_a g_a(\lambda) = 1$ follows by feasibility. Thus, we increase the weights of those agents that were allocated less than they could afford. Clearly, any fixed point of the map $g : \Lambda \to \Lambda$ will satisfy condition (3.74), and thus yield an Arrow–Debreu equilibrium.

Proof of Theorem 3.62. (a): The set Λ is convex and compact. Thus, the existence of a fixed point of the map $g : \Lambda \to \Lambda$ follows from Brouwer's fixed-point theorem as soon as we can verify that g is *continuous*; see, for instance, Corollary 16.52 in [3] for a proof of Brouwer's fixed-point theorem. Suppose that the sequence (λ_n) in Λ converges to $\lambda \in \Lambda$.

In part (c), we show that $X_n := X^{\lambda_n}$ and $\varphi_n := \varphi^{\lambda_n}$ converge P-a. s. to X^λ and φ^λ, respectively. We will show next that we may apply the dominated convergence theorem, so that

$$\lim_{n\uparrow\infty} E[\,\varphi_n W_a\,] = E[\,\varphi^\lambda W_a\,]$$

and

$$\lim_{n\uparrow\infty} E[\,\varphi_n X_n\,] = E[\,\varphi^\lambda X^\lambda\,]$$

and this will prove the continuity of g. To verify the assumptions of the dominated convergence theorem, note that

$$W_a\,\varphi_n \le W\,\varphi_n \le WF,$$

where F is as in (3.73). Moreover,

$$WF \le |\mathscr{A}|F\,\mathbb{1}_{\{W\le|\mathscr{A}|\}} + \max_{a\in\mathscr{A}} u_a'(1)\cdot W \in L^1(P).$$

Thus, $\varphi_n W_a$ and $\varphi_n X_n$ are bounded by $WF \in L^1(P)$.

(b): By our convention (3.26), the map $f : \Lambda\times[0,\infty] \to [0,\infty]$ defined by

$$f(\lambda, y) = \sum_{a\in\mathscr{A}} I_a^+(\lambda_a^{-1}y)$$

is continuous. If we fix $\lambda \in \Lambda$, then the function $f(\lambda,\cdot)$ is continuous on $[0,\infty]$ and strictly decreasing on $(a(\lambda), b(\lambda))$ where

$$a(\lambda) := \max_{a\in\mathscr{A}} \lim_{x\uparrow\infty} \lambda_a\,u_a'(x) \ge 0 \quad\text{and}\quad b(\lambda) = \max_{a\in\mathscr{A}} \lambda_a\,u_a'(0+) \le +\infty.$$

Moreover, $f(\lambda,y) = \infty$ for $y \le a(\lambda)$ and $f(\lambda,y) = 0$ for $y \ge b(\lambda)$. Hence, for each $w \in (0,\infty)$ there exists exactly one solution $y^\lambda \in (a(\lambda), b(\lambda))$ of the equation

$$f(\lambda, y^\lambda) = w.$$

Recall that $[0,\infty]$ can be regarded as a compact topological space. To see that y^λ depends continuously on $\lambda \in \Lambda$, take a sequence $\lambda_n \to \lambda$ and a subsequence (λ_{n_k}) such that the solutions $y_k = y^{\lambda_{n_k}}$ of $f(\lambda_{n_k}, y) = w$ converge to some limit $y_\infty \in [a(\lambda), b(\lambda)]$. By continuity of f,

$$f(\lambda, y_\infty) = \lim_{k\uparrow\infty} f(\lambda_{n_k}, y_k) = w,$$

and so y_∞ must coincide with y^λ.

(c): Recall from (3.71) that

$$X_a^\lambda = I_a^+(\lambda_a^{-1}\varphi^\lambda) \tag{3.75}$$

for any $a \in \mathscr{A}$. By feasibility,

$$W = \sum_{a\in\mathscr{A}} X_a^\lambda = f(\lambda, \varphi^\lambda).$$

Thus, φ^{λ_n} converges P-a. s. to φ^λ as $\lambda_n \to \lambda$ due to part (b), and so X^{λ_n} converges P-a. s. to X^λ due to (3.75). This completes the proof in (a) that the map g is continuous. □

Remark 3.67. In order to simplify the exposition, we have restricted the discussion of equilibrium prices to contingent claims with payoff at time $t = 1$. We have argued in terms of discounted payoffs, and so we have implicitly assumed that the interest rate r has already been fixed. From an economic point of view, also the interest rate should be determined by an equilibrium argument. This requires an intertemporal extension of our setting, which distinguishes between deterministic payoffs y at time $t = 0$ and nominal contingent payoffs Y at time $t = 1$. Thus, we replace $\mathscr{X} = L_+^0$ by the space

$$\overline{\mathscr{Y}} := \{\overline{Y} = (y, Y) \mid y \in [0,\infty),\ Y \in L_+^0\}.$$

A pricing rule is given by a linear functional on $\overline{\mathscr{Y}}$ of the form

$$\overline{\Phi}(\overline{Y}) := \varphi_0 \cdot y + E[\,\varphi Y\,],$$

where $\varphi_0 \in (0,\infty)$ and φ is a price density as before. Any such price system specifies an interest rate for transferring income from time $t = 0$ to time $t = 1$. Indeed, comparing the *forward price* $c \cdot E[\,\varphi\,]$ for the fixed amount c to be delivered at time 1 with the *spot price* $c \cdot \varphi_0$ for the amount c made available at time 0, we see that the implicit interest rate is given by

$$1 + r = \frac{\varphi_0}{E[\,\varphi\,]}.$$

If we describe the preferences of agent $a \in \mathscr{A}$ by a utility functional of the form,

$$U_a(\overline{Y}) = u_{a,0}(y) + E[u_{a,1}(Y)]$$

with smooth utility functions $u_{a,0}$ and $u_{a,1}$, then we can show along the lines of the preceding discussion that an Arrow–Debreu equilibrium exists in this extended setting. Thus, we obtain an equilibrium allocation $(\overline{Y}_a^*)_{a\in\mathscr{A}}$ and an equilibrium price system $\overline{\Phi}^* = (\varphi_0^*, \varphi^*)$ such that each $\overline{Y}_a^*$ maximizes the functional U_a in the agent's budget set determined by an initial endowment in $\overline{\mathscr{Y}}$ and by the pricing rule $\overline{\Phi}^*$. In particular, we have then specified an equilibrium interest rate r^*. Normalizing the price system to

$\varphi_0^* = 1$ and defining P^* as a probability measure with density $\varphi^*/E[\varphi^*]$, we see that the price at time $t = 0$ of a contingent claim with nominal payoff $Y \geq 0$ at time $t = 1$ is given as the expectation

$$E^*\left[\frac{Y}{1+r^*}\right]$$

of the discounted claim with respect to the measure P^*. ◇

Let us now extend the discussion to situations where agents are heterogeneous not only in their utility functions but also in their expectations. Thus, we assume that the preferences of agent $a \in \mathscr{A}$ are described by a Savage functional of the form

$$U_a(X) := E_{Q_a}[\, u_a(X)\,],$$

where Q_a is a probability measure on $(\Omega, \mathscr{F})$ that is equivalent to P. In addition to our assumption

$$\limsup_{x\downarrow 0} x\, u_a'(x) < \infty, \quad a \in \mathscr{A}, \tag{3.76}$$

we assume that

$$E_{Q_a}[\, W\,] < \infty \quad \text{and} \quad E_{Q_a}\left[u_a'\left(\frac{W}{|\mathscr{A}|}\right)\right] < \infty, \quad a \in \mathscr{A}. \tag{3.77}$$

As before, a feasible allocation $(X_a^*)_{a\in\mathscr{A}}$ together with a price density φ^* is called an Arrow–Debreu equilibrium if each X_a^* maximizes the functional U_a on the budget set of agent $a \in \mathscr{A}$, which is determined by φ^*.

Theorem 3.68. *Under assumptions* (3.55), (3.76), *and* (3.77), *there exists an Arrow–Debreu equilibrium.*

Proof. For any $\lambda \in \Lambda$, the general argument of Remark 3.31 yields the existence of a λ-efficient allocation $(X_a^\lambda)_{a\in\mathscr{A}}$, that is, of a feasible allocation that maximizes the functional

$$U^\lambda(X) := \sum_{a\in\mathscr{A}} \lambda_a U_a(X_a)$$

over all feasible allocations $X = (X_a)_{a\in\mathscr{A}}$. Since

$$U_a(X_a^\lambda) = E[\,\varphi_a\, u_a(X_a^\lambda)\,],$$

$(X_a^\lambda)_{a\in\mathscr{A}}$ can be viewed as a λ-efficient allocation in the model where agents have random utility functions of the form

$$\widetilde{u}_a(x,\omega) = u_a(x)\,\varphi_a(\omega),$$

while their expectations are homogeneous and given by P. In view of Corollary 3.37, it follows as before that X^λ satisfies the first-order conditions

$$X_a^\lambda = I_a^+(\lambda_a^{-1}\varphi_a^{-1}\,\varphi^\lambda), \quad a \in \mathscr{A},$$

with

$$\varphi^\lambda = \max_{a\in\mathscr{A}} \lambda_a\, u_a'(X_a^\lambda)\,\varphi_a,$$

and that X_a^λ satisfies

$$U_a(X_a^\lambda) \geq E[\,u_a(Y_a)\,\varphi_a\,] = U_a(Y_a)$$

for all Y_a in the budget set of agent $a \in \mathscr{A}$. The remaining arguments are essentially the same as in the proof of Theorem 3.62. □

4 Monetary measures of risk

In this chapter, we discuss the problem of quantifying the *risk* of a financial position. As in Chapter 2, such a position will be described by the corresponding payoff profile, that is, by a real-valued function X on some set of possible scenarios. In a probabilistic model, specified by a probability measure on scenarios, we could focus on the resulting distribution of X and try to measure the risk in terms of moments or quantiles. Note that a classical measure of risk such as the variance does not capture a basic asymmetry in the financial interpretation of X: Here, it is the *downside risk* that matters. This asymmetry is taken into account by measures such as Value at Risk, which are based on quantiles for the lower tail of the distribution; see Section 4.5 below. Value at Risk, however, fails to satisfy some natural consistency requirements. Such observations have motivated the systematic investigation of measures of risk that satisfy certain basic axioms.

From the point of view of an investor, we could simply turn around the discussion of Chapter 2 and measure the risk of a position X in terms of the loss functional

$$L(X) = -U(X).$$

Here, U is a utility functional representing a given preference relation $\succ$ on financial positions. Assuming robust preferences, we are led to the notion of *robust shortfall risk* defined by

$$L(X) = \sup_{Q \in \mathcal{Q}} E_Q[\, \ell(-X)\,],$$

where $\ell(x) := -u(-x)$ is a convex increasing loss function and $\mathcal{Q}$ is a class of probability measures. The results of Section 2.5 show how such loss functionals can be characterized in terms of convexity and monotonicity properties of the preference relation.

From the point of view of a supervising agency, however, a specific monetary purpose comes into play. In this perspective a risk measure is viewed as a capital requirement: We are looking for the minimal amount of capital that, if added to the position and invested in a risk-free manner, makes the position acceptable. For example, a financial position could be viewed as being acceptable if the robust shortfall risk of X does not exceed a given bound. This monetary interpretation is captured by an additional axiom of *cash invariance*. Together with convexity and monotonicity, it singles out the class of *convex risk measures*. These measures can be represented in the form

$$\rho(X) = \sup_{Q} (E_Q[\, -X\,] - \alpha(Q)),$$

where α is a penalty function defined on probability measures on Ω. Under the additional condition of *positive homogeneity*, we obtain the class of *coherent* risk measures. Here, we are back to the situation in Proposition 2.84, and the representation takes the form

https://doi.org/10.1515/9783111045283-005

$$\rho(X) = \sup_{Q \in \mathcal{Q}} E_Q[\,-X\,],$$

where $\mathcal{Q}$ is some class of probability measures on Ω.

The axiomatic approach to such monetary risk measures was initiated by P. Artzner, F. Delbaen, J. Eber, and D. Heath [14], and it will be developed in the first three sections. Our presentation will be self-contained and can thus be read without relying on the results of the preceding chapters. In Section 4.5, we discuss some coherent risk measures related to Value at Risk. These risk measures only involve the distribution of a position under a given probability measure. In Sections 4.6 and 4.7, we analyze the class of convex risk measures that share this property of law-invariance. Section 4.8 discusses the role of concave distortions, and in Section 4.9 the resulting risk measures are characterized by a property of comonotonicity. In Section 4.10, we discuss risk measures that arise naturally in the context of a financial market model, if acceptability is defined in terms of hedging opportunities. In Section 4.11, we analyze the structure of utility-based risk measures, which are induced by our notion of robust shortfall risk.

4.1 Risk measures and their acceptance sets

Let Ω be a fixed set of scenarios. A *financial position* is described by a mapping $X : \Omega \longrightarrow \mathbb{R}$ where $X(\omega)$ describes the discounted profits and losses (P&L or PnL) of the position at the end of the trading period if the scenario $\omega \in \Omega$ is realized. Our aim is to quantify the risk of X by some number $\rho(X)$, where X belongs to a given class $\mathcal{X}$ of financial positions. Throughout this section, $\mathcal{X}$ will be a linear space of bounded functions containing the constants. We do not assume that a probability measure is given on Ω.

Definition 4.1. A mapping $\rho : \mathcal{X} \to \mathbb{R}$ is called a *monetary risk measure* if it satisfies the following conditions for all $X, Y \in \mathcal{X}$:

- *Monotonicity:* If $X \le Y$, then $\rho(X) \ge \rho(Y)$.
- *Cash invariance:* If $m \in \mathbb{R}$, then $\rho(X + m) = \rho(X) - m$. △

The financial meaning of monotonicity is clear. The downside risk of a position is reduced if the payoff profile is increased. Cash invariance is also called *translation invariance* or *translation property*. It is motivated by the interpretation of $\rho(X)$ as a capital requirement, that is, $\rho(X)$ is the amount that should be added to the position X in order to make it acceptable from the point of view of a supervising agency. Thus, if the amount m is added to the position and invested in a risk-free manner, the capital requirement is reduced by the same amount. In particular, cash invariance implies

$$\rho(X + \rho(X)) = 0, \tag{4.1}$$

and

$$\rho(m) = \rho(0) - m \quad \text{for all } m \in \mathbb{R}.$$

For most purposes, it would be no loss of generality to assume that a given monetary risk measure satisfies the condition of

- *Normalization:* $\rho(0) = 0$.

For a normalized risk measure, cash invariance is equivalent to *cash additivity*, that is, to $\rho(X+m) = \rho(X)+\rho(m)$. In some situations, however, it will be convenient not to insist on normalization.

Remark 4.2. We are using the convention that X describes the P&L of a financial position after discounting. For instance, the discounting factor can be chosen as $1/(1+r)$ where r is the return of a risk-free investment. Instead of measuring the risk of the discounted position X, one could consider directly the nominal profits and losses,

$$\widetilde{X} = (1+r)X.$$

The corresponding risk measure $\widetilde{\rho}(\widetilde{X}) := \rho(X)$ is again monotone. Cash invariance is replaced by the following property:

$$\widetilde{\rho}(\widetilde{X} + (1+r)m) = \widetilde{\rho}(\widetilde{X}) - m, \tag{4.2}$$

that is, the risk is reduced by m if an additional amount m is invested in a risk-free manner. Conversely, any $\widetilde{\rho} : \mathscr{X} \to \mathbb{R}$ that is monotone and satisfies (4.2) defines a monetary measure of risk via $\rho(X) := \widetilde{\rho}((1+r)X)$. ◇

Lemma 4.3. *Any monetary risk measure ρ is Lipschitz continuous with respect to the supremum norm $\|\cdot\|$,*

$$|\rho(X) - \rho(Y)| \leq \|X - Y\|.$$

Proof. Clearly, $X \leq Y + \|X - Y\|$, and so $\rho(Y) - \|X - Y\| \leq \rho(X)$ by monotonicity and cash invariance. Reversing the roles of X and Y yields the assertion. □

From now on, we concentrate on monetary risk measures that have an additional convexity property.

Definition 4.4. A monetary risk measure $\rho : \mathscr{X} \to \mathbb{R}$ is called a *convex risk measure* if it satisfies

- *Convexity:* $\rho(\lambda X + (1-\lambda)Y) \leq \lambda\rho(X) + (1-\lambda)\rho(Y)$, for $0 \leq \lambda \leq 1$. △

Consider the collection of possible future outcomes that can be generated with the resources available to an investor: One investment strategy leads to X, while a second strategy leads to Y. If one *diversifies*, spending only the fraction λ of the resources on the first possibility and using the remaining part for the second alternative, one obtains $\lambda X + (1-\lambda)Y$. Thus, the axiom of convexity gives a precise meaning to the idea that diversification should not increase the risk. This idea becomes even clearer when we note that, for a monetary risk measure, convexity is in fact equivalent to the weaker requirement of

- *Quasi-Convexity:* $\rho(\lambda X + (1-\lambda)Y) \le \rho(X) \vee \rho(Y)$ for $0 \le \lambda \le 1$.

Exercise 4.1.1. Prove that a monetary risk measure is quasi-convex if and only if it is convex. ◇

Exercise 4.1.2. Show that if ρ is convex and normalized, then

$$\begin{aligned} \rho(\lambda X) &\le \lambda\rho(X) \quad \text{for } 0 \le \lambda \le 1, \\ \rho(\lambda X) &\ge \lambda\rho(X) \quad \text{for } \lambda \ge 1. \end{aligned}$$

◇

Definition 4.5. A convex risk measure ρ is called a *coherent risk measure* if it satisfies

- *Positive homogeneity:* If $\lambda \ge 0$, then $\rho(\lambda X) = \lambda\rho(X)$. △

If a monetary risk measure ρ is positively homogeneous, then it is normalized in the sense that $\rho(0) = 0$. Under the assumption of positive homogeneity, convexity is equivalent to

- *Subadditivity:* $\rho(X + Y) \le \rho(X) + \rho(Y)$.

This property allows to decentralize the task of managing the risk arising from a collection of different positions: If separate risk limits are given to different "desks," then the risk of the aggregate position is bounded by the sum of the individual risk limits.

In many situations, however, risk may grow in a *nonlinear* way as the size of the position increases. For this reason, we will not insist on positive homogeneity. Instead, our focus will be mainly on convex measures of risk.

Exercise 4.1.3. Let ρ be a normalized monetary risk measure on $\mathscr{X}$. Show that any two of the following properties imply the remaining third:

- *Convexity*
- *Positive homogeneity*
- *Subadditivity* ◇

A monetary risk measure ρ induces the class

$$\mathscr{A}_\rho := \{X \in \mathscr{X} \mid \rho(X) \le 0\}$$

of positions that are acceptable in the sense that they do not require additional capital. The class $\mathscr{A}_\rho$ will be called the *acceptance set* of ρ. The following two propositions summarize the relations between monetary risk measures and their acceptance sets.

Proposition 4.6. *Suppose that ρ is a monetary risk measure with acceptance set $\mathscr{A} := \mathscr{A}_\rho$.*

(a) *$\mathscr{A}$ is nonempty, closed in $\mathscr{X}$ with respect to the supremum norm $\|\cdot\|$, and satisfies the following two conditions:*

$$\inf\{m \in \mathbb{R} \mid m \in \mathscr{A}\} > -\infty, \tag{4.3}$$

$$X \in \mathscr{A}, Y \in \mathscr{X}, Y \geq X \quad \Longrightarrow \quad Y \in \mathscr{A}. \tag{4.4}$$

(b) *ρ can be recovered from $\mathscr{A}$:*

$$\rho(X) = \inf\{m \in \mathbb{R} \mid m + X \in \mathscr{A}\}. \tag{4.5}$$

(c) *ρ is a* convex *risk measure if and only if $\mathscr{A}$ is convex.*
(d) *ρ is positively homogeneous if and only if $\mathscr{A}$ is a cone. In particular, ρ is coherent if and only if $\mathscr{A}$ is a convex cone.*

Proof. (a): Properties (4.3) and (4.4) are straightforward, and closedness follows from Lemma 4.3.

(b): Cash invariance implies that for $X \in \mathscr{X}$,

$$\begin{aligned}\inf\{m \in \mathbb{R} \mid m + X \in \mathscr{A}_\rho\} &= \inf\{m \in \mathbb{R} \mid \rho(m + X) \leq 0\}\\ &= \inf\{m \in \mathbb{R} \mid \rho(X) \leq m\}\\ &= \rho(X).\end{aligned}$$

(c): $\mathscr{A}$ is clearly convex if ρ is a convex measure of risk. The converse will follow from Proposition 4.7 together with (4.7).

(d): Clearly, positive homogeneity of ρ implies that $\mathscr{A}$ is a cone. The converse follows as in (c). □

Conversely, one can take a given class $\mathscr{A} \subset \mathscr{X}$ of acceptable positions as the primary object. For a position $X \in \mathscr{X}$, we can then define the capital requirement as the minimal amount m for which $m + X$ becomes acceptable:

$$\rho_{\mathscr{A}}(X) := \inf\{m \in \mathbb{R} \mid m + X \in \mathscr{A}\}. \tag{4.6}$$

Note that, with this notation, (4.5) takes the form

$$\rho_{\mathscr{A}_\rho} = \rho. \tag{4.7}$$

Proposition 4.7. *Assume that $\mathscr{A}$ is a nonempty subset of $\mathscr{X}$ that satisfies* (4.3) *and* (4.4). *Then the functional $\rho_{\mathscr{A}}$ has the following properties:*
(a) *$\rho_{\mathscr{A}}$ is a monetary risk measure.*
(b) *If $\mathscr{A}$ is a convex set, then $\rho_{\mathscr{A}}$ is a convex risk measure.*
(c) *If $\mathscr{A}$ is a cone, then $\rho_{\mathscr{A}}$ is positively homogeneous. In particular, $\rho_{\mathscr{A}}$ is a coherent risk measure if $\mathscr{A}$ is a convex cone.*
(d) *$\mathscr{A}_{\rho_{\mathscr{A}}}$ is equal to the closure of $\mathscr{A}$ with respect to the supremum norm $\|\cdot\|$. In particular, $\mathscr{A}_{\rho_{\mathscr{A}}} = \mathscr{A}$ holds if and only if $\mathscr{A}$ is $\|\cdot\|$-closed.*

Proof. (a): It is straightforward to verify that $\rho_{\mathscr{A}}$ satisfies cash invariance and monotonicity. We show next that $\rho_{\mathscr{A}}$ takes only finite values. To this end, fix some Y in the nonempty set $\mathscr{A}$. For $X \in \mathscr{X}$ given, there exists a finite number m with $m + X > Y$, because X and Y are both bounded. Then

$$\rho_{\mathscr{A}}(X) - m = \rho_{\mathscr{A}}(m + X) \le \rho_{\mathscr{A}}(Y) \le 0,$$

and hence $\rho_{\mathscr{A}}(X) \le m < \infty$. Note that (4.3) is equivalent to $\rho_{\mathscr{A}}(0) > -\infty$. To show that $\rho_{\mathscr{A}}(X) > -\infty$ for arbitrary $X \in \mathscr{X}$, we take m' such that $X + m' \le 0$ and conclude by monotonicity and cash invariance that $\rho_{\mathscr{A}}(X) \ge \rho_{\mathscr{A}}(0) + m' > -\infty$.

(b): Suppose that $X_1, X_2 \in \mathscr{X}$ and that $m_1, m_2 \in \mathbb{R}$ are such that $m_i + X_i \in \mathscr{A}$. If $\lambda \in [0,1]$, then the convexity of $\mathscr{A}$ implies that $\lambda(m_1 + X_1) + (1-\lambda)(m_2 + X_2) \in \mathscr{A}$. Thus, by the cash invariance of $\rho_{\mathscr{A}}$,

$$\begin{aligned} 0 &\ge \rho_{\mathscr{A}}(\lambda(m_1 + X_1) + (1-\lambda)(m_2 + X_2)) \\ &= \rho_{\mathscr{A}}(\lambda X_1 + (1-\lambda)X_2) - (\lambda m_1 + (1-\lambda)m_2). \end{aligned}$$

It follows that $\rho_{\mathscr{A}}(\lambda X_1 + (1-\lambda)X_2) \le \lambda m_1 + (1-\lambda)m_2$. Taking $m_i \downarrow \rho(X_i)$ yields the convexity of $\rho_{\mathscr{A}}$.

(c): As in the proof of convexity, we obtain that $\rho_{\mathscr{A}}(\lambda X) \le \lambda \rho_{\mathscr{A}}(X)$ for $\lambda \ge 0$ if $\mathscr{A}$ is a cone. To prove the converse inequality, let $m < \rho_{\mathscr{A}}(X)$. Then $m + X \notin \mathscr{A}$, and hence $\lambda m + \lambda X \notin \mathscr{A}$ for $\lambda > 0$. Thus, $\lambda m < \rho_{\mathscr{A}}(\lambda X)$, and (c) follows.

(d): Let $\overline{\mathscr{A}}$ denote the closure of $\mathscr{A}$ with respect to $\|\cdot\|$. We first show $\overline{\mathscr{A}} \subseteq \mathscr{A}_{\rho_{\mathscr{A}}}$. If $X \in \overline{\mathscr{A}}$, there exists a sequence (X_n) in $\mathscr{A}$ such that $\|X_n - X\| \to 0$. It follows from Lemma 4.3 that $\rho_{\mathscr{A}}(X_n) \to \rho_{\mathscr{A}}(X)$, and so $\rho_{\mathscr{A}}(X) \le 0$. Hence, $X \in \mathscr{A}_{\rho_{\mathscr{A}}}$ and in turn $\overline{\mathscr{A}} \subseteq \mathscr{A}_{\rho_{\mathscr{A}}}$.

For the proof of the converse inclusion, take $X \in \mathscr{A}_{\rho_{\mathscr{A}}}$. Then

$$0 \ge \rho_{\mathscr{A}}(X) = \inf\{m \in \mathbb{R} \mid m + X \in \mathscr{A}\},$$

and so there exists a sequence (m_n) such that $m_n \downarrow \rho_{\mathscr{A}}(X)$ and $m_n + X \in \mathscr{A}$. Moreover, $X_n := m_n + X - \rho_{\mathscr{A}}(X) \ge m_n + X \in \mathscr{A}$, and so $X_n \in \mathscr{A}$ by (4.4). Since $\|X_n - X\| = m_n - \rho_{\mathscr{A}}(X) \downarrow 0$, we obtain $X \in \overline{\mathscr{A}}$ and in turn $\mathscr{A}_{\rho_{\mathscr{A}}} \subseteq \overline{\mathscr{A}}$. □

Exercise 4.1.4. Let $\mathscr{A}$ be a nonempty subset of $\mathscr{X}$ that satisfies (4.3) and (4.4). Show that $\mathscr{A}$ is $\|\cdot\|$-closed if and only if $\mathscr{A}$ satisfies the following property: if $X + m \in \mathscr{A}$ for all $m > 0$ then $X \in \mathscr{A}$. ◇

Exercise 4.1.5. Let I be a nonempty index set. For each $i \in I$, let ρ_i a monetary risk measures on $\mathscr{X}$ with associated acceptance sets $\mathscr{A}_i$, and assume that $\sup_{i \in I} \rho_i(0) < \infty$. Show that

$$\rho(X) := \sup_{i \in I} \rho_i(X)$$

is a monetary risk measure on $\mathscr{X}$ with acceptance set $\mathscr{A} = \bigcap_{i \in I} \mathscr{A}_i$. Show moreover that ρ is a convex risk measure if each ρ_i is a convex risk measure. ◇

In the following examples, we take $\mathscr{X}$ as the linear space of all bounded measurable functions on some measurable space $(\Omega, \mathscr{F})$, and we denote by $\mathscr{M}_1 = \mathscr{M}_1(\Omega, \mathscr{F})$ the class of all probability measures on $(\Omega, \mathscr{F})$.

Example 4.8. Consider the *worst-case risk measure* $\rho_{\max}$ defined by

$$\rho_{\max}(X) = -\inf_{\omega \in \Omega} X(\omega) \quad \text{for all } X \in \mathscr{X}.$$

The value $\rho_{\max}(X)$ is the least upper bound on the potential loss that can occur in any scenario. The corresponding acceptance set $\mathscr{A}$ is given by the convex cone of all nonnegative functions in $\mathscr{X}$. Thus, $\rho_{\max}$ is a coherent risk measure. It is the most conservative measure of risk in the sense that any normalized monetary risk measure ρ on $\mathscr{X}$ satisfies

$$\rho(X) \le \rho\Big(\inf_{\omega \in \Omega} X(\omega)\Big) = \rho_{\max}(X).$$

Note that $\rho_{\max}$ can be represented in the form

$$\rho_{\max}(X) = \sup_{Q \in \mathcal{Q}} E_Q[\,-X\,], \tag{4.8}$$

where $\mathcal{Q}$ is the class $\mathscr{M}_1$ of all probability measures on $(\Omega, \mathscr{F})$. ◇

Example 4.9. Let $\mathcal{Q}$ be a set of probability measures on $(\Omega, \mathscr{F})$, and consider a mapping $\gamma : \mathcal{Q} \to \mathbb{R}$ with $\sup_Q \gamma(Q) < \infty$, which specifies for each $Q \in \mathcal{Q}$ some "floor" $\gamma(Q)$. Suppose that a position X is acceptable if

$$E_Q[\,X\,] \ge \gamma(Q) \quad \text{for all } Q \in \mathcal{Q}.$$

The set $\mathscr{A}$ of such positions satisfies (4.3) and (4.4), and it is convex. Thus, the associated monetary risk measure $\rho = \rho_{\mathscr{A}}$ is *convex*, and it takes the form

$$\rho(X) = \sup_{Q \in \mathcal{Q}} (\gamma(Q) - E_Q[\,X\,]).$$

This can for instance be seen via Exercise 4.1.5 by taking $\rho_Q(X) := \gamma(Q) - E_Q[X]$ for $Q \in I := \mathcal{Q}$. Alternatively, we can write

$$\rho(X) = \sup_{Q \in \mathscr{M}_1} (E_Q[\,-X\,] - \alpha(Q)), \tag{4.9}$$

where the *penalty function* $\alpha : \mathscr{M}_1 \to (-\infty, \infty]$ is defined by $\alpha(Q) = -\gamma(Q)$ for $Q \in \mathcal{Q}$ and $\alpha(Q) = +\infty$ otherwise. Note that ρ is a *coherent* risk measure if $\gamma(Q) = 0$ for all $Q \in \mathcal{Q}$. ◇

Example 4.10. Consider a utility function u on $\mathbb{R}$, a probability measure $Q \in \mathcal{M}_1$, and fix some threshold $c \in \mathbb{R}$. Let us call a position X acceptable if its certainty equivalent is at least c, that is, if its expected utility $E_Q[\,u(X)\,]$ is bounded from below by $u(c)$. Clearly, the set

$$\mathcal{A} := \{X \in \mathcal{X} \mid E_Q[\,u(X)\,] \geq u(c)\},$$

is nonempty, convex, and satisfies (4.3) and (4.4). Thus, $\rho_{\mathcal{A}}$ is a convex risk measure. The corresponding risk measures are called *utility-based shortfall risk measures*. They will be studied in more detail in Section 4.11. As an obvious robust extension, we can define acceptability in terms of a whole class $\mathcal{Q}$ of probability measures on $(\Omega, \mathcal{F})$, that is,

$$\mathcal{A} := \bigcap_{Q \in \mathcal{Q}} \{X \in \mathcal{X} \mid E_Q[\,u(X)\,] \geq u(c_Q)\},$$

with constants c_Q such that $\sup_{Q \in \mathcal{Q}} c_Q < \infty$. This is yet another class of risk measures fitting into the context of Exercise 4.1.5. ◇

Example 4.11. Suppose now that we have specified a probabilistic model, i. e., a probability measure P on $(\Omega, \mathcal{F})$. In this context, a position X is often considered to be acceptable if the probability of a loss is bounded by a given level $\lambda \in (0, 1)$, that is, if

$$P[\,X < 0\,] \leq \lambda.$$

The corresponding monetary risk measure $\mathrm{V@R}_\lambda$, defined by

$$\mathrm{V@R}_\lambda(X) = \inf\{m \in \mathbb{R} \mid P[\,m + X < 0\,] \leq \lambda\},$$

is called *Value at Risk* at level λ. Note that it is well-defined on the space $\mathcal{L}^0(\Omega, \mathcal{F}, P)$ of all random variables that are P-a. s. finite. If X is a Gaussian random variable with variance $\sigma^2(X)$, then

$$\mathrm{V@R}_\lambda(X) = E[\,-X\,] + \Phi^{-1}(1 - \lambda)\sigma(X), \tag{4.10}$$

where Φ^{-1} denotes the inverse of the cumulative distribution function Φ of $N(0, 1)$. Clearly, $\mathrm{V@R}_\lambda$ is positively homogeneous, but in general it is not convex, as shown by Example 4.46 below. In Section 4.5, Value at Risk will be discussed in detail. In particular, we will study some closely related coherent and convex risk measures. ◇

Exercise 4.1.6. Compute $\mathrm{V@R}_\lambda(-X)$ if X is
(a) uniformly distributed on $[0, 1]$,
(b) log-normally distributed, that is, $X = e^{\sigma Z + \mu}$ with $Z \sim N(0, 1)$ and $\mu, \sigma \in \mathbb{R}$,
(c) an indicator function, that is, $X = \mathbb{1}_A$ for some event $A \in \mathcal{F}$. ◇

Example 4.12. As in Example 4.11, we fix a probability measure P on $(\Omega, \mathscr{F})$. For an asset with payoff $\widetilde{X} \in \mathscr{L}^2 = \mathscr{L}^2(\Omega, \mathscr{F}, P)$, price $\pi(\widetilde{X})$, return $R(\widetilde{X}) = (\widetilde{X} - \pi(\widetilde{X}))/\pi(\widetilde{X})$, and variance $\sigma^2(\widetilde{X}) \neq 0$, the *Sharpe ratio* is defined as

$$\frac{E[R(\widetilde{X}) - r]}{\sigma(R(\widetilde{X}) - r)} = \frac{E[\widetilde{X}\,] - \pi(\widetilde{X})(1+r)}{\sigma(\widetilde{X})} = \frac{E[\,X\,]}{\sigma(X)},$$

where $X := \widetilde{X}(1+r)^{-1} - \pi(\widetilde{X})$ is the corresponding discounted P&L. Suppose that we find the position X acceptable if the Sharpe ratio is bounded from below by some constant $c > 0$. The resulting functional ρ_c on $\mathscr{L}^2$ defined by (4.6) for the class

$$\mathscr{A}_c := \{X \in \mathscr{L}^2 \mid E[\,X\,] \geq c \cdot \sigma(X)\}$$

is given by

$$\rho_c(X) = E[\,-X\,] + c \cdot \sigma(X). \tag{4.11}$$

It is sometimes called the *mean-standard deviation risk measure*. It is cash invariant and positively homogeneous, and it is convex since $\sigma(\,\cdot\,)$ is a convex functional on $\mathscr{L}^2$. But ρ_c is not a monetary risk measure, because it is not monotone. Indeed, if $X = e^Z$ and Z is a random variable with normal distribution $N(0, \sigma^2)$, then $X \geq 0$, and monotonicity would imply $\rho_c(X) \leq \rho_c(0) = 0$. But

$$\rho_c(X) = -e^{\sigma^2/2} + ce^{\sigma^2/2}\sqrt{e^{\sigma^2} - 1}$$

becomes positive for sufficiently large σ. Note, however, that (4.10) shows that $\rho_c(X)$ coincides with $\mathrm{V@R}_\lambda(X)$ if X is Gaussian and if $c = \Phi^{-1}(1-\lambda)$ with $0 < \lambda \leq 1/2$. Since $\mathrm{V@R}_\lambda$ is monotone, both ρ_c and $\mathrm{V@R}_\lambda$ have all the properties of a coherent risk measure if restricted to a Gaussian subspace of $\mathscr{L}^2$, that is, any linear space consisting of normally distributed random variables. But neither ρ_c nor $\mathrm{V@R}_\lambda$ can be coherent on $\mathscr{X}$, and hence on the full space $\mathscr{L}^2$, since the existence of normal random variables on $(\Omega, \mathscr{F}, P)$ implies that $\mathscr{X}$ will also contain random variables as considered in Example 4.46. ◇

Example 4.13. Let $u : \mathbb{R} \to \mathbb{R}$ be a strictly increasing continuous function. For $X \in \mathscr{X}$, we consider the *certainty equivalent* of the law of X under P as a functional of X by setting

$$c(X) := u^{-1}(E[\,u(X)\,]).$$

Then $\rho(X) := -c(X)$ is monotone,

$$X \leq Y \Longrightarrow \rho(X) \geq \rho(Y).$$

If ρ is also cash invariant, and hence a monetary risk measure, then Proposition 2.46 shows that u is either linear or a function of exponential form: $u(x) = a + be^{\beta x}$ or $u(x) = a - be^{-\beta x}$ for constants $a \in \mathbb{R}$ and $b, \beta > 0$. In the linear case, we have

$$\rho(X) = E[\,-X\,].$$

In the first exponential case, ρ is of the form

$$\rho(X) = -\frac{1}{\beta}\log E[\,e^{\beta X}\,].$$

In the second exponential case, ρ is given by

$$\rho(X) = \frac{1}{\beta}\log E[\,e^{-\beta X}\,] \tag{4.12}$$

and called the *entropic risk measure* for reasons that will become clear in Example 4.34. There (and in Exercise 4.1.7 below) we will also see that ρ is a convex risk measure. ◇

Exercise 4.1.7. Let $u : \mathbb{R} \to \mathbb{R}$ be a strictly increasing continuous function and let ρ be defined as in Example 4.13. Show that ρ is quasi-convex,

$$\rho(\lambda X + (1-\lambda)Y) \le \rho(X) \vee \rho(Y) \quad \text{for } 0 \le \lambda \le 1,$$

if u is concave. Conclude that the entropic risk measure in (4.12) is a convex risk measure.

◇

Exercise 4.1.8. For any $X \in \mathscr{L}^1$, define

$$\Delta(X) := E[\,|X - \widetilde{X}|\,],$$

where $\widetilde{X}$ is an independent copy of X. Note that $\Delta(\cdot)$, often called *Gini's mean difference*, satisfies $\Delta(X) \in [0, 2E[\,|X|\,]]$, with $\Delta(X) = 0$ if and only if X is P-a. s. constant. In analogy to the mean-standard deviation risk measure in (4.11), consider the functional on $\mathscr{L}^1$ defined by

$$\rho_\lambda(X) := E[\,-X\,] + \lambda\Delta(X)$$

for some constant $\lambda \ge 0$.

(a) Show that $\rho_\lambda(X)$ is a coherent risk measure for any $\lambda \in [0, \frac{1}{2}]$. Note that a nonconstant position X is acceptable if and only if $E[\,X\,] > 0$ and the *Gini coefficient* of X, defined as

$$G(X) := \frac{\Delta(X)}{2E[\,X\,]},$$

satisfies $G(X) \ge (2\lambda)^{-1}$.

(b) Show that for $\lambda > \frac{1}{2}$ there are nonnegative random variables X such that $\rho_\lambda(X) > 0$. In particular, ρ_λ cannot be monotone for $\lambda > \frac{1}{2}$.

(c) Show that

$$\Delta(X) = 2 \int_{-\infty}^{\infty} F(x)(1 - F(x))\, dx,$$

where F denotes the cumulative distribution function of X, and that

$$\Delta(X) = 4 \operatorname{cov}(X, F(X)) \tag{4.13}$$

whenever X has a continuous distribution.

(d) Use (4.13) to show that $\Delta(X) \le 2\sigma(X)/\sqrt{3}$ for all $X \in \mathscr{L}^2$ and conclude that ρ_λ is dominated by the mean-standard deviation risk measure ρ_c in (4.11) if $\lambda \le c\sqrt{3}/2$. As in Example 4.12, we denote here by $\sigma(X)$ the square root of the variance of X.

(e) Compute $\Delta(X)$ and $G(X)$ if X has a Pareto distribution with shape parameter $\alpha > 1$ and minimum 1, that is, $\log X$ is exponentially distributed with parameter α.

(f) Consider a log-normally distributed random variable $X = \exp(m + \sigma Z)$, where Z has a standard normal law $N(0, 1)$. Show that $G(X) = \operatorname{erf}(\sigma/2)$, where $\operatorname{erf}(\cdot)$ is the *Gaussian error function*, that is, $\operatorname{erf}(z) := P[\,|Z| \le z\sqrt{2}\,] = 2\Phi(z\sqrt{2}) - 1$. ◇

Example 4.14. Let $c : \mathscr{F} \to [0, 1]$ be any set function that is normalized and monotone in the sense that $c(\emptyset) = 0$, $c(\Omega) = 1$, and $c(A) \le c(B)$ if $A \subseteq B$. For instance, c can be given by $c(A) := \psi(P[\,A\,])$ for some probability measure P and an increasing function $\psi : [0, 1] \to [0, 1]$ such that $\psi(0) = 0$ and $\psi(1) = 1$. The *Choquet integral* of a bounded measurable function $X \ge 0$ with respect to c is defined as

$$\int X\, dc := \int_0^\infty c(X > x)\, dx.$$

If c is a probability measure, Fubini's theorem implies that $\int X\, dc$ coincides with the usual integral. In the general case, the Choquet integral is a *nonlinear* functional of X, but we still have $\int \lambda X\, dc = \lambda \int X\, dc$ and $\int (X + m)\, dc = \int X\, dc + m$ for constants $\lambda, m \ge 0$. If $X \in \mathscr{X}$ is arbitrary, we take $m \in \mathbb{R}$ such that $X + m \ge 0$ and get

$$\int (X + m)\, dc - m = \int_{-m}^{0} (c(X > x) - 1)\, dx + \int_0^\infty c(X > x)\, dx.$$

The right-hand side is independent of $m \ge -\inf X$, and so it makes sense to extend the definition of the Choquet integral by putting

$$\int X\,dc := \int_{-\infty}^{0} (c(X > x) - 1)\,dx + \int_{0}^{\infty} c(X > x)\,dx$$

for all $X \in \mathscr{X}$. It follows that

$$\int \lambda X\,dc = \lambda \int X\,dc \quad \text{and} \quad \int (X + m)\,dc = \int X\,dc + m$$

for all $\lambda \geq 0$ and $m \in \mathbb{R}$. Moreover, we have

$$\int Y\,dc \geq \int X\,dc \quad \text{for } Y \geq X.$$

Thus, the Choquet integral of the loss,

$$\rho(X) := \int (-X)\,dc, \tag{4.14}$$

is a positively homogeneous monetary risk measure on $\mathscr{X}$. In Section 4.9, we will characterize these risk measures in terms of a property called "comonotonicity." We will also show that ρ is convex, and hence coherent, if and only if c is *submodular* or *strongly subadditive*, that is,

$$c(A \cap B) + c(A \cup B) \leq c(A) + c(B) \quad \text{for } A, B \in \mathscr{F}.$$

In this case, ρ admits the representation

$$\rho(X) = \max_{Q \in \mathscr{Q}_c} E_Q[\,-X\,], \tag{4.15}$$

where $\mathscr{Q}_c$ is the *core* of c, defined as the class of all finitely additive and normalized set functions $Q : \mathscr{F} \to [0,1]$ such that $Q[\,A\,] \leq c(A)$ for all $A \in \mathscr{F}$; see Theorem 4.100. ◇

Exercise 4.1.9. Let P be a probability measure on $(\Omega, \mathscr{F})$ and fix $n \in \mathbb{N}$. For $X \in \mathscr{X}$, let $X_1, \ldots, X_n$ be independent copies of X and set

$$\rho(X) := -E[\,\min(X_1, \ldots, X_n)\,].$$

The functional $\rho : \mathscr{X} \to \mathbb{R}$ is sometimes called MINVAR. Show that ρ is a coherent risk measure on $\mathscr{X}$. Show next that ρ fits into the framework of Example 4.14. More precisely, show that ρ can be represented as a Choquet integral,

$$\rho(X) = \int (-X)\,dc,$$

where the set function c is of the form $c(A) = \psi(P[A])$ for a concave increasing function $\psi : [0,1] \to [0,1]$ such that $\psi(0) = 0$ and $\psi(1) = 1$. ◇

In the next two sections, we are going to show how representations of the form (4.8), (4.15), (4.9), or (4.14) for coherent or convex risk measures arise in a systematic manner.

4.2 Robust representation of convex risk measures

In this section, we consider a situation of Knightian uncertainty, where no probability measure P is fixed on the measurable space $(\Omega, \mathscr{F})$. Let $\mathscr{X}$ denote the space of all bounded measurable functions on $(\Omega, \mathscr{F})$. Recall that $\mathscr{X}$ is a Banach space if endowed with the supremum norm $\|\cdot\|$. As in Section 2.5, we denote by $\mathscr{M}_1 := \mathscr{M}_1(\Omega, \mathscr{F})$ the set of all probability measures on $(\Omega, \mathscr{F})$ and by $\mathscr{M}_{1,f} := \mathscr{M}_{1,f}(\Omega, \mathscr{F})$ the set of all finitely additive set functions $Q : \mathscr{F} \to [0,1]$ that are normalized to $Q[\Omega] = 1$. By $E_Q[X]$, we denote the integral of X with respect to $Q \in \mathscr{M}_{1,f}$; see (G.4) and Remark G.22.

If ρ is a *coherent* risk measure on $\mathscr{X}$, then we are in the context of Proposition 2.84, that is, the functional ϕ defined by $\phi(X) := -\rho(X)$ satisfies the four properties listed in Proposition 2.83. Hence, we have the following result.

Proposition 4.15. *A functional $\rho : \mathscr{X} \to \mathbb{R}$ is a coherent risk measure if and only if there exists a subset $\mathscr{Q}$ of $\mathscr{M}_{1,f}$ such that*

$$\rho(X) = \sup_{Q \in \mathscr{Q}} E_Q[-X], \quad X \in \mathscr{X}. \tag{4.16}$$

Moreover, $\mathscr{Q}$ can be chosen as a convex set for which the supremum in (4.16) *is attained.*

Our first goal in this section is to obtain an analogue of this result for *convex* risk measures. Applied to a coherent risk measure, it will yield an alternative proof of Proposition 4.15, which does not depend on the discussion in Chapter 2. Moreover, it will provide a description of the maximal set $\mathscr{Q}$ in (4.16). Our second goal will be to obtain criteria that guarantee that a risk measure can be represented in terms of σ-additive probability measures.

Let $\alpha : \mathscr{M}_{1,f} \to \mathbb{R} \cup \{+\infty\}$ be any functional such that

$$\inf_{Q \in \mathscr{M}_{1,f}} \alpha(Q) \in \mathbb{R}.$$

For each $Q \in \mathscr{M}_{1,f}$, the functional $X \mapsto E_Q[-X] - \alpha(Q)$ is convex, monotone, and cash invariant on $\mathscr{X}$, and these three properties are preserved when taking the supremum over $Q \in \mathscr{M}_{1,f}$. Hence,

$$\rho(X) := \sup_{Q \in \mathscr{M}_{1,f}} \left(E_Q[-X] - \alpha(Q)\right) \tag{4.17}$$

defines a convex risk measure on $\mathscr{X}$ such that

$$\rho(0) = -\inf_{Q \in \mathscr{M}_{1,f}} \alpha(Q).$$

The functional α will be called a *penalty function* for ρ on $\mathcal{M}_{1,f}$, and we will say that ρ is *represented* by α on $\mathcal{M}_{1,f}$.

Theorem 4.16. *Any convex risk measure ρ on $\mathcal{X}$ is of the form*

$$\rho(X) = \max_{Q\in\mathcal{M}_{1,f}} (E_Q[-X] - \alpha^{\min}(Q)), \quad X \in \mathcal{X}, \tag{4.18}$$

where the penalty function $\alpha^{\min}$ is given by

$$\alpha^{\min}(Q) := \sup_{X\in\mathcal{A}_\rho} E_Q[-X] \quad \textit{for } Q \in \mathcal{M}_{1,f}.$$

Moreover, $\alpha^{\min}$ is the minimal penalty function that represents ρ, that is, any penalty function α for which (4.17) *holds satisfies $\alpha(Q) \geq \alpha^{\min}(Q)$ for all $Q \in \mathcal{M}_{1,f}$.*

Proof. First, we argue that we may assume without loss of generality that $\rho(0) = 0$. Otherwise, we consider the convex risk measure $\widetilde{\rho}(X) = \rho(X) - \rho(0)$. One checks that the acceptance set of $\widetilde{\rho}$ satisfies $\mathcal{A}_{\widetilde{\rho}} = \{X - \rho(0) \mid X \in \mathcal{A}_\rho\}$, so that the minimal penalty function of $\widetilde{\rho}$ is given by

$$\widetilde{\alpha}^{\min}(Q) = \sup_{X\in\mathcal{A}_{\widetilde{\rho}}} E_Q[-X] = \sup_{X\in\mathcal{A}_\rho} E_Q[-X] + \rho(0) = \alpha^{\min}(Q) + \rho(0).$$

It follows that the representation (4.18) holds for ρ if and only if the corresponding representation holds for $\widetilde{\rho}$.

In the next step, we show that

$$\rho(X) \geq \sup_{Q\in\mathcal{M}_{1,f}} (E_Q[-X] - \alpha^{\min}(Q)) \quad \text{for all } X \in \mathcal{X}.$$

To this end, recall that $X' := \rho(X) + X \in \mathcal{A}_\rho$ by (4.1). Thus, for all $Q \in \mathcal{M}_{1,f}$,

$$\alpha^{\min}(Q) \geq E_Q[-X'] = E_Q[-X] - \rho(X).$$

From here, our claim follows.

For X given, we will now construct some $Q_X \in \mathcal{M}_{1,f}$ such that

$$\rho(X) \leq E_{Q_X}[-X] - \alpha^{\min}(Q_X),$$

which, in view of the previous step, will prove our representation (4.18). By cash invariance, it suffices to prove this claim for $X \in \mathcal{X}$ with $\rho(X) = 0$. Due to our assumption $\rho(0) = 0$, X is not contained in the nonempty convex set

$$\mathcal{B} := \{Y \in \mathcal{X} \mid \rho(Y) < 0\}.$$

Since $\mathscr{B}$ is open due to Lemma 4.3, we may apply the separation argument in the form of Theorem H.2. It yields a nonzero continuous linear functional ℓ on $\mathscr{X}$ such that

$$\ell(X) \leq \inf_{Y \in \mathscr{B}} \ell(Y) =: b.$$

We claim that $\ell(Y) \geq 0$ if $Y \geq 0$. Monotonicity and cash invariance of ρ imply that $1 + \lambda Y \in \mathscr{B}$ for any $\lambda > 0$. Hence,

$$\ell(X) \leq \ell(1 + \lambda Y) = \ell(1) + \lambda \ell(Y) \quad \text{for all } \lambda > 0,$$

which could not be true if $\ell(Y) < 0$.

Our next claim is that $\ell(1) > 0$. Since ℓ does not vanish identically, there must be some Y such that $0 < \ell(Y) = \ell(Y^+) - \ell(Y^-)$. We may assume without loss of generality that $\|Y\| < 1$. Positivity of ℓ implies $\ell(Y^+) > 0$ and $\ell(1 - Y^+) \geq 0$. Hence, $\ell(1) = \ell(1 - Y^+) + \ell(Y^+) > 0$.

By the two preceding steps and Theorem G.21, we conclude that there exists some $Q_X \in \mathscr{M}_{1,f}$ such that

$$E_{Q_X}[\,Y\,] = \frac{\ell(Y)}{\ell(1)} \quad \text{for all } Y \in \mathscr{X}.$$

Note that $\mathscr{B} \subset \mathscr{A}_\rho$, and so

$$\alpha^{\min}(Q_X) = \sup_{Y \in \mathscr{A}_\rho} E_{Q_X}[\,-Y\,] \geq \sup_{Y \in \mathscr{B}} E_{Q_X}[\,-Y\,] = -\frac{b}{\ell(1)}.$$

On the other hand, $Y + \varepsilon \in \mathscr{B}$ for any $Y \in \mathscr{A}_\rho$ and each $\varepsilon > 0$. This shows that $\alpha^{\min}(Q_X)$ is in fact equal to $-b/\ell(1)$. It follows that

$$E_{Q_X}[\,-X\,] - \alpha^{\min}(Q_X) = \frac{1}{\ell(1)}(b - \ell(X)) \geq 0 = \rho(X).$$

Thus, Q_X is as desired, and the proof of the representation (4.18) is complete.

Finally, let α be any penalty function for ρ. Then, for all $Q \in \mathscr{M}_{1,f}$ and $X \in \mathscr{X}$,

$$\rho(X) \geq E_Q[\,-X\,] - \alpha(Q),$$

and hence

$$\begin{aligned} \alpha(Q) &\geq \sup_{X \in \mathscr{X}} (E_Q[\,-X\,] - \rho(X)) \\ &\geq \sup_{X \in \mathscr{A}_\rho} (E_Q[\,-X\,] - \rho(X)) \\ &\geq \alpha^{\min}(Q). \end{aligned} \tag{4.19}$$

Thus, α dominates $\alpha^{\min}$. □

Remark 4.17.

(a) If we take $\alpha = \alpha^{\min}$ in (4.19), then all inequalities in (4.19) must be identities. Thus, we obtain an alternative formula for the minimal penalty function $\alpha^{\min}$:

$$\alpha^{\min}(Q) = \sup_{X \in \mathscr{X}} (E_Q[-X] - \rho(X)). \tag{4.20}$$

(b) Note that $\alpha^{\min}$ is convex and lower semicontinuous for the *total variation distance* on $\mathscr{M}_{1,f}$ as defined in Definition G.20, since it is the supremum of affine continuous functions on $\mathscr{M}_{1,f}$.

(c) Suppose ρ is defined via $\rho := \rho_{\mathscr{A}}$ for a given acceptance set $\mathscr{A} \subset \mathscr{X}$. Then $\mathscr{A}$ determines $\alpha^{\min}$:

$$\alpha^{\min}(Q) = \sup_{X \in \mathscr{A}} E_Q[-X] \quad \text{for all } Q \in \mathscr{M}_{1,f}.$$

This follows from part (d) of Proposition 4.7, which states that $\mathscr{A}_\rho$ is the closure of $\mathscr{A}$ with respect to the supremum norm $\|\cdot\|$, together with the $\|\cdot\|$-continuity of the map $X \mapsto E_Q[-X]$ for $Q \in \mathscr{M}_{1,f}$. ◇

Remark 4.18. Equation (4.20) shows that the penalty function $\alpha^{\min}$ may be viewed as the *Fenchel–Legendre transform*, or *conjugate function*, of the convex function ρ on the Banach space $\mathscr{X}$. More precisely,

$$\alpha^{\min}(Q) = \rho^*(\ell_Q), \tag{4.21}$$

where $\rho^* : \mathscr{X}' \to \mathbb{R} \cup \{+\infty\}$ is defined on the dual $\mathscr{X}'$ of $\mathscr{X}$ by

$$\rho^*(\ell) = \sup_{X \in \mathscr{X}} (\ell(X) - \rho(X)),$$

and where $\ell_Q \in \mathscr{X}'$ is given by $\ell_Q(X) = E_Q[-X]$ for $Q \in \mathscr{M}_{1,f}$. This suggests an alternative proof of Theorem 4.16. First, note that, by Theorem G.21, $\mathscr{X}'$ can be identified with the space $ba := ba(\Omega, \mathscr{F})$ of finitely additive set functions with finite total variation. Moreover, ρ is lower semicontinuous with respect to the weak topology $\sigma(\mathscr{X}, \mathscr{X}')$, since any set $\{\rho \le c\}$ is convex, strongly closed due to Lemma 4.3, and hence weakly closed by Theorem H.7. Thus, the general duality theorem for conjugate functions as stated in Theorem H.9 yields

$$\rho^{**} = \rho,$$

where ρ^{**} denotes the conjugate function of ρ^*, that is,

$$\rho(X) = \sup_{\ell \in ba} (\ell(X) - \rho^*(\ell)). \tag{4.22}$$

In a second step, using the arguments in the second part of the proof of Theorem 4.16, we can now check that monotonicity and cash invariance of ρ imply that $\ell \leq 0$ and $\ell(1) = -1$ for any $\ell \in \mathscr{X}' = ba$ such that $\rho^*(\ell) < \infty$. Identifying $-\ell$ with $Q \in \mathcal{M}_{1,f}$ and using equation (4.21), we see that (4.22) reduces to the representation

$$\rho(X) = \sup_{Q \in \mathcal{M}_{1,f}} (E_Q[-X] - \alpha^{\min}(Q)).$$

Moreover, the supremum is actually attained: $\mathcal{M}_{1,f}$ is weak* compact in $\mathscr{X}' = ba$ due to the Banach–Alaoglu theorem stated in Theorem H.10, and so the upper semicontinuous functional $Q \mapsto E_Q[-X] - \alpha^{\min}(Q)$ attains its maximum on $\mathcal{M}_{1,f}$. Thus, the robust representation (4.18) of a convex risk measure ρ follows from the general theory of convex duality. For this reason, it is also called the *dual representation* of ρ. ◇

The representation

$$\rho(X) = \sup_{Q \in \mathcal{Q}} E_Q[-X], \quad X \in \mathscr{X}, \tag{4.23}$$

of a *coherent* risk measure ρ via some set $\mathcal{Q} \subseteq \mathcal{M}_{1,f}$, as formulated in Proposition 4.15, is a particular case of the representation theorem for convex risk measures, since it corresponds to the penalty function

$$\alpha(Q) = \begin{cases} 0 & \text{if } Q \in \mathcal{Q}, \\ +\infty & \text{otherwise.} \end{cases}$$

The following corollary shows that the minimal penalty function of a coherent risk measure is always of this type.

Corollary 4.19. *The minimal penalty function $\alpha^{\min}$ of a coherent risk measure ρ takes only the values 0 and $+\infty$. In particular,*

$$\rho(X) = \max_{Q \in \mathcal{Q}^{\max}} E_Q[-X], \quad X \in \mathscr{X},$$

for the convex set

$$\mathcal{Q}^{\max} := \{Q \in \mathcal{M}_{1,f} \mid \alpha^{\min}(Q) = 0\},$$

and $\mathcal{Q}^{\max}$ is the largest set for which a representation of the form (4.23) *holds.*

Proof. Recall from Proposition 4.6 that the acceptance set $\mathcal{A}_\rho$ of a coherent risk measure is a cone. Thus, the minimal penalty function satisfies

$$\alpha^{\min}(Q) = \sup_{X \in \mathcal{A}_\rho} E_Q[-X] = \sup_{\lambda X \in \mathcal{A}_\rho} E_Q[-\lambda X] = \lambda\, \alpha^{\min}(Q)$$

for all $Q \in \mathcal{M}_{1,f}$ and $\lambda > 0$. Hence, $\alpha^{\min}$ can take only the values 0 and $+\infty$. □

Exercise 4.2.1. Let ρ be a coherent risk measure on $\mathscr{X}$ and assume that ρ admits a representation

$$\rho(X) = \sup_{Q \in \mathcal{Q}} E_Q[-X]$$

with some class $\mathcal{Q} \subseteq \mathscr{M}_{1,f}$.
(a) Show that $\rho(X) + \rho(-X) \geq 0$ for all $X \in \mathscr{X}$.
(b) Show that the following conditions are equivalent:
   (i) ρ is *additive*, that is, $\rho(X+Y) = \rho(X) + \rho(Y)$ for all $X, Y \in \mathscr{X}$.
   (ii) $\rho(X) + \rho(-X) = 0$ for all $X \in \mathscr{X}$.
   (iii) The class $\mathcal{Q}$ reduces to a single element Q, that is, ρ is simply the expected loss with respect to Q. ◇

The preceding exercise can be generalized as follows.

Exercise 4.2.2. Consider a normalized convex risk measure on $\mathscr{X}$ with representation

$$\rho(X) = \sup_{Q \in \mathscr{M}_{1,f}} \left(E_Q[-X] - \alpha(Q) \right)$$

for some penalty function α. Suppose that ρ is *additive* on some linear subspace $\mathscr{Y} \subseteq \mathscr{X}$, that is, $\rho(X+Y) = \rho(X) + \rho(Y)$ for $X, Y \in \mathscr{Y}$.
(a) Show that

$$\alpha(Q) < \infty \implies E_Q[-Y] = \rho(Y) \text{ for all } Y \in \mathscr{Y}.$$

(b) Conclude that ρ is *$\mathscr{Y}$-additive on $\mathscr{X}$*, that is,

$$\rho(X+Y) = \rho(X) + \rho(Y) \quad \text{for } X \in \mathscr{X} \text{ and } Y \in \mathscr{Y}.$$ ◇

The penalty function α arising in the representation (4.17) is not unique, and it is often convenient to represent a convex risk measure by a penalty function that is not the minimal one. For instance, the minimal penalty function may be finite for certain finitely additive set functions while another α is concentrated only on probability measures as in the case of Example 4.8. Another situation of this type occurs for risk measures that are constructed as the supremum of a family of convex risk measures:

Proposition 4.20. *Suppose that for every i in some index set I we are given a convex risk measure ρ_i on $\mathscr{X}$ with associated penalty function α_i. If $\sup_{i \in I} \rho_i(0) < \infty$, then*

$$\rho(X) := \sup_{i \in I} \rho_i(X), \quad X \in \mathscr{X},$$

is a convex risk measure that can be represented with the penalty function

$$\alpha(Q) := \inf_{i\in I} \alpha_i(Q), \quad Q \in \mathcal{M}_{1,f}.$$

Proof. It was shown in Exercise 4.1.5 that ρ is a convex risk measure. So, it remains to show that it can be represented with the penalty function α. Indeed,

$$\begin{aligned}\rho(X) &= \sup_{i\in I} \sup_{Q\in\mathcal{M}_{1,f}} (E_Q[-X] - \alpha_i(Q)) \\ &= \sup_{Q\in\mathcal{M}_{1,f}} \Big(E_Q[-X] - \inf_{i\in I} \alpha_i(Q)\Big),\end{aligned}$$

and the assertion follows. □

In the sequel, we are particularly interested in those convex measures of risk that admit a representation in terms of *σ-additive probability measures*. Such a risk measure ρ can be represented by a penalty function α that is infinite outside the set $\mathcal{M}_1 := \mathcal{M}_1(\Omega, \mathcal{F})$:

$$\rho(X) = \sup_{Q\in\mathcal{M}_1} (E_Q[-X] - \alpha(Q)). \tag{4.24}$$

In this case, one can no longer expect that the supremum above is attained. This is illustrated by Example 4.8 if X does not attain its infimum.

A representation (4.24) in terms of probability measures is closely related to certain continuity properties of ρ. We first examine a necessary condition of "continuity from above."

Lemma 4.21. *A convex risk measure ρ that admits a representation* (4.24) *on $\mathcal{M}_1$ is continuous from above in the sense that*

$$X_n \searrow X \quad \Longrightarrow \quad \rho(X_n) \nearrow \rho(X). \tag{4.25}$$

Moreover, continuity from above is equivalent to the "Fatou property" of lower semicontinuity with respect to bounded pointwise convergence: If (X_n) is a bounded sequence in $\mathcal{X}$ that converges pointwise to $X \in \mathcal{X}$, then

$$\rho(X) \le \liminf_{n\uparrow\infty} \rho(X_n). \tag{4.26}$$

Proof. First, we show (4.26) under the assumption that ρ has a representation in terms of probability measures. Dominated convergence implies that $E_Q[X_n] \to E_Q[X]$ for each $Q \in \mathcal{M}_1$. Hence,

$$\begin{aligned}\rho(X) &= \sup_{Q\in\mathcal{M}_1} (\lim_{n\uparrow\infty} E_Q[-X_n] - \alpha(Q)) \\ &\le \liminf_{n\uparrow\infty} \sup_{Q\in\mathcal{M}_1} (E_Q[-X_n] - \alpha(Q)) \\ &= \liminf_{n\uparrow\infty} \rho(X_n).\end{aligned}$$

In order to show the equivalence of (4.26) and (4.25), let us first assume (4.26). By monotonicity, $\rho(X_n) \le \rho(X)$ for each n if $X_n \searrow X$, and so $\rho(X_n) \nearrow \rho(X)$ follows.

Now we assume continuity from above. Let (X_n) be a bounded sequence in $\mathscr{X}$ that converges pointwise to X. Define $Y_m := \sup_{n\ge m} X_n \in \mathscr{X}$. Then Y_m decreases to X. Since $\rho(X_n) \ge \rho(Y_n)$ by monotonicity, condition (4.25) yields that

$$\liminf_{n\uparrow\infty} \rho(X_n) \ge \lim_{n\uparrow\infty} \rho(Y_n) = \rho(X). \qquad \square$$

The following theorem gives a strong sufficient condition, which guarantees that any penalty function for ρ is concentrated on the set $\mathscr{M}_1$ of probability measures. This condition is "continuity from below" rather than from above; we will see several examples in the next sections.

Theorem 4.22. *For a convex risk measure ρ on $\mathscr{X}$, the following two conditions are equivalent:*

(a) *ρ is* continuous from below *in the sense that*

$$X_n \nearrow X \text{ pointwise on } \Omega \quad \Longrightarrow \quad \rho(X_n) \searrow \rho(X).$$

(b) *The minimal penalty function $\alpha^{\min}$ (and hence every other penalty function representing ρ) is concentrated on the class $\mathscr{M}_1$ of probability measures, that is,*

$$\alpha^{\min}(Q) < \infty \quad \Longrightarrow \quad Q \text{ is } \sigma\text{-additive.}$$

In particular, we have

$$\rho(X) = \max_{Q\in\mathscr{M}_1} \big(E_Q[\,-X\,] - \alpha^{\min}(Q)\big), \quad X \in \mathscr{X},$$

whenever one of these two equivalent conditions is satisfied.

For the proof of this theorem, we need the following lemma.

Lemma 4.23. *Let ρ be a convex risk measure on $\mathscr{X}$ that is represented by the penalty function α on $\mathscr{M}_{1,f}$, and consider the level sets*

$$\Lambda_c := \{Q \in \mathscr{M}_{1,f} \mid \alpha(Q) \le c\}, \quad \text{for } c > -\rho(0) = \inf_{Q\in\mathscr{M}_{1,f}} \alpha(Q).$$

For any sequence (X_n) in $\mathscr{X}$ such that $0 \le X_n \le 1$, the following two conditions are equivalent:

(a) *$\rho(\lambda X_n) \longrightarrow \rho(\lambda)$ for each $\lambda \ge 1$.*

(b) *$\inf_{Q\in\Lambda_c} E_Q[\,X_n\,] \longrightarrow 1$ for all $c > -\rho(0)$.*

Proof. (a)⇒(b): In a first step, we show that for all $Y \in \mathscr{X}$,

$$\inf_{Q\in\Lambda_c} E_Q[\,Y\,] \geq -\frac{c+\rho(\lambda Y)}{\lambda} \quad \text{for all } \lambda > 0. \tag{4.27}$$

Indeed, since α represents ρ, we have for $Q \in \Lambda_c$,

$$c \geq \alpha(Q) \geq E_Q[\,-\lambda Y\,] - \rho(\lambda Y),$$

and dividing by $-\lambda$ yields (4.27).

Now consider a sequence (X_n) that satisfies (a). Then (4.27) shows that for all $\lambda \geq 1$,

$$\liminf_{n\uparrow\infty} \inf_{Q\in\Lambda_c} E_Q[\,X_n\,] \geq -\lim_{n\uparrow\infty} \frac{c+\rho(\lambda X_n)}{\lambda} = 1 - \frac{c+\rho(0)}{\lambda}.$$

Taking $\lambda \uparrow \infty$ and assuming $X_n \leq 1$ proves (b).

(b)⇒(a): Clearly, for all $Y \in \mathscr{X}$ with $0 \leq Y \leq 1$,

$$\rho(\lambda) \leq \rho(\lambda Y) = \sup_{Q\in\mathscr{M}_{1,f}} (E_Q[\,-\lambda Y\,] - \alpha(Q)).$$

Since $E_Q[\,-\lambda Y\,] \leq 0$ for all Q, only those Q can contribute to the supremum on the right-hand side for which

$$\alpha(Q) \leq \delta - \rho(\lambda) = \delta + \lambda - \rho(0) =: c,$$

where $\delta > 0$ is an arbitrary but fixed constant. Hence,

$$\rho(\lambda Y) = \sup_{Q\in\Lambda_c} (E_Q[\,-\lambda Y\,] - \alpha(Q)), \tag{4.28}$$

and in particular for all n,

$$\rho(\lambda X_n) = \sup_{Q\in\Lambda_c} (E_Q[\,-\lambda X_n\,] - \alpha(Q)).$$

But condition (b) implies that $E_Q[\,-\lambda X_n\,]$ converges to $-\lambda$ uniformly in $Q \in \Lambda_c$. Therefore, (4.28) yields that

$$\rho(\lambda X_n) = \sup_{Q\in\Lambda_c} (E_Q[\,-\lambda X_n\,] - \alpha(Q)) \longrightarrow \sup_{Q\in\Lambda_c} (-\lambda - \alpha(Q)) = \rho(\lambda),$$

as $n \uparrow \infty$. □

The proof of our theorem will also rely on *Dini's lemma*, which we recall here for the convenience of the reader.

Lemma 4.24. *On a compact set, a sequence of continuous functions f_n increasing to a continuous function f converges even uniformly.*

Proof. For $\varepsilon > 0$, the compact sets $K_n := \{f_n \le f - \varepsilon\}$ satisfy $\bigcap_n K_n = \emptyset$. Hence, there must be some $n_0 \in \mathbb{N}$ such that $K_n = \emptyset$ for all $n \ge n_0$. □

Proof of Theorem 4.22. To prove the implication (a)⇒(b), recall that $Q \in \mathcal{M}_{1,f}$ is σ-additive if and only if $Q[A_n] \nearrow 1$ for any increasing sequence of events $A_n \in \mathcal{F}$ such that $\bigcup_n A_n = \Omega$. Thus, our claim is implied by the implication (a)⇒(b) of Lemma 4.23 if we take $X_n := \mathbb{1}_{A_n}$.

We now prove the implication (b)⇒(a) of our theorem. Suppose $X_n \nearrow X$ pointwise on Ω. We need to show that $\rho(X_n) \searrow \rho(X)$. By cash invariance, we may assume without loss of generality that $X_n \ge 0$ for all n. As in (4.28), we see that

$$\rho(X_n) = \max_{Q \in \mathcal{M}_{1,f}} (E_Q[-X_n] - \alpha^{\min}(Q)) = \max_{Q \in \Lambda_c}(E_Q[-X_n] - \alpha^{\min}(Q)), \tag{4.29}$$

where $c > 1 - \rho(X)$ and $\Lambda_c = \{Q \in \mathcal{M}_{1,f} \mid \alpha^{\min}(Q) \le c\}$. We will show below that $\Lambda_c \subseteq \mathcal{M}_1$ implies that

$$E_Q[-X_n] \longrightarrow E_Q[-X] \quad \text{uniformly in } Q \in \Lambda_c. \tag{4.30}$$

Together with the representation (4.29), this will imply the desired convergence $\rho(X_n) \to \rho(X)$.

To prove (4.30), recall first from Appendix G, and in particular, from Definition G.20, that $\mathcal{M}_{1,f}$ belongs to the larger vector space $ba := ba(\Omega, \mathcal{F})$ of all finitely additive set functions $\mu : \mathcal{F} \to \mathbb{R}$ with finite total variation $\|\mu\|_{\text{var}}$. In fact, ba can be identified with the topological dual of the Banach space $\mathcal{X}$ with respect to $\|\cdot\|$; see Theorem G.21. Since we can write

$$\mathcal{M}_{1,f} = \{\mu \in ba \mid \mu(\Omega) = 1 \text{ and } \mu(A) \ge 0 \text{ for all } A \in \mathcal{F}\},$$

it is clear that $\mathcal{M}_{1,f}$ is a bounded and weak* closed set in ba. Hence, $\mathcal{M}_{1,f}$ is weak* compact by the Banach–Alaoglu theorem (see Theorem H.10). Since $\alpha^{\min}$ is weak* lower semicontinuous on $\mathcal{M}_{1,f}$ as supremum of the weak* continuous maps $Q \mapsto E_Q[-Y]$ with $Y \in \mathcal{A}_\rho$, the level set Λ_c is also weak* compact.

After these preparations, we can now prove (4.30). Clearly, the functions $\ell_n(Q) := E_Q[-X_n]$ form a decreasing sequence of weak* continuous functions on Λ_c. Moreover, if $\Lambda_c \subseteq \mathcal{M}_1$, we even have $\ell_n(Q) \searrow \ell(Q) := E_Q[-X]$ for each $Q \in \Lambda_c$, due to the monotone convergence theorem. By the established compactness of Λ_c, (4.30) thus follows from Dini' s lemma. □

Remark 4.25. Let ρ be a convex risk measure that is continuous from below. Then ρ is also continuous from above, as can be seen by combining Theorem 4.22 and Lemma 4.21. ◇

Exercise 4.2.3. Show that for a convex risk measure ρ on $\mathcal{X}$, the following two conditions are equivalent:

(a) ρ is continuous from below.
(b) ρ satisfies the following *Lebesgue property*: $\rho(X_n) \longrightarrow \rho(X)$ whenever (X_n) is a bounded sequence in $\mathscr{X}$ that converges pointwise to X. ◇

Exercise 4.2.4. Show that for a convex risk measure ρ on $\mathscr{X}$, the following two conditions are equivalent:
(a) ρ is continuous from below.
(b) For every $c > -\rho(0)$, the coherent risk measures

$$\rho_c(X) := \sup_{Q\in\Lambda_c} E_Q[\,-X\,]$$

are continuous from below, where $\Lambda_c = \{Q \in \mathscr{M}_{1,f} \mid \alpha^{\min}(Q) \leq c\}$. ◇

Example 4.26. Let us consider a utility function u on $\mathbb{R}$, a probability measure $Q \in \mathscr{M}_1(\Omega, \mathscr{F})$, and fix some threshold $c \in \mathbb{R}$. As in Example 4.10, we suppose that a position X is acceptable if its expected utility $E_Q[\,u(X)\,]$ is bounded from below by $u(c)$. Alternatively, we can introduce the convex increasing loss function $\ell(x) = -u(-x)$ and define the convex set of acceptable positions

$$\mathscr{A} := \{X \in \mathscr{X} \mid E_Q[\,\ell(-X)\,] \leq x_0\},$$

where $x_0 := -u(c)$. Let $\rho := \rho_{\mathscr{A}}$ denote the convex risk measure induced by $\mathscr{A}$. In Section 4.11, we will show that ρ is continuous from below, and we will derive a formula for its minimal penalty function. ◇

4.3 Convex risk measures on $C_b(\Omega)$

In this section, we discuss the robust representation of convex risk measures in a *topological* setting. To this end, we will assume throughout this section that Ω is a separable metric space and that $\mathscr{F}$ is the σ-field of Borel sets. As before, $\mathscr{X}$ is the linear space of all bounded measurable functions on $(\Omega, \mathscr{F})$. We denote by $C_b(\Omega)$ the subspace of bounded continuous functions on Ω, and we focus on the representation of convex risk measures viewed as functionals on $C_b(\Omega)$.

Proposition 4.27. *Let ρ be a convex risk measure on $\mathscr{X}$ such that*

$$\rho(X_n) \searrow \rho(\lambda) \quad \textit{for any sequence } (X_n) \textit{ in } C_b(\Omega) \textit{ that increases to a constant } \lambda > 0. \tag{4.31}$$

Then there exists a penalty function α on $\mathscr{M}_1$ such that

$$\rho(X) = \max_{Q\in\mathscr{M}_1} (E_Q[\,-X\,] - \alpha(Q)) \quad \textit{for } X \in C_b(\Omega). \tag{4.32}$$

In fact, one can take

$$\alpha(Q) := \inf\{\alpha^{\min}(\widetilde{Q}) \mid E_{\widetilde{Q}}[\cdot] = E_Q[\cdot] \text{ on } C_b(\Omega)\}. \tag{4.33}$$

Proof. Let $\alpha^{\min}$ be the minimal penalty function of ρ on $\mathcal{M}_{1,f}$. We show that for any $\widetilde{Q}$ with $\alpha^{\min}(\widetilde{Q}) < \infty$ there exists $Q \in \mathcal{M}_1$ such that $E_{\widetilde{Q}}[X] = E_Q[X]$ for all $X \in C_b(\Omega)$. Take a sequence (Y_n) in $C_b(\Omega)$ that increases to some $Y \in C_b(\Omega)$, and choose $\delta > 0$ such that $X_n := 1 + \delta(Y_n - Y) \geq 0$ for all n. Clearly, (X_n) satisfies condition (a) of Lemma 4.23. This lemma yields that $E_{\widetilde{Q}}[X_n] \to 1$, and in turn,

$$E_{\widetilde{Q}}[Y_n] \nearrow E_{\widetilde{Q}}[Y].$$

This continuity property of the linear functional $E_{\widetilde{Q}}[\cdot]$ on $C_b(\Omega)$ implies, via the Daniell–Stone representation theorem as stated in Theorem G.19 that $E_{\widetilde{Q}}[\cdot]$ coincides on $C_b(\Omega)$ with the integral with respect to a σ-additive measure Q. Taking α as in (4.33) gives the result. □

Remark 4.28. If Ω is *compact*, then *any* convex risk measure admits a representation (4.32) on the space $C_b(\Omega) = C(\Omega)$. Indeed, if (X_n) is a sequence in $C_b(\Omega)$ that increases to a constant λ, then this convergence is even uniform by Lemma 4.24. Since ρ is Lipschitz continuous on $C(\Omega)$ by Lemma 4.3, it satisfies condition (4.31).

Alternatively, we could argue as in Remark 4.18 and apply the general duality theorem for the Fenchel–Legendre transform to the convex functional ρ on the Banach space $C(\Omega)$. Just note that any continuous functional ℓ on $C(\Omega)$ that is positive and normalized is of the form $\ell(X) = E_Q[X]$ for some probability measure $Q \in \mathcal{M}_1$; see Theorem G.18. ◇

Definition 4.29. A convex risk measure ρ on $\mathcal{X}$ is called *tight* if there exists an increasing sequence $K_1 \subseteq K_2 \subseteq \cdots$ of compact subsets of Ω such that

$$\rho(\lambda \mathbb{1}_{K_n}) \longrightarrow \rho(\lambda) \quad \text{for all } \lambda \geq 1.$$ △

Note that every convex risk measure is tight if Ω is compact.

Proposition 4.30. *Suppose that the convex risk measure ρ on $\mathcal{X}$ is tight. Then* (4.31) *holds and the conclusion of Proposition* 4.27 *is valid. Moreover, if Ω is a Polish space and α is a penalty function on $\mathcal{M}_1$ such that*

$$\rho(X) = \sup_{Q \in \mathcal{M}_1} (E_Q[-X] - \alpha(Q)) \quad \text{for } X \in C_b(\Omega),$$

then the level sets $\Lambda_c = \{Q \in \mathcal{M}_1 \mid \alpha(Q) \leq c\}$ are relatively compact for the weak topology on $\mathcal{M}_1$.

Proof. We may assume without loss of generality that ρ is normalized. First, we show (4.31). Suppose $X_n \in C_b(\Omega)$ are such that $X_n \nearrow \lambda > 0$.

To exploit the tightness of ρ, we first consider the case in which $\lambda > 1$. Due to the cash invariance of ρ, there is no loss of generality in assuming $X_n \geq 0$ for all n. We will show that $\rho(X_n) \leq \rho(\lambda) + 2\varepsilon$ eventually, where $\varepsilon \in (0, \lambda - 1)$ is given. The tightness of ρ yields a compact set K_N such that

$$\rho((\lambda - \varepsilon)\mathbb{1}_{K_N}) \leq \rho(\lambda - \varepsilon) + \varepsilon = \rho(\lambda) + 2\varepsilon.$$

By Dini's lemma, as recalled in Lemma 4.24, there exists some $n_0 \in \mathbb{N}$ such that $\lambda - \varepsilon \leq X_n$ on K_N, and hence $X_n \geq (\lambda - \varepsilon)\mathbb{1}_{K_N}$, for all $n \geq n_0$. Thus, monotonicity implies

$$\rho(X_n) \leq \rho((\lambda - \varepsilon)\mathbb{1}_{K_N}) \leq \rho(\lambda) + 2\varepsilon.$$

This proves (4.31) for $\lambda > 1$.

Now we use the normalization, monotonicity, and convexity of ρ to establish (4.31) for $0 < \lambda \leq 1$. To this end, we take $\lambda_1 > 1$ and let $\gamma = \lambda/\lambda_1 \in (0, 1)$. Since $Y_n := \gamma^{-1} X_n \nearrow \lambda_1$, we know from the first part of this proof that $\rho(Y_n) \searrow \rho(\lambda_1)$. Using that ρ is normalized, monotone, and convex gives

$$\begin{aligned} -\lambda = \rho(\lambda) &\leq \rho(X_n) = \rho(\gamma Y_n + (1-\gamma)0) \leq \gamma\rho(Y_n) + (1-\gamma)\rho(0) \\ &= \gamma\rho(Y_n) \searrow \gamma\rho(\lambda_1) = -\gamma\lambda_1 = -\lambda. \end{aligned}$$

This concludes the proof of (4.31).

To prove the relative compactness of Λ_c, we will show that for any $\varepsilon > 0$ there exists a compact set $K_\varepsilon \subseteq \Omega$ such that for all $c > 0$,

$$\inf_{Q \in \Lambda_c} Q[\, K_\varepsilon\,] \geq 1 - \varepsilon(c + 1).$$

The relative compactness of Λ_c will then be an immediate consequence of Prohorov's characterization of weakly compact sets in $\mathcal{M}_1$, as stated in Theorem G.9. We fix a countable dense set $\{\omega_1, \omega_2, \dots\}$ in Ω and a complete metric δ that generates the topology of Ω. For $r > 0$, we define continuous functions $\Delta_i^r : \Omega \to [0, 1]$ by

$$\Delta_i^r(\omega) := 1 - \frac{\delta(\omega, \omega_i) \wedge r}{r}.$$

The function Δ_i^r is dominated by the indicator function of the closed metric ball

$$\overline{B}_r(\omega_i) := \{\omega \in \Omega \mid \delta(\omega, \omega_i) \leq r\}.$$

Let

$$X_n^r(\omega) := \max_{i \leq n} \Delta_i^r(\omega).$$

Clearly, X_n^r is continuous and satisfies $0 \leq X_n^r \leq 1$ as well as $X_n^r \nearrow 1$ for $n \uparrow \infty$.

According to (4.27), we have for all $c > 0$ and $\lambda > 0$,

$$\inf_{Q\in\Lambda_c} Q\left[\bigcup_{i=1}^{n}\overline{B}_r(\omega_i)\right] \geq \inf_{Q\in\Lambda_c} E_Q[X_n^r] \geq -\frac{c+\rho(\lambda X_n^r)}{\lambda}.$$

Now we take $\lambda_k := 2^k/\varepsilon$ and $r_k := 1/k$. Using (4.31), as established in the first part of this proof, yields $n_k \in \mathbb{N}$ such that

$$\rho(\lambda_k X_{n_k}^{r_k}) \leq \rho(\lambda_k) + 1 = -\lambda_k + 1,$$

and thus

$$\sup_{Q\in\Lambda_c} Q\left[\bigcap_{i=1}^{n_k}\Omega\setminus\overline{B}_{r_k}(\omega_i)\right] = 1 - \inf_{Q\in\Lambda_c} Q\left[\bigcup_{i=1}^{n_k}\overline{B}_{r_k}(\omega_i)\right] \leq 1 + \frac{c+\rho(\lambda_k X_{n_k}^{r_k})}{\lambda_k} \leq \varepsilon 2^{-k}(c+1).$$

We let

$$K_\varepsilon := \bigcap_{k=1}^{\infty}\bigcup_{i=1}^{n_k}\overline{B}_{r_k}(\omega_i).$$

Then, for each $Q \in \Lambda_c$,

$$\begin{aligned} Q[K_\varepsilon] &= 1 - Q\left[\bigcup_{k=1}^{\infty}\bigcap_{i=1}^{n_k}\Omega\setminus\overline{B}_{r_k}(\omega_i)\right] \\ &\geq 1 - \sum_{k=1}^{\infty}\varepsilon 2^{-k}(c+1) \\ &= 1 - \varepsilon(c+1). \end{aligned}$$

The reader may notice that K_ε is closed, totally bounded, and hence compact. A short proof of this fact goes as follows: Let (x_j) be a sequence in K_ε. We must show that (x_j) has a convergent subsequence. Since K_ε is covered by $\overline{B}_{r_k}(\omega_1), \dots, \overline{B}_{r_k}(\omega_{n_k})$ for each k, there exists some $i_k \leq n_k$ such that infinitely many x_j are contained in $\overline{B}_{r_k}(\omega_{i_k})$. A diagonalization argument yields a single subsequence $(x_{j'})$ that for each k is contained in some $\overline{B}_{r_k}(\omega_{i_k})$. Thus, $(x_{j'})$ is a Cauchy sequence with respect to the complete metric δ, and hence converging to some element $\omega \in \Omega$. □

Remark 4.31. Note that the representation (4.32) does not necessarily extend from $C_b(\Omega)$ to the space $\mathscr{X}$ of all bounded measurable functions. Suppose in fact that Ω is compact but not finite. Then condition (4.31) holds as explained in Remark 4.28. Moreover, there is a finitely additive $Q_0 \in \mathscr{M}_{1,f}$ that does not belong to $\mathscr{M}_1$; see Example G.23. The proof of Proposition 4.27 shows that there is some $\widetilde{Q} \in \mathscr{M}_1$ such that the coherent risk measure ρ defined by $\rho(X) := E_{Q_0}[-X]$ coincides with $E_{\widetilde{Q}}[-X]$ for $X \in C_b(\Omega)$. But ρ does not admit a representation of the form

$$\rho(X) = \sup_{Q \in \mathcal{M}_1} (E_Q[-X] - \alpha(Q)) \quad \text{for all } X \in \mathcal{X}.$$

In fact, this would imply

$$\alpha(Q) \geq E_{Q_0}[X] - E_Q[X]$$

for $Q \in \mathcal{M}_1$ and any $X \in \mathcal{X}$, hence $\alpha(Q) = \infty$ for any $Q \in \mathcal{M}_1$. ◇

4.4 Convex risk measures on L^∞

For the next five sections, we fix a probability measure P on $(\Omega, \mathcal{F})$ and consider risk measures ρ such that

$$\rho(X) = \rho(Y) \quad \text{if } X = Y \text{ } P\text{-a. s.} \tag{4.34}$$

In the present section, only the null sets of P will matter.

Lemma 4.32. *Let ρ be a convex risk measure that satisfies* (4.34) *and that is represented by a penalty function α as in* (4.17). *Then $\alpha(Q) = +\infty$ for any $Q \in \mathcal{M}_{1,f}(\Omega, \mathcal{F})$ that is not absolutely continuous with respect to P.*

Proof. If $Q \in \mathcal{M}_{1,f}(\Omega, \mathcal{F})$ is not absolutely continuous with respect to P, then there exists $A \in \mathcal{F}$ such that $Q[A] > 0$ but $P[A] = 0$. Take any $X \in \mathcal{A}_\rho$, and define $X_n := X - n\mathbb{1}_A$. Then $\rho(X_n) = \rho(X)$, that is, X_n is again contained in $\mathcal{A}_\rho$. Hence,

$$\alpha(Q) \geq \alpha^{\min}(Q) \geq E_Q[-X_n] = E_Q[-X] + n\,Q[A] \longrightarrow \infty$$

as $n \uparrow \infty$. □

In view of (4.34), we can identify $\mathcal{X}$ with the Banach space $L^\infty := L^\infty(\Omega, \mathcal{F}, P)$. Let us denote by

$$\mathcal{M}_1(P) := \mathcal{M}_1(\Omega, \mathcal{F}, P)$$

the set of all probability measures on $(\Omega, \mathcal{F})$ that are absolutely continuous with respect to P. The following theorem characterizes those convex risk measures on L^∞ that can be represented by a penalty function concentrated on probability measures, and hence on $\mathcal{M}_1(P)$, due to Lemma 4.32.

Theorem 4.33. *Suppose $\rho : L^\infty \to \mathbb{R}$ is a convex risk measure. Then the following conditions are equivalent:*

(a) *ρ can be represented by some penalty function on $\mathcal{M}_1(P)$.*

(b) *ρ can be represented by the restriction of the minimal penalty function $\alpha^{\min}$ to $\mathcal{M}_1(P)$:*

$$\rho(X) = \sup_{Q\in\mathcal{M}_1(P)} (E_Q[-X] - \alpha^{\min}(Q)), \quad X \in L^\infty. \tag{4.35}$$

(c) *ρ is continuous from above: If $X_n \searrow X$ P-a. s., then $\rho(X_n) \nearrow \rho(X)$.*

(d) *ρ has the following Fatou property: for any uniformly bounded sequence (X_n) in L^∞ that converges P-a. s. to some X,*

$$\rho(X) \le \liminf_{n\uparrow\infty} \rho(X_n).$$

(e) *ρ is lower semicontinuous for the weak* topology $\sigma(L^\infty, L^1)$.*

(f) *The acceptance set $\mathcal{A}_\rho$ of ρ is weak* closed in L^∞, that is, $\mathcal{A}_\rho$ is closed with respect to the topology $\sigma(L^\infty, L^1)$.*

Proof. The implication s (b)⇒(a) and (e)⇒(f) are obvious, and (a)⇒(c)⇔(d) follows as in Lemma 4.21, replacing pointwise convergence by P-a. s. convergence.

(c)⇒(e): We have to show that $\mathcal{C} := \{\rho \le c\}$ is weak* closed for $c \in \mathbb{R}$. To this end, let

$$\mathcal{C}_r := \mathcal{C} \cap \{X \in L^\infty \mid \|X\|_\infty \le r\}$$

for $r > 0$. If (X_n) is a sequence in $\mathcal{C}_r$ converging in L^1 to some random variable X, then there is a subsequence that converges P-a. s., and the Fatou property of ρ implies that $X \in \mathcal{C}_r$. Hence, $\mathcal{C}_r$ is closed in L^1, and Lemma H.12 implies that $\mathcal{C} := \{\rho \le c\}$ is weak* closed.

(f)⇒(b): We fix some $X \in L^\infty$ and let

$$m = \sup_{Q\in\mathcal{M}_1(P)} (E_Q[-X] - \alpha^{\min}(Q)). \tag{4.36}$$

In view of Theorem 4.16, it suffices to show that $m \ge \rho(X)$, or equivalently that $m + X \in \mathcal{A}_\rho$. Suppose by way of contradiction that $m + X \notin \mathcal{A}_\rho$. Since the nonempty convex set $\mathcal{A}_\rho$ is weak* closed by assumption, we may apply the Hahn–Banach separation theorem in the form of Theorem H.4 in the locally convex space $(L^\infty, \sigma(L^\infty, L^1))$ with $\mathcal{C} := \mathcal{A}_\rho$ and $\mathcal{B} := \{m+X\}$. We obtain a continuous linear functional ℓ on $(L^\infty, \sigma(L^\infty, L^1))$ such that

$$\beta := \inf_{Y\in\mathcal{A}_\rho} \ell(Y) > \ell(m + X) =: \gamma > -\infty. \tag{4.37}$$

By Proposition H.6, ℓ is of the form $\ell(Y) = E[\,YZ\,]$ for some $Z \in L^1$. We claim that $Z \ge 0$ P-a. s. To show this, fix $Y \ge 0$ and note that $\rho(\lambda Y) \le \rho(0)$ for $\lambda \ge 0$, by monotonicity. Hence, $\lambda Y + \rho(0) \in \mathcal{A}_\rho$ for all $\lambda \ge 0$. It follows that

$$-\infty < \gamma < \ell(\lambda Y + \rho(0)) = \lambda\ell(Y) + \ell(\rho(0)).$$

Taking $\lambda \uparrow \infty$ yields that $\ell(Y) \geq 0$ and in turn that $Z \geq 0$ P-almost surely. Moreover, $P[Z > 0] > 0$ since ℓ is nonzero. Thus,

$$\frac{dQ_0}{dP} := \frac{Z}{E[Z]}$$

defines a probability measure $Q_0 \in \mathcal{M}_1(P)$. By (4.37), we see that

$$\alpha^{\min}(Q_0) = \sup_{Y \in \mathcal{A}_\rho} E_{Q_0}[-Y] = -\frac{\beta}{E[Z]}.$$

However,

$$E_{Q_0}[X] + m = \frac{\ell(m+X)}{E[Z]} = \frac{\gamma}{E[Z]} < \frac{\beta}{E[Z]} = -\alpha^{\min}(Q_0),$$

in contradiction to (4.36). Hence, $m + X$ must belong to $\mathcal{A}_\rho$, and thus $m \geq \rho(X)$. □

The theorem shows that any convex risk measure on L^∞ that is continuous from above arises in the following manner. We consider any probabilistic model $Q \in \mathcal{M}_1(P)$, but these models are taken more or less seriously as described by the penalty function. Thus, the value $\rho(X)$ is computed as the worst case, over all models $Q \in \mathcal{M}_1(P)$, of the expected loss $E_Q[-X]$, but reduced by $\alpha(Q)$. In the following example, the given model P is the one that is taken most seriously, and the penalty function $\alpha(Q)$ is proportional to the deviation of Q from P, measured by the relative entropy, which is sometimes also called Kullback–Leibler divergence.

Example 4.34. Consider the penalty function $\alpha : \mathcal{M}_1(P) \to (0, \infty]$ defined by

$$\alpha(Q) := \frac{1}{\beta} H(Q|P),$$

where $\beta > 0$ is a given constant and

$$H(Q|P) = E_Q\left[\log \frac{dQ}{dP}\right]$$

is the relative entropy of $Q \in \mathcal{M}_1(P)$ with respect to P; see Definition C.1. The corresponding *entropic risk measure* ρ_β is given by

$$\rho_\beta(X) = \sup_{Q \in \mathcal{M}_1(P)} \left(E_Q[-X] - \frac{1}{\beta} H(Q|P) \right). \tag{4.38}$$

The variational principle for the relative entropy as stated in Theorem C.5 shows that

$$E_Q[-X] - \frac{1}{\beta} H(Q|P) \leq \frac{1}{\beta} \log E[e^{-\beta X}],$$

and that the upper bound is attained by the measure with the density $e^{-\beta X}/E[\,e^{-\beta X}\,]$. Thus, the entropic risk measure takes the form

$$\rho_\beta(X) = \frac{1}{\beta}\log E[\,e^{-\beta X}\,].$$

Note that α is in fact the minimal penalty function representing ρ_β, since Theorem C.5 implies

$$\alpha^{\min}(Q) = \sup_{X\in L^\infty}\Big(E_Q[\,-X\,] - \frac{1}{\beta}\log E[\,e^{-\beta X}\,]\Big) = \frac{1}{\beta}H(Q|P).$$

A financial interpretation of the entropic risk measure in terms of shortfall risk will be discussed in Example 4.125. ◇

Exercise 4.4.1. Show that the entropic risk measure ρ_β converges to the worst-case risk measure for $\beta \uparrow \infty$ and to the expected loss under P for $\beta \downarrow 0$. ◇

Exercise 4.4.2. Show that the entropic risk measure ρ_β is not coherent for any $\beta \in (0,\infty)$. To this end, take any $X \in \mathscr{X}$ that has positive expectation but carries some downside risk, that is, $E[\,X\,] > 0$ and $P[\,X < 0\,] > 0$. Show that λX is acceptable for sufficiently small $\lambda > 0$, but not acceptable for large λ. Thus, the acceptance set $\mathscr{A}_{\rho_\beta}$ is not a cone. ◇

The following corollary characterizes those convex risk measures on L^∞ that satisfy the property of continuity from below, which is stronger than continuity from above.

Corollary 4.35. *For a convex risk measure ρ on L^∞, the following conditions are equivalent:*

(a) *ρ is continuous from below: $X_n \nearrow X \Longrightarrow \rho(X_n) \searrow \rho(X)$.*
(b) *ρ satisfies the Lebesgue property: $\rho(X_n) \longrightarrow \rho(X)$ whenever (X_n) is a bounded sequence in L^∞ that converges P-a. s. to X.*
(c) *The minimal penalty function $\alpha^{\min}$ is concentrated on $\mathscr{M}_1(P)$, that is, $\alpha^{\min}(Q) < \infty$ implies $Q \in \mathscr{M}_1(P)$.*

In particular, we have

$$\rho(X) = \max_{Q\in\mathscr{M}_1(P)}\big(E_Q[\,-X\,] - \alpha^{\min}(Q)\big), \quad X \in L^\infty, \tag{4.39}$$

whenever one of these three equivalent conditions is satisfied.

Proof. The equivalence of conditions (a) and (b) was shown in Exercise 4.2.3. The equivalence of conditions (a) and (c) as well as (4.39) follow from Theorem 4.22 and Lemma 4.32. □

Exercise 4.4.3. Show that the three conditions in Corollary 4.35 are equivalent to the following fourth condition:

(d) For each $c \in \mathbb{R}$, the level set $\Lambda_c := \{Q \mid \alpha^{\min}(Q) \le c\}$ is contained in $\mathcal{M}_1(P)$ and the corresponding set of densities,

$$\left\{\frac{dQ}{dP} \;\middle|\; Q \in \Lambda_c\right\},$$

is weakly compact in $L^1(\Omega, \mathcal{F}, P)$.

Hint: Use Lemma 4.23 and the Dunford–Pettis theorem (Theorem H.19). ◇

Example 4.36. Let $g : [0, \infty) \to \mathbb{R} \cup \{+\infty\}$ be a lower semicontinuous convex function satisfying $g(1) < \infty$ and the superlinear growth condition $g(x)/x \to +\infty$ as $x \uparrow \infty$. Associated with it is the *g-divergence*,

$$I_g(Q|P) := E\left[g\left(\frac{dQ}{dP}\right)\right], \quad Q \in \mathcal{M}_1(P). \tag{4.40}$$

The g-divergence $I_g(Q|P)$ quantifies the deviation of the hypothetical model Q from the reference measure P. Thus,

$$\alpha_g(Q) := I_g(Q|P), \quad Q \in \mathcal{M}_1(P),$$

is a natural choice for a penalty function. The resulting risk measure

$$\rho_g(X) := \sup_{Q \ll P}(E_Q[-X] - I_g(Q|P)) \tag{4.41}$$

is sometimes called a *divergence risk measure*. Note that, for $g(x) = \frac{1}{\beta}x \log x$, ρ_g is just the entropic risk measure discussed in Example 4.34. Divergence risk measures will be discussed in more detail in Section 4.11. ◇

Exercise 4.4.4. Show that the risk measure in (4.41) is continuous from below and that $\alpha_g(Q) = I_g(Q|P)$, $Q \in \mathcal{M}_1(P)$, is its minimal penalty function. In particular, the supremum in (4.41) is in fact a maximum.

Hint: It can be helpful to use Exercise 4.4.3 and the criterion of de la Vallée–Poussin as stated in Proposition H.16. ◇

Theorem 4.33 takes the following form for *coherent* risk measures; the proof is the same as the one for Corollary 4.19.

Corollary 4.37. *A coherent risk measure on L^∞ can be represented by a set $\mathcal{Q} \subseteq \mathcal{M}_1(P)$ if and only if the equivalent conditions of Theorem 4.33 are satisfied. In this case, the maximal representing subset of $\mathcal{M}_1(P)$ is given by*

$$\mathcal{Q}^{\max} \cap \mathcal{M}_1(P) = \{Q \in \mathcal{M}_1(P) \mid \alpha^{\min}(Q) = 0\}.$$

Let us also state a characterization of those coherent risk measures on L^∞ that are continuous from below. It follows by combining Corollaries 4.35 and 4.37 with Exercise 4.4.3.

Corollary 4.38. *For a coherent risk measure ρ on L^∞, the following properties are equivalent:*

(a) *ρ is continuous from below: $X_n \nearrow X \Longrightarrow \rho(X_n) \searrow \rho(X)$.*
(b) *ρ satisfies the Lebesgue property: $\rho(X_n) \longrightarrow \rho(X)$ whenever (X_n) is a bounded sequence in L^∞ that converges P-a. s. to X.*
(c) *We have $\mathcal{Q}^{\max} \subseteq \mathcal{M}_1(P)$.*
(d) *The set of densities*

$$\left\{ \frac{dQ}{dP} \;\middle|\; Q \in \mathcal{Q}^{\max} \cap \mathcal{M}_1(P) \right\}$$

is weakly compact in $L^1(\Omega, \mathcal{F}, P)$.

If these equivalent conditions hold, then the representation

$$\rho(X) = \max_{Q \in \mathcal{Q}^{\max}} E_Q[-X], \quad X \in L^\infty,$$

involves only σ-additive probability measures.

We now give three examples of coherent risk measures, which will be studied in more detail in Section 4.5.

Example 4.39. In our present context, where we require condition (4.34), the *worst-case risk measure* takes the form

$$\rho_{\max}(X) := -\operatorname{ess\,inf} X = \inf\{m \in \mathbb{R} \mid X + m \geq 0 \; P\text{-a. s.}\}.$$

One can easily check that $\rho_{\max}$ is coherent and satisfies the Fatou property. Moreover, the acceptance set of $\rho_{\max}$ is equal to the positive cone L^∞_+ in L^∞, and this implies $\alpha^{\min}(Q) = 0$ for any $Q \in \mathcal{M}_1(P)$. Thus,

$$\rho_{\max}(X) = \sup_{Q \in \mathcal{M}_1(P)} E_Q[-X].$$

Note, however, that the supremum on the right cannot be replaced by a maximum as soon as $(\Omega, \mathcal{F}, P)$ cannot be reduced to a finite model. Indeed, in that case there exists $X \in L^\infty$ such that X does not attain its essential infimum, and so there can be no $Q \in \mathcal{M}_1(P)$ such that $E_Q[X] = \operatorname{ess\,inf} X = -\rho_{\max}(X)$. In this case, Corollary 4.38 shows that $\rho_{\max}$ is not continuous from below. ◇

Example 4.40. Let $\mathcal{Q}_\lambda$ be the class of all $Q \in \mathcal{M}_1(P)$ whose density dQ/dP is bounded by $1/\lambda$ for some fixed parameter $\lambda \in (0,1)$. The corresponding coherent risk measure

$$\mathrm{AV@R}_\lambda(X) := \sup_{Q\in\mathcal{Q}_\lambda} E_Q[\,-X\,] \tag{4.42}$$

will be called the *Average Value at Risk* at level λ. This terminology will become clear in Section 4.5, which contains a detailed study of $\mathrm{AV@R}_\lambda$. By taking $g(x) := 0$ for $x \le \frac{1}{\lambda}$ and $g(x) := +\infty$ for $x > \frac{1}{\lambda}$, one sees that $\mathrm{AV@R}_\lambda$ falls into the class of divergence risk measures as introduced in Example 4.36. It follows from Exercise 4.4.4 that $\mathcal{Q}_\lambda$ is equal to the maximal representing subset of $\mathrm{AV@R}_\lambda$, that $\mathrm{AV@R}_\lambda$ is continuous from below, and that the supremum in (4.42) is actually attained. An explicit construction of the maximizing measure will be given in the proof of Theorem 4.52. ◇

Example 4.41. We take for $\mathcal{Q}$ the class of all conditional distributions $P[\,\cdot\mid A\,]$ such that $A \in \mathcal{F}$ has $P[\,A\,] > \lambda$ for some fixed level $\lambda \in (0,1)$. The coherent risk measure induced by $\mathcal{Q}$,

$$\mathrm{WCE}_\lambda(X) := \sup\{\, E[\,-X \mid A\,] \mid A \in \mathcal{F},\ P[\,A\,] > \lambda\,\}, \tag{4.43}$$

is called the *worst conditional expectation* at level λ. We will show in Section 4.5 that it coincides with the Average Value at Risk of Example 4.40 if the underlying probability space is rich enough. ◇

Let ρ be a convex risk measure with the Fatou property. We now consider the situation in which ρ admits a representation in terms of *equivalent* probability measures $Q \approx P$, that is,

$$\rho(X) = \sup_{Q\approx P}(E_Q[\,-X\,] - \alpha^{\min}(Q)), \quad X \in L^\infty. \tag{4.44}$$

We will show in the next theorem that this property can be characterized by the following concept of *sensitivity*, which is sometimes also called *relevance*. It formalizes the idea that ρ should react to every nontrivial loss at a sufficiently high level.

Definition 4.42. A convex risk measure ρ on L^∞ is called *sensitive* with respect to P if for every nonconstant $X \in L^\infty$ with $X \ge 0$ there exists $\lambda > 0$ such that $\rho(-\lambda X) > \rho(0)$. △

Theorem 4.43. *For a convex risk measure with the Fatou property, the following conditions are equivalent:*

(a) *ρ admits the representation* (4.44) *in terms of equivalent probability measures.*
(b) *ρ is sensitive with respect to P.*
(c) *For every $A \in \mathcal{F}$ with $P[\,A\,] > 0$, there exists $\lambda > 0$ such that $\rho(-\lambda\mathbb{1}_A) > \rho(0)$.*
(d) *For every $A \in \mathcal{F}$ with $P[\,A\,] > 0$, there exists $Q \in \mathcal{M}_1(P)$ with $\alpha^{\min}(Q) < \infty$ and $Q[\,A\,] > 0$.*
(e) *There exists $Q \approx P$ with $\alpha^{\min}(Q) < \infty$.*

Proof. Throughout the proof, we will assume for simplicity that ρ is normalized in the sense that $\rho(0) = 0$. This can be done without loss of generality.

(a)⇒(b): Take $X \in L_+^\infty$ with $E[\,X\,] > 0$. By (4.44), there exists $Q \approx P$ with $\alpha^{\min}(Q) < \infty$ and

$$\rho(-\lambda X) \geq \lambda E_Q[\,X\,] - \alpha^{\min}(Q).$$

The right-hand side is strictly positive as soon as

$$\lambda > \frac{\alpha^{\min}(Q)}{E_Q[\,X\,]},$$

which is a finite number since $E_Q[\,X\,] > 0$ due to $Q \approx P$.

The implications (b)⇒(c) and (c)⇒(d) are both obvious.

(d)⇒(e): For every $c > 0$, the level set $\Lambda_c := \{Q \in \mathcal{M}_1(P) \mid \alpha^{\min}(Q) \leq c\}$ is nonempty due to our assumption $\rho(0) = 0$. We will show that Λ_c contains some $Q \approx P$. We show first the following auxiliary claim:

$$\text{For any } A \in \mathcal{F} \text{ with } P[\,A\,] > 0 \text{ there exists } Q \in \Lambda_c \text{ with } Q[\,A\,] > 0. \tag{4.45}$$

Indeed, (d) implies that there is $Q_1 \in \mathcal{M}_1(P)$ with $Q_1[\,A\,] > 0$ and $\alpha^{\min}(Q_1) < \infty$. Now we take $Q_0 \in \Lambda_{c/2}$ and let $Q_\varepsilon := \varepsilon Q_1 + (1-\varepsilon)Q_0$ for $0 < \varepsilon < 1$. We clearly have $Q_\varepsilon[\,A\,] > 0$ and

$$\alpha^{\min}(Q_\varepsilon) \leq \varepsilon\alpha^{\min}(Q_1) + (1-\varepsilon)\frac{c}{2} < c$$

for sufficiently small $\varepsilon > 0$. This implies (4.45).

We now apply the Halmos–Savage theorem in the form of Theorem 1.63. It yields the existence of $Q \in \Lambda_c$ with $Q \approx P$ if we can show that Λ_c is countably convex. To show countable convexity, let $(\beta_k)_{k\in\mathbb{N}}$ be a sequence of nonnegative real numbers summing up to 1 and take $Q^k \in \Lambda_c$ for $k \in \mathbb{N}$. We define $Q := \sum_{k=1}^\infty \beta_k Q^k$. Then $Q \in \mathcal{M}_1(P)$ and

$$\begin{aligned}\alpha^{\min}(Q) &= \sup_{X\in\mathcal{A}_\rho} E_Q[\,-X\,] = \sup_{X\in\mathcal{A}_\rho} \sum_{k=1}^\infty \beta_k E_{Q^k}[\,-X\,] \leq \sum_{k=1}^\infty \beta_k \sup_{X\in\mathcal{A}_\rho} E_{Q^k}[\,-X\,]\\ &= \sum_{k=1}^\infty \beta_k \alpha^{\min}(Q^k) \leq c.\end{aligned}$$

Thus, Q belongs to Λ_c, and (e) follows.

(e)⇒(a): By the representation (4.35) in Theorem 4.33, it is sufficient to show that

$$\rho(X) = \sup_{Q\in\mathcal{M}_1(P)} (E_Q[\,-X\,] - \alpha^{\min}(Q)) \leq \sup_{Q\approx P}(E_Q[\,-X\,] - \alpha^{\min}(Q))$$

for any given $X \in L^\infty$. To this end, we take $\delta > 0$ and choose $Q_0 \in \mathcal{M}_1(P)$ such that

$$E_{Q_0}[\,-X\,] - \alpha^{\min}(Q_0) > \rho(X) - \delta.$$

Then we take $Q_1 \approx P$ with $\alpha^{\min}(Q_1) < \infty$, which exists due to (e). When letting $Q_\varepsilon := \varepsilon Q_1 + (1-\varepsilon)Q_0$ we have $Q_\varepsilon \approx P$ for all $\varepsilon \in (0,1)$ and $\alpha^{\min}(Q_\varepsilon) \le \varepsilon\alpha^{\min}(Q_1) + (1-\varepsilon)\alpha^{\min}(Q_0)$. Hence,

$$E_{Q_\varepsilon}[-X] - \alpha^{\min}(Q_\varepsilon) \ge \varepsilon(E_{Q_1}[-X] - \alpha^{\min}(Q_1)) + (1-\varepsilon)(E_{Q_0}[-X] - \alpha^{\min}(Q_0)),$$

and the right-hand side is larger than $\rho(X) - \delta$ if ε is sufficiently small. □

Condition (e) in Theorem 4.43 is of course satisfied if $\alpha^{\min}(P) < \infty$. This is the case for the worst conditional expectation WCE (Example 4.41), Average Value at Risk AV@R (Example 4.40), the divergence risk measures (Example 4.36), and in particular the entropic risk measure (Example 4.34). These, and many other risk measures are therefore sensitive, and admit a representation (4.44) in terms of equivalent risk measures.

Remark 4.44. In analogy to Remark 4.18, the implication (e)⇒(a) in the Representation Theorem 4.33 can be viewed as a special case of the general duality in Theorem H.9 for the *Fenchel–Legendre transform* of the convex function ρ on L^∞, combined with the properties of a monetary risk measure. From this general point of view, it is now clear how to state representation theorems for convex risk measures on the Banach spaces $L^p(\Omega, \mathscr{F}, P)$ for $1 \le p < \infty$. More precisely, let $q \in (1,\infty]$ be such that $\frac{1}{p} + \frac{1}{q} = 1$, and define

$$\mathscr{M}_1^q(P) := \left\{ Q \in \mathscr{M}_1(P) \;\middle|\; \frac{dQ}{dP} \in L^q \right\}.$$

A convex risk measure ρ on L^p is of the form

$$\rho(X) = \sup_{Q \in \mathscr{M}_1^q(P)} (E_Q[-X] - \alpha(Q))$$

if and only if it is lower semicontinuous on L^p, that is, the Fatou property holds in the form

$$X_n \longrightarrow X \text{ in } L^p \quad \Longrightarrow \quad \rho(X) \le \liminf_{n\uparrow\infty} \rho(X_n).$$ ◇

4.5 Value at Risk

A common approach to the problem of measuring the risk of a financial position X consists in specifying a quantile of the distribution of X under the given probability measure P. For $\lambda \in (0,1)$, a *λ-quantile* of a random variable X on $(\Omega, \mathscr{F}, P)$ is any real number q with the property

$$P[X \le q] \ge \lambda \quad \text{and} \quad P[X < q] \le \lambda.$$

Exercise 4.5.1. For $X \in L^1(\Omega, \mathscr{F}, P)$ and $\lambda \in (0,1)$, a real number q is a λ-quantile for X if and only if it minimizes the convex function

$$f_\lambda(c) := \lambda E[\,(X-c)^+\,] + (1-\lambda)E[\,(X-c)^-\,]. \tag{4.46}$$

Hint: A minimizer q of the convex function f_λ is characterized by the conditions that $(f_\lambda)'_+(q) \geq 0$ and $(f_\lambda)'_-(q) \leq 0$. ◇

The set of all λ-quantiles of X is an interval $[q_X^-(\lambda), q_X^+(\lambda)]$, where

$$q_X^-(t) = \sup\{x \mid P[\,X < x\,] < t\} = \inf\{x \mid P[\,X \leq x\,] \geq t\}$$

is the lower and

$$q_X^+(t) = \inf\{x \mid P[\,X \leq x\,] > t\} = \sup\{x \mid P[\,X < x\,] \leq t\}$$

is the upper quantile function of X; see Appendix D and in particular Lemma D.6. In this section, we will focus on the properties of $q_X^+(\lambda)$, viewed as a functional on a space of financial positions X.

Definition 4.45. Fix some level $\lambda \in (0,1)$. For a financial position X, we define its *Value at Risk at level* λ as

$$\mathrm{V@R}_\lambda(X) := -q_X^+(\lambda) = q_{-X}^-(1-\lambda) = \inf\{m \mid P[\,X+m<0\,] \leq \lambda\}. \qquad \triangle$$

In financial terms, $\mathrm{V@R}_\lambda(X)$ is the smallest amount of capital that, if added to X and invested in the risk-free asset, keeps the probability of a negative outcome below the level λ. In other words, the loss $-X$ is below the threshold $\mathrm{V@R}_\lambda(X)$ with a probability of at least $1-\lambda$. Sometimes, this latter interpretation is emphasized in the literature, and Value at Risk is then parameterized by $1-\lambda$ instead of λ. Note that Value at Risk only controls the probability of a loss; it does not capture the size of such a loss if it occurs.

Clearly, $\mathrm{V@R}_\lambda$ is a monetary risk measure on $\mathscr{X} = L^0$, which is positively homogeneous; see also Example 4.11. The following example and the subsequent exercise both show that the acceptance set of $\mathrm{V@R}_\lambda$ is typically not convex, and so $\mathrm{V@R}_\lambda$ is *not* a convex risk measure. Thus, $\mathrm{V@R}_\lambda$ may penalize diversification instead of encouraging it. As a matter of fact, $\mathrm{V@R}_\lambda$ creates an incentive to concentrate risk on an event of small probability.

Example 4.46. Consider two defaultable corporate bonds, each with interest $\tilde{r} > r$, where $r \geq 0$ is the interest rate on a riskless investment. Assuming that in case of a default the entire principal of the bond is wiped out, the discounted P&L of an investment $w > 0$ in the ith bond is given by

$$X_i = \begin{cases} -w & \text{in case of default,} \\ \dfrac{w(\tilde{r} - r)}{1 + r} & \text{otherwise.} \end{cases}$$

If a default of the first bond occurs with probability $p \le \lambda < 1/2$, then

$$P\left[X_1 - \frac{w(\tilde{r} - r)}{1 + r} < 0 \right] = P[\,\text{bond 1 defaults}\,] = p \le \lambda$$

and

$$P\left[X_1 - \frac{w(\tilde{r} - r)}{1 + r} \ge 0 \right] = P[\,\text{bond 1 does not default}\,] = 1 - p > \lambda.$$

Hence,

$$\mathrm{V@R}_\lambda(X_1) = -\frac{w(\tilde{r} - r)}{1 + r} < 0.$$

This means that the position X_1 is acceptable in the sense that it does not carry a positive Value at Risk, regardless of the possible loss of the entire investment w.

Diversifying the portfolio by investing the amount $w/2$ into each of the two bonds leads to the position $Y := (X_1 + X_2)/2$. Let us assume that the two bonds default independently of each other, each of them with probability p. For realistic $\tilde{r}$, the probability that Y is negative is equal to the probability that at least one of the two bonds defaults: $P[\,Y < 0\,] = p(2-p)$. If, for instance, $p = 0.009$ and $\lambda = 0.01$ then we have $p < \lambda < p(2-p)$, hence

$$\mathrm{V@R}_\lambda(Y) = \frac{w}{2} \cdot \left(1 - \frac{\tilde{r} - r}{1 + r}\right).$$

Typically, this value is close to one half of the invested capital w. In particular, the acceptance set of $\mathrm{V@R}_\lambda$ is not convex. This example also shows that V@R may strongly discourage diversification: It penalizes quite drastically the increase of the probability that something goes wrong, without rewarding the significant reduction of the expected loss conditional on the event of default. Thus, optimizing a portfolio with respect to $\mathrm{V@R}_\lambda$ may lead to a concentration of the portfolio in one single asset with a sufficiently small default probability, but with an exposure to large losses; see [127] for further analysis. ◇

Exercise 4.5.2. Let (Y_n) be a sequence of independent and identically distributed random variables in $\mathscr{L}^1(\Omega, \mathscr{F}, P)$. Show that

$$\mathrm{V@R}_\lambda\left(\frac{1}{n}\sum_{i=1}^{n} Y_i\right) \longrightarrow -E[\,Y_1\,] \quad \text{as } n \uparrow \infty$$

for any $\lambda \in (0,1)$. Then choose the common distribution of the Y_i in such a way that convexity is violated for large n, that is,

$$\mathrm{V@R}_\lambda\left(\frac{1}{n}\sum_{i=1}^{n} Y_i\right) > \mathrm{V@R}_\lambda(Y_1). \qquad \diamond$$

In the remainder of this section, we will focus on monetary measures of risk that, in contrast to $\mathrm{V@R}_\lambda$, are convex or even coherent on $\mathscr{X} := L^\infty$. In particular, we are looking for convex risk measures that come close to $\mathrm{V@R}_\lambda$. A first guess might be that one should take the smallest convex measure of risk, continuous from above, that dominates $\mathrm{V@R}_\lambda$. However, since $\mathrm{V@R}_\lambda$ itself is not convex, the following proposition shows that such a smallest $\mathrm{V@R}_\lambda$-dominating convex risk measure does not exist.

Proposition 4.47. *For each $X \in \mathscr{X}$ and each $\lambda \in (0,1)$,*

$$\mathrm{V@R}_\lambda(X) = \min\{\rho(X) \mid \rho \text{ is convex, continuous from above, and} \geq \mathrm{V@R}_\lambda\}.$$

Proof. Let $q := -\mathrm{V@R}_\lambda(X) = q_X^+(\lambda)$ so that $P[\,X < q\,] \leq \lambda$. If $A \in \mathscr{F}$ satisfies $P[\,A\,] > \lambda$, then $P[\,A \cap \{X \geq q\}\,] > 0$. Thus, we may define a measure Q_A by

$$Q_A := P[\,\cdot \mid A \cap \{X \geq q\}\,].$$

It follows that $E_{Q_A}[\,-X\,] \leq -q = \mathrm{V@R}_\lambda(X)$.

Let

$$\mathcal{Q} := \{Q_A \mid A \in \mathscr{F} \text{ such that } P[\,A\,] > \lambda\}$$

and use this set to define a coherent risk measure ρ via

$$\rho(Y) := \sup_{Q \in \mathcal{Q}} E_Q[\,-Y\,].$$

Then $\rho(X) \leq -q = \mathrm{V@R}_\lambda(X)$. Hence, the assertion will follow if we can show that $\rho(Y) \geq \mathrm{V@R}_\lambda(Y)$ for each $Y \in \mathscr{X}$. Let $\varepsilon > 0$ and

$$A := \{Y \leq -\mathrm{V@R}_\lambda(Y) + \varepsilon\} = \{Y \leq q_Y^+(\lambda) + \varepsilon\}.$$

Then $P[\,A\,] > \lambda$, and so $Q_A \in \mathcal{Q}$. Moreover, $Q_A[\,A\,] = 1$, and we obtain

$$\rho(Y) \geq E_{Q_A}[\,-Y\,] \geq \mathrm{V@R}_\lambda(Y) - \varepsilon.$$

Since $\varepsilon > 0$ is arbitrary, it follows that $\rho(Y) \geq \mathrm{V@R}_\lambda(Y)$, as desired. □

For the remainder of this section, we concentrate on the following risk measure, which is defined in terms of Value at Risk but does satisfy the axioms of a coherent risk measure.

Definition 4.48. The *Average Value at Risk at level* $\lambda \in (0,1]$ of a position $X \in \mathscr{X}$ is given by

$$\mathrm{AV@R}_\lambda(X) = \frac{1}{\lambda}\int_0^\lambda \mathrm{V@R}_\gamma(X)\,d\gamma. \qquad \triangle$$

Clearly, Average Value at Risk is a positively homogeneous monetary risk measure, since it inherits these properties from Value at Risk. Sometimes, the Average Value at Risk is also called the “Conditional Value at Risk” or the “Expected Shortfall,” and one writes $CV@R_\lambda(X)$ or $ES_\lambda(X)$. These terms are motivated by formulas (4.49) and (4.47) below, but they are potentially misleading: “Conditional Value at Risk” might also be used to denote the Value at Risk with respect to a conditional distribution, and “Expected Shortfall” might be understood as the expectation of the shortfall X^-. For these reasons, we prefer the term Average Value at Risk. Note that

$$\mathrm{AV@R}_\lambda(X) = -\frac{1}{\lambda}\int_0^\lambda q_X^+(t)\,dt$$

by Definition 4.45. In particular, the definition of $\mathrm{AV@R}_\lambda(X)$ makes sense for any $X \in L^1(\Omega, \mathscr{F}, P)$ and we have, in view of Lemma D.7,

$$\mathrm{AV@R}_1(X) = -\int_0^1 q_X^+(t)\,dt = E[\,-X\,].$$

Lemma D.7 also implies that, in the two preceding identities, q_X^+ can be replaced by any other quantile function of X.

Exercise 4.5.3. Recalling the results from Example 4.11 and Exercise 4.1.6, compute $\mathrm{AV@R}_\lambda(-X)$ if X is
(a) uniformly distributed on $[0,1]$,
(b) normally distributed,
(c) log-normally distributed, that is, $X = e^{\sigma Z+\mu}$ with $Z \sim N(0,1)$ and $\mu, \sigma \in \mathbb{R}$,
(d) an indicator function, that is, $X = \mathbb{1}_A$ for some event $A \in \mathscr{F}$.
Then compare the behavior of $\mathrm{V@R}_\lambda(X)$ and $\mathrm{AV@R}_\lambda(X)$ as the parameter λ increases from 0 to 1. ◇

Remark 4.49. Theorem 2.57(e) shows that the increasing concave order $\succcurlyeq_{\mathrm{icv}}$ can be characterized in terms of Average Value at Risk:

$$\mu \succcurlyeq_{\mathrm{icv}} \nu \iff \mathrm{AV@R}_\lambda(X_\mu) \le \mathrm{AV@R}_\lambda(X_\nu) \quad \text{for all } \lambda \in (0,1],$$

where X_μ and X_ν are random variables with distributions μ and ν that have finite mean. ◇

Remark 4.50. For $X \in L^\infty$, we have

$$\lim_{\lambda \downarrow 0} \mathrm{V@R}_\lambda(X) = -\operatorname{ess\,inf} X = \inf\{m \mid P[\,X + m < 0\,] \le 0\};$$

see Lemma D.3. Hence, it makes sense to define

$$\mathrm{AV@R}_0(X) := \mathrm{V@R}_0(X) := -\operatorname{ess\,inf} X,$$

which is the worst-case risk measure on L^∞ introduced in Example 4.39. Recall that it is continuous from above but in general not from below. ◇

Proposition 4.51. *For $\lambda \in (0,1)$ and any λ-quantile q of X,*

$$\mathrm{AV@R}_\lambda(X) = \frac{1}{\lambda} E[\,(q - X)^+\,] - q = \frac{1}{\lambda} \inf_{r \in \mathbb{R}} (E[\,(r - X)^+\,] - \lambda r). \tag{4.47}$$

Proof. Let q_X be a quantile function with $q_X(\lambda) = q$. By Lemma D.7,

$$\frac{1}{\lambda} E[\,(q - X)^+\,] - q = \frac{1}{\lambda} \int_0^1 (q - q_X(t))^+ \, dt - q = -\frac{1}{\lambda} \int_0^\lambda q_X(t)\, dt = \mathrm{AV@R}_\lambda(X).$$

This proves the first identity. The second one follows from Lemma D.11. It also follows from Exercise 4.5.1, since

$$\begin{aligned} E[\,(r - X)^+\,] - \lambda r &= \lambda(E[\,(r - X)^+\,] - r) + (1 - \lambda)E[\,(X - r)^-\,] \\ &= -\lambda E[\,X\,] + \lambda E[\,(X - r)^+\,] + (1 - \lambda)E[\,(X - r)^-\,] = -\lambda E[\,X\,] + f_\lambda(r), \end{aligned}$$

where we have used the put-call parity from Remark 1.22 in the second step, and f_λ is as in (4.46). □

Exercise 4.5.4. Use the representation (4.47) to show that $\mathrm{AV@R}_\lambda$ is subadditive. ◇

Since we already observed that $\mathrm{AV@R}_\lambda$ is a positively homogeneous monetary risk measure, the preceding exercise, combined with Exercise 4.1.3, implies that $\mathrm{AV@R}_\lambda$ is a coherent risk measure. Alternatively, coherence is implied by the following representation (4.48).

Theorem 4.52. *For $\lambda \in (0,1]$, $\mathrm{AV@R}_\lambda$ is a coherent risk measure that is continuous from below. It has the representation*

$$\mathrm{AV@R}_\lambda(X) = \max_{Q \in \mathcal{Q}_\lambda} E_Q[\,-X\,], \quad X \in \mathscr{X}, \tag{4.48}$$

where $\mathcal{Q}_\lambda$ is the set of all probability measures $Q \ll P$ whose density dQ/dP is P-a. s. bounded by $1/\lambda$. Moreover, $\mathcal{Q}_\lambda$ is equal to the maximal set $\mathcal{Q}^{\max}$ of Corollary 4.38.

Proof. Since $\mathcal{Q}_1 = \{P\}$, the assertion is obvious for $\lambda = 1$. For $0 < \lambda < 1$, consider the coherent risk measure $\rho_\lambda(X) := \sup_{Q\in\mathcal{Q}_\lambda} E_Q[-X]$. First, we assume that we are given some $X < 0$. We define a measure $\widetilde{P} \approx P$ by $d\widetilde{P}/dP = X/E[X]$. Then

$$\rho_\lambda(X) = \frac{E[-X]}{\lambda} \sup\{\widetilde{E}[\varphi] \mid 0 \le \varphi \le 1,\, E[\varphi] = \lambda\}.$$

Clearly, the condition $E[\varphi] = \lambda$ on the right can be replaced by $E[\varphi] \le \lambda$. Thus, we can apply the Neyman–Pearson lemma in the form of Theorem E.3 and conclude that the supremum is attained by

$$\varphi_0 = \mathbb{1}_{\{X<q\}} + \kappa \mathbb{1}_{\{X=q\}}$$

for a λ-quantile q of X and a constant $\kappa \in [0,1]$ that is determined by the condition $E[\varphi_0] = \lambda$. Hence,

$$\rho_\lambda(X) = \frac{E[-X]}{\lambda} \cdot \widetilde{E}[\varphi_0] = \frac{1}{\lambda} E[-X\varphi_0].$$

Since $dQ_0 = \lambda^{-1}\varphi_0\, dP$ defines a probability measure in $\mathcal{Q}_\lambda$, we conclude that

$$\begin{aligned}
\rho_\lambda(X) &= \max_{Q\in\mathcal{Q}_\lambda} E_Q[-X] = E_{Q_0}[-X] \\
&= \frac{1}{\lambda}\big(E[-X;\, X < q] - q\lambda + qP[X < q]\big) \\
&= \frac{1}{\lambda} E[(q-X)^+] - q \\
&= \mathrm{AV@R}_\lambda(X),
\end{aligned}$$

where we have used (4.47) in the last step. This proves (4.48) for $X < 0$. For arbitrary $X \in L^\infty$, we use the cash invariance of both ρ_λ and $\mathrm{AV@R}_\lambda$. In particular, the representation (4.48) shows that $\mathrm{AV@R}_\lambda$ is a coherent risk measure.

It remains to prove that $\mathcal{Q}_\lambda$ is the maximal set of Corollary 4.37. This follows from Exercise 4.4.4 and Example 4.40, but we also give a different argument here. To this end, we show that

$$\sup_{X\in\mathcal{X}} \big(E_Q[-X] - \mathrm{AV@R}_\lambda(X)\big) = +\infty$$

for $Q \notin \mathcal{Q}_\lambda$. We denote by φ the density dQ/dP. There exist $\lambda' \in (0,\lambda)$ and $k > 1/\lambda'$ such that $P[\varphi \wedge k \ge 1/\lambda'] > 0$. For $c > 0$, define $X^{(c)} \in \mathcal{X}$ by

$$X^{(c)} := -c(\varphi \wedge k)\mathbb{1}_{\{\varphi \ge 1/\lambda'\}}.$$

Since, by Markov's inequality,

$$P[\,X^{(c)} < 0\,] = P\Big[\,\varphi \ge \frac{1}{\lambda'}\,\Big] \le \frac{E[\varphi]}{1/\lambda'} = \lambda' < \lambda,$$

we have $\mathrm{V@R}_\lambda(X^{(c)}) = 0$, and (4.47) yields that

$$\mathrm{AV@R}_\lambda(X^{(c)}) = \frac{1}{\lambda}E[\,-X^{(c)}\,] = \frac{c}{\lambda}E\Big[\,\varphi \wedge k;\, \varphi \ge \frac{1}{\lambda'}\,\Big] > 0.$$

On the other hand,

$$E_Q[\,-X^{(c)}\,] = c \cdot E\Big[\,\varphi \cdot (\varphi \wedge k);\, \varphi \ge \frac{1}{\lambda'}\,\Big] \ge \frac{c}{\lambda'}E\Big[\,\varphi \wedge k;\, \varphi \ge \frac{1}{\lambda'}\,\Big].$$

Thus, the difference between $E_Q[\,-X^{(c)}\,]$ and $\mathrm{AV@R}_\lambda(X^{(c)})$ becomes arbitrarily large as $c \uparrow \infty$.

Finally, to show continuity from below, let q be a λ-quantile of $X \in \mathscr{X}$. For any sequence (X_n) in $\mathscr{X}$ that increases to X, dominated convergence implies that

$$\begin{aligned}\mathrm{AV@R}_\lambda(X) &= \frac{1}{\lambda}E[\,(q-X)^+\,] - q = \lim_{n\uparrow\infty}\Big(\frac{1}{\lambda}E[\,(q-X_n)^+\,] - q\Big)\\ &\ge \limsup_{n\uparrow\infty} \mathrm{AV@R}_\lambda(X_n),\end{aligned}$$

using the first identity in (4.47) in the first step and the second identity in (4.47) in the third step. Since $\mathrm{AV@R}_\lambda(X) \le \mathrm{AV@R}_\lambda(X_n)$ by monotonicity, we have shown continuity from below. □

Exercise 4.5.5. Use Corollary 4.38 and the Dunford–Pettis theorem (Theorem H.19) to give an alternative proof of the fact that $\mathrm{AV@R}_\lambda$ is continuous from below. ◇

Remark 4.53. The proof of Theorem 4.52 shows that for $\lambda \in (0,1)$ the maximum in (4.48) is attained by the measure $Q_0 \in \mathcal{Q}_\lambda$, whose density is given by

$$\frac{dQ_0}{dP} = \frac{1}{\lambda}\big(\mathbb{1}_{\{X<q\}} + \kappa\mathbb{1}_{\{X=q\}}\big),$$

where q is a λ-quantile of X, and where κ is defined as

$$\kappa := \begin{cases} 0 & \text{if } P[\,X = q\,] = 0,\\ \dfrac{\lambda - P[\,X < q\,]}{P[\,X = q\,]} & \text{otherwise.}\end{cases}$$
◇

Corollary 4.54. *For all $X \in \mathscr{X}$,*

$$\begin{aligned}\mathrm{AV@R}_\lambda(X) &\ge \mathrm{WCE}_\lambda(X)\\ &\ge E[\,-X \mid -X \ge \mathrm{V@R}_\lambda(X)\,]\\ &\ge \mathrm{V@R}_\lambda(X),\end{aligned} \tag{4.49}$$

where WCE_λ *is the coherent risk measure defined in* (4.43). *Moreover, the first two inequalities are in fact identities if*

$$P[\,X \le q_X^+(\lambda)\,] = \lambda, \tag{4.50}$$

which is the case if X has a continuous distribution.

Proof. If $P[\,A\,] \ge \lambda$, then the density of $P[\,\cdot \mid A\,]$ with respect to P is bounded by $1/\lambda$. Therefore, Theorem 4.52 implies that $\mathrm{AV@R}_\lambda$ dominates WCE_λ. Since

$$P[\,-X \ge \mathrm{V@R}_\lambda(X) - \varepsilon\,] > \lambda,$$

we have

$$\mathrm{WCE}_\lambda(X) \ge E[\,-X \mid -X \ge \mathrm{V@R}_\lambda(X) - \varepsilon\,],$$

and the second inequality follows by taking the limit $\varepsilon \downarrow 0$. Moreover, (4.47) shows that

$$\mathrm{AV@R}_\lambda(X) = E[\,-X \mid -X \ge \mathrm{V@R}_\lambda(X)\,]$$

as soon as (4.50) holds. □

Remark 4.55. We will see in Corollary 4.74 that the two coherent risk measures $\mathrm{AV@R}_\lambda$ and WCE_λ coincide if the underlying probability space is rich enough. If this is not the case, then the first inequality in (4.49) may be strict for some X; see [2]. Moreover, the functional

$$\mathrm{TCE}_\lambda(X) := E[\,-X \mid -X \ge \mathrm{V@R}_\lambda(X)\,],$$

which is sometimes called the *tail-conditional expectation* of X, does not define a convex risk measure as will be shown in the following Exercise 4.5.6. Hence, the second inequality in (4.49) cannot reduce to an identity in general. ◇

Exercise 4.5.6. In this exercise, we will show in particular that TCE_λ, as defined in Remark 4.55, is typically not a coherent risk measure.

(a) Compute $\mathrm{TCE}_\lambda(X)$ if X is the negative of an indicator function, that is, $X = -\mathbb{1}_A$ for some event $A \in \mathscr{F}$. Then compare the result with Exercise 4.5.3(d) and Exercise 4.1.6(c).

(b) Now let $A, B \in \mathscr{F}$ be two disjoint sets with probability $p := P[\,A\,] = P[\,B\,] \in (0, 1/2)$. Show that for $X := -\mathbb{1}_A$, $Y := -\mathbb{1}_B$, and $p \le \lambda < 2p$,

$$\mathrm{TCE}_\lambda(X + Y) > \mathrm{TCE}_\lambda(X) + \mathrm{TCE}_\lambda(Y).$$ ◇

Remark 4.56. We have seen in Proposition 4.47 that there is no smallest convex risk measure dominating $\mathrm{V@R}_\lambda$. But if we restrict our attention to the class of convex risk measures that dominate $\mathrm{V@R}_\lambda$ and only depend on the distribution of a random variable, then the situation is different. In fact, we will see in Theorem 4.73 that $\mathrm{AV@R}_\lambda$ is the smallest risk measure in this class, provided that the underlying probability space is rich enough. In this sense, Average Value at Risk can be regarded as the best conservative approximation to Value at Risk. ◇

4.6 Law-invariant risk measures and the Fatou property

Clearly, $\mathrm{V@R}_\lambda$ and $\mathrm{AV@R}_\lambda$ only involve the distribution of a position under the given probability measure P. In this section and in the section that follows, we study the class of all risk measures that share this property.

Definition 4.57. A monetary risk measure ρ on $\mathscr{X} = L^\infty(\Omega, \mathscr{F}, P)$ is called *law-invariant* if $\rho(X) = \rho(Y)$ whenever X and Y have the same distribution under P. △

Law-invariant risk measure are sometimes also called *law-determined, law-based,* or *law-equivariant.* Throughout this section, we assume that the probability space $(\Omega, \mathscr{F}, P)$ is rich enough in the sense that it supports a random variable with a continuous distribution. This condition is satisfied if and only if $(\Omega, \mathscr{F}, P)$ is atomless; see Proposition D.16. Under this condition, the following result states that every law-invariant convex risk measure is automatically continuous from above. Recall from Theorem 4.33 that, for a convex risk measure ρ on L^∞, continuity from above is equivalent to the $\sigma(L^\infty, L^1)$-lower semicontinuity of ρ and to the Fatou property:

$$\rho(X) \leq \liminf_{n\uparrow\infty} \rho(X_n)$$

for every uniformly bounded sequence (X_n) in L^∞ that converges P-a. s. to $X \in L^\infty$.

Theorem 4.58. *Every law-invariant convex risk measure on L^∞ has the Fatou property.*

The proof of Theorem 4.58 will rely on the following proposition from [93], which is of independent interest.

Proposition 4.59. *On an atomless probability space, let $X_1, X_2, \ldots$ be a uniformly bounded sequence of random variables that converges in probability to the random variable X. Then for each $k \in \mathbb{N}$ there exist $n_k \in \mathbb{N}$ and random variables $X_{k,1}, \ldots, X_{k,n_k}$ such that each $X_{k,j}$ has the same law as X_k and*

$$\frac{1}{n_k}\sum_{j=1}^{n_k} X_{k,j} \longrightarrow X \quad \text{in } L^\infty \text{ as } k \uparrow \infty. \tag{4.51}$$

Once Proposition 4.59 is established, the proof of Theorem 4.58 follows easily.

Proof of Theorem 4.58. Let ρ be a law-invariant convex risk measure and (X_k) in L^∞ a uniformly bounded sequence that converges P-a. s. to $X \in L^\infty$. An application of Proposition 4.59 yields $n_k \in \mathbb{N}$ and random variables $X_{k,1}, \dots, X_{k,n_k}$ with the same law as X_k such that

$$\widetilde{X}_k := \frac{1}{n_k} \sum_{j=1}^{n_k} X_{k,j}$$

converges to X in L^∞. Since ρ is Lipschitz continuous on L^∞ by Lemma 4.3, we have

$$\rho(X) = \lim_{k\uparrow\infty} \rho(\widetilde{X}_k) \le \liminf_{k\uparrow\infty} \frac{1}{n_k} \sum_{j=1}^{n_k} \rho(X_{k,j}) = \liminf_{k\uparrow\infty} \rho(X_k),$$

where we have used the convexity of ρ in the inequality and then the law-invariance of ρ. □

In the remainder of this section, we will prove Proposition 4.59. We start by introducing the notion of a sequence of real numbers that is uniformly distributed modulo 1 and then derive some properties of such sequences. For a real number x, we denote by $\{x\}$ its *fractional part*, defined as $\{x\} = x - \lfloor x \rfloor$, where $\lfloor x \rfloor$ is the greatest integer less than or equal to x. The fractional part $\{x\}$ is sometimes also called the *residue of x modulo 1.*

Definition 4.60. A sequence $(x_n)_{n\in\mathbb{N}}$ of real numbers is called *uniformly distributed modulo 1*, if for all $a, b \in [0, 1]$ with $a < b$,

$$\lim_{n\uparrow\infty} \frac{1}{n} \sum_{k=1}^{n} \mathbb{1}_{[a,b)}(\{x_k\}) = b - a. \qquad \triangle$$

A celebrated theorem by H. Weyl states that the sequence $(n\alpha)_{n\in\mathbb{N}}$ is uniformly distributed modulo 1 if and only if α is irrational; see, for example, Example 2.1 in Chapter 1 of [225]. That chapter contains also many other examples. The following exercise provides a particularly easy example of a sequence that is uniformly distributed modulo 1.

Exercise 4.6.1. Prove that the sequence

$$(x_n)_{n\in\mathbb{N}} = \left(\frac{0}{1}, \frac{0}{2}, \frac{1}{2}, \frac{0}{3}, \frac{1}{3}, \frac{2}{3}, \frac{0}{4}, \dots, \frac{0}{k}, \frac{1}{k}, \dots, \frac{k-1}{k}, \dots \right)$$

is uniformly distributed modulo 1. ◇

Lemma 4.61. *If the sequence $(x_n)_{n\in\mathbb{N}}$ is uniformly distributed modulo 1 and y is a real number, then the sequence $(y + x_n)_{n\in\mathbb{N}}$ is also uniformly distributed modulo 1.*

Proof. Note that congruence modulo 1, that is, $\{x\} = \{z\}$, is an equivalence relation $\sim$ on $\mathbb{R}$ and that the unit interval $[0, 1)$ can be identified with the quotient set $\mathbb{R}/\!\sim$ of all equivalence classes. Thus, $\{x + y\}$ belongs to some interval $[a, b) \subseteq [0, 1)$, if and only if the shifted interval $[a - y, b - y)$ contains some z with $z \sim x$. That is,

$$\{x + y\} \in [a, b) \quad \Longleftrightarrow \quad \{x\} \in \{\{z\} \mid z \in [a - y, b - y)\} =: A.$$

Depending on whether the shifted interval $[a - y, b - y)$ contains an integer or not, the set A is either itself an interval or the union of two intervals, whose (combined) length is equal to $b - a$. The assertion now follows from Definition 4.60. □

Due to Exercise G.1.1, a sequence $(x_n)_{n\in\mathbb{N}}$ is uniformly distributed modulo 1 if and only if the empirical distributions of the fractional parts,

$$\hat{\mu}_n = \frac{1}{n} \sum_{k=1}^{n} \delta_{\{x_k\}}, \quad n \in \mathbb{N}, \tag{4.52}$$

converge weakly to the Lebesgue measure λ on $[0, 1)$. Thus,

$$\frac{1}{n} \sum_{k=1}^{n} f(\{x_k\}) = \int f \, d\hat{\mu}_n \longrightarrow \int_0^1 f(x) \, dx \tag{4.53}$$

as $n \uparrow \infty$ for every continuous function $f : [0, 1] \to \mathbb{R}$.

Proposition 4.62. *If the sequence $(x_n)_{n\in\mathbb{N}}$ is uniformly distributed modulo 1 and $f : [0, 1] \to \mathbb{R}$ is Riemann integrable, then the convergence*

$$\frac{1}{n} \sum_{k=1}^{n} f(\{y + x_k\}) \longrightarrow \int_0^1 f(x) \, dx \tag{4.54}$$

holds uniformly in $y \geq 0$.

Proof. In the first step, we show that the assertion holds if f is a continuous function on $[0, 1]$ with $f(0) = f(1)$. It follows from Lemma 4.61 and (4.53) that the convergence (4.54) holds pointwise for each fixed $y \in \mathbb{R}$.

Note that the function $y \mapsto f(\{y + x\})$ is continuous for each x, due to the assumption $f(0) = f(1)$. Moreover, for every $\varepsilon > 0$ there exists $\delta > 0$ such that for all $x \in \mathbb{R}$,

$$|f(\{y + x\}) - f(\{z + x\})| < \varepsilon \quad \text{if } |y - z| < \delta.$$

It follows that the functions

$$y \longmapsto \frac{1}{n} \sum_{k=1}^{n} f(\{y + x_k\}), \quad n \in \mathbb{N},$$

are uniformly equicontinuous. The Arzelà–Ascoli theorem thus implies that the convergence (4.54) holds uniformly in $y \in I$, where I is any compact interval. Due to the periodicity of $z \mapsto f(\{z\})$, we may conclude that (4.54) holds uniformly in all $y \in \mathbb{R}$.

Now let f be an arbitrary Riemann integrable function on $[0,1]$. We may assume without loss of generality that $f(0) \le f(1)$. Since f is Riemann integrable, there exist numbers $0 = t_0 < t_1 < \cdots < t_m = 1$ and step functions

$$u = \sum_{j=1}^{m} \alpha_j \mathbb{1}_{[t_{j-1},t_j)} + \alpha_m \mathbb{1}_{\{1\}} \quad \text{and} \quad v = \sum_{j=1}^{m} \beta_j \mathbb{1}_{[t_{j-1},t_j)} + \beta_m \mathbb{1}_{\{1\}}$$

such that $u \le f \le v$ and $0 \le \int_0^1 (v(x) - u(x))\,dx < \varepsilon/4$; see, for example, Section 11.3 in [313]. We now claim that there are continuous functions g and h on $[0,1]$ such that $g(0) = g(1)$, $h(0) = h(1)$, $g \le u$, $h \ge v$, and $\int_0^1 (h(x) - g(x))\,dx < \varepsilon/2$. To construct the function g, we choose $\delta < \min_j |t_j - t_{j-1}|/2$. If $\alpha_2 > \alpha_1$, then we let $g_\delta = \alpha_1$ on $[t_0, t_1]$, $g_\delta = \alpha_2$ on $[t_1 + \delta, t_2 - \delta]$, and linearly interpolated on $(t_1, t_1 + \delta)$. If $\alpha_2 \le \alpha_1$, we let $g_\delta = \alpha_1$ on $[t_0, t_1 - \delta]$, $g_\delta = \alpha_2$ on $[t_1, t_2 - \delta]$, and linearly interpolated on $(t_1 - \delta, t_2)$. Like this, we proceed for each interval until we reach the final interval $[t_{m-1}, t_m]$. Here, we set $g_\delta = \alpha_m$ on $[t_{m-1} + \delta, t_m - \delta]$, $g_\delta(1) = g_\delta(0) = \alpha_1$, and interpolate linearly on $(t_m - \delta, t_m)$. Then g_δ is continuous. Moreover, $g_\delta(x) \nearrow u(x)$ for each $x \in [0,1] \setminus \{t_0, t_1, \dots, t_m\}$, and so $\int_0^1 g_\delta(x)\,dx \to \int_0^1 u(x)\,dx$, as $\delta \downarrow 0$. Functions $h_\delta \ge v$ can be constructed in a similar manner. Finally, we let $g := g_\delta$ and $h := h_\delta$ for sufficiently small δ.

Due to the first step of the proof, there exists $n_0 \in \mathbb{N}$ such that for all $y \ge 0$ and all $n \ge n_0$,

$$\left| \frac{1}{n} \sum_{k=1}^{n} g(\{y + x_k\}) - \int_0^1 g(x)\,dx \right| < \frac{\varepsilon}{2} \quad \text{and} \quad \left| \frac{1}{n} \sum_{k=1}^{n} h(\{y + x_k\}) - \int_0^1 h(x)\,dx \right| < \frac{\varepsilon}{2}.$$

Hence, we conclude that

$$\begin{aligned} \frac{1}{n} \sum_{k=1}^{n} f(\{y + x_k\}) - \int_0^1 f(x)\,dx &\le \frac{1}{n} \sum_{k=1}^{n} h(\{y + x_k\}) - \int_0^1 g(x)\,dx \\ &< \frac{1}{n} \sum_{k=1}^{n} h(\{y + x_k\}) - \int_0^1 h(x)\,dx + \frac{\varepsilon}{2} \\ &< \varepsilon \end{aligned}$$

holds for all $y \ge 0$ and all $n \ge n_0$. Replacing h with g in the preceding argument yields a corresponding lower bound, thus completing the proof. □

Via quantile functions, we can apply Proposition 4.62 to general random variables on our atomless probability space $(\Omega, \mathscr{F}, P)$. This is the content of the following lemma, which can be seen as some kind of "law of large numbers." In comparison to classical results, such as Kolmogorov's law of large numbers, it yields stronger convergence, namely convergence in L^∞, but requires a specific choice of random variables Y_j.

Lemma 4.63. *Suppose that Y is a bounded random variable. Then there exists a sequence $Y_1, Y_2, \dots$ of random variables such that each Y_j has the same law as Y and*

$$\frac{1}{n}\sum_{j=1}^{n} Y_j \longrightarrow E[Y] \quad \text{in } L^\infty \text{ as } n \uparrow \infty. \tag{4.55}$$

Proof. Let q_Y denote a quantile function for Y. Since the underlying probability space is atomless, it supports a random variables U with a uniform distribution on $(0, 1)$; see Proposition D.16. Now we take a sequence $(x_j)_{j\in\mathbb{N}}$ that is uniformly distributed modulo 1. For instance, we can take the sequence $(j\alpha)_{j\in\mathbb{N}}$ for irrational α or the sequence in Exercise 4.6.1. We define

$$Y_j := q_Y(\{U + x_j\}).$$

The argument used in the proof of Lemma 4.61 shows that each random variable $\{U + x_j\}$ is again uniformly distributed on the unit interval, and so by Lemma D.7, each Y_j has the same law as Y.

Since Y is bounded, we can extend q_Y to a bounded increasing function on the closed interval $[0, 1]$ by putting $q_Y(0) := \operatorname{ess\,inf} Y$ and $q_Y(1) := \operatorname{ess\,sup} Y$. Note that all bounded and monotone functions on $[0, 1]$ are Riemann integrable; see, e.g., Proposition 11.6.1 in [313]. Hence we may apply Proposition 4.62 to the function $f := q_Y$. It yields that

$$\frac{1}{n}\sum_{j=1}^{n} Y_j(\omega) = \frac{1}{n}\sum_{j=1}^{n} q_Y(\{U(\omega) + x_j\}) \longrightarrow \int_0^1 q_Y(x)\,dx = E[Y], \quad \text{as } n \uparrow \infty,$$

uniformly in $\omega \in \Omega$. □

Now we extend Lemma 4.63 to the case in which the expectation of Y is replaced by a conditional expectation with respect to a finite σ-algebra.

Lemma 4.64. *Suppose that Y is a bounded random variable and $\mathscr{G} \subset \mathscr{F}$ is a finite σ-algebra. Then there exists a sequence $Y_1, Y_2, \dots$ of random variables such that each Y_j has the same law as Y and*

$$\frac{1}{n}\sum_{j=1}^{n} Y_j \longrightarrow E[Y \mid \mathscr{G}] \qquad \text{in } L^\infty \text{ as } n \uparrow \infty. \tag{4.56}$$

Proof. Since $\mathcal{G}$ is a finite σ-algebra, it is generated by a partition $A_1, \dots, A_\ell$ of Ω into disjoint sets.[1] We may assume that, after re-ordering, there is an index $m \in \{1, \dots, \ell\}$ such that $P[A_i] > 0$ for $i \le m$ and $P[A_i] = 0$ for $i > m$. With this convention, the conditional expectation $E[Y \mid \mathcal{G}]$ is given by

$$E[Y \mid \mathcal{G}] = \sum_{i=1}^{m} E[Y \mid A_i]\mathbb{1}_{A_i}, \quad P\text{-a. s.} \tag{4.57}$$

On each of the sets A_i, we can then apply Lemma 4.63 under the conditional probability measure $P[\,\cdot \mid A_i]$. Hence, there are random variables $Y_{1,i}, Y_{2,i}, \dots$ that all have the same distribution as Y under $P[\,\cdot \mid A_i]$ and such that for each $\varepsilon > 0$ there exists n_i such that for all $n \ge n_i$,

$$\left| \frac{1}{n} \sum_{j=1}^{n} Y_{j,i} - E[X \mid A_i] \right| < \varepsilon \quad P[\,\cdot \mid A_i]\text{-a. s.} \tag{4.58}$$

Now we let

$$Y_j := Y_{j,1}\mathbb{1}_{A_1} + \cdots + Y_{j,m}\mathbb{1}_{A_m}.$$

We claim that each Y_j has the same law as Y. To prove this claim, we apply the law of the total probability as follows. For any Borel set $B \subseteq \mathbb{R}$,

$$P[Y_j \in B] = \sum_{i=1}^{m} P[Y_{j,i} \in B \mid A_i] \cdot P[A_i] = \sum_{i=1}^{m} P[Y \in B \mid A_i] \cdot P[A_i] = P[Y \in B].$$

This shows that Y_j has the same law as Y. Finally, combining the law of the total probability with (4.57) and (4.58) yields that for each $n \ge n_1 \vee \cdots \vee n_m$,

$$\left| \frac{1}{n} \sum_{j=1}^{n} Y_j - E[X \mid \mathcal{G}] \right| < \varepsilon \quad P\text{-a. s.},$$

which completes the proof. □

After these preparations, we can now prove Proposition 4.59.

Proof of Proposition 4.59. We may assume without loss of generality that $0 \le X < 1$. For any $\ell \in \mathbb{N}$, we define $\mathcal{G}_\ell$ to be the σ-algebra generated by the events

1 To construct such a partition, let $B_1, \dots, B_N$ denote the elements of the finite σ-algebra $\mathcal{G}$. Then consider the collection $\mathcal{C}$ of all sets of the form $C_1 \cap \cdots \cap C_N$, where each C_i is either B_i or B_i^c. After removing the empty set from $\mathcal{C}$, it will consist of the desired partition.

$$A_{\ell,i} := \left\{\frac{i-1}{\ell} \leq X < \frac{i}{\ell}\right\}, \quad \text{for } i = 1, \dots, \ell,$$

which form a partition of Ω. Then

$$E[X \mid \mathcal{G}_\ell] = \sum_{i=1}^{\ell} E[X \mid A_{\ell,i}] \mathbb{1}_{A_{\ell,i}}, \quad P\text{-a. s.},$$

where we define $E[X \mid A_{\ell,i}] := 0$ if $P[A_{\ell,i}] = 0$. Hence,

$$\|E[X \mid \mathcal{G}_\ell] - X\|_\infty \leq \frac{1}{\ell}. \tag{4.59}$$

Next, dominated convergence implies that $X_k \to X$ in L^1. Therefore, for each fixed ℓ,

$$a_{k,\ell} := \|E[X_k \mid \mathcal{G}_\ell] - E[X \mid \mathcal{G}_\ell]\|_\infty = \max_{i=1,\dots,\ell} |E[X_k - X \mid A_{\ell,i}]| \longrightarrow 0 \quad \text{as } k \uparrow \infty. \tag{4.60}$$

It follows that for each ℓ, there is k_ℓ such that $a_{k,\ell} \leq 1/\ell$ for all $k \geq k_\ell$. We may assume $k_\ell < k_{\ell+1}$ for $\ell = 1, 2, \dots$, and we define $k_0 := 0$. Then, for each k, there exists a unique ℓ, denoted by $\ell(k)$, such that

$$k_\ell \leq k < k_{\ell+1}.$$

Then $\ell(1) \leq \ell(2) \leq \cdots$, and we claim that $\ell(k) \to \infty$ as $k \uparrow \infty$. Indeed, otherwise there are $K, L \in \mathbb{N}$ such that $\ell(k) = L$ for all $k \geq K$, hence $k < k_{L+1}$ for all $k \geq K$, which is impossible.

Finally, for each k, Lemma 4.64 yields random variables $X_{k,1}, X_{k,2}, \dots$ that have the same law as X_k and a number $n_k \in \mathbb{N}$ such that

$$\left\| \frac{1}{n_k} \sum_{j=1}^{n_k} X_{k,j} - E[X_k \mid \mathcal{G}_{\ell(k)}] \right\|_\infty < \frac{1}{\ell(k)}. \tag{4.61}$$

We claim that this choice is as desired. Indeed, by (4.61), (4.60), and (4.59),

$$\begin{aligned}
&\left\| \frac{1}{n_k} \sum_{j=1}^{n_k} X_{k,j} - X \right\|_\infty \\
&\quad \leq \left\| \frac{1}{n_k} \sum_{j=1}^{n_k} X_{k,j} - E[X_k \mid \mathcal{G}_{\ell(k)}] \right\|_\infty + \|E[X_k - X \mid \mathcal{G}_{\ell(k)}]\|_\infty + \|E[X \mid \mathcal{G}_{\ell(k)}] - X\|_\infty \\
&\quad \leq \frac{1}{\ell(k)} + a_{k,\ell(k)} + \frac{1}{\ell(k)} \leq \frac{3}{\ell(k)},
\end{aligned}$$

which converges to zero as $k \uparrow \infty$. □

4.7 Robust representation of law-invariant risk measures

As in Section 4.6, we assume in the sequel that the underlying probability space $(\Omega, \mathscr{F}, P)$ is atomless. Under this condition, Theorem 4.58 states that every law-invariant convex risk measure on L^∞ has the Fatou property or, equivalently, is continuous from above. By Theorem 4.33, ρ therefore admits a robust representation of the form

$$\rho(X) = \sup_{Q \in \mathscr{M}_1(P)} (E_Q[-X] - \alpha^{\min}(Q)), \quad X \in L^\infty,$$

where $\alpha^{\min}$ is the minimal penalty function of ρ. In this section, we analyze additional aspects of this robust representation that arise from the law-invariance of ρ. We start with the following structure theorem for law-invariant convex risk measures.

Theorem 4.65. *Let ρ be a convex risk measure. Then ρ is law-invariant if and only if it is continuous from above and its minimal penalty function $\alpha^{\min}(Q)$ depends only on the law of $\varphi_Q := \frac{dQ}{dP}$ under P for $Q \in \mathscr{M}_1(P)$. In this case, ρ has the representation*

$$\rho(X) = \sup_{Q \in \mathscr{M}_1(P)} \left(\int_0^1 q_{-X}(t) q_{\varphi_Q}(t)\, dt - \alpha^{\min}(Q) \right), \tag{4.62}$$

and the minimal penalty function satisfies

$$\begin{aligned} \alpha^{\min}(Q) &= \sup_{X \in \mathscr{A}_\rho} \int_0^1 q_{-X}(t) q_{\varphi_Q}(t)\, dt \\ &= \sup_{X \in L^\infty} \left(\int_0^1 q_{-X}(t) q_{\varphi_Q}(t)\, dt - \rho(X) \right). \end{aligned} \tag{4.63}$$

For the proof, we will need the following proposition. Here and in the sequel, we will write $X \sim \widetilde{X}$ to indicate that the two random variables X and $\widetilde{X}$ have the same law.

Proposition 4.66. *For two random variables X and Y,*

$$\int_0^1 q_X(t) q_Y(t)\, dt = \max_{\widetilde{X} \sim X} E[\widetilde{X} Y],$$

provided that all occurring integrals are well-defined.

Proof. The upper Hardy–Littlewood inequality in Theorem D.13 yields "≥." To prove the converse inequality, take a random variable U with a uniform distribution on $(0, 1)$ such that $Y = q_Y(U)$ P-almost surely. Such a random variable U exists by Lemma D.17 and our assumption that the underlying probability space is atomless. Since $\widetilde{X} := q_X(U) \sim X$ by Lemma D.7, we obtain

$$E[\widetilde{X}Y] = E[q_X(U)q_Y(U)] = \int_0^1 q_X(t)q_Y(t)\,dt,$$

and hence "≤." □

Proof of Theorem 4.65. Suppose first that ρ is law-invariant. Then ρ is continuous from above by Theorem 4.58. Moreover, $X \in \mathscr{A}_\rho$ implies that $\widetilde{X} \in \mathscr{A}_\rho$ for all $\widetilde{X} \sim X$. Hence,

$$\alpha^{\min}(Q) = \sup_{X\in\mathscr{A}_\rho} E[-X\varphi_Q] = \sup_{X\in\mathscr{A}_\rho}\sup_{\widetilde{X}\sim X} E[-\widetilde{X}\varphi_Q] = \sup_{X\in\mathscr{A}_\rho}\int_0^1 q_{-X}(t)q_{\varphi_Q}(t)\,dt,$$

where we have used Proposition 4.66 in the final step. It follows that $\alpha^{\min}(Q)$ depends only on the law of φ_Q. We have thus also established the first identity in (4.63). For the second one, just note that $\widetilde{X} := X + \rho(X)$ belongs to $\mathscr{A}_\rho$ for any $X \in L^\infty$ and that $q_{-X} - \rho(X)$ is a quantile function for $-\widetilde{X}$.

Conversely, let us assume that $\alpha^{\min}(Q)$ depends only on the law of φ_Q. For the moment, let us write $\widetilde{Q} \sim Q$ to indicate that φ_Q and $\varphi_{\widetilde{Q}}$ have the same law. Then Proposition 4.66 yields

$$\begin{aligned}\rho(X) &= \sup_{Q\in\mathscr{M}_1(P)}(E_Q[-X] - \alpha^{\min}(Q))\\ &= \sup_{Q\in\mathscr{M}_1(P)}\sup_{\widetilde{Q}\sim Q}(E[-X\varphi_{\widetilde{Q}}] - \alpha^{\min}(Q))\\ &= \sup_{Q\in\mathscr{M}_1(P)}\left(\int_0^1 q_{-X}(t)q_{\varphi_Q}(t)\,dt - \alpha^{\min}(Q)\right).\end{aligned}$$

We thus have established (4.62), and the law-invariance of ρ follows. □

Exercise 4.7.1. For $0 < \lambda < 1$, start from the representation of Average Value at Risk as

$$\mathrm{AV@R}_\lambda(X) := \sup\left\{E_Q[-X] \,\middle|\, Q \in \mathscr{M}_1(P) \text{ and } \frac{dQ}{dP} \le \frac{1}{\lambda}\ P\text{-a.s.}\right\},$$

check that the conditions of Theorem 4.65 are satisfied, and deduce from (4.62) that the representation

$$\mathrm{AV@R}_\lambda(X) = \frac{1}{\lambda}\int_0^\lambda \mathrm{V@R}_\gamma(X)\,d\gamma$$

holds. ◇

Example 4.67. Let $u : \mathbb{R} \to \mathbb{R}$ be an increasing concave function, and suppose that a position $X \in L^\infty$ is acceptable if $E[u(X)] \ge c$, where c is a given constant in the interior of $u(\mathbb{R})$. We have seen in Example 4.10 that the corresponding acceptance set induces a convex risk measure ρ. Clearly, ρ is law-invariant, and it will be shown in Proposition 4.124 that ρ is continuous from below, and hence from above. Moreover, the corresponding minimal penalty function can be computed as

$$\alpha^{\min}(Q) = \inf_{\lambda>0} \frac{1}{\lambda}\left(-c + \int_0^1 \ell^*(\lambda \cdot q_{\varphi_Q}(t))\, dt \right),$$

where

$$\ell^*(y) = \sup_{x\in\mathbb{R}}(xy + u(-x)) = \sup_{x\in\mathbb{R}}(xy - \ell(x))$$

is the Fenchel–Legendre transform of the convex increasing loss function $\ell(x) := -u(-x)$; see Theorem 4.126. ◇

The following theorem clarifies the crucial role of the risk measures $\mathrm{AV@R}_\lambda$: they can be viewed as the building blocks for law-invariant convex risk measures on L^∞. Recall that we assume that $(\Omega, \mathscr{F}, P)$ is atomless.

Theorem 4.68. *A convex risk measure ρ is law-invariant if and only if*

$$\rho(X) = \sup_{\mu\in\mathscr{M}_1((0,1])} \left(\int_{(0,1]} \mathrm{AV@R}_\lambda(X)\, \mu(d\lambda) - \beta^{\min}(\mu) \right), \tag{4.64}$$

where

$$\beta^{\min}(\mu) = \sup_{X\in\mathscr{A}_\rho} \int_{(0,1]} \mathrm{AV@R}_\lambda(X)\, \mu(d\lambda).$$

Proof. Clearly, the right-hand side of (4.64) defines a law-invariant convex risk measure that is continuous from above. Conversely, let ρ be law-invariant. Then ρ is continuous from above by Theorem 4.65. We will show that for $Q \in \mathscr{M}_1(P)$ there exists a probability measure $\mu \in \mathscr{M}_1((0,1])$ such that

$$\int_0^1 q_{-X}(t) q_\varphi(t)\, dt = \int_{(0,1]} \mathrm{AV@R}_s(X)\, \mu(ds),$$

where $\varphi := \varphi_Q = \frac{dQ}{dP}$. Then the assertion will follow from Theorem 4.65. Since $q_{-X}(t) = \mathrm{V@R}_{1-t}(X)$ and $q_\varphi(t) = q_\varphi^+(t)$ for a. e. $t \in (0,1)$,

$$\int_0^1 q_{-X}(t) q_\varphi(t)\, dt = \int_0^1 \mathrm{V@R}_t(X) q_\varphi^+(1-t)\, dt. \tag{4.65}$$

Since q_φ^+ is nonnegative, increasing, and right-continuous, we can write $q_\varphi^+(t) = \nu([1-t,1])$ for some positive Radon measure ν on $(0,1]$; see Remark D.8. Moreover, the measure μ given by $\mu(dt) = t\,\nu(dt)$ is a probability measure on $(0,1]$:

$$\begin{aligned}\int_{(0,1]} t\,\nu(dt) &= \int_{(0,1]}\int_0^1 \mathbb{1}_{\{s\le t\}}\,ds\,\nu(dt) = \int_0^1\int_{(0,1]} \mathbb{1}_{\{s\le t\}}\,\nu(dt)\,ds \\ &= \int_0^1 \nu([s,1])\,ds = \int_0^1 q_\varphi^+(s)\,ds = E[\,\varphi\,] = 1.\end{aligned}$$

Thus, from (4.65),

$$\begin{aligned}\int_0^1 q_{-X}(t)q_\varphi(t)\,dt &= \int_0^1 \mathrm{V@R}_t(X)\int_{(t,1]}\frac{1}{s}\,\mu(ds)\,dt \\ &= \int_{(0,1]}\frac{1}{s}\int_0^s \mathrm{V@R}_t(X)\,dt\,\mu(ds) \\ &= \int_{(0,1]} \mathrm{AV@R}_s(X)\,\mu(ds).\end{aligned} \tag{4.66}$$

Conversely, for any probability measure μ on $(0,1]$, the function q defined by $q(t) := \int_{(1-t,1]} s^{-1}\,\mu(ds)$ can be viewed as a quantile function of the density $\varphi := q(U)$ of a measure $Q \in \mathcal{M}_1(P)$, where U has a uniform distribution on $(0,1)$. Indeed, φ is nonnegative, and

$$E[\varphi] = \int_0^1 q(t)\,dt = \int_0^1\int_{(1-t,1]}\frac{1}{s}\,\mu(ds)\,dt = \int_{(0,1]}\int_0^1 \mathbb{1}_{\{t>1-s\}}\,dt\,\frac{1}{s}\,\mu(ds) = \mu((0,1]) = 1.$$

Altogether, we obtain a one-to-one correspondence between laws of densities φ and probability measures μ on $(0,1]$. □

Theorem 4.68 takes the following form for coherent risk measures.

Corollary 4.69. *A coherent risk measure ρ is law-invariant if and only if*

$$\rho(X) = \sup_{\mu\in\mathcal{M}}\int_{(0,1]} \mathrm{AV@R}_\lambda(X)\,\mu(d\lambda)$$

for some set $\mathcal{M} \subseteq \mathcal{M}_1((0,1])$.

Remark 4.70. Any law-invariant monetary risk measure ρ is monotone with respect to the increasing order $\succcurlyeq_{\mathrm{i}}$, which was introduced in Definition 2.67 and is sometimes also called first-order stochastic dominance. More precisely,

$$\mu \succcurlyeq_{\mathrm{i}} \nu \quad \Longrightarrow \quad \rho(X_\mu) \le \rho(X_\nu),$$

if X_μ and X_ν are random variables with distributions μ and ν. To prove this, let q_μ and q_ν be quantile functions for μ and ν and take a random variable U with a uniform distribution on $(0,1)$. Then $\widetilde{X}_\mu := q_\mu(U) \ge q_\nu(U) =: \widetilde{X}_\nu$ by Theorem 2.68, and $\widetilde{X}_\mu$ and $\widetilde{X}_\nu$ have the same distribution as X_μ and X_ν by Lemma D.7. Hence, law-invariance and monotonicity of ρ imply $\rho(X_\mu) = \rho(\widetilde{X}_\mu) \le \rho(\widetilde{X}_\nu) = \rho(X_\nu)$. ◇

Randomness of a position is reduced in terms of P if we replace the position by its conditional expectation with respect to some σ-algebra $\mathscr{G} \subseteq \mathscr{F}$. Such a reduction of randomness is reflected by a convex risk measure if it is law-invariant.

Corollary 4.71. *A convex risk measure ρ is law-invariant if and only if ρ is monotone with respect to the increasing concave order $\succcurlyeq_{\mathrm{icv}}$ introduced in* (3.30)*: For $Y, X \in L^\infty$,*

$$Y \succcurlyeq_{\mathrm{icv}} X \quad \Longrightarrow \quad \rho(Y) \le \rho(X). \tag{4.67}$$

Thus, a law-invariant convex risk measure ρ is dilatation-monotone *in the sense that*

$$\rho(E[X \mid \mathscr{G}]) \le \rho(X), \tag{4.68}$$

for $X \in \mathscr{X}$ and any σ-algebra $\mathscr{G} \subseteq \mathscr{F}$. In particular,

$$\rho(E[X]) = \rho(0) - E[X] \le \rho(X).$$

Proof. If ρ is law-invariant, then (4.67) follows from Theorem 4.68 combined with Remark 4.49. Conversely, if X and Y have the same law, then both $X \succcurlyeq_{\mathrm{icv}} Y$ and $Y \succcurlyeq_{\mathrm{icv}} X$ hold. Hence, (4.67) implies $\rho(X) = \rho(Y)$ and in turn the law-invariance of ρ.

The inequality (4.68) is a special case of (4.67), since $E[X \mid \mathscr{G}] \succcurlyeq_{\mathrm{icv}} X$ according to Theorem 2.57. The inequality for $\rho(E[X])$ follows from (4.68) by taking $\mathscr{G} = \{\emptyset, \Omega\}$. □

It was shown in [60] that the conditions (4.67) and (4.68) are actually equivalent. In other words, a convex risk measure that is dilatation-invariant is also law-invariant. We refer to [242] for additional monotonicity conditions that are equivalent to the law-invariance of a convex risk measure. Recall from Theorem 2.68 that $\mu \succcurlyeq_{\mathrm{i}} \nu$ implies $\mu \succcurlyeq_{\mathrm{icv}} \nu$. Thus, the conclusion of Corollary 4.71 is stronger than the one of Remark 4.70.

Remark 4.72. Let ρ be a law-invariant convex risk measure. If $\mathscr{G}_1 \subseteq \mathscr{G}_2 \subseteq \cdots \subseteq \mathscr{F}$ are σ-algebras and $\mathscr{G}_\infty = \sigma(\bigcup_n \mathscr{G}_n)$, then

$$\rho(E[X \mid \mathscr{G}_n]) \longrightarrow \rho(E[X \mid \mathscr{G}_\infty]) \quad \text{as } n \uparrow \infty,$$

for any $X \in L^\infty$. Indeed, Doob's martingale convergence theorem (see, e. g., Theorem 19.1 in [21]) states that $E[X \mid \mathscr{G}_n] \to E[X \mid \mathscr{G}_\infty]$ P-a. s. as $n \uparrow \infty$. Since the random variables $E[X \mid \mathscr{G}_n]$ are uniformly bounded, the Fatou property and Corollary 4.71 show that

$$\begin{aligned}\rho(E[X \mid \mathscr{G}_\infty]) &= \rho\Big(\lim_{n\uparrow\infty} E[X \mid \mathscr{G}_n]\Big) \\ &\le \liminf_{n\uparrow\infty} \rho(E[X \mid \mathscr{G}_n]) \\ &\le \rho(E[X \mid \mathscr{G}_\infty]).\end{aligned}$$ ◇

Recall from Proposition 4.47 that there exists no smallest convex risk measure ρ that is continuous from above and dominates $\mathrm{V@R}_\lambda$. If, however, we require in addition that ρ is law-invariant, then a best conservative approximation to $\mathrm{V@R}_\lambda$ exists, and it is given by $\mathrm{AV@R}_\lambda$. This is the content of the following theorem.

Theorem 4.73. $\mathrm{AV@R}_\lambda$ *is the smallest law-invariant convex measure of risk that dominates* $\mathrm{V@R}_\lambda$.

Proof. That $\mathrm{AV@R}_\lambda$ dominates $\mathrm{V@R}_\lambda$ was already stated in (4.49). Suppose now that ρ is another law-invariant convex risk measure that dominates $\mathrm{V@R}_\lambda$ and is continuous from above. We must show that for a given $X \in \mathscr{X}$,

$$\rho(X) \ge \mathrm{AV@R}_\lambda(X). \tag{4.69}$$

By cash invariance, we may assume without loss of generality that $X > 0$. Take $\varepsilon > 0$, and let $A := \{-X \ge \mathrm{V@R}_\lambda(X) - \varepsilon\}$ and

$$Y := E[X \mid X\mathbb{1}_{A^c}] = X \cdot \mathbb{1}_{A^c} + E[X \mid A] \cdot \mathbb{1}_A.$$

Since $X > q_X^+(\lambda) + \varepsilon \ge E[X \mid A]$ on A^c, we get $P[Y < E[X \mid A]] = 0$. On the other hand, $P[Y \le E[X \mid A]] \ge P[A] > \lambda$, and this implies that $\mathrm{V@R}_\lambda(Y) = E[-X \mid A]$. Since ρ dominates $\mathrm{V@R}_\lambda$, we have $\rho(Y) \ge E[-X \mid A]$. Thus, by Corollary 4.71,

$$\rho(X) \ge \rho(Y) \ge E[-X \mid -X \ge \mathrm{V@R}_\lambda(X) - \varepsilon].$$

Taking $\varepsilon \downarrow 0$ yields

$$\rho(X) \ge E[-X \mid -X \ge \mathrm{V@R}_\lambda(X)].$$

If the distribution of X is continuous, Corollary 4.54 states that the conditional expectation on the right equals $\mathrm{AV@R}_\lambda(X)$, and we obtain (4.69). If the distribution of X is not continuous, we denote by D the set of all points x such that $P[X = x] > 0$ and take any bounded random variable $Z \ge 0$ with a continuous distribution. Such a random variable exists due to our assumption that $(\Omega, \mathscr{F}, P)$ is atomless. Note that $X_n := X + \frac{1}{n} Z\mathbb{1}_{\{X\in D\}}$ has a continuous distribution. Indeed, for any x,

$$P[X_n = x] = P[X = x, X \notin D] + \sum_{y\in D} P[X = y, Z = n(x-y)] = 0.$$

Moreover, X_n decreases to X. The inequality (4.69) holds for each X_n and extends to X by continuity from above. □

Corollary 4.74. $\mathrm{AV@R}_\lambda$ *and* WCE_λ *coincide under our assumption that the probability space is atomless.*

Proof. We know from Corollary 4.54 that $\mathrm{WCE}_\lambda(X) = \mathrm{AV@R}_\lambda(X)$ if X has a continuous distribution. Repeating the approximation argument at the end of the preceding proof yields $\mathrm{WCE}_\lambda(X) = \mathrm{AV@R}_\lambda(X)$ for each $X \in \mathscr{X}$. □

4.8 Concave distortions

Let us now have a closer look at the coherent risk measures

$$\rho_\mu(X) := \int \mathrm{AV@R}_\lambda(X)\,\mu(d\lambda), \tag{4.70}$$

which appear in the Representation Theorem 4.68 for law-invariant convex risk measures. We are going to characterize these risk measures ρ_μ in two ways, first as Choquet integrals with respect to some concave distortion of the underlying probability measure P, and then in the next section, by a property of comonotonicity.

Again, we will assume throughout this section that the underlying probability space $(\Omega, \mathscr{F}, P)$ is atomless. Since $\mathrm{AV@R}_\lambda$ is coherent, continuous from below, and law-invariant, any mixture ρ_μ for some probability measure μ on $(0,1]$ has the same properties. According to Remark 4.50, we may set $\mathrm{AV@R}_0(X) = -\operatorname{ess\,inf} X$ so that we can extend the definition (4.70) to probability measures μ on the closed interval $[0,1]$. However, ρ_μ will only be continuous from above and not from below if $\mu(\{0\}) > 0$, because $\mathrm{AV@R}_0$ is not continuous from below.

Our first goal is to show that $\rho_\mu(X)$ can be identified with the Choquet integral of the loss $-X$ with respect to the set function $c_\psi(A) := \psi(P[A])$, where ψ is the concave function defined in the following lemma. Choquet integrals were introduced in Example 4.14, and the risk measure MINVAR of Exercise 4.1.9 provides a first example for a risk measure arising as the Choquet integral of a set function c_ψ. Recall that every concave function ψ admits a right-continuous, right-hand derivative ψ'_+ and that $\psi(t) = \psi(0+) + \int_0^t \psi'_+(s)\,ds$ for $t \in (0,1]$; see Proposition A.7.

Lemma 4.75. *The identity*

$$\psi'_+(t) = \int_{(t,1]} s^{-1}\,\mu(ds), \quad 0 < t < 1, \tag{4.71}$$

defines a one-to-one correspondence between probability measures μ *on* $[0,1]$ *and increasing concave functions* $\psi : [0,1] \to [0,1]$ *with* $\psi(0) = 0$ *and* $\psi(1) = 1$*. Moreover, we have* $\psi(0+) = \mu(\{0\})$.

Proof. Suppose first that μ is given and ψ is defined by $\psi(1) = 1$ and (4.71). Then ψ is concave and increasing on $(0,1]$. Moreover,

$$1 - \psi(0+) = \int_0^1 \psi'(t)\,dt = \int_{(0,1]} \frac{1}{s} \int_0^1 \mathbb{1}_{\{t<s\leq 1\}}\,dt\,\mu(ds) = \mu((0,1]) \leq 1.$$

Hence, we may set $\psi(0) := 0$ and obtain an increasing concave function on $[0,1]$. Furthermore, $\psi(0+) = 1 - \mu((0,1]) = \mu(\{0\})$.

Conversely, if ψ is given, then $\psi'_+(t)$ is a decreasing right-continuous function on $(0,1)$ and can be written as $\psi'_+(t) = \nu((t,1])$ for some positive Radon measure ν on $(0,1]$; see Remark D.8. We first define μ on $(0,1]$ by $\mu(dt) = t\,\nu(dt)$. Then (4.71) holds and, by Fubini's theorem,

$$\mu((0,1]) = \int_0^1 \int_{(0,1]} \mathbb{1}_{\{t<s\}}\,\nu(ds)\,dt = 1 - \psi(0+) \leq 1.$$

Hence, setting $\mu(\{0\}) := \psi(0+)$ defines a probability measure μ on $[0,1]$. □

Theorem 4.76. *For a probability measure μ on $[0,1]$, let ψ be the concave function defined in Lemma 4.75. Then, for $X \in \mathscr{X}$,*

$$\begin{aligned}\rho_\mu(-X) &= \psi(0+)\,\mathrm{AV@R}_0(-X) + \int_0^1 q_X(t)\psi'(1-t)\,dt \\ &= \int_{-\infty}^0 (\psi(P[X>x]) - 1)\,dx + \int_0^\infty \psi(P[X>x])\,dx.\end{aligned} \tag{4.72}$$

Proof. Using the fact that $\mathrm{V@R}_\lambda(-X) = q_X^-(1-\lambda)$, we get as in (4.66) that

$$\int_{(0,1]} \mathrm{AV@R}_\lambda(-X)\,\mu(d\lambda) = \int_0^1 q_X(t)\psi'(1-t)\,dt.$$

Hence, we obtain the first identity. For the second one, we assume first $X \geq 0$. Then

$$q_X^+(t) = \sup\{x \geq 0 \mid F_X(x) \leq t\} = \int_0^\infty \mathbb{1}_{\{F_X(x)\leq t\}}\,dx,$$

where F_X is the cumulative distribution function of X. Using Fubini's theorem and the fact that $\int_0^y \psi'(t)\,dt = (\psi(y) - \psi(0+))\mathbb{1}_{\{y>0\}}$, we obtain

$$\int_0^1 q_X(t)\psi'(1-t)\,dt = \int_0^\infty\int_0^1 \mathbb{1}_{\{F_X(x)\le 1-t\}}\psi'(t)\,dt\,dx$$
$$= \int_0^\infty \psi(1-F_X(x))\,dx - \psi(0+)\operatorname{ess\,sup} X.$$

This proves the second identity for $X \ge 0$, since $\psi(0+) = \mu(\{0\})$ and $\operatorname{ess\,sup} X = \mathrm{AV@R}_0(-X)$. If $X \in L^\infty$ is arbitrary, we consider $X + C$, where $C := -\operatorname{ess\,inf} X$. The cash invariance of ρ_μ yields

$$C + \rho_\mu(-X) = \int_0^\infty \psi(P[\,X > x - C\,])\,dx$$
$$= \int_{-C}^0 \psi(P[\,X > x\,])\,dx + \int_0^\infty \psi(P[\,X > x\,])\,dx$$
$$= C + \int_{-\infty}^0 (\psi(P[\,X > x\,]) - 1)\,dx + \int_0^\infty \psi(P[\,X > x\,])\,dx. \qquad \square$$

Example 4.77. Clearly, the risk measure $\mathrm{AV@R}_\lambda$ is itself of the form ρ_μ where $\mu = \delta_\lambda$. For $\lambda > 0$, the corresponding concave distortion function is given by

$$\psi(t) = \left(\frac{t}{\lambda}\right) \wedge 1 = \frac{1}{\lambda}(t \wedge \lambda).$$

Thus, we obtain yet another representation of $\mathrm{AV@R}_\lambda$:

$$\mathrm{AV@R}_\lambda(-X) = \frac{1}{\lambda}\int_0^\infty P[\,X > x\,] \wedge \lambda\,dx \quad \text{for } X \in L^\infty_+. \qquad \diamond$$

Exercise 4.8.1. Find the probability measure μ on $[0,1]$ such that the coherent risk measure MINVAR introduced in Exercise 4.1.9 is of the form (4.70). ◇

Exercise 4.8.2. Suppose that there exists a set $A \in \mathscr{F}$ with $0 < P[\,A\,] < 1$ such that $\rho_\mu(-\mathbb{1}_A) = -\rho_\mu(\mathbb{1}_A)$. Use the representation (4.72) to deduce that $\mu = \delta_1$, that is, $\rho_\mu(X) = E[\,-X\,]$ for all $X \in L^\infty$. More generally, show that $\mu = \delta_1$ if there exists a nonconstant $X \in L^\infty$ such that $\rho_\mu(-X) = -\rho_\mu(X)$. ◇

Corollary 4.78. *If $\mu(\{0\}) = 0$ in Theorem 4.76, then*

$$\rho_\mu(X) = -\int_0^1 q_X(\varphi(t))\,dt,$$

where φ is an inverse function of ψ, taken in the sense of Definition D.1.

Proof. Due to Lemma D.2, the distribution of φ under the Lebesgue measure has the concave distribution function ψ. By Proposition A.7(d), $\psi(t) = \psi(0) + \int_0^t \psi'(s)\,ds$ holds for any $t \in [0,1]$. Hence, the law of φ under the Lebesgue measure has the density ψ'. Therefore,

$$\int_0^1 q_X(\varphi(t))\,dt = \int_0^1 q_X(t)\psi'(t)\,dt = -\int_0^1 q_{-X}(1-t)\psi'(t)\,dt,$$

where we have used Lemma D.12 in the last step. An application of Theorem 4.76 concludes the proof. □

Let us continue with a brief discussion of the set function $c_\psi(A) = \psi(P[A])$.

Definition 4.79. Let $\psi : [0,1] \to [0,1]$ be an increasing function such that $\psi(0) = 0$ and $\psi(1) = 1$. The set function

$$c_\psi(A) := \psi(P[A]), \quad A \in \mathscr{F},$$

is called the *distortion* of the probability measure P with respect to the *distortion function* ψ. △

Definition 4.80. A set function $c : \mathscr{F} \to [0,1]$ is called *monotone* if

$$c(A) \le c(B) \quad \text{for } A \subseteq B$$

and *normalized* if

$$c(\emptyset) = 0 \quad \text{and} \quad c(\Omega) = 1.$$

A monotone set function is called *submodular* or *strongly subadditive* if

$$c(A \cup B) + c(A \cap B) \le c(A) + c(B). \qquad △$$

Clearly, any distortion c_ψ is normalized and monotone. Recall that $(\Omega, \mathscr{F}, P)$ is atomless.

Proposition 4.81. *Let c_ψ be the distortion of P with respect to the distortion function ψ. Then ψ is concave if and only if c_ψ is submodular.*

Proof. Suppose first that ψ is concave. Take $A, B \in \mathscr{F}$ with $P[A] \le P[B]$. We must show that $c := c_\psi$ satisfies

$$c(A) - c(A \cap B) \ge c(A \cup B) - c(B).$$

This is trivial if $r = 0$, where

$$r := P[A] - P[A \cap B] = P[A \cup B] - P[B].$$

For $r > 0$, the concavity of ψ yields via (A.5) that

$$\frac{c(A) - c(A \cap B)}{P[A] - P[A \cap B]} \geq \frac{c(A \cup B) - c(B)}{P[A \cup B] - P[B]}.$$

Multiplying both sides with r gives the result.

Now suppose that $c = c_\psi$ is submodular. By Exercise A.2.2 below, it is sufficient to show that $\psi(y) \geq (\psi(x) + \psi(z))/2$ whenever $0 \leq x \leq z \leq 1$ and $y = (x + z)/2$. To this end, we will construct two sets $A, B \in \mathscr{F}$ such that $P[A] = P[B] = y$, $P[A \cap B] = x$, and $P[A \cup B] = z$. Submodularity then gives $\psi(x) + \psi(z) \leq 2\psi(y)$, and in turn, the concavity of ψ.

In order to construct the two sets A and B, take a random variable U with a uniform distribution on $[0, 1]$, which exists by Proposition D.16 and our assumption that $(\Omega, \mathscr{F}, P)$ is atomless. Then

$$A := \{0 \leq U \leq y\} \quad \text{and} \quad B := \{z - y \leq U \leq z\}$$

are as desired. □

Let us now recall the notion of a Choquet integral, which was introduced in Example 4.14.

Definition 4.82. Let $c : \mathscr{F} \to [0, 1]$ be any set function that is normalized and monotone. The *Choquet integral* of a bounded measurable function X on $(\Omega, \mathscr{F})$ with respect to c is defined as

$$\int X \, dc := \int_{-\infty}^{0} (c(X > x) - 1) \, dx + \int_{0}^{\infty} c(X > x) \, dx. \qquad \triangle$$

Note that the Choquet integral coincides with the usual integral as soon as c is a σ-additive probability measure; see also Lemma 4.103 below.

With this definition, Theorem 4.76 allows us to identify the risk measure ρ_μ as the Choquet integral of the loss with respect to a concave distortion c_ψ of the underlying probability measure P:

Corollary 4.83. *For a probability measure μ on $[0, 1]$, let ψ be the concave distortion function defined in Lemma 4.75, and let c_ψ denote the distortion of P with respect to ψ. Then, for $X \in L^\infty$,*

$$\rho_\mu(X) = \int (-X) \, dc_\psi.$$

Combining Corollary 4.83 with Theorem 4.68, we obtain the following characterization of law-invariant convex risk measures in terms of concave distortions.

Corollary 4.84. *A convex risk measure ρ is law-invariant and continuous from above if and only if*

$$\rho(X) = \sup_{\psi}\left(\int (-X)\, dc_\psi - \gamma^{\min}(\psi) \right),$$

where the supremum is taken over the class of all concave distortion functions ψ and

$$\gamma^{\min}(\psi) := \sup_{X \in \mathscr{A}_\rho} \int (-X)\, dc_\psi.$$

Exercise 4.8.3. Let $c : \mathscr{F} \to [0,1]$ be a normalized and monotone set function and define the dual set function by $\widetilde{c}(A) := 1 - c(\Omega \setminus A)$. Show that the Choquet integrals with respect to c and $\widetilde{c}$ satisfy

$$\int (-X)\, dc = -\int X\, d\widetilde{c}$$

for bounded and measurable X. ◇

Exercise 4.8.4. For $\lambda \in \mathbb{R}$, consider the functional

$$\rho_\lambda(X) := E[\,-X\,] + \lambda E[\,|X - \widetilde{X}|\,], \quad X \in L^\infty,$$

where $\widetilde{X}$ is an independent copy of X. Recall from Exercise 4.1.8 that $\rho_\lambda(X)$ is a coherent risk measure if and only if $0 \le \lambda \le 1/2$. Show that $\rho_\lambda(X)$ can be represented in the form

$$\rho_\lambda(X) = \int_0^1 q_{-X}(t) f_\lambda(t)\, dt$$

for a certain function f_λ and use this representation to give another proof of the fact that ρ_λ is a law-invariant coherent risk measure if and only if $0 \le \lambda \le 1/2$. Then identify the corresponding distortion function in the representation of ρ_λ as a Choquet integral for $0 \le \lambda \le 1/2$. ◇

The following series of exercises should be compared with Exercises 4.1.9 and 4.8.1.

Exercise 4.8.5. For $\beta \ge 1$, consider the concave distortion function $\psi_\beta(x) := x^{\frac{1}{\beta}}$. Show that for $\beta \in \mathbb{N}$ the corresponding risk measure

$$\mathrm{MAXVAR}_\beta(X) := \int (-X)\, dc_{\psi_\beta}$$

has the property that $\mathrm{MAXVAR}_\beta(X) = -E[\,Y_1\,]$, if $Y_1, \dots, Y_\beta$ are independent and identically distributed (i. i. d.) random variables for which $\max(Y_1, \dots, Y_\beta) \sim X$. ◇

Exercise 4.8.6. For $\beta \geq 1$, consider the distortion function $\widetilde{\psi}_\beta(x) := (1-(1-x)^\beta)^{\frac{1}{\beta}}$. Show that for $\beta \in \mathbb{N}$ the corresponding risk measure

$$\text{MAXMINVAR}_\beta(X) := \int (-X)\, dc_{\widetilde{\psi}_\beta}$$

has the property that $\text{MAXMINVAR}_\beta(X) = -E[\,Y_1\,]$, if $Y_1, \dots, Y_\beta$ are i. i. d. random variables for which $\max(Y_1, \dots, Y_\beta) \sim \min(X_1, \dots, X_\beta)$, where $X_1, \dots, X_\beta$ are independent copies of X. ◇

Exercise 4.8.7. For $\beta \geq 1$, consider the distortion function $\hat{\psi}_\beta(x) := 1-(1-x^{\frac{1}{\beta}})^\beta$. Show that for $\beta \in \mathbb{N}$ the corresponding risk measure

$$\text{MINMAXVAR}_\beta(X) := \int (-X)\, dc_{\hat{\psi}_\beta}$$

has the property that $\text{MINMAXVAR}_\beta(X) = -E[\min(Y_1, \dots, Y_\beta)\,]$, if $Y_1, \dots, Y_\beta$ are i. i. d. random variables for which $\max(Y_1, \dots, Y_\beta) \sim X$. ◇

As another consequence of Theorem 4.76, we obtain an explicit description of the maximal representing set $\mathcal{Q}_\mu \subseteq \mathcal{M}_1(P)$ for the coherent risk measure ρ_μ.

Theorem 4.85. *Let μ be a probability measure on $[0,1]$, and let ψ be the corresponding concave function defined in Lemma 4.75. Then ρ_μ can be represented as*

$$\rho_\mu(X) = \sup_{Q \in \mathcal{Q}_\mu} E_Q[\,-X\,],$$

where the set $\mathcal{Q}_\mu$ is given by

$$\mathcal{Q}_\mu := \left\{ Q \in \mathcal{M}_1(P) \;\middle|\; Z := \frac{dQ}{dP} \text{ satisfies } \int_t^1 q_Z(s)\, ds \leq \psi(1-t) \text{ for } t \in (0,1) \right\}.$$

Moreover, $\mathcal{Q}_\mu$ is the maximal subset of $\mathcal{M}_1(P)$ that represents ρ_μ.

Proof. The risk measure ρ_μ is coherent and continuous from above. By Corollary 4.37, it can be represented by taking the supremum of expectations over the set $\mathcal{Q}^{\max} = \{Q \in \mathcal{M}_1(P) \mid \alpha^{\min}(Q) = 0\}$. Using (4.63) and Theorem 4.76, we see that a measure $Q \in \mathcal{M}_1(P)$ with density $Z = dQ/dP$ belongs to $\mathcal{Q}^{\max}$ if and only if

$$\begin{aligned}\int_0^1 q_X(s) q_Z(s)\, ds &\leq \rho_\mu(-X) \\ &= \psi(0+)\,\text{AV@R}_0(-X) + \int_0^1 q_X(s)\psi'(1-s)\, ds\end{aligned} \tag{4.73}$$

for all $X \in L^\infty$. Let X be a Bernoulli-distributed random variable with parameter $1-t$, that is, X takes the two values 0 and 1 and $P[X=1]=1-t$. Then we have $q_X = \mathbb{1}_{[t,1]}$ a. e., and so we obtain

$$\int_t^1 q_Z(s)\,ds \le \psi(0+) + \int_t^1 \psi'(1-s)\,ds = \psi(1-t)$$

for all $t \in (0,1)$. Hence, $\mathcal{Q}^{\max} \subseteq \mathcal{Q}_\mu$. For the proof of the converse inclusion, we show that the density Z of a fixed measure $Q \in \mathcal{Q}_\mu$ satisfies (4.73) for any given $X \in L^\infty$. We may assume without loss of generality that $X \ge 0$. Let ν be the positive finite measure on $[0,1]$ such that $q_X^+(s) = \nu([0,s])$; see Remark D.8. Then, in particular,

$$\nu([0,1]) = q_X(1-) = \operatorname{ess\,sup} X = \mathrm{AV@R}_0(-X).$$

Using Fubini's theorem and the definition of $\mathcal{Q}_\mu$, we get

$$\begin{aligned}\int_0^1 q_X(s)q_Z(s)\,ds &= \int_0^1 q_Z(s)\int_{[0,1]} \mathbb{1}_{\{t\le s\}}\,\nu(dt)\,ds = \int_{[0,1]}\int_t^1 q_Z(s)\,ds\,\nu(dt)\\ &\le \int_{[0,1]} \psi(1-t)\,\nu(dt) = \int_{[0,1]}\left(\psi(0+) + \int_0^{1-t}\psi'(s)\,ds\right)\nu(dt)\\ &= \psi(0+)\nu([0,1]) + \int_{[0,1]}\int_0^1 \mathbb{1}_{\{s\le 1-t\}}\psi'(s)\,ds\,\nu(dt)\\ &= \psi'(0+)\,\mathrm{AV@R}_0(-X) + \int_0^1 \psi'(1-s)\int_{[0,s]}\nu(dt)\,ds,\end{aligned}$$

which coincides with the right-hand side of (4.73). □

Corollary 4.86. *In the context of Theorem 4.85, the following conditions are equivalent:*

(a) ρ_μ *is continuous from below.*

(b) $\mu(\{0\}) = 0$.

If these equivalent conditions are satisfied, then, for any $X \in L^\infty$, the supremum in the representation

$$\rho_\mu(X) = \sup_{Q\in\mathcal{Q}_\mu} E_Q[-X]$$

is attained by the measure $Q_X \in \mathcal{Q}_\mu$ with density $dQ_X/dP = f(X)$, where f is a decreasing function such that

$$f(x) = \begin{cases} \psi'(F_X(x)) & \text{if } F_X(x) \in (0,1) \text{ and } F_X(x-) = F_X(x), \\ \dfrac{1}{F_X(x) - F_X(x-)} \displaystyle\int_{F_X(x-)}^{F_X(x)} \psi'(t)\,dt & \text{if } F_X(x-) < F_X(x). \end{cases}$$

Moreover, with λ denoting the Lebesgue measure on $(0,1)$,

$$\mathcal{Q}_\mu = \left\{ Q \ll P \;\middle|\; P \circ \left(\frac{dQ}{dP}\right)^{-1} \succcurlyeq_{\text{icv}} \lambda \circ (\psi')^{-1} \right\}. \tag{4.74}$$

Proof. If (b) holds, then (a) follows from Theorem 4.52 and monotone convergence. For the proof of the converse implication, we suppose by way of contradiction that $\delta := \mu(\{0\}) > 0$. In this case, we can write

$$\rho_\mu = \delta \,\text{AV@R}_0 + (1-\delta)\rho_{\mu'},$$

where $\mu' := \mu(\cdot \,|\, (0,1])$. Then $\rho_{\mu'}$ is continuous from below since $\mu'(\{0\}) = 0$, but on our atomless probability space, AV@R_0 is not, and so ρ_μ does not satisfy (a); see Remark 4.50.

Let us now prove the remaining assertions. Since $\psi(0+) = \mu(\{0\}) = 0$, a measure Q with density $Z = dQ/dP$ belongs to $\mathcal{Q}_\mu$ if and only if $\int_t^1 q_Z(s)\,ds \le \int_t^1 \psi'(1-s)\,ds$ for all t. Since $\psi'(1-t)$ is a quantile function for the law of ψ' under λ, part (e) of Theorem 2.57 implies (4.74). The problem of identifying the maximizing measure Q_X is hence equivalent to minimizing $E[\,ZX\,]$ under the constraint that Z is a density function such that $P \circ Z^{-1} \succcurlyeq_{\text{icv}} \lambda \circ (\psi')^{-1}$. Let us first assume that $X \ge 0$. Then it follows from Theorem 3.38 that $f(X)$ minimizes $E[\,YX\,]$ over all $Y \in L^1_+$ such that $P \circ Y^{-1} \succcurlyeq_{\text{icv}} \lambda \circ (\psi')^{-1}$. Moreover, equation (3.34) shows that $E[\,f(X)\,] = \int_0^1 \psi'(t)\,dt = 1$, and so $Z_X := f(X) \ge 0$ is the density of an optimal probability measure $Q_X \in \mathcal{Q}_\mu$. If X is not positive, then we may take a constant c such that $X + c \ge 0$ and apply the preceding argument. The formula for f then follows from the fact that $F_{X+c}(x+c) = F_X(x)$. □

Exercise 4.8.8. For a concave distortion function ψ on $[0,1]$ define its convex conjugate function as

$$\varphi(x) := \sup_{y \in [0,1]} (\psi(y) - xy), \quad x \ge 0.$$

Show that the set $\mathcal{Q}_\mu$ in Theorem 4.85 can be represented as

$$\mathcal{Q}_\mu = \left\{ Q \in \mathcal{M}_1(P) \;\middle|\; Z := \frac{dQ}{dP} \text{ satisfies } E[\,(Z-x)^+\,] \le \varphi(x) \text{ for } x \ge 0 \right\}. \qquad \diamond$$

Remark 4.87. As long as we are interested in a law-invariant risk assessment, we can represent a financial position $X \in L^\infty$ by its cumulative distribution function F_X or, equivalently, by the function

$$G_X(t) := 1 - F_X(t) = P[\,X > t\,].$$

If we only consider positions X with values in $[0,1]$ then their proxies G_X vary in the class of right-continuous decreasing functions G on $[0,1]$ such that $G(1) = 0$ and $G(0) \leq 1$. Due to Theorem 4.76, a law-invariant coherent risk measure ρ_μ induces a functional U on the class of proxies via

$$U(G_X) := \rho_\mu(-X) = \int_0^1 \psi(G_X(t))\,dt.$$

Since ψ is increasing and concave, the functional U has the form of a von Neumann–Morgenstern utility functional on the probability space given by Lebesgue measure on the unit interval $[0,1]$. As such, it can be characterized by the axioms in Section 2.3, and this is the approach taken in Yaari's "dual theory of choice" [322]. More generally, we can introduce a utility function u on $[0,1]$ with $u(0) = 0$ and consider the functional

$$U(G_X) := \int_0^1 \psi(G_X(t))\,du(t)$$

introduced by Quiggin [268]. For $u(x) = x$, this reduces to the "dual theory of choice," for $\psi(x) = x$ we recover the classical utility functionals

$$\begin{aligned} U(G_X) &= \int_0^1 G_X(t)\,du(t) \\ &= -\int_0^1 u(t)\,dG_X(t) \\ &= E[\,u(X)\,] \end{aligned}$$

discussed in Section 2.3. ◇

4.9 Comonotonic risk measures

In many situations, the risk of a combined position $X + Y$ will be strictly lower than the sum of the individual risks, because one position serves as a hedge against adverse changes in the other position. If, on the other hand, there is no way for X to work as a hedge for Y then we may want the risk simply to add up. In order to make this idea precise, we introduce the notion of comonotonicity. Our main goal in this section is to characterize the class of all convex risk measures that share this property of comonotonicity.

As in the first two sections of this chapter, we will denote by $\mathscr{X}$ the linear space of all bounded measurable functions on the measurable space $(\Omega, \mathscr{F})$.

Definition 4.88. Two measurable functions X and Y on $(\Omega, \mathscr{F})$ are called *comonotone* if

$$(X(\omega) - X(\omega'))(Y(\omega) - Y(\omega')) \geq 0 \quad \text{for all } (\omega, \omega') \in \Omega \times \Omega. \tag{4.75}$$

A monetary risk measure ρ on $\mathscr{X}$ is called *comonotonic* if

$$\rho(X + Y) = \rho(X) + \rho(Y)$$

whenever $X, Y \in \mathscr{X}$ are comonotone. △

Lemma 4.89. *If ρ is a comonotonic monetary risk measure on $\mathscr{X}$, then ρ is positively homogeneous.*

Proof. Note that (X, X) is a comonotone pair. Hence, $\rho(2X) = 2\rho(X)$. An iteration of this argument yields $\rho(rX) = r\rho(X)$ for all rational numbers $r \geq 0$. Positive homogeneity now follows from the Lipschitz continuity of ρ; see Lemma 4.3. □

We will see below that every comonotonic monetary risk measure on $\mathscr{X}$ arises as the Choquet integral with respect to a certain set function on $(\Omega, \mathscr{F})$. In the sequel, $c : \mathscr{F} \to [0, 1]$ will always denote a set function that is normalized and monotone; see Definition 4.80. Unless otherwise mentioned, we will not assume that c enjoys any additivity properties. Recall from Definition 4.82 that the Choquet integral of $X \in \mathscr{X}$ with respect to c is defined as

$$\int X \, dc = \int_{-\infty}^{0} (c(X > x) - 1) \, dx + \int_{0}^{\infty} c(X > x) \, dx.$$

The proof of the following proposition was already given in Example 4.14.

Proposition 4.90. *The Choquet integral of the loss,*

$$\rho(X) := \int (-X) \, dc,$$

is a monetary risk measure on $\mathscr{X}$ that is positively homogeneous.

Definition 4.91. Let X be a measurable function on $(\Omega, \mathscr{F})$. An inverse function $r_X : (0, 1) \to \mathbb{R}$ of the increasing function $G_X(x) := 1 - c(X > x)$, taken in the sense of Definition D.1, is called a *quantile function* for X with respect to c. △

If c is a probability measure, then $G_X(x) = c(X \leq x)$. Hence, the preceding definition extends the notion of a quantile function given in Definition D.9. The following proposition yields an alternative representation of the Choquet integral in terms of quantile functions with respect to c.

Proposition 4.92. *Let r_X be a quantile function with respect to c for $X \in \mathscr{X}$. Then*

$$\int X\,dc = \int_0^1 r_X(t)\,dt.$$

Proof. We have $\int (X+m)\,dc = \int X\,dc + m$, and one easily checks that $r_{X+m} = r_X + m$ a. e. for all $m \in \mathbb{R}$ and each quantile function r_{X+m} of $X + m$. Thus, we may assume without loss of generality that $X \geq 0$. In this case, Remark D.4 and Lemma D.2 imply that the largest quantile function r_X^+ is given by

$$r_X^+(t) = \sup\{x \geq 0 \mid G_X(x) \leq t\} = \int_0^\infty \mathbb{1}_{\{G_X(x) \leq t\}}\,dx.$$

Since $r_X = r_X^+$ a. e. on $(0,1)$, Fubini's theorem implies

$$\begin{aligned} \int_0^1 r_X(t)\,dt &= \int_0^1 \int_0^\infty \mathbb{1}_{\{G_X(x)\leq t\}}\,dx\,dt \\ &= \int_0^\infty (1 - G_X(x))\,dx \\ &= \int X\,dc. \end{aligned}$$

□

The preceding proposition yields the following generalization of Corollary 4.78 if applied to a continuous distortion of a probability measure as defined in Definition 4.79.

Corollary 4.93. *Let $c_\psi(A) = \psi(P[\,A\,])$ be the distortion of the probability measure P with respect to the continuous distortion function ψ. If φ is an inverse function for the increasing function ψ in the sense of Definition* D.1, *then the Choquet integral with respect to c_ψ satisfies*

$$\int X\,dc_\psi = \int_0^1 q_X(1 - \varphi(t))\,dt,$$

where q_X is a quantile function for $X \in \mathscr{X}$, taken with respect to P.

Proof. Due to the continuity of ψ, we have $\psi(a) \leq t$ if and only if $a \leq \varphi^+(t) = \inf\{x \mid \psi(x) > t\}$. Thus, we can compute the lower quantile function of X with respect to c_ψ:

$$\begin{aligned} r_X^-(t) &= \inf\{x \in \mathbb{R} \mid 1 - c_\psi(X > x) \geq t\} \\ &= \inf\{x \in \mathbb{R} \mid \psi(P[\,X > x\,]) \leq 1 - t\} \end{aligned}$$

$$= \inf\{x \in \mathbb{R} \mid P[X > x] \le \varphi^+(1-t)\}$$
$$= q_X^-(1 - \varphi^+(1-t)).$$

Next, note that $\varphi^+(t) = \varphi(t)$ for a. e. t. Moreover, φ has the continuous cumulative distribution function ψ under the Lebesgue measure, and so we can replace q_X^- by the arbitrary quantile function q_X. □

Theorem 4.94. *A monetary risk measure ρ on $\mathcal{X}$ is comonotonic if and only if there exists a normalized monotone set function c on $(\Omega, \mathcal{F})$ such that*

$$\rho(X) = \int (-X)\, dc, \quad X \in \mathcal{X}.$$

In this case, c is given by $c(A) = \rho(-\mathbb{1}_A)$.

The preceding theorem implies in view of Corollary 4.83 that all mixtures

$$\rho_\mu = \int_{[0,1]} \mathrm{AV@R}_\lambda \; \mu(d\lambda)$$

are comonotonic. We will see in Theorem 4.99 below that these are in fact all convex risk measures that are law-invariant and comonotonic. The proof of Theorem 4.94 requires a further analysis of comonotone random variables.

Lemma 4.95. *For two measurable functions X and Y on $(\Omega, \mathcal{F})$, the following conditions are equivalent:*
(a) *X and Y are comonotone.*
(b) *There exists a measurable function Z on $(\Omega, \mathcal{F})$ and increasing functions f and g on $\mathbb{R}$ such that $X = f(Z)$ and $Y = g(Z)$.*
(c) *There exist increasing and Lipschitz continuous functions f and g on $\mathbb{R}$ such that $X = f(X+Y)$ and $Y = g(X+Y)$.*

Proof. The implications (c)⇒(b) and (b)⇒(a) are obvious. To prove (a)⇒(c), suppose that X and Y are comonotone and define Z by $Z := X + Y$. We show that $z := Z(\omega)$ has a unique decomposition as $z = x + y$, where $(x,y) = (X(\omega'), Y(\omega'))$ for some $\omega' \in \Omega$. After having established this, we can put $f(z) := x$ and $g(z) := y$. We will show later on that these functions are indeed increasing and Lipschitz continuous. The existence of the decomposition $z = x + y$ follows by taking $x := X(\omega)$ and $y := Y(\omega)$, so it remains to show that these are the only possible values x and y. To this end, let us suppose that $X(\omega) + Y(\omega) = z = X(\omega') + Y(\omega')$ for some $\omega' \in \Omega$. Then

$$X(\omega) - X(\omega') = -(Y(\omega) - Y(\omega')),$$

and comonotonicity implies that this expression vanishes. Hence, $x = X(\omega')$ and $y = Y(\omega')$.

Next, we check that both f and g are increasing and Lipschitz continuous functions on $Z(\Omega)$. So, let us suppose that

$$X(\omega_1) + Y(\omega_1) = z_1 \le z_2 = X(\omega_2) + Y(\omega_2). \tag{4.76}$$

This implies

$$X(\omega_2) - X(\omega_1) \ge -(Y(\omega_2) - Y(\omega_1)). \tag{4.77}$$

Comonotonicity thus yields that $X(\omega_2) - X(\omega_1) \ge 0$ and $Y(\omega_2) - Y(\omega_1) \ge 0$, whence $f(z_1) \le f(z_2)$ and $g(z_1) \le g(z_2)$. Thus, f and g are increasing on $Z(\Omega)$. Moreover, (4.76) and (4.77) imply that

$$0 \le f(z_2) - f(z_1) = X(\omega_2) - X(\omega_1) \le X(\omega_2) - X(\omega_1) + Y(\omega_2) - Y(\omega_1) = z_2 - z_1.$$

Hence, $|f(z_2) - f(z_1)| \le |z_2 - z_1|$. In the same way, we get the Lipschitz continuity of g.

We still need to extend the functions $f, g : Z(\Omega) \to \mathbb{R}$ to increasing Lipschitz continuous functions on all of $\mathbb{R}$. To this end, we note first that Lipschitz continuity implies uniform continuity. Hence, f and g can be extended to continuous functions on the closure A of $Z(\Omega)$. Clearly, these extensions, denoted again by f and g, are also increasing and Lipschitz continuous. Next, let us fix $x \in \mathbb{R} \setminus A$. If $x < z$ for all $z \in A$, then $z_1 := \min A$ belongs to the closed set A, and we can define $f(x) := f(z_1)$. Similarly, if $x > z$ for all $z \in A$, we let $f(x) := f(\max A)$. If neither of the preceding two cases holds, then we let $z_1 := \max\{z \in A \mid z < x\}$ and $z_2 := \min\{z \in A \mid z > x\}$ and note that both z_1 and z_2 belong to the closed set A. We may thus define

$$f(x) := \frac{x - z_1}{z_2 - z_1} f(z_2) + \frac{z_2 - x}{z_2 - z_1} f(z_1).$$

It is easy to see that this construction yields the desired extension of f to an increasing and Lipschitz continuous function on $\mathbb{R}$. □

Lemma 4.96. *If $X, Y \in \mathcal{X}$ is a pair of comonotone functions, and r_X, r_Y, r_{X+Y} are quantile functions with respect to c, then*

$$r_{X+Y}(t) = r_X(t) + r_Y(t) \quad \textit{for a. e. } t.$$

Proof. By way of Lemma 4.95(c), we may write $Z = X + Y$, $X = f(Z)$, and $Y = g(Z)$ for two increasing and continuous functions f and g. Using the continuity of f and g, the same argument as in the proof of Lemma D.12 shows that $f(r_Z)$ and $g(r_Z)$ are quantile functions for X and Y under c if r_Z is a quantile function for Z. Here, the continuity of f and g is needed, because we did not make any continuity assumptions on the capacity c. An identical argument applied to the continuous and increasing function $h := f + g$ shows that $h(r_Z) = f(r_Z) + g(r_Z)$ is a quantile function for $X + Y$. The assertion now follows from the fact that all quantile functions of a random variable coincide almost everywhere, due to Lemma D.2. □

Remark 4.97. Applied to the special case of quantile function with respect to a probability measure, the preceding lemma yields that $\mathrm{V@R}_\lambda$ and $\mathrm{AV@R}_\lambda$ are comonotonic. ◇

Proof of Theorem 4.94. We already know from Proposition 4.90 that the Choquet integral of the loss is a monetary risk measure. Comonotonicity follows by combining Proposition 4.92 with Lemma 4.96.

Conversely, suppose now that ρ is comonotonic. Then ρ is positively homogeneous according to Lemma 4.89. In particular, we have $\rho(-m) = \rho(0) + m = m$ for $m \geq 0$. Thus, we obtain a normalized monotone set function by letting $c(A) := \rho(-\mathbb{1}_A)$. Moreover, $\rho_c(X) := \int(-X)\,dc$ is a comonotonic monetary risk measure on $\mathscr{X}$ that coincides with ρ on indicator functions: $\rho(-\mathbb{1}_A) = c(A) = \rho_c(-\mathbb{1}_A)$. Let us now show that ρ and ρ_c coincide on simple random variables of the form

$$X = \sum_{i=1}^{n} x_i \mathbb{1}_{A_i}, \qquad x_i \in \mathbb{R},\ A_i \in \mathscr{F}.$$

Since these random variables are dense in L^∞, Lemma 4.3 will then imply that $\rho = \rho_c$. In order to show that $\rho_c(X) = \rho(X)$ for X as above, we may assume without loss of generality that $x_1 \geq x_2 \geq \cdots \geq x_n$ and that the sets A_i are disjoint. By cash invariance, we may also assume $X \geq 0$, that is, $x_n \geq 0$. Thus, we can write $X = \sum_{i=1}^{n} b_i \mathbb{1}_{B_i}$, where $b_i := x_i - x_{i+1} \geq 0$, $x_{n+1} := 0$, and $B_i := \bigcup_{k=1}^{i} A_k$. Note that $b_i \mathbb{1}_{B_i}$ and $b_k \mathbb{1}_{B_k}$ is a pair of comonotone functions. Hence, also $\sum_{i=1}^{k-1} b_i \mathbb{1}_{B_i}$ and $b_k \mathbb{1}_{B_k}$ are comonotone, and we get inductively

$$\rho(-X) = \sum_{i=1}^{n} b_i \rho(-\mathbb{1}_{B_i}) = \sum_{i=1}^{n} b_i \rho_c(-\mathbb{1}_{B_i}) = \rho_c(-X). \qquad \square$$

Remark 4.98. The argument at the end of the preceding proof shows that the Choquet integral of a simple random variable

$$X = \sum_{i=1}^{n} x_i \mathbb{1}_{A_i} \quad \text{with } x_1 \geq \cdots \geq x_n \geq x_{n+1} := 0$$

and disjoint sets $A_1, \ldots, A_n$ can be computed as

$$\int X\,dc = \sum_{i=1}^{n} (x_i - x_{i+1}) c(B_i) = \sum_{i=1}^{n} x_i \big(c(B_i) - c(B_{i-1})\big),$$

where $B_0 := \emptyset$ and $B_i := \bigcup_{k=1}^{i} A_k$ for $i = 1, \ldots, n$. ◇

Exercise 4.9.1. For $X, Y \in \mathscr{X}$, and a uniform random variable U, define the comonotone copies $\widetilde{X} := q_X(U)$ and $\widetilde{Y} := q_Y(U)$ and show that the law of $X + Y$ dominates the law of $\widetilde{X} + \widetilde{Y}$ with respect to the increasing concave order $\succcurlyeq_{\text{icv}}$.
Hint: Use Remark 4.49 and the comonotonicity of AV@R$_\lambda$. ◇

So far, we have shown that comonotonic monetary risk measures can be identified with Choquet integrals of normalized monotone set functions. Our next goal is to characterize those set functions that induce risk measures with the additional property of *convexity*. To this end, we will first consider law-invariant risk measures. The following result shows that the risk measures AV@R$_\lambda$ may be viewed as the extreme points in the convex class of all law-invariant convex risk measures on L^∞ that are comonotonic.

Theorem 4.99. *On an atomless probability space, the class of risk measures*

$$\rho_\mu(X) := \int \mathrm{AV@R}_\lambda(X)\,\mu(d\lambda), \qquad \mu \in \mathscr{M}_1([0,1]),$$

is precisely the class of all law-invariant convex risk measures on L^∞ that are comonotonic. In particular, any convex risk measure that is law-invariant and comonotonic is also coherent.

Proof. Comonotonicity of ρ_μ follows from Corollary 4.83 and Theorem 4.94. Conversely, let us assume that ρ is a law-invariant convex risk measure that is also comonotonic. By Theorem 4.94, $\rho(X) = \int(-X)\,dc$ for $c(A) := \rho(-\mathbb{1}_A)$. The law-invariance of ρ implies that $c(A)$ is a function of the probability $P[\,A\,]$, that is, there exists an increasing function ψ on $[0,1]$ such that $\psi(0) = 0$, $\psi(1) = 1$, and $c(A) = \psi(P[\,A\,])$. Note that $\mathbb{1}_{A\cup B}$ and $\mathbb{1}_{A\cap B}$ is a pair of comonotone functions for all $A, B \in \mathscr{F}$. Hence, comonotonicity and subadditivity of ρ imply

$$\begin{aligned} c(A\cap B) + c(A\cup B) &= \rho(-\mathbb{1}_{A\cap B}) + \rho(-\mathbb{1}_{A\cup B}) = \rho(-\mathbb{1}_{A\cap B} - \mathbb{1}_{A\cup B}) \\ &= \rho(-\mathbb{1}_A - \mathbb{1}_B) \\ &\le c(A) + c(B). \end{aligned} \tag{4.78}$$

Proposition 4.81 thus implies that ψ is concave. Corollary 4.83 finally shows that the Choquet integral with respect to c can be identified with a risk measure ρ_μ, where μ is obtained from ψ via Lemma 4.75. □

Now we turn to the characterization of *all* comonotonic convex risk measures on $\mathscr{X}$. Recall that, for a positively homogeneous monetary risk measure, convexity is equivalent to subadditivity. Also, recall that $\mathscr{M}_{1,f} := \mathscr{M}_{1,f}(\Omega, \mathscr{F})$ denotes the set of all finitely additive normalized set functions $Q : \mathscr{F} \to [0,1]$, and that $E_Q[\,X\,]$ denotes the integral of $X \in \mathscr{X}$ with respect to $Q \in \mathscr{M}_{1,f}$, as constructed in Theorem G.21.

Theorem 4.100. *For the Choquet integral with respect to a normalized monotone set function c, the following conditions are equivalent:*

(a) $\rho(X) := \int(-X)\,dc$ *is a convex risk measure on* $\mathscr{X}$.
(b) $\rho(X) := \int(-X)\,dc$ *is a coherent risk measure on* $\mathscr{X}$.
(c) *For* $\mathcal{Q}_c := \{Q \in \mathscr{M}_{1,f} \mid Q[A] \le c(A)$ *for all* $A \in \mathscr{F}\}$,

$$\int X\,dc = \max_{Q\in\mathcal{Q}_c} E_Q[X] \quad \textit{for } X \in \mathscr{X}.$$

(d) *The set function c is submodular.*
In this case, $\mathcal{Q}_c$ *is equal to the maximal representing set* $\mathcal{Q}^{\max}$ *for* ρ.

Before giving the proof of this theorem, let us state the following corollary, which gives a complete characterization of all comonotonic convex risk measures, and a remark concerning the set $\mathcal{Q}_c$ in part (c), which is usually called the *core* of c.

Corollary 4.101. *A convex risk measure on* $\mathscr{X}$ *is comonotonic if and only if it arises as the Choquet integral of the loss with respect to a submodular, normalized, and monotone set function c. In this case, c is given by* $c(A) = \rho(-\mathbb{1}_A)$, *and* ρ *has the representation*

$$\rho(X) = \max_{Q\in\mathcal{Q}_c} E_Q[-X],$$

where $\mathcal{Q}_c = \{Q \in \mathscr{M}_{1,f} \mid Q[A] \le c(A)$ *for all* $A \in \mathscr{F}\}$ *is equal to the maximal representing set* $\mathcal{Q}^{\max}$.

Proof. Theorems 4.94 and 4.100 state that $\rho(X) := \int(-X)\,dc$ is a comonotonic coherent risk measure that can be represented as in the assertion, as soon as c is a submodular, normalized, and monotone set function. Conversely, any comonotonic convex risk measure ρ is coherent and arises as the Choquet integral of $c(A) := \rho(-\mathbb{1}_A)$, due to Theorem 4.94. Theorem 4.100 then gives the submodularity of c. □

Remark 4.102. Let c be a normalized, monotone, submodular set function. Theorem 4.100 implies in particular that the core $\mathcal{Q}_c$ of c is nonempty. Moreover, c can be recovered from $\mathcal{Q}_c$:

$$c(A) = \max_{Q\in\mathcal{Q}_c} Q[A] \quad \text{for all } A \in \mathscr{F}.$$

If c has the additional continuity property that $c(A_n) \to 0$ for any decreasing sequence (A_n) of events such that $\bigcap_n A_n = \emptyset$, then this property is shared by any $Q \in \mathcal{Q}_c$, and it follows that Q is σ-additive. Thus, the corresponding coherent risk measure $\rho(X) = \int(-X)\,dc$ admits a representation in terms of σ-additive probability measures. It follows by Lemma 4.21 that ρ is continuous from above. ◇

The proof of Theorem 4.100 requires some preparation. The assertion of the following lemma is not entirely obvious, since Fubini's theorem may fail if $Q \in \mathscr{M}_{1,f}$ is not σ-additive.

Lemma 4.103. *For $X \in \mathscr{X}$ and $Q \in \mathscr{M}_{1,f}$, the integral $E_Q[X]$ is equal to the Choquet integral $\int X\,dQ$.*

Proof. It is enough to prove the result for $X \geq 0$. Suppose first that $X = \sum_{i=1}^n x_i \mathbb{1}_{A_i}$ is as in Remark 4.98. Then

$$\int X\,dQ = \sum_{i=1}^n (x_i - x_{i+1}) Q\left[\bigcup_{k=1}^i A_k\right] = \sum_{i=1}^n x_i Q[A_i] = E_Q[X].$$

The result for general $X \in \mathscr{X}$ follows by approximating X uniformly with X_n that take only finitely many values, and by using the Lipschitz continuity of both $E_Q[\,\cdot\,]$ and $\int \cdot\, dQ$ with respect to the supremum norm. □

Lemma 4.104. *Let $A_1, \dots, A_n$ be a partition of Ω into disjoint measurable sets, and suppose that the normalized monotone set function c is submodular. Let Q be the probability measure on $\mathscr{F}_0 := \sigma(A_1, \dots, A_n)$ with weights*

$$Q[A_k] := c(B_k) - c(B_{k-1}) \quad \textit{for } B_0 := \emptyset \textit{ and } B_k := \bigcup_{j=1}^k A_j, \; k \geq 1. \tag{4.79}$$

Then $\int X\,dc \geq E_Q[X]$ for all $\mathscr{F}_0$-measurable $X = \sum_{i=1}^n x_i \mathbb{1}_{A_i}$, and equality holds if the values of X are arranged in decreasing order: $x_1 \geq \dots \geq x_n$.

Proof. Clearly, it suffices to consider only the case $X \geq 0$. Then Remark 4.98 implies $\int X\,dc = E_Q[X]$ as soon as the values of X are arranged in decreasing order.

Now we prove $\int X\,dc \geq E_Q[X]$ for arbitrary $\mathscr{F}_0$-measurable $X \geq 0$. To this end, note that any permutation σ of $\{1, \dots, n\}$ induces a probability measure Q_σ on $\mathscr{F}_0$ by applying the definition of Q to the relabeled partition $A_{\sigma(1)} \dots, A_{\sigma(n)}$. If σ is a permutation such that $x_{\sigma(1)} \geq \dots \geq x_{\sigma(n)}$, then we have $\int X\,dc = E_{Q_\sigma}[X]$, and so the assertion will follow if we can prove that $E_{Q_\sigma}[X] \geq E_Q[X]$. To this end, it is enough to show that $E_{Q_\tau}[X] \geq E_Q[X]$ if τ is the transposition of two indices i and $i+1$ that are such that $x_i < x_{i+1}$, because σ can be represented as a finite composition of such transpositions.

Note next that

$$E_{Q_\tau}[X] - E_Q[X] = x_i(Q_\tau[A_i] - Q[A_i]) + x_{i+1}(Q_\tau[A_{i+1}] - Q[A_{i+1}]). \tag{4.80}$$

To compute the probabilities $Q_\tau[A_k]$, let us introduce

$$B_0^\tau := \emptyset \quad \text{and} \quad B_k^\tau := \bigcup_{j=1}^k A_{\tau(j)}, \quad k = 1, \dots, n.$$

Then $B_k^\tau = B_k$ for $k \neq i$. Hence,

$$\begin{aligned} Q_\tau[A_i] + Q_\tau[A_{i+1}] &= Q_\tau[A_{\tau(i)}] + Q_\tau[A_{\tau(i+1)}] = c(B^\tau_{i+1}) - c(B^\tau_{i-1}) \\ &= c(B_{i+1}) - c(B_{i-1}) = Q[A_i] + Q[A_{i+1}]. \end{aligned} \tag{4.81}$$

Moreover, $B^\tau_i \cap B_i = B_{i-1}$, $B^\tau_i \cup B_i = B_{i+1}$, and hence $c(B_{i-1}) + c(B_{i+1}) \le c(B^\tau_i) + c(B_i)$, due to the submodularity of c. Thus,

$$Q[A_{i+1}] = c(B_{i+1}) - c(B_i) \le c(B^\tau_i) - c(B^\tau_{i-1}) = Q_\tau[A_{\tau(i)}] = Q_\tau[A_{i+1}].$$

Using (4.80), (4.81), and our assumption $x_i < x_{i+1}$ thus yields $E_{Q_\tau}[X] \ge E_Q[X]$. □

Proof of Theorem 4.100. (a)⇔(b): According to Proposition 4.90, the property of positive homogeneity is shared by all Choquet integrals, and so (a)⇒(b) follows. The converse implication is obvious.

(b)⇒(c): By Corollary 4.19, $\rho(-X) = \max_{Q\in\mathcal{Q}^{\max}} E_Q[X]$, where $Q \in \mathcal{M}_{1,f}$ belongs to $\mathcal{Q}^{\max}$ if and only if

$$E_Q[X] \le \rho(-X) = \int X\,dc \quad \text{for all } X \in \mathcal{X}. \tag{4.82}$$

We will now show that this set $\mathcal{Q}^{\max}$ coincides with the set $\mathcal{Q}_c$. If $Q \in \mathcal{Q}^{\max}$ then, in particular, $Q[A] \le \int \mathbb{1}_A\,dc = c(A)$ for all $A \in \mathcal{F}$. Hence, $Q \in \mathcal{Q}_c$. Conversely, suppose $Q \in \mathcal{Q}_c$. If $X \ge 0$, then

$$\int X\,dc = \int_0^\infty c(X > x)\,dx \ge \int_0^\infty Q[X > x]\,dx = E_Q[X],$$

where we have used Lemma 4.103. Cash invariance yields (4.82).

(c)⇒(b) is obvious.

(b)⇒(d): This follows precisely as in (4.78).

(d)⇒(b): We have to show that the Choquet integral is subadditive. By Lemma 4.3, it is enough to prove this for random variables that take only finitely many values. Thus, let $A_1,\dots,A_n$ be a partition of Ω into finitely many disjoint measurable sets. Let us write $X = \sum_i x_i \mathbb{1}_{A_i}$, $Y = \sum_i y_i \mathbb{1}_{A_i}$, and let us assume that the indices $i = 1,\dots,n$ are arranged such that $x_1 + y_1 \ge \cdots \ge x_n + y_n$. Then the probability measure Q constructed in Lemma 4.104 is such that

$$\int (X+Y)\,dc = E_Q[X+Y] = E_Q[X] + E_Q[Y] \le \int X\,dc + \int Y\,dc.$$

But this is the required subadditivity of the Choquet integral. □

4.10 Measures of risk in a financial market

In this section, we will consider risk measures that arise in the financial market model of Section 1.1. In this model, $d+1$ assets are priced at times $t = 0$ and $t = 1$. Prices at time 1 are modeled as nonnegative random variables $S^0, S^1, \dots, S^d$ on some probability space $(\Omega, \mathscr{F}, P)$, with $S^0 \equiv 1 + r > 0$. Prices at time 0 are given by a vector $\overline{\pi} = (1, \pi)$, with $\pi = (\pi^1, \dots, \pi^d)$. The discounted net gain of a trading strategy $\overline{\xi} = (\xi^0, \xi)$ is given by $\xi \cdot Y$, where the random vector $Y = (Y^1, \dots, Y^d)$ is defined by

$$Y^i = \frac{S^i}{1+r} - \pi^i \quad \text{for } i = 1, \dots, d.$$

As in Sections 4.4–4.8, risk measures will be defined on the space $L^\infty = L^\infty(\Omega, \mathscr{F}, P)$. A financial position X can be viewed as riskless if $X \geq 0$ or, more generally, if X can be hedged without additional costs, that is, if there exists a trading strategy $\overline{\xi} = (\xi^0, \xi)$ such that $\overline{\pi} \cdot \overline{\xi} = 0$ and

$$X + \frac{\overline{\xi} \cdot \overline{S}}{1+r} = X + \xi \cdot Y \geq 0 \quad P\text{-a. s.} \tag{4.83}$$

Thus, we define the following set of acceptable positions in L^∞:

$$\mathscr{A}_0 := \{X \in L^\infty \mid \exists\, \xi \in \mathbb{R}^d \text{ with } X + \xi \cdot Y \geq 0\ P\text{-a. s.}\}.$$

Proposition 4.105. *Suppose that* $\inf\{m \in \mathbb{R} \mid m \in \mathscr{A}_0\} > -\infty$. *Then*

$$\rho_0(X) := \rho_{\mathscr{A}_0}(X) = \inf\{m \in \mathbb{R} \mid m + X \in \mathscr{A}_0\}, \quad X \in L^\infty,$$

is a coherent risk measure. Moreover, ρ_0 *is sensitive in the sense of Definition* 4.42 *if and only if the market model is arbitrage-free. In this case,* ρ_0 *is continuous from above and can be represented in terms of the set* $\mathscr{P}$ *of equivalent risk-neutral measures:*

$$\rho_0(X) = \sup_{P^* \in \mathscr{P}} E^*[\,-X\,]. \tag{4.84}$$

Proof. The fact that ρ_0 is a coherent risk measure follows from Proposition 4.7, because $\mathscr{A}_0$ is a convex cone. If the model is arbitrage-free, then the arbitrage bound in Theorem 1.34 yields the representation (4.84), and it follows that ρ_0 is sensitive and continuous from above.

Conversely, suppose by way of contradiction that ρ_0 is sensitive, but the market model admits an arbitrage opportunity. Then there are $\xi \in \mathbb{R}^d$ and $\varepsilon > 0$ such that $0 \leq \xi \cdot Y$ P-a. s. and $A := \{\xi \cdot Y \geq \varepsilon\}$ satisfies $P[\,A\,] > 0$. It follows that $\xi \cdot Y - \varepsilon \mathbb{1}_A \geq 0$, that is, $-\varepsilon \mathbb{1}_A$ is acceptable. Thus, the positive homogeneity of ρ_0, which follows from the fact that $\mathscr{A}_0$ is a cone, implies that $\rho_0(-\lambda \mathbb{1}_A) \leq 0$ for any $\lambda > 0$. On the other hand, Theorem 4.43 implies that $\rho_0(-\lambda \mathbb{1}_A) > \rho_0(0) = 0$ for some (and hence all) $\lambda > 0$. This is the desired contradiction. □

There are several reasons why it may make sense to allow in (4.83) only strategies ξ that belong to a proper subset $\mathscr{S}$ of the class $\mathbb{R}^d$ of all strategies. For instance, if the resources available to an investor are limited, only those strategies should be considered for which the initial investment in risky assets is below a certain amount. Such a restriction corresponds to an upper bound on $\xi \cdot \pi$. There may be other constraints. For instance, *short sales constraints* are lower bounds on the number of shares in the portfolio. In view of *market illiquidity*, the investor may also wish to avoid holding too many shares of one single asset, since the market capacity may not suffice to resell the shares. Such constraints will be taken into account by assuming throughout the remainder of this section that $\mathcal{S}$ has the following properties:

- $0 \in \mathcal{S}$.
- $\mathcal{S}$ is convex.
- Each $\xi \in \mathcal{S}$ is *admissible* in the sense that $\xi \cdot Y$ is P-a. s. bounded from below.

Under these conditions, the set

$$\mathcal{A}^{\mathcal{S}} := \{X \in L^\infty \mid \exists\, \xi \in \mathcal{S} \text{ with } X + \xi \cdot Y \geq 0 \ P\text{-a. s.}\} \tag{4.85}$$

is nonempty, convex, and contains all $X \in \mathcal{X}$ that dominate some $Z \in \mathcal{A}^{\mathcal{S}}$. Moreover, we will assume from now on that

$$\inf\{m \in \mathbb{R} \mid m \in \mathcal{A}^{\mathcal{S}}\} > -\infty. \tag{4.86}$$

Proposition 4.7 then guarantees that the induced risk measure

$$\rho^{\mathcal{S}}(X) := \rho_{\mathcal{A}^{\mathcal{S}}}(X) = \inf\{m \in \mathbb{R} \mid m + X \in \mathcal{A}^{\mathcal{S}}\}$$

is a convex risk measure on L^∞. Note that (4.86) holds, in particular, if $\mathcal{S}$ does not contain arbitrage opportunities in the sense that $\xi \cdot Y \geq 0$ P-a. s. for $\xi \in \mathcal{S}$ implies $P[\,\xi \cdot Y = 0\,] = 1$.

Remark 4.106. Admissibility of portfolios is a serious restriction; in particular, it prevents unhedged short sales of any unbounded asset. Note, however, that it is consistent with our notion of acceptability for bounded claims in (4.85), since $X + \xi \cdot Y \geq 0$ implies $\xi \cdot Y \geq -\|X\|$. ◇

Two questions arise: Which conditions guarantee that $\rho^{\mathcal{S}}$ is continuous from above, and thus admits a representation (4.35) in terms of probability measures? And, if such a representation exists, how can we identify the minimal penalty function $\alpha_{\mathcal{S}}^{\min}$ on $\mathcal{M}_1(P)$? In the case $\mathcal{S} = \mathbb{R}^d$, both questions were addressed in Proposition 4.105. For general $\mathcal{S}$, only the second question has a straightforward answer, which will be given in Proposition 4.108. As can be seen from the proof of Proposition 4.105, an analysis of the first question requires an extension of the arbitrage theory in Chapter 1 for the case of portfolio constraints. Such a theory will be developed in Chapter 9 in a more general dynamic setting, and we will address both questions for the corresponding risk measures in Corollary 9.32. This result implies the following theorem for the simple one-period model of the present section.

Theorem 4.107. *In addition to the above assumptions, suppose that the market model is nonredundant in the sense of Definition* 1.15 *and that $\mathscr{S}$ is a closed subset of $\mathbb{R}^d$ such that the cone $\{\lambda\xi \mid \xi \in \mathscr{S}, \lambda \geq 0\}$ is closed. Then $\rho^{\mathscr{S}}$ is sensitive if and only if $\mathscr{S}$ contains no arbitrage opportunities. In this case, $\rho^{\mathscr{S}}$ is continuous from above and admits the representation*

$$\rho^{\mathscr{S}}(X) = \sup_{Q\in\mathcal{M}_1(P)} \Big(E_Q[\,-X\,] - \sup_{\xi\in\mathscr{S}} E_Q[\,\xi\cdot Y\,]\Big). \tag{4.87}$$

In the following proposition, we will explain the specific form of the penalty function in (4.87). This result will not require the additional assumptions of Theorem 4.107.

Proposition 4.108. *For $Q \in \mathcal{M}_1(P)$, the minimal penalty function $\alpha_{\mathscr{S}}^{\min}$ of $\rho^{\mathscr{S}}$ is given by*

$$\alpha_{\mathscr{S}}^{\min}(Q) = \sup_{\xi\in\mathscr{S}} E_Q[\,\xi\cdot Y\,].$$

In particular, $\rho^{\mathscr{S}}$ can be represented as in (4.87) *if $\rho^{\mathscr{S}}$ is continuous from above.*

Proof. Fix $Q \in \mathcal{M}_1(P)$. Clearly, the expectation $E_Q[\,\xi\cdot Y\,]$ is well-defined for each $\xi \in \mathscr{S}$ by admissibility. If $X \in \mathscr{A}^{\mathscr{S}}$, there exists $\eta \in \mathscr{S}$ such that $-X \leq \eta\cdot Y$ P-almost surely. Thus,

$$E_Q[\,-X\,] \leq E_Q[\,\eta\cdot Y\,] \leq \sup_{\xi\in\mathscr{S}} E_Q[\,\xi\cdot Y\,]$$

for any $Q \in \mathcal{M}_1(P)$. Hence, the definition of the minimal penalty function yields

$$\alpha_{\mathscr{S}}^{\min}(Q) \leq \sup_{\xi\in\mathscr{S}} E_Q[\,\xi\cdot Y\,]. \tag{4.88}$$

To prove the converse inequality, take $\xi \in \mathscr{S}$. Note that $X_k := -((\xi\cdot Y)\wedge k)$ is bounded since ξ is admissible. Moreover,

$$X_k + \xi\cdot Y = (\xi\cdot Y - k)\,\mathbb{1}_{\{\xi\cdot Y\geq k\}} \geq 0,$$

so that $X_k \in \mathscr{A}^{\mathscr{S}}$. Hence,

$$\alpha_{\mathscr{S}}^{\min}(Q) \geq E_Q[\,-X_k\,] = E_Q[\,(\xi\cdot Y)\wedge k\,],$$

and so $\alpha_{\mathscr{S}}^{\min}(Q) \geq E_Q[\,\xi\cdot Y\,]$ by monotone convergence. □

Exercise 4.10.1. Show that the identity

$$\alpha_{\mathscr{S}}^{\min}(Q) = \sup_{\xi\in\mathscr{S}} E_Q[\,\xi\cdot Y\,]$$

in Proposition 4.108 remains true even for $Q \in \mathcal{M}_{1,f}(P)$ if we assume in addition that Y is P-a. s. bounded. We thus obtain the representation

$$\rho^{\mathcal{S}}(X) = \max_{Q \in \mathcal{M}_{1,f}(P)} \Big(E_Q[\,-X\,] - \sup_{\xi \in \mathcal{S}} E_Q[\,\xi \cdot Y\,] \Big)$$

without assuming continuity from above. ◇

Remark 4.109. Suppose that $\mathcal{S}$ is a cone. Then the acceptance set $\mathcal{A}^{\mathcal{S}}$ is also a cone, and $\rho^{\mathcal{S}}$ is a coherent measure of risk. If $\rho^{\mathcal{S}}$ is continuous from above, then Corollary 4.37 yields the representation

$$\rho^{\mathcal{S}}(X) = \sup_{Q \in \mathcal{Q}_{\mathcal{S}}} E_Q[\,-X\,]$$

in terms of the nonempty set $\mathcal{Q}_{\mathcal{S}} = \{Q \in \mathcal{M}_1(P) \mid \alpha^{\min}_{\mathcal{S}}(Q) = 0\}$. It follows from Proposition 4.108 that for $Q \in \mathcal{M}_1(P)$,

$$Q \in \mathcal{Q}_{\mathcal{S}} \text{ if and only if } E_Q[\,\xi \cdot Y\,] \le 0 \text{ for all } \xi \in \mathcal{S}.$$

If $\rho^{\mathcal{S}}$ is sensitive, then the set $\mathcal{S}$ cannot contain any arbitrage opportunities, and $\mathcal{Q}_{\mathcal{S}}$ contains the set $\mathcal{P}$ of all equivalent martingale measures whenever such measures exist. More precisely, $\mathcal{Q}_{\mathcal{S}}$ can be described as the set of absolutely continuous supermartingale measures with respect to $\mathcal{S}$; this will be discussed in more detail in the dynamical setting of Chapter 9. ◇

Let us now relax the condition of acceptability in (4.85). We no longer insist that the final outcome of an acceptable position, suitably hedged, should always be nonnegative. Instead, we only require that the hedged position is acceptable in terms of a given convex risk measure $\rho_{\mathcal{A}}$ with acceptance set $\mathcal{A}$. Thus, we define

$$\tilde{\mathcal{A}} := \big\{X \in L^\infty \mid \exists\, \xi \in \mathcal{S},\, A \in \mathcal{A} \text{ with } X + \xi \cdot Y \ge A \ P\text{-a. s.}\big\}. \tag{4.89}$$

Clearly, $\mathcal{A} \subseteq \tilde{\mathcal{A}}$, and hence

$$\rho_{\mathcal{A}} \ge \overline{\rho} := \rho_{\tilde{\mathcal{A}}}.$$

From now on, we assume that

$$\overline{\rho}(0) > -\infty, \tag{4.90}$$

which also implies our assumption (4.86) for $\mathcal{A}^{\mathcal{S}}$.

Proposition 4.110. *The minimal penalty function $\overline{\alpha}^{\min}$ for $\overline{\rho}$ is given by*

$$\overline{\alpha}^{\min}(Q) = \alpha^{\min}_{\mathcal{S}}(Q) + \alpha^{\min}(Q), \quad Q \in \mathcal{M}_{1,f}, \tag{4.91}$$

where $\alpha_{\mathscr{S}}^{\min}$ is the minimal penalty function for $\rho^{\mathscr{S}}$ and $\alpha^{\min}$ is the minimal penalty function for $\rho_{\mathscr{A}}$.

Proof. We claim that

$$\{X \in L^\infty \mid \overline{\rho}(X) < 0\} \subseteq \{X^{\mathscr{S}} + A \mid X^{\mathscr{S}} \in \mathscr{A}^{\mathscr{S}}, A \in \mathscr{A}\} \subseteq \bar{\mathscr{A}}. \tag{4.92}$$

If $\overline{\rho}(X) < 0$, then $X \in \bar{\mathscr{A}}$. Hence, there exists $A \in \mathscr{A}$ and $\xi \in \mathscr{S}$ such that $X + \xi \cdot Y \geq A$. Therefore, $X^{\mathscr{S}} := X - A \in \mathscr{A}^{\mathscr{S}}$, and we obtain the first inclusion in (4.92). To prove the second inclusion, take $X^{\mathscr{S}} \in \mathscr{A}^{\mathscr{S}}$. Then there exists $\xi \in \mathscr{S}$ such that $X^{\mathscr{S}} + \xi \cdot Y \geq 0$. Hence, for any $A \in \mathscr{A}$, we get $X^{\mathscr{S}} + A + \xi \cdot Y \geq A \in \mathscr{A}$, that is, $X := X^{\mathscr{S}} + A \in \bar{\mathscr{A}}$. This establishes the second inclusion in (4.92).

Now let $Q \in \mathscr{M}_{1,f}$ be given. In view of (4.92), we have

$$\sup_{X:\overline{\rho}(X)<0} E_Q[\,-X\,] \leq \sup_{X^{\mathscr{S}} \in \mathscr{A}^{\mathscr{S}}} \sup_{A\in\mathscr{A}} E_Q[\,-X^{\mathscr{S}} - A\,] \leq \sup_{X\in\bar{\mathscr{A}}} E_Q[\,-X\,] \leq \sup_{X\in\mathscr{A}_{\overline{\rho}}} E_Q[\,-X\,].$$

But the left- and rightmost terms in the preceding chain of inequalities are equal, and so we get that

$$\begin{aligned}\overline{\alpha}^{\min}(Q) &= \sup_{X\in\bar{\mathscr{A}}} E_Q[\,-X\,]\\ &= \sup_{X^{\mathscr{S}} \in \mathscr{A}^{\mathscr{S}}} \sup_{A\in\mathscr{A}} E_Q[\,-X^{\mathscr{S}} - A\,]\\ &= \alpha_{\mathscr{S}}^{\min}(Q) + \alpha^{\min}(Q).\end{aligned}$$

□

Remark 4.111. It can happen that the sum of two penalty functions as on the right-hand side of (4.91) is infinite for every $Q \in \mathscr{M}_{1,f}(P)$. In this case, the condition (4.90) will be violated and the risk measure $\overline{\rho}$ does not exist. Consider, for example, the situation of a complete market model without trading constraints. Then $\rho^{\mathscr{S}}(X) = E^*[\,-X\,]$, where P^* is the unique equivalent risk-neutral measure. That is,

$$\alpha_{\mathscr{S}}^{\min}(Q) = \begin{cases} 0 & \text{if } Q = P^*,\\ +\infty & \text{otherwise.}\end{cases}$$

For $\rho_{\mathscr{A}}$, we take $\mathrm{AV@R}_\lambda$. Then $\alpha^{\min}(Q) + \alpha_{\mathscr{S}}^{\min}(Q)$ is infinite for $Q \neq P^*$, and $\alpha^{\min}(P^*) + \alpha_{\mathscr{S}}^{\min}(P^*)$ is finite (and in fact zero) if and only if

$$\frac{dP^*}{dP} \leq \frac{1}{\lambda} \quad P\text{-a.s.}$$

But the latter condition is violated as soon as $P^* \neq P$ and λ is close enough to 1. ◇

Exercise 4.10.2. Suppose that condition (4.90) is satisfied. Show that the risk measure $\overline{\rho}$ can also be obtained as

$$\overline{\rho}(X) = \inf\{\rho(X - X^{\mathscr{S}}) \mid X^{\mathscr{S}} \in \mathscr{A}^{\mathscr{S}}\}$$

and as

$$\overline{\rho}(X) = \inf_{Z \in L^\infty} (\rho(X - Z) + \rho^{\mathscr{S}}(Z)). \qquad \diamond$$

The preceding exercise shows that $\overline{\rho}$ is a special case of an *inf-convolution*, which we now define in a general context.

Definition 4.112. For two convex risk measures ρ_1 and ρ_2 on L^∞,

$$\rho_1 \,\square\, \rho_2(X) := \inf_{Z \in L^\infty} (\rho_1(X - Z) + \rho_2(Z)), \quad X \in L^\infty,$$

is called the *inf-convolution* of ρ_1 and ρ_2. △

Exercise 4.10.3. Let ρ_1 and ρ_2 be two convex risk measures on L^∞ and assume that

$$\rho_1 \,\square\, \rho_2(0) = \inf_{Z \in L^\infty} (\rho_1(-Z) + \rho_2(Z)) > -\infty.$$

(a) Show that $\rho_1 \,\square\, \rho_2 = \rho_2 \,\square\, \rho_1$.
(b) Show that $\rho := \rho_1 \,\square\, \rho_2$ is a convex risk measure on L^∞.
(c) Show that the minimal penalty function $\alpha^{\min}$ of ρ is equal to the sum of the respective minimal penalty functions $\alpha_1^{\min}$ and $\alpha_2^{\min}$ of ρ_1 and ρ_2:

$$\alpha^{\min}(Q) = \alpha_1^{\min}(Q) + \alpha_2^{\min}(Q) \quad \text{for } Q \in \mathscr{M}_{1,f}.$$

(d) Show that ρ is continuous from below as soon as ρ_1 is continuous from below. ◇

For the remainder of this section, we consider the following case study, which is based on [52]. Let us fix a finite class

$$\mathcal{Q}_0 = \{Q_1, \ldots, Q_n\}$$

of equivalent probability measures $Q_i \approx P$ such that $|Y| \in L^1(Q_i)$; as in [52], we call the measures in $\mathcal{Q}_0$ *valuation measures*. Define the sets

$$\mathscr{B} := \{X \in L^0 \mid E_{Q_i}[X] \text{ exists and is } \geq 0,\ i = 1, \ldots, n\} \tag{4.93}$$

and

$$\mathscr{B}_0 := \{X \in \mathscr{B} \mid E_{Q_i}[X] = 0 \text{ for } i = 1, \ldots, n\}.$$

Note that

$$\mathscr{B}_0 \cap L^0_+ = \{0\}, \tag{4.94}$$

since $X = 0$ P-a. s. as soon as $X \geq 0$ P-a. s. and $E_{Q_i}[X] = 0$, due to the equivalence $Q_i \approx P$.

As the initial acceptance set, we take the convex cone

$$\mathscr{A} := \mathscr{B} \cap L^\infty. \tag{4.95}$$

The corresponding set $\tilde{\mathscr{A}}$ of positions that become acceptable if combined with a suitable hedge, is defined as in (4.89):

$$\tilde{\mathscr{A}} := \{X \in L^\infty \mid \exists\, \xi \in \mathbb{R}^d \text{ with } X + \xi \cdot Y \in \mathscr{B}\}. \tag{4.96}$$

Let us now introduce the following stronger version of the no-arbitrage condition $\mathscr{K} \cap L^0_+ = \{0\}$, where $\mathscr{K} := \{\xi \cdot Y \mid \xi \in \mathbb{R}^d\}$:

$$\mathscr{K} \cap \mathscr{B} = \mathscr{K} \cap \mathscr{B}_0. \tag{4.97}$$

In other words, there is no portfolio $\xi \in \mathbb{R}^d$ such that the result satisfies the valuation inequalities in (4.93) and is strictly favorable in the sense that at least one of the inequalities is strict.

Note that (4.97) implies the absence of arbitrage opportunities:

$$\mathscr{K} \cap L^0_+ = \mathscr{K} \cap \mathscr{B} \cap L^0_+ = \mathscr{K} \cap \mathscr{B}_0 \cap L^0_+ = \{0\},$$

where we have used (4.94) and $\mathscr{B} \cap L^0_+ = L^0_+$. Thus, (4.97) implies, in particular, the existence of an equivalent martingale measure, that is, $\mathscr{P} \neq \emptyset$. The following proposition may be viewed as an extension of the "fundamental theorem of asset pricing." Let us denote by

$$\mathscr{R} := \left\{ \sum_{i=1}^n \lambda_i Q_i \;\middle|\; \lambda_i > 0,\ \sum_{i=1}^n \lambda_i = 1 \right\}$$

the class of all "representative" models for the class $\mathcal{Q}_0$, that is, all mixtures such that each $Q \in \mathcal{Q}_0$ appears with a positive weight.

Proposition 4.113. *The following two properties are equivalent:*
(a) $\mathscr{K} \cap \mathscr{B} = \mathscr{K} \cap \mathscr{B}_0$.
(b) $\mathscr{P} \cap \mathscr{R} \neq \emptyset$.

Proof. (b)⇒(a): For $V \in \mathscr{K} \cap \mathscr{B}$ and $R \in \mathscr{R}$, we have $E_R[\,V\,] \geq 0$. If we can choose $R \in \mathscr{P} \cap \mathscr{R}$, then we get $E_R[\,V\,] = 0$, and hence $V \in \mathscr{B}_0$.

(a)⇒(b): Consider the convex set

$$\mathscr{C} := \{E_R[\,Y\,] \mid R \in \mathscr{R}\} \subseteq \mathbb{R}^d\,;$$

we have to show that $\mathscr{C}$ contains the origin. If this is not the case, then there exists $\xi \in \mathbb{R}^d$ such that

$$\xi \cdot x \geq 0 \quad \text{for } x \in \mathscr{C}, \tag{4.98}$$

and

$$\xi \cdot x^* > 0 \quad \text{for some } x^* \in \mathscr{C};$$

see Proposition A.5. Define $V := \xi \cdot Y \in \mathscr{K}$. Condition (4.98) implies

$$E_R[\, V\,] \geq 0 \quad \text{for all } R \in \mathscr{R},$$

hence $V \in \mathscr{K} \cap \mathscr{B}$. Let $R^* \in \mathscr{R}$ be such that $x^* = E_{R^*}[\, Y\,]$. Then V satisfies $E_{R^*}[\, V\,] > 0$, hence $V \notin \mathscr{K} \cap \mathscr{B}_0$, in contradiction to our assumption (a). □

We can now state a representation theorem for the coherent risk measure $\overline{\rho}$ corresponding to the convex cone $\bar{\mathscr{A}}$. It is a special case of Theorem 4.116 that will be proved below.

Theorem 4.114. *Under assumption* (4.97), *the coherent risk measure* $\overline{\rho} := \rho_{\bar{\mathscr{A}}}$ *corresponding to the acceptance set* $\bar{\mathscr{A}}$ *in* (4.96) *is given by*

$$\overline{\rho}(X) = \sup_{P^* \in \mathscr{P} \cap \mathscr{R}} E^*[\, -X\,].$$

Let us now introduce a second finite set $\mathcal{Q}_1 \subseteq \mathscr{M}_1(P)$ of probability measures $Q \ll P$ with $|Y| \in \mathscr{L}^1(Q)$; as in [52], we call them *stress test measures*. In addition to the valuation inequalities in (4.93), we require that an admissible position passes a stress test specified by a "floor"

$$\gamma(Q) < 0 \quad \text{for each } Q \in \mathcal{Q}_1.$$

Thus, the convex cone $\mathscr{A}$ in (4.95) is reduced to the convex set

$$\mathscr{A}_1 := \mathscr{A} \cap \mathscr{B}_1 = L^\infty \cap (\mathscr{B} \cap \mathscr{B}_1),$$

where

$$\mathscr{B}_1 := \{X \in L^0 \mid E_Q[\, X\,] \geq \gamma(Q) \text{ for } Q \in \mathcal{Q}_1\}.$$

Let

$$\bar{\mathscr{A}}_1 := \{X \in L^\infty \mid \exists\, \xi \in \mathbb{R}^d \text{ with } X + \xi \cdot Y \in \mathscr{B} \cap \mathscr{B}_1\}$$

denote the resulting acceptance set for positions combined with a suitable hedge.

Remark 4.115. The analogue

$$\mathcal{K} \cap (\mathcal{B} \cap \mathcal{B}_1) = \mathcal{K} \cap (\mathcal{B}_0 \cap \mathcal{B}_1) \tag{4.99}$$

of our condition (4.97) looks weaker, but it is in fact equivalent to (4.97). Indeed, for $X \in \mathcal{K} \cap \mathcal{B}$ we can find $\varepsilon > 0$ such that $X_1 := \varepsilon X$ satisfies the additional constraints

$$E_Q[X_1] \geq \gamma(Q) \quad \text{for } Q \in \mathcal{Q}_1.$$

Since $X_1 \in \mathcal{K} \cap \mathcal{B} \cap \mathcal{B}_1$, condition (4.99) implies $X_1 \in \mathcal{K} \cap \mathcal{B}_0$, hence $X = \frac{1}{\varepsilon} X_1 \in \mathcal{K} \cap \mathcal{B}_0$, since $\mathcal{K} \cap \mathcal{B}_0$ is a cone. Thus, (4.99) implies the inclusion $\mathcal{K} \cap \mathcal{B} \subseteq \mathcal{K} \cap \mathcal{B}_0$, which is clearly the same as (4.97). The converse implication (4.97)$\Rightarrow$ (4.99) is obvious. ◇

Let us now identify the convex risk measure $\overline{\rho}_1$ induced by the convex acceptance set $\bar{\mathcal{A}}_1$. Define

$$\mathcal{R}_1 := \left\{ \sum_{Q \in \mathcal{Q}} \lambda(Q) \cdot Q \,\middle|\, \lambda(Q) \geq 0,\ \sum_{Q \in \mathcal{Q}} \lambda(Q) = 1 \right\} \supseteq \mathcal{R}$$

as the convex hull of $\mathcal{Q} := \mathcal{Q}_0 \cup \mathcal{Q}_1$, and define

$$\gamma(R) := \sum_{Q \in \mathcal{Q}} \lambda(Q) \gamma(Q)$$

for $R = \sum_Q \lambda(Q) Q \in \mathcal{R}$ with $\gamma(Q) := 0$ for $Q \in \mathcal{Q}_0$.

Theorem 4.116. *Under assumption* (4.97), *the convex risk measure $\overline{\rho}_1$ induced by the acceptance set $\bar{\mathcal{A}}_1$ is given by*

$$\overline{\rho}_1(X) = \sup_{P^* \in \mathcal{P} \cap \mathcal{R}_1} \left(E^*[-X] + \gamma(P^*) \right), \tag{4.100}$$

that is, $\overline{\rho}_1$ is determined by the penalty function

$$\overline{\alpha}_1(Q) := \begin{cases} +\infty & \text{for } Q \notin \mathcal{P} \cap \mathcal{R}_1, \\ -\gamma(Q) & \text{for } Q \in \mathcal{P} \cap \mathcal{R}_1. \end{cases}$$

Proof. Let ρ^* denote the convex risk measure defined by the right-hand side of (4.100), and let $\mathcal{A}^*$ denote the corresponding acceptance set:

$$\mathcal{A}^* := \{ X \in L^\infty \mid E^*[X] \geq \gamma(P^*) \text{ for all } P^* \in \mathcal{P} \cap \mathcal{R}_1 \}.$$

It is enough to show $\mathcal{A}^* = \bar{\mathcal{A}}_1$.

(a): In order to show $\bar{\mathcal{A}}_1 \subseteq \mathcal{A}^*$, take $X \in \bar{\mathcal{A}}_1$ and $P^* \in \mathcal{P} \cap \mathcal{R}_1$. There exists $\xi \in \mathbb{R}^d$ and $A_1 \in \mathcal{A}_1$ such that $X + \xi \cdot Y \geq A_1$. Thus,

$$E^*[X + \xi \cdot Y] \geq E^*[A_1] \geq \gamma(P^*),$$

due to $P^* \in \mathscr{R}_1$. Since $E^*[\xi \cdot Y] = 0$ due to $P^* \in \mathscr{P}$, we obtain $E^*[X] \geq \gamma(P^*)$, and hence $X \in \mathscr{A}^*$.

(b): In order to show $\mathscr{A}^* \subseteq \bar{\mathscr{A}_1}$, we take $X \in \mathscr{A}^*$ and assume that $X \notin \bar{\mathscr{A}_1}$. This means that the vector $x^* = (x_1^*, \dots, x_N^*)$ with components

$$x_i^* := E_{Q_i}[X] - \gamma(Q_i)$$

does *not* belong to the convex cone

$$\mathscr{C} := \{(E_{Q_i}[\xi \cdot Y])_{i=1,\dots,N} + y \mid \xi \in \mathbb{R}^d,\ y \in \mathbb{R}_+^N\} \subset \mathbb{R}^N,$$

where $\mathcal{Q}_0 \cup \mathcal{Q}_1 = \{Q_1, \dots, Q_N\}$ with $N \geq n$. In part (c) of this proof, we will show that $\mathscr{C}$ is closed. Thus, there exists $\lambda \in \mathbb{R}^N$ such that

$$\lambda \cdot x^* < \inf_{x \in \mathscr{C}} \lambda \cdot x; \tag{4.101}$$

see Proposition A.5. Since $\mathscr{C} \supseteq \mathbb{R}_+^N$, we obtain $\lambda_i \geq 0$ for $i = 1, \dots, N$, and we may assume $\sum_i \lambda_i = 1$ since $\lambda \neq 0$. Define

$$R := \sum_{i=1}^N \lambda_i Q_i \in \mathscr{R}_1.$$

Since $\mathscr{C}$ contains the linear space of vectors $(E_{Q_i}[V])_{i=1,\dots,N}$ with $V \in \mathscr{K}$, (4.101) implies

$$E_R[V] = 0 \quad \text{for } V \in \mathscr{K},$$

hence $R \in \mathscr{P}$. Moreover, the right-hand side of (4.101) must be zero, and the condition $\lambda \cdot x^* < 0$ translates into

$$E_R[X] < \gamma(R),$$

contradicting our assumption $X \in \mathscr{A}^*$.

(c): It remains to show that $\mathscr{C}$ is *closed*. For $\xi \in \mathbb{R}^d$, we define $y(\xi)$ as the vector in $\mathbb{R}^N$ with coordinates $y_i(\xi) = E_{Q_i}[\xi \cdot Y]$. Any $x \in \mathscr{C}$ admits a representation

$$x = y(\xi) + z$$

with $z \in \mathbb{R}_+^N$ and $\xi \in N^\perp$, where

$$N := \{\eta \in \mathbb{R}^d \mid E_{Q_i}[\eta \cdot Y] = 0 \text{ for } i = 1, \dots, N\},$$

and

$$N^{\perp} := \{\xi \in \mathbb{R}^d \mid \xi \cdot \eta = 0 \text{ for all } \eta \in N\}.$$

Take a sequence

$$x_n = y(\xi_n) + z_n, \quad n = 1, 2, \ldots,$$

with $\xi_n \in N^{\perp}$ and $z_n \in \mathbb{R}^N_+$, such that x_n converges to $x \in \mathbb{R}^N$. If $\liminf_n |\xi_n| < \infty$, then we may assume, passing to a subsequence if necessary that ξ_n converges to $\xi \in \mathbb{R}^d$. In this case, z_n must converge to some $z \in \mathbb{R}^N_+$, and we have $x = y(\xi) + z \in \mathscr{C}$. Let us now show that the case $\lim_n |\xi_n| = \infty$ is in fact excluded. In that case, $\alpha_n := (1 + |\xi_n|)^{-1}$ converges to 0, and the vectors $\zeta_n := \alpha_n \xi_n$ stay bounded. Thus, we may assume that ζ_n converges to $\zeta \in N^{\perp}$. This implies

$$y(\zeta) = \lim_{n\uparrow\infty} y(\zeta_n) = -\lim_{n\uparrow\infty} \alpha_n z_n \in -\mathbb{R}^N_+.$$

Since $\zeta \in N^{\perp}$ and $|\zeta| = \lim_n |\zeta_n| = 1$, we obtain $y(\zeta) \neq 0$. Thus, the inequality

$$E_{Q_i}[\,(-\zeta) \cdot Y\,] = -y_i(\zeta) \geq 0$$

holds for all i and is strict for some i, in contradiction to our assumption (4.97). □

4.11 Utility-based shortfall risk measures

In this section, we will establish a connection between convex risk measures and the expected utility theory of Chapter 2. Suppose that a risk-averse investor assesses the downside risk of a financial position $X \in \mathscr{X}$ by taking the expected utility $E[\,u(-X^-)\,]$ derived from the *shortfall* X^-, or by considering the expected utility $E[\,u(X)\,]$ of the position itself. If the focus is on the downside risk, then it is natural to change the sign and to replace u by the function $\ell(x) := -u(-x)$. Then ℓ is a strictly convex and increasing function, and the maximization of expected utility is equivalent to minimizing the expected loss $E[\,\ell(-X)\,]$ or the *shortfall risk* $E[\,\ell(X^-)\,]$. In order to unify the discussion of both cases, we do not insist on strict convexity. In particular, ℓ may vanish on $(-\infty, 0]$, and in this case the shortfall risk takes the form

$$E[\,\ell(X^-)\,] = E[\,\ell(-X)\,].$$

Definition 4.117. A function $\ell : \mathbb{R} \to \mathbb{R}$ is called a *loss function* if it is increasing and not identically constant. △

Let us return to the setting where we consider monetary risk measures defined on the class $\mathscr{X}$ of all bounded measurable functions on some given measurable space $(\Omega, \mathscr{F})$. Let us fix a probability measure P on $(\Omega, \mathscr{F})$. For a given loss function ℓ and

$$x_0 \in (\inf \ell, \sup \ell),$$

we define the following acceptance set:

$$\mathcal{A} := \{X \in \mathcal{X} \mid E[\,\ell(-X)\,] \le x_0\}. \tag{4.102}$$

Alternatively, we can write

$$\mathcal{A} = \{X \in \mathcal{X} \mid E[\,u(X)\,] \ge y_0\}, \tag{4.103}$$

where $u(x) = -\ell(-x)$ and $y_0 = -x_0$.

Lemma 4.118. *The acceptance set $\mathcal{A}$ is nonempty and induces a monetary risk measure ρ on L^∞, which is given by*

$$\begin{aligned} \rho(X) &= \inf\{m \in \mathbb{R} \mid E[\,\ell(-X-m)] \le x_0\} \\ &= \inf\{m \in \mathbb{R} \mid E[\,u(X+m)] \ge y_0\}. \end{aligned} \tag{4.104}$$

If ℓ is convex or u concave, then ρ is a convex risk measure. It is normalized if $x_0 = \ell(0)$.

Proof. The set $\mathcal{A}$ clearly contains all constants $m \in \mathbb{R}$ such that $\ell(-m) \le x_0$. Since $x_0 > \inf \ell$, it follows that $\mathcal{A}$ is nonempty. Moreover, since $x_0 < \sup \ell$, we also have $\inf\{m \in \mathbb{R} \mid m \in \mathcal{A}\} > -\infty$. That is, $\mathcal{A}$ satisfies (4.3). It also satisfies (4.4), and so part (a) of Proposition 4.7 yields that $\mathcal{A}$ induces a monetary risk measure ρ given by

$$\rho(X) = \inf\{m \in \mathbb{R} \mid X + m \in \mathcal{A}\} = \inf\{m \in \mathbb{R} \mid E[\,u(X+m)] \ge y_0\}.$$

This risk measure satisfies (4.34), and hence can be regarded as a monetary risk measure on L^∞. □

Example 4.119. Value at Risk at level $\lambda \in (0,1)$,

$$\mathrm{V@R}_\lambda(X) = \inf\{m \in \mathbb{R} \mid P[\,m + X < 0\,] \le \lambda\} = \inf\{m \in \mathbb{R} \mid E[\,\mathbb{1}_{(0,\infty)}(-m-X)\,] < 0\,] \le \lambda\},$$

is of the form (4.104) with $\ell = \mathbb{1}_{(0,\infty)}$ and $x_0 = \lambda$. ◇

Exercise 4.11.1. Let ℓ be a strictly increasing continuous loss function and suppose that the risk measure ρ associated to ℓ via (4.104) satisfies

$$\rho(X) \ge \rho(Y) \quad \Longleftrightarrow \quad E[\,\ell(-X)\,] \ge E[\,\ell(-Y)\,]$$

for any $X, Y \in L^\infty$. Show that ℓ is either affine or exponential.
Hint: Apply Proposition 2.46. ◇

For the remainder of this section, we will only consider convex loss functions $\ell : \mathbb{R} \to \mathbb{R}$. By Proposition A.7, any such loss function ℓ is continuous. Moreover, the risk measure defined via (4.104) is convex.

Definition 4.120. The convex risk measure in (4.104) is called *utility-based shortfall risk measure.* △

Proposition 4.121. *For every $X \in L^\infty$, the number $r := \rho(X)$ is the unique solution to the equation*

$$E[\,\ell(-X-r)\,] = x_0. \tag{4.105}$$

Proof. Since X is bounded and the function ℓ is continuous, dominated convergence implies that the function $m \mapsto E[\ell(-X-m)]$ is continuous. Now let (m_n) be a sequence such that $E[\ell(-X-m_n)] \le x_0$ for each n and

$$m_n \searrow \inf\{m \in \mathbb{R} \mid E[\,\ell(-X-m)] \le x_0\} = \rho(X) =: r.$$

Then the continuity of $m \mapsto E[\ell(-X-m)]$ implies that $E[\ell(-X-r)] \le x_0$. Moreover, if $E[\ell(-X-r)] < x_0$, then we can find $y < r$ such that $E[\ell(-X-y)] \le x_0$, in contradiction to the definition of $r = \rho(X)$. Therefore, r solves (4.105).

To show the uniqueness of solutions to (4.105), we suppose by way of contradiction that $\tilde{r}$ is another solution. We may assume without loss of generality that $\tilde{r} < r$. Clearly, $\ell(-X-r) \le \ell(-X-\tilde{r})$ P-almost surely. We claim moreover that $\ell(-X-r) < \ell(-X-\tilde{r})$ with strictly positive probability. Once that claim is established, we will obtain

$$E[\,\ell(-X-r)\,] < E[\,\ell(-X-\tilde{r})\,],$$

which is a contradiction to our assumption that both $\tilde{r}$ and r solve (4.105). To prove our claim that $\ell(-X-r) < \ell(-X-\tilde{r})$ with strictly positive probability, we note first that the increasing convex function ℓ is strictly increasing on $\ell^{-1}((\inf\ell, \sup\ell))$ by (A.5) and that $\sup\ell = \infty$. Since $P[\ell(-X-\tilde{r}) \ge x_0]$ must be strictly positive and since $x_0 > \inf\ell$, it follows that both $-X-r$ and $-X-\tilde{r}$ belong to $\ell^{-1}((\inf\ell, \infty))$ with strictly positive probability. Hence, $\ell(-X-r) < \ell(-X-\tilde{r})$ with strictly positive probability. □

Remark 4.122. In the preceding proof, we did not use the assumption $x_0 \in (\inf\ell, \sup\ell)$ for showing that $r = \rho(X)$ solves the equation $E[\,\ell(-X-r)\,] = x_0$. Hence, this fact remains true also in the case where $x_0 = \inf\ell$ and ℓ attains its infimum. This observation will be used in the proof of Theorem 4.137 in Section 4.12. However, the assumption $x_0 \in (\inf\ell, \sup\ell)$ was used to derive the uniqueness of the solution r to the equation $E[\,\ell(-X-r)\,] = x_0$. ◇

Remark 4.123. Every law-invariant convex risk measure ρ on L^∞ gives rise to a functional R on a subset of the space $\mathcal{M}_b(\mathbb{R})$ of all Borel probability measures on $\mathbb{R}$ with bounded support:

$$R(\mu) := \rho(X) \quad \text{whenever } P \circ X^{-1} = \mu.$$

This subset is equal to all of $\mathscr{M}_b(\mathbb{R})$ if the underlying probability space $(\Omega, \mathscr{F}, P)$ is atomless; see Appendix D.4. If ρ is a utility-based shortfall risk measure, then Proposition 4.121 shows that $R(\mu)$ is given by the unique solution r to the equation

$$\int \ell(-x-r)\,\mu(dx) = x_0.$$

Since the left-hand side of this equation is an affine function of μ, the functional R has the following property:

$$\text{If } R(\mu) = R(\nu) = r, \text{ then } R(\lambda\mu + (1-\lambda)\nu) = r \text{ for all } \lambda \in [0,1]. \tag{4.106}$$

This property is sometimes called the *convexity of the level sets at the level of distributions*, or *CxLS*. It is connected to the so-called elicitability of risk measures, and under certain conditions, characterizes the shortfall risk measures of the form (4.104); see [320] and [94] and the references therein. ◇

Proposition 4.124. *The utility-based shortfall risk measure ρ is continuous from below. In particular, ρ can be represented in the form*

$$\rho(X) = \max_{Q \in \mathscr{M}_1(P)} \big(E_Q[-X] - \alpha^{\min}(Q)\big). \tag{4.107}$$

Proof. Suppose that (X_n) is a sequence in L^∞ that increases P-a.s. to some $X \in L^\infty$. Then $\rho(X_n)$ decreases to some finite limit r. Using the continuity of ℓ and dominated convergence, it follows that

$$E[\,\ell(-\rho(X_n) - X_n)] \longrightarrow E[\,\ell(-r - X)\,].$$

By Proposition 4.121, each of the approximating expectations equals x_0, and so r is a solution to (4.105). Hence, $r = \rho(X)$, and this proves continuity from below. The representation (4.107) now follows from Corollary 4.38. □

The following exercise can be compared with Exercise 4.5.1.

Exercise 4.11.2. Let $\beta \in (0,1)$ be given.

(a) Show that the following conditions are equivalent for $X \in L^2$ and $\zeta \in \mathbb{R}$:

(i) The number ζ minimizes the function

$$f_\beta(z) := \beta E\big[\,((X-z)^+)^2\,\big] + (1-\beta)E\big[\,((X-z)^-)^2\,\big].$$

(ii) We have

$$\beta E[\,(X-\zeta)^+\,] = (1-\beta)E[\,(X-\zeta)^-\,].$$

A number ζ satisfying the preceding two equivalent conditions is called an *expectile* for X at level β.

(b) Now suppose that $0 < \beta \le 1/2$, so that the loss function $\ell(x) = (1-\beta)x^+ - \beta x^-$ is convex. Show that, in this case, the corresponding shortfall risk measure

$$\rho(X) = \inf\{m \in \mathbb{R} \mid E[\,\ell(-X-m)] \le 0\}$$

is such that $-\rho(X)$ is an expectile for X at level β. ◇

Let us now compute the minimal penalty function $\alpha^{\min}$ of a utility-based shortfall risk measure. We start by analyzing the following special case.

Example 4.125. For an exponential loss function $\ell(x) = e^{\beta x}$, the minimal penalty function can be described in terms of relative entropy, and the resulting risk measure coincides, up to an additive constant, with the entropic risk measure introduced in Example 4.34. In fact,

$$\rho(X) = \inf\{m \in \mathbb{R} \mid E[\,e^{-\beta(m+X)}\,] \le x_0\} = \frac{1}{\beta}(\log E[\,e^{-\beta X}\,] - \log x_0).$$

In this special case, the general formula (4.20) for $\alpha^{\min}$ reduces to the variational formula for the relative entropy $H(Q|P)$ of Q with respect to P:

$$\begin{aligned}\alpha^{\min}(Q) &= \sup_{X\in L^\infty}\left(E_Q[\,-X\,] - \frac{1}{\beta}\log E[\,e^{-\beta X}\,]\right) + \frac{\log x_0}{\beta}\\ &= \frac{1}{\beta}(H(Q|P) + \log x_0),\end{aligned} \tag{4.108}$$

where the second identity follows from Theorem C.5. Thus, the representation (4.107) of ρ is equivalent to the following dual variational identity:

$$\log E[\,e^X\,] = \max_{Q\in\mathcal{M}_1(P)}(E_Q[\,X\,] - H(Q|P));$$

see also Remark C.8. ◇

In general, the minimal penalty function $\alpha^{\min}$ on $\mathcal{M}_1(P)$ can be expressed in terms of the *Fenchel–Legendre transform* or *conjugate function* ℓ^* of the convex function ℓ defined by

$$\ell^*(z) := \sup_{x\in\mathbb{R}}(zx - \ell(x)).$$

Theorem 4.126. *For any convex loss function ℓ, the minimal penalty function in the representation* (4.107) *is given by*

$$\alpha^{\min}(Q) = \inf_{\lambda>0}\frac{1}{\lambda}\left(x_0 + E\left[\,\ell^*\left(\lambda\frac{dQ}{dP}\right)\right]\right),\quad Q\in\mathcal{M}_1(P). \tag{4.109}$$

In particular,

$$\rho(X) = \max_{Q \in \mathcal{M}_1(P)} \left(E_Q[-X] - \inf_{\lambda>0} \frac{1}{\lambda}\left(x_0 + E\left[\ell^*\left(\lambda \frac{dQ}{dP}\right)\right]\right)\right), \quad X \in L^\infty. \tag{4.110}$$

Let us now discuss some examples and applications of Theorem 4.126. The proof of Theorem 4.126 will start after Exercise 4.11.4. After the proof is complete, we will analyze what can still be said if we drop our assumption $x_0 \in (\inf \ell, \sup \ell) = (\inf \ell, \infty)$.

Example 4.127. Take

$$\ell(x) := \begin{cases} \frac{1}{p} x^p & \text{if } x \geq 0, \\ 0 & \text{otherwise,} \end{cases}$$

where $p > 1$. Then

$$\ell^*(z) := \begin{cases} \frac{1}{q} z^q & \text{if } z \geq 0, \\ +\infty & \text{otherwise,} \end{cases}$$

where $q = p/(p-1)$ is the usual dual coefficient. We may apply Theorem 4.126 for any $x_0 > 0$. Let $Q \in \mathcal{M}_1(P)$ with density $\varphi := dQ/dP$. Clearly, $\alpha^{\min}(Q) = +\infty$ if $\varphi \notin L^q(\Omega, \mathcal{F}, P)$. Otherwise, the infimum in (4.109) is attained for

$$\lambda_Q = \left(\frac{p x_0}{E[\varphi^q]} \right)^{1/q}.$$

Hence, we can identify $\alpha^{\min}(Q)$ for any $Q \ll P$ as

$$\alpha_p^{\min}(Q) = (p x_0)^{1/p} \cdot E\left[\left(\frac{dQ}{dP} \right)^q \right]^{1/q}.$$

Taking the limit $p \downarrow 1$, we obtain the case $\ell(x) = x^+$, where we measure the risk in terms of the expectation of the shortfall $(-X)^+$. Here, we have

$$\ell^*(z) = \begin{cases} 0 & \text{if } 0 \leq z \leq 1, \\ +\infty & \text{otherwise,} \end{cases}$$

and hence

$$\alpha_1^{\min}(Q) = x_0 \cdot \left\| \frac{dQ}{dP} \right\|_\infty.$$

◇

Exercise 4.11.3. For an exponential loss function $\ell(x) = e^{\beta x}$, we are in the context of Example 4.125 and ρ is the entropic risk measure. Compute ℓ^* and use Theorem 4.126

to derive the formula (4.108) for the minimal penalty function of the entropic risk measure ρ. ◇

Example 4.128. As in Exercise 4.11.2, we fix $0 < \beta \le 1/2$ and consider the loss function $\ell(x) = (1-\beta)x^+ - \beta x^-$ and the expectile risk measure

$$\rho(X) = \inf\{m \in \mathbb{R} \mid E[\,\ell(-X-m)] \le 0\}.$$

Note that ρ is a coherent risk measure, because the function ℓ is positively homogeneous. The conjugate function of ℓ is given by

$$\ell^*(z) = \begin{cases} 0 & \text{if } \beta \le z \le 1-\beta, \\ +\infty & \text{otherwise.} \end{cases}$$

By Theorem 4.126, ρ can hence be represented as

$$\rho(-X) = \min_{Q \in \mathcal{Q}} E_Q[-X],$$

where

$$\mathcal{Q} = \left\{ Q \in \mathcal{M}_1(P) \;\middle|\; \text{there is } \lambda > 0 \text{ such that } \beta \le \lambda \frac{dQ}{dP} \le (1-\beta)\ P\text{-a.s.} \right\}. \qquad ◇$$

When combined with Proposition 4.20, Theorem 4.126 yields the following result for risk measures that are defined in terms of a robust notion of bounded shortfall risk. Here, it is convenient to define $\ell^*(\infty) := \infty$.

Corollary 4.129. *Suppose that $\mathcal{Q}$ is a family of probability measures on $(\Omega, \mathcal{F})$, and that ℓ, ℓ^*, and x_0 are as in Theorem 4.126. We define a set of acceptable positions by*

$$\mathcal{A} := \{X \in \mathcal{X} \mid E_P[\,\ell(-X)\,] \le x_0 \text{ for all } P \in \mathcal{Q}\}.$$

Then the corresponding convex risk measure can be represented in terms of the penalty function

$$\alpha(Q) = \inf_{\lambda > 0} \frac{1}{\lambda}\left(x_0 + \inf_{P \in \mathcal{Q}} E_P\left[\ell^*\left(\lambda \frac{dQ}{dP} \right) \right] \right), \qquad Q \in \mathcal{M}_1(\Omega, \mathcal{F}),$$

where dQ/dP is the density appearing in the Lebesgue decomposition of Q with respect to P as in Theorem B.9.

Example 4.130. In the case of Example 4.125, the corresponding robust problem in Corollary 4.129 leads to the following entropy minimization problem: For a given Q and a set $\mathcal{Q}$ of probability measures, find

$$\inf_{P \in \mathcal{Q}} H(Q|P).$$

Note that this problem is different from the standard problem of minimizing $H(Q|P)$ with respect to the first variable Q as it appears in Section 3.2 and Appendix C. ◇

Example 4.131. Take $x_0 = 0$ in (4.102) and $\ell(x) := x$. Then

$$\ell^*(z) := \begin{cases} 0 & \text{if } z = 1, \\ +\infty & \text{otherwise.} \end{cases}$$

Therefore, $\alpha(Q) = \infty$ if $Q \neq P$, and $\rho(X) = E[\,-X\,]$. If $\mathcal{Q}$ is a set of probability measures, the risk measure ρ of Corollary 4.129 is coherent, and it is given by

$$\rho(X) = \sup_{P \in \mathcal{Q}} E_P[\,-X\,]. \qquad \diamond$$

Exercise 4.11.4. Consider a situation in which model uncertainty is described by a parametric family P_θ for $\theta \in \Theta$. In each model P_θ, the expected utility of a random variable $X \in L^\infty$ is $E_\theta[\,u(X)\,]$, where $u : \mathbb{R} \to \mathbb{R}$ is a given utility function. In a Bayesian approach, one would choose a prior distribution μ on Θ. In terms of F. Knight's distinction between risk and uncertainty, we would then be in a situation of *model risk*. Risk neutrality with respect to this model risk would be described by the utility functional

$$U(X) = \int E_\theta[\,u(X)\,]\,\mu(d\theta);$$

here we assume that Θ is endowed with the σ-field generated by the maps $\theta \mapsto E_\theta[X]$ for $X \in L^\infty$. In order to capture *model risk aversion*, we could choose another utility function $\hat{u} : \mathbb{R} \to \mathbb{R}$ and consider the functional $\hat{U}(X)$ defined by

$$\hat{u}(\hat{U}(X)) = \int \hat{u}(E_\theta[\,u(X)\,])\,\mu(d\theta).$$

Show that $\hat{U}$ is quasi-concave, that is,

$$\hat{U}(\lambda X + (1-\lambda)Y) \geq \hat{U}(X) \wedge \hat{U}(Y) \quad \text{for } X, Y \in L^\infty \text{ and } 0 \leq \lambda \leq 1.$$

Show next that for $\hat{u}(x) = 1 - e^{-\beta x}$, the functional $\hat{U}(X)$ takes the form

$$\hat{U}(X) = -\rho(u(X))$$

for a convex risk measure ρ. Then compute the minimal penalty function in the robust representation of ρ (here you may assume that Θ is a finite set). Discuss the limit $\beta \uparrow \infty$.
◇

In the remainder of this section, we will prove Theorem 4.126. To prepare for the proof, we summarize some properties of the functions ℓ and ℓ^* as stated in Appendix A.1.

First, note that ℓ^* is a proper convex function, that is, it is convex and takes some finite value. This follows, for example, from Proposition A.9. We denote by

$$J(z) := (\ell^*)'_+(z)$$

its right-hand derivative, which is defined for z in the interior of the effective domain $\operatorname{dom}\ell^*$ of the convex function ℓ^* (see Proposition A.7). Recall from Proposition A.9 that $\ell^{**} = \ell$ and that, for $x, z \in \mathbb{R}$,

$$xz \le \ell(x) + \ell^*(z) \quad \text{with equality if } z \in \operatorname{dom}\ell^* \text{ and } x = J(z). \tag{4.111}$$

Lemma 4.132. *Let (ℓ_n) be a sequence of convex loss functions that decrease pointwise to the convex loss function ℓ. Then the corresponding conjugate functions ℓ_n^* increase pointwise to ℓ^*.*

Proof. It follows immediately from the definition of the Fenchel–Legendre transform that each ℓ_n^* is dominated by ℓ^*, and that $\ell_n^*(z)$ increases to some limit $\ell_\infty^*(z)$. We have to prove that $\ell_\infty^* = \ell^*$.

The function $z \mapsto \ell_\infty^*(z)$ is a lower semicontinuous convex function as the increasing limit of such functions. Moreover, ℓ_∞^* is a proper convex function, since

$$-\infty < -\ell_1(0) \le \ell_1^* \le \ell_\infty^* \le \ell^*,$$

and ℓ^* is proper. Consider the conjugate function $\ell_\infty^{**} := (\ell_\infty^*)^*$ of ℓ_∞^*. Since $\ell_\infty^* \le \ell^*$, we have $\ell_\infty^{**} \ge \ell^{**}$, and since $\ell^{**} = \ell$ by Proposition A.9, we get $\ell_\infty^{**} \ge \ell$. On the other hand, an analogous argument yields $\ell_\infty^{**} \le \ell_n$ for each n. By taking $n \uparrow \infty$, this shows $\ell_\infty^{**} = \ell$, which in turn gives $\ell_\infty^* = \ell^*$. □

By ℓ'_- and ℓ'_+, we denote the respective left- and right-hand derivatives of the convex function ℓ; see Proposition A.7.

Lemma 4.133. *The functions ℓ and ℓ^* have the following properties:*

(a) *We have $\ell^*(0) = -\inf_{x\in\mathbb{R}} \ell(x)$ and $\ell^*(z) \ge -\ell(0)$ for all z. Moreover, $\ell^*(z)$ is finite for all z with*

$$\inf_x \ell'_-(x) < z < \sup_x \ell'_+(x).$$

(b) *There exists some $z_1 \in [0, \infty)$ such that*

$$\ell^*(z) = \sup_{x\ge 0}(xz - \ell(x)) \quad \text{for } z > z_1.$$

In particular, ℓ^* is increasing on (z_1, ∞).

(c) $\frac{\ell^*(z)}{z} \longrightarrow \infty$ *as* $z \uparrow \infty$.

(d) *If ℓ^* is finite on (z_1, ∞), then $J(z) \nearrow \infty$ as $z \uparrow \infty$.*

(e) *Suppose that ℓ attains its infimum and that κ is equal to the largest x such that $\ell(x) = \inf \ell$. Then ℓ^* is continuous on $[0, \sup \ell'_+)$ and*

$$J(0) := J(0+) = \lim_{z \downarrow 0} J(z) = \kappa.$$

Proof. (a): It follows immediately from the definition of the Fenchel–Legendre transform that $\ell^*(0) = -\inf_{x \in \mathbb{R}} \ell(x)$ and $\ell^*(z) \geq -\ell(0)$ for all z. Now suppose that $\inf \ell'_- < z < \sup \ell'_+$. Then there are real numbers y_- and y_+ such that

$$\inf \ell'_- < \ell'_-(y_-) < z < \ell'_+(y_+) < \sup \ell'_+.$$

We can write

$$\ell^*(z) \leq \sup_{x<y_-}(xz - \ell(x)) + \sup_{y_- \leq x \leq y_-}(xz - \ell(x)) + \sup_{x>y_+}(xz - \ell(x)). \tag{4.112}$$

The center term on the right-hand side is finite, due to the continuity of ℓ. To show that the rightmost term is also finite, we use that for all $x > y_+$,

$$\ell(x) = \ell(y_+) + \int_{y_+}^{x} \ell'_+(y)\,dy \geq \ell(y_+) + \ell'_+(y_+)(x - y_+) \tag{4.113}$$

by Proposition A.7. This fact gives us

$$\sup_{x>y_+}(xz - \ell(x)) \leq \sup_{x>y_+} x(z - \ell'_+(y_+)) - \ell(y_+) + \ell'_+(y_+)y_+.$$

Since $z - \ell'_+(y_+) < 0$, the supremum on the right is equal to $y_+(z - \ell'_+(y_+))$, and so the entire right-hand side is finite. To see that also the leftmost supremum on the right-hand side of (4.112) is finite, we use the fact that $\ell(x) \geq \ell(y_-) + \ell'_-(y_-)(x - y_-)$ for all $x < y_-$, which can be obtained as in (4.113). It yields

$$\sup_{x<y_-}(xz-\ell(x)) \leq \sup_{x<y_-} x(z-\ell'_-(y_-))-\ell(y_-)+\ell'_-(y_-)y_- = y_-(z-\ell'_-(y_-))-\ell(y_-)+\ell'_-(y_-)y_- < \infty.$$

Thus, $-\ell(0) \leq \ell^*(z) < \infty$ whenever $\inf \ell'_- < z < \sup \ell'_+$.

(b): We show in a first step that the set S of all real numbers z for which $zx \leq \ell(x) - \ell(0)$ for all $x \in \mathbb{R}$ is nonempty. To this end, recall from the proof of Proposition A.7(b) that

$$d(x) := \frac{\ell(x) - \ell(0)}{x}$$

is an increasing function of $x \in \mathbb{R} \setminus \{0\}$. Hence, S is equal to the closed and nonempty real interval with endpoints

$$\sup_{x<0} d(x) \quad \text{and} \quad \inf_{x>0} d(x).$$

In the next step, we show that the set

$$N := \{z \in \mathbb{R} \mid \ell^*(z) = -\ell(0)\} = \{z \in \mathbb{R} \mid \ell^*(z) \leq -\ell(0)\}$$

contains S. Indeed, for $z \in S$ we have $\ell^*(z) \leq -\ell(0)$ while $\ell^*(z) \geq -\ell(0)$ holds for all z by (a). This shows that $S \subseteq N$.

Now we let $z_1 := \sup N$. Since $\ell^*(z) \geq z - \ell(1)$, we have $\ell^*(z) \to \infty$ as $z \uparrow \infty$. Hence, we get $z_1 < \infty$. Moreover, since ℓ is increasing, we have $z \geq 0$ for every $z \in S$, and hence get $z_1 \geq 0$. If $z > z_1$ and $x < 0$, then

$$xz - \ell(x) \leq xz_1 - \ell(x) \leq \ell^*(z_1) \leq -\ell(0),$$

where the rightmost inequality follows from the lower semicontinuity of the Fenchel–Legendre transform ℓ^* and the definition of N. But $z \notin N$, and so $\ell^*(z) > -\ell(0)$. Hence,

$$\sup_{x<0}(xz - \ell(x)) < \ell^*(z).$$

(c): For $z > z_1$,

$$\frac{\ell^*(z)}{z} = \sup_{x\geq 0}(x - \ell(x)/z)$$

by (b). Hence,

$$\frac{\ell^*(z)}{z} \geq x_z - 1,$$

where $x_z := \sup\{x \mid \ell(x) \leq z\}$. Since ℓ is convex, increasing, and takes only finite values, we have $x_z \to \infty$ as $z \uparrow \infty$.

(d): By Proposition A.7(d) and the fact that J is increasing, we have that for $z > z_2 > z_1$,

$$\ell^*(z) - \ell^*(z_2) = \int_{z_2}^{z} J(y)\,dy \leq J(z)(z - z_2).$$

If moreover $\ell^*(z) \geq 0$, which by part (c) holds for all sufficiently large z, then

$$J(z) \geq \frac{\ell^*(z) - \ell^*(z_2)}{z - z_2} \geq \frac{\ell^*(z)}{z} - \frac{\ell^*(z_2)}{z - z_2}.$$

By part (c), the right-hand side converges to infinity as $z \uparrow \infty$.

(e): We note first that the fact that the increasing function ℓ attains its infimum implies that $\inf \ell'_- = 0$. Hence, part (a) yields that ℓ^* and J are both well-defined and finite on the nonempty open interval $I := (0, \sup \ell'_+)$. Moreover, the Fenchel–Legendre transform ℓ^* is lower semicontinuous as the supremum of affine functions. By Proposition A.7, it is also upper semicontinuous, and hence continuous, on its effective domain, which contains $I \cup \{0\}$. Now we show that $J(z) \geq \kappa$ for all $z \in I$. Indeed, (4.111) yields that for $z \in I$,

$$J(z)z = \ell(J(z)) + \ell^*(z) \geq \ell(J(z)) + z\kappa - \ell(\kappa) = \ell(J(z)) + z\kappa - \inf \ell, \tag{4.114}$$

and so

$$J(z) \geq \frac{1}{z}(\ell(J(z)) - \inf \ell) + \kappa \geq \kappa.$$

In particular, we have $\kappa \leq J(0) \leq J(z) < \infty$ for an arbitrary $z \in I$. Now we show that $J(0)$ is actually equal to κ. We have already seen that $J(0)$ is finite. Thus, we have $J(z)z \to 0$ as $z \downarrow 0$. Hence, the leftmost identity in (4.114) yields that $\ell(J(z)) + \ell^*(z) \to 0$ as $z \downarrow 0$. As noted above, ℓ^* is continuous on its effective domain. Part (a) of this lemma hence yields that $\ell^*(z) \to -\inf \ell$ as $z \downarrow 0$. Therefore, we must have that $\ell(J(z)) \to \inf \ell$, which is only possible if $J(z) \searrow k$ as $z \downarrow 0$. □

Proof of Theorem 4.126. Fix $Q \in \mathcal{M}_1(P)$, and denote by $\varphi := dQ/dP$ its density. First, we show that it suffices to prove the claim for $x_0 > \ell(0)$. Otherwise, we can find some $a \in \mathbb{R}$ such that $\ell(-a) < x_0$, since $x_0 > \inf \ell$. Let $\widetilde{\ell}(x) := \ell(x - a)$, and

$$\widetilde{\mathcal{A}} := \{\widetilde{X} \in L^\infty \mid E[\,\widetilde{\ell}(-\widetilde{X})\,] \leq x_0\,\}.$$

Then $\widetilde{\mathcal{A}} = \{X - a \mid X \in \mathcal{A}\}$, and hence

$$\sup_{\widetilde{X} \in \widetilde{\mathcal{A}}} E_Q[\,-\widetilde{X}\,] = \sup_{X \in \mathcal{A}} E_Q[\,-X\,] + a. \tag{4.115}$$

The convex loss function $\widetilde{\ell}$ satisfies the requirement $\widetilde{\ell}(0) < x_0$. So, if the assertion is established in this case, we find that

$$\sup_{\widetilde{X} \in \widetilde{\mathcal{A}}} E_Q[\,-\widetilde{X}\,] = \inf_{\lambda > 0} \frac{1}{\lambda}(x_0 + E[\,\widetilde{\ell}^*(\lambda\varphi)\,]) = \inf_{\lambda > 0} \frac{1}{\lambda}(x_0 + E[\,\ell^*(\lambda\varphi)\,]) + a\,;$$

here we have used the fact that the Fenchel–Legendre transform $\widetilde{\ell}^*$ of $\widetilde{\ell}$ satisfies $\widetilde{\ell}^*(z) = \ell^*(z) + az$. Together with (4.115), this proves that the reduction to the case $x_0 > \ell(0)$ is indeed justified.

For any $\lambda > 0$ and $X \in \mathscr{A}$, (4.111) implies

$$-X\varphi = \frac{1}{\lambda}(-X)(\lambda\varphi) \leq \frac{1}{\lambda}(\ell(-X) + \ell^*(\lambda\varphi)).$$

Hence, for any $\lambda > 0$

$$\alpha^{\min}(Q) \leq \sup_{X \in \mathscr{A}} \frac{1}{\lambda}(E[\,\ell(-X)\,] + E[\,\ell^*(\lambda\varphi)]) \leq \frac{1}{\lambda}(x_0 + E[\,\ell^*(\lambda\varphi)]).$$

Thus, it remains to prove that

$$\alpha^{\min}(Q) \geq \inf_{\lambda > 0} \frac{1}{\lambda}(x_0 + E[\,\ell^*(\lambda\varphi)]) \quad \text{if } \alpha^{\min}(Q) < \infty. \tag{4.116}$$

This will be done first under the following extra conditions on ℓ, its Fenchel–Legendre transform ℓ^*, and the right-hand derivative J of ℓ^*:

$$\ell \text{ attains its infimum,} \tag{4.117}$$

$$\ell^* \text{ is finite on } (0, \infty), \tag{4.118}$$

$$J \text{ is continuous on } (0, \infty). \tag{4.119}$$

Note that (4.118) and Lemma 4.133 (d) imply that $J(z)$ increases to $+\infty$ as $z \uparrow \infty$. Hence, so does $\ell(J(z))$. Moreover, assumption (4.117) implies by way of Lemma 4.133(a) that $\ell^*(0) = -\inf \ell$ is finite. Since

$$\ell^*(z) \geq -\ell(0) > -x_0 \quad \text{for all } z, \tag{4.120}$$

it thus follows from (4.111) and part (e) of Lemma 4.133 that

$$\lim_{z \downarrow 0} \ell(J(z)) - x_0 < \lim_{z \downarrow 0}(\ell(J(z)) + \ell^*(z)) = \lim_{z \downarrow 0} zJ(z) = 0.$$

These facts, the continuity of J, and parts (d) and (e) of Lemma 4.133 imply that for all sufficiently large n there exists some $\lambda_n > 0$ such that

$$E[\,\ell(J(\lambda_n\varphi)\mathbb{1}_{\{\varphi \leq n\}})\,] = x_0.$$

Let us define

$$X_n := -J(\lambda_n\varphi)\mathbb{1}_{\{\varphi \leq n\}}.$$

Then X_n is bounded and belongs to $\mathscr{A}$. Hence, it follows from (4.111) and (4.120) that

$$\begin{aligned}\alpha^{\min}(Q) &\geq E_Q[\,-X_n\,]\\ &= \frac{1}{\lambda_n}E[\,\lambda_n\varphi J(\lambda_n\varphi)\mathbb{1}_{\{\varphi\leq n\}}\,]\\ &= \frac{1}{\lambda_n}E[\,(\ell(-X_n)+\ell^*(\lambda_n\varphi))\mathbb{1}_{\{\varphi\leq n\}}\,]\\ &= \frac{1}{\lambda_n}(x_0-\ell(0)\cdot P[\,\varphi>n\,]+E[\,\ell^*(\lambda_n\varphi)\mathbb{1}_{\{\varphi\leq n\}}\,])\\ &\geq \frac{x_0-\ell(0)}{\lambda_n}.\end{aligned}$$

Since we assumed that $\alpha^{\min}(Q)<\infty$, it follows that the decreasing limit λ_∞ of λ_n must be strictly positive. The fact that the lower semicontinuous function ℓ^* is bounded from below allows us to apply Fatou's lemma:

$$\begin{aligned}\alpha^{\min}(Q) &\geq \liminf_{n\uparrow\infty}\frac{1}{\lambda_n}(x_0-\ell(0)\cdot P[\,\varphi>n\,]+E[\,\ell^*(\lambda_n\varphi)\mathbb{1}_{\{\varphi\leq n\}}\,])\\ &\geq \frac{1}{\lambda_\infty}(x_0+E[\,\ell^*(\lambda_\infty\varphi)\,]).\end{aligned}$$

This proves (4.116) under the assumptions (4.117), (4.118), and (4.119). In the remainder of this proof, we will remove these assumption step by step.

In the next step, we will remove condition (4.119). That is, we will establish (4.116) if (4.117) and (4.118) are retained but J is no longer assumed to be continuous. To this end, we will show that for given $\varepsilon>0$,

$$(1+\varepsilon)\alpha^{\min}(Q)\geq\inf_{\lambda>0}\frac{1}{\lambda}\Big(x_0+E\Big[\,\ell^*\Big(\lambda\frac{dQ}{dP}\Big)\Big]\Big). \tag{4.121}$$

To prove (4.121), we start by defining the function

$$J_\varepsilon(z)=\frac{1}{\varepsilon z}\int_z^{(1+\varepsilon)z}J(y)\,dy,\qquad z>0.$$

Then J_ε is continuous on $(0,\infty)$, and the fact that J is increasing implies that $J_\varepsilon\geq J$ and that J_ε is increasing as well. Moreover, part (e) of Lemma 4.133 yields that $J_\varepsilon(0):=J_\varepsilon(0+)$ satisfies $J_\varepsilon(0)\geq J(0)>-\infty$ and $J_\varepsilon(0)\leq J(1)<\infty$. In particular, $J_\varepsilon(0)$ is finite and $\int_0^z J_\varepsilon(y)\,dy$ exists and is finite for all $z\geq0$. Now we define the function

$$\ell_\varepsilon^*(z):=\begin{cases}\ell^*(0)+\displaystyle\int_0^z J_\varepsilon(y)\,dy & \text{if } z\geq0,\\ +\infty & \text{if } z<0.\end{cases} \tag{4.122}$$

Then ℓ_ε^* is convex and lower semicontinuous. It is clear that ℓ_ε^* and its derivative J_ε satisfy (4.118) and (4.119). Now let

$$\ell_\varepsilon(x) := \sup_{z\in\mathbb{R}}(xz - \ell_\varepsilon^*(z)) = \sup_{z\geq 0}(xz - \ell_\varepsilon^*(z)) \tag{4.123}$$

be the Fenchel–Legendre transform of ℓ_ε^*. We claim that ℓ_ε is a convex loss function satisfying (4.117). Since the Fenchel–Legendre transform of ℓ_ε is equal to ℓ_ε^* by Proposition A.9(b), we will then be within the context of the preceding part of the proof, where all three conditions (4.117), (4.118), and (4.119) hold. First, the convexity of ℓ_ε is clear. Next, the function ℓ_ε^* has superlinear growth, because $J_\varepsilon \geq J$ and $J(x) \nearrow \infty$ by (4.119) and part (d) of Lemma 4.133. Hence, the rightmost supremum in (4.123) is attained in some $z_x \geq 0$ for every $x \in \mathbb{R}$. In particular, ℓ_ε takes only finite values. Moreover, for $0 \leq x \leq y$,

$$\ell_\varepsilon(x) = xz_x - \ell_\varepsilon^*(z_x) \leq yz_x - \ell_\varepsilon^*(z_x) \leq \sup_{z\geq 0}(yz - \ell_\varepsilon^*(z)) = \ell_\varepsilon(y),$$

and so ℓ_ε is increasing. Therefore, ℓ_ε is indeed a loss function. We will show below that the functions ℓ^* and ℓ_ε^* are related as follows:

$$\ell^*(z) \leq \ell_\varepsilon^*(z) \leq \ell^*((1+\varepsilon)z) \quad \text{for } z \geq 0. \tag{4.124}$$

Since $\ell_\varepsilon = (\ell_\varepsilon^*)^*$ and $\ell^{**} = \ell$ hold by Proposition A.9, we obtain from (4.124) that ℓ and ℓ_ε satisfy the following relation:

$$\ell\left(\frac{x}{1+\varepsilon}\right) \leq \ell_\varepsilon(x) \leq \ell(x) \quad \text{for all } x \in \mathbb{R}. \tag{4.125}$$

Hence, since ℓ attains its infimum according to our assumption (4.117), so does ℓ_ε.

Now we prove (4.124). Proposition A.7 (d) yields that $\ell^*(z) = \ell^*(\delta) + \int_\delta^z J(y)\,dy$ for all $\delta, z > 0$. Sending $\delta \downarrow 0$ and using the fact that $\ell^*(\delta) \to \ell^*(0)$ by Lemma 4.133(e) hence yields

$$\ell^*(z) = \ell^*(0) + \int_0^z J(y)\,dy \quad \text{for all } z > 0. \tag{4.126}$$

In view of (4.122) and the fact that $J_\varepsilon \geq J$, this yields the first inequality in (4.124). The second inequality can be obtained as follows:

$$\begin{aligned}\ell_\varepsilon^*(z) - \ell^*(0) = \ell_\varepsilon^*(z) - \ell_\varepsilon^*(0) &= \int_0^z J_\varepsilon(y)\,dy = \int_0^z \frac{1}{\varepsilon y}\int_y^{(1+\varepsilon)y} J(x)\,dx\,dy \\ &= \int_0^z \frac{1}{\varepsilon y}\int_0^{(1+\varepsilon)z} \mathbb{1}_{\{y\leq x\leq(1+\varepsilon)y\}} J(x)\,dx\,dy\end{aligned}$$

$$
\begin{aligned}
&= \int_0^{(1+\varepsilon)z} J(x)\frac{1}{\varepsilon}\int_{\frac{x}{1+\varepsilon}}^{x\wedge z}\frac{1}{y}\,dy\,dx \le \int_0^{(1+\varepsilon)z} J(x)\frac{1}{\varepsilon}\int_{\frac{x}{1+\varepsilon}}^{x}\frac{1}{y}\,dy\,dx \\
&= \int_0^{(1+\varepsilon)z} J(x)\frac{\log x - \log\frac{x}{1+\varepsilon}}{\varepsilon}\,dx = \frac{\log(1+\varepsilon)}{\varepsilon}\int_0^{(1+\varepsilon)z} J(x)\,dx \\
&\le \int_0^{(1+\varepsilon)z} J(x)\,dx = \ell^*((1+\varepsilon)z) - \ell^*(0).
\end{aligned}
$$

After these preparations, we can now prove (4.121). Using (4.125), we get

$$
\mathscr{A}_\varepsilon := \{X \in L^\infty \mid E[\,\ell_\varepsilon(-X)\,] \le x_0\} \subseteq \{(1+\varepsilon)X \mid E[\,\ell(-X)\,] \le x_0\} = (1+\varepsilon)\mathscr{A}.
$$

Since ℓ_ε satisfies the conditions (4.117), (4.118), and (4.119), the first step of this proof yields that

$$
\sup_{X\in\mathscr{A}_\varepsilon} E_Q[\,-X\,] = \inf_{\lambda>0}\frac{1}{\lambda}\left(x_0 + E\left[\,\ell_\varepsilon^*\left(\lambda\frac{dQ}{dP}\right)\right]\right).
$$

Using (4.124), we now get

$$
\begin{aligned}
\inf_{\lambda>0}\frac{1}{\lambda}\left(x_0 + E\left[\,\ell^*\left(\lambda\frac{dQ}{dP}\right)\right]\right) &\le \inf_{\lambda>0}\frac{1}{\lambda}\left(x_0 + E\left[\,\ell_\varepsilon^*\left(\lambda\frac{dQ}{dP}\right)\right]\right) \\
&= \sup_{X\in\mathscr{A}_\varepsilon} E_Q[\,-X\,] \\
&\le \sup_{X\in(1+\varepsilon)\mathscr{A}} E_Q[\,-X\,] \\
&= (1+\varepsilon)\alpha^{\min}(Q).
\end{aligned}
$$

This establishes (4.121).

Finally, we remove conditions (4.117) and (4.118). If $\ell^*(z) = +\infty$ for some z, then z must be an upper bound for the slope of ℓ. So, we will approximate ℓ by a sequence (ℓ_n) of convex loss functions whose slope is unbounded. Simultaneously, we can handle the case where ℓ does not attain its infimum. To this end, we choose a sequence $y_n \downarrow \inf \ell$ such that $y_n \le \ell(0) < x_0$. We can define, for instance,

$$
\ell_n(x) := \ell(x) \vee y_n + \frac{1}{n}(e^x - 1)^+.
$$

Then ℓ_n decreases pointwise to ℓ. Each loss function ℓ_n satisfies (4.117) and (4.118). Hence, for any $\varepsilon > 0$ there are λ_n^ε such that

$$
\infty > \alpha^{\min}(Q) \ge \alpha_n^{\min}(Q) \ge \frac{1}{\lambda_n^\varepsilon}(x_0 + E[\,\ell_n^*(\lambda_n^\varepsilon\varphi)\,]) - \varepsilon \quad \text{for each } n, \tag{4.127}
$$

where $\alpha_n^{\min}(Q)$ is the penalty function arising from ℓ_n. Note that

$$\ell_n^*(z) \nearrow \ell^*(z) \quad \text{for all } z \in \mathbb{R} \tag{4.128}$$

by Lemma 4.132. Our assumption $\alpha^{\min}(Q) < \infty$, the fact that

$$\inf_{z\in\mathbb{R}} \ell_n^*(z) \geq -\ell_n(0) = -\ell(0) > -x_0, \tag{4.129}$$

and (4.128) in combination with part (c) of Lemma 4.133 show that the sequence $(\lambda_n^\varepsilon)_{n\in\mathbb{N}}$ must be bounded away from zero and from infinity. Therefore, by passing to a subsequence if necessary, we may assume that λ_n^ε converges to some $\lambda^\varepsilon \in (0,\infty)$. Now let $\delta > 0$ be given. By (4.128) and (4.129), we may apply monotone convergence to find some $n_0 \in \mathbb{N}$ such that for all $n \geq n_0$,

$$E[\,\ell_n^*(\lambda^\varepsilon\varphi)\,] \geq \begin{cases} E[\,\ell^*(\lambda^\varepsilon\varphi)\,] - \delta & \text{if } E[\,\ell^*(\lambda^\varepsilon\varphi)\,] < \infty, \\ 1/\delta & \text{otherwise.} \end{cases}$$

Using (4.127), (4.128), (4.129), the lower semicontinuity of $\ell_{n_0}^*$, and Fatou's lemma yields that

$$\begin{aligned}
\alpha^{\min}(Q) + \varepsilon &\geq \liminf_{n\uparrow\infty} \frac{1}{\lambda_n^\varepsilon}\big(x_0 + E[\,\ell_n^*(\lambda_n^\varepsilon\varphi)\,]\big) \\
&\geq \liminf_{n\uparrow\infty} \frac{1}{\lambda_n^\varepsilon}\big(x_0 + E[\,\ell_{n_0}^*(\lambda_n^\varepsilon\varphi)\,]\big) \\
&\geq \frac{1}{\lambda^\varepsilon}\big(x_0 + E[\,\ell_{n_0}^*(\lambda^\varepsilon\varphi)\,]\big) \\
&\geq \begin{cases} \dfrac{1}{\lambda^\varepsilon}\big(x_0 + E[\,\ell^*(\lambda^\varepsilon\varphi)\,]\big) - \dfrac{\delta}{\lambda^\varepsilon} & \text{if } E[\,\ell^*(\lambda^\varepsilon\varphi)\,] < \infty, \\ \dfrac{x_0}{\lambda^\varepsilon} + \dfrac{1}{\lambda^\varepsilon\delta} & \text{if } E[\,\ell^*(\lambda^\varepsilon\varphi)\,] = \infty. \end{cases}
\end{aligned}$$

Sending $\delta \downarrow 0$ shows that the case $E[\,\ell^*(\lambda^\varepsilon\varphi)\,] = \infty$ cannot occur and that

$$\alpha^{\min}(Q) + \varepsilon \geq \frac{1}{\lambda^\varepsilon}\big(x_0 + E[\,\ell^*(\lambda^\varepsilon\varphi)\,]\big).$$

This completes the proof of Theorem 4.126. □

In this section, we have so far assumed that x_0 belongs to $(\inf\ell, \sup\ell) = (\inf\ell, \infty)$. Let us now relax this assumption. We start with the case in which ℓ attains its infimum and $x_0 = \inf\ell = \min\ell$. Then there exists $\kappa \in \mathbb{R}$ such that $\ell(x) = x_0$ for $x \leq \kappa$ and $\ell(x) > x_0$ for $x > \kappa$. It follows that

$$\rho(X) = \inf\{m \in \mathbb{R} \mid E[\,\ell(-X-m)] \leq x_0\} = -\operatorname{ess\,inf} X - \kappa, \tag{4.130}$$

and so ρ is a translation of the worst-case measure on L^∞, which was introduced in Example 4.39. If $(\Omega, \mathscr{F}, P)$ cannot be reduced to a finite probability space, then ρ is not continuous from below, and the conclusions of Propositions 4.121 and 4.124 are no longer valid. Nevertheless, the following proposition shows that ρ still admits a dual representation terms of the penalty function:

$$\alpha(Q) = \inf_{\lambda>0} \frac{1}{\lambda}\left(x_0 + E\left[\ell^*\left(\lambda \frac{dQ}{dP}\right)\right]\right), \quad Q \ll P, \tag{4.131}$$

even though α may no longer be the minimal penalty function; see Example 4.135.

Proposition 4.134. *Suppose that $\ell(\mathbb{R}) = [x_0, \infty)$ and let κ denote the largest x such that $\ell(x) = \min \ell = x_0$. Then the penalty function $\alpha(Q)$ defined in (4.131) satisfies $\alpha(Q) \geq \kappa$ for all $Q \ll P$ and $\alpha(Q) = \kappa$ if dQ/dP is P-a. s. bounded. In particular, we have*

$$\rho(X) = \sup_{Q \ll P}(E_Q[-X] - \alpha(Q)), \quad X \in L^\infty. \tag{4.132}$$

Proof. We have for all $z \geq 0$,

$$\ell^*(z) = \sup_x (xz - \ell(x)) \geq \kappa z - \ell(\kappa) = \kappa z - x_0.$$

Hence, for all $Q \ll P$ and each $\lambda > 0$,

$$\frac{1}{\lambda}\left(x_0 + E\left[\ell^*\left(\lambda \frac{dQ}{dP}\right)\right]\right) \geq \kappa E\left[\frac{dQ}{dP}\right] = \kappa. \tag{4.133}$$

Therefore, $\alpha(Q) \geq \kappa$.

Now let $Q \ll P$ be such that dQ/dP is P-a. s. bounded by a constant $C > 0$. To show that $\alpha(Q) = \kappa$, we use (A.5) and the fact that $\ell^*(0) = -x_0$ to see that the function

$$\frac{1}{\lambda}(x_0 + \ell^*(\lambda y)) = y \cdot \frac{\ell^*(\lambda y) - \ell^*(0)}{\lambda y} \tag{4.134}$$

is increasing in $\lambda > 0$ for each $y > 0$. Hence, this function approaches its infimum in the limit $\lambda \downarrow 0$. Therefore, and in view of (4.133), we will have $\alpha(Q) = \kappa$ if we can show that

$$\lim_{\lambda \downarrow 0} \frac{1}{\lambda}\left(x_0 + E\left[\ell^*\left(\lambda \frac{dQ}{dP}\right)\right]\right) \leq \kappa. \tag{4.135}$$

The dual representation (4.132) will then result from (4.130) and the following chain of inequalities:

$$\begin{aligned}\sup_{Q\ll P}(E_Q[-X]-\alpha(Q)) &\le \sup_{Q\ll P}(E_Q[-X]-\kappa) = -\operatorname{ess\,inf} X-\kappa\\ &= \sup_{\substack{Q\ll P\\ dQ/dP\in L^\infty}}(E_Q[-X]-\kappa) = \sup_{\substack{Q\ll P\\ dQ/dP\in L^\infty}}(E_Q[-X]-\alpha(Q))\\ &\le \sup_{Q\ll P}(E_Q[-X]-\alpha(Q)).\end{aligned}$$

Let us now establish (4.135). We have seen in Lemma 4.133(a) that ℓ^* is finite on an interval of the form $[0,a)$ for some $a > 0$ and that $\ell^*(z) \ge -\ell(0)$ for all z. In part (e) of Lemma 4.133, we have seen moreover that the right-continuous derivative $J(z) := (\ell^*)'_+(z)$ of ℓ^* is finite and increasing on $[0,a)$ and satisfies $J(0) := J(0+) = \kappa$. By Proposition A.7(d), we have for $0 < \delta < z < a$ that $\ell^*(z) = \ell^*(\delta) + \int_\delta^z J(y)\,dy$. Sending $\delta \downarrow 0$ and using the fact that ℓ^* is continuous on $[0,a)$ by Lemma 4.133(e) therefore yields that

$$\ell^*(z) = \ell^*(0) + \int_0^z J(y)\,dy = -x_0 + \int_0^z J(y)\,dy.$$

It follows that

$$\frac{x_0+\ell^*(z)}{z} = \frac{1}{z}\int_0^z J(y)\,dy \searrow J(0) = \kappa \quad \text{as } z \downarrow 0.$$

Recall that $dQ/dP \le C$ P-almost surely. Hence, if $\lambda < a/C$, the expectation $E[\ell^*(\lambda dQ/dP)]$ is finite, and dominated convergence yields that

$$\lim_{\lambda\downarrow 0}\frac{1}{\lambda}\left(x_0 + E\left[\ell^*\left(\lambda\frac{dQ}{dP}\right)\right]\right) = \kappa E\left[\frac{dQ}{dP}\right] = \kappa.$$

Thus, we arrive at (4.133). □

Example 4.135. Consider the situation of Example 4.127, where $\ell(x) = (x^+)^p/p$ for some $p > 1$. This time, however, we take $x_0 = 0 = \inf \ell$. Then ρ is equal to the worst-case risk measure

$$\rho(X) = -\operatorname{ess\,inf} X = \sup_{Q\ll P} E_Q[-X].$$

Since $\ell^*(z) = z^q/q$ for $z \ge 0$ and $q = p/(p-1)$, the penalty function $\alpha(Q)$ defined in (4.131) is given by

$$\alpha(Q) = \inf_{\lambda>0}\frac{1}{\lambda}\left(x_0 + E\left[\ell^*\left(\lambda\frac{dQ}{dP}\right)\right]\right) = \inf_{\lambda>0}\lambda^{q-1}E\left[\left(\frac{dQ}{dP}\right)^q\right] = \begin{cases}0 & \text{if } dQ/dP \in L^q,\\ \infty & \text{otherwise.}\end{cases}$$

Thus, while α is a penalty function for ρ, it is not the minimal penalty function, which is zero for all $Q \ll P$. ◇

For fixed $X \in L^\infty$, we now consider the function

$$R_X(x) = \inf\{m \in \mathbb{R} \mid E[\ell(-m - X)] \le x\}, \quad x \in \mathbb{R}. \tag{4.136}$$

With the convention $\inf \emptyset = +\infty$, the function $R_X(x)$ is finite if and only if x belongs to the range $\ell(\mathbb{R})$ of ℓ. Moreover, Theorem 4.126, applied for $x > \inf \ell$, and Proposition 4.134, applied for $x = \min \ell$, yield the duality relation

$$R_X(x) = \sup_{Q \ll P}\Big(E_Q[-X] - \inf_{\lambda>0} \frac{1}{\lambda}\Big(x + E\Big[\ell^*\Big(\lambda \frac{dQ}{dP}\Big)\Big]\Big)\Big) \tag{4.137}$$

for all $x \in \ell(\mathbb{R})$. The following result shows that (4.137) remains true for $x \notin \ell(\mathbb{R})$.

Corollary 4.136. *For any $X \in L^\infty$, the duality relation* (4.137) *holds for all $x \in \mathbb{R}$.*

Proof. We already know that (4.137) holds for all $x \in \ell(\mathbb{R})$ and that the left-hand side of (4.137) is infinite if $x \notin \ell(\mathbb{R})$. Thus, it remains to show that the right-hand side of (4.137) is infinite for $x \notin \ell(\mathbb{R})$. By considering the special case $Q = P$, this will follow as soon as we can show that

$$\inf_{\lambda>0} \frac{1}{\lambda}\Big(x + E\Big[\ell^*\Big(\lambda \frac{dP}{dP}\Big)\Big]\Big) = \inf_{\lambda>0} \frac{1}{\lambda}(x + \ell^*(\lambda)) = -\infty. \tag{4.138}$$

Since the existence of $x \notin \ell(\mathbb{R})$ implies that $\inf \ell$ must be finite, we must have $\inf \ell'_- = 0$. Thus, Lemma 4.133(a) implies that there exists $\lambda_0 > 0$ such that $-\inf \ell = \ell^*(0) \le \ell^*(\lambda) < \infty$ for all $\lambda \in (0, \lambda_0)$.

Let us now consider the case $x < \inf \ell$. Then, for $\lambda \in (0, \lambda_0)$,

$$\frac{1}{\lambda}(x + \ell^*(\lambda)) = \frac{1}{\lambda}(x - \inf \ell) + \frac{1}{\lambda}(\ell^*(\lambda) - \ell^*(0)).$$

The first term on the right-hand side clearly converges to $-\infty$ as $\lambda \downarrow 0$. As already seen in (4.134), the rightmost term is an increasing function of λ, and hence bounded by a finite constant for all $\lambda \in (0, \lambda_0/2]$. Thus, we obtain (4.138) for $x < \inf \ell$.

Finally, we consider the case in which $x = \inf \ell$ and ℓ does not attain its infimum. Then

$$\frac{1}{\lambda}(x + \ell^*(\lambda)) = \frac{\ell^*(\lambda) - \ell^*(0)}{\lambda}$$

is an increasing function of $\lambda \in (0, \lambda_0)$, and so the infimum will be attained in the limit $\lambda \downarrow 0$. If we let $\lambda_n := \ell'_-(-n)$, then $\lambda_n \searrow 0$ as $n \uparrow \infty$. Moreover, Proposition A.9 implies that

$$\ell(-n) = \ell'_-(-n)(-n) - \ell^*(\ell'_-(-n)) = -n\lambda_n - \ell^*(\lambda_n) \quad \text{for all } n.$$

It follows that

$$\frac{\ell^*(\lambda_n) - \ell^*(0)}{\lambda_n} = \frac{\ell^*(\lambda_n) + \inf \ell}{\lambda_n} < \frac{\ell^*(\lambda_n) + \ell(-n)}{\lambda_n} = -n.$$

Sending $n \uparrow \infty$ now yields (4.138). □

4.12 Divergence risk measures

In this section, we explore the relations between shortfall risk and the *divergence risk measures* introduced in Example 4.36. To this end, let $g : [0, \infty) \to \mathbb{R} \cup \{+\infty\}$ be a lower semicontinuous convex function satisfying $g(1) < \infty$ and the superlinear growth condition $g(x)/x \to +\infty$ as $x \uparrow \infty$. For notational convenience, we extend g to all of $\mathbb{R}$ by letting $g(x) := +\infty$ for $x < 0$. Recall the definitions of the *g-divergence*,

$$I_g(Q|P) := E\left[g\left(\frac{dQ}{dP} \right) \right], \quad Q \in \mathcal{M}_1(P), \tag{4.139}$$

and of the corresponding *divergence risk measure*

$$\rho_g(X) := \sup_{Q \ll P} (E_Q[\,-X\,] - I_g(Q|P)), \quad X \in L^\infty. \tag{4.140}$$

We have seen in Exercise 4.4.4 that ρ is continuous from below and that $I_g(\cdot\,|P)$ is its minimal penalty function, so the supremum in (4.140) is actually a maximum.

Theorem 4.137. *Let $g^*(y) = \sup_{x \geq 0}(xy - g(x))$ be the Fenchel–Legendre transform of g. Then*

$$\rho_g(X) = \inf_{z \in \mathbb{R}} (E[\,g^*(z - X)\,] - z), \quad X \in L^\infty. \tag{4.141}$$

Before turning to the proof of Theorem 4.137, let us take a look at two special cases of the representation (4.141).

Example 4.138. If we let $g = \infty \cdot \mathbb{1}_{(1/\lambda, \infty)}$ for some $\lambda \in (0, 1)$, then the representation (4.140) is given by

$$\rho_g(X) = \sup_{Q \in \mathcal{Q}_\lambda} E_Q[-X],$$

where

$$\mathcal{Q}_\lambda = \left\{ Q \ll P \;\middle|\; \frac{dQ}{dP} \leq \frac{1}{\lambda} \text{ } P\text{-a.s.} \right\}.$$

By Theorem 4.52, the divergence risk measure ρ_g hence coincides with Average Value at Risk $\mathrm{AV@R}_\lambda$. One checks that the Fenchel–Legendre transform g^* of g is given by

$g^*(z) = z^+/\lambda$, and so the representation for AV@R$_\lambda$ obtained in Proposition 4.51 is a special case of (4.141):

$$\mathrm{AV@R}_\lambda(X) = \frac{1}{\lambda}\inf_{z\in\mathbb{R}}(E[\,(z-X)^+\,] - \lambda z). \qquad \diamond$$

Exercise 4.12.1. For given $\beta > 0$, let $g(x) := \frac{1}{\beta}x\log x$. Then

$$I_g(Q|P) = \frac{1}{\beta}H(Q|P),$$

where $H(Q|P)$ is the relative entropy of Q with respect to P, and so ρ_g is the entropic risk measure. Compute the Fenchel–Legendre transform g^* of g and show that the representation (4.141) is equivalent to the dual representation (4.38) given in Example 4.34. ◇

The proof of Theorem 4.137 is based on Corollary 4.136. In fact, we will see that, in some sense, the two representations (4.110) and (4.141) are dual with respect to each other. We prepare for the proof with the following exercises and lemmas.

Exercise 4.12.2. If h is a lower semicontinuous and proper convex function on $[0,\infty)$, then

$$(x,y) \longmapsto xh\left(\frac{y}{x}\right)$$

is a convex function of $(x,y) \in (0,\infty)\times[0,\infty)$. ◇

Exercise 4.12.3. The goal of this exercise is to illustrate the purpose of our assumption $g(1) < \infty$. So, let $\overline{g} : [0,\infty) \to \mathbb{R}\cup\{+\infty\}$ be any lower semicontinuous convex function. Show that there exists a probability measure $Q \ll P$ such that

$$I_{\overline{g}}(Q|P) := E\left[\,\overline{g}\left(\frac{dQ}{dP}\right)\,\right] < \infty,$$

if and only if $\overline{g}(1) < \infty$. ◇

For $\lambda > 0$, we now define $g_\lambda(x) := \lambda g(x/\lambda)$ and

$$\gamma_\lambda(Q) = I_{g_\lambda}(Q|P) = E\left[\,g_\lambda\left(\frac{dQ}{dP}\right)\right], \quad Q \ll P.$$

Lemma 4.139. *The map $(\lambda, Q) \mapsto \gamma_\lambda(Q)$ is a proper convex function on $(0,\infty)\times\mathcal{M}_1(P)$. Moreover, $\gamma_\lambda(Q) < \infty$ for some Q if and only if $g(1/\lambda) < \infty$.*

Proof. Exercise 4.12.3 states that $\gamma_\lambda(Q) < \infty$ for some Q if and only if $g(1/\lambda) < \infty$. Next, our assumptions on g imply that g is bounded from below. Hence, $\gamma_\lambda(Q)$ does not take the value $-\infty$. Exercise 4.12.2 implies moreover that $(\lambda, Q) \mapsto \gamma_\lambda(Q)$ is convex. Finally,

$g_1(P) = g(1) < \infty$ by assumption. Therefore, the convex function $(\lambda, Q) \mapsto \gamma_\lambda(Q)$ is proper. □

For given $X \in L^\infty$ and $\lambda \in \mathbb{R}$, we now define

$$h_X(\lambda) := \begin{cases} \inf_{Q \in \mathcal{M}_1(P)} (E_Q[X] + \gamma_\lambda(Q)) & \text{if } \lambda > 0, \\ +\infty & \text{otherwise.} \end{cases} \tag{4.142}$$

Then $h_X(1) = -\rho_g(X)$.

Lemma 4.140. *Let $X \in L^\infty$ be given.*

(a) *For $\lambda > 0$, we have $h_X(\lambda) < \infty$ if and only if $g(1/\lambda) < \infty$.*
(b) *h_X is a proper convex function on $\mathbb{R}$.*
(c) *h_X is lower semicontinuous.*

Proof. (a) If $h_X(\lambda) < \infty$, then there exists $Q \ll P$ such that $\gamma_\lambda(Q) < \infty$. Now the convexity of g_λ and Jensen's inequality yield that

$$\infty > \gamma_\lambda(Q) = E\left[g_\lambda\left(\frac{dQ}{dP} \right) \right] \geq g_\lambda\left(E\left[\frac{dQ}{dP} \right] \right) = g_\lambda(1) = \lambda g(1/\lambda). \tag{4.143}$$

Conversely, if $g(1/\lambda) < \infty$, then

$$\gamma_\lambda(P) = g_\lambda(1) = \lambda g(1/\lambda) < \infty,$$

and in turn $h_X(\lambda) < \infty$.

(b) To prove convexity, we will show that

$$h_X(\theta\lambda_1 + (1-\theta)\lambda_0) \leq \theta h_X(\lambda_1) + (1-\theta) h_X(\lambda_0)$$

whenever $\lambda_0, \lambda_1 \in \mathbb{R}$ and $\theta \in [0,1]$. To this end, we may assume that the right-hand side is finite so that $\lambda_0, \lambda_1 > 0$. It follows from Exercise 4.4.4 that there are $Q_i \ll P$ such that $h_X(\lambda_i) = E_{Q_i}[X] + \gamma_{\lambda_i}(Q_i)$ for $i = 0, 1$. Lemma 4.139 now yields that

$$\begin{aligned} h_X(\theta\lambda_1 + (1-\theta)\lambda_0) &\leq E_{\theta Q_{\lambda_1} + (1-\theta)Q_{\lambda_0}}[X] + \gamma_{\theta\lambda_1 + (1-\theta)\lambda_0}(\theta Q_{\lambda_1} + (1-\theta)Q_{\lambda_0}) \\ &\leq \theta E_{Q_{\lambda_1}}[X] + (1-\theta)E_{Q_{\lambda_0}}[X] + \theta\gamma_{\lambda_1}(Q_{\lambda_1}) + (1-\theta)\gamma_{\lambda_0}(Q_{\lambda_0}) \\ &= \theta h_X(\lambda_1) + (1-\theta)h_X(\lambda_0). \end{aligned}$$

This proves the convexity of h_X. To see that h_X is proper, it suffices to note that $h_X(1) = -\rho_g(X) < \infty$ and that $h_X(\lambda) > -\infty$ for all $\lambda \in \mathbb{R}$ due to (4.143).

(c) Let Λ denote the set of all $\lambda > 0$ such that $1/\lambda \in \operatorname{dom} g$. Then Λ is an interval with endpoints $a \geq 0$ and $b \leq \infty$. Part (a) implies that $h_X(\lambda)$ is finite if and only if $\lambda \in \Lambda$. By part (b) and Proposition A.7, the function h_X is continuous on the open interval (a, b)

and infinite on Λ^c. It therefore suffices to verify the lower semicontinuity of h_X in the endpoints of Λ.

We first consider the case in which the lower endpoint a is equal to zero. By (4.143), we have for any $Q \ll P$,

$$E_Q[X] + \gamma_\lambda(Q) \geq \operatorname{ess\,inf} X + \lambda g(1/\lambda) \longrightarrow \infty \quad \text{as } \lambda \downarrow 0,$$

due to the superlinear growth condition on g. Thus, $h_X(\lambda) \to \infty$ as $\lambda \downarrow 0$, which proves the lower semicontinuity of h_X in $a = 0$.

Now let c be an endpoint of Λ such that $0 < c < \infty$. Let (λ_n) be a sequence in Λ such that $\lim_n \lambda_n = c$. If $h_X(\lambda_n) \to \infty$, then the convexity of h_X implies that $h_X(c) = \infty$, and we are done. Otherwise, the values $h_X(\lambda_n)$ are uniformly bounded in n. As in the proof of part (b), we see that for every n there exists $Q_n \ll P$ such that $h_X(\lambda_n) = E_{Q_n}[X] + \gamma_{\lambda_n}(Q_n)$. The boundedness of X and the fact that $\lim_n \lambda_n = c > 0$ yield that for $\varphi_n := dQ_n/dP$,

$$\infty > \sup_{n \in \mathbb{N}} \frac{1}{\lambda_n} \gamma_{\lambda_n}(Q_n) = \sup_{n \in \mathbb{N}} E\left[g\left(\frac{\varphi_n}{\lambda_n} \right) \right].$$

The criterion of de la Vallée Poussin in Proposition H.16 hence states that the sequence $(\varphi_n/\lambda_n)_{n \in \mathbb{N}}$ is uniformly integrable. The Dunford–Pettis theorem in Theorem H.19 and the Eberlein–Šmulian theorem in Theorem H.13 now yield that, after passing to a subsequence if necessary, the random variables φ_n/λ_n converge weakly in L^1 to a random variable ψ. Since weak convergence preserves nonnegativity, we have $\psi \geq 0$. Moreover, $E[\psi] = \lim_n E[\varphi_n/\lambda_n] = 1/c$, and so $\varphi_c := c\psi$ is the density of a probability measure $Q_c \ll P$. Clearly, $\varphi_n \to \varphi_c$ weakly in L^1, and so the fact that $X \in L^\infty$ yields

$$E_{Q_c}[X] = E[\varphi_c X] = \lim_{n \uparrow \infty} E[\varphi_n X] = \lim_{n \uparrow \infty} E_{Q_n}[X]. \tag{4.144}$$

Since g is bounded from below, we may apply Fatou's lemma to conclude that the level sets

$$L_C := \{Y \in L^1 \mid E[g(Y)] \leq C\}$$

are closed in L^1 for all $C \in \mathbb{R}$. Moreover, each L_C is convex, because g is convex. Hence, Theorem H.7 yields each level set L_C is weakly closed in L^1. That is, the map $Y \mapsto E[g(Y)]$ is weakly lower semicontinuous. The weak convergence $\varphi_n/\lambda_n \to \varphi_c/c$ hence implies that

$$\gamma_c(Q_c) = E[g_c(\varphi_c)] = cE\left[g\left(\frac{\varphi_c}{c} \right) \right] \leq \liminf_{n \uparrow \infty} \lambda_n E\left[g\left(\frac{\varphi_n}{\lambda_n} \right) \right] = \liminf_{n \uparrow \infty} \gamma_{\lambda_n}(Q_n). \tag{4.145}$$

Combining (4.144) with (4.145) hence gives

$$h(c) \le E_{Q_c}[X] + \gamma_c(Q_c) \le \liminf_{n\uparrow\infty}(E_{Q_n}[X] + \gamma_{\lambda_n}(Q)) = \liminf_{n\uparrow\infty} h(\lambda_n).$$

This establishes the lower semicontinuity of h at c. □

Proof of Theorem 4.137. Let g_λ and h be as in Lemma 4.140. Our aim is to establish a duality formula for $h(1)$. The idea is to use Corollary 4.136 to identify the Fenchel–Legendre transform h^* of h. To this end, we first claim that the convex function

$$\ell(x) := g^*(x) = \sup_{z\ge 0}(xz - g(z)) \tag{4.146}$$

is a loss function in the sense of Definition 4.117. That is, ℓ is finite, increasing, and not identically constant. To prove this claim, we note that the superlinear growth condition on g implies that the supremum in (4.146) is attained in some number z_x for each $x \in \mathbb{R}$. In particular, ℓ takes only finite values. Moreover, if $x \le y$, then

$$\ell(x) = xz_x - g(z_x) \le yz_x - g(z_x) \le \ell(y),$$

and so ℓ is increasing. Finally, ℓ is not constant, because $\ell(x)$ is bounded from below by the affine function $x - g(1)$. Therefore, ℓ is indeed a loss function in the sense of Definition 4.117.

For the next step, we let

$$R_X(x) = \inf\{m \in \mathbb{R} \mid E[\,\ell(-m - X)\,] \le x\}, \quad x \in \mathbb{R},$$

be as in (4.136). Note also that $\ell^* = g^{**} = g$ by Proposition A.9. Corollary 4.136 now yields that for all $x \in \mathbb{R}$,

$$\begin{aligned} R_X(x) &= \sup_{Q\in\mathcal{M}_1(P)}\left(E_Q[\,-X\,] - \inf_{\lambda>0}\left(\lambda x + E\left[\,g_\lambda\left(\frac{dQ}{dP}\right)\right]\right)\right) \\ &= -\inf_{\lambda>0}\inf_{Q\in\mathcal{M}_1(P)}(E_Q[\,X\,] + \lambda x + \gamma_\lambda(Q)) \\ &= -\inf_{\lambda>0}(\lambda x + h_X(\lambda)) = h_X^*(-x). \end{aligned}$$

Lemma 4.140 and Proposition A.9 hence yield that

$$h_X(1) = h_X^{**}(1) = \sup_{x\in\mathbb{R}}(x - h_X^*(x)) = \sup_{x\in\mathbb{R}}(x - R_X(-x)) = \sup_{x\in\ell(\mathbb{R})}(-x - R_X(x)),$$

where in the final step we have used the fact that $R_X(x)$ is finite if $x \in \ell(\mathbb{R})$ and $R_X(x) = \inf\emptyset = +\infty$ otherwise. Proposition 4.121 and Remark 4.122 state that $x = E[\,\ell(-X - R_X(x))\,]$. Hence,

$$
\begin{aligned}
h_X(1) &= \sup_{x\in\ell(\mathbb{R})} (-x - R_X(x)) = \sup_{x\in\ell(\mathbb{R})} (-E[\,\ell(-X - R_X(x))\,] - R_X(x)) \\
&= -\inf_{y\in R_X(\ell(\mathbb{R}))} (E[\,\ell(-y - X)\,] + y) = -\inf_{z\in -R_X(\ell(\mathbb{R}))} (E[\,\ell(z - X)\,] - z).
\end{aligned}
$$

If ℓ does not attain its infimum, then the range $R_X(\ell(\mathbb{R}))$ is clearly equal to the entire real line $\mathbb{R}$, and the assertion follows by recalling the definition (4.142) of h and the fact that $\ell = g^*$. Now consider the case in which ℓ attains its infimum. We denote by κ the largest x such that $\ell(x) = \min \ell$. Then (4.130) states that $R_X(\min \ell) = -\operatorname{ess\,inf} X - \kappa := -y$. Thus, $-R_X(\ell(\mathbb{R})) = [y, \infty)$. If $z \le y$, then

$$
E[\,\ell(z - X)\,] - z = \min \ell - z \ge \min \ell - y = E[\,\ell(y - X)\,] - y,
$$

and so

$$
\inf_{z\in[y,\infty)} (E[\,\ell(z - X)\,] - z) = \inf_{z\in\mathbb{R}} (E[\,\ell(z - X)\,] - z).
$$

This completes the proof. □

Part II: **Dynamic hedging**

5 Dynamic arbitrage theory

In this chapter, we develop a dynamic version of the arbitrage theory of Chapter 1. Here, we will work in a multi-period setting, where the stochastic price fluctuation of a financial asset is described as a stochastic process in discrete time. Portfolios will be successively readjusted, taking into account the information available at each time. Such a dynamic trading strategy may produce a positive expected gain. If it does so without any downside risk, then it is called an arbitrage opportunity. In its weakest form, market efficiency requires that such arbitrage opportunities are excluded.

In Section 5.2, we show that an arbitrage-free model is characterized by the existence of an equivalent martingale measure. Under such a measure, the discounted price processes of the traded assets are martingales, that is, they have the mathematical structure of a fair game. In Section 5.3, we introduce European contingent claims. These are financial instruments whose payoff at the expiration date depends on the behavior of the underlying primary assets, and possibly on other factors. We discuss the problem of pricing such contingent claims in a manner that does not create new arbitrage opportunities. The pricing problem is closely related to the problem of hedging a given claim by using a dynamic trading strategy based on the primary assets.

An ideal situation occurs if any contingent claim can be perfectly replicated by the final outcome of such a strategy. In such a *complete* model, the equivalent martingale measure P^* is unique, and derivatives are priced in a canonical manner by taking the expectation of the discounted payoff with respect to the measure P^*. Section 5.5 contains a simple case study for completeness, the binomial model introduced by Cox, Ross, and Rubinstein. In this context, it is possible to obtain explicit pricing formulas for a number of exotic options, as explained in Section 5.6. In Section 5.7, we pass to the limiting diffusion model of geometric Brownian motion. Using a suitable version of the central limit theorem, we are led to the general Black–Scholes formula for European contingent claims and to explicit pricing formulas for some exotic options such as lookback options and the up-and-in and up-and-out calls.

The general structure of complete models is described in Section 5.4. There it will become clear that completeness is the exception rather than the rule: Typical market models in discrete time are *incomplete*.

5.1 The multi-period market model

Throughout this chapter, we consider a market model in which $d + 1$ assets are priced at times $t = 0, 1, \dots, T$. The price of the ith asset at time t is modeled as a nonnegative random variable S_t^i on a given probability space $(\Omega, \mathscr{F}, P)$. The random vector $\overline{S}_t = (S_t^0, S_t) = (S_t^0, S_t^1, \dots, S_t^d)$ is assumed to be measurable with respect to a σ-algebra $\mathscr{F}_t \subseteq \mathscr{F}$. One should think of $\mathscr{F}_t$ as the class of all events that are observable up to time t. Thus, it is natural to assume that

https://doi.org/10.1515/9783111045283-007

$$\mathscr{F}_0 \subseteq \mathscr{F}_1 \subseteq \cdots \subseteq \mathscr{F}_T. \tag{5.1}$$

Definition 5.1. A family $(\mathscr{F}_t)_{t=0,\dots,T}$ of σ-algebras satisfying (5.1) is called a *filtration*. In this case, $(\Omega, \mathscr{F}, (\mathscr{F}_t)_{t=0,\dots,T}, P)$ is also called a *filtered probability space*. △

To simplify the presentation, we will assume from now on that

$$\mathscr{F}_0 = \{\emptyset, \Omega\} \quad \text{and} \quad \mathscr{F} = \mathscr{F}_T. \tag{5.2}$$

Let $(E, \mathscr{E})$ be a measurable space. A *stochastic process* with state space $(E, \mathscr{E})$ is given by a family of E-valued random variables on $(\Omega, \mathscr{F}, P)$ indexed by time. In our context, the typical parameter sets will be $\{0, \dots, T\}$ or $\{1, \dots, T\}$, and the state space will be some Euclidean space.

Definition 5.2. A stochastic process $Y = (Y_t)_{t=0,\dots,T}$ is called *adapted* with respect to the filtration $(\mathscr{F}_t)_{t=0,\dots,T}$ if each Y_t is $\mathscr{F}_t$-measurable. A stochastic process $Z = (Z_t)_{t=1,\dots,T}$ is called *predictable* with respect to $(\mathscr{F}_t)_{t=0,\dots,T}$ if each Z_t is $\mathscr{F}_{t-1}$-measurable. △

Note that in our definition predictable processes start at $t = 1$ while adapted processes are also defined at $t = 0$. In particular, the asset prices $\overline{S} = (\overline{S}_t)_{t=0,\dots,T}$ form an adapted stochastic process with values in $\mathbb{R}^{d+1}$.

Definition 5.3. A *trading strategy* is a predictable $\mathbb{R}^{d+1}$-valued process

$$\overline{\xi} = (\xi^0, \xi) = (\xi_t^0, \xi_t^1, \dots, \xi_t^d)_{t=1,\dots,T}.$$ △

The value ξ_t^i of a trading strategy $\overline{\xi}$ corresponds to the quantity of shares of the ith asset held during the tth trading period between $t-1$ and t. Thus, $\xi_t^i S_{t-1}^i$ is the amount invested into the ith asset at time $t-1$, while $\xi_t^i S_t^i$ is the resulting value at time t. The total value of the portfolio $\overline{\xi}_t$ at time $t-1$ is

$$\overline{\xi}_t \cdot \overline{S}_{t-1} = \sum_{i=0}^{d} \xi_t^i S_{t-1}^i.$$

By time t, the value of the portfolio $\overline{\xi}_t$ has changed to

$$\overline{\xi}_t \cdot \overline{S}_t = \sum_{i=0}^{d} \xi_t^i S_t^i.$$

The predictability of $\overline{\xi}$ expresses the fact that investments must be allocated at the beginning of each trading period, without anticipating future price increments.

Definition 5.4. A trading strategy $\overline{\xi}$ is called *self-financing* if

$$\overline{\xi}_t \cdot \overline{S}_t = \overline{\xi}_{t+1} \cdot \overline{S}_t \quad \text{for } t = 1, \dots, T-1.$$ △

Intuitively, a trading strategy is self-financing if its portfolio is always rearranged in such a way that its present value is preserved. One may think of a brokerage account that is funded at time $t = 0$ and whose portfolio is subsequently rebalanced without adding or withdrawing any funds. It follows that the accumulated gains and losses resulting from the asset price fluctuations are the only source of variations of the portfolio value:

$$\overline{\xi}_{t+1} \cdot \overline{S}_{t+1} - \overline{\xi}_t \cdot \overline{S}_t = \overline{\xi}_{t+1} \cdot (\overline{S}_{t+1} - \overline{S}_t). \tag{5.3}$$

In fact, $\overline{\xi}$ is self-financing if and only if (5.3) holds for $t = 1, \dots, T-1$. It follows through summation over (5.3) that

$$\overline{\xi}_t \cdot \overline{S}_t = \overline{\xi}_1 \cdot \overline{S}_0 + \sum_{k=1}^{t} \overline{\xi}_k \cdot (\overline{S}_k - \overline{S}_{k-1}) \quad \text{for } t = 1, \dots, T.$$

Here, the constant $\overline{\xi}_1 \cdot \overline{S}_0$ can be interpreted as the initial investment for the purchase of the portfolio $\overline{\xi}_1$.

Example 5.5. Often it is assumed that the 0th asset plays the role of a *locally riskless bond*. In this case, one takes $S_0^0 \equiv 1$ and one lets S_t^0 evolve according to a *spot rate* $r_t \geq 0$: At time t, an investment x made at time $t-1$ yields the payoff $x(1 + r_t)$. Thus, a unit investment at time 0 produces the value

$$S_t^0 = \prod_{k=1}^{t} (1 + r_k)$$

at time t. An investment in S^0 is "locally riskless" if the spot rate r_t is known beforehand at time $t-1$. This idea can be made precise by assuming that the process r is *predictable*. ◇

Without assuming predictability as in the preceding example, we assume from now on that

$$S_t^0 > 0 \quad P\text{-a. s. for all } t.$$

This assumption allows us to use the 0th asset as a *numéraire* and to form the *discounted* price processes

$$X_t^i := \frac{S_t^i}{S_t^0}, \quad t = 0, \dots, T, \ i = 0, \dots, d.$$

Then $X_t^0 \equiv 1$, and $X_t = (X_t^1, \dots, X_t^d)$ expresses the value of the remaining assets in units of the numéraire. As explained in Remark 1.11, discounting allows comparison of asset prices that are quoted at different times.

Definition 5.6. The (discounted) *value process* $V = (V_t)_{t=0,\dots,T}$ associated with a trading strategy $\overline{\xi}$ is given by

$$V_0 := \overline{\xi}_1 \cdot \overline{X}_0 \quad \text{and} \quad V_t := \overline{\xi}_t \cdot \overline{X}_t \quad \text{for } t = 1, \dots, T.$$

The *gains process* associated with $\overline{\xi}$ is defined as

$$G_0 := 0 \quad \text{and} \quad G_t := \sum_{k=1}^{t} \xi_k \cdot (X_k - X_{k-1}) \quad \text{for } t = 1, \dots, T. \qquad \triangle$$

Clearly,

$$V_t = \overline{\xi}_t \cdot \overline{X}_t = \frac{\overline{\xi}_t \cdot \overline{S}_t}{S_t^0},$$

so V_t can be interpreted as the portfolio value at the end of the tth trading period expressed in units of the numéraire asset. The gains process

$$G_t = \sum_{k=1}^{t} \xi_k \cdot (X_k - X_{k-1})$$

reflects, in terms of the numéraire, the net gains that have accumulated through the trading strategy $\overline{\xi}$ up to time t. For a self-financing trading strategy $\overline{\xi}$, the identity

$$\overline{\xi}_t \cdot \overline{S}_t = \overline{\xi}_1 \cdot \overline{S}_0 + \sum_{k=1}^{t} \overline{\xi}_k \cdot (\overline{S}_k - \overline{S}_{k-1}) \tag{5.4}$$

remains true if all relevant quantities are computed in units of the numéraire. This is the content of the following simple proposition.

Proposition 5.7. *For a trading strategy $\overline{\xi}$, the following conditions are equivalent:*
(a) *$\overline{\xi}$ is self-financing.*
(b) *$\overline{\xi}_t \cdot \overline{X}_t = \overline{\xi}_{t+1} \cdot \overline{X}_t$ for $t = 1, \dots, T-1$.*
(c) *$V_t = V_0 + G_t = \overline{\xi}_1 \cdot \overline{X}_0 + \sum_{k=1}^{t} \xi_k \cdot (X_k - X_{k-1})$ for all t.*

Proof. By dividing both sides of the self-financing condition $\overline{\xi}_t \cdot \overline{S}_t = \overline{\xi}_{t+1} \cdot \overline{S}_t$ by S_t^0, it is seen that condition (b) is a reformulation of Definition 5.4. Moreover, (b) holds if and only if

$$\overline{\xi}_{t+1} \cdot \overline{X}_{t+1} - \overline{\xi}_t \cdot \overline{X}_t = \overline{\xi}_{t+1} \cdot (\overline{X}_{t+1} - \overline{X}_t) = \xi_{t+1} \cdot (X_{t+1} - X_t)$$

for $t = 1, \dots, T-1$, and this identity is equivalent to (c). $\square$

Remark 5.8. The numéraire component of a self-financing trading strategy $\overline{\xi}$ satisfies

$$\xi_{t+1}^0 - \xi_t^0 = -(\xi_{t+1} - \xi_t) \cdot X_t \quad \text{for } t = 1, \ldots, T-1. \tag{5.5}$$

Since

$$\xi_1^0 = V_0 - \xi_1 \cdot X_0, \tag{5.6}$$

the entire process ξ^0 is determined by the initial investment V_0 and the d-dimensional process ξ. Consequently, if a constant V_0 and an arbitrary d-dimensional predictable process ξ are given, then we can use (5.6) and (5.5) as the definition of a predictable process ξ^0, and this construction yields a self-financing trading strategy $\overline{\xi} := (\xi^0, \xi)$. In dealing with self-financing strategies $\overline{\xi}$, it is thus sufficient to focus on the initial investment V_0 and the d-dimensional processes X and ξ. ◇

Remark 5.9. Different economic agents investing into the same market may choose different numéraires. For example, consider the following simple market model in which all prices are quoted in euros (€) as the domestic currency. Let S^0 be a locally riskless €-bond with the predictable spot rate process r^0, that is,

$$S_t^0 = \prod_{k=1}^{t} (1 + r_k^0),$$

and let S^1 describe the price of a locally riskless investment into US dollars ($). Since the price of this $-bond is quoted in €, the asset S^1 is modeled as

$$S_t^1 = U_t \cdot \prod_{k=1}^{t} (1 + r_k^1),$$

where r^1 is the spot rate for a $-investment, and U_t denotes the price of 1$ in terms of €, that is, U_t is the exchange rate of the $ versus the €, which is often denoted by USDEUR. While it may be natural for European investors to take S^0 as their numéraire, it may be reasonable for an American investor to choose S^1. This simple example explains why it may be relevant to check which concepts and results of our theory are invariant under a *change of numéraire*; see, for example, the discussion at the end of Section 5.2. ◇

Exercise 5.1.1. Consider a market model with two assets that are modeled as usual by the stochastic process $\overline{S} = (S^0, S^1)$ that is adapted to the filtration $(\mathscr{F}_t)_{t=0,\ldots,T}$. Decide which of the following processes ξ are predictable and which in general are not:

(i) $\xi_t = \mathbb{1}_{\{S_t^1 > S_{t-1}^1\}}$;
(ii) $\xi_1 = 1$ and $\xi_t = \mathbb{1}_{\{S_{t-1}^1 > S_{t-2}^1\}}$ for $t \geq 2$;
(iii) $\xi_t = \mathbb{1}_A \cdot \mathbb{1}_{\{t > t_0\}}$, where $t_0 \in \{0, \ldots, T\}$ and $A \in \mathscr{F}_{t_0}$;
(iv) $\xi_t = \mathbb{1}_{\{S_t^1 > S_0^1\}}$;
(v) $\xi_1 = 1$ and $\xi_t = 2\xi_{t-1}\mathbb{1}_{\{S_{t-1}^1 < S_0^1\}}$ for $t \geq 2$. ◇

5.2 Arbitrage opportunities and martingale measures

Intuitively, an arbitrage opportunity is an investment strategy that yields a positive profit with positive probability but without any downside risk.

Definition 5.10. A self-financing trading strategy is called an *arbitrage opportunity* if its value process V satisfies

$$V_0 \leq 0, \quad V_T \geq 0 \;\; P\text{-a.s.}, \quad \text{and} \quad P[\, V_T > 0\,] > 0. \qquad \triangle$$

The existence of such an arbitrage opportunity may be regarded as a market inefficiency in the sense that certain assets are not priced in a reasonable way. In this section, we will characterize those market models, which do not allow for arbitrage opportunities. Such models will be called *arbitrage-free*. The following proposition shows that the market model is arbitrage-free if and only if there are no arbitrage opportunities for each single trading period. Later on, this fact will allow us to apply the results of Section 1.6 to our multi-period model.

Proposition 5.11. *The market model admits an arbitrage opportunity if and only if there exist $t \in \{1, \dots, T\}$ and $\eta \in \mathscr{L}^0(\Omega, \mathscr{F}_{t-1}, P; \mathbb{R}^d)$ such that*

$$\eta \cdot (X_t - X_{t-1}) \geq 0 \;\; P\text{-a.s.}, \quad \text{and} \quad P[\, \eta \cdot (X_t - X_{t-1}) > 0\,] > 0. \tag{5.7}$$

Proof. To prove necessity, take an arbitrage opportunity $\bar{\xi} = (\xi^0, \xi)$ with value process V, and let

$$t := \min\{k \mid V_k \geq 0 \; P\text{-a.s., and } P[\, V_k > 0\,] > 0\,\}.$$

Then $t \leq T$ by assumption, and either $V_{t-1} = 0$ P-a.s. or $P[\, V_{t-1} < 0\,] > 0$. In the first case, it follows that

$$\xi_t \cdot (X_t - X_{t-1}) = V_t - V_{t-1} = V_t \quad P\text{-a.s.}$$

Thus, $\eta := \xi_t$ satisfies (5.7). In the second case, we let $\eta := \xi_t \, \mathbb{1}_{\{V_{t-1}<0\}}$. Then η is $\mathscr{F}_{t-1}$-measurable, and

$$\eta \cdot (X_t - X_{t-1}) = (V_t - V_{t-1})\, \mathbb{1}_{\{V_{t-1}<0\}} \geq -V_{t-1}\, \mathbb{1}_{\{V_{t-1}<0\}}.$$

The expression on the right-hand side is nonnegative and strictly positive with a positive probability, so (5.7) holds.

Now we prove sufficiency. For t and η as in (5.7), define a d-dimensional predictable process ξ by

$$\xi_s := \begin{cases} \eta & \text{if } s = t, \\ 0 & \text{otherwise.} \end{cases}$$

Via (5.6) and (5.5), ξ uniquely defines a self-financing trading strategy $\overline{\xi} = (\xi^0, \xi)$ with initial investment $V_0 = 0$. Since the corresponding value process satisfies $V_T = \eta \cdot (X_t - X_{t-1})$, the strategy $\overline{\xi}$ is an arbitrage opportunity. □

Exercise 5.2.1. Let V be the value process of a self-financing strategy in an arbitrage-free market model. Prove that the following two implications hold for all $t \in \{0, \dots, T-1\}$ and $A \in \mathscr{F}_t$ with $P[A] > 0$:

$$P[\, V_{t+1} - V_t \geq 0 \mid A\,] = 1 \Longrightarrow P[\, V_{t+1} - V_t = 0 \mid A\,] = 1,$$
$$P[\, V_{t+1} - V_t \leq 0 \mid A\,] = 1 \Longrightarrow P[\, V_{t+1} - V_t = 0 \mid A\,] = 1. \qquad \diamond$$

Definition 5.12. A stochastic process $M = (M_t)_{t=0,\dots,T}$ on a filtered probability space $(\Omega, \mathscr{F}, (\mathscr{F}_t), Q)$ is called a *martingale* if M is adapted, satisfies $E_Q[\,|M_t|\,] < \infty$ for all t, and if

$$M_s = E_Q[\, M_t \mid \mathscr{F}_s\,] \quad \text{for } 0 \leq s \leq t \leq T. \qquad \triangle$$

A martingale can be regarded as the mathematical formalization of a "fair game": For each time s and for each horizon $t > s$, the conditional expectation of the future gain $M_t - M_s$ is zero, given the information available at s.

Exercise 5.2.2. Let $M = (M_t)_{t=0,\dots,T}$ be an adapted process on $(\Omega, \mathscr{F}, (\mathscr{F}_t), Q)$ such that $E_Q[\,|M_t|\,] < \infty$ for all t. Show that the following conditions are equivalent:
(a) M is a martingale.
(b) $M_t = E_Q[\, M_{t+1} \mid \mathscr{F}_t\,]$ for $0 \leq t \leq T-1$.
(c) There exists $F \in \mathscr{L}^1(\Omega, \mathscr{F}_T, Q)$ such that $M_t = E_Q[\, F \mid \mathscr{F}_t\,]$ for $t = 0, \dots, T$, that is, M arises as a sequence of successive conditional expectations. ◇

Exercise 5.2.3. Let $\widetilde{Q}$ be a probability measure on $(\Omega, \mathscr{F}, (\mathscr{F}_t)_{0\leq t\leq T})$ that is absolutely continuous with respect to Q. Show that the *density process*

$$Z_t := \left.\frac{d\widetilde{Q}}{dQ}\right|_{\mathscr{F}_t}, \qquad t = 0, \dots, T,$$

is a martingale with respect to Q. ◇

Whether or not a given process M is a martingale depends on the underlying probability measure Q. If we wish to emphasize the dependence of the martingale property of M on a particular measure Q, we will say that M is a *Q-martingale* or that M is a *martingale under the measure Q*.

Definition 5.13. A probability measure Q on $(\Omega, \mathscr{F}_T)$ is called a *martingale measure* if the discounted price process X is a (d-dimensional) Q-martingale, that is,

$$E_Q[\, X_t^i\,] < \infty \quad \text{and} \quad X_s^i = E_Q[\, X_t^i \mid \mathscr{F}_s\,], \quad 0 \leq s \leq t \leq T,\ i = 1, \dots, d.$$

A martingale measure P^* is called an *equivalent martingale measure* if it is equivalent to the original measure P on $\mathscr{F}_T$. The set of all equivalent martingale measures is denoted by $\mathscr{P}$. △

The following result is a version of Doob's fundamental "systems theorem" for martingales. It states that a fair game admits no realistic gambling system that produces a positive expected gain. Recall that Y^- denotes the negative part $-(Y \wedge 0)$ of a random variable Y.

Theorem 5.14. *For a probability measure Q, the following conditions are equivalent:*
(a) *Q is a martingale measure.*
(b) *If $\overline{\xi} = (\xi^0, \xi)$ is self-financing and ξ is bounded, then the value process V of $\overline{\xi}$ is a Q-martingale.*
(c) *If $\overline{\xi} = (\xi^0, \xi)$ is self-financing and its value process V satisfies $E_Q[\, V_T^- \,] < \infty$, then V is a Q-martingale.*
(d) *If $\overline{\xi} = (\xi^0, \xi)$ is self-financing and its value process V satisfies $V_T \geq 0$ Q-a. s., then $E_Q[\, V_T \,] = V_0$.*

Proof. (a)⇒(b): Let V be the value process of a self-financing trading strategy $\overline{\xi} = (\xi^0, \xi)$ such that there is a constant c such that $|\xi^i| \leq c$ for all i. Then, for each t,

$$|V_t| \leq |V_0| + \sum_{k=1}^{t} c(|X_k| + |X_{k-1}|).$$

By Definition 5.13, each $|X_k|$ belongs to $\mathscr{L}^1(Q)$, and so it follows that $E_Q[\, |V_t| \,] < \infty$. Moreover, for $0 \leq t \leq T-1$,

$$\begin{aligned} E_Q[\, V_{t+1} \mid \mathscr{F}_t \,] &= E_Q[\, V_t + \xi_{t+1} \cdot (X_{t+1} - X_t) \mid \mathscr{F}_t \,] \\ &= V_t + \xi_{t+1} \cdot E_Q[\, X_{t+1} - X_t \mid \mathscr{F}_t \,] \\ &= V_t, \end{aligned}$$

where we have used that ξ_{t+1} is $\mathscr{F}_t$-measurable and bounded.

(b)⇒(c): We will show the following implication:

$$\text{If } E_Q[\, V_t^- \,] < \infty \text{ then } E_Q[\, V_t \mid \mathscr{F}_{t-1} \,] = V_{t-1}. \tag{5.8}$$

Since $E_Q[\, V_T^- \,] < \infty$ by assumption, we will then get

$$E_Q[\, V_{T-1}^- \,] = E_Q[\, E_Q[\, V_T \mid \mathscr{F}_{T-1} \,]^- \,] \leq E_Q[\, V_T^- \,] < \infty,$$

due to Jensen's inequality for conditional expectations. Repeating this argument will yield $E_Q[\, V_t^- \,] < \infty$ and $E_Q[\, V_t \mid \mathscr{F}_{t-1} \,] = V_{t-1}$ for all t. By induction, we will also get

$E_Q[V_t] = V_0 \in \mathbb{R}$, which together with the fact that $E_Q[V_t^-] < \infty$ implies $V_t \in \mathscr{L}^1(Q)$ for all t. Thus, the martingale property of V will follow.

To prove (5.8), note first that $E_Q[V_t \mid \mathscr{F}_{t-1}]$ is well-defined due to our assumption $E_Q[V_t^-] < \infty$. Next, let $\xi_t^{(a)} := \xi_t \mathbb{1}_{\{|\xi_t| \le a\}}$ for $a > 0$. Then $\xi_t^{(a)} \cdot (X_t - X_{t-1})$ is a martingale increment due to (b). In particular, $\xi_t^{(a)} \cdot (X_t - X_{t-1}) \in \mathscr{L}^1(Q)$ and $E_Q[\xi_t^{(a)} \cdot (X_t - X_{t-1}) \mid \mathscr{F}_{t-1}] = 0$. Hence,

$$\begin{aligned} E_Q[V_t \mid \mathscr{F}_{t-1}] \mathbb{1}_{\{|\xi_t| \le a\}} &= E_Q[V_t \mathbb{1}_{\{|\xi_t| \le a\}} \mid \mathscr{F}_{t-1}] - E_Q[\xi_t^{(a)} \cdot (X_t - X_{t-1}) \mid \mathscr{F}_{t-1}] \\ &= E_Q[V_t \mathbb{1}_{\{|\xi_t| \le a\}} - \xi_t^{(a)} \cdot (X_t - X_{t-1}) \mid \mathscr{F}_{t-1}] \\ &= E_Q[V_{t-1} \mathbb{1}_{\{|\xi_t| \le a\}} \mid \mathscr{F}_{t-1}] \\ &= V_{t-1} \mathbb{1}_{\{|\xi_t| \le a\}}. \end{aligned}$$

By sending $a \uparrow \infty$, we obtain (5.8).

(c)⇒(d): Since $V_T \ge 0$, it follows from (c) that V is a Q-martingale. Thus, by (5.2),

$$V_0 = E_Q[V_T \mid \mathscr{F}_0] = E_Q[V_T].$$

(d)⇒(a): To prove that $X_t^i \in \mathscr{L}^1(Q)$ for given i and t, consider the deterministic process ξ defined by $\xi_s^i := \mathbb{1}_{\{s \le t\}}$ and $\xi_s^j := 0$ for $j \ne i$. By Remark 5.8, ξ can be complemented with a predictable process ξ^0 such that $\overline{\xi} = (\xi^0, \xi)$ is a self-financing strategy with initial investment $V_0 = X_0^i$. The corresponding value process satisfies

$$V_T = V_0 + \sum_{s=1}^{T} \xi_s \cdot (X_s - X_{s-1}) = X_t^i \ge 0.$$

From (d), we get

$$E_Q[X_t^i] = E_Q[V_T] = V_0 = X_0^i, \tag{5.9}$$

which yields $X_t^i \in \mathscr{L}^1(Q)$.

Condition (a) will follow if we can show that $E_Q[X_t^i; A] = E_Q[X_{t-1}^i; A]$ for given t, i, and $A \in \mathscr{F}_{t-1}$. To this end, we define a d-dimensional predictable process η by $\eta_s^i := \mathbb{1}_{\{s<t\}} + \mathbb{1}_{A^c} \mathbb{1}_{\{s=t\}}$ and $\eta_s^j := 0$ for $j \ne i$. As above, we take a predictable process η^0 such that $\overline{\eta} = (\eta^0, \eta)$ is a self-financing strategy with initial investment $\widetilde{V}_0 = X_0^i$. Its terminal value is given by

$$\widetilde{V}_T = \widetilde{V}_0 + \sum_{s=1}^{T} \eta_s \cdot (X_s - X_{s-1}) = X_t^i \mathbb{1}_{A^c} + X_{t-1}^i \mathbb{1}_A \ge 0.$$

Using (d) yields

$$X_0^i = \widetilde{V}_0 = E_Q[\,\widetilde{V}_T\,] = E_Q[\,X_t^i;\, A^c\,] + E_Q[\,X_{t-1}^i;\, A\,].$$

By comparing this identity with (5.9), we conclude that $E_Q[\,X_t^i;\, A\,] = E_Q[\,X_{t-1}^i;\, A\,]$. □

Remark 5.15.

(a) Suppose that the "objective" measure P is itself a martingale measure, so that the fluctuation of prices may be viewed as a fair game. In this case, the preceding theorem shows that there are no realistic self-financing strategies that would generate a positive expected gain. Thus, the assumption $P \in \mathscr{P}$ is a strong version of the so-called *efficient market hypothesis*. For a market model containing a locally riskless bond, this strong hypothesis would imply that risk-averse investors would not be attracted toward investing into the risky assets if their expectations are consistent with P; see Example 2.40.

(b) The strong assumption $P \in \mathscr{P}$ implies, in particular, that there is no arbitrage opportunity, that is, no self-financing strategy with positive expected gain and without any downside risk. Indeed, Theorem 5.14 implies that the value process of any self-financing strategy with $V_0 \leq 0$ and $V_T \geq 0$ satisfies $E[\,V_T\,] = V_0$, hence $V_T = 0$ P-almost surely. The assumption that the market model is arbitrage-free may be viewed as a much milder, and hence more flexible form of the *efficient market hypothesis*. ◇

We can now state the following dynamic version of the "fundamental theorem of asset pricing," which relates the absence of arbitrage opportunities to the existence of equivalent martingale measures.

Theorem 5.16. *The market model is arbitrage-free if and only if the set $\mathscr{P}$ of all equivalent martingale measures is nonempty. In this case, there exists a $P^* \in \mathscr{P}$ with bounded density dP^*/dP.*

Proof. Suppose first that there exists an equivalent martingale measure P^*. Then it follows as in Remark 5.15(b) that the market model in which the probability measure P is replaced by P^* is arbitrage-free. Since the notion of an arbitrage opportunity depends on the underlying measure only through its null sets and since these are common for the two equivalent measures P and P^*, it follows that also the original market model is arbitrage-free.

Let us turn to the proof of the converse assertion. For $t \in \{1, \dots, T\}$, we define

$$\mathscr{K}_t := \{\eta \cdot (X_t - X_{t-1}) \mid \eta \in L^0(\Omega, \mathscr{F}_{t-1}, P; \mathbb{R}^d)\,\}. \tag{5.10}$$

By Proposition 5.11, the market model is arbitrage-free if and only if

$$\mathscr{K}_t \cap L_+^0(\Omega, \mathscr{F}_t, P) = \{0\} \tag{5.11}$$

holds for all t. Note that (5.11) depends on the measure P only through its null sets.

Condition (5.11) allows us to apply Theorem 1.57 to the tth trading period. For $t = T$, we obtain a probability measure $\widetilde{P}_T \approx P$ that has a bounded density $d\widetilde{P}_T/dP$ and that satisfies

$$\widetilde{E}_T[\,X_T - X_{T-1} \mid \mathscr{F}_{T-1}\,] = 0.$$

Now suppose that we already have a probability measure $\widetilde{P}_{t+1} \approx P$ with a bounded density $d\widetilde{P}_{t+1}/dP$ such that

$$\widetilde{E}_{t+1}[\,X_k - X_{k-1} \mid \mathscr{F}_{k-1}\,] = 0 \quad \text{for } t+1 \le k \le T. \tag{5.12}$$

The equivalence of $\widetilde{P}_{t+1}$ and P implies that (5.11) also holds with P replaced by $\widetilde{P}_{t+1}$. Applying Theorem 1.57 to the tth trading period yields a probability measure $\widetilde{P}_t$ with a bounded $\mathscr{F}_t$-measurable density $Z_t := d\widetilde{P}_t/d\widetilde{P}_{t+1} > 0$ such that

$$\widetilde{E}_t[\,X_t - X_{t-1} \mid \mathscr{F}_{t-1}\,] = 0.$$

Clearly, $\widetilde{P}_t$ is equivalent to P and has a bounded density, since by Proposition B.4,

$$\frac{d\widetilde{P}_t}{dP} = \frac{d\widetilde{P}_t}{d\widetilde{P}_{t+1}} \cdot \frac{d\widetilde{P}_{t+1}}{dP}$$

is the product of two bounded densities. Moreover, if $t+1 \le k \le T$, Proposition B.8 and the $\mathscr{F}_t$-measurability of $Z_t = d\widetilde{P}_t/d\widetilde{P}_{t+1}$ imply

$$\begin{aligned}
\widetilde{E}_t[\,X_k - X_{k-1} \mid \mathscr{F}_{k-1}\,] &= \frac{\widetilde{E}_{t+1}[\,(X_k - X_{k-1})Z_t \mid \mathscr{F}_{k-1}\,]}{\widetilde{E}_{t+1}[\,Z_t \mid \mathscr{F}_{k-1}\,]} \\
&= \widetilde{E}_{t+1}[\,X_k - X_{k-1} \mid \mathscr{F}_{k-1}\,] \\
&= 0.
\end{aligned}$$

Hence, (5.12) carries over from $\widetilde{P}_{t+1}$ to $\widetilde{P}_t$. We can repeat this recursion until finally $P^* := \widetilde{P}_1$ yields the desired equivalent martingale measure. □

Clearly, the absence of arbitrage in the market is independent of the choice of the numéraire, while the set $\mathscr{P}$ of equivalent martingale measures generally does depend on the numéraire. In order to investigate the structure of this dependence, suppose that the first asset S^1 is P-a. s. strictly positive, so that it can serve as an alternative numéraire. The price process discounted by S^1 is denoted by

$$\overline{Y}_t = (Y_t^0, Y_t^1, \dots, Y_t^d) := \left(\frac{S_t^0}{S_t^1}, 1, \frac{S_t^2}{S_t^1}, \dots, \frac{S_t^d}{S_t^1}\right) = \frac{S_t^0}{S_t^1}\,\overline{X}_t = \frac{\overline{X}_t}{X_t^1}, \quad t = 0, \dots, T. \tag{5.13}$$

Let $\widetilde{\mathcal{P}}$ be the set of equivalent martingale measures for $\overline{Y}$. Then $\widetilde{\mathcal{P}} \neq \emptyset$ if and only if $\mathcal{P} \neq \emptyset$, according to Theorem 5.16 and the fact that the existence of arbitrage opportunities is independent of the choice of the numéraire.

Proposition 5.17. *The two sets $\mathcal{P}$ and $\widetilde{\mathcal{P}}$ are related via the identity*

$$\widetilde{\mathcal{P}} = \left\{ \widetilde{P}^* \;\middle|\; \frac{d\widetilde{P}^*}{dP^*} = \frac{X_T^1}{X_0^1} \text{ for some } P^* \in \mathcal{P} \right\}.$$

Proof. By assumption, the process X_t^1/X_0^1 is a strictly positive P^*-martingale for any $P^* \in \mathcal{P}$. In particular, $E^*[\,X_T^1/X_0^1\,] = 1$, and the formula

$$\frac{d\widetilde{P}^*}{dP^*} = \frac{X_T^1}{X_0^1}$$

defines a probability measure $\widetilde{P}^*$ that is equivalent to P. Moreover, by Proposition B.8 and (5.13),

$$\begin{aligned}\widetilde{E}^*[\,\overline{Y}_t \mid \mathcal{F}_s\,] &= \frac{1}{X_s^1} \cdot E^*[\,\overline{Y}_t \cdot X_t^1 \mid \mathcal{F}_s\,] \\ &= \frac{1}{X_s^1} \cdot E^*[\,\overline{X}_t \mid \mathcal{F}_s\,] \\ &= \overline{Y}_s.\end{aligned}$$

Hence, $\widetilde{P}^*$ is an equivalent martingale measure for $\overline{Y}$, and it follows that

$$\widetilde{\mathcal{P}} \supseteq \left\{ \widetilde{P}^* \;\middle|\; \frac{d\widetilde{P}^*}{dP^*} = \frac{X_T^1}{X_0^1} \text{ for some } P^* \in \mathcal{P} \right\}.$$

Reversing the roles of $\overline{X}$ and $\overline{Y}$ yields the identity of the two sets. □

Remark 5.18. Unless X_T^1 is P-a. s. constant, the two sets $\mathcal{P}$ and $\widetilde{\mathcal{P}}$ satisfy

$$\mathcal{P} \cap \widetilde{\mathcal{P}} = \emptyset.$$

This can be proved as in Remark 1.12. ◇

Exercise 5.2.4. Let $X_t := X_t^1$ be the P-a. s. strictly positive discounted price process of a risky asset. The corresponding returns are

$$\widetilde{R}_t := \frac{X_t - X_{t-1}}{X_{t-1}}, \qquad t = 1, \dots, T,$$

so that

$$X_t = X_0 \prod_{k=1}^{t} (1 + \widetilde{R}_k).$$

We take as filtration $\mathscr{F}_t = \sigma(X_0, \ldots, X_t)$.

(a) Show that X is a P-martingale if the $(\widetilde{R}_t)$ are independent and integrable random variables with $E[\widetilde{R}_t] = 0$.

(b) Now give necessary and sufficient conditions on the $(\widetilde{R}_t)$ such that X is a P-martingale.

(c) Construct an example in which X is a martingale but the $(\widetilde{R}_t)$ are not independent.

◇

Exercise 5.2.5. Let $Z_1, \ldots, Z_T$ be independent standard normal random variables on $(\Omega, \mathscr{F}, P)$, and let $\mathscr{F}_t$ be the σ-field generated by $Z_1, \ldots, Z_t$, where $t = 1, \ldots, T$. We also let $\mathscr{F}_0 := \{\emptyset, \Omega\}$. For constants $X_0^1 > 0$, $\sigma_i > 0$, and $m_i \in \mathbb{R}$, we now define the discounted price process of a risky asset as the following sequence of log-normally distributed random variables:

$$X_t^1 := X_0^1 \prod_{i=1}^{t} e^{\sigma_i Z_i + m_i}, \quad t = 0, \ldots, T. \tag{5.14}$$

Construct an equivalent martingale measure for X^1 under which the random variables X_t^1 have still a log-normal distribution. ◇

Exercise 5.2.6. For a square-integrable random variable X on $(\Omega, \mathscr{F}, P)$ and a σ-algebra $\mathscr{F}_0 \subseteq \mathscr{F}$, the *conditional variance* of X given $\mathscr{F}_0$ is defined as

$$\operatorname{var}(X|\mathscr{F}_0) := E[\,(X - E[\,X \mid \mathscr{F}_0\,])^2 \mid \mathscr{F}_0\,].$$

Show that

$$\operatorname{var}(X|\mathscr{F}_0) = E[\,X^2 \mid \mathscr{F}_0\,] - (E[\,X \mid \mathscr{F}_0\,])^2$$

and that

$$\operatorname{var}(X) = E[\,\operatorname{var}(X|\mathscr{F}_0)\,] + \operatorname{var}(E[\,X \mid \mathscr{F}_0\,]).$$ ◇

Exercise 5.2.7. Let Y_1 and Y_2 be jointly normal random variables with mean 0, variance 1, and correlation $\varrho \in (-1, 1)$. That is, the joint distribution of (Y_1, Y_2) has the density

$$\varphi(y_1, y_2) = \frac{1}{2\pi\sqrt{1-\varrho^2}} e^{-\frac{1}{2(1-\varrho^2)}(y_1^2 + y_2^2 - 2\varrho y_1 y_2)}, \quad (y_1, y_2) \in \mathbb{R}^2.$$

(a) Compute the conditional expectation $E[\, Y_2 \mid Y_1 \,]$.
(b) Compute the conditional variance $\operatorname{var}(Y_2|Y_1)$.
(c) For constants $m, \sigma \in \mathbb{R}$ compute $E[\, e^{\sigma Y_2 + m} \mid Y_1 \,]$. ◇

Exercise 5.2.8. Let Y_1 and Y_2 be as in Exercise 5.2.7. We use the Y_i to construct a log-normal price process in analogy to (5.14):

$$X_t^1 := X_0^1 \prod_{i=1}^{t} e^{\sigma_i Y_i + m_i}, \quad t = 0, \dots, 2, \tag{5.15}$$

for constants $X_0^1, \sigma_i > 0$ and $m_i \in \mathbb{R}$ $(i = 1, 2)$.
(a) Compute the conditional expectation $E[\, X_2^1 \mid X_1^1 \,]$.
(b) Construct an equivalent martingale measure for the price process in (5.15) if the filtration is the one generated by the process X^1. ◇

Exercise 5.2.9. Let $X_0, X_1, \dots$ describe the discounted prices of a risky asset in a market model with infinite time horizon that is modeled on a filtered probability space $(\Omega, (\mathcal{F}_t)_{t=0,1,\dots}, P)$. Suppose that every market model $X_0, \dots, X_T$ with finite time horizon $T \in \mathbb{N}$ is arbitrage-free.
(a) Show that there exists a sequence $(P_T^*)_{T=1,2\dots}$ of probability measures such that P_T^* is defined on $(\Omega, \mathcal{F}_T)$, is equivalent to P on $\mathcal{F}_T$, and such that the restriction of P_T^* to $\mathcal{F}_{T-1}$ equals P_{T-1}^*, that is, $P_T^*[\, A \,] = P_{T-1}^*[\, A \,]$ for all $A \in \mathcal{F}_{T-1}$.
(b) Can you give conditions under which P_T^* arises as the restriction to $\mathcal{F}_T$ of a measure P^* that is defined on $\mathcal{F}_\infty := \sigma(\bigcup_{t \geq 0} \mathcal{F}_t)$?
Hint: You may choose a setting in which one can apply the Kolmogorov extension theorem. ◇

We have assumed so far that all assets can be traded in arbitrary fractions. This assumption, however, is an idealization of reality, where most assets can only be traded in integer quantities. The following two exercises explore the absence of arbitrage in a market model in which only integer quantities of the risky assets can be traded and asset prices are quoted in rational numbers. To this end, we first formulate the following auxiliary exercise.

Exercise 5.2.10. Let $(a_{ij})_{i=1,\dots,m,\, j=1,\dots,n}$ be a matrix with rational entries $a_{ij} \in \mathbb{Q}$. Suppose there exists $(x_1, \dots, x_n) \in \mathbb{R}^n$ such that

$$\sum_{j=1}^{n} a_{ij} x_j \geq 0 \quad \text{for all } i \text{, with at least one inequality being strict.} \tag{5.16}$$

Then there exists $(x_1, \dots, x_n) \in \mathbb{Q}^n$ satisfying (5.16). ◇

Exercise 5.2.11. Suppose that Ω is finite, $\mathscr{F}$ is the power set, and P satisfies $P[\{\omega\}] > 0$ for all $\omega \in \Omega$. Suppose moreover that $S_t^0 = (1+r)^t$ for some $r \in \mathbb{Q} \cap (-1, \infty)$ and that all risky assets take only rational values in the sense that $S_t^i(\omega) \in \mathbb{Q} \cap [0, \infty)$ for $i = 1, \dots, d$, $t = 0, \dots, T$, and $\omega \in \Omega$. Use Exercise 5.2.10 to prove that the model is arbitrage-free if and only if there exists no arbitrage strategy $\overline{\xi} = (\xi^0, \xi)$ for which ξ is integer-valued, that is, $\xi_t^i(\omega) \in \mathbb{Z}$ for $i = 1, \dots, d$, $t = 1, \dots, T$, and $\omega \in \Omega$. ◇

5.3 European contingent claims

A key topic of mathematical finance is the analysis of *derivative securities* or *contingent claims*, that is, of certain assets whose payoff depends on the behavior of the primary assets $S^0, S^1, \dots, S^d$ and, in some cases, also on other factors.

Definition 5.19. A nonnegative random variable C on $(\Omega, \mathscr{F}_T, P)$ is called a *European contingent claim*. A European contingent claim C is called a *derivative* of the underlying assets $S^0, S^1, \dots, S^d$ if C is measurable with respect to the σ-algebra generated by the price process $(\overline{S}_t)_{t=0,\dots,T}$. △

A European contingent claim has the interpretation of an asset that yields at time T the amount $C(\omega)$, depending on the scenario ω of the market evolution. T is called the *expiration date* or the *maturity* of C. Of course, maturities prior to the final trading period T of our model are also possible, but unless it is otherwise mentioned, we will assume that our European contingent claims expire at T. In Chapter 6, we will meet another class of derivative securities, the so-called *American contingent claims*. As long as there is no risk of confusion between European and American contingent claims, we will use the term “contingent claim” to refer to a European contingent claim.

Example 5.20. The owner of a *European call option* has the right, but not the obligation, to buy an asset at time T for a fixed price K, called the *strike price*. This corresponds to a contingent claim of the form

$$C^{\text{call}} = (S_T^i - K)^+.$$

Conversely, a *European put option* gives the right, but not the obligation, to sell the asset at time T for a strike price K. This corresponds to the contingent claim

$$C^{\text{put}} = (K - S_T^i)^+.$$

As discussed in Example 1.21, put and call options are traded on a vast variety of financial instruments such as stocks, currencies, futures, and indices. ◇

Example 5.21. The payoff of an *Asian option* depends on the average price

$$S_{\text{av}}^i := \frac{1}{|\mathbb{T}|} \sum_{t \in \mathbb{T}} S_t^i$$

of the underlying asset during a predetermined set of periods $\mathbb{T} \subseteq \{0, \ldots, T\}$. For instance, an *average price call* with strike K corresponds to the contingent claim

$$C_{\text{av}}^{\text{call}} := (S_{\text{av}}^i - K)^+,$$

and an *average price put* has the payoff

$$C_{\text{av}}^{\text{put}} := (K - S_{\text{av}}^i)^+.$$

Average price options can be used, for instance, to secure regular cash streams against exchange rate fluctuations. For example, assume that an economic agent receives at each time $t \in \mathbb{T}$ a fixed amount of a foreign currency with exchange rates S_t^i. In this case, an average price put option may be an efficient instrument for securing the incoming cash stream against the risk of unfavorable exchange rates.

An *average strike call* corresponds to the contingent claim

$$(S_T^i - S_{\text{av}}^i)^+,$$

while an *average strike put* pays off the amount

$$(S_{\text{av}}^i - S_T^i)^+.$$

An average strike put can be used, for example, to secure the risk from selling at time T a quantity of an asset that was bought at successive times over the period $\mathbb{T}$. ◇

Example 5.22. The payoff of a *barrier option* depends on whether the price of the underlying asset reaches a certain level before maturity. Most barrier options are either *knock-out* or *knock-in options.* A knock-in option pays off only if the barrier B is reached. The simplest example is a *digital option*

$$C^{\text{dig}} := \begin{cases} 1 & \text{if } \max\limits_{0 \le t \le T} S_t^i \ge B, \\ 0 & \text{otherwise,} \end{cases}$$

which has a unit payoff if the price process reaches a given upper barrier $B > S_0^i$. Another example is the *down-and-in put* with strike price K and lower barrier $\widetilde{B} < S_0^i$, which pays off

$$C_{\text{d\&i}}^{\text{put}} := \begin{cases} (K - S_T^i)^+ & \text{if } \min\limits_{0 \le t \le T} S_t^i \le \widetilde{B}, \\ 0 & \text{otherwise.} \end{cases}$$

A knock-out barrier option has a zero payoff once the price of the underlying asset reaches the predetermined barrier. For instance, an *up-and-out call* corresponds to the contingent claim

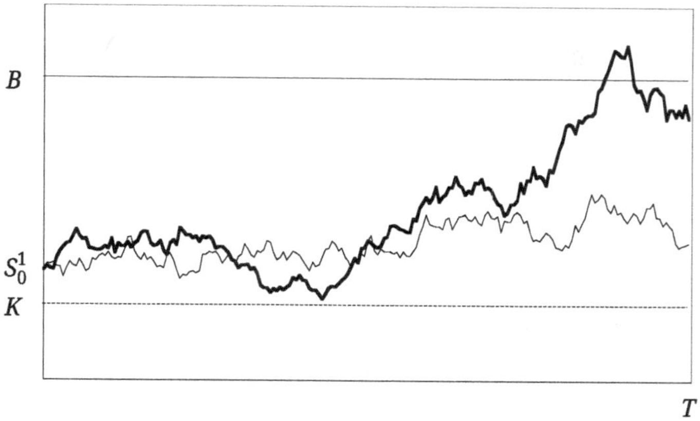

Figure 5.1: In one scenario, the payoff of the up-and-out call becomes zero because the stock price hits the barrier B before time T. In the other scenario, the payoff is given by $(S_T - K)^+$.

$$C^{\text{call}}_{\text{u\&o}} := \begin{cases} (S_T^i - K)^+ & \text{if } \max\limits_{0\le t\le T} S_t^i < B, \\ 0 & \text{otherwise;} \end{cases}$$

see Figure 5.1. *Down-and-out* and *up-and-in options* are defined analogously. ◇

Example 5.23. Using a *lookback option*, one can trade the underlying asset at the maximal or minimal price that occurred during the life of the option. A *lookback call* has the payoff

$$S_T^i - \min_{0\le t\le T} S_t^i,$$

while a *lookback put* corresponds to the contingent claim

$$\max_{0\le t\le T} S_t^i - S_T^i. \qquad \diamond$$

The discounted value of a contingent claim C when using the numéraire S^0 is given by

$$H := \frac{C}{S_T^0}.$$

We will call H the *discounted European claim* or just the *discounted claim* associated with C. In the remainder of this text, "H" will be the generic notation for the discounted payoff of any type of contingent claim.

The reader may wonder why we work simultaneously with the notions of a contingent claim and a discounted claim. From a purely mathematical point of view, there would be no loss of generality in assuming that the numéraire asset is identically equal to one. In fact, the entire theory to be developed in Part II can be seen as a discrete-time "stochastic analysis" for the d-dimensional process $X = (X^1, \dots, X^d)$ and its "stochastic integrals"

$$\sum_{k=1}^{t} \xi_k \cdot (X_k - X_{k-1})$$

of predictable d-dimensional processes ξ. However, some of the economic intuition would be lost if we limited the discussion to this level. For instance, we have already seen the economic relevance of the particular choice of the numéraire, even though this choice may be irrelevant from the mathematician's point of view. As a compromise between the mathematician's preference for conciseness and the economist's concern for keeping track explicitly of economically relevant quantities, we develop the mathematics on the level of discounted prices, but we will continue to discuss definitions and results in terms of undiscounted prices whenever it seems appropriate.

From now on, we will assume that our market model is arbitrage-free or, equivalently, that

$$\mathcal{P} \neq \emptyset.$$

Definition 5.24. A contingent claim C is called *attainable* (*replicable, redundant*) if there exists a self-financing trading strategy $\overline{\xi}$ whose terminal portfolio value coincides with C, that is,

$$C = \overline{\xi}_T \cdot \overline{S}_T \quad P\text{-a. s.}$$

Such a trading strategy $\overline{\xi}$ is called a *replicating strategy* for C. △

Clearly, a contingent claim C is attainable if and only if the corresponding discounted claim $H = C/S_T^0$ is of the form

$$H = \overline{\xi}_T \cdot \overline{X}_T = V_T = V_0 + \sum_{t=1}^{T} \xi_t \cdot (X_t - X_{t-1}),$$

for a self-financing trading strategy $\overline{\xi} = (\xi^0, \xi)$ with value process V. In this case, we will say that the discounted claim H is *attainable*, and we will call $\overline{\xi}$ a replicating strategy for H. The following theorem yields the surprising result that an attainable discounted claim is automatically integrable with respect to every equivalent martingale measure. Note, however, that integrability may not hold for an attainable contingent claim *prior* to discounting.

Theorem 5.25. *Any attainable discounted claim H is integrable with respect to each equivalent martingale measure, that is,*

$$E^*[\,H\,] < \infty \quad \textit{for all } P^* \in \mathcal{P}.$$

Moreover, for each $P^* \in \mathcal{P}$ the value process of any replicating strategy satisfies

$$V_t = E^*[\,H \mid \mathcal{F}_t\,] \quad P\text{-a. s. for } t = 0, \ldots, T.$$

In particular, V is a nonnegative P^*-martingale.

Proof. This follows from $V_T = H \geq 0$ and the systems theorem in the form of Theorem 5.14. □

Remark 5.26. The identity

$$V_t = E^*[\,H \mid \mathcal{F}_t\,], \quad t = 0, \ldots, T,$$

appearing in Theorem 5.25 has two remarkable implications. Since its right-hand side is independent of the particular replicating strategy, all such strategies must have the same value process. Moreover, the left-hand side does not depend on the choice of $P^* \in \mathcal{P}$. Hence, V_t is a version of the conditional expectation $E^*[\,H \mid \mathcal{F}_t\,]$ for every $P^* \in \mathcal{P}$. In particular, $E^*[\,H\,]$ is the same for all $P^* \in \mathcal{P}$. ◇

Remark 5.27. When applied to an attainable contingent claim C prior to discounting, Theorem 5.25 states that

$$\overline{\xi}_t \cdot \overline{S}_t = S_t^0 E^*\left[\,\frac{C}{S_T^0} \;\Big|\; \mathcal{F}_t\,\right], \quad t = 0, \ldots, T,$$

P-a. s. for all $P^* \in \mathcal{P}$ and for every replicating strategy $\overline{\xi}$. In particular, the initial investment that is needed for a replication of C, is given by

$$\overline{\xi}_1 \cdot \overline{S}_0 = S_0^0 E^*\left[\,\frac{C}{S_T^0}\,\right]. \qquad ◇$$

Let us now turn to the problem of *pricing* a contingent claim. Consider first a discounted claim H that is attainable. Then the (discounted) initial investment

$$\overline{\xi}_1 \cdot \overline{X}_0 = V_0 = E^*[\,H\,] \tag{5.17}$$

needed for the replication of H can be interpreted as the unique (discounted) "*fair price*" of H. In fact, a different price for H would create an arbitrage opportunity. For instance, if H could be sold at time 0 for a price $\tilde{\pi}$ that is higher than (5.17), then selling H and buying the replicating portfolio $\overline{\xi}$ yields the profit

$$\tilde{\pi} - \overline{\xi}_1 \cdot \overline{X}_0 > 0$$

at time 0, although the terminal portfolio value $V_T = \overline{\xi}_T \cdot \overline{X}_T$ suffices for settling the claim H at maturity T. In order to make this idea precise, let us formalize the idea of an "arbitrage-free price" of a general discounted claim H.

Definition 5.28. A real number $\pi^H \geq 0$ is called an *arbitrage-free price* of a discounted claim H, if there exists an adapted stochastic process X^{d+1} such that

$$\begin{aligned} X_0^{d+1} &= \pi^H, \\ X_t^{d+1} &\geq 0 \quad \text{for } t = 1, \dots, T-1, \quad \text{and} \\ X_T^{d+1} &= H, \end{aligned} \tag{5.18}$$

and such that the enlarged market model with price process $(X^0, X^1, \dots, X^d, X^{d+1})$ is arbitrage-free. The set of all arbitrage-free prices of H is denoted by $\Pi(H)$. The lower and upper bounds of $\Pi(H)$ are denoted by

$$\pi_{\inf}(H) := \inf \Pi(H) \quad \text{and} \quad \pi_{\sup}(H) := \sup \Pi(H). \qquad \triangle$$

Thus, an arbitrage-free price π^H of a discounted claim H is by definition a price at which H can be traded at time 0 without introducing arbitrage opportunities into the market model: If H is sold for π^H, then neither buyer nor seller can find an investment strategy that both eliminates all the risk and yields an opportunity to make a positive profit. Our aim in this section is to characterize the set of all arbitrage-free prices of a discounted claim H.

Note that an arbitrage-free price π^H is quoted in units of the numéraire asset. The amount that corresponds to π^H in terms of currency units prior to discounting is equal to

$$\pi^C := S_0^0 \pi^H,$$

and π^C is an (undiscounted) arbitrage-free price of the contingent claim $C := S_T^0 H$.

Theorem 5.29. *The set of arbitrage-free prices of a discounted claim H is nonempty and given by*

$$\Pi(H) = \{E^*[\,H\,] \mid P^* \in \mathcal{P} \text{ and } E^*[\,H\,] < \infty\}. \tag{5.19}$$

Moreover, the lower and upper bounds of $\Pi(H)$ *are given by*

$$\pi_{\inf}(H) = \inf_{P^* \in \mathcal{P}} E^*[\,H\,] \quad \text{and} \quad \pi_{\sup}(H) = \sup_{P^* \in \mathcal{P}} E^*[\,H\,].$$

Proof. By Theorem 5.16, π^H is an arbitrage-free price for H if and only if we can find an equivalent martingale measure $\hat{P}$ for the market model extended via (5.18). $\hat{P}$ must satisfy

$$X_t^i = \hat{E}[\,X_T^i \mid \mathscr{F}_t] \quad \text{for } t = 0, \dots, T \text{ and } i = 1, \dots, d+1.$$

In particular, $\hat{P}$ belongs to $\mathscr{P}$ and satisfies $\pi^H = \hat{E}[\,H\,]$. Thus, we obtain the inclusion $\subseteq$ in (5.19).

Conversely, if $\pi^H = E^*[\,H\,]$ for some $P^* \in \mathscr{P}$, then we can define the stochastic process

$$X_t^{d+1} := E^*[\,H \mid \mathscr{F}_t], \quad t = 0, \dots, T,$$

which satisfies all the requirements of (5.18). Moreover, the same measure P^* is clearly an equivalent martingale measure for the extended market model, which hence is arbitrage-free. Thus, we obtain the identity of the two sets in (5.19).

To show that $\Pi(H)$ is nonempty, we first fix some measure $\widetilde{P} \approx P$ such that $\widetilde{E}[\,H\,] < \infty$. For instance, we can take $d\widetilde{P} = c(1+H)^{-1}\,dP$, where c is the normalizing constant. Under $\widetilde{P}$, the market model is arbitrage-free. Hence, Theorem 5.16 yields $P^* \in \mathscr{P}$ such that $dP^*/d\widetilde{P}$ is bounded. In particular, $E^*[\,H\,] < \infty$, and hence $E^*[\,H\,] \in \Pi(H)$.

The formula for $\pi_{\inf}(H)$ follows immediately from (5.19) and the fact that $\Pi(H) \neq \emptyset$. The one for $\pi_{\sup}(H)$ needs an additional argument. Suppose there exists $P^\infty \in \mathscr{P}$ such that $E^\infty[\,H\,] = \infty$. We must show that for any $c > 0$ there exists some $\pi \in \Pi(H)$ with $\pi > c$. To this end, let n be such that $\widetilde{\pi} := E^\infty[\,H \wedge n\,] > c$, and define

$$X_t^{d+1} := E^\infty[\,H \wedge n \mid \mathscr{F}_t\,], \quad t = 0, \dots, T.$$

Then P^∞ is an equivalent martingale measure for the extended market model $(X^0, \dots, X^d, X^{d+1})$, which hence is arbitrage-free. Applying the already established fact that the set of arbitrage-free prices of any contingent claim is nonempty in the extended market model yields an equivalent martingale measure P^* for $(X^0, \dots, X^d, X^{d+1})$ such that $E^*[\,H\,] < \infty$. Since P^* is also a martingale measure for the original market model, the first part of this proof implies that $\pi := E^*[\,H\,] \in \Pi(H)$. Finally, note that

$$\pi = E^*[\,H\,] \geq E^*[\,H \wedge n\,] = E^*[\,X_T^{d+1}\,] = X_0^{d+1} = \widetilde{\pi} > c.$$

Hence, the formula for $\pi_{\sup}(H)$ is proved. □

Example 5.30. In an arbitrage-free market model, we consider a European call option $C^{\text{call}} = (S_T^1 - K)^+$ with strike $K > 0$ and with maturity T. We assume that the numéraire S^0 is the predictable price process of a locally riskless bond with nonnegative interest rates as in Example 5.5. Then S_t^0 is increasing in t and satisfies $S_0^0 \equiv 1$. For any $P^* \in \mathscr{P}$, Theorem 5.29 yields an arbitrage-free price π^{call} of C^{call} that is given by

$$\pi^{\text{call}} = E^*\left[\frac{C^{\text{call}}}{S_T^0}\right] = E^*\left[\left(X_T^1 - \frac{K}{S_T^0}\right)^+\right].$$

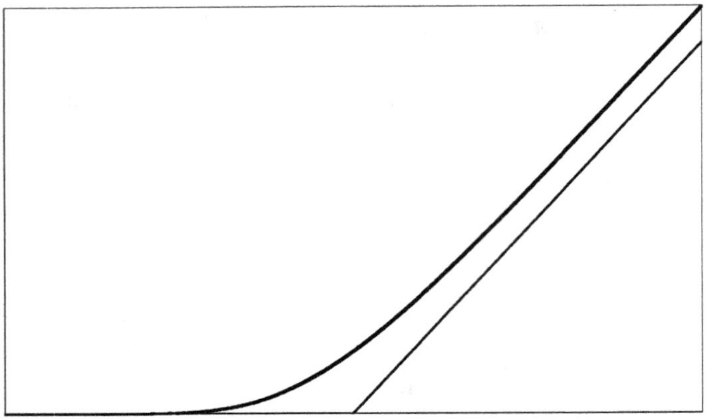

Figure 5.2: In a regime with nonnegative interest rates, the typical price of a European call option as a function of S_0^1 is always above the option's intrinsic value $(S_0^1 - K)^+$.

Due to the convexity of the function $x \mapsto x^+ = x \vee 0$ and our assumptions on S^0, π^{call} can be bounded from below as follows:

$$\begin{aligned} \pi^{\text{call}} &\geq \left(E^*\left[X_T^1 - \frac{K}{S_T^0}\right]\right)^+ \\ &= \left(S_0^1 - E^*\left[\frac{K}{S_T^0}\right]\right)^+ \\ &\geq (S_0^1 - K)^+. \end{aligned} \tag{5.20}$$

In financial language, this fact is usually expressed by saying that the value of the option is higher than its *intrinsic value* $(S_0^1 - K)^+$, that is, the payoff if the option were exercised immediately. The difference of the price π^{call} of an option and its intrinsic value is often called the *time-value* of the European call option; see Figure 5.2. ◇

Example 5.31. For a European put option $C^{\text{put}} = (K - S_T^1)^+$, the situation is more complicated. If we consider the same situation as in Example 5.30, then the analogue of (5.20) fails unless the numéraire S_t^0 is nonincreasing in t, which happens in an environment with vanishing or negative interest rates. For positive interest rates, the time value of a put option that is "in the money" can become negative; see Figure 5.3. This fact can be derived as a consequence of the put-call parity discussed in Exercise 5.3.1 below. ◇

Exercise 5.3.1 (Put-call parity in a multi-period model). Consider an arbitrage-free market model with a single risky asset, S^1, and a riskless numéraire, $S_t^0 = (1 + r)^t$, for some $r > -1$. Suppose that an arbitrage-free price π^{call} has been fixed for the discounted claim $H^{\text{call}} = (S_T^1 - K)^+/S_T^0$ of a European call option with strike $K \geq 0$. Then there exists a nonnegative adapted process X^2 with $X_0^2 = \pi^{\text{call}}$ and $X_T^2 = H^{\text{call}}$ such that the extended

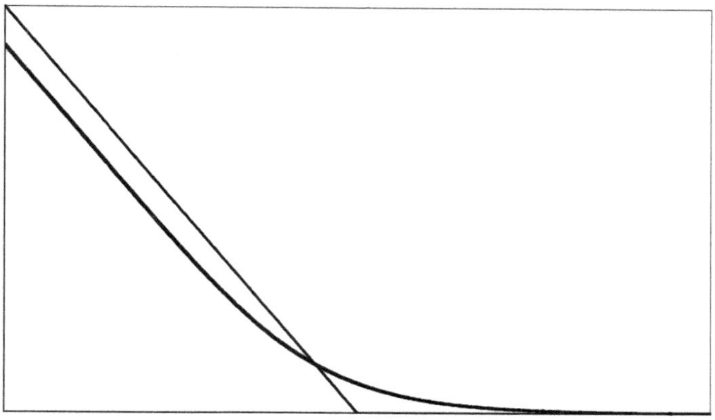

Figure 5.3: The typical price of a European put option as a function of S_0^1 compared with the option's intrinsic value $(K - S_0^1)^+$.

market model with discounted price process (X^0, X^1, X^2) is arbitrage-free. Show that the discounted claim $H^{\text{put}} = (K - S_T^1)^+/S_T^0$ is attainable in the extended model, and that its unique arbitrage-free price is given by

$$\pi^{\text{put}} = \frac{K}{(1+r)^T} - S_0^1 + \pi^{\text{call}}. \qquad \diamond$$

Our next aim is to characterize the structure of the set of arbitrage-free prices of a discounted claim H. It follows from Theorem 5.29 that every arbitrage-free price π^H of H must lie between the two numbers

$$\pi_{\text{inf}}(H) = \inf_{P^* \in \mathcal{P}} E^*[\, H\,] \quad \text{and} \quad \pi_{\text{sup}}(H) = \sup_{P^* \in \mathcal{P}} E^*[\, H\,].$$

We also know that $\pi_{\text{inf}}(H)$ and $\pi_{\text{sup}}(H)$ are equal if H is attainable. The following theorem shows that also the converse implication holds. Thus, H is attainable if and only if $\pi_{\text{inf}}(H) = \pi_{\text{sup}}(H)$.

Theorem 5.32. *Let H be a discounted claim.*

(a) *If H is attainable, then the set $\Pi(H)$ of arbitrage-free prices for H consists of the single element V_0, where V is the value process of any replicating strategy for H.*

(b) *If H is not attainable, then $\pi_{\text{inf}}(H) < \pi_{\text{sup}}(H)$ and*

$$\Pi(H) = \big(\pi_{\text{inf}}(H), \pi_{\text{sup}}(H)\big).$$

Proof. The first assertion follows from Remark 5.26 and Theorem 5.29.

To prove (b), note first that

$$\Pi(H) = \{E^*[\,H\,] \mid P^* \in \mathcal{P},\, E^*[\,H\,] < \infty\}$$

is an interval because $\mathcal{P}$ is a convex set. We will show that $\Pi(H)$ is open by constructing for any $\pi \in \Pi(H)$ two arbitrage-free prices $\breve{\pi}$ and $\hat{\pi}$ for H such that $\breve{\pi} < \pi < \hat{\pi}$. To this end, take $P^* \in \mathcal{P}$ such that $\pi = E^*[\,H\,]$. We will first use a separation argument to construct an equivalent martingale measure $\hat{P} \in \mathcal{P}$ such that $\hat{E}[\,H\,] > E^*[\,H\,]$. Let

$$U_t := E^*[\,H \mid \mathcal{F}_t\,], \quad t = 0, \dots, T,$$

and represent H by the following telescopic sum:

$$H = U_0 + \sum_{t=1}^{T} (U_t - U_{t-1}).$$

Since H is not attainable, there must be some $t \in \{1, \dots, T\}$ such that $U_t - U_{t-1} \notin \mathcal{K}_t \cap L^1(P^*)$, where

$$\mathcal{K}_t := \{\eta \cdot (X_t - X_{t-1}) \mid \eta \in L^0(\Omega, \mathcal{F}_{t-1}, P; \mathbb{R}^d)\,\}.$$

By Lemma 1.71, $\mathcal{K}_t \cap L^1(P^*)$ is a closed linear subspace of $L^1(\Omega, \mathcal{F}_t, P^*)$. Therefore, the Hahn–Banach separation theorem in the form of Theorem H.4 applied with $\mathcal{B} := \{U_t - U_{t-1}\}$ and $\mathcal{C} := \mathcal{K}_t \cap L^1(P^*)$ yields some $Z \in L^\infty(\Omega, \mathcal{F}_t, P^*)$ such that

$$\sup\{E^*[\,W Z\,] \mid W \in \mathcal{K}_t \cap L^1(P^*)\} < E^*[\,(U_t - U_{t-1}) Z\,] < \infty.$$

From the linearity of $\mathcal{K}_t \cap L^1(P^*)$, we deduce that

$$E^*[\,W Z\,] = 0 \quad \text{for all } W \in \mathcal{K}_t \cap L^1(P^*), \tag{5.21}$$

and hence that

$$E^*[\,(U_t - U_{t-1}) Z\,] > 0. \tag{5.22}$$

There is no loss of generality in assuming that $|Z| \le 1/3$, so that

$$\hat{Z} := 1 + Z - E^*[\,Z \mid \mathcal{F}_{t-1}\,]$$

can be taken as the density $d\hat{P}/dP^* = \hat{Z}$ of a new probability measure $\hat{P} \approx P$. Since Z is $\mathcal{F}_t$-measurable, the expectation of H under $\hat{P}$ satisfies

$$\begin{aligned}\hat{E}[\,H\,] &= E^*[\,H \hat{Z}\,]\\ &= E^*[\,H\,] + E^*[\,E^*[\,H \mid \mathcal{F}_t] Z\,] - E^*[\,E^*[\,H \mid \mathcal{F}_{t-1}]\, E^*[\,Z \mid \mathcal{F}_{t-1}]\,]\end{aligned}$$

$$= E^*[H] + E^*[U_t Z] - E^*[U_{t-1} Z]$$
$$> E^*[H],$$

where we have used (5.22) in the final step. On the other hand, $\hat{E}[H] \le \frac{5}{3}E^*[H] < \infty$. Thus, $\hat{\pi} := \hat{E}[H]$ will yield the desired arbitrage-free price larger than π if we have $\hat{P} \in \mathcal{P}$.

Let us prove that $\hat{P} \in \mathcal{P}$. For $k > t$, the $\mathcal{F}_t$-measurability of $\hat{Z}$ and Proposition B.8 yield that

$$\hat{E}[X_k - X_{k-1} \mid \mathcal{F}_{k-1}] = E^*[X_k - X_{k-1} \mid \mathcal{F}_{k-1}] = 0.$$

For $k = t$, (5.21) yields $E^*[(X_t - X_{t-1}) Z \mid \mathcal{F}_{t-1}] = 0$. Thus, it follows from $E^*[\hat{Z} \mid \mathcal{F}_{t-1}] = 1$ that

$$\begin{aligned}
&\hat{E}[X_t - X_{t-1} \mid \mathcal{F}_{t-1}] \\
&\quad = E^*[(X_t - X_{t-1})(1 - E^*[Z \mid \mathcal{F}_{t-1}]) \mid \mathcal{F}_{t-1}] + E^*[(X_t - X_{t-1}) Z \mid \mathcal{F}_{t-1}] \\
&\quad = 0.
\end{aligned}$$

Finally, if $k < t$ then P^* and $\hat{P}$ coincide on $\mathcal{F}_k$. Hence,

$$\hat{E}[X_k - X_{k-1} \mid \mathcal{F}_{k-1}] = E^*[X_k - X_{k-1} \mid \mathcal{F}_{k-1}] = 0,$$

and we may conclude that $\hat{P} \in \mathcal{P}$.

It remains to construct another equivalent martingale measure $\check{P}$ such that

$$\check{\pi} := \check{E}[H] < E^*[H] = \pi. \tag{5.23}$$

But this is simply achieved by letting

$$\frac{d\check{P}}{dP^*} := 2 - \frac{d\hat{P}}{dP^*},$$

which defines a probability measure $\check{P} \approx P$, because the density $d\hat{P}/dP^*$ is bounded from above by 5/3 and below by 1/3. $\check{P} \in \mathcal{P}$ is then obvious as is (5.23). □

Remark 5.33. So far, we have assumed that a contingent claim is settled at the terminal time T. A natural way of dealing with an $\mathcal{F}_{T_0}$-measurable payoff $C_0 \ge 0$ maturing at some time $T_0 < T$ is to apply our results to the corresponding discounted claim

$$H_0 := \frac{C_0}{S^0_{T_0}}$$

in the market model with the restricted time horizon T_0. Clearly, this restricted model is arbitrage-free. An alternative approach is to invest the payoff C_0 at time T_0 into the numéraire asset S^0. At time T, this yields the contingent claim

$$C := C_0 \cdot \frac{S_T^0}{S_{T_0}^0},$$

whose discounted claim

$$H = \frac{C}{S_T^0} = \frac{C_0}{S_{T_0}^0}$$

is formally identical to H_0. Moreover, our results can be directly applied to H. It is intuitively clear that these two approaches for determining the arbitrage-free prices of C_0 should be equivalent. A formal proof must show that the set $\Pi(H)$ is equal to the set

$$\Pi(H_0) := \{E_0^*[\,H_0\,] \mid P_0^* \in \mathcal{P}_0 \text{ and } E_0^*[\,H_0\,] < \infty\}$$

of arbitrage-free prices of H_0 in the market model whose time horizon is T_0. Here, $\mathcal{P}_0$ denotes the set of measures P_0^* on $(\Omega, \mathcal{F}_{T_0})$ that are equivalent to P on $\mathcal{F}_{T_0}$ and that are martingale measures for the restricted price process $(X_t)_{t=0,\dots,T_0}$. Clearly, each $P^* \in \mathcal{P}$ defines an element of $\mathcal{P}_0$ by restricting P^* to the σ-algebra $\mathcal{F}_{T_0}$. In fact, Proposition 5.34 below shows that every element in $\mathcal{P}_0$ arises in this way. Thus, the two sets of arbitrage-free prices for H and H_0 coincide, that is,

$$\Pi(H) = \Pi(H_0).$$

It follows, in particular, that H_0 is attainable if and only if H is attainable. ◇

Proposition 5.34. *Consider the situation described in Remark 5.33 and let $P_0^* \in \mathcal{P}_0$ be given. Then there exists some $P^* \in \mathcal{P}$ whose restriction to $\mathcal{F}_{T_0}$ is equal to P_0^*.*

Proof. Let $\hat{P} \in \mathcal{P}$ be arbitrary, and denote by Z_{T_0} the density of P_0^* with respect to the restriction of $\hat{P}$ to the σ-algebra $\mathcal{F}_{T_0}$. Then Z_{T_0} is $\mathcal{F}_{T_0}$-measurable, and

$$dP^* := Z_{T_0}\, d\hat{P}$$

defines a probability measure on $\mathcal{F}$. Clearly, P^* is equivalent to $\hat{P}$ and to P, and it coincides with P_0^* on $\mathcal{F}_{T_0}$. To check that $P^* \in \mathcal{P}$, it suffices to show that $X_t - X_{t-1}$ is a martingale increment under P^* for $t > T_0$. For these t, the density Z_{T_0} is $\mathcal{F}_{t-1}$-measurable, so Proposition B.8 implies that

$$E^*[\,X_t - X_{t-1} \mid \mathcal{F}_{t-1}\,] = \hat{E}[\,X_t - X_{t-1} \mid \mathcal{F}_{t-1}\,] = 0.$$ □

Example 5.35. Let us consider the situation of Example 5.30, where the numéraire S^0 is a locally riskless bond with nonnegative interest rates. Remark 5.33 allows us to compare the arbitrage-free prices of two European call options $C_0 = (S_{T_0}^1 - K)^+$ and $C = (S_T^1 - K)^+$ with the same strikes and underlyings but with different maturities $T_0 < T$. As in Example 5.30, we get that for $P^* \in \mathcal{P}$,

$$E^*\left[\frac{C}{S_T^0}\,\Big|\,\mathscr{F}_{T_0}\right] \geq \frac{1}{S_{T_0}^0}\left(S_{T_0}^1 - K\,E^*\left[\frac{S_{T_0}^0}{S_T^0}\,\Big|\,\mathscr{F}_{T_0}\right]\right)^+ \qquad (5.24)$$
$$\geq \frac{C_0}{S_{T_0}^0}.$$

Hence, if P^* is used to calculate arbitrage-free prices for C_0 and C, the resulting price of C_0 is lower than the price of C:

$$E^*\left[\frac{C}{S_T^0}\right] \geq E^*\left[\frac{C_0}{S_{T_0}^0}\right].$$

This argument suggests that the price of a European call option should be an increasing function of the maturity. ◇

Exercise 5.3.2. Let Y_k $(k = 1,\dots,T)$ be independent identically distributed random variables in $\mathscr{L}^1(\Omega,\mathscr{F},P)$, and suppose that the Y_k are not P-a. s. constant and satisfy $E[\,Y_k\,] = 0$. Let furthermore

$$X_t := X_0 + \sum_{k=1}^{t} Y_k, \quad t = 0,\dots,T.$$

Then X is a P-martingale if we consider the filtration given by $\mathscr{F}_0 = \{\emptyset,\Omega\}$ and $\mathscr{F}_t = \sigma(Y_1,\dots,Y_t)$ for $t = 1,\dots,T$. We now enlarge the filtration by adding "insider information" of the terminal value X_T. That is, we consider the enlarged filtration

$$\widetilde{\mathscr{F}}_n := \sigma(\mathscr{F}_n \cup \sigma(X_T)).$$

(a) Show that X is no longer a P-martingale with respect to $(\widetilde{\mathscr{F}}_t)$.
(b) Prove that the process

$$\widetilde{X}_t := X_t - \sum_{k=0}^{t-1} \frac{1}{T-k}\,(X_T - X_k),$$

is a P-martingale with respect to the enlarged filtration $(\widetilde{\mathscr{F}}_t)$.
(c) The insider information of the terminal value X_T implies the existence of self-financing strategies with positive expected profit. Construct a strategy ξ^* that maximizes the expected profit

$$E\left[\sum_{t=1}^{T} \xi_t(X_t - X_{t-1})\right]$$

within the class of all $(\widetilde{\mathscr{F}}_t)$-predictable strategies ξ with $|\xi_t| \leq 1$ P-a. s. for all t. ◇

5.4 Complete markets

We have seen in Theorem 5.32 that any attainable claim in an arbitrage-free market model has a unique arbitrage-free price. Thus, the situation becomes particularly transparent if all contingent claims are attainable.

Definition 5.36. An arbitrage-free market model is called *complete* if every contingent claim is attainable. △

Complete market models are precisely those models in which every contingent claim has a unique and unambiguous arbitrage-free price. However, in discrete time, only a very limited class of models enjoys this property. The following characterization of market completeness is sometimes called the "second fundamental theorem of asset pricing." It uses the notion of an *atom* of the probability space $(\Omega, \mathscr{F}_T, P)$ as provided in Definition D.15.

Theorem 5.37. *An arbitrage-free market model is complete if and only if there exists exactly one equivalent martingale measure. In this case, the probability space $(\Omega, \mathscr{F}_T, P)$ can be decomposed into finitely many atoms, whose number is bounded from above by $(d+1)^T$.*

Proof. If the model is complete, then $H := \mathbb{1}_A$ is an attainable discounted claim for every $A \in \mathscr{F}_T$. It follows from the results of Section 5.3 that the mapping $P^* \mapsto E^*[\,H\,] = P^*[\,A\,]$ is constant over the set $\mathscr{P}$. Hence, there can be only one equivalent martingale measure.

Conversely, if $|\mathscr{P}| = 1$, then the set $\Pi(H)$ of arbitrage-free prices of every discounted claim H has exactly one element. Hence, Theorem 5.32 implies that H is attainable.

To prove the second assertion, note first that the asserted bound on the number of atoms in $\mathscr{F}_T$ holds for $T = 1$ by Corollary 1.44. We proceed by induction on T. Suppose that the assertion holds for $T - 1$. By assumption, any bounded $\mathscr{F}_T$-measurable random variable $H \geq 0$ can be written as

$$H = V_{T-1} + \xi_T \cdot (X_T - X_{T-1}),$$

where both V_{T-1} and ξ_T are $\mathscr{F}_{T-1}$-measurable, and hence constant on each atom A of $(\Omega, \mathscr{F}_{T-1}, P)$. It follows that the dimension of the linear space $L^\infty(\Omega, \mathscr{F}_T, P[\,\cdot\,|A])$ is less than or equal to $d+1$. Thus, Proposition 1.43 implies that $(\Omega, \mathscr{F}_T, P[\,\cdot\,|A])$ has at most $d+1$ atoms. Applying the induction hypothesis concludes the proof. □

Below we state additional characterizations of market completeness. Denote by $\mathcal{Q}$ the set of all martingale measures in the sense of Definition 5.13. Then both $\mathscr{P}$ and $\mathcal{Q}$ are convex sets. Recall that an element of a convex set is called an *extreme point* of this set if it cannot be written as a nontrivial convex combination of members of this set.

Property (d) in the following theorem is usually called the *predictable representation property*, or the *martingale representation property*, of the P^*-martingale X.

Theorem 5.38. *For $P^* \in \mathscr{P}$, the following conditions are equivalent:*

(a) $\mathcal{P} = \{P^*\}$.
(b) P^* *is an extreme point of* $\mathcal{P}$
(c) P^* *is an extreme point of* $\mathcal{Q}$.
(d) *Every* P^**-martingale* M *can be represented as a "stochastic integral" of a d-dimensional predictable process* ξ:

$$M_t = M_0 + \sum_{k=1}^{t} \xi_k \cdot (X_k - X_{k-1}) \quad \textit{for } t = 0, \dots, T.$$

Proof. (a)⇒(c): If P^* can be written as $P^* = \alpha Q_1 + (1-\alpha)Q_2$ for $\alpha \in (0,1)$ and $Q_1, Q_2 \in \mathcal{Q}$, then Q_1 and Q_2 are both absolutely continuous with respect to P^*. By defining

$$P_i := \frac{1}{2}(Q_i + P^*), \quad i = 1, 2,$$

we thus obtain two martingale measures P_1 and P_2, which are equivalent to P^*. Hence, (a) yields that $P_1 = P_2 = P^*$ and, in turn, $Q_1 = Q_2 = P^*$.

(c)⇒(b): This is obvious since $\mathcal{P} \subseteq \mathcal{Q}$.

(b)⇒(a): Suppose that there exists a $\hat{P} \in \mathcal{P}$ that is different from P^*. We will show below that in this case $\hat{P}$ can be chosen such that the density $d\hat{P}/dP^*$ is bounded by some constant $c > 0$. Then, if $\varepsilon > 0$ is less than $1/c$,

$$\frac{dP'}{dP^*} := 1 + \varepsilon - \varepsilon \frac{d\hat{P}}{dP^*}$$

defines another measure $P' \in \mathcal{P}$ different from P^*. Moreover, P^* can be represented as the convex combination

$$P^* = \frac{\varepsilon}{1+\varepsilon}\hat{P} + \frac{1}{1+\varepsilon}P',$$

which contradicts condition (b). Hence, P^* must be the unique equivalent martingale measure.

It remains to prove the existence of $\hat{P} \in \mathcal{P}$ with a bounded density $d\hat{P}/dP^*$ if there exists some $\widetilde{P} \in \mathcal{P}$ that is different from P^*. Since $\widetilde{P} \neq P^*$, there exists a set $A \in \mathcal{F}_T$ such that $P^*[\,A\,] \neq \widetilde{P}[\,A\,]$. We enlarge our market model by introducing the additional asset

$$X_t^{d+1} := \widetilde{P}[\,A \mid \mathcal{F}_t\,], \quad t = 0, \dots, T,$$

and we take P^* instead of P as our reference measure. By definition, $\widetilde{P}$ is an equivalent martingale measure for $(X^0, X^1, \dots, X^d, X^{d+1})$. Hence, the extended market model is arbitrage-free, and Theorem 5.16 guarantees the existence of an equivalent martingale measure $\hat{P}$ such that the density $d\hat{P}/dP^*$ is bounded. Moreover, $\hat{P}$ must be different from P^*, since P^* is not a martingale measure for X^{d+1}:

$$X_0^{d+1} = \widetilde{P}[A] \neq P^*[A] = E^*[X_T^{d+1}].$$

(a)⇒(d): The terminal value M_T of a P^*-martingale M can be decomposed into the difference of its positive and negative parts:

$$M_T = M_T^+ - M_T^-.$$

M_T^+ and M_T^- can be regarded as two discounted claims, which are attainable by Theorem 5.37. Hence, there exist two d-dimensional predictable process ξ^+ and ξ^- such that

$$M_T^\pm = V_0^\pm + \sum_{k=1}^T \xi_k^\pm \cdot (X_k - X_{k-1}) \quad P^*\text{-a. s.}$$

for two nonnegative constants V_0^+ and V_0^-. Since the value processes

$$V_t^\pm := V_0^\pm + \sum_{k=1}^t \xi_k^\pm \cdot (X_k - X_{k-1})$$

are P^*-martingales by Theorem 5.25, we get that

$$M_t = E^*[M_T^+ - M_T^- \mid \mathscr{F}_t] = V_t^+ - V_t^-.$$

This proves that the desired representation of M holds in terms of the d-dimensional predictable process $\xi := \xi^+ - \xi^-$.

(d)⇒(a): Applying our assumption to the martingale $M_t := P^*[A \mid \mathscr{F}_t]$ shows that $H = \mathbb{1}_A$ is an attainable contingent claim for every $A \in \mathscr{F}_T$. Hence, it follows as in the proof of Theorem 5.37 that the mapping $P^* \mapsto P^*[A]$ is constant over the set $\mathscr{P}$. Thus, there can be only one equivalent martingale measure. □

Exercise 5.4.1. Consider the sample space

$$\Omega := \{-1, +1\}^T = \{\omega = (y_1, \dots, y_T) \mid y_i \in \{-1, +1\}\}$$

and denote by $Y_t(\omega) := y_t$, for $\omega = (y_1, \dots, y_T)$, the projection on the tth coordinate. As filtration, we take $\mathscr{F}_0 := \{\emptyset, \Omega\}$ and $\mathscr{F}_t := \sigma(Y_1, \dots, Y_t)$ for $t = 1, \dots, T$. We consider a financial market model with two assets such that the discounted price process $X_t := X_t^1 = S_t^1 / S_t^0$ is of the form

$$X_t = X_0 \exp\left(\sum_{k=1}^t (\sigma_k Y_k + m_k) \right)$$

for a constant $X_0 > 0$ and two predictable processes (σ_t) and (m_t). We suppose that $0 \leq |m_t| < \sigma_t$ for all t.

(a) Show that if P is a probability measure on Ω for which $P[\,\{\omega\}\,] > 0$ for all ω, then there exists a unique equivalent martingale measure P^*.

(b) By using the binary structure of the model, and without using Theorem 5.37, prove the following martingale representation result. If P^* is as in (a), every P^*-martingale M can be represented as

$$M_t = M_0 + \sum_{k=1}^{t} \xi_k (X_k - X_{k-1}), \quad t = 0, \dots, T,$$

where the predictable process ξ is given by

$$\xi_k = \frac{M_k - M_{k-1}}{X_k - X_{k-1}}.$$

◇

Exercise 5.4.2. Consider a complete market model with $T = 2$ and discounted price process $(X_t)_{t=0,1,2}$, defined on some filtered probability space $(\Omega, \mathscr{F}, (\mathscr{F}_t)_{t=0,1,2}, P)$. By P^* we denote the unique equivalent martingale measure in this complete model. Now we enlarge the model by adding two external states ω^+ and ω^-, corresponding to additional information revealed at time $t = 2$. That is, we define $\hat{\Omega} := \Omega \times \{\omega^+, \omega^-\}$, $\hat{\mathscr{F}} := \sigma(A \times \{\omega^+\} \mid A \in \mathscr{F})$, and

$$\hat{P}[\,\{(\omega_0, \omega^\pm)\}\,] := \frac{1}{2} P[\,\{\omega\}\,], \quad \omega \in \Omega.$$

The fact that the additional information is revealed at time $t = 2$ means that we choose the price process

$$\hat{X}_t(\omega, \omega^\pm) := X_t(\omega)$$

and the following filtration:

$$\hat{\mathscr{F}}_0 := \{\emptyset, \hat{\Omega}\}, \quad \hat{\mathscr{F}}_1 := \sigma(A \times \{\omega^+, \omega^-\} \mid A \in \mathscr{F}_1), \quad \text{and} \quad \hat{\mathscr{F}}_2 := \hat{\mathscr{F}}.$$

For $0 < p < 1$, we furthermore define the probability measure $\hat{P}^*_p$ on $\hat{\mathscr{F}}$ via

$$\hat{P}^*_p[\,A \times \{\omega^+\}\,] := p \cdot P^*[\,A\,], \quad A \in \mathscr{F}.$$

(a) Show that each measure $\hat{P}^*_p$ is an equivalent martingale measure for the extended model.

(b) Argue that the extended model is incomplete and find a nonattainable contingent claim.

(c) Does every equivalent martingale measure belong to the set $\{\hat{P}^*_p \mid 0 < p < 1\}$? If yes, give a proof, if not, find a counterexample.

◇

5.5 The binomial model

A complete financial market model with only one risky asset must have a binary tree structure, as we have seen in Theorem 5.37. Under an additional homogeneity assumption, this reduces to the following particularly simple model, which was introduced by Cox, Ross, and Rubinstein in [69]. It involves the riskless bond

$$S_t^0 := (1+r)^t, \quad t = 0, \dots, T,$$

with $r > -1$ and one risky asset $S^1 = S$, whose return

$$R_t := \frac{S_t - S_{t-1}}{S_{t-1}}$$

in the tth trading period can only take two possible values $a, b \in \mathbb{R}$ such that

$$-1 < a < b.$$

Thus, the stock price jumps from S_{t-1} either to the higher value $S_t = S_{t-1}(1+b)$ or to the lower value $S_t = S_{t-1}(1+a)$. In this context, we are going to derive explicit formulas for the arbitrage-free prices and replicating strategies of various contingent claims.

Let us construct the model on the sample space

$$\Omega := \{-1, +1\}^T = \{\omega = (y_1, \dots, y_T) \mid y_i \in \{-1, +1\}\}.$$

Denote by

$$Y_t(\omega) := y_t \quad \text{for } \omega = (y_1, \dots, y_T) \tag{5.25}$$

the projection on the tth coordinate, and let

$$R_t(\omega) := a\,\frac{1 - Y_t(\omega)}{2} + b\,\frac{1 + Y_t(\omega)}{2} = \begin{cases} a & \text{if } Y_t(\omega) = -1, \\ b & \text{if } Y_t(\omega) = +1. \end{cases}$$

The price process of the risky asset is modeled as

$$S_t := S_0 \prod_{k=1}^{t} (1 + R_k),$$

where the initial value $S_0 > 0$ is a given constant. The discounted price process takes the form

$$X_t = \frac{S_t}{S_t^0} = S_0 \prod_{k=1}^{t} \frac{1 + R_k}{1 + r}.$$

As filtration, we take

$$\mathscr{F}_t := \sigma(S_0, \dots, S_t) = \sigma(X_0, \dots, X_t), \quad t = 0, \dots, T.$$

Note that $\mathscr{F}_0 = \{\emptyset, \Omega\}$, and

$$\mathscr{F}_t = \sigma(Y_1, \dots, Y_t) = \sigma(R_1, \dots, R_t) \quad \text{for } t = 1, \dots, T;$$

$\mathscr{F} := \mathscr{F}_T$ coincides with the power set of Ω. Let us fix any probability measure P on $(\Omega, \mathscr{F})$ such that

$$P[\,\{\omega\}\,] > 0 \quad \text{for all } \omega \in \Omega. \tag{5.26}$$

Such a model will be called a *binomial model* or a *CRR model*. The following theorem characterizes those parameter values a, b, r for which the model is arbitrage-free.

Theorem 5.39. *The CRR model is arbitrage-free if and only if $a < r < b$. In this case, the CRR model is complete, and there is a unique martingale measure P^*. The martingale measure is characterized by the fact that the random variables $R_1, \dots, R_T$ are independent under P^* with common distribution*

$$P^*[\,R_t = b\,] = p^* := \frac{r - a}{b - a}, \quad t = 1, \dots, T.$$

Proof. A measure Q on $(\Omega, \mathscr{F})$ is a martingale measure if and only if the discounted price process is a martingale under Q, that is,

$$X_t = E_Q[\,X_{t+1} \mid \mathscr{F}_t\,] = X_t\, E_Q\Big[\,\frac{1 + R_{t+1}}{1 + r} \,\Big|\, \mathscr{F}_t\,\Big] \quad Q\text{-a.s.}$$

for all $t \le T - 1$. This identity is equivalent to the equation

$$r = E_Q[\,R_{t+1} \mid \mathscr{F}_t\,] = b \cdot Q[\,R_{t+1} = b \mid \mathscr{F}_t\,] + a \cdot (1 - Q[\,R_{t+1} = b \mid \mathscr{F}_t\,]),$$

that is, to the condition

$$Q[\,R_{t+1} = b \mid \mathscr{F}_t\,](\omega) = p^* = \frac{r - a}{b - a} \quad \text{for } Q\text{-a.e. } \omega \in \Omega.$$

Since the right-hand side is deterministic, this identity holds if and only if the random variables $R_1, \dots, R_T$ are independent under Q with common distribution $Q[\,R_t = b\,] = p^*$. In particular, there can be at most one martingale measure for X.

If the market model is arbitrage-free, then there exists an *equivalent* martingale measure P^*. The condition $P^* \approx P$ implies

$$p^* = P^*[\,R_1 = b\,] \in (0, 1),$$

which holds if and only if $a < r < b$.

Conversely, if $a < r < b$ then we can define a measure $P^* \approx P$ on $(\Omega, \mathscr{F})$ by

$$P^*[\{\omega\}] := (p^*)^k \cdot (1 - p^*)^{T-k} > 0$$

where k denotes the number of occurrences of $+1$ in $\omega = (y_1, \dots, y_T)$. Under P^*, $Y_1, \dots, Y_T$, and hence $R_1, \dots, R_T$, are independent random variables with common distribution $P^*[Y_t = 1] = P^*[R_t = b] = p^*$, and so P^* is an equivalent martingale measure. □

From now on, we consider only CRR models that are arbitrage-free, and we denote by P^* the unique equivalent martingale measure.

Remark 5.40. Note that the unique martingale measure P^*, and hence the valuation of any contingent claim, is completely independent of the initial choice of the "objective" measure P within the class of measures satisfying (5.26). ◇

Let us now turn to the problem of *pricing* and *hedging* a given contingent claim C. The discounted claim $H = C/S_T^0$ can be written as

$$H = h(S_0, \dots, S_T)$$

for a suitable function h.

Proposition 5.41. *The value process*

$$V_t = E^*[H \mid \mathscr{F}_t], \quad t = 0, \dots, T,$$

of a replicating strategy for H is of the form

$$V_t(\omega) = v_t(S_0, S_1(\omega), \dots, S_t(\omega)), \tag{5.27}$$

where the function v_t is given by

$$v_t(x_0, \dots, x_t) = E^*\left[h\left(x_0, \dots, x_t, x_t \frac{S_1}{S_0}, \dots, x_t \frac{S_{T-t}}{S_0} \right) \right]. \tag{5.28}$$

Proof. Clearly,

$$V_t = E^*\left[h\left(S_0, S_1, \dots, S_t, S_t \frac{S_{t+1}}{S_t}, \dots, S_t \frac{S_T}{S_t} \right) \Bigm| \mathscr{F}_t \right].$$

Each ratio S_{t+s}/S_t is independent of $\mathscr{F}_t$ and has under P^* the same distribution as

$$\frac{S_s}{S_0} = \prod_{k=1}^{s} (1 + R_k).$$

Hence, (5.28) follows from the standard properties of conditional expectations. □

Since V is characterized by the recursion

$$V_T := H \quad \text{and} \quad V_t = E^*[\, V_{t+1} \mid \mathscr{F}_t\,], \quad t = T-1, \dots, 0,$$

we obtain a recursive formula for the functions v_t defined in (5.28):

$$\begin{aligned} v_T(x_0,\dots,x_T) &= h(x_0,\dots,x_T), \\ v_t(x_0,\dots,x_t) &= p^* \cdot v_{t+1}(x_0,\dots,x_t,x_t\hat{b}) + (1-p^*)\cdot v_{t+1}(x_0,\dots,x_t,x_t\hat{a}), \end{aligned} \tag{5.29}$$

where

$$\hat{a} := 1+a \quad \text{and} \quad \hat{b} := 1+b.$$

In the literature on the binomial model, the quantities $\hat{a}$ and $\hat{b}$ are sometimes also denoted by d and u, respectively, where d stands for "down" and u stands for "up."

Example 5.42. If $H = h(S_T)$ depends only on the terminal value S_T of the stock price, then V_t depends only on the value S_t of the current stock price:

$$V_t(\omega) = v_t(S_t(\omega)).$$

Moreover, the formula (5.28) for v_t reduces to an expectation with respect to the binomial distribution with parameter p^*:

$$v_t(x_t) = \sum_{k=0}^{T-t} h(x_t\,\hat{a}^{T-t-k}\,\hat{b}^k)\binom{T-t}{k}(p^*)^k(1-p^*)^{T-t-k}.$$

In particular, the unique arbitrage-free price of H is given by

$$\pi(H) = v_0(S_0) = \sum_{k=0}^{T} h(S_0\,\hat{a}^{T-k}\,\hat{b}^k)\binom{T}{k}(p^*)^k(1-p^*)^{T-k}.$$

For $h(x) = (x-K)^+/(1+r)^T$ or $h(x) = (K-x)^+/(1+r)^T$, we obtain explicit formulas for the arbitrage-free prices of European call or put options with strike price K. For instance, the price of $H^{\text{call}} := (S_T-K)^+/(1+r)^T$ is given by

$$\pi(H^{\text{call}}) = \frac{1}{(1+r)^T}\sum_{k=0}^{T}(S_0\,\hat{a}^{T-k}\,\hat{b}^k - K)^+\binom{T}{k}(p^*)^k(1-p^*)^{T-k}. \qquad \diamond$$

Example 5.43. Denote by

$$M_t := \max_{0\le s\le t} S_s, \quad 0 \le t \le T,$$

the *running maximum* of S, and consider a discounted claim of the form $H = h(S_T, M_T)$. For instance, H can be an up-and-in or up-and-out barrier option or a lookback put. Then the value process of H is of the form

$$V_t = v_t(S_t, M_t),$$

where

$$v_t(x_t, m_t) = E^*\left[h\left(x_t \frac{S_{T-t}}{S_0}, m_t \vee \left(x_t \frac{M_{T-t}}{S_0}\right)\right)\right].$$

This follows from (5.28) or directly from the fact that

$$M_T = M_t \vee \left(S_t \max_{t \leq u \leq T} \frac{S_u}{S_t}\right),$$

where $\max_{t \leq u \leq T} S_u/S_t$ is independent of $\mathscr{F}_t$ and has the same law as M_{T-t}/S_0 under P^*. The same argument works for options that depend on the minimum of the stock price such as lookback calls or down-and-in barrier options. ◇

Exercise 5.5.1. For an *Asian option* depending on the average price

$$S_{\mathrm{av}} := \frac{1}{|\mathbb{T}|} \sum_{t \in \mathbb{T}} S_t$$

during a predetermined set of periods $\mathbb{T} \subseteq \{0, \dots, T\}$, we introduce the process

$$A_t := \sum_{s \in \mathbb{T},\, s \leq t} S_s.$$

Show that the value process V_t of the Asian option is a function of S_t, A_t, and t. ◇

Let us now derive a formula for the hedging strategy $\bar{\xi} = (\xi^0, \xi)$ of our discounted claim $H = h(S_0, \dots, S_T)$.

Proposition 5.44. *The hedging strategy is given by*

$$\xi_t(\omega) = \Delta_t(S_0, S_1(\omega), \dots, S_{t-1}(\omega)),$$

where

$$\Delta_t(x_0, \dots, x_{t-1}) := (1+r)^t \frac{v_t(x_0, \dots, x_{t-1}, x_{t-1}\hat{b}) - v_t(x_0, \dots, x_{t-1}, x_{t-1}\hat{a})}{x_{t-1}\hat{b} - x_{t-1}\hat{a}}.$$

Proof. For each $\omega = (y_1, \dots, y_T)$, ξ_t must satisfy

$$\xi_t(\omega)(X_t(\omega) - X_{t-1}(\omega)) = V_t(\omega) - V_{t-1}(\omega). \tag{5.30}$$

In this equation, the random variables ξ_t, X_{t-1}, and V_{t-1} depend only on the first $t-1$ components of ω. For a fixed t, let us define ω^+ and ω^- by

$$\omega^{\pm} := (y_1, \dots, y_{t-1}, \pm 1, y_{t+1}, \dots, y_T).$$

Plugging ω^+ and ω^- into (5.30) shows

$$\xi_t(\omega) \cdot (X_{t-1}(\omega)\,\hat{b}(1+r)^{-1} - X_{t-1}(\omega)) = V_t(\omega^+) - V_{t-1}(\omega),$$
$$\xi_t(\omega) \cdot (X_{t-1}(\omega)\,\hat{a}(1+r)^{-1} - X_{t-1}(\omega)) = V_t(\omega^-) - V_{t-1}(\omega).$$

Solving for $\xi_t(\omega)$ and using our formula (5.27) for V_t, we obtain

$$\xi_t(\omega) = (1+r)\frac{V_t(\omega^+) - V_t(\omega^-)}{X_{t-1}(\omega)(\hat{b} - \hat{a})} = \Delta_t(S_0, S_1(\omega), \ldots, S_{t-1}(\omega)). \qquad \square$$

Remark 5.45. The term Δ_t may be viewed as a discrete "derivative" of the value function v_t with respect to the possible stock price changes. In financial language, a hedging strategy based on a derivative of the value process is often called a *Delta hedge*. ◇

Remark 5.46. Let $H = h(S_T)$ be a discounted claim that depends on the terminal value of S by way of an *increasing* function h. For instance, h can be the discounted payoff function $h(x) = (x-K)^+/(1+r)^T$ of a European call option. Then

$$v_t(x) = E^*[\,h(x\,S_{T-t}/S_0)\,]$$

is also increasing in x, and so the hedging strategy satisfies

$$\xi_t(\omega) = (1+r)^t\frac{v_t(S_{t-1}(\omega)\,\hat{b}) - v_t(S_{t-1}(\omega)\,\hat{a})}{S_{t-1}(\omega)\,\hat{b} - S_{t-1}(\omega)\,\hat{a}} \geq 0.$$

In other words, the hedging strategy for H does not involve *short sales* of the risky asset. ◇

Exercise 5.5.2. Let $T_0 \in \{1, \ldots, T-1\}$ and $K > 0$. The payoff of *forward starting call option* has the form

$$\left(\frac{S_T}{S_{T_0}} - K\right)^+.$$

Determine its arbitrage-free price and hedging strategy. ◇

5.6 Exotic derivatives

The recursion formula (5.29) can be used for the numeric computation of the value process of any contingent claim. For the value processes of certain *exotic derivatives* that depend on the *maximum* of the stock price, it is even possible to obtain simple closed-form solutions if we make the additional assumption that

$$\hat{a} = \frac{1}{\hat{b}}, \quad \text{where } \hat{a} = 1 + a \text{ and } \hat{b} = 1 + b.$$

Under this assumption, the price process of the risky asset is of the form

$$S_t(\omega) = S_0 \hat{b}^{Z_t(\omega)}$$

where, for Y_k as in (5.25),

$$Z_0 := 0 \quad \text{and} \quad Z_t := Y_1 + \cdots + Y_t, \quad t = 1, \ldots, T.$$

Let $\mathbb{P}$ denote the uniform distribution

$$\mathbb{P}[\,\{\omega\}\,] := \frac{1}{|\Omega|} = 2^{-T}, \qquad \omega \in \Omega.$$

Under the measure $\mathbb{P}$, the random variables Y_t are independent with common distribution $\mathbb{P}[\,Y_t = +1\,] = \frac{1}{2}$. Thus, the stochastic process Z becomes a *simple random walk* under $\mathbb{P}$. Therefore,

$$\mathbb{P}[\,Z_t = k\,] = \begin{cases} 2^{-t}\binom{t}{\frac{t+k}{2}} & \text{if } t + k \text{ is even,} \\ 0 & \text{otherwise.} \end{cases} \tag{5.31}$$

The following lemma is the key to numerous explicit results on the distribution of Z under the measure $\mathbb{P}$; see, for example, Chapter III of [129]. We denote by

$$M_t := \max_{0 \le s \le t} Z_s$$

the *running maximum* of Z.

Lemma 5.47 (Reflection principle). *For all $k \in \mathbb{N}$ and $l \in \mathbb{N}_0$,*

$$\mathbb{P}[\,M_T \ge k \text{ and } Z_T = k - l\,] = \mathbb{P}[\,Z_T = k + l\,],$$

and

$$\mathbb{P}[\,M_T = k \text{ and } Z_T = k - l\,] = \mathbb{P}[\,Z_T = k + l\,] - \mathbb{P}[\,Z_T = k + l + 2\,]. \tag{5.32}$$

Proof. Let

$$\tau(\omega) := \inf\{t \ge 0 \mid Z_t(\omega) = k\} \wedge T.$$

For $\omega = (y_1, \ldots, y_T) \in \Omega$ we define $\phi(\omega)$ by $\phi(\omega) = \omega$ if $\tau(\omega) = T$ and by

$$\phi(\omega) = (y_1, \ldots, y_{\tau(\omega)}, -y_{\tau(\omega)+1}, \ldots, -y_T)$$

otherwise, that is, if the level k is reached before the deadline T. Intuitively, the two trajectories $(Z_t(\omega))_{t=0,\ldots,T}$ and $(Z_t(\phi(\omega)))_{t=0,\ldots,T}$ coincide up to $\tau(\omega)$, but from then on the latter path is obtained by reflecting the original one on the horizontal axis at level k; see Figure 5.4.

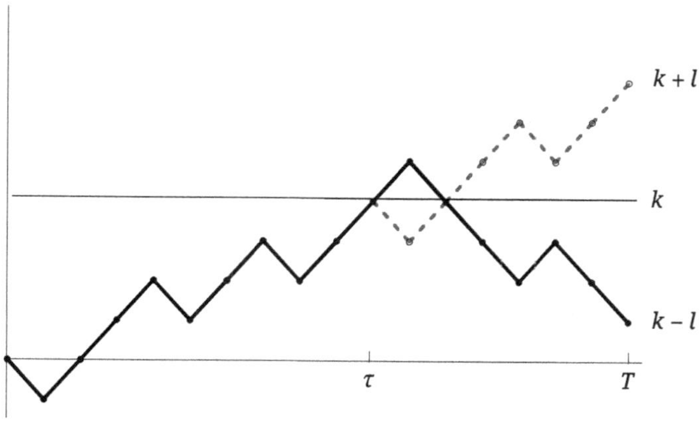

Figure 5.4: The reflection principle.

Let $A_{k,l}$ denote the set of all $\omega \in \Omega$ such that $Z_T(\omega) = k - l$ and $M_T \geq k$. Then ϕ is a bijection between $A_{k,l}$ and the set

$$\{M_T \geq k \text{ and } Z_T = k + l\},$$

which coincides with $\{Z_T = k + l\}$, due to our assumption $l \geq 0$. Hence, the uniform distribution $\mathbb{P}$ must assign the same probability to $A_{k,l}$ and $\{Z_T = k + l\}$, and we obtain our first formula. The second one follows by part one of this lemma:

$$\begin{aligned}
&\mathbb{P}[\,M_T = k,\, Z_T = k - l\,] \\
&\quad = \mathbb{P}[\,M_T \geq k,\, Z_T = k - l\,] - \mathbb{P}[\,M_T \geq k + 1,\, Z_T = k - l\,] \qquad (5.33) \\
&\quad = \mathbb{P}[\,M_T \geq k,\, Z_T = k - l\,] - \mathbb{P}[\,M_T \geq k + 1,\, Z_T = (k + 1) - (l + 1)\,] \\
&\quad = \mathbb{P}[\,Z_T = k + l\,] - \mathbb{P}[\,Z_T = k + l + 2\,].
\end{aligned}$$

□

Exercise 5.6.1. Use the reflection principle to deduce the following formulas for the distribution of the maximum of the simple random walk:

$$\begin{aligned}
\mathbb{P}[M_T \geq k] &= \mathbb{P}[Z_T = k] + 2\mathbb{P}[Z_T > k] \quad \text{and} \\
\mathbb{P}[M_T = k] &= \mathbb{P}[Z_T = k] + \mathbb{P}[Z_T = k + 1], \quad k \in \mathbb{N}_0.
\end{aligned}$$

Note that at least one of the two probabilities $\mathbb{P}[Z_T = k]$ and $\mathbb{P}[Z_T = k + 1]$ must be zero.

◇

Remark 5.48. Suppose that the random walk Z is defined even up to time $T+1$; this can always be achieved by enlarging the probability space $(\Omega, \mathscr{F}, \mathbb{P})$. Then the probability in (5.32) can be rewritten as

$$\mathbb{P}[\,M_T = k \text{ and } Z_T = k-l\,] = 2\frac{k+l+1}{T+1}\,\mathbb{P}[\,Z_{T+1} = 1+k+l\,]. \tag{5.34}$$

Indeed, (5.34) is trivial in case $T+k+l$ is not even; otherwise, we let $j := (T+k+l)/2$ and apply (5.31) to get

$$\begin{aligned}&\mathbb{P}[\,Z_T = k+l\,] - \mathbb{P}[\,Z_T = k+l+2\,]\\&\quad = 2^{-T}\binom{T}{j} - 2^{-T}\binom{T}{j+1} = 2^{-T}\binom{T+1}{j+1}\frac{2j+1-T}{T+1},\end{aligned}$$

and the latter expression is equal to the right-hand side of (5.34). ◇

Formula (5.31) will change if we replace the uniform distribution $\mathbb{P}$ by our martingale measure P^*, described in Theorem 5.39:

$$P^*[\,Z_t = k\,] = \begin{cases}(p^*)^{\frac{t+k}{2}}(1-p^*)^{\frac{t-k}{2}}\binom{t}{\frac{t+k}{2}} & \text{if } t+k \text{ is even,}\\ 0 & \text{otherwise.}\end{cases}$$

Let us now show how the reflection principle carries over to P^*.

Lemma 5.49 (Reflection principle for P^*). *For all $k \in \mathbb{N}$ and $l \in \mathbb{N}_0$,*

$$\begin{aligned}P^*[\,M_T \ge k,\, Z_T = k-l\,] &= \left(\frac{1-p^*}{p^*}\right)^l P^*[\,Z_T = k+l\,]\\ &= \left(\frac{p^*}{1-p^*}\right)^k P^*[\,Z_T = -k-l\,],\end{aligned}$$

and

$$\begin{aligned}&P^*[\,M_T = k,\, Z_T = k-l\,]\\&\quad = \left(\frac{1-p^*}{p^*}\right)^l P^*[\,Z_T = k+l\,] - \left(\frac{1-p^*}{p^*}\right)^{l+1} P^*[\,Z_T = k+l+2\,]\\&\quad = \left(\frac{p^*}{1-p^*}\right)^k P^*[\,Z_T = -k-l\,] - \left(\frac{p^*}{1-p^*}\right)^{k+1} P^*[\,Z_T = -k-l-2\,].\end{aligned}$$

Proof. We show first that the density of P^* with respect to $\mathbb{P}$ is given by

$$\frac{dP^*}{d\mathbb{P}} = 2^T \cdot (p^*)^{\frac{T+Z_T}{2}}(1-p^*)^{\frac{T-Z_T}{2}}. \tag{5.35}$$

Indeed, P^* puts the weight

$$P^*[\{\omega\}] = (p^*)^k (1-p^*)^{T-k}$$

to each $\omega = (y_1, \dots, y_T) \in \Omega$ that contains exactly k components with $y_i = +1$. But for such an ω, we have $Z_T(\omega) = k - (T-k) = 2k - T$, and our formula follows.

From the density formula (5.35), we get

$$P^*[M_T \geq k \text{ and } Z_T = k - l] = 2^T (p^*)^{\frac{T+k-l}{2}} (1-p^*)^{\frac{T+l-k}{2}} \mathbb{P}[M_T \geq k \text{ and } Z_T = k - l].$$

Applying the reflection principle and using again the density formula, we see that the probability term on the right is equal to

$$\mathbb{P}[Z_T = k + l] = 2^{-T} (p^*)^{-\frac{T+k+l}{2}} (1-p^*)^{-\frac{T-k-l}{2}} P^*[Z_T = k + l],$$

which gives the first identity. The proof of the second one is analogous. The second set of identities now follows as in (5.33). □

Remark 5.50. If we assume once again that Z is defined up to time $T+1$, then the arguments from the proof of Lemma 5.49 applied to (5.34) yield the alternative representation

$$\begin{aligned} P^*[M_T = k, Z_T = k - l] &= \frac{1}{p^*} \cdot \left(\frac{1-p^*}{p^*}\right)^l \cdot \frac{k+l+1}{T+1} P^*[Z_{T+1} = 1 + k + l] \\ &= \frac{1}{1-p^*} \cdot \left(\frac{p^*}{1-p^*}\right)^k \cdot \frac{k+l+1}{T+1} P^*[Z_{T+1} = -1 - k - l]. \end{aligned} \tag{5.36}$$

◇

Example 5.51 (Up-and-in call option). Consider an up-and-in call option with payoff

$$C_{\text{u\&i}}^{\text{call}} = \begin{cases} (S_T - K)^+ & \text{if } \max_{0 \leq t \leq T} S_t \geq B, \\ 0 & \text{otherwise,} \end{cases}$$

where $B > S_0 \vee K$ denotes a given barrier, and where $K > 0$ is the strike price. Our aim is to compute the arbitrage-free price

$$\pi(C_{\text{u\&i}}^{\text{call}}) = \frac{1}{(1+r)^T} E^*[C_{\text{u\&i}}^{\text{call}}].$$

Clearly,

$$\begin{aligned} E^*[C_{\text{u\&i}}^{\text{call}}] &= E^*\Big[(S_T - K)^+;\ \max_{0 \leq t \leq T} S_t \geq B\Big] \\ &= E^*[(S_T - K)^+;\ S_T \geq B] + E^*\Big[(S_T - K)^+;\ \max_{0 \leq t \leq T} S_t \geq B,\ S_T < B\Big]. \end{aligned}$$

The first expectation on the right can be computed explicitly in terms of the binomial distribution. Thus, it remains to compute the second expectation, which we denote by I. To this end, we may assume without loss of generality that B lies within the range of possible asset prices, that is, there exists some $k \in \mathbb{N}$ such that $B = S_0\hat{b}^k$. Then, by Lemma 5.49,

$$\begin{aligned}
I &= \sum_{l\geq 1} E^*[\,(S_T - K)^+;\, M_T \geq k,\, Z_T = k - l\,] \\
&= \sum_{l\geq 1} (S_0\hat{b}^{k-l} - K)^+ P^*[\,M_T \geq k,\, Z_T = k - l\,] \\
&= \sum_{l\geq 1} (S_0\hat{b}^{k-l} - K)^+ \left(\frac{p^*}{1-p^*}\right)^k P^*[\,Z_T = -k - l\,] \\
&= \left(\frac{p^*}{1-p^*}\right)^k \hat{b}^{2k} \sum_{l\geq 1} (S_0\hat{b}^{-k-l} - \widetilde{K})^+ P^*[\,Z_T = -k - l\,] \\
&= \left(\frac{p^*}{1-p^*}\right)^k \left(\frac{B}{S_0}\right)^2 E^*[\,(S_T - \widetilde{K})^+;\, S_T < \widetilde{B}\,],
\end{aligned}$$

where

$$\widetilde{K} = K\hat{b}^{-2k} = K\left(\frac{S_0}{B}\right)^2 \quad \text{and} \quad \widetilde{B} := \frac{S_0^2}{B}.$$

Hence, we obtain the formula

$$\pi(C^{\text{call}}_{\text{u\&i}}) = \frac{1}{(1+r)^T}\left(E^*[\,(S_T - K)^+;\, S_T \geq B\,] + \left(\frac{p^*}{1-p^*}\right)^k \left(\frac{B}{S_0}\right)^2 E^*[\,(S_T - \widetilde{K})^+;\, S_T < \widetilde{B}\,]\right).$$

Both expectations on the right now only involve the binomial distribution with parameters p^* and T. They can be computed as in Example 5.42, and so we get the explicit formula

$$\begin{aligned}
\pi(C^{\text{call}}_{\text{u\&i}}) = \frac{1}{(1+r)^T}\Bigg[&\sum_{n=0}^{n_k} (S_0\hat{b}^{T-2n} - K)^+ (p^*)^{T-n}(1-p^*)^n \binom{T}{T-n} + \\
&+ \left(\frac{p^*}{1-p^*}\right)^k \left(\frac{B}{S_0}\right)^2 \sum_{n=m_k}^{T} (S_0\hat{b}^{T-2n} - \widetilde{K})^+ (p^*)^{T-n}(1-p^*)^n \binom{T}{T-n}\Bigg],
\end{aligned}$$

where n_k is the largest integer n such that $T - 2n \geq k$ and m_k is the smallest integer m such that $2m > T + k$. ◇

Example 5.52 (Up-and-out call option). Consider an up-and-out call option with payoff

$$C^{\text{call}}_{\text{u\&o}} = \begin{cases} 0 & \text{if } \max\limits_{0\leq t\leq T} S_t \geq B, \\ (S_T - K)^+ & \text{otherwise,} \end{cases}$$

where $K > 0$ is the strike price and $B > S_0 \vee K$ is an upper barrier for the stock price. As in the preceding example, we assume that $B = S_0(1+b)^k$ for some $k \in \mathbb{N}$. Let

$$C^{\text{call}} := (S_T - K)^+$$

denote the corresponding "plain vanilla call," whose arbitrage-free price is given by

$$\pi(C^{\text{call}}) = \frac{1}{(1+r)^T} E^*[\,(S_T - K)^+\,].$$

Since $C^{\text{call}} = C^{\text{call}}_{\text{u\&o}} + C^{\text{call}}_{\text{u\&i}}$, we get from Example 5.51 that

$$\begin{aligned}\pi(C^{\text{call}}_{\text{u\&o}}) &= \pi(C^{\text{call}}) - \pi(C^{\text{call}}_{\text{u\&i}})\\ &= \frac{1}{(1+r)^T}\left[E^*[\,(S_T - K)^+;\, S_T < B\,] - \left(\frac{p^*}{1-p^*}\right)^k \left(\frac{B}{S_0}\right)^2 E^*[\,(S_T - \widetilde{K})^+;\, S_T < \widetilde{B}\,]\right],\end{aligned}$$

where $\widetilde{K} = KS_0^2/B^2$ and $\widetilde{B} := S_0^2/B$. These expectations can be computed as in Example 5.51. ◇

Exercise 5.6.2. Derive a formula for the arbitrage-free price of a down-and-in put option with payoff

$$C^{\text{put}}_{\text{d\&i}} = \begin{cases} 0 & \text{if } \min\limits_{0\le t\le T} S_t > B,\\ (K - S_T)^+ & \text{otherwise,}\end{cases}$$

where $K > 0$ is the strike price and $B < S_0$ is a lower barrier for the stock price. Then compute the price of the option for the following specific parameter values:

$$T = 3, \quad S_0 = 100, \quad a = -0.2, \quad r = 0.05, \quad B = 81, \quad K = 90.$$ ◇

In the following example, we compute the price of a lookback put option.

Example 5.53 (Lookback put option). A lookback put option corresponds to the contingent claim

$$C^{\text{put}}_{\max} := \max_{0\le t\le T} S_t - S_T\,;$$

see Example 5.23. In the CRR model, the discounted arbitrage-free price of $C^{\text{put}}_{\max}$ is given by

$$\pi(C^{\text{put}}_{\max}) = \frac{1}{(1+r)^T} E^*\Big[\max_{0\le t\le T} S_t\Big] - S_0.$$

The expectation of the maximum can be computed as

$$E^*\Big[\max_{0\le t\le T} S_t\Big] = S_0 \sum_{k=0}^{T} \hat{b}^k\, P^*[\,M_T = k\,].$$

The formula (5.36) yields

$$\begin{aligned} P^*[M_T = k] &= \sum_{l\geq 0} P^*[M_T = k, Z_T = k-l] \\ &= \sum_{l\geq 0} \frac{1}{1-p^*}\left(\frac{p^*}{1-p^*}\right)^k \frac{k+l+1}{T+1} P^*[Z_{T+1} = -1-k-l] \\ &= \frac{1}{1-p^*}\left(\frac{p^*}{1-p^*}\right)^k \frac{1}{T+1} E^*[-Z_{T+1}; Z_{T+1} \leq -1-k]. \end{aligned}$$

Thus, we arrive at the expression

$$\begin{aligned} &\pi(C^{\text{put}}_{\text{max}}) + S_0 \\ &= \frac{S_0}{(1+r)^T(1-p^*)(T+1)} \sum_{k=0}^{T} \hat{b}^k \left(\frac{p^*}{1-p^*}\right)^k E^*[-Z_{T+1}; Z_{T+1} \leq -1-k]. \end{aligned}$$

As before, one can give explicit formulas for the expectations occurring on the right. ◇

Exercise 5.6.3. Derive a formula for the price of a lookback call option with payoff

$$S_T - \min_{0\leq t\leq T} S_t.$$ ◇

5.7 Convergence to the Black–Scholes price

In practice, a huge number of trading periods may occur between the current time $t = 0$ and the maturity T of a European contingent claim. Thus, the computation of option prices in terms of some martingale measure may become rather elaborate. On the other hand, one can hope that the pricing formulas in discrete time converge to a transparent limit as the number of intermediate trading periods grows larger and larger. In this section, we will formulate conditions under which such a convergence occurs.

Throughout this section, T will *not* denote the *number* of trading periods in a fixed discrete-time market model but rather a physical *date*. The time interval $[0, T]$ will be divided into N equidistant time steps $\frac{T}{N}, \frac{2T}{N}, \dots, \frac{NT}{N}$, and the date $\frac{kT}{N}$ will correspond to the kth trading period of an arbitrage-free market model. For simplicity, we will assume that each market model contains a riskless bond and just one risky asset. In the Nth approximation, the risky asset will be denoted by $S^{(N)}$, and the riskless bond will be defined by a constant interest rate $r_N > -1$.

The question is whether the prices of contingent claims in the approximating market models converge as N tends to infinity. Since the terminal values of the riskless bonds should converge, we assume that

$$\lim_{N\uparrow\infty} (1 + r_N)^N = e^{rT}, \tag{5.37}$$

where r is a finite constant. By taking logarithms on both sides of (5.37), one sees that this condition is in fact equivalent to the following one:

$$\lim_{N\uparrow\infty} N\, r_N = r\, T.$$

Let us now consider the risky assets. We assume that the initial prices $S_0^{(N)}$ do not depend on N, that is, $S_0^{(N)} = S_0$ for some constant $S_0 > 0$. The prices $S_k^{(N)}$ are random variables on some probability space $(\Omega_N, \mathscr{F}^{(N)}, P_N^*)$, where P_N^* is a risk-neutral measure for each approximating market model, that is, the discounted price process

$$X_k^{(N)} := \frac{S_k^{(N)}}{(1+r_N)^k}, \qquad k = 0, \dots, N,$$

is a P_N^*-martingale with respect to the filtration $\mathscr{F}_k^{(N)} := \sigma(S_1^{(N)}, \dots, S_k^{(N)})$. Our remaining conditions will be stated in terms of the returns

$$R_k^{(N)} := \frac{S_k^{(N)} - S_{k-1}^{(N)}}{S_{k-1}^{(N)}}, \qquad k = 1, \dots, N.$$

First, we assume that, for each N, the random variables $R_1^{(N)}, \dots, R_N^{(N)}$ are independent under P_N^* and satisfy

$$-1 < \alpha_N \le R_k^{(N)} \le \beta_N, \quad k = 1, \dots, N,$$

for constants α_N and β_N such that

$$\lim_{N\uparrow\infty} \alpha_N = \lim_{N\uparrow\infty} \beta_N = 0.$$

Second, we assume that the variances $\mathrm{var}_N(R_k^{(N)})$ under P_N^* are such that

$$\sigma_N^2 := \frac{1}{T} \sum_{k=1}^N \mathrm{var}_N(R_k^{(N)}) \longrightarrow \sigma^2 \in (0, \infty).$$

The following result can be regarded as a multiplicative version of the *central limit theorem.*

Theorem 5.54. *Under the above assumptions, the distributions of $S_N^{(N)}$ under P_N^* converge weakly to the log-normal distribution with parameters $\log S_0 + rT - \frac{1}{2}\sigma^2 T$ and $\sigma\sqrt{T}$, that is, to the distribution of*

$$S_T := S_0 \exp\Big(\sigma\, W_T + \Big(r - \frac{1}{2}\sigma^2\Big) T\Big), \tag{5.38}$$

where W_T has a centered normal law $N(0, T)$ with variance T.

Proof. We may assume without loss of generality that $S_0 = 1$. Consider the Taylor expansion

$$\log(1+x) = x - \frac{1}{2}x^2 + \rho(x)\,x^2, \tag{5.39}$$

where the remainder term ρ is such that

$$|\rho(x)| \le \delta(\alpha,\beta) \quad \text{for } -1 < \alpha \le x \le \beta,$$

and where $\delta(\alpha,\beta) \to 0$ for $\alpha,\beta \to 0$. Applied to

$$S_N^{(N)} = \prod_{k=1}^{N}(1 + R_k^{(N)}),$$

this yields

$$\log S_N^{(N)} = \sum_{k=1}^{N}\Big(R_k^{(N)} - \frac{1}{2}(R_k^{(N)})^2\Big) + \Delta_N,$$

where

$$|\Delta_N| \le \delta(\alpha_N,\beta_N)\sum_{k=1}^{N}(R_k^{(N)})^2.$$

Since P_N^* is a martingale measure, we have $E_N^*[\,R_k^{(N)}\,] = r_N$, and it follows that

$$E_N^*[\,|\Delta_N|\,] \le \delta(\alpha_N,\beta_N)\sum_{k=1}^{N}(\mathrm{var}_N(R_k^{(N)}) + r_N^2) \longrightarrow 0.$$

In particular, $\Delta_N \to 0$ in probability, and the corresponding laws converge weakly to the Dirac measure δ_0. In view of Slutsky's theorem, as stated in Theorem G.8, it suffices to show that the distributions of

$$Z_N := \sum_{k=1}^{N}\Big(R_k^{(N)} - \frac{1}{2}(R_k^{(N)})^2\Big) =: \sum_{k=1}^{N} Y_k^{(N)}$$

under P_N^* converge weakly to the normal law $N(rT - \frac{1}{2}\sigma^2 T, \sigma^2 T)$. To this end, we will check that the conditions of the central limit theorem in the form of Theorem G.2 are satisfied.

Note that

$$\max_{1\le k\le N}|Y_k^{(N)}| \le \gamma_N + \frac{1}{2}\gamma_N^2 \longrightarrow 0$$

for $\gamma_N := |\alpha_N| \vee |\beta_N|$, and that

$$E_N^*[Z_N] = Nr_N - \frac{1}{2}(\sigma_N^2 T + Nr_N^2) \longrightarrow rT - \frac{1}{2}\sigma^2 T.$$

Finally,

$$\mathrm{var}_N(Z_N) \longrightarrow \sigma^2 T,$$

since for $p > 2$

$$\sum_{k=1}^{N} E_N^*[|R_k^{(N)}|^p] \le \gamma_N^{p-2} \sum_{k=1}^{N} E_N^*[(R_k^{(N)})^2] \longrightarrow 0.$$

Thus, the conditions of Theorem G.2 are satisfied. □

Remark 5.55. The assumption of independent returns in Theorem 5.54 can be relaxed. Instead of Theorem G.2, we can apply a *central limit theorem for martingales* under suitable assumptions on the behavior of the *conditional* variances

$$\mathrm{var}_N(R_k^{(n)} \mid \mathscr{F}_{k-1});$$

for details see, for example, Section 9.3 of [64]. ◇

Example 5.56. Suppose the approximating model in the Nth stage is a CRR model with interest rate

$$r_N = \frac{rT}{N},$$

and with returns $R_k^{(N)}$ that can take the two possible values a_N and b_N; see Section 5.5. We assume that

$$\hat{a}_N = 1 + a_N = e^{-\sigma\sqrt{T/N}} \quad \text{and} \quad \hat{b}_N = 1 + b_N = e^{\sigma\sqrt{T/N}}$$

for some given $\sigma > 0$. Since

$$\sqrt{N} r_N \longrightarrow 0, \quad \sqrt{N} a_N \longrightarrow -\sigma\sqrt{T}, \quad \sqrt{N} b_N \longrightarrow \sigma\sqrt{T} \quad \text{as } N \uparrow \infty, \tag{5.40}$$

we have $a_N < r_N < b_N$ for large enough N. Theorem 5.39 yields that the Nth model is arbitrage-free and admits a unique equivalent martingale measure P_N^*. The measure P_N^* is characterized by

$$P_N^*[R_k^{(N)} = b_N] =: p_N^* = \frac{r_N - a_N}{b_N - a_N},$$

and we obtain from (5.40) that

$$\lim_{N\uparrow\infty} p_N^* = \frac{1}{2}.$$

Moreover, $E_N^*[\,R_k^{(N)}\,] = r_N$, and we get

$$\sum_{k=1}^{N} \operatorname{var}_N(R_k^{(N)}) = N(p_N^* b_N^2 + (1-p_N^*)a_N^2 - r_N^2) \longrightarrow \sigma^2 T$$

as $N \uparrow \infty$. Hence, the assumptions of Theorem 5.54 are satisfied. ◇

Let us consider a derivative that is defined in terms of a function $f \geq 0$ of the risky asset's terminal value. In each approximating model, this corresponds to a contingent claim

$$C^{(N)} = f(S_N^{(N)}).$$

Corollary 5.57. *If f is bounded and continuous, the arbitrage-free prices of $C^{(N)}$ calculated under P_N^* converge to a discounted expectation with respect to a log-normal distribution, which is often called the Black–Scholes price. More precisely,*

$$\begin{aligned}\lim_{N\uparrow\infty} E_N^*\Big[\,\frac{C^{(N)}}{(1+r_N)^N}\,\Big] &= \frac{e^{-rT}}{\sqrt{2\pi}} \int_{-\infty}^{\infty} f(S_0 e^{\sigma\sqrt{T}y+rT-\sigma^2T/2}) e^{-y^2/2}\,dy \\ &= e^{-rT} E^*[\,f(S_T)],\end{aligned} \tag{5.41}$$

where S_T is a random variable on a probability space $(\Omega, \mathscr{F}, P^*)$ that has the form (5.38).

We will see in Remark 5.65 how the probability measure P^* in (5.41) can be chosen in such a way that it becomes the unique equivalent martingale measure in a continuous-time financial market model. Corollary 5.57 applies in particular to the choice $f(x) = (K-x)^+$ corresponding to a European put option with strike K. Since the put-call parity

$$E_N^*\Big[\,\frac{(S_N^{(N)}-K)^+}{(1+r_N)^N}\,\Big] = E_N^*\Big[\,\frac{(K-S_N^{(N)})^+}{(1+r_N)^N}\,\Big] + S_0 - \frac{K}{(1+r_N)^N}$$

holds for each N and in the limit $N \uparrow \infty$, the convergence (5.41) is also true for a European call option with the unbounded payoff profile $f(x) = (x-K)^+$.

Example 5.58 (Black–Scholes formula for the price of a call option). The limit of the arbitrage-free prices of $C^{(N)} = (S_N^{(N)} - K)^+$ is given by $v(S_0, T)$, where

$$v(x,T) = \frac{e^{-rT}}{\sqrt{2\pi}} \int_{-\infty}^{\infty} (xe^{\sigma\sqrt{T}y+rT-\sigma^2T/2} - K)^+ e^{-y^2/2}\,dy.$$

The integrand on the right vanishes for

$$y \leq -\frac{\log \frac{x}{K} + (r - \frac{1}{2}\sigma^2)T}{\sigma\sqrt{T}} =: -d_-(x,T) =: -d_-.$$

Let us also define

$$d_+ := d_+(x,T) := d_-(x,T) + \sigma\sqrt{T} = \frac{\log \frac{x}{K} + (r + \frac{1}{2}\sigma^2)T}{\sigma\sqrt{T}},$$

and let us denote by $\Phi(x) = (2\pi)^{-1/2}\int_{-\infty}^{x} e^{-y^2/2}\,dy$ the cumulative distribution function of the standard normal distribution. Then

$$v(x,T) = \frac{x}{\sqrt{2\pi}} \int_{-d_-}^{+\infty} e^{-(y-\sigma\sqrt{T})^2/2}\,dy - e^{-rT}K(1 - \Phi(-d_-)),$$

and we arrive at the *Black–Scholes formula* for the price of a European call option with strike K and maturity T:

$$v(x,T) = x\,\Phi(d_+(x,T)) - e^{-rT}K\Phi(d_-(x,T)). \tag{5.42}$$

See Figure 5.5 for the plot of the function $v(x,t)$. ◇

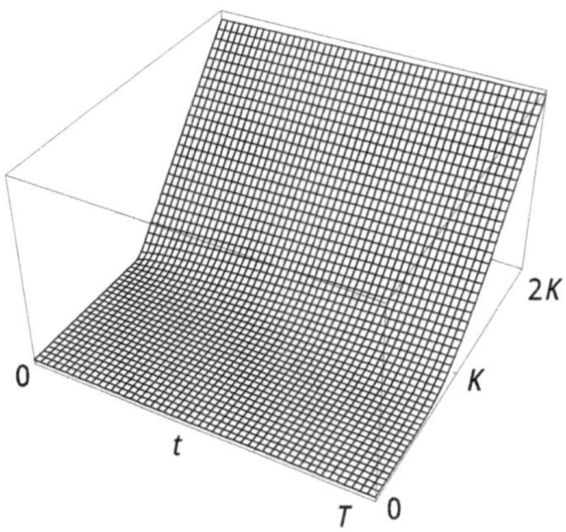

Figure 5.5: The Black–Scholes price $v(x,t)$ of a European call option $(S_T - K)^+$ plotted as a function of the initial spot price $x = S_0$ and the time to maturity t.

Remark 5.59. For fixed x and T, the Black–Scholes price of a European call option increases to the upper arbitrage bound x as $\sigma \uparrow \infty$. In the limit $\sigma \downarrow 0$, we obtain the lower arbitrage bound $(x - e^{-rT}K)^+$; see Remark 1.39. ◇

The following proposition gives a criterion for the convergence (5.41) in case f is not necessarily bounded and continuous. It applies in particular to $f(x) = (x-K)^+$, and so we get an alternative proof for the convergence of call option prices to the Black–Scholes price.

Proposition 5.60. *Let $f : (0,\infty) \to \mathbb{R}$ be measurable, continuous a. e., and such that $|f(x)| \le c\,(1+x)^q$ for some $c \ge 0$ and $0 \le q < 2$. Then*

$$E_N^*[\, f(S_N^{(N)})\,] \longrightarrow E^*[\, f(S_T)],$$

where S_T has the form (5.38) *under P^*.*

Proof. Let us note first that by the Taylor expansion (5.39)

$$\begin{aligned}\log E_N^*\Big[\Big(\frac{S_N^{(N)}}{S_0}\Big)^2\Big] &= \log \prod_{k=1}^N \big(\operatorname{var}_N(1+R_k^{(N)}) + E_N^*[\,1+R_k^{(N)}]^2\big)\\ &= \sum_{k=1}^N \log\big(\operatorname{var}_N(R_k^{(N)}) + (1+r_N)^2\big)\\ &\le \sigma_N^2 T + 2N\, r_N + N\, r_N^2 + \widetilde{c}\,\sum_{k=1}^N \big(\operatorname{var}_N(R_k^{(N)}) + 2|r_N| + r_N^2\big)^2\end{aligned}$$

for a finite constant $\widetilde{c}$. Thus,

$$\sup_N E_N^*[\,(S_N^{(N)})^2\,] < \infty.$$

With this property established, the assertion follows immediately from Theorem 5.54 and the Corollaries G.13 and G.14, but we also give the following more elementary proof. To this end, we may assume that $q > 0$, and we define $p := 2/q > 1$. Then

$$\sup_N E_N^*[\,|f(S_N^{(N)})|^p\,] \le c^p \sup_N E_N^*[\,(1+S_N^{(N)})^2\,] < \infty,$$

and the assertion follows from Lemma 5.61 below. □

Lemma 5.61. *Suppose $(\mu_N)_{N\in\mathbb{N}}$ is a sequence of probability measures on $\mathbb{R}$ converging weakly to μ. If f is a measurable and μ-a. e. continuous function on $\mathbb{R}$ such that*

$$c := \sup_{N\in\mathbb{N}} \int |f|^p \, d\mu_N < \infty \quad \textit{for some } p > 1,$$

then

$$\int f\, d\mu_N \longrightarrow \int f\, d\mu.$$

Proof. We may assume without loss of generality that $f \ge 0$. Then $f_k := f \wedge k$ is a bounded and μ-a. e. continuous function for each $k > 0$. Clearly,

$$\int f\,d\mu_N = \int f_k\,d\mu_N + \int (f-k)^+\,d\mu_N.$$

Due to part (e) of the portmanteau theorem in the form of Theorem G.4, the first integral on the right converges to $\int f_k\,d\mu$ as $N \uparrow \infty$. Let us consider the second term on the right:

$$\int (f-k)^+\,d\mu_N \le \int_{\{f>k\}} f\,d\mu_N \le \frac{1}{k^{p-1}} \int f^{p-1} f\,d\mu_N \le \frac{c}{k^{p-1}},$$

uniformly in N. Hence,

$$\begin{aligned}\int f_k\,d\mu &= \lim_{N\uparrow\infty} \int f_k\,d\mu_N \le \liminf_{N\uparrow\infty} \int f\,d\mu_N \\ &\le \limsup_{N\uparrow\infty} \int f\,d\mu_N \le \int f_k\,d\mu + \frac{c}{k^{p-1}}.\end{aligned}$$

Letting $k \uparrow \infty$, we have $\int f_k\,d\mu \nearrow \int f\,d\mu$, and convergence follows. □

Let us now continue the discussion of the Black–Scholes price of a European call option where $f(x) = (x-K)^+$. We are particularly interested how it depends on the various model parameters. The dependence on the spot price $S_0 = x$ can be analyzed via the x-derivatives of the function $v(t,x)$ appearing in the Black–Scholes formula (5.42). The first derivative

$$\Delta(x,t) := \frac{\partial}{\partial x} v(x,t) = \Phi\big(d_+(x,t)\big) \tag{5.43}$$

is called the option's *Delta*; see Figure 5.6. In analogy to the formula for the hedging strategy in the binomial model obtained in Proposition 5.44, $\Delta(x,t)$ determines the "Delta hedging portfolio" needed for a replication of the call option in continuous time, as explained in (5.52) below.

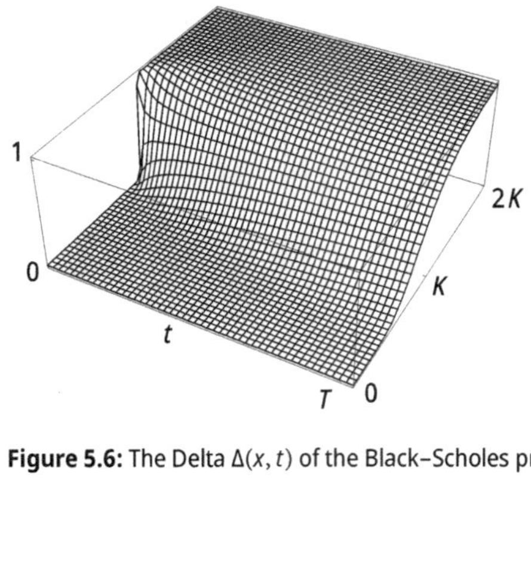

Figure 5.6: The Delta $\Delta(x,t)$ of the Black–Scholes price of a European call option.

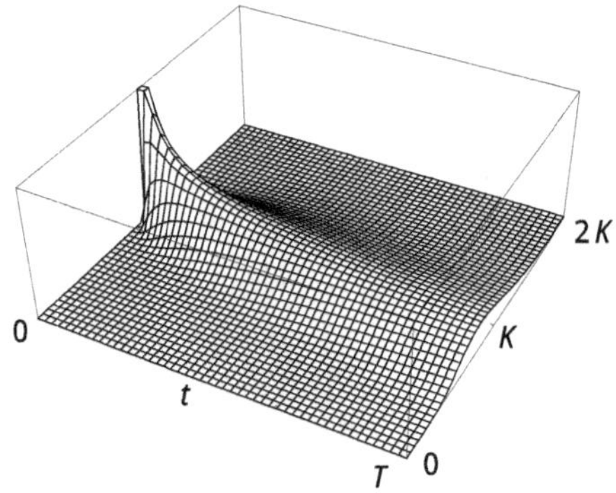

Figure 5.7: The option's Gamma $\Gamma(x,t)$.

The *Gamma* of the call option is given by

$$\Gamma(x,t) := \frac{\partial}{\partial x}\Delta(x,t) = \frac{\partial^2}{\partial x^2}v(x,t) = \varphi(d_+(x,t))\frac{1}{x\sigma\sqrt{t}}; \tag{5.44}$$

see Figure 5.7. Here, $\varphi(x) = \Phi'(x) = e^{-x^2/2}/\sqrt{2\pi}$ stands as usual for the density of the standard normal distribution. Large Gamma values occur in regions where the Delta changes rapidly, corresponding to the need for frequent readjustments of the Delta hedging portfolio. Note that Γ is always strictly positive. It follows that $v(x,t)$ is a strictly convex function of its first argument.

Exercise 5.7.1. Prove the formulas (5.43) and (5.44) for Delta and Gamma of a European call option. ◇

Remark 5.62. On the one hand, $0 \le \Delta(x,t) \le 1$ implies that

$$\big|v(x,t) - v(y,t)\big| \le |x-y|.$$

Thus, the *total change* of the option values is always less than a corresponding change in the asset prices. On the other hand, the strict convexity of $x \mapsto v(x,t)$ together with (A.5) yields that for $t > 0$ and $z > y$,

$$\frac{v(z,t)-v(y,t)}{z-y} > \frac{v(y,t)-v(0,t)}{y-0} = \frac{v(y,t)}{y},$$

and hence

$$\frac{v(z,t)-v(y,t)}{v(y,t)} > \frac{z-y}{y}.$$

Similarly, one obtains

$$\frac{v(x,t)-v(y,t)}{v(y,t)} < \frac{x-y}{y}$$

for $x < y$. Thus, the *relative change* of option prices is larger in absolute value than the relative change of asset values. This fact can be interpreted as the *leverage effect* for call options; see also Example 1.45. ◇

Another important parameter is the *Theta*

$$\Theta(x,t) := \frac{\partial}{\partial t}v(x,t) = \frac{x\sigma}{2\sqrt{t}}\varphi(d_+(x,t)) + Kr\,e^{-rt}\Phi(d_-(x,t))\,; \tag{5.45}$$

see Figure 5.8. The fact $\Theta > 0$ corresponds to our general observation, made in Example 5.35 that arbitrage-free prices of European call options are typically increasing functions of the maturity.

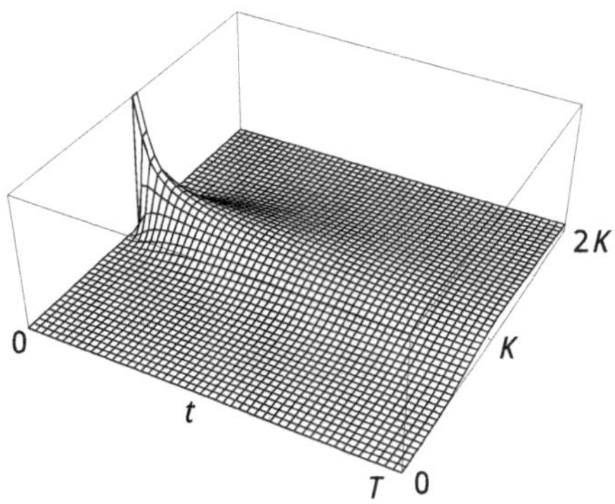

Figure 5.8: The Theta $\Theta(x,t)$.

Exercise 5.7.2. Prove the formula (5.45) for the Theta of a European call option. Then show that the parameters Δ, Γ, and Θ are related by the equation

$$\Theta(x,t) = rx\,\Delta(x,t) + \frac{1}{2}\sigma^2x^2\,\Gamma(x,t) - r\,v(x,t) \tag{5.46}$$

if $t > 0$. ◇

Equation (5.46) implies that, for $(x,t) \in (0,\infty)\times(0,\infty)$, the function v solves the partial differential equation

$$\frac{\partial v}{\partial t} = rx\frac{\partial v}{\partial x} + \frac{1}{2}\sigma^2x^2\frac{\partial^2 v}{\partial x^2} - rv, \tag{5.47}$$

often called the *Black–Scholes equation*. Since

$$v(x,t) \longrightarrow f(x) = (x-K)^+ \quad \text{as } t \downarrow 0, \tag{5.48}$$

$v(x,t)$ is a solution to the Cauchy problem defined via (5.47) and (5.48). This fact is not limited to call options, it remains valid for all reasonable payoff profiles f.

Proposition 5.63. *Let f be a continuous function on $(0, \infty)$ such that $|f(x)| \le c(1+x)^p$ for some c, $p \ge 0$, and define*

$$u(x,t) := e^{-rt} E^*[\,f(S_t)\,] = \frac{e^{-rt}}{\sqrt{2\pi}} \int_{-\infty}^{\infty} f(xe^{\sigma\sqrt{t}y + rt - \sigma^2 t/2}) e^{-y^2/2}\, dy,$$

where $S_t = x\exp(\sigma W_t + rt - \sigma^2 t/2)$ and W_t has law $N(0,t)$ under P^*. Then u solves the Cauchy problem defined by the Black–Scholes equation (5.47) *and the initial condition $\lim_{t\downarrow 0} u(x,t) = f(x)$, locally uniformly in x.*

The proof of Proposition 5.63 is the content of the next exercise.

Exercise 5.7.3. In the context Proposition 5.63, use the formula (2.31) for the density of a log-normally distributed random variable to show that

$$E^*[\,f(S_t)\,] = \int_0^{\infty} \frac{1}{y\sigma\sqrt{t}} \varphi\Big(\frac{\log y - rt + \sigma^2 t/2 - \log x}{\sigma\sqrt{t}}\Big) f(y)\, dy,$$

where $\varphi(x) = e^{-x^2/2}/\sqrt{2\pi}$. Then verify the validity of (5.47) by differentiating under the integral. Use the bound $|f(x)| \le c(1+x)^p$ for some c, $p \ge 0$ to justify the interchange of differentiation and integration and to verify the initial condition $\lim_{t\downarrow 0} u(x,t) = f(x)$. ◇

Recall that the Black–Scholes price $v(S_0, T)$ was obtained as the expectation of the *discounted* payoff $e^{-rT}(S_T - K)^+$ under the measure P^*. Thus, at a first glance, it may come as a surprise that the *Rho* of the option,

$$\varrho(x,t) := \frac{\partial}{\partial r} v(x,t) = Kt\, e^{-rt}\, \Phi(d_-(x,t)), \tag{5.49}$$

is strictly positive, that is, the price is *increasing* in r; see Figure 5.9. Note, however, that the measure P^* depends itself on the interest rate r, since $E^*[\,e^{-rT} S_T\,] = S_0$. In a simple

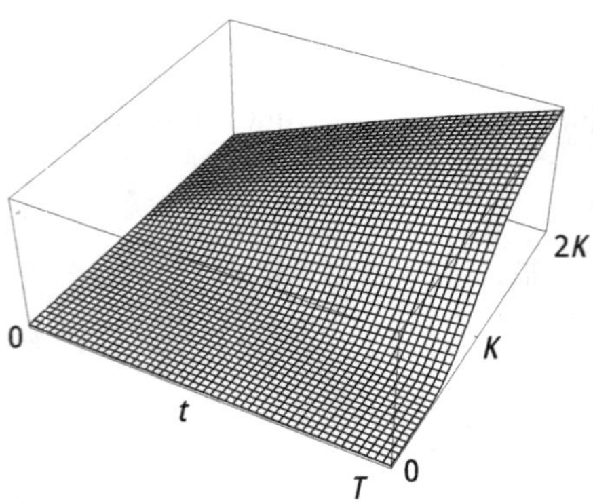

Figure 5.9: The Rho $\varrho(x,t)$ of a call option.

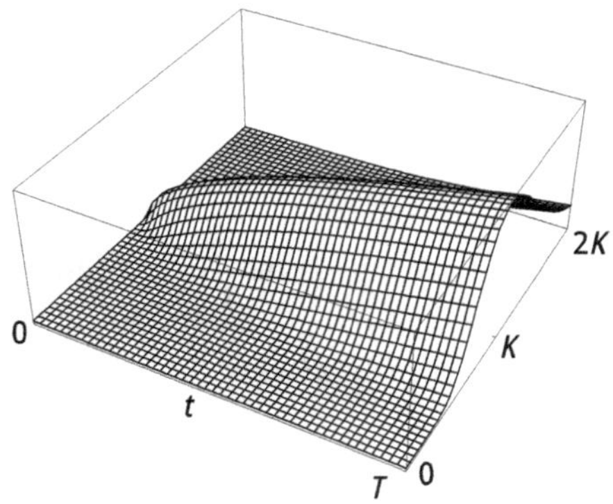

Figure 5.10: The Vega $\mathscr{V}(x,t)$.

one-period model, we have already seen this effect in Example 1.45. Intuitively, it is due to the fact that the replicating strategy involves a long position in the risky asset, which is financed by taking out a loan in the riskless bond. This loan becomes more expensive when interest rates rise, thus increasing the Black–Scholes price of the call option.

The parameter σ is called the *volatility*. As we have seen, the Black–Scholes price of a European call option is an increasing function of the volatility, and this is reflected in the strict positivity of

$$\mathscr{V}(x,t) := \frac{\partial}{\partial\sigma}v(x,t) = x\sqrt{t}\,\varphi(d_+(x,t)); \tag{5.50}$$

see Figure 5.10. The function $\mathscr{V}$ is often called the *Vega* of the call option price, and the functions Δ, Γ, Θ, ϱ, and $\mathscr{V}$ are usually called the *Greeks* (although "vega" does not correspond to a letter of the Greek alphabet).

Exercise 5.7.4. Prove the respective formulas (5.49) and (5.50) for Rho and Vega of a European call option. Then derive formulas for the option's *Vanna*,

$$\frac{\partial^2 v}{\partial x\partial\sigma} = \frac{\partial\Delta}{\partial\sigma} = \frac{\partial\mathscr{V}}{\partial x},$$

Charm,

$$\frac{\partial^2 v}{\partial x\partial\tau} = \frac{\partial\Delta}{\partial t} = \frac{\partial\Theta}{\partial x},$$

and the option's *Volga*, which is also called *Vomma*,

$$\frac{\partial^2 v}{\partial\sigma^2} = \frac{\partial\mathscr{V}}{\partial\sigma}. \qquad \diamond$$

Let us conclude this section with some informal comments on the dynamic picture in continuous time behind the convergence result in Theorem 5.54 and the pricing formulas in Example 5.58 and Proposition 5.60. The constant r is viewed as the interest rate of a risk-free savings account

$$S_t^0 = e^{rt}, \quad 0 \le t \le T.$$

The prices of the risky asset in each discrete-time model are considered as a *continuous* process $\widetilde{S}^{(N)} = (\widetilde{S}_t^{(N)})_{0\le t\le T}$, defined as $\widetilde{S}_t^{(N)} := S_k^{(N)}$ at the dates $t = \frac{kT}{N}$, and by linear interpolation in between. Theorem 5.54 shows that the distributions of $\widetilde{S}_t^{(N)}$ converge for each fixed t weakly to the distribution of

$$S_t = S_0 \exp\left(\sigma W_t + \left(r - \frac{1}{2}\sigma^2\right)t\right), \tag{5.51}$$

where W_t has a centered normal distribution with variance t. In fact, one can prove convergence in the much stronger sense of a functional central limit theorem: The laws of the processes $\widetilde{S}^{(N)}$, considered as $C[0,T]$-valued random variables on $(\Omega_N, \mathscr{F}^{(N)}, P_N^*)$, converge weakly to the law of a *geometric Brownian motion* $S = (S_t)_{0\le t\le T}$, where each S_t is of the form (5.51), and where the process $W = (W_t)_{0\le t\le T}$ is a standard *Brownian motion* or *Wiener process*. A Wiener process is characterized by the following properties:

- $W_0 = 0$ almost surely,
- $t \mapsto W_t$ is continuous,
- for each sequence $0 = t_0 < t_1 < \cdots < t_n = T$, the increments

$$W_{t_1} - W_{t_0}, \dots, W_{t_n} - W_{t_{n-1}}$$

are independent and have normal distributions $N(0, t_i - t_{i-1})$;

see, for example, [205]. This multiplicative version of a functional central limit theorem follows as above if we replace the classical central limit theorem by Donsker's invariance principle; for details, see, for example, [114]. Sample paths of Brownian motion and geometric Brownian motion can be found in Figures 5.11 and 5.12.

Geometric Brownian motion is the classical reference model in continuous-time mathematical finance. In order to describe the model more explicitly, we denote by

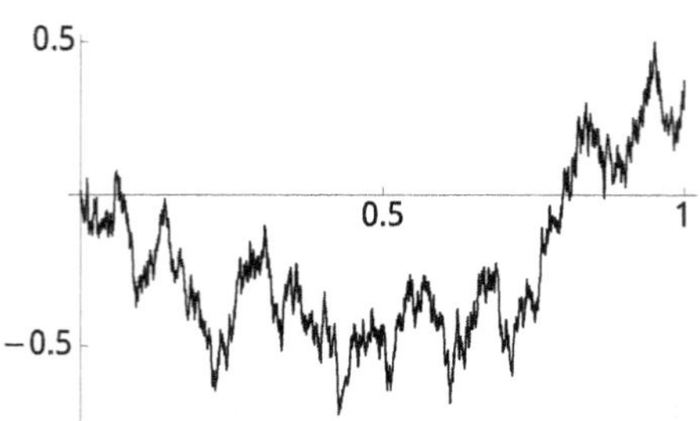

Figure 5.11: A sample path of Brownian motion.

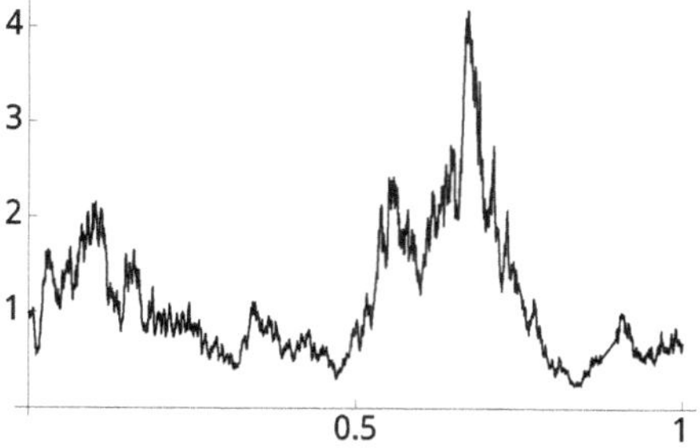

Figure 5.12: A sample path of geometric Brownian motion.

$W = (W_t)_{0 \le t \le T}$ the coordinate process on the canonical path space $\Omega = C[0, T]$, defined by $W_t(\omega) = \omega(t)$, and furthermore by $(\mathscr{F}_t)_{0 \le t \le T}$ the filtration given by $\mathscr{F}_t = \sigma(W_s;\ s \le t)$. There is exactly one probability measure $\mathbb{P}$ on $(\Omega, \mathscr{F}_T)$ such that W is a Wiener process under $\mathbb{P}$, and it is called the *Wiener measure*. Let us now model the price process of a risky asset as a geometric Brownian motion S defined by (5.51). The discounted price process

$$X_t := \frac{S_t}{e^{rt}} = S_0 e^{\sigma W_t - \sigma^2 t/2}, \quad 0 \le t \le T,$$

is a martingale under $\mathbb{P}$, since

$$\mathbb{E}[\, X_t \mid \mathscr{F}_s \,] = X_s \mathbb{E}[\, e^{\sigma(W_t - W_s) - \sigma^2 (t-s)/2} \,] = X_s$$

for $0 \le s \le t \le T$. In fact, $\mathbb{P}$ is the only probability measure equivalent to $\mathbb{P}$ with that property.

As in discrete time, uniqueness of the equivalent martingale measure implies completeness of the model. Let us sketch the construction of the replicating strategy for a given European option with reasonable payoff profile $f(S_T)$, for example a call option with strike K. At time t, the price of the asset is $S_t(\omega)$, the remaining time to maturity is $T - t$, and the discounted price of the option is given by

$$V_t(\omega) = e^{-rt} u(S_t(\omega), T - t),$$

where u is the function defined in Proposition 5.63. The process $V = (V_t)_{0 \le t \le T}$ can be viewed as the value process of the trading strategy $\overline{\xi} = (\xi^0, \xi)$ defined by

$$\xi_t = \Delta(S_t, T - t), \quad \xi_t^0 = e^{-rt} u(S_t, T - t) - \xi_t X_t, \tag{5.52}$$

where $\Delta = \partial u / \partial x$ is the option's Delta. Indeed, if we view ξ as the number of shares in the risky asset S and ξ^0 as the number of shares in the risk-free savings account $S_t^0 = e^{rt}$, then the value of the resulting portfolio in units of the numéraire is given by

$$V_t = \xi_t \cdot X_t + \xi_t^0 = e^{-rt}(\xi_t \cdot S_t + \xi_t^0 \cdot S_t^0).$$

The strategy replicates the option since

$$V_T := \lim_{t \uparrow T} e^{-rt} u(S_t, T-t) = e^{-rT} f(S_T) = \frac{f(S_T)}{S_T^0},$$

due to Proposition 5.63. Moreover, its initial cost is given by the Black–Scholes price

$$V_0 = u(S_0, T) = e^{-rT} \mathbb{E}[f(S_T)] = \frac{e^{-rT}}{\sqrt{2\pi}} \int_{-\infty}^{\infty} f(xe^{\sigma\sqrt{T}y + rT - \sigma^2 T/2}) e^{-y^2/2}\, dy.$$

It remains to show that the strategy is *self-financing* in the sense that changes in the portfolio value are only due to price changes in the underlying assets and do not require any additional capital. To this end, we use *Itô's formula*

$$dF(W_t, t) = \frac{\partial F}{\partial x}(W_t, t)\, dW_t + \left(\frac{1}{2}\frac{\partial^2 F}{\partial x^2} + \frac{\partial F}{\partial t}\right)(W_t, t)\, dt;$$

for a smooth function F; see, for example, [205] or, for a strictly pathwise approach, [138]. Applied to the function $F(x,t) = \exp(\sigma x + rt - \sigma^2 t/2)$, it shows that the price process S satisfies the *stochastic differential equation*

$$dS_t = \sigma S_t\, dW_t + r S_t\, dt. \tag{5.53}$$

Thus, the infinitesimal return dS_t/S_t is the sum of the safe return $r\,dt$ and an additional noise term with zero expectation under P^*. The strength of the noise is measured by the *volatility* parameter σ. Similarly, we obtain

$$dX_t = \sigma X_t\, dW_t = e^{-rt}(dS_t - rS_t\, dt). \tag{5.54}$$

Applying Itô's formula to the function

$$F(x,t) = e^{-rt} u(\exp(x + rt - \sigma^2 t/2), T-t)$$

and using (5.53), we obtain

$$dV_t = e^{-rt}\frac{\partial u}{\partial x}(S_t, t)\, dS_t + e^{-rt}\left(\frac{1}{2}\sigma^2 S_t^2 \frac{\partial^2 u}{\partial x^2} - \frac{\partial u}{\partial t} - ru\right)(S_t, t)\, dt.$$

The Black–Scholes partial differential equation (5.47) shows that the term in parenthesis is equal to $-rS_t \partial u/\partial x$, and we obtain from (5.54) that

$$dV_t = \frac{\partial u}{\partial x}(S_t, t)\, dX_t = \xi_t\, dX_t.$$

More precisely,

$$V_t = V_0 + \int_0^t \xi_s \, dX_s,$$

where the integral with respect to X is defined as an *Itô integral*, that is, as the limit of nonanticipating Riemann sums

$$\sum_{t_i \in D_n, \, t_i \le t} \xi_{t_i} (X_{t_{i+1}} - X_{t_i})$$

along an increasing sequence (D_n) of partitions of the interval $[0, T]$; see, for example, [138]. Thus, the Itô integral can be interpreted in financial terms as the cumulative net gain generated by dynamic hedging in the discounted risky asset as described by the hedging strategy ξ. This fact is an analogue of property (c) in Proposition 5.7, and in this sense $\overline{\xi} = (\xi^0, \xi)$ is a self-financing trading strategy in continuous time. Similarly, we obtain the following continuous-time analogue of (5.4), which describes the undiscounted value of the portfolio as a result of dynamic trading both in the undiscounted risky asset and the risk-free asset:

$$e^{rt} V_t = V_0 + \int_0^t \xi_s \, dS_s + \int_0^t \xi_s^0 \, dS_s^0.$$

Perfect replication also works for exotic options $C(S)$ defined by reasonable functionals C on the path space $C[0, T]$, due to a general representation theorem for such functionals as Itô integrals of the underlying Brownian motion W, or via (5.54) of the process X. Weak convergence on path space implies, in analogy to Proposition 5.63 that the arbitrage-free prices of the options $C(S^{(N)})$, computed as discounted expectations under the measure P_N^*, converge to the discounted expectation

$$e^{-rT} \mathbb{E}[\, C(S) \,]$$

under the Wiener measure $\mathbb{P}$.

On the other hand, the discussion in Section 5.6 suggests that the prices of certain exotic contingent claims, such as barrier options, can be computed in closed form as the Black–Scholes price for some corresponding payoff profile of the form $f(S_T)$. This is illustrated by the following example, where the price of an up-and-in call is computed in terms of the distribution of the terminal stock price under the equivalent martingale measure.

Example 5.64 (Black–Scholes price of an up-and-in call option). Consider an up-and-in call option

$$C^{\text{call}}_{\text{u\&i}}(S) = \begin{cases} (S_T - K)^+ & \text{if } \max\limits_{0\le t\le T} S_t \ge B, \\ 0 & \text{otherwise,} \end{cases}$$

where $B > S_0 \vee K$ denotes a given barrier, and where $K > 0$ is the strike price. As approximating models, we choose the CRR models of Example 5.56. That is, we have interest rates

$$r_N = \frac{r\,T}{N}$$

and parameters a_N and b_N defined by

$$\hat{a}_N = 1 + a_N = e^{-\sigma\sqrt{T/N}} \quad \text{and} \quad \hat{b}_N = 1 + b_N = e^{\sigma\sqrt{T/N}}$$

for some given $\sigma > 0$. Applying the formula obtained in Example 5.51 yields

$$\begin{aligned} &E^*_N[\, C^{\text{call}}_{\text{u\&i}}(S^{(N)})\,] \\ &= E^*_N[\,(S^{(N)}_N - K)^+;\, S^{(N)}_N \ge B_N\,] + \Big(\frac{p^*_N}{1-p^*_N}\Big)^{k_N}\Big(\frac{B_N}{S_0}\Big)^2 E^*_N[\,(S^{(N)}_N - \widetilde{K}_N)^+;\, S^{(N)}_N < \widetilde{B}_N\,], \end{aligned}$$

where

$$k_N = \Big\lceil \frac{\sqrt{N}}{\sigma\sqrt{T}} \log\frac{B}{S_0} \Big\rceil$$

is the smallest integer k such that $B \le S_0\hat{b}^k_N$, $B_N := S_0\hat{b}^{k_N}_N$, $\widetilde{B}_N := S_0^2/B_N = S_0\hat{a}^{k_N}_N$, and

$$\widetilde{K}_N = K\hat{b}^{-2k_N}_N = K\Big(\frac{S_0}{B_N}\Big)^2.$$

Then we have

$$B_N \searrow B, \quad \widetilde{B}_N \nearrow \widetilde{B} = \frac{S_0^2}{B}, \quad \text{and} \quad \widetilde{K}_N \nearrow \widetilde{K} = K\Big(\frac{S_0}{B}\Big)^2.$$

Since $f(x) = (x-K)^+\mathbb{1}_{\{x\ge B\}}$ is continuous a. e., we obtain

$$E^*_N[\,(S^{(N)}_N - K)^+;\, S^{(N)}_N \ge B_N\,] = E^*_N[\,(S^{(N)}_N - K)^+;\, S^{(N)}_N \ge B\,] \longrightarrow \mathbb{E}[\,(S_T - K)^+;\, S_T \ge B\,],$$

due to Proposition 5.60. Combining the preceding argument with the fact that

$$P^*_N[\,\widetilde{K}_N \le S^{(N)}_N \le \widetilde{K}\,] \longrightarrow 0$$

also gives the convergence of the second expectation:

$$E^*_N[\,(S^{(N)}_N - \widetilde{K}_N)^+;\, S^{(N)}_N < \widetilde{B}_N\,] = E^*_N[\,(S^{(N)}_N - \widetilde{K}_N)^+;\, S^{(N)}_N < \widetilde{B}\,] \longrightarrow \mathbb{E}[\,(S_T - \widetilde{K})^+;\, S_T < \widetilde{B}\,].$$

Next, we note that for constants $c, d > 0$,

$$\lim_{x\downarrow 0} \frac{1}{x} \log \frac{cx^2 + 1 - e^{-dx}}{e^{dx} - 1 - cx^2} = \frac{2c}{d} - d,$$

due to l'Hôpital's rule. From this fact, one deduces that

$$\left(\frac{p_N^*}{1-p_N^*}\right)^{k_N} \longrightarrow \left(\frac{B}{S_0}\right)^{\frac{2r}{\sigma^2}-1}.$$

Thus, we may conclude that the arbitrage-free prices

$$\frac{1}{(1+r_N)^N} E_N^*[\, C_{\text{u\&i}}^{\text{call}}(\widetilde{S}^{(N)})\,]$$

in the Nth approximating model converge to

$$e^{-rT}\left(\mathbb{E}[\,(S_T - K)^+;\, S_T \ge B\,] + \left(\frac{B}{S_0}\right)^{\frac{2r}{\sigma^2}+1} \mathbb{E}[\,(S_T - \widetilde{K})^+;\, S_T < \widetilde{B}\,]\right).$$

The expectations occurring in this formula are integrals with respect to a log-normal distribution and can be explicitly computed as in Example 5.58. Moreover, our limit is in fact equal to the *Black–Scholes price of the up-and-in call option*: The functional $C_{\text{u\&i}}^{\text{call}}(\,\cdot\,)$ is continuous in each path in $C[0, T]$ whose maximum is different from the value B, and one can show that these paths have full measure for the law of S under $\mathbb{P}$. Hence, $C_{\text{u\&i}}^{\text{call}}(\,\cdot\,)$ is continuous $\mathbb{P} \circ S^{-1}$-a. e., and the functional version of Proposition 5.60 yields

$$E_N^*[\, C_{\text{u\&i}}^{\text{call}}(\widetilde{S}^{(N)})\,] \longrightarrow \mathbb{E}[\, C_{\text{u\&i}}^{\text{call}}(S)\,],$$

so that our limiting price must coincide with the discounted expectation on the right. ◇

Remark 5.65. Let us assume more generally that the price process S is defined by

$$S_t = S_0 e^{\sigma W_t + \alpha t}, \quad 0 \le t \le T,$$

for some $\alpha \in \mathbb{R}$. Applying Itô's formula as in (5.53), we see that S is governed by the stochastic differential equation

$$dS_t = \sigma S_t \, dW_t + b S_t \, dt$$

with $b = \alpha + \frac{1}{2}\sigma^2$. The discounted price process is given by

$$X_t = S_0 e^{\sigma W_t + (\alpha - r)t} = S_0 e^{\sigma W_t^* - \sigma^2 t/2}$$

with $W_t^* = W_t + \lambda t$ for $\lambda = (b - r)/\sigma$. The process W^* is a Wiener process under the measure $P^* \approx \mathbb{P}$ defined by the density

$$\frac{dP^*}{d\mathbb{P}} = e^{-\lambda W_T - \lambda^2 T/2}.$$

In fact, P^* is the unique equivalent martingale measure for X. We can now repeat the arguments above to conclude that the cost of perfect replication for a contingent claim $C(S)$ is given by

$$e^{-rT} E^* [\, C(S) \,].$$ ◇

Even in the context of simple diffusion models such as geometric Brownian motion, however, completeness is lost as soon as the future behavior of the volatility parameter σ is unknown. If, for instance, volatility itself is modeled as a stochastic process, we are facing incompleteness. Thus, the problems of pricing and hedging in discrete-time incomplete markets as discussed in this book reappear in continuous time. Other versions of the invariance principle may lead to other classes of continuous-time models with discontinuous paths, for instance to geometric Poisson or Lévy processes. Discontinuity of paths is another important source of incompleteness. In fact, this has already been illustrated in this book, since discrete-time models can be regarded as stochastic processes in continuous time, where jumps occur at predictable dates.

6 American contingent claims

So far, we have studied European contingent claims whose payoff is due at a fixed maturity date. In the case of American options, the buyer can exercise the claim at any time up to the expiration of the contract.

First, we take the point of view of the *seller*, whose aim is to hedge against all possible claims of the buyer. In Section 6.1, this problem is solved under the assumption of market completeness, using the Snell envelope of the contingent claim. The *buyer* tries to choose the best date for exercising the claim, contingent on the information available up to that time. Since future prices are usually unknown, a formulation of this problem will typically involve subjective preferences. If preferences are expressed in terms of expected utility, the choice of the best exercise date amounts to solving an *optimal stopping problem*. In the special case of a complete market model, any exercise strategy that maximizes the expected payoff under the unique equivalent martingale measure, turns out to be optimal even in an almost sure sense.

In Section 6.3, we characterize the set of all arbitrage-free prices of an American contingent claim in an incomplete market model. This involves a lower Snell envelope of the claim, which is analyzed in Section 6.5, using the fact that the class of equivalent martingale measures is stable under pasting. This notion of stability under pasting is discussed in Section 6.4 in a general context, and in Section 6.5 we point out its connection with the time-consistency of dynamic risk measures. This connection will be discussed systematically in Chapter 11. The results on lower Snell envelopes can also be regarded as a solution to the buyer's optimal stopping problem in the case where preferences are described by robust Savage functionals. Moreover, these results will be used in the theory of superhedging of Chapter 7.

6.1 Hedging strategies for the seller

Throughout this chapter, we will continue to use the setting described in Section 5.1. We start by introducing the Doob decomposition of an adapted process and the notion of a supermartingale.

Proposition 6.1. *Let Q be a probability measure on $(\Omega, \mathscr{F}_T)$, and suppose that Y is a stochastic process that is adapted to the filtration $(\mathscr{F}_t)_{t=0,\dots,T}$ and satisfies $Y_t \in \mathscr{L}^1(Q)$ for all t. Then there exists a unique decomposition*

$$Y = M - A, \tag{6.1}$$

where M is a Q-martingale and A is a process such that $A_0 = 0$ and $(A_t)_{t=1,\dots,T}$ is predictable. The decomposition (6.1) *is called the* Doob decomposition *of Y with respect to the probability measure Q.*

https://doi.org/10.1515/9783111045283-008

Proof. Define A by

$$A_t - A_{t-1} := -E_Q[\, Y_t - Y_{t-1} \mid \mathscr{F}_{t-1} \,] \quad \text{for } t = 1, \dots, T. \tag{6.2}$$

Then A is predictable and $M_t := Y_t + A_t$ is a Q-martingale. Clearly, any process A with the required properties must satisfy (6.2), so the uniqueness of the decomposition follows. □

Definition 6.2. Let Q be a probability measure on $(\Omega, \mathscr{F}_T)$ and suppose that Y is an adapted process such that $Y_t \in \mathscr{L}^1(Q)$ for all t. Denote by $Y = M - A$ the Doob decomposition of Y.
(a) Y is called a *Q-supermartingale* if A is increasing.
(b) Y is called a *Q-submartingale* if A is decreasing. △

Clearly, a process is a martingale if and only if it is both a supermartingale and a submartingale, that is, if and only if $A \equiv 0$. The following exercise gives equivalent characterizations of the supermartingale property of a process Y.

Exercise 6.1.1. Let Y be an adapted process with $Y_t \in \mathscr{L}^1(Q)$ for all t. Show that the following conditions are equivalent:
(a) Y is a Q-supermartingale.
(b) $Y_s \geq E_Q[\, Y_t \mid \mathscr{F}_s \,]$ for $0 \leq s \leq t \leq T$.
(c) $Y_{t-1} \geq E_Q[\, Y_t \mid \mathscr{F}_{t-1} \,]$ for $t = 1, \dots, T$.
(d) $-Y$ is a Q-submartingale. ◇

Exercise 6.1.2. Let Y be a nonnegative Q-supermartingale. Show that for $0 \leq s \leq T - t$ we have $Y_{t+s} = 0$ Q-a. s. on $\{Y_t = 0\}$. ◇

We now return to the market model introduced in Section 5.1. An American option, or American contingent claim, corresponds to a contract that is issued at time 0 and that obliges the seller to pay a certain amount $C_\tau \geq 0$ if the buyer decides at time τ to exercise the option. The choice of the exercise time τ is entirely up to the buyer, except that the claim is automatically exercised at the expiration date of the claim. The American contingent claim can be exercised only once: It becomes invalid as soon as the payoff has been claimed by the buyer. This concept is formalized as follows.

Definition 6.3. An *American contingent claim* is a nonnegative adapted process $C = (C_t)_{t=0,\dots,T}$ on the filtered space $(\Omega, (\mathscr{F}_t)_{t=0,\dots,T})$. △

For each t, the random variable C_t is interpreted as the payoff of the American contingent claim if the claim is exercised at time t. The time horizon T plays the role of the expiration date of the claim. The possible exercise times for C are not limited to fixed deterministic times $t \in \{0. \dots, T\}$; the buyer may exercise the claim in a way that depends on the scenario $\omega \in \Omega$ of the market evolution.

Definition 6.4. An *exercise strategy* for an American contingent claim C is an $\mathscr{F}_T$-measurable random variable τ taking values in $\{0, \dots, T\}$. The payoff obtained by using τ is equal to

$$C_\tau(\omega) := C_{\tau(\omega)}(\omega), \quad \omega \in \Omega. \qquad \triangle$$

Example 6.5. An *American put option* on the ith asset and with strike $K > 0$ gives the owner the right, but not the obligation, to sell the underlying asset for the price K at any time until expiration. Thus, exercising the option at time t corresponds to the payoff

$$C_t^{\text{put}} := (K - S_t^i)^+.$$

An *American call option* gives the right to purchase the asset at price K. Hence, if exercised at time t, its payoff is

$$C_t^{\text{call}} := (S_t^i - K)^+.$$

Clearly, the American call option is "out of the money" (i. e., has zero payoff) if the corresponding American put is "in the money" (i. e., has nonzero payoff). It is therefore a priori clear that the respective owners of C^{put} and C^{call} will usually exercise their claims at different times. In particular, there will be no put-call parity for American options. ◇

Similarly, one can define American versions of most options mentioned in the examples of Section 5.3. Clearly, the value of an American option is at least as high as the value of the corresponding European option with maturity T.

Remark 6.6. It should be emphasized that the concept of American contingent claims can be regarded as a generalization of European contingent claims: If C^E is a European contingent claim, then we can define a corresponding American claim C^A by

$$C_t^A = \begin{cases} 0 & \text{if } t < T, \\ C^E & \text{if } t = T. \end{cases} \qquad \diamond$$

Example 6.7. A *Bermuda option* can be exercised by its buyer at each time of a predetermined subset $\mathbb{T} \subseteq \{0, \dots, T\}$. For instance, a Bermuda call option pays the amount $(S_t^i - K)^+$ if it is exercised at some time $t \in \mathbb{T}$. Thus, a Bermuda option is a financial instrument "between" an American option with $\mathbb{T} = \{0, \dots, T\}$ and a European option with $\mathbb{T} = \{T\}$, just as Bermuda lies between America and Europe; hence, the name "Bermuda option." A Bermuda option can be regarded as a particular American option C that pays the amount $C_t = 0$ for $t \notin \mathbb{T}$. ◇

The process

$$H_t = \frac{C_t}{S_t^0}, \quad t = 0, \dots, T,$$

of discounted payoffs of C will be called the *discounted American claim* associated with C. As far as the mathematical theory is concerned, the discounted American claim H will be the primary object. For certain examples, it will be helpful to keep track of the numéraire, and thus of the payoffs C_t prior to discounting.

In this section, we will analyze the theory of hedging American claims in a *complete* market model. That is, we assume that

$$\mathcal{P} = \{P^*\}. \tag{6.3}$$

Under this assumption, we will construct a suitable trading strategy that permits the seller of an American claim to hedge against the buyer's discounted claim H_τ. Let us first try to characterize the minimal amount of capital U_t that will be needed at time $t \in \{0, \dots, T\}$. Since the choice of the exercise time τ is entirely up to the buyer, the seller must be prepared to pay at any time t the current payoff H_t of the option. This amounts to the condition $U_t \geq H_t$. Moreover, the amount U_t must suffice to cover the purchase of the hedging portfolio for the possible payoffs H_u for $u > t$. Since the latter condition is void at maturity, we require

$$U_T = H_T.$$

At time $T-1$, our first requirement on U_{T-1} reads $U_{T-1} \geq H_{T-1}$. The second requirement states that the amount U_{T-1} must suffice for hedging the claim H_T in case the option is not exercised before time T. Due to our assumption of market completeness, the latter amount equals

$$E^*[\, H_T \mid \mathcal{F}_{T-1}\,] = E^*[\, U_T \mid \mathcal{F}_{T-1}\,].$$

Thus,

$$U_{T-1} := H_{T-1} \vee E^*[\, U_T \mid \mathcal{F}_{T-1}\,]$$

is the minimal amount that fulfills both requirements. Iterating this argument leads to the following recursive scheme for U_t:

$$U_T := H_T, \quad U_t := H_t \vee E^*[\, U_{t+1} \mid \mathcal{F}_t\,] \quad \text{for } t = T-1, \dots, 0. \tag{6.4}$$

Definition 6.8. The process $U^{P^*} := U$ defined by the recursion (6.4) is called the *Snell envelope* of the process H with respect to the measure P^*. △

Example 6.9. Let H^E be a discounted European claim. Then the Snell envelope with respect to P^* of the discounted American claim H^A associated with H^E as in Remark 6.6 satisfies

$$U_t^{P^*} = E^*[\, H_T^A \mid \mathcal{F}_t\,] = E^*[\, H^E \mid \mathcal{F}_t\,].$$

Thus, U is equal to the value process of a replicating strategy for H^E. ◇

Clearly, a Snell envelope U^Q can be defined for any probability measure Q on $(\Omega, \mathscr{F}_T)$ and for any adapted process H that satisfies the following integrability condition:

$$H_t \in \mathscr{L}^1(Q) \quad \text{for } t = 0, \dots, T. \tag{6.5}$$

In our finite-time setting, this condition is equivalent to

$$E_Q\Big[\max_{t \le T} |H_t|\Big] < \infty.$$

If $Q = P^*$, then assumption (6.3) and Theorem 5.37 imply that the probability space $(\Omega, \mathscr{F}, Q)$ has a finite structure so that the integrability condition (6.5) is satisfied automatically. For later applications, the following proposition is stated for a general measure Q, and we do not assume (6.3) here.

Proposition 6.10. *Let H be an adapted process such that* (6.5) *holds. Then the Snell envelope U^Q of H with respect to Q is the smallest Q-supermartingale dominating H: If $\widetilde{U}$ is another Q-supermartingale such that $\widetilde{U}_t \ge H_t$ Q-a. s. for all t, then $\widetilde{U}_t \ge U_t^Q$ Q-a. s. for all t.*

Proof. Since $|U_t^Q| \le |H_t| + E_Q[\,|U_{t+1}^Q| \mid \mathscr{F}_t\,]$, backward induction yields that $U_t^Q \in \mathscr{L}^1(Q)$ for all t. Moreover, it follows from the definition of U^Q that $U_{t-1}^Q \ge E_Q[\,U_t^Q \mid \mathscr{F}_{t-1}\,]$ so that U^Q is indeed a supermartingale. If $\widetilde{U}$ is another supermartingale dominating H, then $\widetilde{U}_T \ge H_T = U_T^Q$. We now proceed by backward induction on t. If we already know that $\widetilde{U}_t \ge U_t^Q$, then

$$\widetilde{U}_{t-1} \ge E_Q[\,\widetilde{U}_t \mid \mathscr{F}_{t-1}\,] \ge E_Q[\,U_t^Q \mid \mathscr{F}_{t-1}\,].$$

Adding our assumption $\widetilde{U}_{t-1} \ge H_{t-1}$ yields that

$$\widetilde{U}_{t-1} \ge H_{t-1} \vee E_Q[\,U_t^Q \mid \mathscr{F}_{t-1}\,] = U_{t-1}^Q,$$

and the result follows. □

Let us now return to the setting of a complete market model satisfying $\mathscr{P} = \{P^*\}$. We show next how Proposition 6.10 yields a (super-) hedge for the seller of a discounted American claim H. To this end, we start from the Doob decomposition

$$U_t^{P^*} = M_t - A_t, \quad t = 0, \dots, T,$$

of the Snell envelope U^{P^*} with respect to P^*. Thus, M is a P^*-martingale, A is increasing, and $(A_t)_{t=1,\dots,T}$ is predictable. Since we assume the completeness of the market model, Theorem 5.38 yields the representation of the martingale M as the discrete-time "stochastic integral" of a suitable d-dimensional predictable process ξ:

$$M_t = U_0^{P^*} + \sum_{k=1}^{t} \xi_k \cdot (X_k - X_{k-1}), \quad t = 0, \dots, T. \tag{6.6}$$

It follows that

$$M_t \geq U_t^{P^*} \geq H_t \quad \text{for all } t.$$

By adding a numéraire component ξ^0 such that $\overline{\xi} = (\xi^0, \xi)$ becomes a self-financing trading strategy with initial investment $U_0^{P^*}$, we obtain a (super-) hedge for H, namely a self-financing trading strategy whose value process V satisfies

$$V_t \geq H_t \quad \text{for all } t. \tag{6.7}$$

Thus, $U_t^{P^*}$ may be viewed as the resulting capital at each time t if we use the self-financing strategy $\overline{\xi}$, combined with a refunding scheme where we withdraw successively the amounts defined by the increments of A. In fact, $U_t^{P^*}$ is the minimal investment at time t for which one can purchase a hedging strategy such that (6.7) holds. This follows from our next result.

Theorem 6.11. *Let H be a discounted American claim with Snell envelope U^{P^*}. Then there exists a d-dimensional predictable process ξ such that*

$$U_t^{P^*} + \sum_{k=t+1}^{u} \xi_k \cdot (X_k - X_{k-1}) \geq H_u \quad \textit{for all } u \geq t \ P\textit{-a. s.} \tag{6.8}$$

Moreover, any $\mathscr{F}_t$-measurable random variable $\widetilde{U}_t$ that, for some predictable ξ, satisfies (6.8) *in place of $U_t^{P^*}$ is such that*

$$\widetilde{U}_t \geq U_t^{P^*} \quad P\text{-a. s.}$$

Thus, $U_t^{P^*}$ is the minimal amount of capital necessary to hedge H from time t up to maturity.

Proof. Due to the argument preceding the statement of this theorem, U^{P^*} satisfies (6.8) for ξ as in (6.6). Now suppose that $\widetilde{U}_t$ is $\mathscr{F}_t$-measurable, that $\widetilde{\xi}$ is predictable, and that

$$V_u := \widetilde{U}_t + \sum_{k=t+1}^{u} \widetilde{\xi}_k \cdot (X_k - X_{k-1}) \geq H_u \quad \text{for all } u \geq t \ P\text{-a. s.}$$

For $t = 0$, one can therefore use Doob's systems theorem, Theorem 5.14, to conclude that V is a martingale dominating H. Proposition 6.10 thus yields $V_u \geq U_u^{P^*}$ for all u and, in particular, $\widetilde{U}_0 = V_0 \geq U_0^{P^*}$.

The following alternative argument works for all $t \geq 0$. We proceed by backward induction on u. First, $V_T \geq H_T = U_T^{P^*}$ holds by assumption. Next, we assume that $V_{u+1} \geq U_{u+1}^{P^*}$ for some $u \geq t$. Since our market model is complete, Theorem 5.37 implies that $\widetilde{\xi}$ is bounded. Hence, we get

$$E^*[\, V_{u+1} - V_u \mid \mathscr{F}_u \,] = E^*[\, \widetilde{\xi}_{u+1} \cdot (X_{u+1} - X_u) \mid \mathscr{F}_u \,] = 0 \quad P\text{-a. s.}$$

It follows that

$$V_u = E^*[\, V_{u+1} \mid \mathscr{F}_u \,] \geq H_u \vee E^*[\, U_{u+1}^{P^*} \mid \mathscr{F}_u \,] = U_u^{P^*}. \qquad \square$$

6.2 Stopping strategies for the buyer

In this section, we take the point of view of the buyer of an American contingent claim. Thus, our aim is to optimize the exercise strategy. It is natural to assume that the decision to exercise the claim at a particular time t depends only on the market information available at t. This constraint can be formulated as follows.

Definition 6.12. A function $\tau : \Omega \to \{0, 1, \ldots, T\} \cup \{+\infty\}$ is called a *stopping time* if $\{\tau = t\} \in \mathscr{F}_t$ for $t = 0, \ldots, T$. △

In particular, the constant function $\tau \equiv t$ is a stopping time for fixed $t \in \{0, \ldots, T\}$.

Exercise 6.2.1. Show that a function $\tau : \Omega \to \{0, 1, \ldots, T\} \cup \{+\infty\}$ is a stopping time if and only if $\{\tau \leq t\} \in \mathscr{F}_t$ for each t. Show next that, if τ and σ are two stopping times, then the following functions are also stopping times:

$$\tau \wedge \sigma, \qquad \tau \vee \sigma, \qquad (\tau + \sigma) \wedge T. \qquad \diamond$$

Example 6.13. A typical example of a nontrivial stopping time is the first time at which an adapted process Y exceeds a certain level c:

$$\tau(\omega) := \inf\{t \geq 0 \mid Y_t(\omega) \geq c\}.$$

Indeed,

$$\{\tau \leq t\} = \bigcup_{s=0}^{t} \{Y_s \geq c\} \in \mathscr{F}_t$$

for $t = 0, \ldots, T$. This example also illustrates the role of the value $+\infty$ in Definition 6.12: We have $\tau(\omega) = +\infty$ if, for this particular ω, the criterion that triggers τ is not met for any $t \in \{0, \ldots, T\}$. ◇

Definition 6.14. For any stochastic process Y and each stopping time τ, we denote by Y^τ the *process stopped in* τ:

$$Y_t^\tau(\omega) := Y_{t \wedge \tau(\omega)}(\omega) \quad \text{for } \omega \in \Omega \text{ and for all } t \in \{0, \ldots, T\}. \qquad \triangle$$

It follows from the definition of a stopping time that Y^τ is an adapted process if Y is. Informally, the following basic theorem states that a martingale cannot be turned into a favorable game by using a clever stopping strategy. This result is often called *Doob's stopping theorem* or the *optional sampling theorem*. Recall that we assume $\mathscr{F}_0 = \{\emptyset, \Omega\}$.

Theorem 6.15. *Let M be an adapted process such that $M_t \in \mathscr{L}^1(Q)$ for each t. Then the following conditions are equivalent:*
(a) *M is a Q-martingale.*
(b) *For any stopping time τ the stopped process M^τ is a Q-martingale.*
(c) *$E_Q[M_{\tau\wedge T}] = M_0$ for any stopping time τ.*

Proof. (a)⇒(b): Note that

$$M^\tau_{t+1} - M^\tau_t = (M_{t+1} - M_t)\,\mathbb{1}_{\{\tau>t\}}.$$

We can thus use Doob's systems theorem, Theorem 5.14, to conclude that M^τ is a martingale. A direct argument goes as follows: Since $\{\tau > t\} \in \mathscr{F}_t$, we obtain that

$$E_Q[M^\tau_{t+1} - M^\tau_t \mid \mathscr{F}_t] = E_Q[M_{t+1} - M_t \mid \mathscr{F}_t]\mathbb{1}_{\{\tau>t\}} = 0.$$

(b)⇒(c): This follows simply from the fact that the expectation of M^τ_t is constant in t.

(c)⇒(a): We need to show that if $t < T$, then

$$E_Q[M_T; A] = E_Q[M_t; A] \tag{6.9}$$

for each $A \in \mathscr{F}_t$, where we use again the notation $E_Q[M_t; A] = E_Q[M_t\mathbb{1}_A]$. Fix $A \in \mathscr{F}_t$ and define a stopping time τ as

$$\tau(\omega) := \begin{cases} t & \text{if } \omega \in A, \\ T & \text{if } \omega \notin A. \end{cases}$$

We obtain that

$$M_0 = E_Q[M_{T\wedge\tau}] = E_Q[M_t; A] + E_Q[M_T; A^c].$$

Using the constant stopping time T instead of τ yields that

$$M_0 = E_Q[M_T] = E_Q[M_T; A] + E_Q[M_T; A^c].$$

Subtracting the latter identity from the previous one yields (6.9). □

Exercise 6.2.2. Let $Y = M - A$ be the Doob decomposition with respect to Q of an adapted process Y with $Y_t \in \mathscr{L}^1(Q)$ $(t = 0, \dots, T)$, and let τ be a stopping time. Show that $Y^\tau = M^\tau - A^\tau$ is the Doob decomposition of Y^τ. ◇

Corollary 6.16. *Let U be an adapted process such that $U_t \in \mathscr{L}^1(Q)$ for each t. Then the following conditions are equivalent:*

(a) *U is a Q-supermartingale.*

(b) *For any stopping time τ, the stopped process U^τ is a Q-supermartingale.*

Moreover, in this case we have $E_Q[\, U_{\tau\wedge T}\,] \le U_0$ for every stopping time τ.

Proof. If $U = M - A$ is the Doob decomposition of U, then Exercise 6.2.2 implies that $U^\tau = M^\tau - A^\tau$ is the Doob decomposition of U^τ. Since M^τ is a martingale by Theorem 6.15 and A is increasing if and only if A^τ is increasing for all stopping times τ, we obtain the equivalence of (a) and (b). Next, if U is a Q-supermartingale, then $E_Q[\, U_{\tau\wedge T}\,] \le E_Q[\, M_{\tau\wedge T}\,] = M_0 = U_0$, where we have once again used Theorem 6.15 for the martingale M. □

Let us return to the problem of finding an optimal exercise time τ for a discounted American claim H. We assume that the buyer chooses the possible exercise times from the set

$$\mathcal{T} := \{\tau \mid \tau \text{ is a stopping time with } \tau \le T\} \tag{6.10}$$

of all stopping times that do not take the value $+\infty$. Assume that the aim of the buyer is to choose a payoff from the class $\{H_\tau \mid \tau \in \mathcal{T}\}$ that is optimal in the sense that it has maximal expectation. Thus, the problem is:

$$\text{Maximize } E[\, H_\tau\,] \text{ over all } \tau \in \mathcal{T}. \tag{6.11}$$

The analysis of the *optimal stopping problem* (6.11) does not require any properties of the underlying market model, not even the absence of arbitrage. Thus, we make no assumptions on the structure of the underlying probability space $(\Omega, \mathcal{F}, P)$, before we resume our discussion of a complete financial market model, following the proof of Theorem 6.21. We may also drop the positivity assumption on H: All we have to assume is that H is an adapted process that satisfies

$$H_t \in \mathcal{L}^1(\Omega, \mathcal{F}_t, P) \quad \text{for all } t. \tag{6.12}$$

This relaxed assumption will be useful in Chapter 9, and it allows us to include the interpretation of the optimal stopping problem in terms of the following utility maximization problem.

Remark 6.17. Suppose the buyer uses a preference relation on $\mathcal{X} := \{H_\tau \mid \tau \in \mathcal{T}\}$ that can be represented in terms of a *Savage representation*

$$\mathcal{U}(H_\tau) = E_Q[\, u(H_\tau)\,]$$

where Q is a probability measure on $(\Omega, \mathcal{F})$, and u is a measurable or continuous function; see Section 2.5. Then a natural goal is to maximize the utility $\mathcal{U}(H_\tau)$ over all $\tau \in \mathcal{T}$. This is equivalent to the optimal stopping problem (6.11) for the transformed process

$\widetilde{H}_t := u(H_t)$, and with respect to the measure Q instead of P. This utility maximization problem is covered by the discussion in this section as long as $\widetilde{H}_t \in \mathscr{L}^1(Q)$ for all t. In Remark 6.50, we will discuss the problem of maximizing the more general utility functionals that appear in a *robust* Savage representation. ◇

Under the assumption (6.12), we can construct the Snell envelope $U := U^P$ of H with respect to P, that is, U is defined via the recursive formula

$$U_T := H_T \quad \text{and} \quad U_t := H_t \vee E[\, U_{t+1} \mid \mathscr{F}_t\,], \quad t = T-1, \dots, 0.$$

Let us define a stopping time $\tau_{\min}$ by

$$\tau_{\min} := \min\{t \geq 0 \mid U_t = H_t\}.$$

Note that $\tau_{\min} \leq T$ since $U_T = H_T$. As we will see in the following theorem, $\tau_{\min}$ maximizes the expectation of H_τ among all $\tau \in \mathcal{T}$. In other words, $\tau_{\min}$ is a solution to our optimal stopping problem (6.11). Similarly, we let

$$\tau_{\min}^{(t)} := \min\{u \geq t \mid U_u = H_u\},$$

which is a member of the set

$$\mathcal{T}_t := \{\tau \in \mathcal{T} \mid \tau \geq t\}.$$

The following theorem uses the *essential supremum* of a family of random variables as explained in Appendix F.

Theorem 6.18. *The Snell envelope U of H satisfies*

$$U_t = E[\, H_{\tau_{\min}^{(t)}} \mid \mathscr{F}_t\,] = \operatorname*{ess\,sup}_{\tau \in \mathcal{T}_t} E[\, H_\tau \mid \mathscr{F}_t\,].$$

In particular,

$$U_0 = E[\, H_{\tau_{\min}}\,] = \sup_{\tau \in \mathcal{T}} E[\, H_\tau\,].$$

Proof. Since U is a supermartingale under P, Corollary 6.16 shows that for $\tau \in \mathcal{T}_t$,

$$U_t \geq E[\, U_\tau \mid \mathscr{F}_t\,] \geq E[\, H_\tau \mid \mathscr{F}_t\,].$$

Therefore,

$$U_t \geq \operatorname*{ess\,sup}_{\tau \in \mathcal{T}_t} E[\, H_\tau \mid \mathscr{F}_t\,].$$

Hence, the theorem will be proved if we can show that $U_t = E[\,H_{\tau^{(t)}_{\min}} \mid \mathscr{F}_t\,]$, which is in turn implied by the identity

$$U_t = E[\,U_{\tau^{(t)}_{\min}} \mid \mathscr{F}_t\,], \tag{6.13}$$

because $U_{\tau^{(t)}_{\min}} = H_{\tau^{(t)}_{\min}}$. In order to prove (6.13), let $U^{(t)}$ denote the stopped process

$$U^{(t)}_s := U_{s\wedge\tau^{(t)}_{\min}},$$

and fix some s between t and T. Then $U_s > H_s$ on $\{\tau^{(t)}_{\min} > s\}$. Hence, P-a. s. on $\{\tau^{(t)}_{\min} > s\}$,

$$U^{(t)}_s = U_s = H_s \vee E[\,U_{s+1} \mid \mathscr{F}_s\,] = E[\,U_{s+1} \mid \mathscr{F}_s\,] = E[\,U^{(t)}_{s+1} \mid \mathscr{F}_s\,].$$

On the set $\{\tau^{(t)}_{\min} \le s\}$, one has $U^{(t)}_{s+1} = U_{\tau^{(t)}_{\min}} = U^{(t)}_s$, hence $U^{(t)}_s = E[\,U^{(t)}_{s+1} \mid \mathscr{F}_s\,]$. Thus, $U^{(t)}$ is a martingale from time t on

$$U^{(t)}_s = E[\,U^{(t)}_{s+1} \mid \mathscr{F}_s\,] \quad \text{for all } s \in \{t, t+1, \dots, T-1\}.$$

It follows that

$$E[\,U_{\tau^{(t)}_{\min}} \mid \mathscr{F}_t\,] = E[\,U^{(t)}_T \mid \mathscr{F}_t\,] = U^{(t)}_t = U_t.$$

This proves the claim (6.13). □

Definition 6.19. A stopping time $\tau^* \in \mathcal{T}$ is called *optimal* (*with respect to P*) if

$$E[\,H_{\tau^*}\,] = \sup_{\tau\in\mathcal{T}} E[\,H_\tau\,]. \qquad \triangle$$

In particular, $\tau_{\min}$ is an optimal stopping time in the sense of this definition. The following result implies that $\tau_{\min}$ is in fact the minimal optimal stopping time.

Proposition 6.20. *A stopping time $\tau \in \mathcal{T}$ is optimal if and only if $H_\tau = U_\tau$ P-a. s., and if the stopped process U^τ is a martingale. In particular, any optimal stopping time τ satisfies $\tau \ge \tau_{\min}$.*

Proof. First, note that $\tau \in \mathcal{T}$ is optimal if it satisfies the two conditions of the assertion, because then Theorem 6.18 implies that

$$\sup_{\sigma\in\mathcal{T}} E[\,H_\sigma\,] = U_0 = U^\tau_0 = E[\,U^\tau_T\,] = E[\,U_\tau\,] = E[\,H_\tau\,].$$

For the converse implication, we apply the assumption of optimality, the fact that $H_\tau \le U_\tau$, and the stopping theorem for supermartingales, Corollary 6.16, to obtain that

$$U_0 = E[\,H_\tau\,] \le E[\,U_\tau\,] \le U_0,$$

so that all inequalities are in fact equalities. Since $U_\tau \geq H_\tau$, it follows in particular that $H_\tau = U_\tau$ P-almost surely. Moreover, the identity $E[U_\tau] = U_0$ implies that the stopped process U^τ is a supermartingale with constant expectation U_0, and hence is a martingale. □

In general, there can be many different optimal stopping times. The largest optimal stopping time admits an explicit description: It is the first time before T for which the Snell envelope U loses the martingale property:

$$\begin{aligned}\tau_{\max} &:= \inf\{t \geq 0 \mid E[U_{t+1} - U_t \mid \mathcal{F}_t] \neq 0\} \wedge T \\ &= \inf\{t \geq 0 \mid A_{t+1} \neq 0\} \wedge T.\end{aligned}$$

Here, A denotes the increasing process obtained from the Doob decomposition of U under P.

Theorem 6.21. *A stopping time τ is optimal if and only if P-a. s. $\tau \leq \tau_{\max}$ and $U_\tau = H_\tau$. Moreover, the stopping time $\tau_{\max}$ is the largest optimal stopping time.*

Proof. Let $U = M - A$ be the Doob decomposition of U. Recall from Exercise 6.2.2 that $U^\tau = M^\tau - A^\tau$ is the Doob decomposition of U^τ for any stopping time τ. Thus, U^τ is a martingale if and only if $A_\tau = 0$, because A is increasing. Therefore, U^τ is a martingale if and only if $\tau \leq \tau_{\max}$, and so the first part of the assertion follows from Proposition 6.20.

It remains to prove that $\tau_{\max}$ itself is optimal, that is, that $U_{\tau_{\max}} = H_{\tau_{\max}}$. This is clear on the set $\{\tau_{\max} = T\}$. On the set $\{\tau_{\max} = t\}$ for $t < T$ one has $A_t = 0$ and $A_{t+1} > 0$. Hence,

$$E[U_{t+1} - U_t \mid \mathcal{F}_t] = -(A_{t+1} - A_t) = -A_{t+1} < 0 \quad \text{on } \{\tau_{\max} = t\}.$$

Thus, $U_t > E[U_{t+1} \mid \mathcal{F}_t]$ and the definition of the Snell envelope yields that $U_t = H_t \vee E[U_{t+1} \mid \mathcal{F}_t] = H_t$ on $\{\tau_{\max} = t\}$. □

Let us now return to our complete financial market model, where H_t is the discounted payoff of an American contingent claim. Thus, an optimal stopping strategy for H maximizes the *expected* payoff $E[H_\tau]$. But a stopping time turns out to be the best choice even in a *pathwise* sense, provided that it is optimal with respect to the unique equivalent martingale measure P^*. In order to explain this fact, let us first recall from Section 6.1 the construction of a perfect hedge of H from the seller's perspective. Let

$$U^{P^*} = M - A$$

denote the Doob decomposition of the Snell envelope U^{P^*} of H with respect to P^*. Since P^* is the unique equivalent martingale measure in our model, the martingale M has the representation

$$M_t = U_0^{P^*} + \sum_{k=1}^{t} \xi_k \cdot (X_k - X_{k-1}), \quad t = 0, \ldots, T,$$

for a d-dimensional predictable process ξ. Clearly, M is equal to the value process of the self-financing strategy constructed from ξ and the initial investment $U_0^{P^*}$. Since M dominates H, this yields a perfect hedge of H from the perspective of the seller: If the buyer exercises the option at some stopping time τ, then the seller makes a profit $M_\tau - H_\tau \geq 0$. The following corollary states that the buyer can in fact meet the value of the seller's hedging portfolio, and that this happens if and only if the option is exercised at an optimal stopping time with respect to P^*. In this sense, $U_0^{P^*}$ can be regarded as the unique *arbitrage-free price* of the discounted American claim H.

Corollary 6.22. *With the above notation,*

$$H_\tau \leq M_\tau = U_0^{P^*} + \sum_{k=1}^{\tau} \xi_k \cdot (X_k - X_{k-1}), \quad P^*\text{-a. s. for all } \tau \in \mathcal{T},$$

and equality holds P^*-almost surely if and only if τ is optimal with respect to P^*.

Proof. At time τ,

$$H_\tau \leq U_\tau^{P^*} = M_\tau - A_\tau \leq M_\tau.$$

Moreover, by Theorem 6.21, both $H_\tau = U_\tau^{P^*}$ and $A_\tau = 0$ hold P^*-a. s. if and only if τ is optimal with respect to P^*. □

Let us now compare a discounted American claim H to the corresponding discounted European claim H_T, that is, to the contract that is obtained from H by restricting the exercise time to be T. In particular, we are interested in the relation between American and European put or call options. Let

$$V_t := E^*[\, H_T \mid \mathcal{F}_t \,]$$

denote the amount needed at time t to hedge H_T. Since our market model is complete, V_t can also be regarded as the unique arbitrage-free price of the discounted claim H_T at time t. From the seller's perspective, $U_t^{P^*}$ plays a similar role for the American option. It is intuitively clear that an American claim should be more expensive than the corresponding European one. This is made mathematically precise in the following statement.

Proposition 6.23. *With the above notation, $U_t^{P^*} \geq V_t$ for all t. Moreover, if V dominates H, then U^{P^*} and V coincide.*

Proof. The first statement follows immediately from the supermartingale property of U^{P^*}:

$$U_t^{P^*} \geq E^*[\, U_T^{P^*} \mid \mathcal{F}_t \,] = E^*[\, H_T \mid \mathcal{F}_t \,] = V_t.$$

Next, if the P^*-martingale V dominates H, then it also dominates the corresponding Snell envelope U^{P^*} by Proposition 6.10. Thus, V and U^{P^*} must coincide. □

Remark 6.24. The situation in which V dominates H occurs, in particular, if the process H is a P^*-submartingale. This happens, for instance, if H is obtained by applying a convex function $f : \mathbb{R}^d \to [0,\infty)$ to the discounted price process X. Indeed, in this case, Jensen's inequality for conditional expectations implies that

$$E^*[f(X_{t+1}) \mid \mathscr{F}_t] \ge f(E^*[X_{t+1} \mid \mathscr{F}_t]) = f(X_t). \qquad \diamond$$

Example 6.25. The discounted payoff of an American call option $C_t^{\text{call}} = (S_t^1 - K)^+$ is given by

$$H_t^{\text{call}} = \left(X_t^1 - \frac{K}{S_t^0} \right)^+.$$

Under the hypothesis that S_t^0 is increasing in t, (5.24) states that

$$E^*[H_{t+1}^{\text{call}} \mid \mathscr{F}_t] \ge H_t^{\text{call}} \quad P^*\text{-a. s. for } t = 0, \dots, T-1.$$

In other words, H^{call} is a submartingale, and the Snell envelope U^{P^*} of H^{call} coincides with the value process

$$V_t = E^*\left[\left(X_T^1 - \frac{K}{S_T^0} \right)^+ \,\Big|\, \mathscr{F}_t \right]$$

of the corresponding European call option with maturity T. In particular, we have $U_0^{P^*} = V_0$, that is, the unique arbitrage-free price of the American call option is equal to its European counterpart. Moreover, Theorem 6.21 implies that the maximal optimal stopping time with respect to P^* is given by $\tau_{\max} \equiv T$. This suggests that, in a complete model with nonnegative interest rates, an American call should not be exercised before maturity. $\diamond$

Example 6.26. For an *American put option* $C_t^{\text{put}} := (K - S_t^1)^+$ the situation is different, because the argument in (5.24) fails unless S^0 is decreasing, which happens, for instance, if S^0 is a riskless bond in an environment with negative interest rates. If S^0 is increasing and nonconstant, then the time value

$$W_t := S_t^0 E^*\left[\frac{(K - S_T^1)^+}{S_T^0} \,\Big|\, \mathscr{F}_t \right] - (K - S_t^1)^+$$

of a European put $(K - S_T^1)^+$ typically becomes negative at a certain time t, corresponding to an *early exercise premium* $-W_t$; see Figure 5.3. Thus, the early exercise premium is the surplus that an owner of the American put option would have over the value of the European put $(K - S_T^1)^+$.

The relation between the price of a put option and its intrinsic value can be illustrated in the context of the CRR model. With the notation of Section 5.5, the price process of the risky asset $S_t = S_t^1$ can be written as

$$S_t = S_0 \Lambda_t \quad \text{for } \Lambda_t := \prod_{k=1}^{t}(1 + R_k)$$

and with the constant $S_0 \geq 0$. Recall that the returns R_k can take only two possible values a and b with $-1 < a < b$, and that the market model is arbitrage-free if and only if the riskless interest rate r satisfies $a < r < b$. In this case, the model is complete, and the unique equivalent martingale measure P^* is characterized by the fact that it makes $R_1, \ldots, R_T$ independent with common distribution

$$P^*[\, R_k = b \,] = p^* = \frac{r - a}{b - a}. \tag{6.14}$$

Let

$$\pi(x) := \sup_{\tau \in \mathcal{T}} E^*\left[\frac{(K - x\Lambda_\tau)^+}{(1 + r)^\tau} \right]$$

denote the price of C^{put} regarded as a function of $x := S_0$. Clearly, $\pi(x)$ is a convex and decreasing function in x. Let us assume that $r > 0$ and that the parameter a is strictly negative. A trivial situation occurs if the option is "far out of the money" in the sense that

$$x \geq \frac{K}{(1 + a)^T},$$

because then $S_t = x\Lambda_t \geq K$ for all t, and the payoff of C^{put} is always zero. In particular, $\pi(x) = 0$. If

$$x \leq \frac{K}{(1 + b)^T} \tag{6.15}$$

then $S_t = x\Lambda_t \leq K$ for all t, and hence

$$\pi(x) = \sup_{\tau \in \mathcal{T}} \left(E^*\left[\frac{K}{(1 + r)^\tau} \right] - x \right) = K - x.$$

In this case, the price of the American put option is equal to its intrinsic value $(K - x)^+$ at time $t = 0$, and an optimal strategy for the owner would simply consist in exercising the option immediately, that is, there is no demand for the option in the regime (6.15).

Now consider the case

$$K \leq x < \frac{K}{(1 + a)^T}$$

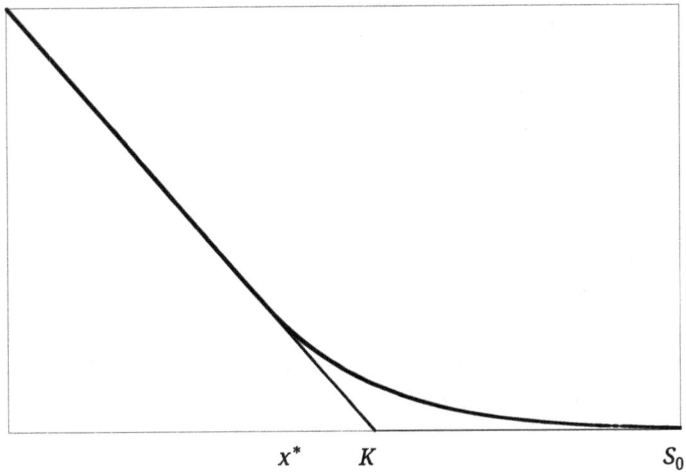

Figure 6.1: The price $\pi(S_0)$ of an American put option as a function of S_0 compared with the option's intrinsic value $(K - S_0)^+$.

of a put option that is "at the money" or "not too far out of the money." For large enough $t > 0$, the probability $P^*[\,C_t^{\text{put}} > 0\,]$ of a nonzero payoff is strictly positive, while the intrinsic value $(K - x)^+$ vanishes. It follows that the price $\pi(x)$ is strictly higher than the intrinsic value, and so it is not optimal for the buyer to exercise the option immediately.

Summarizing our observations, we can say that there exists a value x^* with

$$\frac{K}{(1+b)^T} \leq x^* < K$$

such that

$$\begin{aligned} \pi(x) &= (K-x)^+ && \text{for } x \leq x^*, \\ \pi(x) &> (K-x)^+ && \text{for } x^* < x < K/(1+a)^T, \quad \text{and} \\ \pi(x) &= 0 && \text{for } x \geq K/(1+a)^T; \end{aligned}$$

see Figure 6.1. ◇

Remark 6.27. In the context of an arbitrage-free CRR model, we consider a discounted American claim H whose payoff is determined by a function of time and of the current spot price, that is,

$$H_t = h_t(S_t) \quad \text{for all } t.$$

Clearly, this setting includes American call and put options as special cases. By using the same arguments as in the derivation of (5.29), we get that the Snell envelope U^{P^*} of H is of the form

$$U_t^{P^*} = u_t(S_t), \quad t = 0, \ldots, T,$$

where the functions u_t are determined by the recursion

$$u_T(x) = h_T(x) \quad \text{and} \quad u_t(x) = h_t(x) \vee \big(u_{t+1}(x\,\hat{b}\,)\,p^* + u_{t+1}(x\,\hat{a})\,(1 - p^*)\big).$$

Here, p^* is defined as in (6.14), and the parameters $\hat{a}$ and $\hat{b}$ are given by $\hat{a} = 1 + a$ and $\hat{b} = 1 + b$. Thus, the space $[0, T] \times [0, \infty)$ can be decomposed into the two regions

$$\mathscr{R}_c := \{(t, x) \mid u_t(x) > h_t(x)\} \quad \text{and} \quad \mathscr{R}_s := \{(t, x) \mid u_t(x) = h_t(x)\},$$

and the minimal optimal stopping time $\tau_{\min}$ can be described as the first exit time of the space time process (t, S_t) from the *continuation region* $\mathscr{R}_c$ or, equivalently, as the first entrance time into the *stopping region* $\mathscr{R}_s$:

$$\tau_{\min} = \min\{t \geq 0 \mid (t, S_t) \notin \mathscr{R}_c\} = \min\{t \geq 0 \mid (t, S_t) \in \mathscr{R}_s\}. \qquad \diamond$$

Exercise 6.2.3. Consider a market model with two assets and an American contingent claim. The development of the discounted price process $X := X^1$ and the discounted American claim is described by the following diagram.

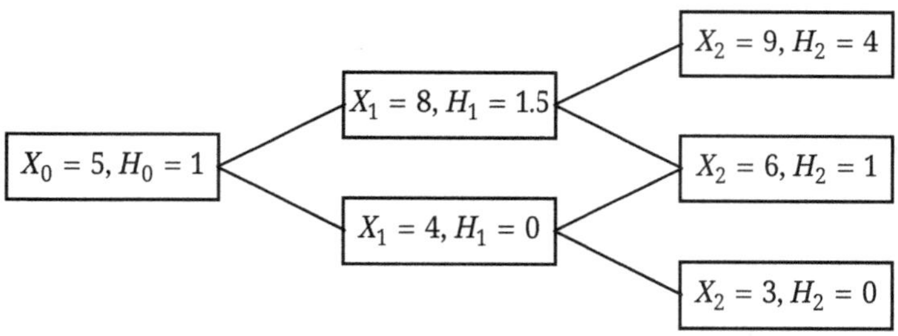

The buyer of the American claim uses a probability measure P that assigns equal probability to each of the possible scenarios. Find an optimal stopping strategy that maximizes $E[\,H_\tau\,]$ over $\tau \in \mathcal{T}$. What would be an optimal stopping time if the buyer uses the utility function $u(x) = \sqrt{x}$, and thus aims at maximizing $E[\,\sqrt{H_\tau}\,]$? Then show that the market model admits a unique risk-neutral measure P^* and compute the corresponding Snell envelope U^{P^*}. $\diamond$

Exercise 6.2.4. Let

$$H_t^K := \frac{(K - S_t)^+}{(1 + r)^t}$$

be the discounted payoff of an American put option with strike K in a market model with one risky asset $S = (S_t)_{t=0,\ldots T}$ and a riskless asset $S_t^0 = (1 + r)^t$, where $r > 0$. We denote by $\tau^K_{\min}$ the minimal optimal stopping time of the buyer's problem to maximize $E[\,H^K_\tau\,]$ over $\tau \in \mathcal{T}$.

(a) Show that $\tau^K_{\min} \geq \tau^{K'}_{\min}$ P-a.s. if $K \leq K'$.

(b) Show that $\operatorname{ess\,inf}_{K\geq 0} \tau^K_{\min} = 0$ P-a.s.

(c) Use (b) and the fact that $\mathcal{F}_0 = \{\emptyset, \Omega\}$ to conclude that there exists $K_0 \geq 0$ such that $\tau^K_{\min} = 0$ P-a.s. for all $K \geq K_0$. ◇

6.3 Arbitrage-free prices

In this section, we drop the condition of market completeness, and we develop the notion of an arbitrage-free price π for a discounted American claim H in a general incomplete framework. The basic idea consists in reducing the problem to the determination of the arbitrage-free price for the payoff H_τ that arises from H by fixing the exercise strategy τ. The following remark explains that H_τ can be treated like the discounted payoff of a European contingent claim, whose set of arbitrage-free prices is given by

$$\Pi(H_\tau) = \{E^*[\,H_\tau\,] \mid P^* \in \mathcal{P},\ E^*[\,H_\tau\,] < \infty\}. \tag{6.16}$$

Remark 6.28. As observed in Remark 5.33, a discounted payoff $\widetilde{H}_t$ that is received at time $t < T$ can be regarded as a discounted European claim $\widetilde{H}^E$ maturing at T. $\widetilde{H}^E$ is obtained from $\widetilde{H}_t$ by investing at time t the payoff $S^0_t\widetilde{H}_t$ into the numéraire, that is, by buying $\widetilde{H}_t$ shares of the 0th asset, and by considering the discounted terminal value of this investment:

$$\widetilde{H}^E = \frac{1}{S^0_T}\,(S^0_T\,\widetilde{H}_t) = \widetilde{H}_t.$$

In the case of our discounted American claim H that is paid off at the random time τ, we can either apply this argument to each payoff

$$\widetilde{H}_t := H_\tau\,\mathbb{1}_{\{\tau=t\}} = H_t\,\mathbb{1}_{\{\tau=t\}}, \qquad t = 0, 1, \dots, T,$$

or directly use a stopping time version of this argument. We conclude that H_τ can be regarded as a discounted European claim, whose arbitrage-free prices are given by (6.16). ◇

Now suppose that H is offered at time $t = 0$ for a price $\pi \geq 0$. From the buyer's point of view, there should be at least one exercise strategy τ such that the proposed price π is not too high in the sense that $\pi \leq \pi'$ for some $\pi' \in \Pi(H_\tau)$. From the seller's point of view, the situation looks different: There should be no exercise strategy τ' such that the proposed price π is too low in the sense that $\pi < \pi'$ for all $\pi' \in \Pi(H_{\tau'})$. By adding the assumption that the buyer only uses stopping times in exercising the option, we obtain the following formal definition, where we use again $\mathcal{T}$ to denote the set of all stopping times $\tau \le T$; see (6.10).

Definition 6.29. A real number π is called an *arbitrage-free price* of a discounted American claim H if the following two conditions are satisfied.

- The price π is not too high in the sense that there exist $\tau \in \mathcal{T}$ and $\pi' \in \Pi(H_\tau)$ such that $\pi \le \pi'$.
- The price π is not too low in the sense that there exists no $\tau' \in \mathcal{T}$ such that $\pi < \pi'$ for all $\pi' \in \Pi(H_{\tau'})$.

The set of all arbitrage-free prices of H is denoted $\Pi(H)$, and we define

$$\pi_{\inf}(H) := \inf \Pi(H) \quad \text{and} \quad \pi_{\sup}(H) := \sup \Pi(H). \qquad \triangle$$

Recall from Remark 6.6 that every discounted European claim H^E can be regarded as a discounted American claim H^A whose payoff is zero if H^A is exercised before T, and whose payoff at T equals H^E. Clearly, the two sets $\Pi(H^E)$ and $\Pi(H^A)$ coincide, and so the two Definitions 5.28 and 6.29 are consistent with each other.

Remark 6.30. It follows from Definition 6.29 that any arbitrage-free price π for H must be an arbitrage-free price for H_τ for some $\tau \in \mathcal{T}$. Hence, (6.16) implies that $\pi = E^*[\,H_\tau\,]$ for some $P^* \in \mathcal{P}$. The second condition in Definition 6.29 also yields that $\pi \ge \inf_{P^* \in \mathcal{P}} E^*[\,H_{\tau'}\,]$ for all $\tau' \in \mathcal{T}$. It follows that

$$\sup_{\tau \in \mathcal{T}} \inf_{P^* \in \mathcal{P}} E^*[\,H_\tau\,] \le \pi \le \sup_{\tau \in \mathcal{T}} \sup_{P^* \in \mathcal{P}} E^*[\,H_\tau\,] \quad \text{for all } \pi \in \Pi(H). \tag{6.17}$$

In particular,

$$\sup_{\tau \in \mathcal{T}} E^*[\,H_\tau\,]$$

is the unique arbitrage-free price of H if P^* is the unique equivalent martingale measure in a complete market model, and so Definition 6.29 is consistent with the results of the Sections 6.1 and 6.2. $\diamond$

Exercise 6.3.1. Show that in every arbitrage-free market model and for any discounted American claim H,

$$\inf_{P^* \in \mathcal{P}} \sup_{\tau \in \mathcal{T}} E^*[\,H_\tau\,] < \infty, \tag{6.18}$$

and that the set $\Pi(H)$ of arbitrage-free prices is nonempty. $\diamond$

Our main goal in this section is to characterize the set $\Pi(H)$, and to identify the upper and lower bounds in (6.17) with the quantities $\pi_{\sup}(H)$ and $\pi_{\inf}(H)$. We will work under the simplifying assumption that

$$H_t \in \mathcal{L}^1(P^*) \quad \text{for all } t \text{ and each } P^* \in \mathcal{P}. \tag{6.19}$$

For each $P^* \in \mathcal{P}$, we denote by U^{P^*} the corresponding Snell envelope of H, that is,

$$U_t^{P^*} = \operatorname*{ess\,sup}_{\tau \in \mathcal{T}_t} E^*[\, H_\tau \mid \mathcal{F}_t\,].$$

With this notation, the right-hand bound in (6.17) can be written as

$$\sup_{\tau \in \mathcal{T}} \sup_{P^* \in \mathcal{P}} E^*[\, H_\tau\,] = \sup_{P^* \in \mathcal{P}} \sup_{\tau \in \mathcal{T}} E^*[\, H_\tau\,] = \sup_{P^* \in \mathcal{P}} U_0^{P^*}.$$

In fact, a similar relation also holds for the lower bound in (6.17):

$$\sup_{\tau \in \mathcal{T}} \inf_{P^* \in \mathcal{P}} E^*[\, H_\tau\,] = \inf_{P^* \in \mathcal{P}} \sup_{\tau \in \mathcal{T}} E^*[\, H_\tau\,] = \inf_{P^* \in \mathcal{P}} U_0^{P^*}. \tag{6.20}$$

The proof that the above interchange of infimum and supremum is indeed justified under assumption (6.19) is postponed to Section 6.5; see Theorem 6.46. We will now apply it in the following result, which characterizes the set of arbitrage-free prices of a discounted American claim in an arbitrage-free market model that is not necessarily complete.

Theorem 6.31. *Under condition* (6.19), *the set of arbitrage-free prices for H is a real interval with endpoints*

$$\pi_{\inf}(H) = \sup_{\tau \in \mathcal{T}} \inf_{P^* \in \mathcal{P}} E^*[\, H_\tau\,] = \inf_{P^* \in \mathcal{P}} \sup_{\tau \in \mathcal{T}} E^*[\, H_\tau\,] = \inf_{P^* \in \mathcal{P}} U_0^{P^*}$$

and

$$\pi_{\sup}(H) = \sup_{\tau \in \mathcal{T}} \sup_{P^* \in \mathcal{P}} E^*[\, H_\tau\,] = \sup_{P^* \in \mathcal{P}} \sup_{\tau \in \mathcal{T}} E^*[\, H_\tau\,] = \sup_{P^* \in \mathcal{P}} U_0^{P^*}.$$

Moreover, $\Pi(H)$ either consists of one single point or does not contain its upper endpoint $\pi_{\sup}(H)$.

Proof. Let τ^* be a stopping time that is optimal with respect to a given $P^* \in \mathcal{P}$. Then $U_0^{P^*} = E^*[\, H_{\tau^*}\,] = \sup_{\tau' \in \mathcal{T}} E^*[\, H_{\tau'}\,]$, and consequently $U_0^{P^*} \in \Pi(H)$. Together with the a priori bounds (6.17), we obtain the inclusions

$$\{U_0^{P^*} \mid P^* \in \mathcal{P}\} \subseteq \Pi(H) \subseteq [a, b], \tag{6.21}$$

where

$$a := \sup_{\tau \in \mathcal{T}} \inf_{P^* \in \mathcal{P}} E^*[\, H_\tau\,] \quad \text{and} \quad b := \sup_{\tau \in \mathcal{T}} \sup_{P^* \in \mathcal{P}} E^*[\, H_\tau\,].$$

Moreover, the minimax identity (6.20) shows that

$$a = \inf_{P^* \in \mathcal{P}} \sup_{\tau \in \mathcal{T}} E^*[\, H_\tau\,] = \inf_{P^* \in \mathcal{P}} U_0^{P^*} \quad \text{and} \quad b = \sup_{P^* \in \mathcal{P}} U_0^{P^*}.$$

Together with (6.21), this yields the identification of $\pi_{\inf}(H)$ and $\pi_{\sup}(H)$ as a and b.

Now we claim that $\{U_0^{P^*} \mid P^* \in \mathcal{P}\}$ is an interval, which, in view of the preceding step, will prove that $\Pi(H)$ is also an interval. Take $P_0, P_1 \in \mathcal{P}$ and define $P_\alpha \in \mathcal{P}$ by $P_\alpha := \alpha P_1 + (1-\alpha)P_0$ for $0 \le \alpha \le 1$. By Theorem 6.18, $f(\alpha) := U_0^{P_\alpha}$ is the supremum of the affine functions

$$\alpha \mapsto E_\alpha[\,H_\tau\,] = \alpha E_1[\,H_\tau\,] + (1-\alpha)E_0[\,H_\tau\,], \quad \tau \in \mathcal{T}.$$

Thus, f is finite, convex, and lower semicontinuous on $[0,1]$, hence continuous; see part (a) of Proposition A.7. Since $\mathcal{P}$ is convex, this proves our claim.

It remains to exclude the possibility that b belongs to $\Pi(H)$ in case $a < b$. Suppose by way of contradiction that $b \in \Pi(H)$. Then the first condition in Definition 6.29 implies the existence of $\hat{\tau} \in \mathcal{T}$ and $\hat{P} \in \mathcal{P}$ such that $b \le \hat{E}[\,H_{\hat{\tau}}\,]$. It follows that

$$\hat{E}[\,H_{\hat{\tau}}\,] \ge b = \sup_{\tau\in\mathcal{T}} \sup_{P^*\in\mathcal{P}} E^*[\,H_\tau\,].$$

In particular, $\hat{P}$ attains the supremum of $E^*[\,H_{\hat{\tau}}\,]$ over $P^* \in \mathcal{P}$. Theorem 5.32 implies that the discounted European claim $H_{\hat{\tau}}$ is attainable and that $E^*[\,H_{\hat{\tau}}\,]$ is in fact independent of $P^* \in \mathcal{P}$. Hence,

$$b = \hat{E}[\,H_{\hat{\tau}}\,] = \inf_{P^*\in\mathcal{P}} E^*[\,H_{\hat{\tau}}\,] \le \sup_{\tau\in\mathcal{T}} \inf_{P^*\in\mathcal{P}} E^*[\,H_\tau\,],$$

and we end up with the contradiction $b \le a$. Thus, b cannot belong to $\Pi(H)$. □

Comparing the previous result with Theorem 5.32, one might wonder whether $\Pi(H)$ contains its lower bound if $\pi_{\inf}(H) < \pi_{\sup}(H)$. At a first glance, it may come as a surprise that both cases

$$\pi_{\inf}(H) \in \Pi(H) \quad \text{and} \quad \pi_{\inf}(H) \notin \Pi(H)$$

can occur, as is illustrated by the following simple example.

Example 6.32. Consider the situation of Exercise 5.4.2 in which a complete market model with $T = 2$ was enlarged by adding two external states ω^+ and ω^-, corresponding to additional information revealed at time $t = 2$. That is, the extended model is defined on the probability space $(\hat{\Omega}, \hat{\mathcal{F}}, \hat{P})$, where $\hat{\Omega} := \Omega \times \{\omega^+, \omega^-\}$, $\hat{\mathcal{F}} := \sigma(A \times \{\omega^+\} \mid A \in \mathcal{F})$, and $P[\,\{(\omega_0, \omega^\pm)\}\,] := \frac{1}{2} P[\,\{\omega\}\,]$. It was shown in Exercise 5.4.2 that for $0 < p < 1$ the probability measure

$$\hat{P}_p^*[\,A \times \{\omega^+\}\,] := p \cdot P^*[\,A\,], \quad A \in \mathcal{F},$$

is an equivalent martingale measure for the enlarged model, where P^* is the unique equivalent martingale measure in the original, complete model.

Consider the discounted American claim H defined as

$$H_0 \equiv 0, \quad H_1 \equiv 1, \quad \text{and} \quad H_2(\omega) := \begin{cases} 2 & \text{if } \omega = (\omega_0, \omega^+), \\ 0 & \text{if } \omega = (\omega_0, \omega^-). \end{cases}$$

Clearly, $\Pi(H) \subseteq [1, 2)$. On the other hand, $\tau_2 \equiv 2$ is an optimal stopping time for $\hat{P}_p^*$ if $p > \frac{1}{2}$, while $\tau_1 \equiv 1$ is optimal for $p \leq \frac{1}{2}$. Hence,

$$\Pi(H) = [1, 2),$$

and the lower bound $\pi_{\inf}(H) = 1$ is an arbitrage-free price for H. Now consider the discounted American claim $\widetilde{H}$ defined by $\widetilde{H}_t = H_t$ for $t = 0, 2$ and by $\widetilde{H}_1 \equiv 0$. In this case, we have

$$\Pi(\widetilde{H}) = (0, 2). \qquad \diamond$$

Theorem 6.31 suggests that an American claim H that admits a unique arbitrage-free price should be attainable in an appropriate sense. Corollary 6.22, our hedging result in the case of a complete market, suggests the following definition of attainability.

Definition 6.33. A discounted American claim H is called *attainable* if there exists a stopping time $\tau \in \mathcal{T}$ and a self-financing trading strategy $\overline{\xi}$ whose value process V satisfies P-a. s.,

$$V_t \geq H_t \quad \text{for all } t, \text{ and } V_\tau = H_\tau.$$

The trading strategy $\overline{\xi}$ is called a *hedging strategy* for H. △

If H is attainable, then a hedging strategy protects the seller not only against those claims H_τ that arise from stopping times τ. The seller is on the safe side even if the buyer has full knowledge of future prices and exercises H at an arbitrary $\mathcal{F}_T$-measurable random time σ. For instance, the buyer even could choose σ such that

$$H_\sigma = \max_{0 \leq t \leq T} H_t.$$

In fact, we will see in Remark 7.12 that H is attainable in the sense of Definition 6.33 if and only if $V_t \geq H_t$ for all t and $V_\sigma = H_\sigma$ for some $\mathcal{F}_T$-measurable random time σ.

If the market model is *complete*, then every American claim H is attainable. Moreover, Theorem 6.11 and Corollary 6.22 imply that the minimal initial investment needed for the purchase of a hedging strategy for H is equal to the unique arbitrage-free price of H. In a general, possibly incomplete market model, every *attainable* discounted American claim H satisfies our integrability condition (6.19) and has a unique arbitrage-free price, which is equal to the initial investment of a hedging strategy for H. This follows from Theorem 5.25. In fact, the following stronger result is true.

Theorem 6.34. *For a discounted American claim H satisfying* (6.19), *the following conditions are equivalent:*
(a) *H is attainable.*
(b) *H admits a unique arbitrage-free price $\pi(H)$, that is, $\Pi(H) = \{\pi(H)\}$.*
(c) *$\pi_{\sup}(H) \in \Pi(H)$.*
Moreover, if H is attainable, then $\pi(H)$ is equal to the initial investment of any hedging strategy for H.

The equivalence of (b) and (c) is an immediate consequence of Theorem 6.31. The remainder of the proof of Theorem 6.34 is postponed to Remark 7.10 because it requires the technique of superhedging, which will be introduced in Section 7.

6.4 Stability under pasting

In this section, we define the pasting of two equivalent probability measures at a given stopping time. This operation will play an important role in the analysis of lower and upper Snell envelopes as developed in Section 6.5. In particular, we will prepare for the proof of the minimax identity (6.20), which was used in the characterization of arbitrage-free prices of an American contingent claim. Let us start with a few preparations.

Definition 6.35. Let τ be a stopping time. The σ-algebra of events that are observable up to time τ is defined as

$$\mathscr{F}_\tau := \{A \in \mathscr{F} \mid A \cap \{\tau \le t\} \in \mathscr{F}_t \text{ for all } t\}. \qquad \triangle$$

Exercise 6.4.1. Let τ be a stopping time.
(a) Prove that $\mathscr{F}_\tau$ is indeed a σ-algebra.
(b) Argue that τ is $\mathscr{F}_\tau$-measurable.
(c) Show next that

$$\mathscr{F}_\tau = \{A \in \mathscr{F} \mid A \cap \{\tau = t\} \in \mathscr{F}_t \text{ for all } t\}$$

and conclude that $\mathscr{F}_\tau = \mathscr{F}_t$ if τ is the constant stopping time with $\tau(\omega) = t$ for all $\omega \in \Omega$.
(d) Let σ be another stopping time. Show that $A \cap \{\sigma \le \tau\} \in \mathscr{F}_\tau$ for all $A \in \mathscr{F}_\sigma$.
(e) Conclude that $\mathscr{F}_\sigma \subseteq \mathscr{F}_\tau$ if σ is a stopping time with $\sigma(\omega) \le \tau(\omega)$ for all $\omega \in \Omega$. $\diamond$

Lemma 6.36. *Let τ and σ be two stopping times and suppose that Y is $\mathscr{F}_\tau$-measurable. Then $Y\mathbb{1}_{\{\tau\le\sigma\}}$ and $Y\mathbb{1}_{\{\tau<\sigma\}}$ are both $\mathscr{F}_\tau \cap \mathscr{F}_\sigma$-measurable.*

Proof. On the one hand, $Y\mathbb{1}_{\{\tau\le t\}}$ and $Y\mathbb{1}_{\{\tau<t\}}$ are $\mathscr{F}_t$-measurable, and so

$$Y\mathbb{1}_{\{\tau\le\sigma\}}\mathbb{1}_{\{\sigma=t\}} = Y\mathbb{1}_{\{\tau\le t\}}\mathbb{1}_{\{\sigma=t\}} \quad \text{and} \quad Y\mathbb{1}_{\{\tau<\sigma\}}\mathbb{1}_{\{\sigma=t\}} = Y\mathbb{1}_{\{\tau<t\}}\mathbb{1}_{\{\sigma=t\}}$$

are $\mathscr{F}_t$-measurable for each t. Therefore, $Y\mathbb{1}_{\{\tau\le\sigma\}}$ and $Y\mathbb{1}_{\{\tau<\sigma\}}$ are both $\mathscr{F}_\sigma$-measurable by Exercise 6.4.1(c). On the other hand, $Y\mathbb{1}_{\{\tau=t\}}$ is $\mathscr{F}_t$-measurable, $\{\sigma \ge t\} = \{\sigma < t\}^c$ is $\mathscr{F}_{t-1}$-measurable, $\{\sigma > t\} = \{\sigma \le t\}^c$ is $\mathscr{F}_t$-measurable, and so

$$Y\mathbb{1}_{\{\tau\le\sigma\}}\mathbb{1}_{\{\tau=t\}} = Y\mathbb{1}_{\{\tau=t\}}\mathbb{1}_{\{\sigma\ge t\}} \quad \text{and} \quad Y\mathbb{1}_{\{\tau<\sigma\}}\mathbb{1}_{\{\tau=t\}} = Y\mathbb{1}_{\{\tau=t\}}\mathbb{1}_{\{\sigma>t\}}$$

are $\mathscr{F}_t$-measurable for each t. It follows again from Exercise 6.4.1(c) that $Y\mathbb{1}_{\{\tau\le\sigma\}}$ is also $\mathscr{F}_\tau$-measurable. □

The following result is an addendum to Doob's stopping theorem, Theorem 6.15.

Proposition 6.37. *For an adapted process M in $\mathscr{L}^1(Q)$, the following conditions are equivalent:*
(a) *M is a Q-martingale.*
(b) *$E_Q[\,M_\tau \mid \mathscr{F}_\sigma\,] = M_{\tau\wedge\sigma}$ for all $\tau \in \mathcal{T}$ and all stopping times σ.*

Proof. (a)⇒(b): Take a set $A \in \mathscr{F}_\sigma$ and let us write

$$E_Q[\,M_\tau; A\,] = E_Q[\,M_\tau; A\cap\{\tau\le\sigma\}\,] + E_Q[\,M_\tau; A\cap\{\tau>\sigma\}\,].$$

Condition (b) will follow if we may replace M_τ by M_σ in the rightmost expectation. To this end, note that

$$A\cap\{\sigma=t\}\cap\{\tau>\sigma\} = A\cap\{\sigma=t\}\cap\{\tau>t\} \in \mathscr{F}_t.$$

Thus, since the stopped process M^τ is a martingale by Theorem 6.15,

$$\begin{aligned} E_Q[\,M_\tau; A\cap\{\tau>\sigma\}\,] &= \sum_{t=0}^{T} E_Q[\,M_T^\tau; A\cap\{\sigma=t\}\cap\{\tau>\sigma\}\,] \\ &= \sum_{t=0}^{T} E_Q[\,M_t^\tau; A\cap\{\sigma=t\}\cap\{\tau>\sigma\}\,] \\ &= E_Q[\,M_\sigma; A\cap\{\tau>\sigma\}\,]. \end{aligned}$$

(b)⇒(a): This follows by taking $\tau \equiv t$ and $\sigma \equiv s \le t$. □

Exercise 6.4.2. Let Z be the density process of a probability measure $\widetilde{Q}$ that is absolutely continuous with respect to Q; see Exercise 5.2.3. Show that for a stopping time σ, we have $\widetilde{Q} \ll Q$ on $\mathscr{F}_\sigma$ with density given by

$$\frac{d\widetilde{Q}}{dQ}\Big|_{\mathscr{F}_\sigma} = E_Q[\,Z_T \mid \mathscr{F}_\sigma\,] = Z_{T\wedge\sigma}.$$ ◇

Exercise 6.4.3. Show that for a stopping time τ, a random variable $Y \in \mathscr{L}^1(\Omega, \mathscr{F}, Q)$, and $t \in \{0,\dots,T\}$,

$$E_Q[\, Y \mid \mathscr{F}_\tau \,] = E_Q[\, Y \mid \mathscr{F}_t \,] \quad Q\text{-a. s. on } \{\tau = t\}. \tag{6.22}$$

◇

We next state the following extension of Theorem 6.18. It provides the solution to the optimal stopping problem posed at any stopping time $\tau \le T$.

Proposition 6.38. *Let H be an adapted process in $\mathscr{L}^1(\Omega, \mathscr{F}, Q)$, and define for $\tau \in \mathcal{T}$,*

$$\mathcal{T}_\tau := \{\sigma \in \mathcal{T} \mid \sigma \ge \tau\}.$$

Then the Snell envelope U^Q of H satisfies Q-a. s.

$$U_\tau^Q = \operatorname*{ess\,sup}_{\sigma \in \mathcal{T}_\tau} E_Q[\, H_\sigma \mid \mathscr{F}_\tau \,],$$

and the essential supremum is attained for

$$\sigma_{\min}^{(\tau)} := \min\{t \ge \tau \mid H_t = U_t^Q\}.$$

Exercise 6.4.4. Prove Proposition 6.38 by using the identity (6.22). ◇

Definition 6.39. Let Q_1 and Q_2 be two equivalent probability measures and take $\sigma \in \mathcal{T}$. The probability measure

$$\widetilde{Q}[\, A \,] := E_{Q_1}[\, Q_2[\, A \mid \mathscr{F}_\sigma \,] \,], \quad A \in \mathscr{F}_T,$$

is called the *pasting of Q_1 and Q_2 in σ.* △

The monotone convergence theorem for conditional expectations guarantees that $\widetilde{Q}$ is indeed a probability measure and that

$$E_{\widetilde{Q}}[\, Y \,] = E_{Q_1}[\, E_{Q_2}[\, Y \mid \mathscr{F}_\sigma \,] \,]$$

for all $\mathscr{F}_T$-measurable $Y \ge 0$. Note that $\widetilde{Q}$ coincides with Q_1 on $\mathscr{F}_\sigma$, that is,

$$E_{\widetilde{Q}}[\, Y \,] = E_{Q_1}[\, Y \,] \quad \text{for all } \mathscr{F}_\sigma\text{-measurable } Y \ge 0.$$

Lemma 6.40. *For $Q_1 \approx Q_2$, their pasting in $\sigma \in \mathcal{T}$ is equivalent to Q_1 and satisfies*

$$\frac{d\widetilde{Q}}{dQ_1} = \frac{Z_T}{Z_\sigma},$$

where Z is the density process of Q_2 with respect to Q_1.

Proof. For $Y \ge 0$,

$$
\begin{aligned}
E_{\widetilde{Q}}[\,Y\,] &= E_{Q_1}[\,E_{Q_2}[\,Y \mid \mathscr{F}_\sigma\,]\,] \\
&= E_{Q_1}\Big[\,\frac{1}{Z_\sigma}E_{Q_1}[\,YZ_T \mid \mathscr{F}_\sigma\,]\,\Big] \\
&= E_{Q_1}\Big[\,\frac{Z_T}{Z_\sigma}\,Y\,\Big],
\end{aligned}
$$

where we have used the martingale property of Z, the fact that $Z_\sigma > 0$ Q_1-almost surely, and Proposition B.8. The equivalence of $\widetilde{Q}$ and Q_1 follows from $Z_T > 0$ Q_1-almost surely. □

Lemma 6.41. *For $Q_1 \approx Q_2$, let $\widetilde{Q}$ be their pasting in $\sigma \in \mathcal{T}$. Then, for all stopping times τ and $\mathscr{F}_T$-measurable $Y \geq 0$,*

$$
E_{\widetilde{Q}}[\,Y \mid \mathscr{F}_\tau\,] = E_{Q_1}[\,E_{Q_2}[\,Y \mid \mathscr{F}_{\sigma\vee\tau}\,] \mid \mathscr{F}_\tau\,].
$$

Proof. We first show the claimed identity on the set $\{\tau \leq \sigma\}$. To this end, take an $\mathscr{F}_\tau$-measurable test function $\varphi \geq 0$. Then $\varphi\mathbb{1}_{\{\tau\leq\sigma\}}$ is $\mathscr{F}_\sigma\cap\mathscr{F}_\tau$-measurable by Lemma 6.36. Therefore,

$$
\begin{aligned}
E_{\widetilde{Q}}[\,Y\varphi\mathbb{1}_{\{\tau\leq\sigma\}}\,] &= E_{Q_1}[\,E_{Q_2}[\,Y\varphi\mathbb{1}_{\{\tau\leq\sigma\}} \mid \mathscr{F}_\sigma\,]\,] \\
&= E_{Q_1}[\,E_{Q_2}[\,Y \mid \mathscr{F}_\sigma\,]\varphi\mathbb{1}_{\{\tau\leq\sigma\}}\,] \\
&= E_{Q_1}[\,E_{Q_1}[\,E_{Q_2}[\,Y \mid \mathscr{F}_\sigma\,] \mid \mathscr{F}_\tau\,]\varphi\mathbb{1}_{\{\tau\leq\sigma\}}\,].
\end{aligned}
$$

It follows that

$$
E_{\widetilde{Q}}[\,Y \mid \mathscr{F}_\tau\,] = E_{Q_1}[\,E_{Q_2}[\,Y \mid \mathscr{F}_\sigma\,] \mid \mathscr{F}_\tau\,] \quad \text{a. s. on } \{\tau \leq \sigma\}. \tag{6.23}
$$

Next, Exercise 6.4.3 yields that

$$
E_{Q_2}[\,Y \mid \mathscr{F}_{\sigma\vee\tau}\,]\mathbb{1}_{\{\tau\leq\sigma\}} = E_{Q_2}[\,Y \mid \mathscr{F}_\sigma\,]\mathbb{1}_{\{\tau\leq\sigma\}}. \tag{6.24}
$$

Moreover, $\mathbb{1}_{\{\tau\leq\sigma\}}$ is $\mathscr{F}_\tau$-measurable by Lemma 6.36. Hence,

$$
\begin{aligned}
E_{Q_1}[\,E_{Q_2}[\,Y \mid \mathscr{F}_\sigma\,] \mid \mathscr{F}_\tau\,]\mathbb{1}_{\{\tau\leq\sigma\}} &= E_{Q_1}[\,E_{Q_2}[\,Y \mid \mathscr{F}_\sigma\,]\mathbb{1}_{\{\tau\leq\sigma\}} \mid \mathscr{F}_\tau\,] \\
&= E_{Q_1}[\,E_{Q_2}[\,Y \mid \mathscr{F}_{\sigma\vee\tau}\,]\mathbb{1}_{\{\tau\leq\sigma\}} \mid \mathscr{F}_\tau\,] \\
&= E_{Q_1}[\,E_{Q_2}[\,Y \mid \mathscr{F}_{\sigma\vee\tau}\,] \mid \mathscr{F}_\tau\,]\mathbb{1}_{\{\tau\leq\sigma\}}.
\end{aligned}
$$

Plugging this result into (6.23) shows that our assertion holds on $\{\tau \leq \sigma\}$.

To prove the assertion on $\{\tau > \sigma\}$, we show first a preliminary identity. Lemma 6.36 implies that $\psi\mathbb{1}_{\{\tau>\sigma\}}$ is $\mathscr{F}_\tau$-measurable for any $\mathscr{F}_\sigma$-measurable ψ. Taking first $\psi = 1$ and then $\psi = 1/Z_\sigma$ yields that for any $\mathscr{F}_T$-measurable W,

$$
E_{Q_1}\Big[\,W\frac{Z_T}{Z_\sigma}\,\Big|\,\mathscr{F}_\tau\,\Big]\mathbb{1}_{\{\tau>\sigma\}} = E_{Q_1}\Big[\,W\frac{Z_T}{Z_\sigma}\mathbb{1}_{\{\tau>\sigma\}}\,\Big|\,\mathscr{F}_\tau\,\Big] = E_{Q_1}[\,WZ_T \mid \mathscr{F}_\tau\,]\frac{1}{Z_\sigma}\mathbb{1}_{\{\tau>\sigma\}}. \tag{6.25}
$$

After this preparation, we use Proposition B.8 and Lemma 6.40 to obtain that

$$E_{\widetilde{Q}}[\,Y \mid \mathscr{F}_\tau\,] = \frac{1}{E_{Q_1}[\,\frac{Z_T}{Z_\sigma} \mid \mathscr{F}_\tau\,]} E_{Q_1}\Big[\,Y\frac{Z_T}{Z_\sigma} \;\Big|\; \mathscr{F}_\tau\,\Big]. \tag{6.26}$$

Using (6.25) with $W = 1$ on the denominator and with $W = Y$ on the numerator of (6.26) yields

$$\begin{aligned} E_{\widetilde{Q}}[\,Y \mid \mathscr{F}_\tau\,]\mathbb{1}_{\{\tau>\sigma\}} &= \frac{1}{E_{Q_1}[\,Z_T \mid \mathscr{F}_\tau\,]} E_{Q_1}[\,YZ_T \mid \mathscr{F}_\tau\,]\mathbb{1}_{\{\tau>\sigma\}} \\ &= \frac{1}{Z_\tau} E_{Q_1}[\,YZ_T \mid \mathscr{F}_\tau\,]\mathbb{1}_{\{\tau>\sigma\}} \\ &= E_{Q_2}[\,Y \mid \mathscr{F}_\tau\,]\mathbb{1}_{\{\tau>\sigma\}} \\ &= E_{Q_1}[\,E_{Q_2}[\,Y \mid \mathscr{F}_\tau\,]\mathbb{1}_{\{\tau>\sigma\}} \mid \mathscr{F}_\tau\,] \\ &= E_{Q_1}[\,E_{Q_2}[\,Y \mid \mathscr{F}_{\sigma\vee\tau}\,]\mathbb{1}_{\{\tau>\sigma\}} \mid \mathscr{F}_\tau\,] \\ &= E_{Q_1}[\,E_{Q_2}[\,Y \mid \mathscr{F}_{\sigma\vee\tau}\,] \mid \mathscr{F}_\tau\,]\mathbb{1}_{\{\tau>\sigma\}}. \end{aligned}$$

Here, we have used Proposition B.8 in the third identity, the $\mathscr{F}_\tau$-measurability of $\{\tau > \sigma\}$ in the fourth and sixth steps, and (6.24) with reversed roles of τ and σ in the fifth identity. □

Definition 6.42. A set $\mathcal{Q}$ of equivalent probability measures on $(\Omega, \mathscr{F})$ is called *stable* if, for any $Q_1, Q_2 \in \mathcal{Q}$ and $\sigma \in \mathcal{T}$, the pasting in σ is also contained in $\mathcal{Q}$. △

The condition of stability in the preceding definition is sometimes also called *fork convexity*, *m-stability*, or *stability under pasting*. For the purposes of this book, the most important example of a stable set is the class $\mathscr{P}$ of all equivalent martingale measures. But in Section 6.5 we will also discuss the connection between stable sets and dynamic risk measures.

Proposition 6.43. *$\mathscr{P}$ is stable.*

Proof. Take $P_1, P_2 \in \mathscr{P}$, and denote by $\widetilde{P}$ their pasting in a given $\sigma \in \mathcal{T}$. Doob's stopping theorem in the form of Proposition 6.37 and Lemma 6.41 applied with $Y := X_t^i \geq 0$ and $\tau \equiv s$ yield that for $s \leq t$,

$$\widetilde{E}[\,X_t \mid \mathscr{F}_s\,] = E_1[\,E_2[\,X_t \mid \mathscr{F}_{\sigma\vee s}\,] \mid \mathscr{F}_s\,] = E_1[\,X_{\sigma\vee s} \mid \mathscr{F}_s\,] = X_s.$$

It follows in particular that each component X_t^i is in $\mathscr{L}^1(\widetilde{P})$ since $\widetilde{E}[\,X_t^i\,] = X_0^i < \infty$, concluding the proof of $\widetilde{P} \in \mathscr{P}$. □

We conclude this section by an alternative characterization of stable sets. It will be used in Section 11.2. Suppose that $\sigma \in \mathcal{T}$ takes at most one value $t \in \{0, \dots, T\}$ that is different from T. Then there exists a set $B \in \mathscr{F}_t$ such that $\sigma = t \cdot \mathbb{1}_B + T \cdot \mathbb{1}_{B^c}$. It follows that the pasting $\widetilde{Q}$ of two equivalent probability measures Q_1 and Q_2 in σ is given by

$$\widetilde{Q}[A] := E_{Q_1}[\, Q_2[A \mid \mathscr{F}_t] \cdot \mathbb{1}_B + \mathbb{1}_{A\cap B^c} \,], \quad A \in \mathscr{F}_T. \tag{6.27}$$

This observation can be used to give the following characterization of stable sets.

Proposition 6.44. *A set $\mathcal{Q}$ of equivalent probability measures is stable if and only if for any $t \in \{0,\dots,T\}$ and $B \in \mathscr{F}_t$ the probability measure $\widetilde{Q}$ defined in* (6.27) *belongs again to $\mathcal{Q}$.*

Proof. We have already seen that $\widetilde{Q} \in \mathcal{Q}$ if $\mathcal{Q}$ is stable. For the proof of the converse implication, let $\sigma \in \mathcal{T}$ be a stopping time and take $Q_1, Q_2 \in \mathcal{Q}$. We define recursively $\widetilde{Q}_T := Q_1$ and

$$\widetilde{Q}_{t-1}[A] := E_{\widetilde{Q}_t}[\, Q_2[A \mid \mathscr{F}_{t-1}] \cdot \mathbb{1}_{\{\sigma=t-1\}} + \mathbb{1}_{A\cap\{\sigma\neq t-1\}} \,]$$

for $t = T,\dots,1$. Then $\widetilde{Q}_0 \in \mathcal{Q}$ by assumption. We claim that $\widetilde{Q}_0$ coincides with the pasting of Q_1 and Q_2 in σ, and this will prove the assertion. To verify our claim, note that the densities of $\widetilde{Q}_t$ with respect to Q_1 satisfy the recursion

$$\frac{d\widetilde{Q}_{t-1}}{dQ_1} = \frac{d\widetilde{Q}_t}{dQ_1}\left(\frac{Z_T}{Z_{t-1}}\mathbb{1}_{\{\sigma=t-1\}} + \mathbb{1}_{\{\sigma\neq t-1\}}\right),$$

where (Z_t) is the density process of Q_2 with respect to Q_1. But this implies that

$$\frac{d\widetilde{Q}_{t-1}}{dQ_1} = \frac{Z_T}{Z_\sigma}\mathbb{1}_{\{\sigma\geq t-1\}} + \mathbb{1}_{\{\sigma<t-1\}},$$

and so our claim $\widetilde{Q}_0 = \widetilde{Q}$ follows from Lemma 6.40. □

6.5 Lower and upper Snell envelopes

Our main goal in this section is to provide a proof of the minimax identity (6.20), which was used in the characterization of the set of arbitrage-free prices of an American contingent claim. The techniques and results we develop here will help to characterize the time consistency of dynamic coherent risk measures and they will also be needed in Chapter 7. Moreover, they can be interpreted in terms of an optimal stopping problem for general utility functionals that appear in a robust Savage representation of preferences on payoff profiles. Let us now fix a set $\mathcal{Q}$ of equivalent probability measures and an adapted process H such that

$$H_t \in \mathscr{L}^1(Q) \quad \text{for all } t \text{ and each } Q \in \mathcal{Q}.$$

Recall that this condition implies

$$\inf_{Q\in\mathcal{Q}} \sup_{\tau\in\mathcal{T}} E_Q[\, H_\tau \,] = \inf_{Q\in\mathcal{Q}} U_0^Q < \infty,$$

where U^Q denotes the Snell envelope of H with respect to $Q \in \mathcal{Q}$. Let us also assume that

$$\mathcal{Q} \text{ is stable.}$$

Definition 6.45. The *lower Snell envelope* of H is defined as

$$U_t^{\downarrow} := \operatorname*{ess\,inf}_{Q\in\mathcal{Q}} U_t^Q = \operatorname*{ess\,inf}_{Q\in\mathcal{Q}} \operatorname*{ess\,sup}_{\tau\in\mathcal{T}_t} E_Q[\,H_\tau \mid \mathscr{F}_t\,], \quad t = 0,\dots,T.$$

The *upper Snell envelope* of H is defined as

$$U_t^{\uparrow} := \operatorname*{ess\,sup}_{Q\in\mathcal{Q}} U_t^Q = \operatorname*{ess\,sup}_{\tau\in\mathcal{T}_t} \operatorname*{ess\,sup}_{Q\in\mathcal{Q}} E_Q[\,H_\tau \mid \mathscr{F}_t\,], \quad t = 0,\dots,T. \qquad \triangle$$

We will first study the lower Snell envelope. The following “minimax theorem” states that the essential infimum and the essential supremum occurring in the definition of $U^{\downarrow}$ may be interchanged if $\mathcal{Q}$ is stable. Applied at $t = 0$ and combined with Proposition 6.43, this gives the identity (6.20), which was used in our characterization of the arbitrage-free prices of an American claim H.

Theorem 6.46. *The lower Snell envelope of H satisfies*

$$U_t^{\downarrow} = \operatorname*{ess\,sup}_{\tau\in\mathcal{T}_t} \operatorname*{ess\,inf}_{Q\in\mathcal{Q}} E_Q[\,H_\tau \mid \mathscr{F}_t\,] \quad \textit{for each } t. \tag{6.28}$$

In particular,

$$U_0^{\downarrow} = \inf_{Q\in\mathcal{Q}} \sup_{\tau\in\mathcal{T}} E_Q[\,H_\tau\,] = \sup_{\tau\in\mathcal{T}} \inf_{Q\in\mathcal{Q}} E_Q[\,H_\tau\,].$$

The inequality $\geq$ in (6.28) is obvious. Its converse is an immediate consequence of the next theorem, which solves the following optimal stopping problem that is formulated with respect to the *nonadditive* expectation operator $\inf_{Q\in\mathcal{Q}} E_Q[\,\cdot\,]$:

$$\text{maximize } \inf_{Q\in\mathcal{Q}} E_Q[\,H_\tau\,] \text{ over all } \tau \in \mathcal{T}.$$

Theorem 6.47. *Define a stopping time $\tau_t \in \mathcal{T}_t$ by*

$$\tau_t := \min\{u \geq t \mid U_u^{\downarrow} = H_u\}.$$

Then, P-a. s.,

$$U_t^{\downarrow} = \operatorname*{ess\,inf}_{Q\in\mathcal{Q}} E_Q[\,H_{\tau_t} \mid \mathscr{F}_t\,]. \tag{6.29}$$

In particular,

$$\sup_{\tau\in\mathcal{T}} \inf_{Q\in\mathcal{Q}} E_Q[\,H_\tau\,] = \inf_{Q\in\mathcal{Q}} E_Q[\,H_{\tau_0}\,] = U_0^{\downarrow}.$$

For the proof of Theorem 6.47, we need some preparations.

Lemma 6.48. *Suppose that we are given $Q_1, Q_2 \in \mathcal{Q}$, a stopping time $\tau \in \mathcal{T}$, and a set $B \in \mathcal{F}_\tau$. Let $\widetilde{Q} \in \mathcal{Q}$ be the pasting of Q_1 and Q_2 in the stopping time*

$$\sigma := \tau \mathbb{1}_B + T \mathbb{1}_{B^c}.$$

Then the Snell envelopes associated with these three measures are related as follows:

$$U_\tau^{\widetilde{Q}} = U_\tau^{Q_2} \mathbb{1}_B + U_\tau^{Q_1} \mathbb{1}_{B^c} \quad P\text{-a. s.} \tag{6.30}$$

Proof. With Proposition 6.38 and its notation, we have

$$U_\tau^{\widetilde{Q}} = \operatorname*{ess\,sup}_{\rho \in \mathcal{T}_\tau} E_{\widetilde{Q}}[\, H_\rho \mid \mathcal{F}_\tau \,].$$

To compute the conditional expectation on the right, note first that

$$E_{Q_2}[\, H_\rho \mid \mathcal{F}_{\sigma \vee \tau} \,] = E_{Q_2}[\, H_\rho \mid \mathcal{F}_\tau \,] \mathbb{1}_B + H_\rho \mathbb{1}_{B^c}.$$

Hence, Lemma 6.41 yields that

$$E_{\widetilde{Q}}[\, H_\rho \mid \mathcal{F}_\tau \,] = E_{Q_2}[\, H_\rho \mid \mathcal{F}_\tau \,] \mathbb{1}_B + E_{Q_1}[\, H_\rho \mid \mathcal{F}_\tau \,] \mathbb{1}_{B^c}.$$

Moreover, whenever $\rho_1, \rho_2 \in \mathcal{T}_\tau$, then

$$\widetilde{\rho} := \rho_1 \mathbb{1}_B + \rho_2 \mathbb{1}_{B^c}$$

is also a stopping time in $\mathcal{T}_\tau$. Thus,

$$U_\tau^{\widetilde{Q}} = \operatorname*{ess\,sup}_{\rho \in \mathcal{T}_\tau} E_{Q_2}[\, H_\rho \mid \mathcal{F}_\tau \,] \mathbb{1}_B + \operatorname*{ess\,sup}_{\rho \in \mathcal{T}_\tau} E_{Q_1}[\, H_\rho \mid \mathcal{F}_\tau \,] \mathbb{1}_{B^c},$$

and (6.30) follows. □

Lemma 6.49. *For any $Q \in \mathcal{Q}$ and $\tau \in \mathcal{T}$, there exist $Q_k \in \mathcal{Q}$ such that $Q_k = Q$ on $\mathcal{F}_\tau$ and*

$$U_\tau^{Q_k} \searrow \operatorname*{ess\,inf}_{\hat{Q} \in \mathcal{Q}} U_\tau^{\hat{Q}} = U_\tau^{\downarrow}.$$

Similarly, there exist $Q^k \in \mathcal{Q}$ such that $Q^k = Q$ on $\mathcal{F}_\tau$ and

$$U_\tau^{Q^k} \nearrow \operatorname*{ess\,sup}_{\hat{Q} \in \mathcal{Q}} U_\tau^{\hat{Q}} =: U_\tau^{\uparrow}.$$

Proof. For $Q_1, Q_2 \in \mathcal{Q}$, $B := \{U_\tau^{Q_1} > U_\tau^{Q_2}\}$, take $\widetilde{Q} \in \mathcal{Q}$ as in Lemma 6.48. Then

$$U_\tau^{\widetilde{Q}} = U_\tau^{Q_1} \cdot \mathbb{1}_{B^c} + U_\tau^{Q_2} \cdot \mathbb{1}_B = U_\tau^{Q_1} \wedge U_\tau^{Q_2}. \tag{6.31}$$

Moreover, if $Q_1 = Q$ on $\mathcal{F}_\tau$ then also $\widetilde{Q} = Q$ on $\mathcal{F}_\tau$. Hence, the set

$$\Phi := \{U_\tau^{\hat{Q}} \mid \hat{Q} \in \mathcal{Q} \text{ and } \hat{Q} = Q \text{ on } \mathcal{F}_\tau\}$$

is such that $U_\tau^{\downarrow} = \operatorname{ess\,inf} \Phi$. Moreover, (6.31) implies that Φ is directed downwards, and the second part of Theorem F.1 states the existence of the desired sequence (Q_k) in $\mathcal{Q}$. The proof for the essential supremum is analogous. □

Proof of Theorem 6.47. To prove (6.29), observe first that $U_t^Q \geq E_Q[H_{\tau_t} \mid \mathcal{F}_t]$ for each $Q \in \mathcal{Q}$, so that $\geq$ holds in (6.29). For the proof of the converse inequality, note that

$$\tau_t \leq \min\{u \geq t \mid U_u^Q = H_u\} =: \tau_t^Q \quad \text{for } Q \in \mathcal{Q}.$$

It was shown in Theorem 6.18 that τ_t^Q is the minimal optimal stopping time after time t and with respect to Q. It was also shown in the proof of Theorem 6.18 that the stopped process $(U^Q)^{\tau_t^Q}$ is a Q-martingale from time t on. In particular,

$$U_t^Q = E_Q[\,U^Q_{\tau_t \wedge \tau_t^Q} \mid \mathcal{F}_t\,] = E_Q[\,U_{\tau_t}^Q \mid \mathcal{F}_t\,] \quad \text{for all } Q \in \mathcal{Q}. \tag{6.32}$$

Let us now fix some $Q \in \mathcal{Q}$. Lemma 6.49 yields $Q_k \in \mathcal{Q}$ with $Q_k = Q$ on $\mathcal{F}_{\tau_t}$ such that $U_{\tau_t}^{Q_k}$ decreases to $U_{\tau_t}^{\downarrow}$. We obtain

$$\begin{aligned} E_Q[\,H_{\tau_t} \mid \mathcal{F}_t\,] &= E_Q[\,U_{\tau_t}^{\downarrow} \mid \mathcal{F}_t\,] = E_Q\Big[\lim_{k\uparrow\infty} U_{\tau_t}^{Q_k} \mid \mathcal{F}_t\Big] \\ &= \lim_{k\uparrow\infty} E_Q[\,U_{\tau_t}^{Q_k} \mid \mathcal{F}_t\,] = \lim_{k\uparrow\infty} E_{Q_k}[\,U_{\tau_t}^{Q_k} \mid \mathcal{F}_t\,] \\ &= \lim_{k\uparrow\infty} U_t^{Q_k} \geq U_t^{\downarrow}. \end{aligned}$$

Here, we have used that $H_{\tau_t} \leq U_{\tau_t}^{Q_k} \leq U_{\tau_t}^{Q_1}$ and $E_Q[\,|U_{\tau_t}^{Q_1}|\,] = E_{Q_1}[\,|U_{\tau_t}^{Q_1}|\,] < \infty$ together with dominated convergence in the third step, the fact that $Q_k = Q$ on $\mathcal{F}_{\tau_t} \supseteq \mathcal{F}_t$ in the fourth, and (6.32) in the fifth identity. □

Remark 6.50. Suppose the buyer of an American option uses a utility functional of the form

$$\inf_{Q \in \mathcal{Q}} E_Q[\,u(Z)\,],$$

where $\mathcal{Q}$ is a set of probability measures and u is a measurable function. This may be viewed as a robust Savage representation of a preference relation on discounted asset payoffs; see Section 2.5. Thus, the aim of the buyer is to maximize the utility

$$\inf_{Q\in\mathcal{Q}} E_Q[\,u(H_\tau)\,]$$

of the discounted payoff H_τ over all stopping times $\tau \in \mathcal{T}$. This generalized utility maximization problem can be solved with the results developed in this section, provided that the set $\mathcal{Q}$ is a *stable* set of equivalent probability measures. Indeed, assume

$$\widetilde{H}_t := u(H_t) \in \mathcal{L}^1(Q) \quad \text{for all } t \text{ and each } Q \in \mathcal{Q},$$

and let U^Q be the Snell envelope of $\widetilde{H}_t$ with respect to $Q \in \mathcal{Q}$. Theorem 6.47 states that the generalized optimal stopping problem is solved by the stopping time

$$\tau^* := \min\Big\{t \ge 0 \mid \operatorname*{ess\,inf}_{Q\in\mathcal{Q}} U_t^Q = \widetilde{H}_t\Big\},$$

that is,

$$\inf_{Q\in\mathcal{Q}} \sup_{\tau\in\mathcal{T}} E_Q[\,u(H_\tau)\,] = U_0^{\downarrow} = \inf_{Q\in\mathcal{Q}} E_Q[\,u(H_{\tau^*})\,].$$ ◇

Let us now turn to the analysis of the upper Snell envelope

$$U_t^{\uparrow} := \operatorname*{ess\,sup}_{Q\in\mathcal{Q}} U_t^Q = \operatorname*{ess\,sup}_{\tau\in\mathcal{T}_t} \operatorname*{ess\,sup}_{Q\in\mathcal{Q}} E_Q[\,H_\tau \mid \mathcal{F}_t\,], \quad t = 0,\dots,T.$$

In order to simplify the presentation, we will assume from now on that

$$\sup_{Q\in\mathcal{Q}} E_Q[\,|H_t|\,] < \infty \quad \text{for all } t.$$

This condition implies that

$$U_0^{\uparrow} = \sup_{Q\in\mathcal{Q}} U_0^Q \le \sup_{\tau\in\mathcal{T}} \sup_{Q\in\mathcal{Q}} E_Q[\,|H_\tau|\,] < \infty.$$

Our main result on upper Snell envelopes states that, for stable sets $\mathcal{Q}$, the upper Snell envelope $U^{\uparrow}$ satisfies a recursive scheme that is similar to the one for ordinary Snell envelopes. In contrast to (6.4), however, it involves the *nonadditive conditional expectation operators* $\operatorname{ess\,sup}_Q E_Q[\,\cdot \mid \mathcal{F}_t\,]$.

Theorem 6.51. *$U^{\uparrow}$ satisfies the following recursive scheme:*

$$U_T^{\uparrow} = H_T \quad \textit{and} \quad U_t^{\uparrow} = H_t \vee \operatorname*{ess\,sup}_{Q\in\mathcal{Q}} E_Q[\,U_{t+1}^{\uparrow} \mid \mathcal{F}_t\,], \quad t = T-1\dots,0. \tag{6.33}$$

Proof. The definition of the Snell envelope U^Q implies that

$$U_t^{\uparrow} = \operatorname*{ess\,sup}_{Q\in\mathcal{Q}} U_t^Q = H_t \vee \operatorname*{ess\,sup}_{Q\in\mathcal{Q}} E_Q[\,U_{t+1}^Q \mid \mathcal{F}_t\,]. \tag{6.34}$$

Next, we fix $Q \in \mathcal{Q}$ and denote by $\mathcal{Q}_{t+1}(Q)$ the set of all $\hat{Q} \in \mathcal{Q}$ that coincide with Q on $\mathcal{F}_{t+1}$. According to Lemma 6.49, there are $Q^k \in \mathcal{Q}_{t+1}(Q)$ such that $U^{Q^k}_{t+1} \nearrow U^{\uparrow}_{t+1}$. The fact that $E_Q[\,|U^{Q_1}_{t+1}|\,] = E_{Q_1}[\,|U^{Q_1}_{t+1}|\,] < \infty$ combined with monotone convergence for conditional expectations shows that

$$\begin{aligned}
\operatorname*{ess\,sup}_{Q\in\mathcal{Q}} E_Q[\,U^{\uparrow}_{t+1} \mid \mathcal{F}_t\,] &\geq \operatorname*{ess\,sup}_{Q\in\mathcal{Q}} E_Q[\,U^{Q}_{t+1} \mid \mathcal{F}_t\,] \\
&= \operatorname*{ess\,sup}_{Q\in\mathcal{Q}}\ \operatorname*{ess\,sup}_{\hat{Q}\in\mathcal{Q}_{t+1}(Q)} E_Q[\,U^{\hat{Q}}_{t+1} \mid \mathcal{F}_t\,] \\
&\geq \operatorname*{ess\,sup}_{Q\in\mathcal{Q}} \liminf_{k\uparrow\infty} E_Q[\,U^{Q^k}_{t+1} \mid \mathcal{F}_t\,] \qquad (6.35)\\
&= \operatorname*{ess\,sup}_{Q\in\mathcal{Q}} E_Q[\,U^{\uparrow}_{t+1} \mid \mathcal{F}_t\,].
\end{aligned}$$

In particular, all inequalities are in fact identities. Together with (6.34), we obtain the recursive scheme for $U^{\uparrow}$. □

The following result shows that the nonadditive conditional expectation operators $\operatorname{ess\,sup}_Q E_Q[\,\cdot \mid \mathcal{F}_t\,]$ associated with a stable set $\mathcal{Q}$ enjoy a consistency property that is similar to the martingale property for ordinary conditional expectations.

Theorem 6.52. *Let $\mathcal{Q}$ be a set of equivalent probability measures and*

$$V^{\uparrow}_t := \operatorname*{ess\,sup}_{Q\in\mathcal{Q}} E_Q[\,H \mid \mathcal{F}_t\,], \quad t = 0, \dots, T,$$

for some $\mathcal{F}_T$-measurable $H \geq 0$ such that $V^{\uparrow}_0 < \infty$. If $\mathcal{Q}$ is stable, then

$$V^{\uparrow}_\sigma = \operatorname*{ess\,sup}_{Q\in\mathcal{Q}} E_Q[\,V^{\uparrow}_\tau \mid \mathcal{F}_\sigma\,] \quad \textit{for } \sigma, \tau \in \mathcal{T} \textit{ with } \sigma \leq \tau.$$

Remark 6.53. Note that, for H as in the theorem and $\tau \in \mathcal{T}$,

$$\begin{aligned}
V^{\uparrow}_\tau &= \sum_{t=0}^{T} \operatorname*{ess\,sup}_{Q\in\mathcal{Q}} E_Q[\,H \mid \mathcal{F}_t\,]\mathbb{1}_{\{\tau=t\}} \\
&= \sum_{t=0}^{T} \operatorname*{ess\,sup}_{Q\in\mathcal{Q}} E_Q[\,H \mid \mathcal{F}_\tau\,]\mathbb{1}_{\{\tau=t\}} \\
&= \operatorname*{ess\,sup}_{Q\in\mathcal{Q}} E_Q[\,H \mid \mathcal{F}_\tau\,],
\end{aligned}$$

where we have used (6.22) in the second identity. ◇

Proof of Theorem 6.52. By Remark 6.53,

$$V^{\uparrow}_\sigma = \operatorname*{ess\,sup}_{Q\in\mathcal{Q}} E_Q[\,H \mid \mathcal{F}_\sigma\,] = \operatorname*{ess\,sup}_{Q\in\mathcal{Q}} E_Q[\,E_Q[\,H \mid \mathcal{F}_\tau\,] \mid \mathcal{F}_\sigma\,].$$

The proof that the right-hand side is equal to $\operatorname{ess\,sup}_{Q\in\mathcal{Q}} E_Q[\, V_\tau^\uparrow \mid \mathscr{F}_\sigma\,]$ is done by first noting that $V^\uparrow$ is equal to the upper Snell envelope of the process H_t given by $H_T = H$ and $H_t = 0$ for $t < T$. Then the same argument as in (6.35) applies. All one has to do is to replace $t+1$ by τ. □

Remark 6.54. Let us conclude this section by pointing out the connection between stability under pasting and the time-consistency of dynamic coherent risk measures. Let

$$\rho(Y) := \sup_{Q\in\mathcal{Q}} E_Q[\,-Y\,], \quad Y \in L^\infty(P),$$

be a coherent risk measure on $L^\infty(P)$ defined in terms of a set $\mathcal{Q}$ of probability measures equivalent to P. In the context of a dynamic financial market model, it is natural to update the initial risk assessment at later times $t > 0$. If one continues to use $\mathcal{Q}$ as a basis to compute the risk but takes into account the available information, one is led to consider the *conditional risk measures*

$$\rho_t(Y) = \operatorname*{ess\,sup}_{Q\in\mathcal{Q}} E_Q[\,-Y \mid \mathscr{F}_t\,], \quad t = 0,\dots,T. \tag{6.36}$$

The sequence $\rho_0\ \dots, \rho_T$ can be regarded as a *dynamic coherent risk measure*. Such a dynamic risk measure is called *time-consistent* or *dynamically consistent* if

$$\rho_s(-\rho_t(Y)) = \rho_s(Y) \quad \text{for } 0 \le s \le t \le T. \tag{6.37}$$

If the set $\mathcal{Q}$ in (6.36) is a stable set of equivalent probability measures, then Theorem 6.52 implies immediately the time consistency (6.37). The following converse of this statement, and hence a converse of Theorem 6.52, will be given in Theorem 11.22: if (ρ_t) is a dynamically consistent sequence of conditional coherent risk measures satisfying certain regularity assumptions, then there exists a stable set $\mathcal{Q}$ of equivalent probability measures such that (6.37) holds. An extension of dynamic consistency to dynamic *convex* risk measures will be given in Section 11.2.

Note that Theorem 6.52 shows that in (6.37) the deterministic times s and t can even be replaced by stopping times if $\mathcal{Q}$ is stable. ◇

7 Superhedging

The idea of superhedging is to find a self-financing trading strategy with minimal initial investment that covers any possible future obligation resulting from the sale of a contingent claim. If the contingent claim is not attainable, the proof of the existence of such a "superhedging strategy" requires new techniques, and in particular a new uniform version of the Doob decomposition. We will develop this theory for general American contingent claims. In doing so, we will also obtain new results for European contingent claims. In the first three sections of this chapter, we assume that our market model is arbitrage-free or, equivalently, that the set of equivalent martingale measures satisfies

$$\mathcal{P} \neq \emptyset.$$

In the final Section 7.4, we discuss liquid options in a setting where no probabilistic model is fixed a priori. Such options may be used for the construction of specific martingale measures, and also for the purpose of superhedging illiquid exotic derivatives.

7.1 $\mathcal{P}$-supermartingales

In this section, H denotes a discounted American claim with

$$\sup_{P^* \in \mathcal{P}} E^*[\, H_t \,] < \infty \quad \text{for all } t. \tag{7.1}$$

Our aim in this chapter is to find the minimal amount of capital U_t that will be needed at time t in order to purchase a self-financing trading strategy whose value process satisfies $V_u \geq H_u$ for all $u \geq t$. In analogy to our derivation of the recursive scheme (6.4), we will now heuristically derive a formula for U_t. At time T, the minimal amount needed is clearly given by

$$U_T = H_T.$$

At time $T-1$, a first requirement is to have $U_{T-1} \geq H_{T-1}$. Moreover, the amount U_{T-1} must suffice to purchase an $\mathcal{F}_{T-1}$-measurable portfolio $\overline{\xi}_T$ such that $\overline{\xi}_T \cdot X_T \geq H_T$ almost surely. An informal application of Theorem 1.34, conditional on $\mathcal{F}_{T-1}$, shows that

$$U_{T-1} \geq \operatorname*{ess\,sup}_{P^* \in \mathcal{P}} E^*[\, H_T \mid \mathcal{F}_{T-1} \,].$$

Hence, the minimal amount U_{T-1} is equal to the maximum of H_{T-1} and this essential supremum. An iteration of this argument yields the recursive scheme

$$U_T = H_T \quad \text{and} \quad U_t = H_t \vee \operatorname*{ess\,sup}_{P^* \in \mathcal{P}} E^*[\, U_{t+1} \mid \mathcal{F}_t \,]$$

https://doi.org/10.1515/9783111045283-009

for $t = T-1, \dots, 0$. By combining Proposition 6.43 and Theorem 6.51, we can identify U as the *upper Snell envelope*

$$U_t^{\uparrow} = \operatorname*{ess\,sup}_{P^* \in \mathcal{P}} U_t^{P^*} = \operatorname*{ess\,sup}_{\tau \in \mathcal{T}_t} \operatorname*{ess\,sup}_{P^* \in \mathcal{P}} E^*[\, H_\tau \mid \mathcal{F}_t \,]$$

of H with respect to the stable set $\mathcal{P}$, where U^{P^*} denotes the Snell envelope of H with respect to P^*. In the first three sections of this chapter, we will in particular give a rigorous version of the heuristic argument above.

Note first that condition (7.1) implies that

$$\pi_{\sup}(H) = \sup_{P^* \in \mathcal{P}} U_0^{P^*} = \sup_{P^* \in \mathcal{P}} \sup_{\tau \in \mathcal{T}} E^*[\, H_\tau \,] < \infty,$$

where we have used the identification of the upper bound $\pi_{\sup}(H)$ of the arbitrage-free prices of H given in Theorem 6.31. It will turn out that the following definition applies to the upper Snell envelope if we choose $\mathcal{Q} = \mathcal{P}$.

Definition 7.1. Suppose that $\mathcal{Q}$ is a nonempty set of probability measures on $(\Omega, \mathcal{F}_T)$. An adapted process is called a *$\mathcal{Q}$-supermartingale* if it is a supermartingale with respect to each $Q \in \mathcal{Q}$. Analogously, we define the notions of a *$\mathcal{Q}$-submartingale* and of a *$\mathcal{Q}$-martingale.* △

In Theorem 5.25, we have already encountered an example of a $\mathcal{P}$-martingale, namely the value process of the replicating strategy of an attainable discounted European claim.

Theorem 7.2. *The upper Snell envelope $U^{\uparrow}$ of H is the smallest $\mathcal{P}$-supermartingale that dominates H.*

Proof. For each $P^* \in \mathcal{P}$, the recursive scheme (6.33) implies that P^*-a. s.,

$$U_t^{\uparrow} \geq H_t \vee E^*[\, U_{t+1}^{\uparrow} \mid \mathcal{F}_t \,] \geq E^*[\, U_{t+1}^{\uparrow} \mid \mathcal{F}_t \,].$$

Since $U_0^{\uparrow}$ is a finite constant due to our integrability assumption (7.1) and $U_t^{\uparrow} \geq 0$ for all t, induction on t shows that $U_t^{\uparrow}$ is integrable with respect to each $P^* \in \mathcal{P}$, and hence is a $\mathcal{P}$-supermartingale dominating H.

If $\widetilde{U}$ is another $\mathcal{P}$-supermartingale that dominates H, then $\widetilde{U}_T \geq H_T = U_T^{\uparrow}$. Moreover, if $\widetilde{U}_{t+1} \geq U_{t+1}^{\uparrow}$ for some t, then

$$\widetilde{U}_t \geq H_t \vee E^*[\, \widetilde{U}_{t+1} \mid \mathcal{F}_t \,] \geq H_t \vee E^*[\, U_{t+1}^{\uparrow} \mid \mathcal{F}_t \,].$$

Thus,

$$\widetilde{U}_t \geq H_t \vee \operatorname*{ess\,sup}_{P^* \in \mathcal{P}} E^*[\, U_{t+1}^{\uparrow} \mid \mathcal{F}_t \,] = U_t^{\uparrow},$$

and backward induction shows that $\widetilde{U}$ dominates $U^{\uparrow}$. □

For European claims, Theorem 7.2 takes the following form.

Corollary 7.3. *Let H^E be a discounted European claim such that*

$$\sup_{P^* \in \mathscr{P}} E^*[\, H^E \,] < \infty.$$

Then

$$V_t^{\uparrow} := \operatorname*{ess\,sup}_{P^* \in \mathscr{P}} E^*[\, H^E \mid \mathscr{F}_t \,], \quad t = 0, \dots, T,$$

is the smallest $\mathscr{P}$-supermartingale whose terminal value dominates H^E.

Remark 7.4. Note that the proof of Theorem 7.2 did not use any special properties of the set $\mathscr{P}$. Thus, if $\mathcal{Q}$ is an arbitrary set of equivalent probability measures, the process U defined by the recursion

$$U_T = H_T \quad \text{and} \quad U_t = H_t \vee \operatorname*{ess\,sup}_{Q \in \mathcal{Q}} E_Q[\, U_{t+1} \mid \mathscr{F}_t \,]$$

is the smallest $\mathcal{Q}$-supermartingale dominating the adapted process H. ◇

7.2 Uniform Doob decomposition

The aim of this section is to give a complete characterization of all nonnegative $\mathscr{P}$-supermartingales. It will turn out that an integrable and nonnegative process U is a $\mathscr{P}$-supermartingale if and only if it can be written as the difference of a $\mathscr{P}$-martingale N and an increasing adapted process B satisfying $B_0 = 0$. This decomposition may be viewed as a uniform version of the Doob decomposition since it involves simultaneously the whole class $\mathscr{P}$. It will turn out that the $\mathscr{P}$-martingale N has a special structure: It can be written as a "stochastic integral" of the underlying process X, which defines the class $\mathscr{P}$. On the other hand, the increasing process B is only adapted, not predictable as in the Doob decomposition with respect to a single measure.

Theorem 7.5. *For an adapted, nonnegative process U, the following two statements are equivalent.*

(a) *U is a $\mathscr{P}$-supermartingale.*

(b) *There exists an adapted increasing process B with $B_0 = 0$ and a d-dimensional predictable process ξ such that*

$$U_t = U_0 + \sum_{k=1}^{t} \xi_k \cdot (X_k - X_{k-1}) - B_t \quad P\text{-a.s. for all } t.$$

Proof. First, we prove the easier implication (b)⇒(a). Fix $P^* \in \mathcal{P}$ and note that

$$V_T := U_0 + \sum_{k=1}^{T} \xi_k \cdot (X_k - X_{k-1}) \ge U_T \ge 0.$$

Hence, V is a $\mathcal{P}$-martingale by Theorem 5.14. It follows that $U_t \in \mathcal{L}^1(P^*)$ for all t. Moreover, for $P^* \in \mathcal{P}$,

$$E^*[\, U_{t+1} \mid \mathcal{F}_t \,] = E^*[\, V_{t+1} - B_{t+1} \mid \mathcal{F}_t \,] \le V_t - B_t = U_t,$$

and so U is a $\mathcal{P}$-supermartingale.

The proof of the implication (a)⇒(b) is similar to the proof of Theorem 5.32. We must show that for any given $t \in \{1, \dots, T\}$, there exist $\xi_t \in L^0(\Omega, \mathcal{F}_{t-1}, P; \mathbb{R}^d)$ and $R_t \in L^0_+(\Omega, \mathcal{F}_t, P)$ such that

$$U_t - U_{t-1} = \xi_t \cdot (X_t - X_{t-1}) - R_t.$$

This condition can be written as

$$U_t - U_{t-1} \in \mathcal{K}_t - L^0_+(\Omega, \mathcal{F}_t, P),$$

where $\mathcal{K}_t$ is as in (5.10), that is,

$$\mathcal{K}_t := \{\eta \cdot (X_t - X_{t-1}) \mid \eta \in L^0(\Omega, \mathcal{F}_{t-1}, P; \mathbb{R}^d)\,\}.$$

There is no loss of generality in assuming that P is itself a martingale measure. In this case, $U_t - U_{t-1}$ is contained in $L^1(\Omega, \mathcal{F}_t, P)$ by the definition of a $\mathcal{P}$-supermartingale. Assume that

$$U_t - U_{t-1} \notin \mathcal{C} := \big(\mathcal{K}_t - L^0_+(\Omega, \mathcal{F}_t, P)\big) \cap L^1(P).$$

Absence of arbitrage and Lemma 1.70 imply that $\mathcal{C}$ is closed in $L^1(\Omega, \mathcal{F}_t, P)$. Hence, the Hahn–Banach separation theorem in the form of Theorem H.4 implies the existence of some $Z \in L^\infty(\Omega, \mathcal{F}_t, P)$ such that

$$\alpha := \sup_{W \in \mathcal{C}} E[\, Z\, W\,] < E[\, Z\,(U_t - U_{t-1})\,] =: \delta < \infty. \tag{7.2}$$

In fact, we must have $\alpha = 0$ since $\mathcal{C}$ is a cone containing the constant function 0. Lemma 1.60 implies that such a random variable Z is nonnegative and satisfies

$$E[\,(X_t - X_{t-1})\, Z \mid \mathcal{F}_{t-1}] = 0. \tag{7.3}$$

We claim next that we can modify Z such that it is bounded from below by some $\varepsilon > 0$ and still satisfies (7.2). To see this, note first that every $W \in \mathcal{C}$ is dominated by a term of the form $\xi_t \cdot (X_t - X_{t-1})$. Hence, our assumption $P \in \mathscr{P}$, the integrability of W, and an application of Fatou's lemma yield that

$$E[\,W\,] \le E[\,\xi_t \cdot (X_t - X_{t-1})\,] \le \liminf_{c\uparrow\infty} E[\,\mathbb{1}_{\{|\xi_t|\le c\}}\xi_t \cdot (X_t - X_{t-1})\,] \le 0.$$

Thus, if we let $Z^\varepsilon := \varepsilon + Z$, then Z^ε also satisfies $E[\,Z^\varepsilon\,W\,] \le 0$ for all $W \in \mathscr{C}$. Moreover, if we chose ε small enough, then $E[\,Z^\varepsilon\,(U_t - U_{t-1})\,]$ is still larger than 0; that is, Z^ε also satisfies (7.2) and in turn (7.3). Therefore, we may assume from now on that our Z with (7.2) is bounded from below by some constant $\varepsilon > 0$.

Let

$$Z_{t-1} := E[\,Z \mid \mathscr{F}_{t-1}\,],$$

and define a new measure $\widetilde{P} \approx P$ by

$$\frac{d\widetilde{P}}{dP} := \frac{Z}{Z_{t-1}}.$$

We claim that $\widetilde{P} \in \mathscr{P}$. To prove this, note first that $X_k \in L^1(\widetilde{P})$ for all k, because the density $d\widetilde{P}/dP$ is bounded. Next, let

$$\varphi_k := E\Big[\,\frac{Z}{Z_{t-1}}\,\Big|\,\mathscr{F}_k\,\Big], \qquad k = 0, \dots, T.$$

By Proposition B.7, φ_k is the density of $\widetilde{P}$ with respect to P on $\mathscr{F}_k$. If $k \neq t$, then $\varphi_{k-1} = \varphi_k$; this is clear for $k > t$, and for $k < t$ it follows from

$$\varphi_k = E\Big[\,\frac{E[\,Z \mid \mathscr{F}_{t-1}\,]}{Z_{t-1}}\,\Big|\,\mathscr{F}_k\,\Big] = 1.$$

Thus, Proposition B.8 yields that for $k \neq t$,

$$\begin{aligned}
\widetilde{E}[\,X_k - X_{k-1} \mid \mathscr{F}_{k-1}\,] &= \frac{1}{\varphi_{k-1}} E[\,(X_k - X_{k-1})\,\varphi_k \mid \mathscr{F}_{k-1}\,] \\
&= E[\,X_k - X_{k-1} \mid \mathscr{F}_{k-1}\,] \\
&= 0.
\end{aligned}$$

If $k = t$, then (7.3) and Proposition B.8 yield that

$$\widetilde{E}[\,X_k - X_{k-1} \mid \mathscr{F}_{k-1}\,] = \frac{1}{Z_{t-1}} E[\,(X_t - X_{t-1})\,Z \mid \mathscr{F}_{t-1}\,] = 0.$$

Hence, $\widetilde{P} \in \mathscr{P}$.

Since $\widetilde{P} \in \mathcal{P}$, we have $\widetilde{E}[\, U_t - U_{t-1} \mid \mathcal{F}_{t-1}\,] \le 0$, and we get

$$\begin{aligned} 0 &\ge \widetilde{E}[\,\widetilde{E}[\, U_t - U_{t-1} \mid \mathcal{F}_{t-1}\,] Z_{t-1}\,] \\ &= \widetilde{E}[\,(U_t - U_{t-1}) Z_{t-1}\,] \\ &= E[\,(U_t - U_{t-1}) Z\,] \\ &= \delta. \end{aligned}$$

This, however, contradicts the fact that $\delta > 0$ by (7.2). □

Remark 7.6. The decomposition in part (b) of Theorem 7.5 is sometimes called the *optional decomposition* of the $\mathcal{P}$-supermartingale U. The existence of such a decomposition was first proved by El Karoui and Quenez [121] and D. Kramkov [219] in a continuous-time framework. In this context, the process B is not necessarily predictable, but it is "optional" in the sense of Chapter IV in [101]; this explains the terminology. ◇

7.3 Superhedging of American and European claims

Let H be a discounted American claim such that

$$\sup_{P^* \in \mathcal{P}} E^*[\, H_t\,] < \infty \quad \text{for all } t,$$

which is equivalent to the condition that the upper bound on the arbitrage-free prices of H is finite:

$$\pi_{\sup}(H) = \sup_{P^* \in \mathcal{P}} \sup_{\tau \in \mathcal{T}} E^*[\, H_\tau\,] < \infty.$$

Our aim in this section is to construct self-financing trading strategies that allow the seller of H to always stay on the safe side in the sense that the corresponding portfolio value is always above H.

Definition 7.7. Any self-financing trading strategy $\overline{\xi}$ whose value process V satisfies

$$V_t \ge H_t \quad P\text{-a. s. for all } t$$

is called a *superhedging strategy* for H. △

Sometimes, a superhedging strategy is also called a *superreplication strategy*. According to Definition 6.33, H is attainable if and only if there exist $\tau \in \mathcal{T}$ and a superhedging strategy whose value process satisfies $V_\tau = H_\tau$ P-almost surely.

Lemma 7.8. *If H is not attainable, then the value process V of any superhedging strategy satisfies*

$$P[\, V_t > H_t \textit{ for all } t\,] > 0.$$

Proof. We introduce the stopping time

$$\tau := \inf\{t \geq 0 \mid H_t = V_t\}.$$

Then $P[\, \tau = \infty\,] = P[\, V_t > H_t$ for all $t\,]$. Suppose that $P[\, \tau = \infty\,] = 0$. In this case, $V_\tau = H_\tau$ P-a. s. so that we arrive at the contradiction that H must be an attainable American claim. □

Let us now turn to the question whether superhedging strategies *exist*. Suppose first that P^* is the unique equivalent martingale measure in a complete market model. In Section 6.1, we have already seen how one can use the Doob decomposition of the Snell envelope U^{P^*} of H together with the martingale representation of Theorem 5.38 in order to obtain a superhedging strategy for the price $U_0^{P^*}$. We have also seen that $U_0^{P^*}$ is the minimal amount for which such a superhedging strategy is available, and that $U_0^{P^*}$ is the unique arbitrage-free price of H. The same is true of any *attainable* American claim in an incomplete market model.

In the context of a *nonattainable* American claim H in an *incomplete* financial market model, the P^*-Snell envelope will be replaced by the upper Snell envelope $U^\uparrow$ of H. The uniform Doob decomposition will take over the roles played by the usual Doob decomposition and the martingale representation theorem. Since $U^\uparrow$ is a $\mathcal{P}$-supermartingale by Theorem 7.2, the uniform Doob decomposition states that $U^\uparrow$ takes the form

$$U_t^\uparrow = U_0^\uparrow + \sum_{s=1}^{t} \xi_s \cdot (X_s - X_{s-1}) - B_t \geq H_t \tag{7.4}$$

for some predictable process ξ and some increasing process B. Thus, the self-financing trading strategy $\overline{\xi} = (\xi^0, \xi)$ defined by ξ and the initial capital

$$\overline{\xi}_1 \cdot \overline{X}_0 = U_0^\uparrow = \pi_{\sup}(H)$$

is a superhedging strategy for H. Moreover, if $\widetilde{V}$ is the value process of any superhedging strategy, then Lemma 7.8 implies that $\widetilde{V}_0 > E^*[\, H_\tau\,]$ for all $\tau \in \mathcal{T}$ and each $P^* \in \mathcal{P}$. In particular, $\widetilde{V}_0$ is larger than any arbitrage-free price for H, and it follows that $\widetilde{V}_0 \geq \pi_{\sup}(H)$. Thus, we have proved the following result.

Corollary 7.9. *There exists a superhedging strategy with initial investment $\pi_{\sup}(H)$, and this is the minimal amount needed to implement a superhedging strategy.*

We will call $\pi_{\sup}(H)$ the *cost of superhedging* of H. Sometimes, a superhedging strategy is also called a *superreplication strategy*, and one says that $\pi_{\sup}(H)$ is the *cost of superreplication* or the *upper hedging price* of H. Recall, however, that $\pi_{\sup}(H)$ is typically not an arbitrage-free price for H. In particular, the seller cannot expect to receive the amount $\pi_{\text{sup}}(H)$ for selling H.

On the other hand, the process B in the decomposition (7.4) can be interpreted as a *refunding scheme*: Using the superhedging strategy $\overline{\xi}$, the seller may withdraw successively the amounts defined by the increments of B. With this capital flow, the hedging portfolio at time t has the value $U_t^\uparrow \geq H_t$. Thus, the seller is on the safe side at no matter when the buyer decides to exercise the option. As we are going to show in Theorem 7.13 below, this procedure is optimal in the sense that, if started at any time t, it requires a minimal amount of capital.

Remark 7.10. Suppose $\pi_{\text{sup}}(H)$ belongs to the set $\Pi(H)$ of arbitrage-free prices for H. By Theorem 6.31, this holds if and only if $\pi_{\text{sup}}(H)$ is the only element of $\Pi(H)$. In this case, the definition of $\Pi(H)$ yields a stopping time $\tau \in \mathcal{T}$ and some $P^* \in \mathcal{P}$ such that

$$\pi_{\text{sup}}(H) = E^*[\,H_\tau\,].$$

Now let V be the value process of a superhedging strategy bought at $V_0 = \pi_{\text{sup}}(H)$. It follows that $E^*[\,V_\tau\,] = \pi_{\text{sup}}(H)$. Hence, $V_\tau = H_\tau$ P-a. s., so that H is attainable in the sense of Definition 6.33. This observation completes the proof of Theorem 6.34. ◇

Remark 7.11. If the American claim H is not attainable, then $\pi_{\text{sup}}(H)$ is not an arbitrage-free price of H. Thus, one may expect the existence of arbitrage opportunities if H is traded at the price $\pi_{\text{sup}}(H)$. Indeed, selling H for $\pi_{\text{sup}}(H)$ and buying a superhedging strategy $\overline{\xi}$ creates such an arbitrage opportunity: The balance at $t = 0$ is zero, but Lemma 7.8 implies that the value process V of $\overline{\xi}$ cannot be reached by any exercise strategy σ, that is, we always have

$$V_\sigma \geq H_\sigma \quad \text{and} \quad P[\,V_\sigma > H_\sigma\,] > 0. \tag{7.5}$$

Note that (7.5) is not limited to exercise strategies that are stopping times but holds for arbitrary $\mathcal{F}_T$-measurable random times $\sigma : \Omega \to \{0, \dots, T\}$. In other words, $\pi_{\text{sup}}(H)$ is too expensive even if the buyer of H has full information about the future price evolution. ◇

Remark 7.12. The argument of Remark 7.11 implies that an American claim H is attainable if and only if there exists an $\mathcal{F}_T$-measurable random time $\sigma : \Omega \to \{0, \dots, T\}$ such that $H_\sigma = V_\sigma$, where V the value process of a superhedging strategy. In other words, the notion of attainability of American claims does not need the restriction to stopping times. ◇

We already know that $\pi_{\text{sup}}(H)$ is the smallest amount for which one can buy a superhedging strategy at time 0. The following "superhedging duality theorem" extends this result to times $t > 0$. To this end, denote by $\mathcal{U}_t^\uparrow(H)$ the set of all $\mathcal{F}_t$-measurable random variables $\widetilde{U}_t \geq 0$ for which there exists a d-dimensional predictable process $\widetilde{\xi}$ such that

$$\widetilde{U}_t + \sum_{k=t+1}^{u} \widetilde{\xi}_k \cdot (X_k - X_{k-1}) \geq H_u \quad \text{for all } u \geq t \text{ } P\text{-a.s.} \tag{7.6}$$

Theorem 7.13. *The upper Snell envelope $U_t^{\uparrow}$ of H is the minimal element of $\mathscr{U}_t^{\uparrow}(H)$. More precisely:*

(a) $U_t^{\uparrow} \in \mathscr{U}_t^{\uparrow}(H)$,
(b) $U_t^{\uparrow} = \operatorname{ess\,inf} \mathscr{U}_t^{\uparrow}(H)$.

Proof. Assertion (a) follows immediately from the uniform Doob decomposition of the $\mathscr{P}$-supermartingale $U^{\uparrow}$. As to part (b), we clearly get $U_t^{\uparrow} \geq \operatorname{ess\,inf} \mathscr{U}_t^{\uparrow}(H)$ from (a). For the proof of the converse inequality, take $\widetilde{U}_t \in \mathscr{U}_t^{\uparrow}(H)$ and choose a predictable process $\widetilde{\xi}$ for which (7.6) holds. We must show that the set $B := \{U_t^{\uparrow} \leq \widetilde{U}_t\}$ satisfies $P[\,B\,] = 1$. Let

$$\hat{U}_t := U_t^{\uparrow} \wedge \widetilde{U}_t = U_t^{\uparrow} \cdot \mathbb{1}_B + \widetilde{U}_t \cdot \mathbb{1}_{B^c}.$$

Then $\hat{U}_t \leq U_t^{\uparrow}$, and our claim will follow if we can show that $U_t^{\uparrow} \leq \hat{U}_t$. Let ξ denote the predictable process obtained from the uniform Doob decomposition of the $\mathscr{P}$-supermartingale $U^{\uparrow}$, and define

$$\hat{\xi}_s := \begin{cases} \xi_s & \text{if } s \leq t, \\ \xi_s \cdot \mathbb{1}_B + \widetilde{\xi}_s \cdot \mathbb{1}_{B^c} & \text{if } s > t. \end{cases}$$

With this choice, $\hat{U}_t$ satisfies (7.6), that is, $\hat{U}_t \in \mathscr{U}_t^{\uparrow}(H)$. Let

$$\hat{V}_s := U_0^{\uparrow} + \sum_{k=1}^{s} \hat{\xi}_k \cdot (X_k - X_{k-1}).$$

Then $\hat{V}_s \geq U_s^{\uparrow} \geq H_s$ for all $s \leq t$. In particular $\hat{V}_T \geq \hat{U}_T \geq H_T \geq 0$, which implies that $\hat{V}$ is a $\mathscr{P}$-martingale; see Theorem 5.25. Hence, Doob's stopping theorem implies

$$\begin{aligned} U_t^{\uparrow} &= \operatorname*{ess\,sup}_{P^* \in \mathscr{P}} \operatorname*{ess\,sup}_{\tau \in \mathcal{T}_t} E^*[\,H_\tau \mid \mathscr{F}_t\,] \\ &\leq \operatorname*{ess\,sup}_{P^* \in \mathscr{P}} \operatorname*{ess\,sup}_{\tau \in \mathcal{T}} E^*\left[\,\hat{U}_t + \sum_{k=t+1}^{\tau} \hat{\xi}_k \cdot (X_k - X_{k-1}) \,\middle|\, \mathscr{F}_t\,\right] \\ &= \hat{U}_t + \operatorname*{ess\,sup}_{P^* \in \mathscr{P}} \operatorname*{ess\,sup}_{\tau \in \mathcal{T}} E^*[\,\hat{V}_\tau - \hat{V}_t \mid \mathscr{F}_t\,] \\ &= \hat{U}_t, \end{aligned}$$

which concludes the proof. □

We now take the point of view of the buyer of the American claim H. The buyer allocates an initial investment π to purchase H, and then receives the amount $H_\tau \geq 0$. The objective is to find an exercise strategy and a self-financing trading strategy $\overline{\eta}$ with initial investment $-\pi$, such that the portfolio value is covered by the payoff of the claim.

In other words, find $\tau \in \mathcal{T}$ and a self-financing trading strategy with value process V such that $V_0 = -\pi$ and $V_\tau + H_\tau \geq 0$. As shown below, the maximal π for which this is possible is equal to

$$\pi_{\inf}(H) = \sup_{\tau \in \mathcal{T}} \inf_{P^* \in \mathcal{P}} E^*[\,H_\tau\,] = \inf_{P^* \in \mathcal{P}} \sup_{\tau \in \mathcal{T}} E^*[\,H_\tau\,] = U_0^{\downarrow},$$

where, by Theorem 6.46,

$$\begin{aligned} U_t^{\downarrow} &= \operatorname*{ess\,inf}_{P^* \in \mathcal{P}} U_t^{P^*} \\ &= \operatorname*{ess\,inf}_{P^* \in \mathcal{P}} \operatorname*{ess\,sup}_{\tau \in \mathcal{T}_t} E^*[\,H_\tau \mid \mathcal{F}_t\,] \\ &= \operatorname*{ess\,sup}_{\tau \in \mathcal{T}_t} \operatorname*{ess\,inf}_{P^* \in \mathcal{P}} E^*[\,H_\tau \mid \mathcal{F}_t\,] \end{aligned}$$

is the lower Snell envelope of H with respect to the stable set $\mathcal{P}$. More generally, we will consider the buyer's problem for arbitrary $t \geq 0$. To this end, denote by $\mathcal{U}_t^{\downarrow}(H)$ the set of all $\mathcal{F}_t$-measurable random variables $\widetilde{U}_t \geq 0$ for which there exists a d-dimensional predictable process $\widetilde{\eta}$ and a stopping time $\sigma \in \mathcal{T}_t$ such that

$$\widetilde{U}_t - \sum_{k=t+1}^{\sigma} \widetilde{\eta}_k \cdot (X_k - X_{k-1}) \leq H_\sigma \qquad P\text{-a. s.}$$

Theorem 7.14. *$U_t^{\downarrow}$ is the maximal element of $\mathcal{U}_t^{\downarrow}(H)$. More precisely:*
(a) $U_t^{\downarrow} \in \mathcal{U}_t^{\downarrow}(H)$,
(b) $U_t^{\downarrow} = \operatorname{ess\,sup} \mathcal{U}_t^{\downarrow}(H)$.

Proof. (a): Let $\overline{\xi}$ be a superhedging strategy for H with initial investment $\pi_{\sup}(H)$, and denote by V the value process of $\overline{\xi}$. The main idea of the proof is to use that $V_t - H_t \geq 0$ can be regarded as a new discounted American claim, to which we can apply Theorem 7.13. However, we must take care of the basic asymmetry of the hedging problem for American options: The seller of H must hedge against *all* possible exercise strategies, while the buyer must find only *one* suitable stopping time. It will turn out that a suitable stopping time is given by $\tau_t := \inf\{u \geq t \mid U_u^{\downarrow} = H_u\}$. With this choice, let us define a modified discounted American claim $\widetilde{H}$ by

$$\widetilde{H}_u = (V_u - H_u) \cdot \mathbb{1}_{\{u=\tau_t\}}, \qquad u = 0, \dots, T.$$

Clearly, $\widetilde{H}_\sigma \leq \widetilde{H}_{\tau_t}$ for all $\sigma \in \mathcal{T}_t$. It follows that

$$\begin{aligned} \operatorname*{ess\,sup}_{\sigma \in \mathcal{T}_t} \operatorname*{ess\,sup}_{P^* \in \mathcal{P}} E^*[\,\widetilde{H}_\sigma \mid \mathcal{F}_t\,] &= \operatorname*{ess\,sup}_{P^* \in \mathcal{P}} E^*[\,\widetilde{H}_{\tau_t} \mid \mathcal{F}_t\,] \\ &= V_t - \operatorname*{ess\,inf}_{P^* \in \mathcal{P}} E^*[\,H_{\tau_t} \mid \mathcal{F}_t\,] \\ &= V_t - U_t^{\downarrow}, \end{aligned}$$

where we have used that V is a $\mathscr{P}$-martingale in the second and Theorem 6.47 in the third step. Thus, $V_t - U_t^{\downarrow}$ is equal to the upper Snell envelope $\widetilde{U}^{\uparrow}$ of $\widetilde{H}$ at time t. Let $\widetilde{\xi}$ be the d-dimensional predictable process obtained from the uniform Doob decomposition of $\widetilde{U}^{\uparrow}$. Then, due to part (a) of Theorem 7.13,

$$V_t - U_t^{\downarrow} + \sum_{k=t+1}^{u} \widetilde{\xi}_k \cdot (X_k - X_{k-1}) \geq \widetilde{H}_u = (V_u - H_u) \cdot \mathbb{1}_{\{u=\tau_t\}} \quad \text{for all } u \geq t.$$

Thus, $\eta := \widetilde{\xi} - \xi$ is as desired.

(b): Part (a) implies the inequality $\leq$ in (b). To prove its converse, take $\widetilde{U}_t \in \mathscr{U}_t^{\downarrow}$, a d-dimensional predictable process $\widetilde{\eta}$, and $\sigma \in \mathcal{T}_t$ such that

$$\widetilde{U}_t - \sum_{k=t+1}^{\sigma} \widetilde{\eta}_k \cdot (X_k - X_{k-1}) \leq H_\sigma \quad P\text{-a. s.}$$

We will show below that

$$E^*\left[\sum_{k=t+1}^{\sigma} \widetilde{\eta}_k \cdot (X_k - X_{k-1}) \,\middle|\, \mathscr{F}_t \right] = 0 \quad \text{for all } P^* \in \mathscr{P}. \tag{7.7}$$

Given this fact, we obtain that

$$\widetilde{U}_t \leq E^*[\, H_\sigma \mid \mathscr{F}_t \,] \leq \operatorname*{ess\,sup}_{\tau \in \mathcal{T}_t} E^*[\, H_\tau \mid \mathscr{F}_t \,]$$

for all $P^* \in \mathscr{P}$. Taking the essential infimum over $P^* \in \mathscr{P}$ thus yields $\widetilde{U}_t \leq U_t^{\downarrow}$ and in turn (b).

To prove (7.7), let

$$\widetilde{G}_s := \mathbb{1}_{\{s \geq t+1\}} \sum_{k=t+1}^{s} \mathbb{1}_{\{k \leq \sigma\}} \widetilde{\eta}_k \cdot (X_k - X_{k-1}), \quad s = 0, \ldots, T.$$

Then $\widetilde{G}_T \geq \widetilde{U}_t - H_\sigma \geq -H_\sigma \in \mathscr{L}^1(P^*)$ for all P^*, and Theorem 5.14 implies that $\widetilde{G}$ is a $\mathscr{P}$-martingale. Hence, (7.7) follows. □

We conclude this section by stating explicitly the corresponding results for European claims. Recall from Remark 6.6 that every discounted European claim H^E can be regarded as the discounted American claim. Therefore, the results we have obtained so far include the corresponding "European" counterparts as special cases.

Corollary 7.15. *For any discounted European claim H^E such that*

$$\sup_{P^* \in \mathscr{P}} E^*[\, H^E \,] < \infty,$$

there exist two d-dimensional predictable processes ξ and η such that P-a. s.,

$$\operatorname*{ess\,sup}_{P^* \in \mathcal{P}} E^*[\, H^E \mid \mathcal{F}_t \,] + \sum_{k=t+1}^{T} \xi_k \cdot (X_k - X_{k-1}) \geq H^E, \tag{7.8}$$

$$\operatorname*{ess\,inf}_{P^* \in \mathcal{P}} E^*[\, H^E \mid \mathcal{F}_t \,] - \sum_{k=t+1}^{T} \eta_k \cdot (X_k - X_{k-1}) \leq H^E. \tag{7.9}$$

Remark 7.16. For $t = 0$, (7.8) takes the form

$$\sup_{P^* \in \mathcal{P}} E^*[\, H^E \,] + \sum_{k=1}^{T} \xi_k \cdot (X_k - X_{k-1}) \geq H^E \quad P\text{-a.s.}$$

Thus, the self-financing trading strategy $\overline{\xi}$ arising from ξ and the initial investment $\overline{\xi}_1 \cdot \overline{X}_0 = \sup_{P^* \in \mathcal{P}} E^*[\, H^E \,]$ allow the seller to cover all possible obligations without any downside risk. Similarly, (7.9) yields an interpretation of the self-financing trading strategy $\overline{\eta}$ that arises from η and the initial investment $\overline{\eta}_1 \cdot \overline{X}_0 = -\inf_{P^* \in \mathcal{P}} E^*[\, H^E \,]$. The latter quantity corresponds to the largest loan the buyer can take out and still be sure that, by using the trading strategy $\overline{\eta}$, this debt will be covered by the payoff H^E. ◇

Remark 7.17. Let H^E be a discounted European claim such that

$$\sup_{P^* \in \mathcal{P}} E^*[\, H^E \,] < \infty.$$

Suppose that $\hat{P} \in \mathcal{P}$ is such that $\hat{E}[\, H^E \,] = \pi_{\sup}(H^E)$. If $\overline{\xi} = (\xi^0, \xi)$ is a superhedging strategy for H^E, then

$$\hat{H}^E := \hat{E}[\, H^E \,] + \sum_{k=1}^{T} \xi_k \cdot (X_k - X_{k-1})$$

satisfies $\hat{H}^E \geq H^E \geq 0$. Hence, $\hat{H}^E$ is an attainable discounted claim, and it follows from Theorem 5.25 that

$$\hat{E}[\, \hat{H}^E \,] = \hat{E}[\, H^E \,].$$

This shows that $\hat{H}^E$ and H^E are P-a.s. identical and that H^E is attainable. We have thus obtained another proof of Theorem 5.32. ◇

As the last result in this section, we formulate the following "superhedging duality theorem," which states that the bounds in (7.8) and (7.9) are optimal.

Corollary 7.18. *Suppose that H^E is a discounted European claim with*

$$\sup_{P^* \in \mathcal{P}} E^*[\, H^E \,] < \infty.$$

Denote by $\mathcal{U}_t^{\uparrow}(H^E)$ the set of all $\mathcal{F}_t$ measurable random variables $\widetilde{U}_t$ for which there exists a d-dimensional predictable process $\widetilde{\xi}$ such that

$$\widetilde{U}_t + \sum_{k=t+1}^{T} \widetilde{\xi}_k \cdot (X_k - X_{k-1}) \geq H^E \quad P\text{-a. s.}$$

Then

$$\operatorname*{ess\,sup}_{P^* \in \mathcal{P}} E^*[\, H^E \mid \mathcal{F}_t \,] = \operatorname{ess\,inf} \mathcal{U}_t^{\uparrow}(H^E).$$

By $\mathcal{U}_t^{\downarrow}(H^E)$, we denote the set of all $\mathcal{F}_t$ measurable random variables $\widetilde{U}_t$ for which there exists a d-dimensional predictable process $\widetilde{\eta}$ such that

$$\widetilde{U}_t - \sum_{k=t+1}^{T} \widetilde{\eta}_k \cdot (X_k - X_{k-1}) \leq H^E \quad P\text{-a. s.}$$

Then

$$\operatorname*{ess\,inf}_{P^* \in \mathcal{P}} E^*[\, H^E \mid \mathcal{F}_t \,] = \operatorname{ess\,sup} \mathcal{U}_t^{\downarrow}(H^E).$$

Remark 7.19. Define $\mathcal{A}$ as the set of financial positions $Z \in L^\infty(\Omega, \mathcal{F}_T, P)$ that are acceptable in the sense that there exists a d-dimensional predictable process ξ such that

$$Z + \sum_{k=1}^{T} \xi_k \cdot (X_k - X_{k-1}) \geq 0 \quad P\text{-a. s.}$$

As in Section 4.10, this set $\mathcal{A}$ induces a coherent risk measure ρ on $L^\infty(\Omega, \mathcal{F}_T, P)$:

$$\rho(Z) = \inf\{m \in \mathbb{R} \mid m + Z \in \mathcal{A}\}, \quad Z \in L^\infty(\Omega, \mathcal{F}_T, P).$$

Corollary 7.18 implies that ρ can be represented as

$$\rho(Z) = \sup_{P^* \in \mathcal{P}} E^*[\,-Z\,].$$

We therefore obtain a multiperiod version of Proposition 4.105. ◇

Remark 7.20. Often, the superhedging strategy in a given incomplete model can be identified as the perfect hedge in an associated "extremal" model. As an example, consider a one-period model with d discounted risky assets given by bounded random variables $X^1, \dots, X^d$. Denote by μ the distribution of $X = (X^1, \dots, X^d)$ and by $\Gamma(\mu)$ the convex hull of the support of μ. The closure $K := \overline{\Gamma}(\mu)$ of $\Gamma(\mu)$ is convex and compact. We know from Section 1.5 that the model is arbitrage-free if and only if the price system $\pi = (\pi^1, \dots, \pi^d)$ is contained in the relative interior of $\Gamma(\mu)$, and the equivalent martingale measures can be identified with the measures $\mu^* \approx \mu$ with barycenter π. Consider a derivative $H = h(X)$ given by a *convex* function h on K. The cost of superhedging is given by

$$\sup_{\mu^*} \int h\,d\mu^* = \inf\{\alpha(\pi) \mid \alpha \text{ affine on } K,\ \alpha \geq h\ \mu\text{-a.s.}\},$$

which is a special case of the duality result of Theorem 1.34. For an affine function α on K, the set $\{\alpha \geq h\}$ is convex and closed, and the condition $\mu(\alpha \geq h) = 1$ implies $\alpha \geq h$ on K. Denoting by $\mathscr{M}(\pi)$ the class of all probability measures on K with barycenter π, we obtain

$$\int h\,d\tilde{\mu} \leq \int \alpha\,d\tilde{\mu} = \alpha(\pi)$$

for any $\tilde{\mu} \in \mathscr{M}(\pi)$. Thus,

$$\hat{h}(\pi) = \sup_{\tilde{\mu} \in \mathscr{M}(\pi)} \int h\,d\tilde{\mu} \tag{7.10}$$

where we define for $f \in C(K)$

$$\hat{f} := \inf\{\alpha \mid \alpha \text{ affine on } K,\ \alpha \geq f\ \mu\text{-a.s.}\}.$$

The supremum in (7.10) is attained since $\mathscr{M}(\pi)$ is weakly compact. More precisely, it is attained by any measure $\hat{\mu} \in \mathscr{M}(\pi)$ on K that is maximal with respect to the convex order $\succcurlyeq_{cx}$ defined for measures on K as in (2.29); see Théorème X. 41 in [102]. But such a maximal measure is supported by the set of extreme points of the convex compact set K, that is, by the *Choquet boundary* of K. This follows from a general integral representation theorem of Choquet; see, for example, Théorème X. 43 of [102]. In our finite-dimensional setting, $\hat{\mu}$ can in fact be chosen to have a support consisting of at most $d+1$ points, due to a theorem of Carathéodory and the representation of K as the convex hull of its extreme points; see [273], Theorems 17.1 and 18.5. But this means that $\hat{\mu}$ can be identified with a *complete* model, due to Proposition 1.43. Thus, the cost of superhedging $\hat{h}(\pi)$ can be identified with the canonical price

$$\hat{\pi} := \int h\,d\hat{\mu}$$

of the derivative H, computed in the complete model $\hat{\mu}$. Note that $\hat{\mu}$ sits on the Choquet boundary of $K = \overline{\Gamma}(\mu)$, but typically it will no longer be equivalent or absolutely continuous with respect to the original measure μ. As a simple illustration, consider a one-period model with one risky asset X^1. If X^1 is bounded, then the distribution μ of X^1 has bounded support, and $\Gamma(\mu)$ is of the form $[a, b]$. In this case, the cost of superhedging $H = h(X^1)$ for a convex function h is given by the price

$$p^* h(b) + (1 - p^*)h(a),$$

computed in the binary model in which X^1 takes only the values a and b, and where $p^* \in (0, 1)$ is determined by

$$p^* b + (1 - p^*)a = \pi^1.$$ ◇

The following example illustrates that a superhedging strategy is typically too expensive from a practical point of view. However, we will see in Chapter 8 how superhedging strategies can be used in order to construct other hedging strategies that are efficient in terms of cost and shortfall risk.

Example 7.21. Consider a simple one-period model where S_1^1 has under P a Poisson distribution and where $S^0 \equiv 1$. Let $H := (S_1^1 - K)^+$ be a call option with strike $K > 0$. We have seen in Example 1.40 that $\pi_{\inf}(H)$ and $\pi_{\sup}(H)$ coincide with the universal arbitrage bounds of Remark 1.39:

$$(S_0^1 - K)^+ = \pi_{\inf}(H) \quad \text{and} \quad \pi_{\sup}(H) = S_0^1.$$

Thus, the superhedging strategy for the seller consists in the trivial hedge of buying the asset at time 0, while the corresponding strategy for the buyer is a short-sale of the asset in case the option is in the money, that is, if $S_0^1 > K$. ◇

7.4 Model-free superhedging with liquid options

In practice, some derivatives such as put or call options are traded so frequently that their prices are quoted just like those of the primary assets. The prices of such *liquid options* can be regarded as an additional source of information on the expectations of the market as to the future evolution of asset prices. This information can be exploited in various ways. First, it serves to single out those martingale measures P^* that are compatible with the observed options prices, in the sense that the observed prices coincide with the expectations of the discounted payoff under P^*. Second, liquid options may be used as instruments for hedging more exotic options.

Our aim in this section is to illustrate these ideas in a simple setting. Assume that there is only one risky asset S^1 such that S_0^1 is a positive constant, and that S^0 is a riskless bond with interest rate $r = 0$. Thus, the discounted price process of the risky asset is given by $X_t = S_t^1 \geq 0$ for $t = 0, \dots, T$. As the underlying space of scenarios, we use the product space

$$\Omega := [0, \infty)^T.$$

We define $X_t(\omega) = x_t$ for $\omega = (x_1, \dots, x_T) \in \Omega$, and denote by $\mathscr{F}_t$ the σ-algebra generated by $X_0, \dots, X_t$; note that $\mathscr{F}_0 = \{\emptyset, \Omega\}$. No probability measure P is given a priori. Let us now introduce a linear space $\mathscr{X}$ of $\mathscr{F}_T$-measurable functions as the smallest linear space such that the following conditions are satisfied:

(a) $1 \in \mathscr{X}$.
(b) $(X_t - X_s) \mathbb{1}_A \in \mathscr{X}$ for $0 \leq s < t \leq T$ and $A \in \mathscr{F}_s$.
(c) $(X_t - K)^+ \in \mathscr{X}$ for $K \geq 0$ and $t = 1, \dots, T$.

The functions in the space $\mathscr{X}$ will be interpreted as (discounted) payoffs of liquid derivatives. The constant function 1 in (a) corresponds to a unit investment into the riskless bond. The function $X_t - X_s$ in (b) corresponds to the payoff of a forward contract on the risky asset, issued at time s for the price X_s and expiring at time t. The decision to buy such a forward contract at time s may depend on the market situation at time s; this is taken into account by allowing for payoffs $(X_t - X_s)\,\mathbb{1}_A$ with $A \in \mathscr{F}_s$. Linearity of $\mathscr{X}$ together with conditions (a) and (b) implies that

$$X_t \in \mathscr{X} \qquad \text{for all } t.$$

Finally, condition (c) states that European call options with any possible strike and any maturity up to time T can be used as liquid securities.

Suppose that a linear pricing rule Φ is given on $\mathscr{X}$. The value $\Phi(Y)$ will be interpreted as the market price of the liquid security $Y \in \mathscr{X}$. The price of a liquid European call option with strike K and maturity t will be denoted by

$$C_t(K) := \Phi((X_t - K)^+).$$

Assumption 7.22. We assume that $\Phi : \mathscr{X} \to \mathbb{R}$ is a linear functional that satisfies the following conditions:
(a) $\Phi(1) = 1$.
(b) $\Phi(Y) \geq 0$ if $Y \geq 0$.
(c) $\Phi((X_t - X_s)\,\mathbb{1}_A) = 0$ for all $0 \leq s < t \leq T$ and $A \in \mathscr{F}_s$.
(d) $C_t(K) = \Phi((X_t - K)^+) \to 0$ as $K \uparrow \infty$ for all t.

The first two conditions must clearly be satisfied if the pricing rule Φ shall not create arbitrage opportunities. Condition (c) states that X_s is the fair price for a forward contract issued at time s. This condition is quite natural in view of Theorem 5.29. In our present setting, it can also be justified by the following simple replication argument. At time s, take out a loan $X_s(\omega)$ and use it for buying the asset. At time t, the asset is worth $X_t(\omega)$ and the loan must be paid back, which results in a balance $X_t(\omega) - X_s(\omega)$. Since this replication strategy requires zero initial capital, the price of the corresponding payoff should also be zero. The continuity condition (d) is also quite natural.

Our first goal is to show that any such pricing rule Φ is compatible with the paradigm that arbitrage-free prices can be identified as expectations with respect to some martingale measure for X. More precisely, we are going to construct a martingale measure P^* such that $\Phi(Y) = E^*[\,Y\,]$ for all $Y \in \mathscr{X}$. On the one hand, this will imply regularity properties of Φ. On the other hand, this will yield an extension of our pricing rule Φ to a larger space of payoffs including path-dependent exotic options. As a first step in this direction, we have the following result.

Lemma 7.23. *For each t, there exists a unique probability measure μ_t on $[0, \infty)$ such that for all $K \geq 0$,*

$$C_t(K) = \Phi((X_t - K)^+) = \int (x - K)^+ \, \mu_t(dx).$$

In particular, μ_t has the mean

$$\int x \, \mu_t(dx) = X_0.$$

Proof. Since $K \mapsto (X_t - K)^+$ is convex and decreasing, linearity and positivity of Φ imply that the function $\phi_t(K) := \Phi((X_t - K)^+)$ is convex and decreasing as well. Hence, by Proposition A.7, $C_t(K)$ admits a right-hand derivative, denoted by $-f(K)$ in the sequel. The function f is then right-continuous, decreasing, and such that

$$\begin{aligned} C_t(K) &= C_t(0) - \int_0^K f(x) \, dx \\ &= X_0 - \int_0^K f(x) \, dx. \end{aligned}$$

Our fourth condition on Φ yields

$$\int_0^\infty f(x) \, dx = X_0,$$

so that $f(x) \searrow 0$ as $x \uparrow \infty$. Hence, there exists a positive measure μ_t on $(0, \infty)$ such that

$$f(x) = \mu_t((x, \infty)) \quad \text{for } x > 0;$$

see Remark D.8. Fubini's theorem implies

$$\int_{(0,\infty)} x \, \mu_t(dx) = \int_0^\infty f(y) \, dy = X_0$$

and

$$\begin{aligned} C_t(K) &= C_t(0) - \int_0^K \mu_t((x, \infty)) \, dx \\ &= X_0 - \int_0^K \int_{(0,\infty)} \mathbb{1}_{\{y<x\}} \, \mu_t(dx) \, dy \\ &= X_0 - \int_{(0,\infty)} \int_0^K \mathbb{1}_{\{y<x\}} \, dy \, \mu_t(dx) \\ &= \int_{(0,\infty)} (x - K)^+ \, \mu_t(dx). \end{aligned}$$

It remains to show that μ_t can be extended to a probability measure on $[0,\infty)$, that is, we must show that $\mu_t((0,\infty)) \le 1$, so that we can put $\mu_t(\{0\}) := 1 - \mu_t((0,\infty))$. To this end, we will use the "put-call parity"

$$C_t(K) = X_0 - K + \Phi((K - X_t)^+),$$

which follows from the fact that $(X_t - K)^+ = X_0 - K + (K - X_t)^+$ and our assumptions on Φ. Thus,

$$\begin{aligned}\Phi((K - X_t)^+) &= K - X_0 + C_t(X) = K - \int_0^K f(x)\,dx \\ &= \int_0^K g(x)\,dx,\end{aligned}$$

where $g(x) = 1 - f(x)$. Since $K \mapsto \Phi((K - X_t)^+)$ is increasing, the right-continuous function g must be nonnegative, and so $g(0) \ge 0$. Hence, $1 \ge f(0) = \mu_t((0,\infty))$. □

The following lemma shows that the measures μ_t constructed in Lemma 7.23 are related to each other by the *convex order* $\succcurlyeq_{cx}$, defined by

$$\mu \succcurlyeq_{cx} \nu \quad \iff \quad \int f\,d\mu \ge \int f\,d\nu \quad \text{for all convex functions } f$$

for probability measures with finite expectation; see Remark 2.63.

Lemma 7.24. *The map $t \mapsto \mu_t$ is increasing with respect to the convex order $\succcurlyeq_{cx}$:*

$$\mu_{t+1} \succcurlyeq_{cx} \mu_t \quad \textit{for all } t.$$

Proof. Note that

$$(X_{t+1} - X_t)\,\mathbb{1}_{\{X_t > K\}} + (X_t - K)^+ = (X_{t+1} - K)\,\mathbb{1}_{\{X_t > K\}} \le (X_{t+1} - K)^+\,\mathbb{1}_{\{X_t > K\}}.$$

Hence,

$$\begin{aligned}&(X_{t+1} - K)^+ - (X_{t+1} - X_t)\,\mathbb{1}_{\{X_t > K\}} - (X_t - K)^+ \\ &\qquad \ge (X_{t+1} - K)^+ - (X_{t+1} - K)^+\,\mathbb{1}_{\{X_t > K\}} \\ &\qquad \ge 0.\end{aligned}$$

Since the price of the forward contract $(X_{t+1} - X_t)\,\mathbb{1}_{\{X_t > K\}}$ vanishes under our pricing rule Φ, we must have that for all $K \ge 0$,

$$\begin{aligned}\int (x-K)^+ \,\mu_{t+1}(dx) - \int (x-K)^+ \,\mu_t(dx) \\ &= C_{t+1}(K) - C_t(K) \\ &\geq 0.\end{aligned}$$

An application of Corollary 2.61 now yields that $\mu_{t+1} \succcurlyeq_{\mathrm{cx}} \mu_t$. □

Let us introduce the class

$$\mathscr{P}_\Phi := \{P^* \in \mathscr{M}_1(\Omega, \mathscr{F}) \mid E^*[\,Y\,] = \Phi(Y) \text{ for all } Y \in \mathscr{X}\,\}$$

of all probability measures P^* on $(\Omega, \mathscr{F})$ that coincide with Φ on $\mathscr{X}$. Note that for any $P^* \in \mathscr{P}_\Phi$,

$$E^*[\,(X_t - X_s)\,\mathbb{1}_A\,] = 0 \quad \text{for } s < t \text{ and } A \in \mathscr{F}_s,$$

so that $\mathscr{P}_\Phi$ consists of *martingale measures* for X. Our first main result in this section can be regarded as a version of the "fundamental theorem of asset pricing" without an a priori measure P.

Theorem 7.25. *Under Assumption* 7.22, *the class $\mathscr{P}_\Phi$ is nonempty. Moreover, there exists $P^* \in \mathscr{P}_\Phi$ with the* Markov property: *For $0 \leq s \leq t \leq T$ and each bounded measurable function f,*

$$E^*[\,f(X_t) \mid \mathscr{F}_s\,] = E^*[\,f(X_t) \mid X_s\,]. \tag{7.11}$$

Proof. Since $\mu_{t+1} \succcurlyeq_{\mathrm{cx}} \mu_t$, Corollary 2.61 yields the existence of a stochastic kernel Q_{t+1} such that

$$\int y\, Q_{t+1}(x, dy) = x \tag{7.12}$$

and $\mu_{t+1} = \mu_t Q_{t+1}$. Let us define

$$P^* := \mu_1 \otimes Q_2 \otimes \cdots \otimes Q_T,$$

that is, for each measurable set $A \subseteq \Omega = [0, \infty)^T$,

$$P^*[\,A\,] = \int \mu_1(dx_1) \int Q_2(x_1, dx_2) \cdots \int Q_T(x_{T-1}, dx_T)\, \mathbb{1}_A(x_1, x_2, \ldots, x_T).$$

Clearly, μ_t is the law of X_t under P^*. In particular, all call options are priced correctly by calculating their expectation with respect to P^*. Then one checks that P^*-a. s.,

$$E^*[\,f(X_{t+1}) \mid \mathscr{F}_t\,] = \int f(y)\, Q_{t+1}(X_t, dy) = E^*[\,f(X_{t+1}) \mid X_t\,]. \tag{7.13}$$

Recalling (7.12), this yields $E^*[X_{t+1} \mid \mathscr{F}_t] = X_t$. In particular, P^* is a martingale measure, and the expectation of $(X_t - X_s)\mathbb{1}_A$ vanishes for $s < t$ and $A \in \mathscr{F}_s$. It follows that $E^*[Y] = \Phi(Y)$ for all $Y \in \mathscr{X}$. Finally, an induction argument applied to (7.13) yields the Markov property (7.11) for all $0 \le s \le t \le T$. □

So far, we have assumed that our space $\mathscr{X}$ of liquidly traded derivatives contains call options with *all* possible strike prices and maturities. From now on, we will simplify our setting by assuming that only call options with maturity T are liquidly traded. Thus, we replace $\mathscr{X}$ by the smaller space $\mathscr{X}_T$ that is defined as the linear hull of the constant functions, of all forward contracts

$$(X_t - X_s)\mathbb{1}_A, \quad 0 \le s < t \le T, A \in \mathscr{F}_s,$$

and of all call options with maturity T

$$(X_T - K)^+, \quad K \ge 0.$$

The observed market prices of derivatives in $\mathscr{X}_T$ are as before modeled by a linear pricing rule

$$\Phi_T : \mathscr{X}_T \to \mathbb{R}.$$

We assume that Φ_T satisfies Assumption 7.22 in the sense that its condition (d) is only required for $t = T$:
(d′) $C_T(K) := \Phi_T((X_T - K)^+) \to 0$ as $K \uparrow \infty$.

By

$$\mathscr{P}_{\Phi_T} := \{P^* \in \mathscr{M}_1(\Omega, \mathscr{F}) \mid E^*[Y] = \Phi_T(Y) \text{ for all } Y \in \mathscr{X}_T\}$$

we denote the class of all probability measures P^* on $(\Omega, \mathscr{F})$ that coincide with Φ_T on $\mathscr{X}_T$. As before, it follows from condition (c) of Assumption 7.22 that any $P^* \in \mathscr{P}_{\Phi_T}$ will be a *martingale measure* for the price process X. Obviously, any linear pricing rule Φ that is defined on the full space $\mathscr{X}$ and that satisfies Assumption 7.22 can be restricted to $\mathscr{X}_T$, and this restriction satisfies the above assumptions. Thus, we have $\mathscr{P}_{\Phi_T} \supseteq \mathscr{P}_\Phi \ne \emptyset$. Hence, Theorem 7.25 immediately implies the following "fundamental theorem of asset pricing", which does not involve an a priori measure P.

Proposition 7.26. *Under the above assumptions, $\mathscr{P}_{\Phi_T}$ is nonempty.*

A measure $P^* \in \mathscr{P}_{\Phi_T}$ can be regarded as an extension of the pricing rule Φ_T to the larger space $\mathscr{L}^1(P^*)$, and the expectation $E^*[H]$ of some European claim $H \ge 0$ can be regarded as an *arbitrage-free price* for H. Our aim is to obtain upper and lower bounds for $E^*[H]$ that hold simultaneously for all $P^* \in \mathscr{P}_\Phi$. We will derive such bounds for various exotic options; this will amount to the construction of certain superhedging strategies in terms of liquid securities.

As a first example, we consider the following *digital option:*

$$H^{\text{dig}} := \begin{cases} 1 & \text{if } \max_{0\le t\le T} X_t \ge B, \\ 0 & \text{otherwise,} \end{cases}$$

which has a unit payoff if the price process reaches a given upper barrier $B > X_0$ before or at time T. If we denote by

$$\tau_B := \inf\{t \ge 0 \mid X_t \ge B\}$$

the first hitting time of the barrier B, then the payoff of the digital option can also be described as

$$H^{\text{dig}} = \mathbb{1}_{\{\tau_B \le T\}}.$$

For simplicity, we will assume from now on that

$$C_T(B) > 0, \tag{7.14}$$

so that in particular $\mu_T((B,\infty)) > 0$, where μ_T is as in Lemma 7.23.

Theorem 7.27. *The following upper bound on the arbitrage-free prices of the digital option holds:*

$$\max_{P^* \in \mathscr{P}_{\Phi_T}} E^*[\, H^{\text{dig}} \,] = \min_{0\le K<B} \frac{C_T(K)}{B-K}. \tag{7.15}$$

Proof. For $0 \le K < B$, we have $X_{\tau_B} \ge B$. Hence,

$$H^{\text{dig}} = \mathbb{1}_{\{\tau_B \le T\}} \le \frac{(X_T - K)^+}{B-K} + \frac{X_{\tau_B} - X_T}{B-K} \mathbb{1}_{\{\tau_B \le T\}}.$$

Taking expectations with respect to some $P^* \in \mathscr{P}_{\Phi_T}$ yields

$$E^*[\, H^{\text{dig}} \,] \le \frac{C_T(K)}{B-K} + \frac{1}{B-K} E^*[\, (X_{\tau_B} - X_T)\, \mathbb{1}_{\{\tau_B \le T\}} \,]. \tag{7.16}$$

Since P^* is a martingale measure, the stopping theorem in the form of Proposition 6.37 implies

$$E^*[\, X_T\, \mathbb{1}_{\{\tau_B \le T\}} \,] = E^*[\, X_{\tau_B}\, \mathbb{1}_{\{\tau_B \le T\}} \,],$$

and so the rightmost term in (7.16) vanishes. This shows that

$$\sup_{P^*\in\mathscr{P}_{\Phi_T}} E^*[\,H^{\mathrm{dig}}\,] \le \inf_{0\le K<B} \frac{C_T(K)}{B-K}.$$

The converse inequality as well as the fact that both the supremum and the infimum are attained will be obtained in Lemmas 7.28 and 7.29 below. □

Lemma 7.28. *If we let*

$$\lambda := 1 - \inf_{0\le K<B} \frac{C_T(K)}{B-K}, \tag{7.17}$$

then $0 < \lambda < 1$ *and the infimum on the right-hand side is attained in a number* $K^* \in [0, B)$ *such that* $\mu_T([0, K^*)) \le \lambda \le \mu_T([0, K^*])$*. That is,* K^* *is a* λ*-quantile for* μ_T*.*

Proof. To see that $0 < \lambda < 1$, note first that

$$\inf_{0\le K<B} \frac{C_T(K)}{B-K} \le \frac{C_T(0)}{B} = \frac{X_0}{B},$$

which is strictly less than 1, due to our assumption $B > X_0$. Hence, $\lambda > 0$. On the other hand, the fact that $K \mapsto C_T(K)$ is decreasing yields that for any $K < B$,

$$\frac{C_T(K)}{B-K} \ge \frac{C_T(B)}{B-K} \ge \frac{C_T(B)}{B} > 0,$$

where we have used our assumption (7.14). This yields $\lambda < 1$.

Next, the function $g(K) := C_T(K)/(B-K)$ is continuous on $[0, B)$ and, due to our assumption $C_T(B) > 0$, satisfies $g(K) \to \infty$ as $K \uparrow B$. Hence, it attains its infimum in some number $K^* \in [0, B)$.

Note that by Fubini's theorem,

$$C_T(K) = \int (x-K)^+\,\mu_T(dx) = \int\int_K^\infty \mathbb{1}_{\{x>y\}}\,dy\,\mu_T(dx) = \int_K^\infty \mu_T((y,\infty))\,dy.$$

Hence, at each $K \in \mathbb{R}$, the convex function $C_T(K)$ has the left- and right-hand derivatives

$$(C_T)'_-(K) = -\mu_T([K,\infty)) \quad \text{and} \quad (C_T)'_+(K) = -\mu_T((K,\infty)). \tag{7.18}$$

Thus, at each $K < B$, the function $g(K) := C_T(K)/(B-K)$ admits the left- and right-hand derivatives

$$g'_\pm(K) = \frac{(C_T)'_\pm(K)}{B-K} + \frac{C_T(K)}{(B-K)^2} = \frac{1}{B-K}\big((C_T)'_\pm(K) + g(K)\big). \tag{7.19}$$

Moreover, since K^* is a minimizer of g on $[0, B)$, we have $g(K^*) = 1-\lambda$, $g'_+(K^*) \ge 0$, and $g'_-(K^*) \le 0$. Plugging these conditions into (7.19) and using (7.18) yields

$$\mu_T([0,K^*)) \le \lambda \le \mu_T([0,K^*]).$$

Thus, K^* is a λ-quantile of μ_T. □

Lemma 7.29. *There exists a martingale measure $\hat{P} \in \mathcal{P}_{\Phi_T}$ such that*

$$\hat{P}[\,\tau_B \le T\,] = \min_{0\le K<B} \frac{C_T(K)}{B-K}.$$

Moreover, $\hat{P}$ can be taken such that

$$\tau_B = T-1 \quad \text{and} \quad X_{\tau_B} = B \quad \hat{P}\text{-a. s. on } \{\tau_B \le T\} \tag{7.20}$$

and

$$\{X_T > K^*\} \subseteq \{\tau_B \le T\} \subseteq \{X_T \ge K^*\} \quad \text{modulo } \hat{P}\text{-null sets,} \tag{7.21}$$

where K^* is as in Lemma 7.28.

Proof. Let λ and K^* be as in Lemma 7.28, and let $q = q_{\mu_T}$ be a quantile function for μ_T such that $q(\lambda) = K^*$; see Appendix D. We take an auxiliary probability space $(\widetilde{\Omega}, \widetilde{\mathcal{F}}, \widetilde{P})$ supporting a random variable U that is uniformly distributed on $(0,1)$. By Lemma D.7, $\widetilde{X}_T := q(U)$ has distribution μ_T under $\widetilde{P}$. Hence, Lemma 7.23 yields that

$$C_T(K) = \int (x-K)^+\,\mu_T(dx) = \widetilde{E}[(\widetilde{X}_T - K)^+] \quad \text{and} \quad \widetilde{E}[\widetilde{X}_T] = \int x\,\mu_T(dx) = X_0. \tag{7.22}$$

Now we let $\widetilde{X}_t := X_0$ for $0 \le t \le T-2$. To construct a suitable random variable $\widetilde{X}_{T-1}$, let γ be such that

$$X_0 = \gamma\lambda + B(1-\lambda).$$

Since $B > X_0$ and $0 < \lambda < 1$, we have $0 \le \gamma < X_0$. We define $\widetilde{X}_{T-1}$ by

$$\widetilde{X}_{T-1} := \gamma\,\mathbb{1}_{\{U\le\lambda\}} + B\,\mathbb{1}_{\{U>\lambda\}}.$$

We now prove that $\widetilde{X} = (\widetilde{X}_t)_{t=0,\dots,T}$ is a $\widetilde{P}$-martingale with respect to its natural filtration $\widetilde{\mathcal{F}}_t := \sigma(\widetilde{X}_0,\dots,\widetilde{X}_t)$. We clearly have $\widetilde{E}[\widetilde{X}_t|\widetilde{\mathcal{F}}_s] = X_0 = \widetilde{X}_s$ for $0 \le s \le T-2$. Moreover, $\widetilde{\mathcal{F}}_{T-2} = \{\emptyset, \widetilde{\Omega}\}$, and hence

$$\widetilde{E}[\,\widetilde{X}_{T-1} \mid \widetilde{\mathcal{F}}_{T-2}\,] = \widetilde{E}[\,\widetilde{X}_{T-1}\,] = \gamma\lambda + B(1-\lambda) = X_0 = \widetilde{X}_{T-2}.$$

Furthermore, by (7.22) and since $K^* = q(\lambda)$ for λ as in (7.17),

$$\begin{aligned}\widetilde{E}[\,\widetilde{X}_T;\,\widetilde{X}_{T-1}=B\,] &= \widetilde{E}[\,q(U);\,U>\lambda\,]\\ &= \widetilde{E}[q(U)-q(\lambda);U>\lambda]+q(\lambda)\widetilde{P}[U>\lambda]\\ &= \widetilde{E}[\,(\widetilde{X}_T-K^*)^+\,]+K^*\,\widetilde{P}[\,U>\lambda\,]\\ &= C_T(K^*)+K^*\,(1-\lambda)\\ &= (1-\lambda)(B-K^*)+K^*\,(1-\lambda)\\ &= B\cdot\widetilde{P}[\,\widetilde{X}_{T-1}=B\,].\end{aligned}$$

Hence, by the second identity in (7.22) and since $\widetilde{P}[\,\widetilde{X}_{T-1}=B\,]=1-\lambda=1-\widetilde{P}[\,\widetilde{X}_{T-1}=\gamma\,]$,

$$\begin{aligned}\widetilde{E}[\,\widetilde{X}_T;\,\widetilde{X}_{T-1}=\gamma\,] &= \widetilde{E}[\,\widetilde{X}_T\,]-\widetilde{E}[\,\widetilde{X}_T;\,\widetilde{X}_{T-1}=B\,]\\ &= X_0-B\cdot\widetilde{P}[\,\widetilde{X}_{T-1}=B\,]\\ &= \gamma\,\widetilde{P}[\,\widetilde{X}_{T-1}=\gamma\,].\end{aligned}$$

Since $\widetilde{X}_{T-1}$ takes only the two values γ and B, it follows that

$$\begin{aligned}\widetilde{E}[\,\widetilde{X}_T\mid\mathscr{F}_{T-1}\,] &= \widetilde{E}[\,\widetilde{X}_T\mid\widetilde{X}_{T-1}\,]\\ &= \widetilde{E}[\widetilde{X}_T|\widetilde{X}_{T-1}=\gamma]\mathbb{1}_{\{\widetilde{X}_{T-1}=\gamma\}}+\widetilde{E}[\widetilde{X}_T|\widetilde{X}_{T-1}=B]\mathbb{1}_{\{\widetilde{X}_{T-1}=B\}}\\ &= \widetilde{X}_{T-1},\end{aligned}$$

and so $\widetilde{X}$ is indeed a martingale.

Let us denote the first time at which $\widetilde{X}$ hits the barrier B by

$$\widetilde{\tau}_B := \inf\{t\ge 0\mid \widetilde{X}_t\ge B\}.$$

Since $\widetilde{X}_t=X_0<B$ for $t\le T-2$ by construction, $\widetilde{\tau}_B$ can only take the values $T-1$, T, and $+\infty$. Moreover, since $B>K^*=q(\lambda)$ and q is increasing,

$$\begin{aligned}\{\widetilde{X}_T\ge B\} &\subseteq \{\widetilde{X}_T>K^*\}=\{q(U)>q(\lambda)\}\\ &\subseteq \{U>\lambda\}=\{\widetilde{X}_{T-1}=B\}\\ &\subseteq \{U\ge\lambda\}\subseteq\{q(U)\ge q(\lambda)\}=\{\widetilde{X}_T\ge K^*\}.\end{aligned}\tag{7.23}$$

It follows that

$$\{\widetilde{\tau}_B\le T\}=\{\widetilde{\tau}_B=T-1\}=\{\widetilde{X}_{T-1}=B\}\tag{7.24}$$

and in turn

$$\widetilde{P}[\,\widetilde{\tau}_B\le T\,]=\widetilde{P}[\,\widetilde{X}_{T-1}=B\,]=1-\lambda=\min_{0\le K<B}\frac{C_T(K)}{B-K}.$$

Thus, the law of $(\widetilde{X}_t)_{t=1,\dots,T}$ under $\widetilde{P}$ defines a martingale measure $\hat{P}$ on $\Omega = [0,\infty)^T$ satisfying the properties in the assertion. Specifically, the fact that $\hat{P} \in \mathscr{P}_{\Phi_T}$ follows from (7.22), the property (7.20) follows from (7.24), and (7.21) follows from (7.23). □

Remark 7.30. The inequality

$$H^{\mathrm{dig}} \le \frac{(X_T - K)^+}{B-K} + \frac{X_{\tau_B} - X_T}{B-K} \mathbb{1}_{\{\tau_B \le T\}}$$

appearing in the proof of Theorem 7.27 can be interpreted in terms of a suitable *superhedging strategy* for the claim H^{dig} that uses call options and forward contracts: At time $t = 0$, we buy $(B-K)^{-1}$ call options with strike K, and at the first time the price process passes the barrier B, we sell forward $(B-K)^{-1}$ shares of the asset. This strategy will be optimal if the strike price K is such that it realizes the minimum on the right-hand side of (7.15). ◇

Let us now derive bounds on the arbitrage-free prices of *barrier call options*. More precisely, we will consider an *up-and-in call option* with strike K and barrier B,

$$H^{\mathrm{call}}_{\mathrm{u\&i}} := \begin{cases} (X_T - K)^+ & \text{if } \max\limits_{0\le t\le T} X_t \ge B, \\ 0 & \text{otherwise,} \end{cases}$$

and the corresponding *up-and-out call*

$$H^{\mathrm{call}}_{\mathrm{u\&o}} := \begin{cases} (X_T - K)^+ & \text{if } \max\limits_{0\le t\le T} X_t < B, \\ 0 & \text{otherwise.} \end{cases}$$

If the barrier B is at or below the strike price K, then the up-and-in call is identical to a "plain vanilla call" $(X_T - K)^+$, and the payoff of the up-and-out call is zero. Thus, we assume from now on that

$$K < B.$$

Recall that K^* denotes the minimizer of the function $c \mapsto C_T(c)/(B-c)$ as constructed in Lemma 7.28.

Theorem 7.31. *For an up-and-in call option with strike K and barrier B,*

$$\max_{P^* \in \mathscr{P}_{\Phi_T}} E^*[\,H^{\mathrm{call}}_{\mathrm{u\&i}}\,] = \begin{cases} C_T(K) & \text{if } K^* \le K, \\ \dfrac{B-K}{B-K^*}\, C_T(K^*) & \text{if } K^* > K. \end{cases}$$

Proof. We have for any $K \ge 0$,

$$H^{\mathrm{call}}_{\mathrm{u\&i}} = (X_T - K)^+ \mathbb{1}_{\{\tau_B \le T\}} \le (X_T - K)^+.$$

Taking expectations with respect to any $P^* \in \mathscr{P}_{\Phi_T}$ yields $E^*[H^{\text{call}}_{\text{u\&i}}] \le C_T(K)$. Now we show the corresponding lower bound in case $K^* \le K$. To this end, let $\hat{P}$ be the martingale measure constructed in Lemma 7.29. Then (7.21) in Lemma 7.29 yields that for $K^* \le K$,

$$(X_T - K)^+ \, \mathbb{1}_{\{\tau_B \le T\}} = (X_T - K)^+ \quad \hat{P}\text{-a. s.},$$

and so $\hat{E}[\, H^{\text{call}}_{\text{u\&i}} \,] = C_T(K)$. This proves the assertion in case $K^* \le K$.

Let us now consider the case $K^* > K$. We first prove the lower bound. Using again $\hat{P}$ from Lemma 7.29 and (7.21), one checks that $\hat{P}$-a. s.

$$H^{\text{call}}_{\text{u\&i}} = \frac{B-K}{B-K^*} \, (X_T - K^*)^+ + \frac{K^* - K}{B-K^*} \, (B - X_T) \, \mathbb{1}_{\{\tau_B \le T\}}.$$

Taking expectations with respect to $\hat{P}$ hence yields $\ge$.

Now we prove $\le$ for $K^* > K$. To this end, we claim first that, for any c with $K \le c < B$,

$$H^{\text{call}}_{\text{u\&i}} \le \frac{B-K}{B-c} \, (X_T - c)^+ + \frac{c-K}{B-c} \, (X_{\tau_B} - X_T) \, \mathbb{1}_{\{\tau_B \le T\}}. \tag{7.25}$$

To prove this claim, note first that on $\{X_T \le K\}$ or on $\{\tau_B > T\}$ the payoff of $H^{\text{call}}_{\text{u\&i}}$ is zero, while the right-hand side of (7.25) is nonnegative. On $\{X_T \ge c,\ \tau_B \le T\}$, the right-hand side of (7.25) is equal to

$$\begin{aligned} \frac{B-K}{B-c} \, (X_T - c) + \frac{c-K}{B-c} \, (X_{\tau_B} - X_T) &= X_T - \frac{c(B-K) - X_{\tau_B}(c-K)}{B-c} \\ &\ge X_T - K = H^{\text{call}}_{\text{u\&i}}, \end{aligned}$$

where the inequality follows from the fact that $X_{\tau_B} \ge B$ and our assumption $c \ge K$. On $\{c > X_T > K,\ \tau_B \le T\}$, we get in the same way that the right-hand side of (7.25) is larger than or equal to

$$\frac{c-K}{B-c}(B - X_T) \ge c - K \ge X_T - K = H^{\text{call}}_{\text{u\&i}}.$$

This establishes (7.25).

The expectation of the right-hand side of (7.25) under a martingale measure $P^* \in \mathscr{P}_{\Phi_T}$ is equal to

$$\frac{B-K}{B-c} \, C_T(c) + \frac{c-K}{B-c} \, E^*[\, (X_{\tau_B} - X_T) \, \mathbb{1}_{\{\tau_B \le T\}} \,] = \frac{B-K}{B-c} \, C_T(c),$$

due to the stopping theorem in the form of Proposition 6.37. By Lemma 7.28, and since $K^* > K$, the minimum of the right-hand side over $c \in [K, B)$ is attained in $c = K^*$. This proves the inequality $\le$ in the assertion for the case $K^* > K$, and hence completes the proof. □

Remark 7.32. The inequality

$$H^{\text{call}}_{\text{u\&i}} \leq \frac{B-K}{B-c}\,(X_T - c)^+ + \frac{c-K}{B-c}\,(X_{\tau_B} - X_T)\,\mathbb{1}_{\{\tau_B \leq T\}}$$

appearing in the preceding proof can be interpreted as a *superhedging strategy* for the up-and-in call with liquid derivatives in case $K < K^*$: At time $t = 0$, we purchase $(B-K)/(B-c)$ call options with strike c, and at the first time the stock price passes the barrier B, we sell forward $(c-K)/(B-c)$ shares of the asset. This strategy will be optimal for $c = K^*$. ◇

We now turn to the analysis of the up-and-out call option

$$H^{\text{call}}_{\text{u\&o}} = (X_T - K)^+\,\mathbb{1}_{\{\tau_B > T\}}.$$

Theorem 7.33. *For an up-and-out call,*

$$\max_{P^* \in \mathcal{P}_{\Phi_T}} E^*[\,H^{\text{call}}_{\text{u\&o}}\,] = C_T(K) - C_T(B) - (B-K)\,\mu_T([B,\infty)).$$

Proof. Clearly,

$$\begin{aligned} H^{\text{call}}_{\text{u\&o}} &\leq (X_T - K)^+\,\mathbb{1}_{\{X_T < B\}} \\ &= (X_T - K)^+ - (X_T - B)^+ - (B-K)\,\mathbb{1}_{\{X_T \geq B\}}. \end{aligned} \tag{7.26}$$

Taking expectations yields $\leq$ in the assertion.

Now consider the measure $\widetilde{P}$ on $(\Omega, \mathcal{F})$ defined as

$$\widetilde{P} := \delta_{X_0} \otimes \cdots \otimes \delta_{X_0} \otimes \mu_T,$$

that is, under $\widetilde{P}$ we have $X_t = X_0$ $\widetilde{P}$-a. s. for $t < T$, and the law of X_T is μ_T. Clearly, $\widetilde{P} \in \mathcal{P}_{\Phi_T}$, and (7.26) is $\widetilde{P}$-a. s. an identity. Taking expectations in (7.26) with respect to $\widetilde{P}$ hence yields $\geq$. □

Using the identity

$$(X_T - K)^+ = H^{\text{call}}_{\text{u\&o}} + H^{\text{call}}_{\text{u\&i}},$$

we get the following lower bounds as an immediate corollary.

Corollary 7.34. *We have*

$$\min_{P^* \in \mathcal{P}_{\Phi_T}} E^*[\,H^{\text{call}}_{\text{u\&i}}\,] = C_T(B) + (B-K)\,\mu_T([B,\infty)),$$

and

$$\min_{P^* \in \mathscr{P}_{\Phi_T}} E^*[\, H_{\mathrm{u\&o}}^{\mathrm{call}}\,] = \begin{cases} 0 & \textit{if } K^* \leq K, \\ C_T(K) - \dfrac{B-K}{B-K^*}\, C_T(K^*) & \textit{if } K^* > K. \end{cases}$$

8 Efficient hedging

In an incomplete financial market model, a contingent claim typically will not admit a perfect hedge. Superhedging provides a method for staying on the safe side, but the required cost is usually too high both from a theoretical and from a practical point of view. It is thus natural to relax the requirements.

As a first preliminary step, we consider strategies of *quantile hedging*, which stay on the safe side with high probability. In other words, we maximize the probability for staying on the safe side under a given cost constraint. The main idea consists in reducing the construction of such strategies for a given claim H to a problem of superhedging for a modified claim $\widetilde{H}$, which is the solution to a static optimization problem of Neyman–Pearson type. Typically, $\widetilde{H}$ will have the form of a knock-out option, that is, $\widetilde{H} = H \cdot \mathbb{1}_A$. At this stage, we only focus on the probability that a shortfall occurs; we do not take into account the *size* of the shortfall if it does occur.

In Sections 8.2 and 8.3, we take a more comprehensive view of the downside risk. Our discussion of risk measures in Section 4.10 suggests to quantify the downside risk in terms of an acceptance set for suitably hedged positions. If acceptability is defined in terms of utility-based shortfall risk as in Section 4.11, we are led to the problem of constructing *efficient* strategies that minimize the utility-based shortfall risk under a given cost constraint. As in the case of quantile hedging, this problem can be decomposed into a static optimization problem and the construction of a superhedging strategy for a modified payoff profile $\widetilde{H}$. In Section 8.3, we go even one step further and assess the shortfall risk in terms of a general convex risk measure. For a complete market model and the case of AV@R, we discuss the structure of the modified payoff profile $\widetilde{H}$ and point out the relation to the problem of robust utility maximization discussed in Section 3.6.

8.1 Quantile hedging

Let H be a discounted European claim in an arbitrage-free market model such that $P[\, H = 0\,] < 1$ and

$$\pi_{\text{sup}}(H) = \sup_{P^* \in \mathcal{P}} E^*[\, H\,] < \infty.$$

We saw in Corollary 7.15 that there exists a self-financing trading strategy whose value process $V^\uparrow$ satisfies

$$V_T^\uparrow \geq H \quad P\text{-a.s.}$$

By using such a superhedging strategy, the seller of H can cover almost any possible obligation that may arise from the sale of H, and thus eliminate completely the corresponding risk. The smallest amount for which such a superhedging strategy is available

https://doi.org/10.1515/9783111045283-010

is given by $\pi_{\sup}(H)$. This cost will often be too high from a practical point of view, as illustrated by Example 7.21. Furthermore, if H is not attainable then $\pi_{\sup}(H)$, viewed as a price for H, is too high from a theoretical point of view since it permits arbitrage. Even if H is attainable, a complete elimination of risk by using a replicating strategy for H would consume the entire proceeds from the sale of H, and any opportunity of making a profit would be lost along with the risk.

Let us therefore suppose that the seller is unwilling to put up the initial amount of capital required by a superhedge and is ready to accept some risk. What is the optimal partial hedge that can be achieved with a given smaller amount of capital? In order to make this question precise, we need a criterion expressing the seller's attitude toward risk. Several of such criteria will be studied in the following sections. In this section, our aim is to construct a strategy that *maximizes the probability of a successful hedge* given a constraint on the initial cost.

More precisely, let us fix an initial amount

$$v < \pi_{\sup}(H).$$

We are looking for a self-financing trading strategy whose value process maximizes the probability

$$P[\, V_T \geq H \,]$$

over all those strategies whose initial investment V_0 is bounded by v and that respect the bounds $V_t \geq 0$ for $t = 0, \ldots, T$. In view of Theorem 5.25, the second restriction amounts to admissibility in the following sense.

Definition 8.1. A self-financing trading strategy is called an *admissible strategy* if its value process satisfies $V_T \geq 0$ P-almost surely. △

The problem of *quantile hedging* consists in constructing an admissible strategy $\overline{\xi}^*$ such that its value process V^* satisfies

$$P[\, V_T^* \geq H \,] = \max P[\, V_T \geq H \,] \tag{8.1}$$

where the maximum is taken over all value processes V of admissible strategies subject to the constraint

$$V_0 \leq v. \tag{8.2}$$

Note that this problem would not be well posed if considered without the constraint of admissibility.

Let us emphasize that the idea of quantile hedging corresponds to a *Value at Risk* criterion, and that it invites the same criticism: Only the *probability* of a shortfall is taken into account, not the *size* of the loss if a shortfall occurs. This exclusive focus on the shortfall probability may be reasonable in cases where a loss is to be avoided by any means. But for most applications, other optimality criteria as considered in the next section will usually be more appropriate from an economic point of view. In view of the mathematical techniques, however, some key ideas already appear quite clearly in our present context.

Let us first consider the particularly transparent situation of a complete market model before passing to the general incomplete case. The set

$$\{V_T \geq H\}$$

will be called the *success set* associated with the value process V of an admissible strategy. As a first step, we reduce our problem to the construction of a success set of maximal probability.

Proposition 8.2. *Let P^* denote the unique equivalent martingale measure in a complete market model, and assume that $A^* \in \mathscr{F}_T$ maximizes the probability $P[A]$ over all sets $A \in \mathscr{F}_T$ satisfying the constraint*

$$E^*[H \cdot \mathbb{1}_A] \leq v. \tag{8.3}$$

Then the replicating strategy $\overline{\xi}^*$ of the knock-out option

$$H^* := H \cdot \mathbb{1}_{A^*}$$

solves the optimization problem defined by (8.1) *and* (8.2), *and A^* coincides up to P-null sets with the success set of $\overline{\xi}^*$.*

Proof. As a first step, let V be the value process of any admissible strategy such that $V_0 \leq v$. We denote by $A := \{V_T \geq H\}$ the corresponding success set. Admissibility yields that $V_T \geq H \cdot \mathbb{1}_A$. Moreover, the results of Section 5.3 imply that V is a P^*-martingale. Hence, we obtain that

$$E^*[H \cdot \mathbb{1}_A] \leq E^*[V_T] = V_0 \leq v.$$

Therefore, A fulfills the constraint (8.3) and it follows that

$$P[A] \leq P[A^*].$$

As a second step, we consider the trading strategy $\overline{\xi}^*$ and its value process V^*. Clearly, $\overline{\xi}^*$ is admissible, and its success set satisfies

$$\{V_T^* \geq H\} = \{H \cdot \mathbb{1}_{A^*} \geq H\} \supseteq A^*.$$

On the other hand, the first part of the proof yields that

$$P[\,V_T^* \geq H\,] \leq P[\,A^*\,].$$

It follows that the two sets A^* and $\{V_T^* \geq H\}$ coincide up to P-null sets. In particular, $\overline{\xi}^*$ is an optimal strategy. □

Our next goal is the construction of the optimal success set A^*, whose existence was assumed in Proposition 8.2. This problem is solved by using the Neyman–Pearson lemma. To this end, we introduce the measure Q^* given by

$$\frac{dQ^*}{dP^*} := \frac{H}{E^*[\,H\,]}. \tag{8.4}$$

The constraint (8.3) can be written as

$$Q^*[\,A\,] \leq \alpha := \frac{v}{E^*[\,H\,]}. \tag{8.5}$$

Thus, an optimal success set must maximize the probability $P[\,A\,]$ under the constraint $Q^*[\,A\,] \leq \alpha$. We denote by dP/dQ^* the generalized density of P with respect to Q^* in the sense of the Lebesgue decomposition as constructed in Theorem B.9. Thus, we may define the level

$$c^* := \inf\left\{c \geq 0 \,\middle|\, Q^*\left[\,\frac{dP}{dQ^*} > c \cdot E^*[\,H\,]\,\right] \leq \alpha\right\}, \tag{8.6}$$

and the set

$$A^* := \left\{\frac{dP}{dQ^*} > c^* \cdot E^*[\,H\,]\right\} = \left\{\frac{dP}{dP^*} > c^* \cdot H\right\}. \tag{8.7}$$

Proposition 8.3. *If the set A^* in* (8.7) *satisfies*

$$Q^*[\,A^*\,] = \alpha,$$

then A^* maximizes the probability $P[\,A\,]$ over all $A \in \mathscr{F}_T$ satisfying the constraint

$$E^*[\,H \cdot \mathbb{1}_A\,] \leq v.$$

Proof. The condition $E^*[\,H \cdot \mathbb{1}_A\,] \leq v$ is equivalent to $Q^*[\,A\,] \leq \alpha = Q^*[\,A^*\,]$. Thus, the particular form of the set A^* in (8.7) and the Neyman–Pearson lemma in the form of Proposition E.1 imply that $P[\,A\,] \leq P[\,A^*\,]$. □

By combining the two Propositions 8.2 and 8.3, we obtain the following result.

Corollary 8.4. *Denote by P^* the unique equivalent martingale measure in a complete market model, and assume that the set A^* of* (8.7) *satisfies*

$$Q^*[\,A^*\,] = \alpha.$$

Then the optimal strategy solving (8.1) *and* (8.2) *is given by the replicating strategy of the knock-out option* $H^* = H \cdot \mathbb{1}_{A^*}$.

Our solution to the optimization problem (8.1) and (8.2) still relies on the assumption that the set A^* of (8.7) satisfies $Q^*[\, A^* \,] = \alpha$. This condition is clearly satisfied if

$$P\Big[\, \frac{dP}{dP^*} = c^* \cdot H \,\Big] = 0.$$

However, it may not in general be possible to find any set A whose Q^*-probability is exactly α. In such a situation, the Neyman–Pearson theory suggests replacing the indicator function $\mathbb{1}_{A^*}$ of the "critical region" A^* by a *randomized test*, that is, by an $\mathscr{F}_T$-measurable $[0,1]$-valued function ψ. Let $\mathscr{R}$ denote the class of all randomized tests, and consider the following optimization problem:

$$E[\, \psi^* \,] = \max\{E[\, \psi \,] \mid \psi \in \mathscr{R} \text{ and } E_{Q^*}[\, \psi \,] \le \alpha\},$$

where Q^* is the measure defined in (8.4) and $\alpha = v/E^*[\, H \,]$ as in (8.5). The generalized Neyman–Pearson lemma in the form of Theorem E.3 states that the solution is given by

$$\psi^* = \mathbb{1}_{\{\frac{dP}{dP^*} > c^* \cdot H\}} + \gamma \cdot \mathbb{1}_{\{\frac{dP}{dP^*} = c^* \cdot H\}}, \tag{8.8}$$

where c^* is defined through (8.6) and γ is chosen such that $E_{Q^*}[\, \psi^* \,] = \alpha$, that is,

$$\gamma = \frac{\alpha - Q^*[\, \frac{dP}{dP^*} > c^* \cdot H \,]}{Q^*[\, \frac{dP}{dP^*} = c^* \cdot H \,]} \quad \text{in case} \quad P\Big[\, \frac{dP}{dP^*} = c^* \cdot H \,\Big] \neq 0.$$

Definition 8.5. Let V be the value process of an admissible strategy $\overline{\xi}$. The *success ratio* of $\overline{\xi}$ is defined as the randomized test

$$\psi_V = \mathbb{1}_{\{V_T \ge H\}} + \frac{V_T}{H} \cdot \mathbb{1}_{\{V_T < H\}}. \qquad \triangle$$

Note that the set $\{\psi_V = 1\}$ coincides with the success set $\{V_T \ge H\}$ of V. In the extended version of our original problem, we are now looking for a strategy that maximizes the expected success ratio $E[\, \psi_V \,]$ under the measure P subject to the cost constraint $V_0 \le v$.

Theorem 8.6. *Suppose that* P^* *is the unique equivalent martingale measure in a complete market model. Let* ψ^* *be given by* (8.8)*, and denote by* $\overline{\xi}^*$ *a replicating strategy for the discounted claim* $H^* = H \cdot \psi^*$. *Then the success ratio* ψ_{V^*} *of* $\overline{\xi}^*$ *maximizes the expected success ratio* $E[\, \psi_V \,]$ *among all admissible strategies with initial investment* $V_0 \le v$. *Moreover, the optimal success ratio* ψ_{V^*} *is P-a. s. equal to* ψ^*.

We do not prove this theorem here, as it is a special case of Theorem 8.7 below and its proof is similar to the one of Corollary 8.4, once the optimal randomized test ψ^* has been determined by the generalized Neyman–Pearson lemma. Note that the condition

$$P\left[\frac{dP}{dP^*} = c^* \cdot H\right] = 0$$

implies that $\psi^* = \mathbb{1}_{A^*}$ with A^* as in (8.7), so in this case the strategy $\overline{\xi}^*$ reduces to the one described in Corollary 8.4.

Now we turn to the general case of an arbitrage-free but possibly incomplete market model, that is, we no longer assume that the set $\mathcal{P}$ of equivalent martingale measures consists of a single element, but we assume only that

$$\mathcal{P} \neq \emptyset.$$

In this setting, our aim is to find an admissible strategy whose success ratio ψ_{V^*} satisfies

$$E[\,\psi_{V^*}\,] = \max E[\,\psi_V\,], \tag{8.9}$$

where the maximum on the right-hand side is taken over all admissible strategies whose initial investment satisfies the constraint

$$V_0 \leq v. \tag{8.10}$$

Theorem 8.7. *In any arbitrage-free market model, there exists a randomized test ψ^* such that*

$$\sup_{P^* \in \mathcal{P}} E^*[\,H \cdot \psi^*\,] = v, \tag{8.11}$$

and that maximizes $E[\,\psi\,]$ over all $\psi \in \mathcal{R}$ subject to the constraints

$$E^*[\,H \cdot \psi\,] \leq v \quad \textit{for all } P^* \in \mathcal{P}. \tag{8.12}$$

Moreover, the superhedging strategy for the modified claim

$$H^* = H \cdot \psi^*$$

with initial investment $\pi_{\sup}(H^*)$ solves the problem (8.9) *and* (8.10).

Proof. Denote by $\mathcal{R}_0$ the set of all $\psi \in \mathcal{R}$ that satisfy the constraints (8.12), and take a sequence $\psi_n \in \mathcal{R}_0$ such that

$$\lim_{n\uparrow\infty} E[\,\psi_n\,] = \sup_{\psi\in\mathcal{R}_0} E[\,\psi\,].$$

Lemma 1.72 yields a sequence of convex combinations $\widetilde{\psi}_n \in \text{conv}\{\psi_n, \psi_{n+1}, \dots\}$ converging P-a. s. to a function $\widetilde{\psi} \in \mathscr{R}$. Clearly, $\widetilde{\psi}_n \in \mathscr{R}_0$ for each n. Hence, Fatou's lemma yields that

$$E^*[\,H\,\widetilde{\psi}\,] \le \liminf_{n\uparrow\infty} E^*[\,H\,\widetilde{\psi}_n\,] \le v \quad \text{for all } P^* \in \mathscr{P},$$

and it follows that $\widetilde{\psi} \in \mathscr{R}_0$. Moreover, dominated convergence yields that $E[\,\widetilde{\psi}\,] = \lim_n E[\,\widetilde{\psi}_n\,]$. Next, since each $E[\,\widetilde{\psi}_n\,]$ is a convex combination of $E[\,\psi_k\,]$ for certain $k \in \{n, n+1, \dots\}$, the expectations $E[\,\widetilde{\psi}_n\,]$ must converge to the same limit as $E[\,\psi_k\,]$. That is,

$$E[\,\widetilde{\psi}\,] = \lim_{n\uparrow\infty} E[\,\widetilde{\psi}_n\,] = \lim_{k\uparrow\infty} E[\,\psi_k\,] = \sup_{\psi\in\mathscr{R}_0} E[\,\psi\,].$$

Hence, $\psi^* := \widetilde{\psi}$ is the desired maximizer.

To show that (8.11) holds, we assume by way of contradiction that

$$\sup_{P^*\in\mathscr{P}} E^*[\,H\cdot\psi^*\,] < v.$$

Then there exists $\varepsilon \in (0,1)$ such that $\psi_\varepsilon := \varepsilon + (1-\varepsilon)\psi^* \in \mathscr{R}_0$. Since $P[\,\psi^* = 1\,] = 1$ is impossible due to our assumption $v < \pi_{\text{sup}}(H)$, we have $E[\,\psi_\varepsilon\,] > E[\,\psi^*\,]$, in contradiction to the maximality of $E[\,\psi^*\,]$. This proves (8.11).

Now let $\overline{\xi}$ be any admissible strategy whose value process V satisfies $V_0 \le v$. If ψ_V denotes the corresponding success ratio, then

$$H\cdot\psi_V = H \wedge V_T \le V_T.$$

The $\mathscr{P}$-martingale property of V yields that for all $P^* \in \mathscr{P}$,

$$E^*[\,H\cdot\psi_V\,] \le E^*[\,V_T\,] = V_0 \le v. \tag{8.13}$$

Therefore, ψ_V is contained in $\mathscr{R}_0$ and it follows that

$$E[\,\psi_V\,] \le E[\,\psi^*\,]. \tag{8.14}$$

Consider the superhedging strategy $\overline{\xi}^*$ of $H^* = H\cdot\psi^*$ and denote by V^* its value process. Clearly, $\overline{\xi}^*$ is an admissible strategy. Moreover,

$$V_0^* = \pi_{\text{sup}}(H^*) = \sup_{P^*\in\mathscr{P}} E^*[\,H\cdot\psi^*\,] = v.$$

Thus, (8.14) yields that ψ_{V^*} satisfies

$$E[\,\psi_{V^*}\,] \le E[\,\psi^*\,]. \tag{8.15}$$

On the other hand, V_T^* dominates H^*, and so

$$H \cdot \psi_{V^*} = H \wedge V_T^* \geq H \wedge H^* = H \cdot \psi^*.$$

Therefore, ψ_{V^*} dominates ψ^* on the set $\{H > 0\}$. Moreover, any success ratio is equal to one on $\{H = 0\}$, and we obtain that $\psi_{V^*} \geq \psi^*$ P-almost surely. According to (8.15), this can only happen if the two randomized tests ψ_{V^*} and ψ^* coincide P-almost everywhere. This proves that $\overline{\xi}^*$ solves the hedging problem (8.9) and (8.10). □

8.2 Hedging with minimal shortfall risk

Our starting point in this section is the same as in the previous one: At time T, an investor must pay the discounted random amount $H \geq 0$. A complete elimination of the corresponding risk would involve the cost

$$\pi_{\sup}(H) = \sup_{P^* \in \mathcal{P}} E^*[\, H \,]$$

of superhedging H, but the investor is only willing to put up a smaller amount

$$v \in (0, \pi_{\sup}(H)).$$

This means that the investor is ready to take some risk: Any “partial” hedging strategy whose value process V satisfies the capital constraint $V_0 \leq v$ will generate a nontrivial *shortfall*

$$(H - V_T)^+.$$

In the previous section, we constructed trading strategies that minimize the shortfall probability

$$P[\, V_T < H \,]$$

among the class of trading strategies whose initial investment is bounded by v, and that are admissible in the sense of Definition 8.1, that is, their terminal value V_T is nonnegative. In this section, we assess the shortfall in terms of a *loss function*, that is, an increasing function $\ell : \mathbb{R} \to \mathbb{R}$ that is not identically constant. We assume furthermore that

$$\ell(x) = 0 \text{ for } x \leq 0 \quad \text{and} \quad E[\, \ell(H) \,] < \infty.$$

A particular role will be played by *convex* loss functions, which correspond to *risk aversion* in view of the shortfall; compare the discussion in Section 4.11.

Definition 8.8. Given a loss function ℓ satisfying the above assumptions, the *shortfall risk* of an admissible strategy with value process V is defined as the expectation

$$E[\,\ell(H - V_T)\,] = E[\,\ell((H - V_T)^+)\,]$$

of the shortfall weighted by the loss function ℓ. △

Our aim is to minimize the shortfall risk over all admissible strategies satisfying the capital constraint $V_0 \le v$. Alternatively, we could minimize the cost under a given bound on the shortfall risk. In other words, the problem consists in constructing strategies that are *efficient* with respect to the trade-off between cost and shortfall risk. This generalizes our discussion of quantile hedging in the previous Section 8.1, which corresponds to a minimization of the shortfall risk with respect to the nonconvex loss function

$$\ell(x) = \mathbb{1}_{(0,\infty)}(x).$$

Remark 8.9. Recall our discussion of risk measures in Chapter 4. From this point of view, it is natural to quantify the downside risk in terms of an acceptance set $\mathscr{A}$ for hedged positions. As in Section 4.10, we denote by $\bar{\mathscr{A}}$ the class of all positions X such that there exists an admissible strategy $\bar{\xi}$ with value process V such that

$$V_0 = 0 \quad \text{and} \quad X + V_T \ge A \quad P\text{-a.s.}$$

for some $A \in \mathscr{A}$. Thus, the downside risk of the position $-H$ takes the form

$$\bar{\rho}(-H) = \inf\{m \in \mathbb{R} \mid m - H \in \bar{\mathscr{A}}\}.$$

Suppose that the acceptance set $\mathscr{A}$ is defined in terms of shortfall risk, that is,

$$\mathscr{A} := \{X \in L^\infty \mid E[\,\ell(X^-)\,] \le x_0\},$$

where ℓ is a convex loss function and x_0 is a given threshold. Then $\bar{\rho}(-H)$ is the smallest amount m such that there exists an admissible strategy $\bar{\xi}$ whose value process V satisfies $V_0 = m$ and

$$E[\,\ell((H - V_T)^+)\,] \le x_0.$$

For a given m, we are thus led to the problem of finding a strategy $\bar{\xi}$ that minimizes the shortfall risk under the cost constraint $V_0 \le m$. In this way, the problem of quantifying the downside risk of a contingent claim is reduced to the construction of efficient hedging strategies as discussed in this section. ◇

As in the preceding section, the construction of the optimal hedging strategy is carried out in two steps. The first one is to solve the "static" problem of minimizing

$$E[\,\ell(H-Y)\,]$$

over all $\mathscr{F}_T$-measurable random variables $Y \geq 0$ satisfying the constraints

$$\sup_{P^* \in \mathscr{P}} E^*[\,Y\,] \leq v.$$

If Y^* solves this problem, then so does $\widetilde{Y} := H \wedge Y^*$. Hence, we may assume that $0 \leq Y^* \leq H$ or, equivalently, that $Y^* = H\,\psi^*$ for some *randomized test* ψ^* that belongs to the set $\mathscr{R}$ of all $\mathscr{F}_T$-measurable random variables with values in $[0,1]$. Thus, the static problem can be reformulated as follows: Find a randomized test $\psi^* \in \mathscr{R}$ that minimizes the "shortfall risk"

$$E[\,\ell(H(1-\psi))] \tag{8.16}$$

over all $\psi \in \mathscr{R}$ subject to the constraints

$$E^*[\,H\,\psi\,] \leq v \quad \text{for all } P^* \in \mathscr{P}. \tag{8.17}$$

The next step is to fit the terminal value V_T of an admissible strategy to the optimal profile $H\,\psi^*$. It turns out that this step can be carried out without any further assumptions on our loss function ℓ. Thus, we assume at this point that the optimal ψ^* of step one is granted, and we construct the corresponding optimal strategy.

Theorem 8.10. *Given a randomized test ψ^* that minimizes* (8.16) *subject to* (8.17), *a superhedging strategy $\overline{\xi}^*$ for the modified discounted claim*

$$H^* := H\,\psi^*$$

with initial investment $\pi_{\sup}(H^*)$ has minimal shortfall risk among all admissible strategies $\overline{\xi}$ that satisfy the capital constraint $\overline{\xi}_1 \cdot \overline{X}_0 \leq v$.

Proof. The proof extends the last argument in the proof of Theorem 8.7. As a first step, we take any admissible strategy $\overline{\xi}$ such that the corresponding value process V satisfies the capital constraint $V_0 \leq v$. Denote by

$$\psi_V = \mathbb{1}_{\{V_T \geq H\}} + \frac{V_T}{H} \cdot \mathbb{1}_{\{V_T < H\}}$$

the corresponding success ratio. It follows as in (8.13) that ψ_V satisfies the constraints

$$E^*[\,H\,\psi_V] \leq v \quad \text{for all } P^* \in \mathscr{P}.$$

Thus, the optimality of ψ^* implies the following lower bound on the shortfall risk of $\overline{\xi}$:

$$E[\,\ell(H-V_T)] = E[\,\ell(H(1-\psi_V))] \geq E[\,\ell(H(1-\psi^*))].$$

In the second step, we consider the admissible strategy $\overline{\xi}^*$ and its value process V^*. On the one hand,

$$V_0^* = \pi_{\sup}(H^*) = \sup_{P^* \in \mathcal{P}} E^*[\, H\,\psi^*\,] \leq v,$$

so $\overline{\xi}^*$ satisfies the capital constraint. Hence, the first part of the proof yields

$$E[\,\ell(H - V_T^*)] = E[\,\ell(H(1 - \psi_{V^*}))] \geq E[\,\ell(H(1 - \psi^*))]. \tag{8.18}$$

On the other hand, $V_T^* \geq H^* = H\,\psi^*$ and, therefore,

$$\psi_{V^*} \geq \psi^* \quad P\text{-a. s.}$$

Hence, the inequality in (8.18) is in fact an equality, and the assertion follows. □

Let us now return to the static problem defined by (8.16) and (8.17). We start by considering the special case of risk aversion in view of the shortfall.

Proposition 8.11. *If the loss function ℓ is convex, then there exists a randomized test $\psi^* \in \mathcal{R}$ that minimizes the shortfall risk*

$$E[\,\ell(H(1 - \psi))]$$

over all $\psi \in \mathcal{R}$ subject to the constraints

$$E^*[\,H\,\psi\,] \leq v \quad \textit{for all } P^* \in \mathcal{P}. \tag{8.19}$$

If ℓ is strictly convex on $[0, \infty)$, then ψ^* is uniquely determined on $\{H > 0\}$.

Proof. The proof is similar to the one of Proposition 3.30. Let $\mathcal{R}_0$ denote the set of all randomized tests that satisfy the constraints (8.19). Take $\psi_n \in \mathcal{R}_0$ such that $E[\ell(H(1 - \psi_n))]$ converges to the infimum of the shortfall risk, and use Lemma 1.72 to select convex combinations $\widetilde{\psi}_n \in \operatorname{conv}\{\psi_n, \psi_{n+1}, \dots\}$ that converge P-a. s. to some $\widetilde{\psi} \in \mathcal{R}$. Since ℓ is continuous and nonnegative, Fatou's lemma implies that

$$E[\,\ell(H(1-\widetilde{\psi}))] \leq \liminf_{n\uparrow\infty} E[\,\ell(H(1-\widetilde{\psi}_n))] \leq \liminf_{n\uparrow\infty} E[\ell(H(1-\psi_n))] = \inf_{\psi\in\mathcal{R}_0} E[\,\ell(H(1-\psi))],$$

where we have used the convexity of ℓ in the second step. Fatou's lemma also yields that for all $P^* \in \mathcal{P}$,

$$E^*[\,H\,\widetilde{\psi}\,] \leq \liminf_{n\uparrow\infty} E^*[\,H\,\widetilde{\psi}_n\,] \leq v.$$

Hence, $\widetilde{\psi} \in \mathcal{R}_0$, and we conclude that $\psi^* := \widetilde{\psi}$ is the desired minimizer. The uniqueness part is obvious. □

Remark 8.12. The proof shows that the analogous existence result holds if we use a robust version of the shortfall risk defined as

$$\sup_{Q \in \mathcal{Q}} E_Q[\,\ell(H(1-\psi))],$$

where $\mathcal{Q}$ is a class of equivalent probability measures; see also Remark 3.31 and Sections 8.2 and 8.3. ◇

Combining Proposition 8.11 and Theorem 8.10 yields existence and uniqueness of an optimal hedging strategy under risk aversion in a general arbitrage-free market model.

Corollary 8.13. *Assume that the loss function ℓ is strictly convex on $[0,\infty)$. Then there exists an admissible strategy that is optimal in the sense that it minimizes the shortfall risk over all admissible strategies $\overline{\xi}$ subject to the capital constraint $\overline{\xi}_1 \cdot \overline{X}_0 \leq v$. Moreover, any optimal strategy requires the exact initial investment v, and its success ratio is P-a. s. equal to*

$$\psi^* \cdot \mathbb{1}_{\{H>0\}} + \mathbb{1}_{\{H=0\}},$$

where ψ^* denotes the solution to the static problem constructed in Proposition 8.11.

Proof. The existence of an optimal strategy follows by combining Proposition 8.11 and Theorem 8.10. Strict convexity of ℓ implies that ψ^* is P-a. s. unique on $\{H > 0\}$. In particular, ψ^* and the success ratio ψ_{V^*} of any optimal strategy $\overline{\xi}^*$ must coincide P-a. s. on $\{H > 0\}$. On $\{H = 0\}$, the success ratio ψ_{V^*} is equal to 1 by definition.

Since ℓ is strictly increasing on $[0,\infty)$, we must have that

$$\sup_{P^* \in \mathcal{P}} E^*[\,H\psi^*\,] = v,$$

for otherwise we could find some $\varepsilon > 0$ such that $\psi_\varepsilon := \varepsilon + (1-\varepsilon)\psi^*$ also satisfies the constraints (8.17). Since we have assumed that $v < \pi_{\text{sup}}(H)$, the constraints (8.17) imply that ψ^* is not P-a. s. equal to 1 on $\{H > 0\}$, and hence that

$$E[\,\ell(H(1-\psi_\varepsilon))] < E[\,\ell(H(1-\psi^*))].$$

This, however, contradicts the optimality of ψ^*.

Since the value process V^* of an optimal strategy is a $\mathcal{P}$-martingale, and since

$$V_T^* \geq H\,\psi_{V^*} = H\,\psi^*,$$

we conclude from the above that

$$v \geq V_0^* = \sup_{P^* \in \mathcal{P}} E^*[\,V_T^*\,] \geq \sup_{P^* \in \mathcal{P}} E^*[\,H\psi^*\,] = v.$$

Thus, V_0^* is equal to v. □

Beyond the general existence statement of Proposition 8.11, it is possible to obtain an explicit formula for the optimal solution to the static problem if the market model is complete. Recall that we assume that the loss function $\ell(x)$ vanishes for $x \le 0$. In addition, we will also assume that

$$\ell \text{ is strictly convex and continuously differentiable on } (0, \infty).$$

Then the derivative ℓ' of ℓ is strictly increasing on $(0, \infty)$. Let J denote the inverse function of ℓ' defined on the range of ℓ', that is, on the interval (a, b) where $a := \lim_{x \downarrow 0} \ell'(x)$ and $b := \lim_{x \uparrow \infty} \ell'(x)$. We extend J to a function $J^+ : [0, \infty] \to [0, \infty]$ by setting

$$J^+(y) := \begin{cases} +\infty & \text{for } y \ge b, \\ 0 & \text{for } y \le a. \end{cases}$$

From now on, we assume also that

$$\mathcal{P} = \{P^*\},$$

that is, P^* is the unique equivalent martingale measure in a complete market model. Its density will be denoted by

$$\varphi^* := \frac{dP^*}{dP}.$$

Theorem 8.14. *Under the above assumptions, the solution to the static optimization problem of Proposition* 8.11 *is given by*

$$\psi^* = 1 - \frac{J^+(c\,\varphi^*)}{H} \wedge 1 \quad P\text{-a. s. on } \{H > 0\},$$

where the constant c is determined by the condition $E^*[\,H\,\psi^*\,] = v$.

Proof. The problem is of the same type as those considered in Section 3.3. It can in fact be reduced to Corollary 3.37 by considering the random utility function

$$u(x, \omega) := -\ell(H(\omega) - x), \quad 0 \le x \le H(\omega).$$

Just note that the shortfall risk $E[\,\ell(H - Y)\,]$ coincides with the negative expected utility $-E[u(Y, \cdot)\,]$ for any profile Y such that $0 \le Y \le H$. Moreover, since our market model is complete, it has a finite structure by Theorem 5.37, and so all integrability conditions are automatically satisfied. Thus, Corollary 3.37 states that the optimal profile $H^* := Y^*$ that maximizes the expected utility $E[\,u(Y, \cdot)\,]$ under the constraints $0 \le Y \le H$ and $E^*[\,Y\,] \le v$ is given by

$$H^*(\omega) = I^+(c\,\varphi^*(\omega), \omega) \wedge H(\omega) = \left(H(\omega) - J^+(c\,\varphi^*(\omega))\right)^+.$$

Dividing by H yields the formula for the optimal randomized test ψ^*. □

Corollary 8.15. *In the situation of Theorem* 8.14, *suppose that the objective probability measure P is equal to the martingale measure P^*. Then the modified discounted claim takes the simple form*

$$H^* = H\psi^* = (H - J^+(c))^+.$$

Example 8.16. Consider the discounted payoff H of a European call option $(S^i_T - K)^+$ with strike K under the assumption that the numéraire S^0 is a riskless bond, that is, that $S^0_t = (1+r)^t$ for a certain constant $r \geq 0$. If the assumptions of Corollary 8.15 hold, then the modified profile H^* is the discounted value of the European call option struck at $\widetilde{K} := K + J^+(c^*) \cdot (1+r)^T$, that is,

$$H^* = \frac{(S^i_T - \widetilde{K})^+}{(1+r)^T}.$$ ◇

Example 8.17. Consider an exponential loss function $\ell(x) = (e^{\alpha x} - 1)^+$ for some $\alpha > 0$. In this case,

$$J^+(y) = \left(\frac{1}{\alpha}\log\frac{y}{\alpha}\right)^+, \quad y \geq 0,$$

and the optimal profile is given by

$$H^* = H - \left(\frac{1}{\alpha}\log\frac{c\varphi^*}{\alpha}\right)^+ \wedge H.$$ ◇

Example 8.18. If ℓ is the particular loss function,

$$\ell(x) = \frac{x^p}{p}, \quad x \geq 0,$$

for some $p > 1$, then the problem is to minimize a lower partial moment of the difference $V_T - H$. Theorem 8.14 implies that it is optimal to hedge the modified claim

$$H\psi^*_p = H - (c_p \cdot \varphi^*)^{1/(p-1)} \wedge H \tag{8.20}$$

where the constant c_p is determined by $E^*[H\psi^*_p] = v$. ◇

Let us now consider the limit $p \uparrow \infty$ in (8.20), corresponding to ever increasing risk aversion with respect to large losses.

Proposition 8.19. *Let us consider the loss functions*

$$\ell_p(x) = \frac{x^p}{p}, \quad x \geq 0,$$

for $p > 1$*. As* $p \uparrow \infty$*, the modified claims* $H\,\psi_p^*$ *of* (8.20) *converge P-a. s. and in* $L^1(P^*)$ *to the discounted claim*

$$(H - c_\infty)^+$$

where the constant c_∞ *is determined by*

$$E^*[\,(H - c_\infty)^+\,] = v. \tag{8.21}$$

Proof. Let $\gamma(p)$ be shorthand for $1/(p-1)$ and note that

$$(\varphi^*)^{\gamma(p)} \longrightarrow 1 \quad P\text{-a. s. as } p \uparrow \infty.$$

Hence, if (p_n) is a sequence for which $c_{p_n}^{\gamma(p_n)}$ converges to some $\tilde{c} \in [0, \infty]$, then

$$\lim_{n\uparrow\infty} H\,\psi_{p_n}^* = H - \tilde{c} \wedge H = (H - \tilde{c}\,)^+.$$

Hence,

$$E^*[\,H\,\psi_{p_n}^*\,] \longrightarrow E^*[\,(H - \tilde{c}\,)^+\,].$$

Since each term on the left-hand side equals v, we must have

$$E^*[\,(H - \tilde{c}\,)^+\,] = v,$$

which determines $\tilde{c}$ uniquely as the constant c_∞ of (8.21). □

Example 8.20. If the discounted claim H in Proposition 8.19 is the discounted payoff of a call option with strike K, and the numéraire is a riskless bond as in Example 8.16, then the limiting profile $\lim_{p\uparrow\infty} H\,\psi_p^*$ is equal to the discounted call with the higher strike price $K + c_\infty \cdot S_T^0$. ◇

In the remainder of this section, we consider loss functions corresponding to risk neutrality and to risk-seeking preferences. In the first case, the loss function will be convex but not strictly convex. In the second case, it will not be convex. Let us first consider the risk-neutral case.

Example 8.21. In the case of risk neutrality, the loss function is given by

$$\ell(x) = x \quad \text{for } x \geq 0.$$

Thus, the task is to minimize the expectation of the shortfall,

$$E[\,(H - V_T)^+\,]$$

under the capital constraint $V_0 \leq v$. Let P^* be the unique equivalent martingale measure in a complete market model. Then the static problem corresponding to Proposition 8.11 is to maximize the expectation

$$E[\,H\,\psi\,]$$

under the constraint that $\psi \in \mathscr{R}$ satisfies

$$E^*[\,H\,\psi\,] \leq v.$$

We can define two equivalent measures Q and Q^* by

$$\frac{dQ}{dP} = \frac{H}{E[\,H\,]} \quad \text{and} \quad \frac{dQ^*}{dP^*} = \frac{H}{E^*[\,H\,]}.$$

The problem then becomes the hypothesis testing problem of maximizing $E_Q[\,\psi\,]$ under the side condition

$$E_{Q^*}[\,\psi\,] \leq \alpha := \frac{v}{E^*[\,H\,]}.$$

Since the density dQ/dQ^* is proportional to the inverse of the density $\varphi^* = dP^*/dP$, Theorem E.3 implies that the optimal test takes the form

$$\psi_1^* = \mathbb{1}_{\{\varphi^* < c_1\}} + \gamma \cdot \mathbb{1}_{\{\varphi^* = c_1\}} \qquad P\text{-a.s. on } \{H > 0\},$$

where the constant c_1 is given by

$$c_1 = \sup\{c \in \mathbb{R} \mid E^*[\,H; \varphi^* < c\,] \leq v\},$$

and where the constant $\gamma \in [0,1]$ is chosen such that $E^*[\,H\,\psi_1^*\,] = v$. On $\{H = 0\}$, ψ_1^* can be arbitrary. ◇

Assume now that the shortfall risk is assessed by an investor who, instead of being risk-averse, is in fact inclined to take risk. In our context, this corresponds to a loss function that is concave on $[0,\infty)$ rather than convex. It is not difficult to generalize Theorem 8.14 so that it covers this situation. Here, we limit ourselves to the following explicit case study.

Example 8.22. Consider the loss function

$$\ell(x) = \frac{x^q}{q}, \qquad x \geq 0,$$

for some $q \in (0,1)$. In order to solve our static optimization problem, one could apply the results and techniques of Section 3.3. Here, we will use an approach based on the Neyman–Pearson lemma. Note first that for $\psi \in \mathscr{R}$

$$\ell(H(1-\psi)) = (1-\psi)^q \cdot \ell(H) \geq \ell(H) - \psi \cdot \ell(H).$$

Hence, we get a lower bound on the "shortfall risk" of ψ:

$$E[\,\ell(H(1-\psi))\,] \geq E[\ell(H)\,] - E[\,\psi \cdot \ell(H)\,]. \tag{8.22}$$

The problem of finding a minimizer of the right-hand side is equivalent to maximizing the expectation $E_Q[\,\psi\,]$ under the constraint that $E_{Q^*}[\,\psi\,] \leq v/E^*[\,H\,]$ for the measures Q and Q^* defined via

$$\frac{dQ}{dP} = \frac{H^q}{E[\,H^q\,]} \quad \text{and} \quad \frac{dQ^*}{dP^*} = \frac{H}{E^*[\,H\,]},$$

if we assume again $P[\,H > 0\,] = 1$. As in Example 8.21, we then conclude that the optimal test must be of the form

$$\mathbb{1}_{\{1 > c_q^* \varphi^* H^{1-q}\}} + \gamma \cdot \mathbb{1}_{\{1 = c_q^* \varphi^* H^{1-q}\}}, \tag{8.23}$$

for certain constants c_q^* and γ. Under the simplifying assumption that

$$P[\,1 = c_q^* \varphi^* H^{1-q}\,] = 0, \tag{8.24}$$

the formula (8.23) reduces to

$$\psi_q^* = \begin{cases} 1 & \text{on } \{1 > c_q^* \varphi^* H^{1-q}\}, \\ 0 & \text{otherwise.} \end{cases} \tag{8.25}$$

By taking $\psi = \psi_q^*$, we obtain an identity in (8.22), and so ψ_q^* must be a minimizer for $E[\,\ell(H(1-\psi))\,]$ under the constraint that $E^*[\,H\,\psi\,] \leq v$. ◇

In our last result of this section, we recover the knock-out option

$$H \cdot \mathbb{1}_{\{1 > c_0^* \cdot H\,\varphi^*\}},$$

which was obtained as the solution to the problem of quantile hedging by taking the limit $q \downarrow 0$ in (8.25). Intuitively, decreasing q corresponds to an increasing appetite for risk in view of the shortfall.

Proposition 8.23. *Let us assume for simplicity that* (8.24) *holds for all $q \in (0,1)$, that $P[\,H > 0\,] = 1$, and that there exists a unique constant c_0^* such that*

$$E^*[\,H \cdot \mathbb{1}_{\{1 > c_0^* \cdot H\,\varphi^*\}}\,] = v. \tag{8.26}$$

Then the solutions ψ_q^* of (8.25) *converge P-a. s. to the solution*

$$\psi_0^* = \mathbb{1}_{\{1 > c_0^* \cdot H \varphi^*\}}$$

of the corresponding problem of quantile hedging as constructed in Proposition 8.3.

Proof. Take any sequence $q_n \downarrow 0$ such that $(c_{q_n}^*)^{1/(1-q_n)}$ converges to some $\widetilde{c} \in [0, \infty]$. Then

$$\lim_{n \uparrow \infty} \psi_{q_n}^* = \mathbb{1}_{\{1 > \widetilde{c} \cdot H \varphi^*\}}.$$

Hence,

$$E^*[H \psi_{q_n}^*] \longrightarrow E^*[H \cdot \mathbb{1}_{\{1 > \widetilde{c} \cdot H \varphi^*\}}].$$

Since we assumed (8.24) for all $q \in (0, 1)$, the left-hand terms are all equal to v, and it follows from (8.26) that $\widetilde{c} = c_0^*$. This establishes the desired convergence. □

8.3 Efficient hedging with convex risk measures

As in the previous sections of this chapter, we consider the shortfall

$$(H - V_T)^+$$

arising from hedging the discounted claim H with an admissible trading strategy with initial capital

$$V_0 = v \in (0, \pi_{\text{sup}}(H)).$$

Recall from Definition 8.1 that a self-financing trading strategy is called admissible if its value process satisfies $V_T \geq 0$ P-a. s. In this section, our aim is to minimize the shortfall risk

$$\rho(-(H - V_T)^+),$$

where ρ is a given convex risk measure as discussed in Chapter 4. Here, we assume that ρ is defined on a suitable function space, such as $L^p(\Omega, \mathcal{F}, P)$, so that the shortfall risk is well-defined and finite; cf. Remark 4.44. In particular, we assume that $\rho(Y) = \rho(\widetilde{Y})$ whenever $Y = \widetilde{Y}$ P-a. s.

As in the preceding two sections, the construction of the optimal hedging strategy can be carried out in two steps. The first step is to solve the static problem of minimizing

$$\rho(-(H - Y)^+)$$

over all $\mathcal{F}_T$-measurable random variables $Y \geq 0$ that satisfy the constraint

$$\sup_{P^* \in \mathcal{P}} E^*[\, Y \,] \le v.$$

If Y^* solves this problem, then so does $H \wedge Y^*$. Hence, we may assume that $0 \le Y^* \le H$, and we can reformulate the problem as

$$\text{minimize } \rho(Y - H) \text{ subject to } 0 \le Y \le H \text{ and } \sup_{P^* \in \mathcal{P}} E^*[\, Y \,] \le v. \tag{8.27}$$

The next step is to fit the terminal value V_T of an admissible strategy to the optimal profile Y^*. It turns out that this step can be carried out without any further assumptions on our risk measure ρ. Thus, we assume at this point that the optimal Y^* of step one is granted, and we construct the corresponding optimal strategy.

Proposition 8.24. *A superhedging strategy for a solution Y^* of* (8.27) *with initial investment $\pi_{sup}(Y^*)$ has minimal shortfall risk among all admissible strategies whose value process satisfies the capital constraint $V_0 \le v$.*

Proof. Let V be the value process of any admissible strategy such that $V_0 \le v$. Due to Doob's systems theorem in the form of Theorem 5.14, V is a martingale under any $P^* \in \mathcal{P}$, and so

$$\sup_{P^* \in \mathcal{P}} E^*[\, V_T \,] = V_0 \le v.$$

Thus, $Y := H \wedge V_T$ satisfies the constraints in (8.27), and we get

$$\rho(-(H - V_T)^+) = \rho(Y - H) \ge \rho(Y^* - H).$$

Next, let V^* be the value process of a superhedging strategy for Y^* with initial investment

$$V_0^* = \pi_{\text{sup}}(Y^*) = \sup_{P^* \in \mathcal{P}} E^*[\, Y^* \,].$$

Then we have $V_0^* \le v$ and $V_T^* \ge 0$. Moreover, $V_T^* \ge Y^*$ P-a. s., and thus

$$\rho(Y^* - H) = \rho(-(H - Y^*)^+) \ge \rho(-(H - V_T^*)^+).$$

This concludes the proof. □

Let us now return to the static problem defined by (8.27).

Proposition 8.25. *If the convex risk measure ρ is lower semicontinuous with respect to P-a. s. convergence of random variables in the class $\{-Y \mid 0 \le Y \le H\}$ and $\rho(-Y) < \infty$ for one such Y, then there exists a solution to the static optimization problem* (8.27)*. In particular, there exists a solution if H is bounded and ρ is continuous from above.*

Proof. Take a sequence Y_n with $0 \le Y_n \le H$ and $\sup_{P^* \in \mathcal{P}} E^*[\,Y_n\,] \le v$ such that $\rho(Y_n - H)$ converges to the infimum A of the shortfall risk. We can use Lemma 1.72 to select convex combinations $Z_n \in \operatorname{conv}\{Y_n, Y_{n+1}, \dots\}$ that converge P-a. s. to some random variable Z. Then $0 \le Z \le H$ and Fatou's lemma yields that

$$E^*[\,Z\,] \le \liminf_{n\uparrow\infty} E^*[\,Z_n\,] \le v$$

for all $P^* \in \mathcal{P}$. Lower semicontinuity of ρ implies that

$$\rho(Z - H) \le \liminf_{n\uparrow\infty} \rho(Z_n - H).$$

Moreover, the right-hand side is equal to A, due to the convexity of ρ. Hence, Z is the desired minimizer. □

Combining Proposition 8.25 and Proposition 8.24 yields the existence of a risk-minimizing hedging strategy in a general arbitrage-free market model. So far, all arguments were practically the same as in the preceding two sections.

Beyond the general existence statement of Proposition 8.25, it is sometimes possible to obtain an explicit formula for the optimal solution to the static problem (8.27) if the market model is complete and so $\mathcal{P} = \{P^*\}$. In this case, the static optimization problem simplifies to

$$\text{minimize } \rho(Y - H) \text{ subject to } 0 \le Y \le H \text{ and } E^*[\,Y\,] \le v.$$

By substituting Z for $H - Y$, this is equivalent to the problem

$$\text{minimize } \rho(-Z) \text{ subject to } 0 \le Z \le H \text{ and } E^*[\,Z\,] \ge \widetilde{v}, \tag{8.28}$$

where $\widetilde{v} := E^*[\,H\,] - v$. We will now discuss this problem in the case $\rho = \mathrm{AV@R}_\lambda$. Our approach relies on the general idea that a minimax problem can be transformed into a standard minimization problem by using a duality result for the expression involving the maximum. In the case of $\mathrm{AV@R}_\lambda$, we can use the following representation of $\mathrm{AV@R}_\lambda$ from Proposition 4.51:

$$\mathrm{AV@R}_\lambda(-Z) = \frac{1}{\lambda}\min_{r\in\mathbb{R}}\big(E[\,(Z-r)^+\,] + \lambda r\big) = \frac{1}{\lambda}\min_{r\ge 0}\big(E[\,(Z-r)^+\,] + \lambda r\big) \tag{8.29}$$

for $Z \ge 0$. Our discussion of problem (8.28) will be valid also beyond the setting of a complete discrete-time market model, whose underlying probability space has necessarily a discrete structure by Theorem 5.37. In fact, it applies whenever P and P^* are two equivalent probability measures on a given measurable space $(\Omega, \mathcal{F})$. This is important in view of the application of the next theorem in Example 3.47.

Theorem 8.26. *Suppose that $H \in L^1(P)$ and denote by $\varphi := dP^*/dP$ the price density of P^* with respect to P. Then the problem (8.28) admits a solution for $\rho = \mathrm{AV@R}_\lambda$ that is of the form*

$$Z^* = H\mathbb{1}_{\{\varphi>c\}} + (H \wedge r^*)\mathbb{1}_{\{\varphi<c\}} + (\gamma H + (1-\gamma)(H \wedge r^*))\mathbb{1}_{\{\varphi=c\}} \tag{8.30}$$

for certain constants $c > 0$, $r^* \geq 0$, and $\gamma \in [0,1]$.

Proof. Recall from Section 4.5 that

$$\mathrm{AV@R}_\lambda(-Y) = \sup_{Q\in\mathcal{Q}_\lambda} E_Q[\,Y\,],$$

where $\mathcal{Q}_\lambda$ is the set of all probability measures $Q \ll P$ whose density dQ/dP is P-a. s. bounded by $1/\lambda$. Thus, it follows from Fatou's lemma that $\mathrm{AV@R}_\lambda$ is lower semicontinuous with respect to P-a. s. convergence of random variables in the class $\{-Y \mid 0 \leq Y \leq H\} \subset L^1$ (the same argument also gives upper semicontinuity, and hence continuity, but this fact is not needed here). Proposition 8.25 hence yields the existence of a solution Z^* of the minimization problem (8.28). By (8.29), Z^* must solve

$$\text{minimize } E[\,(Z - r^*)^+\,] \text{ subject to } 0 \leq Z \leq H \text{ and } E^*[\,Z\,] \geq \widetilde{v}, \tag{8.31}$$

where $r^* \geq 0$ is such that

$$\mathrm{AV@R}_\lambda(-Z^*) = \frac{1}{\lambda}E[\,(Z^* - r^*)^+\,] + r^*. \tag{8.32}$$

It is clear that the expectation constraint in (8.31) must be binding. That is, $E^*[\,Z^*\,] = \widetilde{v}$.

Let us now solve (8.31). To this end, we consider first the case in which $r^* = 0$. Then we are in the situation of Example 8.21 and obtain

$$Z^* = H\mathbb{1}_{\{\varphi>c\}} + \gamma H\mathbb{1}_{\{\varphi=c\}}$$

for constants $c > 0$ and $\gamma \in [0,1]$. This representation coincides with (8.30) for $r^* = 0$.

Now we consider the case $r^* > 0$. We claim first that we must have $Z^* \geq H \wedge r^*$ P-almost surely. To see why this is the case, let us assume by way of contradiction that $P[\,Z^* < H\wedge r^*\,] > 0$. Then we must have either $P[\,Z^* \leq H\wedge r^*\,] = 1$ or $P[\,Z^* > H\wedge r^*\,] > 0$. In the first case, we have $P[Z^* \leq r^*] = 1$, and so (8.32) yields that $\mathrm{AV@R}_\lambda(-Z^*) = r^*$. Now we choose r' such that $E^*[H \wedge r'] = \widetilde{v}$ and let $Z' := H \wedge r'$. Due to our assumptions $P[\,Z^* \leq H \wedge r^*\,] = 1$ and $P[\,Z^* < H \wedge r^*\,] > 0$ and the fact that $E^*[Z^*] = \widetilde{v}$, we must have $r' < r^*$. On the other hand, Z' satisfies the constraints in (8.31), and (8.29) implies that

$$\mathrm{AV@R}_\lambda(-Z') \leq \frac{1}{\lambda}E[\,(Z' - r')^+\,] + r' = r' < r^* = \mathrm{AV@R}_\lambda(-Z^*).$$

But this is a contradiction to the optimality of Z^*. Now we consider the second case in which we have both $P[\,Z^* > H \wedge r^*\,] > 0$ and $P[\,Z^* < H \wedge r^*\,] > 0$. Since Z^* satisfies the constraint $0 \le Z^* \le H$, we have $Z^* > r^*$ P-a. s. on $\{Z^* > H \wedge r^*\}$. We define

$$Z_\theta := \mathbb{1}_{\{Z^* < H\wedge r^*\}} H \wedge r^* + \mathbb{1}_{\{Z^* \ge H\wedge r^*\}} (H \wedge r^* + \theta(Z^* - H \wedge r^*)),$$

where we take the smallest $\theta \ge 0$ such that $E^*[Z_\theta] \ge \widetilde{v} = E^*[Z^*]$. Since $Z_\theta > Z^*$ on the set $\{Z^* < H \wedge r^*\}$, which has a strictly positive probability, it follows that θ must be strictly less than 1. Hence, $Z_\theta < Z^*$ on $\{Z^* > H \wedge r^*\}$. Thus,

$$E[(Z_\theta - r^*)^+] < E[(Z^* - r^*)^+],$$

in contradiction to our assumption that Z^* solves (8.31). This establishes our claim that $Z^* \ge H \wedge r^*$ P-a. s.

Now we continue with our proof for the case $r^* > 0$. Since $Z^* \ge H \wedge r^*$ P-a. s., the problem (8.31) is equivalent to minimizing $E[\,(\hat{Z} + H \wedge r^* - r^*)^+\,]$ subject to

$$0 \le \hat{Z} \le H - H \wedge r^* \quad \text{and} \quad E^*[\,\hat{Z}\,] \ge \hat{v} := \widetilde{v} - E^*[\,H \wedge r^*\,].$$

Any $\hat{Z}$ satisfying these constraints must be concentrated on $\{H > r^*\}$, so that the problem is equivalent to

$$\text{minimize } E[\,\hat{Z}\,] \text{ subject to } 0 \le \hat{Z} \le H - H \wedge r^* \text{ and } E^*[\,\hat{Z}\,] \ge \hat{v}. \tag{8.33}$$

But this problem is equivalent to the one for $r^* = 0$ if we replace H by $H - H \wedge r^*$. Hence, it is solved by

$$\hat{Z}^* = (H - H \wedge r^*)\mathbb{1}_{\{\varphi > c\}} + \gamma(H - H \wedge r^*)\mathbb{1}_{\{\varphi = c\}}$$

for some constants $c > 0$ and $\gamma \in [0, 1]$. It follows that

$$Z^* = \hat{Z}^* + H \wedge r^* = H\mathbb{1}_{\{\varphi > c\}} + (H \wedge r^*)\mathbb{1}_{\{\varphi < c\}} + (\gamma H + (1-\gamma)(H \wedge r^*))\mathbb{1}_{\{\varphi = c\}}. \qquad \square$$

We now solve (8.28) in the more specific situation in that $H = 1$. In this case, one sees that r^* and c are functions of the capital $\widetilde{v}$. The next result shows that these functions behave as follows. As long as $\widetilde{v}$ is below some critical threshold v^*, we have $r^* = 0$, and $c = c(\widetilde{v})$ is determined by the requirement $E^*[\,Z^*\,] = \widetilde{v}$. Above the critical threshold v^*, the value of c is always equal to $c(v^*)$, and now $r^* > 0$ is determined by the requirement $E^*[\,Z^*\,] = \widetilde{v}$.

Theorem 8.27. *Consider the setting of Theorem* 8.26. *Assume in addition that* $H = 1$ *and that* φ *has a continuous and strictly increasing quantile function* q_φ *with respect to* P *and satisfies* $\|\varphi\|_\infty > \lambda^{-1}$. *Then the solution* Z^* *of problem* (8.28) *is* P-*a. s. unique. Moreover, there exists a critical capital level* v^* *such that*

$$Z^* = \mathbb{1}_{\{\varphi > q_\varphi(t_0)\}} \quad P\text{-a. s. for } \widetilde{v} \le v^*,$$

where t_0 is determined by the condition $E^*[\,Z^*\,] = \widetilde{v}$, and

$$Z^* = r^* + (1 - r^*)\mathbb{1}_{\{\varphi > q_\varphi(t_\lambda)\}} \quad P\text{-a. s. for } \widetilde{v} > v^*,$$

where $r^* = 1 - \frac{1-\widetilde{v}}{\Phi(t_\lambda)} > 0$ for $\Phi(t) := \int_0^t q_\varphi(s)\,ds$, and the level t_λ is the unique solution to the equation

$$q_\varphi(t_\lambda)(t_\lambda - 1 + \lambda) = \Phi(t_\lambda). \tag{8.34}$$

Finally, the critical capital level v^* is equal to $1 - \Phi(t_\lambda)$.

Proof. We fix $\widetilde{v} \in (0, 1)$. For a constant c, we then let

$$Z_r = r + (1 - r)\mathbb{1}_{\{\varphi > c\}},$$

where $r = r(c) \ge 0$ is such that $E^*[\,Z_r\,] = \widetilde{v}$, that is,

$$r(c) = \frac{\widetilde{v} - E[\,\varphi;\ \varphi > c\,]}{E[\,\varphi;\ \varphi \le c\,]}.$$

This makes sense as long as $c \ge c_0$, where c_0 is defined via $\widetilde{v} = E[\,\varphi;\ \varphi > c_0\,]$. Theorem 8.26 states that a solution to our problem can be found within the class $\{Z_{r(c)} \mid c \ge c_0\}$. Thus, we have to minimize

$$\lambda\,\mathrm{AV@R}_\lambda(-Z_{r(c)}) = \int_{1-\lambda}^{1} q_{Z_{r(c)}}(s)\,ds = \lambda r(c) + (1 - r(c)) \int_{1-\lambda}^{1} \mathbb{1}_{\{q_\varphi(s) > c\}}\,ds$$

over $c \ge c_0$. Here, we have used Lemma D.12 in the second identity. This minimization problem can be simplified further by using the reparameterization $c = q_\varphi(t)$, which is one-to-one according to our assumptions. Indeed, by letting

$$\varrho(t) := r(q_\varphi(t)) = 1 - \frac{1 - \widetilde{v}}{\Phi(t)},$$

we simply have to minimize the function

$$\begin{aligned} R(t) := \lambda\,\mathrm{AV@R}_\lambda(-Z_{\varrho(t)}) &= \lambda\varrho(t) + (1 - \varrho(t)) \int_{1-\lambda}^{1} \mathbb{1}_{(t,1]}(s)\,ds \\ &= \lambda\varrho(t) + (1 - \varrho(t))(\lambda - (t - 1 + \lambda)^+) \\ &= \lambda - (1 - \widetilde{v})\frac{(t - 1 + \lambda)^+}{\Phi(t)} \end{aligned}$$

over $t \geq t_0 := F_\varphi(c_0)$. For $t \leq 1 - \lambda$, we get $R(t) = \lambda$, which cannot be optimal. We show next that the function

$$\Psi(t) := \frac{t - 1 + \lambda}{\Phi(t)}$$

has a unique maximizer $t_\lambda \in (1 - \lambda, 1]$, which will define the solution as soon as $t_\lambda \geq t_0$ and as long as $t = t_0$ does not give a better result. To this end, we note first that

$$\Psi'(t) = \frac{\Phi(t) - (t - 1 + \lambda)q_\varphi(t)}{\Phi(t)^2}. \tag{8.35}$$

The numerator of this expression is strictly larger than zero for $t \leq 1 - \lambda$ and equal to $\Phi(1 - \lambda) > 0$ at $t = 1 - \lambda$. Moreover, for $t > 1 - \lambda$,

$$\Phi(t) - (t - 1 + \lambda)q_\varphi(t) = \int_0^{1-\lambda} q_\varphi(s)\, ds + \int_{1-\lambda}^{t} q_\varphi(s) - q_\varphi(t)\, ds,$$

which is easily seen to be strictly decreasing in t. For $t \uparrow 1$ this expression converges to $1 - \lambda\|\varphi\|_\infty$, which is strictly negative due to our assumption $\|\varphi\|_\infty > \lambda^{-1}$. Hence, the numerator in (8.35) has a unique zero $t_\lambda \in (1 - \lambda, 1)$, which is the unique solution to the equation

$$q_\varphi(t_\lambda)(t_\lambda - 1 + \lambda) = \Phi(t_\lambda),$$

and this solution t_λ is consequently the unique maximizer of Ψ.

If $t_\lambda \leq t_0$, then R has no minimizer on $(t_0, 1]$, and it follows that $t^* = t_0$ is its minimizer. Let us compare $R(t_\lambda)$ with $R(t_0)$ in case $t_\lambda > t_0$. We have

$$R(t_\lambda) = \lambda - (1 - \widetilde{v})\frac{(t_\lambda - 1 + \lambda)^+}{\Phi(t_\lambda)} = \lambda - \Phi(t_0)\frac{(t_\lambda - 1 + \lambda)^+}{\Phi(t_\lambda)}$$

and

$$R(t_0) = \lambda - (t_0 + \lambda - 1)^+ = \lambda - \Phi(t_0)\frac{(t_0 - 1 + \lambda)^+}{\Phi(t_0)}.$$

Since t_λ is the unique maximizer of the function $t \mapsto (t - 1 + \lambda)^+/\Phi(t)$, we thus see that $R(t_\lambda)$ is strictly smaller than $R(t_0)$. Hence, the solution is defined by

$$t^* := t_0 \vee t_\lambda.$$

Clearly, t_λ is independent of $\widetilde{v}$, while t_0 decreases from 1 to 0 as $\widetilde{v}$ increases from 0 to 1. Thus, by taking v^* as the capital level for which $t_\lambda = t_0$, we see that the optimal solution has the form

$$Z^* = \begin{cases} \mathbb{1}_{\{\varphi > q_\varphi(t_0)\}} & \text{for } \tilde{v} \leq v^*, \\ r^* + (1 - r^*)\mathbb{1}_{\{\varphi > q_\varphi(t_\lambda)\}} & \text{for } \tilde{v} > v^*, \end{cases}$$

where $r^* = 1 - \frac{1-\tilde{v}}{\Phi(t_\lambda)} > 0$.

Finally, if $\tilde{v}$ is equal to the critical capital level v^*, we must have that $t_\lambda = t_0$. But t_0 was defined to be the solution to $1 - \Phi(t) = \tilde{v}$, and so $v^* = 1 - \Phi(t_\lambda)$. □

Let us now point out the connections of the preceding theorem with robust statistical test theory as explained in Section 3.6 and in particular in Remark 3.51. To this end, let $\mathscr{R}$ denote the set of all measurable functions $\psi : \Omega \to [0, 1]$, which will be interpreted as randomized statistical tests; see Remark E.4. Problem (8.28) for $H = 1$ can then be rewritten as

$$\text{minimize } \sup_{Q \in \mathcal{Q}_\lambda} E_Q[\,\psi\,] \text{ subject to } \psi \in \mathscr{R} \text{ and } E^*[\,\psi\,] \geq \tilde{v},$$

where $\mathcal{Q}_\lambda$ is the set of all probability measures $Q \ll P$ whose density dQ/dP is P-a. s. bounded by $1/\lambda$. The solution ψ^* to this problem is described in Theorem 8.27. It is easy to see that ψ^* must also solve the following dual problem:

$$\text{maximize } E^*[\,\psi\,] \text{ subject to } \psi \in \mathscr{R} \text{ and } \sup_{Q \in \mathcal{Q}_\lambda} E_Q[\,\psi\,] \leq \alpha,$$

where $\alpha = \sup_{Q \in \mathcal{Q}_\lambda} E_Q[\,\psi^*\,]$. Thus, ψ^* is an optimal randomized test for testing the hypothesis P^* against the composite null hypothesis $\mathcal{Q}_\lambda$; see Remark 3.51. If $\mathcal{Q}_\lambda$ admits a least-favorable measure Q_0 with respect to P^* in the sense of Definition 3.45, then ψ^* must also be a standard Neyman–Pearson test for testing the hypothesis P^* against the null hypothesis Q_0. By Theorem E.3, it must hence be of the form

$$\psi^* = \kappa \mathbb{1}_{\{\pi = c\}} + \mathbb{1}_{\{\pi > c\}},$$

where $c > 0$ and $\kappa \in [0, 1]$ are constants and

$$\pi = \frac{dP^*}{dQ_0}.$$

Under the assumptions of Theorem 8.27, this is the case if $\pi = c(\varphi \vee q_\varphi(t_\lambda))$, where c is a suitable normalization constant. It follows that

$$\frac{dQ_0}{dP} = \frac{1}{c} \cdot \frac{\varphi}{\varphi \vee q_\varphi(t_\lambda)}.$$

We have by (8.34),

$$E\left[\frac{\varphi}{\varphi \vee q_\varphi(t_\lambda)}\right] = P[\,\varphi > q_\varphi(t_\lambda)\,] + \frac{1}{q_\varphi(t_\lambda)} E[\,\varphi;\, \varphi \le q_\varphi(t_\lambda)\,]$$
$$= 1 - t_\lambda + \frac{1}{q_\varphi(t_\lambda)} \Phi(t_\lambda) = \lambda.$$

This yields that $c = \lambda$, and we see that

$$\frac{dQ_0}{dP} = \frac{1}{\lambda} \cdot \frac{\varphi}{\varphi \vee q_\varphi(t_\lambda)} \tag{8.36}$$

defines a probability measure $Q_0 \in \mathcal{Q}_\lambda$ with

$$\pi = \frac{dP^*}{dQ_0} = \lambda(\varphi \vee q_\varphi(t_\lambda)).$$

The following result now follows immediately from Theorem 8.27.

Corollary 8.28. *For a measure $P^* \approx P$ satisfying the assumptions of Theorem* 8.27, *the measure Q_0 in* (8.36) *is a least-favorable measure for*

$$\mathcal{Q}_\lambda = \left\{ Q \in \mathcal{M}_1(P) \;\middle|\; \frac{dQ}{dP} \le \frac{1}{\lambda}\ P\text{-a.s.} \right\}$$

in the sense of Definition 3.45.

9 Hedging under constraints

So far, we have focussed on *frictionless* market models, where asset transactions can be carried out with no limitation. In this chapter, we study the impact of market imperfections generated by *convex trading constraints*. Thus, we develop the theory of dynamic hedging under the condition that only trading strategies from a given convex class $\mathscr{S}$ may be used. In Section 9.1, we characterize those market models for which $\mathscr{S}$ does not contain arbitrage opportunities. Then we take a direct approach to the superhedging duality for American options. To this end, we first derive a uniform Doob decomposition under constraints in Section 9.2. The appropriate upper Snell envelopes are analyzed in Section 9.3. In Section 9.4, we derive a superhedging duality under constraints, and we explain its role in the analysis of convex risk measures in a financial market model.

9.1 Absence of arbitrage opportunities

In practice, it may be reasonable to restrict the class of trading strategies that are admissible for hedging purposes. As discussed in Section 4.10, there may be upper bounds on the capital invested into risky assets, or upper and lower bounds on the number of shares of an asset. Here, we model such portfolio constraints by a set $\mathscr{S}$ of d-dimensional predictable processes, viewed as admissible investment strategies into risky assets. Throughout this chapter, we will assume that $\mathscr{S}$ satisfies the following conditions:

(a) $0 \in \mathscr{S}$.

(b) $\mathscr{S}$ is *predictably convex*: If $\xi, \eta \in \mathscr{S}$ and h is a predictable process with $0 \leq h \leq 1$, then the process

$$h_t\,\xi_t + (1 - h_t)\,\eta_t, \quad t = 1, \ldots, T,$$

belongs to $\mathscr{S}$.

(c) For each $t \in \{1, \ldots, T\}$, the set

$$\mathscr{S}_t := \{\xi_t \mid \xi \in \mathscr{S}\}$$

is closed in $L^0(\Omega, \mathscr{F}_{t-1}, P; \mathbb{R}^d)$.

(d) For all t, $\xi_t \in \mathscr{S}_t$ implies $\xi_t^\perp \in \mathscr{S}_t$.

In order to explain condition (d), let us recall from Lemma 1.68 that each $\xi_t \in L^0(\Omega, \mathscr{F}_{t-1}, P; \mathbb{R}^d)$ can be uniquely decomposed as

$$\xi_t = \eta_t + \xi_t^\perp, \quad \text{where } \eta_t \in N_t \text{ and } \xi_t^\perp \in N_t^\perp,$$

https://doi.org/10.1515/9783111045283-011

and where

$$N_t = \{\eta_t \in L^0(\Omega, \mathscr{F}_{t-1}, P; \mathbb{R}^d) \mid \eta_t \cdot (X_t - X_{t-1}) = 0 \ P\text{-a.s.}\},$$

$$N_t^\perp = \{\xi_t \in L^0(\Omega, \mathscr{F}_{t-1}, P; \mathbb{R}^d) \mid \xi_t \cdot \eta_t = 0 \ P\text{-a.s. for all } \eta_t \in N_t\}.$$

Remark 9.1. Under condition (d), we may replace $\xi_t \cdot (X_t - X_{t-1})$ by $\xi_t^\perp \cdot (X_t - X_{t-1})$, and $\xi_t^\perp \cdot (X_t - X_{t-1}) = 0$ P-a.s. implies $\xi_t^\perp = 0$. Note that condition (d) holds if the price increments satisfy the following *nonredundance condition*: For all $t \in \{1, \dots, T\}$ and $\xi_t \in L^0(\Omega, \mathscr{F}_{t-1}, P; \mathbb{R}^d)$,

$$\xi_t \cdot (X_t - X_{t-1}) = 0 \ P\text{-a.s.} \quad \Longrightarrow \quad \xi_t = 0 \ P\text{-a.s.} \tag{9.1}$$

Indeed, under (9.1) we have $N_t = \{0\}$ and $N_t^\perp = L^0(\Omega, \mathscr{F}_{t-1}, P; \mathbb{R}^d)$. ◇

Example 9.2. For each t, let C_t be a closed convex subset of $\mathbb{R}^d$ such that $0 \in C_t$. Take $\mathscr{S}$ as the class of all d-dimensional predictable processes ξ such that $\xi_t \in C_t$ P-a.s. for all t. If the nonredundance condition (9.1) holds, then $\mathscr{S}$ satisfies conditions (a) through (d). This case includes short sales constraints and restrictions on the size of a long position. ◇

Example 9.3. Let a, b be two constants such that $-\infty \le a < 0 < b \le \infty$, and take $\mathscr{S}$ as the set of all d-dimensional predictable processes such that

$$a \le \xi_t \cdot X_{t-1} \le b \quad P\text{-a.s. for } t = 1, \dots, T.$$

This class $\mathscr{S}$ corresponds to constraints on the capital invested into risky assets. If we assume that the nonredundance condition (9.1) holds, then $\mathscr{S}$ satisfies conditions (a) through (d). More generally, instead of the two constants a and b, one can take dynamic margins defined via two predictable processes (a_t) and (b_t). ◇

Let $\overline{\mathscr{S}}$ denote the set of all self-financing trading strategies $\overline{\xi} = (\xi^0, \xi)$ that arise from an investment strategy $\xi \in \mathscr{S}$, that is,

$$\overline{\mathscr{S}} = \{\overline{\xi} = (\xi^0, \xi) \mid \overline{\xi} \text{ is self-financing and } \xi \in \mathscr{S}\}.$$

In this section, our goal is to characterize the absence of arbitrage opportunities in $\overline{\mathscr{S}}$. The existence of an equivalent martingale measure $P^* \in \mathscr{P}$ is clearly sufficient. Under an additional technical assumption, a condition that is both necessary and sufficient will involve a larger class $\mathscr{P}_{\mathscr{S}} \supseteq \mathscr{P}$. In order to introduce these conditions, we need some preparation.

Definition 9.4. An adapted stochastic process Z on $(\Omega, \mathscr{F}, (\mathscr{F}_t), Q)$ is called a *local Q-martingale* if there exists a sequence of stopping times $(\tau_n)_{n\in\mathbb{N}}$ such that $\tau_n \nearrow \infty$ Q-a.s., and such that the stopped processes Z^{τ_n} are Q-martingales. The sequence $(\tau_n)_{n\in\mathbb{N}}$

is called a *localizing sequence* for Z. In the same way, we define *local supermartingales* and *local submartingales*. △

Remark 9.5. If Q is a martingale measure for the discounted price process X, then the value process V of each self-financing trading strategy $\overline{\xi} = (\xi^0, \xi)$ is a local Q-martingale. To prove this, one can take the sequence

$$\tau_n := \inf\{t \geq 0 \mid |\xi_{t+1}| > n\}$$

as a localizing sequence. With this choice, $|\xi_t| \leq n$ on $\{\tau_n \geq t\}$, and the increments

$$V_t^{\tau_n} - V_{t-1}^{\tau_n} = \mathbb{1}_{\{\tau_n \geq t\}}\, \xi_t \cdot (X_t - X_{t-1}), \quad t = 1, \dots, T,$$

of the stopped process V^{τ_n} are Q-integrable and satisfy

$$E_Q[\, V_t^{\tau_n} - V_{t-1}^{\tau_n} \mid \mathscr{F}_{t-1}\,] = \mathbb{1}_{\{\tau_n \geq t\}}\, \xi_t \cdot E_Q[\, X_t - X_{t-1} \mid \mathscr{F}_{t-1}\,] = 0. \qquad \diamond$$

The following proposition is a generalization of an argument, which we have already used in the proof of Theorem 5.25. We emphasize that throughout this chapter, we will assume that $\mathscr{F}_0 = \{\emptyset, \Omega\}$ and $\mathscr{F}_T = \mathscr{F}$.

Proposition 9.6. *For a local Q-supermartingale Z, the following conditions are equivalent:*

(a) *Z is a Q-supermartingale.*
(b) $E_Q[\, Z_T^- \,] < \infty$.

If, moreover, $Z_T \geq 0$ Q-a. s., then $Z_t \geq 0$ Q-a. s. for all t.

Proof. The implication (a)⇒(b) follows immediately from the definition of a Q-supermartingale, which includes the integrability of Z_t for all t.

Now we prove the implication (b)⇒(a). We show first that the negative parts Z_t^- are integrable for all t. To this end, we proceed by backward induction on t. If $t = T$, our claim holds by assumption. So, let us assume that Z_{t+1}^- is integrable for some $t \in \{0, \dots, T-1\}$. Let (τ_n) be a localizing sequence for Z. Since $\tau_n \nearrow \infty$ Q-a. s., we have

$$Z_{t+1} = \lim_{n \uparrow \infty} Z_{t+1} \mathbb{1}_{\{\tau_n \geq t+1\}} = \lim_{n \uparrow \infty} Z_{t+1}^{\tau_n} \mathbb{1}_{\{\tau_n \geq t+1\}}. \tag{9.2}$$

Since the random variables $Z_{t+1}^{\tau_n} \mathbb{1}_{\{\tau_n \geq t+1\}}$ are uniformly bounded from below by the integrable random variable $-Z_{t+1}^-$, we may apply Fatou's lemma for conditional expectations as follows, and subsequently use the fact that $\{\tau_n \geq t+1\} = \{\tau_n \leq t\}^c \in \mathscr{F}_t$ and the supermartingale property of Z^{τ_n},

$$
\begin{aligned}
E_Q[\, Z_{t+1} \mid \mathscr{F}_t \,] &\le \liminf_{n\uparrow\infty} E_Q[\, Z_{t+1}^{\tau_n} \mathbb{1}_{\{\tau_n \ge t+1\}} \mid \mathscr{F}_t \,] \\
&= \liminf_{n\uparrow\infty} E_Q[\, Z_{t+1}^{\tau_n} \mid \mathscr{F}_t \,] \mathbb{1}_{\{\tau_n \ge t+1\}} \\
&\le \liminf_{n\uparrow\infty} Z_t^{\tau_n} \mathbb{1}_{\{\tau_n \ge t+1\}} \\
&= Z_t.
\end{aligned}
\tag{9.3}
$$

Next, the convexity of $x \mapsto x^-$ and Jensen's inequality for conditional expectations imply that

$$
0 \le (E_Q[\, Z_{t+1} \mid \mathscr{F}_t \,])^- \le E_Q[\, Z_{t+1}^- \mid \mathscr{F}_t \,].
$$

Since the right-hand side is integrable according to the induction hypothesis, the integrability of Z_t^- now follows from (9.3). Note that (9.3) also yields the supermartingale inequality

$$
E_Q[\, Z_{t+1} \mid \mathscr{F}_t \,] \le Z_t. \tag{9.4}
$$

Hence, Z will be a supermartingale if we can show that $Z_t \in L^1$ for all t. To this end, we now apply (9.3) successively to get

$$
E_Q[\, Z_t \,] = E_Q[\, Z_t \mid \mathscr{F}_0 \,] \le Z_0 < \infty.
$$

In view of $E_Q[\, Z_t^- \,] < \infty$, this yields $E_Q[\, Z_t^+ \,] < \infty$ and in turn the integrability of Z_t. If, moreover, $Z_T \ge 0$ Q-a. s., then backward induction and (9.4) yield $Z_t \ge 0$ Q-a. s. for all t. □

Exercise 9.1.1. Use Proposition 9.6 to show that the following conditions are equivalent for a local Q-martingale Z:

(a) Z is a Q-martingale.
(b) $Z_t \in L^1$ for all t.
(c) $Z_T \in L^1$.
(d) $Z_T^- \in L^1$.
(e) $Z_T^+ \in L^1$. ◇

Exercise 9.1.2. Let $Y_1, Y_2 \in L^1$ be two independent, centered, and nonconstant random variables on $(\Omega, \mathscr{F}, P)$. Suppose furthermore that f is a measurable function on $\mathbb{R}$ such that $\xi := f(Y_1)$ is real-valued but not integrable. Provide a concrete example for this situation. Then show that $M_0 := 0$, $M_1 := Y_1$, and $M_2 := Y_1 + \xi Y_2$ is a local martingale, but not a martingale, with respect to the filtration $\mathscr{F}_0 = \{\emptyset, \Omega\}$, $\mathscr{F}_1 = \sigma(Y_1)$, and $\mathscr{F}_2 = \mathscr{F}$. ◇

Definition 9.7. By $\mathscr{P}_{\mathscr{S}}$, we denote the class of all probability measures $\widetilde{P} \approx P$ such that

$$X_t^i \in L^1(\widetilde{P}) \quad \text{for } i = 1, \dots, d \text{ and } t = 1, \dots, T, \tag{9.5}$$

and such that the value process of any trading strategy in $\overline{\mathscr{S}}$ is a local $\widetilde{P}$-supermartingale. △

Remark 9.8. If $\overline{\mathscr{S}}$ contains all self-financing trading strategies $\overline{\xi} = (\xi^0, \xi)$ with bounded ξ, then $\mathscr{P}_{\mathscr{S}}$ coincides with the class $\mathscr{P}$ of all equivalent martingale measures. To prove this, let $\widetilde{P} \in \mathscr{P}_{\mathscr{S}}$, and note that $\xi_t \cdot (X_t - X_{t-1}) \in L^1(\widetilde{P})$ by (9.5) if ξ_t is bounded. Therefore, the value process V of any such $\overline{\xi}$ is a $\widetilde{P}$-supermartingale by Proposition 9.6. The same applies to the strategy $-\overline{\xi}$, so V is in fact a $\widetilde{P}$-martingale, and Theorem 5.14 shows that $\widetilde{P}$ is a martingale measure for X. ◇

Our first goal is to extend the "fundamental theorem of asset pricing" to our present setting; see Theorem 5.16. Let us introduce the positive cone

$$\overline{\mathscr{R}} := \{\lambda \overline{\xi} \mid \overline{\xi} \in \overline{\mathscr{S}}, \lambda \geq 0\}$$

generated by $\overline{\mathscr{S}}$. Accordingly, we define the cones $\mathscr{R}$ and $\mathscr{R}_t$. Clearly, $\overline{\mathscr{R}}$ contains no arbitrage opportunities if and only if $\overline{\mathscr{S}}$ is arbitrage-free. We will need the following condition on the L^0-closure $\hat{\mathscr{R}}_t$ of $\mathscr{R}_t$:

$$\text{for each } t, \text{ we have } \hat{\mathscr{R}}_t \cap L^\infty(\Omega, \mathscr{F}_t, P; \mathbb{R}^d) \subseteq \mathscr{R}_t. \tag{9.6}$$

This condition clearly holds if $\mathscr{R}_t$ itself is closed in L^0 and in particular if $\mathscr{S}_t$ is itself a cone, and hence $\mathscr{S}_t = \mathscr{R}_t$ for all t.

Theorem 9.9. *Under condition* (9.6), *there are no arbitrage opportunities in $\overline{\mathscr{S}}$ if and only if $\mathscr{P}_{\mathscr{S}}$ is nonempty. In this case, there exists a measure $\widetilde{P} \in \mathscr{P}_{\mathscr{S}}$ that has a bounded density $d\widetilde{P}/dP$.*

Example 9.10. Consider the situation of Example 9.2, where $\mathscr{S}$ consists of all predictable processes ξ such that $\xi_t \in C_t$ P-a. s. for closed convex sets $C_t \subseteq \mathbb{R}^d$ with $0 \in C_t$. Here, condition (9.6) will be satisfied as soon as the cones generated by the convex sets C_t are closed in $\mathbb{R}^d$. This case includes short sales constraints and constraints on the size of a long position, which are modeled by taking $C_t = [a_t^1, b_t^1] \times \cdots \times [a_t^d, b_t^d]$ for certain numbers a_t^k, b_t^k such that $-\infty \leq a_t^k \leq 0 \leq b_t^k \leq \infty$. ◇

Example 9.11. Consider now the situation of Example 9.3, where $\mathscr{S}$ consists of all predictable processes ξ with $a \leq \xi_t \cdot X_{t-1} \leq b$ P-a. s. for two constants a, b with $-\infty \leq a < 0 < b \leq +\infty$. We claim that $\mathscr{S}$ does not contain arbitrage opportunities if and only if the unconstrained market is arbitrage-free. To prove this, note that the existence of an arbitrage opportunity in the unconstrained market is equivalent to the existence of some t and some $\mathscr{F}_{t-1}$-measurable ξ_t such that $\xi_t \cdot (X_t - X_{t-1}) \geq 0$ P-a. s. and

$P[\,\xi_t \cdot (X_t - X_{t-1}) > 0\,] > 0$ (see Proposition 5.11). In this case, there exists a constant $c > 0$ such that these properties are shared by $\widetilde{\xi}_t := \xi_t \mathbb{1}_{\{|\xi_t \cdot X_{t-1}| \le c\}}$ and in turn by $\varepsilon \widetilde{\xi}_t$, where $\varepsilon > 0$. But $\varepsilon \widetilde{\xi}_t \in \mathscr{S}_t$ if ε is small enough. This argument shows moreover that $\mathscr{P}_{\mathscr{S}} = \mathscr{P}$. ◇

As to the proof of Theorem 9.9, we will first show that the condition $\mathscr{P}_{\mathscr{S}} \neq \emptyset$ implies the absence of arbitrage opportunities in $\overline{\mathscr{S}}$.

Proof of sufficiency in Theorem 9.9. Suppose $\widetilde{P}$ is a measure in $\mathscr{P}_{\mathscr{S}}$, and V is the value process of a trading strategy in $\overline{\mathscr{S}}$ such that $V_T \ge 0$ P-almost surely. Proposition 9.6 shows that V is a $\widetilde{P}$-supermartingale. Hence, $V_0 \ge \widetilde{E}[\,V_T\,]$, so V cannot be the value process of an arbitrage opportunity. □

Let us now prepare for the proof that the condition $\mathscr{P}_{\mathscr{S}} \neq \emptyset$ is necessary. First, we argue that the absence of arbitrage opportunities in $\overline{\mathscr{S}}$ is equivalent to the absence of arbitrage opportunities in each of the embedded one-period models, that is, to the nonexistence of $\xi_t \in \mathscr{S}_t$ such that $\xi_t \cdot (X_t - X_{t-1})$ amounts to a nontrivial positive gain. This observation will allow us to apply the techniques of Section 1.6. Let us denote

$$\mathscr{S}^\infty := \{\xi \in \mathscr{S} \mid \xi \text{ is bounded}\}.$$

Similarly, we define

$$\mathscr{S}_t^\infty := \{\xi_t \mid \xi \in \mathscr{S}^\infty\} = \mathscr{S}_t \cap L^\infty(\Omega, \mathscr{F}_{t-1}, P; \mathbb{R}^d)$$

and the class $\overline{\mathscr{S}^\infty} \subseteq \overline{\mathscr{S}}$. The following two lemmas show that for many purposes it is sufficient to consider strategies in $\overline{\mathscr{S}^\infty}$. The first lemma concerns arbitrage opportunities, the second probability measures in the class $\mathscr{P}_{\mathscr{S}}$.

Lemma 9.12. *The following conditions are equivalent:*

(a) *There exists an arbitrage opportunity in* $\overline{\mathscr{S}}$.

(b) *There exist* $t \in \{1, \dots, T\}$ *and* $\xi_t \in \mathscr{S}_t$ *such that*

$$\xi_t \cdot (X_t - X_{t-1}) \ge 0 \ P\text{-a. s.}, \quad \textit{and} \quad P[\,\xi_t \cdot (X_t - X_{t-1}) > 0\,] > 0. \tag{9.7}$$

(c) *There exist* $t \in \{1, \dots, T\}$ *and* $\xi_t \in \mathscr{S}_t^\infty$ *that satisfies* (9.7).

(d) *There exists an arbitrage opportunity in* $\overline{\mathscr{S}^\infty}$.

Proof. The proof is essentially the same as the one of Proposition 5.11. □

Lemma 9.13. *For a probability measure* $\widetilde{P} \approx P$, *the following two conditions are equivalent:*

(a) $\widetilde{P} \in \mathscr{P}_{\mathscr{S}}$.

(b) *The value process of every strategy in* $\overline{\mathscr{S}^\infty}$ *is a* $\widetilde{P}$-*supermartingale.*

Proof. The implication (a)⇒(b) follows just as in Remark 9.8. To prove the converse implication, (b)⇒(a), we need to show that the value process V of any given $\overline{\xi} = (\xi^0, \xi) \in \overline{\mathscr{S}}$ is a local $\widetilde{P}$-supermartingale. To this end, we define $\tau_n := \inf\{t \mid |\xi_t| \geq n\}$. Clearly, $\tau_n \nearrow \infty$ as $n \uparrow \infty$. Since $0 \in \mathscr{S}$ and $h_t := \mathbb{1}_{\{\tau_n > t\}}$ is predictable, the predictable process

$$\xi_t^{(n)} := h_t \xi_t = h_t \xi_t + (1 - h_t) 0$$

belongs to $\mathscr{S}$ and in fact to $\mathscr{S}^\infty$. Hence, condition (b) implies that the value process corresponding to $\xi^{(n)}$ and initial investment V_0 is a $\widetilde{P}$-supermartingale. But this value process coincides with V^{τ_n}, and so we conclude that V is a local $\widetilde{P}$-supermartingale. □

In order to apply the results of Section 1.6, we introduce the convex sets

$$\mathscr{K}_t^{\mathscr{S}} := \{\xi_t \cdot (X_t - X_{t-1}) \mid \xi_t \in \mathscr{S}_t\},$$

for $t \in \{1, \ldots, T\}$. Lemma 9.12 shows that $\overline{\mathscr{S}}$ contains no arbitrage opportunities if and only if the condition

$$\mathscr{K}_t^{\mathscr{S}} \cap L_+^0 = \{0\} \tag{9.8}$$

holds for all $t \in \{1, \ldots, T\}$.

Lemma 9.14. *Condition* (9.8) *implies that* $\mathscr{K}_t^{\mathscr{S}} - L_+^0(\Omega, \mathscr{F}_t, P)$ *is a closed convex subset of* $L^0(\Omega, \mathscr{F}_t, P)$.

Proof. The proof is essentially the same as the one of Lemma 1.70. Only the following additional observation is required: If (ξ^n) is sequence in $\mathscr{S}_t$, and if α and σ are two $\mathscr{F}_{t-1}$-measurable random variables such that $0 \leq \alpha \leq 1$ and σ is integer-valued, then $\zeta := \alpha \xi^\sigma \in \mathscr{S}_t$. Indeed, predictable convexity of $\mathscr{S}$ and our assumption that $0 \in \mathscr{S}$ imply that

$$\alpha \sum_{k=1}^{n} \mathbb{1}_{\{\sigma = k\}} \xi^k \in \mathscr{S}_t$$

for each n, and the closedness of $\mathscr{S}_t$ in $L^0(\Omega, \mathscr{F}_{t-1}, P; \mathbb{R}^d)$ yields

$$\zeta = \alpha \sum_{k=1}^{\infty} \mathbb{1}_{\{\sigma = k\}} \xi^k \in \mathscr{S}_t.$$

□

From now on, we will assume that

$$E[\,|X_s|\,] < \infty \quad \text{for all } s. \tag{9.9}$$

For the purpose of proving Theorem 9.9, this can be assumed without loss of generality: If (9.9) does not hold, then we replace P by an equivalent measure P' that has a bounded density dP'/dP and for which the price process X is integrable. For instance, we can take

$$dP' = c \exp\left(- \sum_{s=1}^{T} |X_s| \right) dP,$$

where c denotes the normalizing constant. If there exist a measure $\widetilde{P} \approx P'$ such that each value process for a strategy in $\overline{\mathscr{S}}$ is a local $\widetilde{P}$-supermartingale and such that the density $d\widetilde{P}/dP'$ is bounded, then $\widetilde{P} \in \mathscr{P}_{\mathscr{S}}$, and the density $d\widetilde{P}/dP$ is bounded as well.

Lemma 9.15. *If $\overline{\mathscr{S}}$ contains no arbitrage opportunities and condition* (9.6) *holds, then for each $t \in \{1, \dots, T\}$ there exists some $Z_t \in L^\infty(\Omega, \mathscr{F}_t, P)$ such that $Z_t > 0$ P-a. s. and such that*

$$E[\, Z_t \, \xi_t \cdot (X_t - X_{t-1}) \,] \leq 0 \quad \text{for all } \xi \in \mathscr{S}^\infty. \tag{9.10}$$

Proof. Recall that $\overline{\mathscr{R}}$ does not contain arbitrage opportunities if and only if $\overline{\mathscr{S}}$ is arbitrage-free. Hence, for each t, $\mathscr{K}_t^{\mathscr{R}} \cap L^0_+(\Omega, \mathscr{F}_t, P) = \{0\}$ by Lemma 9.12. By the equivalence of conditions (a) and (c) of that lemma and condition (9.6), we even get

$$\mathscr{K}_t^{\hat{\mathscr{R}}} \cap L^0_+(\Omega, \mathscr{F}_t, P) = \{0\}, \tag{9.11}$$

where $\hat{\mathscr{R}}_t$ denotes again the L^0-closure of $\mathscr{R}_t$. The cone $\hat{\mathscr{R}}_t$ satisfies all conditions required from $\mathscr{S}_t$, and hence Lemma 9.14 implies that

$$\mathscr{C}_t^{\hat{\mathscr{R}}} := (\mathscr{K}_t^{\hat{\mathscr{R}}} - L^0_+(\Omega, \mathscr{F}_t, P)) \cap L^1$$

is a closed convex cone in L^1 that contains $-L^\infty_+(\Omega, \mathscr{F}_t, P)$. Furthermore, it follows from (9.11) and the argument in the proof of (a)⇔(b) of Theorem 1.57 that $\mathscr{C}_t^{\hat{\mathscr{R}}} \cap L^0_+ = \{0\}$. Hence, $\mathscr{C}_t^{\hat{\mathscr{R}}}$ satisfies the assumptions of the Kreps–Yan theorem, as stated in Theorem 1.64. We conclude that there exist $Z_t \in L^\infty(\Omega, \mathscr{F}_t, P)$ such that $P[\, Z_t > 0 \,] = 1$, and such that $E[\, Z_t \, W \,] \leq 0$ for each $W \in \mathscr{C}_t^{\hat{\mathscr{R}}}$. As $\xi_t \cdot (X_t - X_{t-1}) \in \mathscr{C}_t^{\hat{\mathscr{R}}}$ for each $\xi \in \mathscr{S}^\infty$, Z_t has property (9.10). □

Now we can complete the proof of Theorem 9.9 by showing that the absence of arbitrage opportunities in $\overline{\mathscr{S}}$ implies the existence of a measure $\widetilde{P}$ that belongs to the class $\mathscr{P}_{\mathscr{S}}$ and has a bounded density $d\widetilde{P}/dP$.

Proof of necessity in Theorem 9.9. Suppose that $\overline{\mathscr{S}}$ does not contain arbitrage opportunities. We are going to construct the desired measure $\widetilde{P}$ via backward recursion. First, we consider the case $t = T$. Take a bounded random variable $Z_T > 0$ as constructed in Lemma 9.15, and define a probability measure $\widetilde{P}_T$ by

$$\frac{d\widetilde{P}_T}{dP} = \frac{Z_T}{E[\, Z_T \,]}.$$

Clearly, $\widetilde{P}_T$ is equivalent to P, and $X_t \in L^1(\widetilde{P}_T)$ for all t. We claim that

$$\widetilde{E}_T[\,\xi_T \cdot (X_T - X_{T-1}) \mid \mathscr{F}_{T-1}\,] \le 0 \quad P\text{-a. s. for all } \xi \in \mathscr{S}^\infty. \tag{9.12}$$

To prove this claim, consider the family

$$\Phi := \{\widetilde{E}_T[\,\xi_T \cdot (X_T - X_{T-1}) \mid \mathscr{F}_{T-1}\,] \mid \xi \in \mathscr{S}^\infty\}.$$

For $\xi, \widetilde{\xi} \in \mathscr{S}^\infty$, let

$$A := \{\widetilde{E}_T[\,\xi_T \cdot (X_T - X_{T-1}) \mid \mathscr{F}_{T-1}\,] > \widetilde{E}_T[\,\widetilde{\xi}_T \cdot (X_T - X_{T-1}) \mid \mathscr{F}_{T-1}\,]\},$$

and define ξ' by $\xi'_t = 0$ for $t < T$ and

$$\xi'_T := \xi_T \mathbb{1}_A + \widetilde{\xi}_T \mathbb{1}_{A^c}.$$

The predictable convexity of $\mathscr{S}$ implies that $\xi' \in \mathscr{S}^\infty$. Furthermore, we have

$$\begin{aligned}
&\widetilde{E}_T[\,\xi'_T \cdot (X_T - X_{T-1}) \mid \mathscr{F}_{T-1}\,] \\
&\qquad = \widetilde{E}_T[\,\xi_T \cdot (X_T - X_{T-1}) \mid \mathscr{F}_{T-1}\,]\mathbb{1}_A + \widetilde{E}_T[\,\widetilde{\xi}_T \cdot (X_T - X_{T-1}) \mid \mathscr{F}_{T-1}\,]\mathbb{1}_{A^c} \\
&\qquad = \widetilde{E}_T[\,\xi_T \cdot (X_T - X_{T-1}) \mid \mathscr{F}_{T-1}\,] \vee \widetilde{E}_T[\,\widetilde{\xi}_T \cdot (X_T - X_{T-1}) \mid \mathscr{F}_{T-1}\,].
\end{aligned}$$

Hence, the family Φ is directed upwards in the sense of Theorem F.1. By virtue of that theorem, ess sup Φ is the increasing limit of a sequence (φ_n) in Φ. By monotone convergence, we get

$$\begin{aligned}
\widetilde{E}_T\Big[\operatorname*{ess\,sup}_{\xi \in \mathscr{S}^\infty} \widetilde{E}_T[\,\xi_T \cdot (X_T - X_{T-1}) \mid \mathscr{F}_{T-1}\,]\Big]& \\
&= \widetilde{E}_T[\operatorname{ess\,sup} \Phi\,] = \widetilde{E}_T\Big[\lim_{n \uparrow \infty} \varphi_n\Big] \\
&= \lim_{n \uparrow \infty} \widetilde{E}_T[\,\varphi_n\,] \\
&\le \sup_{\xi \in \mathscr{S}^\infty} \widetilde{E}_T[\,\widetilde{E}_T[\,\xi_T \cdot (X_T - X_{T-1}) \mid \mathscr{F}_{T-1}\,]\,] \\
&= \frac{1}{E[\,Z_T\,]} \sup_{\xi \in \mathscr{S}^\infty} E[\,\xi_T \cdot (X_T - X_{T-1})\, Z_T\,] \\
&\le 0,
\end{aligned} \tag{9.13}$$

where we have used (9.10) in the final step. Since $\mathscr{S}^\infty$ contains 0, we must have ess sup $\Phi \ge 0$, and so (9.13) implies that

$$\operatorname*{ess\,sup}_{\xi \in \mathscr{S}^\infty} \widetilde{E}_T[\,\xi_T \cdot (X_T - X_{T-1}) \mid \mathscr{F}_{T-1}\,] = 0 \quad \widetilde{P}_T\text{-a. s.},$$

which yields our claim (9.12).

Now we apply the previous argument inductively: Suppose we already have a probability measure $\widetilde{P}_{t+1} \approx P$ with a bounded density $d\widetilde{P}_{t+1}/dP$ such that

$$\widetilde{E}_{t+1}[\,|X_s|\,] < \infty \quad \text{for all } s,$$

and such that

$$\widetilde{E}_{t+1}[\,\xi_k \cdot (X_k - X_{k-1}) \mid \mathscr{F}_{k-1}\,] \leq 0 \quad P\text{-a.s. for } k \geq t+1 \text{ and } \xi \in \mathscr{S}^\infty. \tag{9.14}$$

Then we may apply Lemma 9.15 with P replaced by $\widetilde{P}_{t+1}$, and we get some strictly positive $\widetilde{Z}_t \in L^\infty(\Omega, \mathscr{F}_t, \widetilde{P}_{t+1})$ satisfying (9.10) with $\widetilde{P}_{t+1}$ in place of P. We now proceed as in the first step by defining a probability measure $\widetilde{P}_t \approx \widetilde{P}_{t+1} \approx P$ as

$$\frac{d\widetilde{P}_t}{d\widetilde{P}_{t+1}} = \frac{\widetilde{Z}_t}{\widetilde{E}_{t+1}[\,\widetilde{Z}_t\,]}.$$

Then $\widetilde{P}_t$ has bounded densities with respect to both $\widetilde{P}_{t+1}$ and P. In particular, $\widetilde{E}_t[\,|X_s|\,] < \infty$ for all s. Moreover, the $\mathscr{F}_t$-measurability of $d\widetilde{P}_t/d\widetilde{P}_{t+1}$ implies that (9.14) is satisfied for $\widetilde{P}_t$ replacing $\widetilde{P}_{t+1}$. Repeating the arguments that led to (9.12) yields

$$\widetilde{E}_t[\,\xi_t \cdot (X_t - X_{t-1}) \mid \mathscr{F}_{t-1}\,] \leq 0 \quad P\text{-a.s. for all } \xi \in \mathscr{S}^\infty.$$

After T steps, we arrive at a measure $\widetilde{P} := \widetilde{P}_1$. Under this measure $\widetilde{P}$, the value process of every strategy in $\overline{\mathscr{S}^\infty}$ is a supermartingale, and so $\widetilde{P} \in \mathscr{P}_{\mathscr{S}}$ by Lemma 9.13. □

9.2 Uniform Doob decomposition

The goal of this section is to characterize those nonnegative adapted processes U that can be decomposed as

$$U_t = U_0 + \sum_{k=1}^{t} \xi_k \cdot (X_k - X_{k-1}) - B_t, \tag{9.15}$$

where the predictable d-dimensional process ξ belongs to $\mathscr{S}$, and where B is an adapted and increasing process such that $B_0 = 0$. In the unconstrained case, where $\mathscr{S}$ consists of all strategies, we have seen in Section 7.2 that such a decomposition exists if and only if U is a supermartingale under each equivalent martingale measure $P^* \in \mathscr{P}$. In our present context, a first guess might be that the role of $\mathscr{P}$ is now played by $\mathscr{P}_{\mathscr{S}}$. Since each value process of a strategy in $\overline{\mathscr{S}}$ is a local $\widetilde{P}$-supermartingale for each $\widetilde{P} \in \mathscr{P}_{\mathscr{S}}$, any process U that has a decomposition (9.15) is also a local $\widetilde{P}$-supermartingale for $\widetilde{P} \in \mathscr{P}_{\mathscr{S}}$. Thus, one might suspect that the latter property would also be sufficient for the existence of a decomposition (9.15). This, however, is not the case, as is illustrated by the following simple example.

Example 9.16. Consider a one-period market model with the riskless bond $S_0^0 \equiv S_1^0 \equiv 1$ and with one risky asset S^1. We assume that $S_0^1 \equiv 1$ and that S_1^1 takes the values $S_1^1(\omega^-) = \frac{1}{2}$

and $S_1^1(\omega^+) = \frac{3}{2}$ on $\Omega := \{\omega^-, \omega^+\}$. We choose any measure P on Ω that assigns positive mass to both ω^+ and ω^-. If we let $\mathscr{S} = [0,1]$, then a measure $\widetilde{P}$ belongs to $\mathscr{P}_{\mathscr{S}}$ if and only if $\widetilde{P}[\,\{\omega^+\}\,] \in (0, \frac{1}{2}]$. Thus, for any positive initial value U_0, the process defined by $U_1(\omega^-) := 0$ and $U_1(\omega^+) := 2U_0$ is a $\mathscr{P}_{\mathscr{S}}$-supermartingale. If U can be decomposed according to (9.15), then we must be able to write

$$2U_0 = U_1(\omega^+) = U_0 + \xi \cdot (S_1^1(\omega^+) - S_0^1(\omega^+)) - B_1(\omega^+)$$

for some $B_1(\omega^+) \geq 0$. This requirement is equivalent to $U_0 \leq \xi/2$. Hence, the decomposition (9.15) fails for $U_0 > 1/2$. ◇

The reason for the failure of the decomposition (9.15) for certain $\mathscr{P}_{\mathscr{S}}$-supermartingales is that $\mathscr{P}_{\mathscr{S}}$ does not reflect the full structure of $\mathscr{S}$; the definition of $\mathscr{P}_{\mathscr{S}}$ depends only on the cone

$$\{\lambda\,\xi \mid \lambda > 0,\ \xi \in \mathscr{S}\}$$

generated by $\mathscr{S}$. In the approach we are going to present here, the structure of $\mathscr{S}$ will be reflected by a stochastic process we associate to any measure $Q \ll P$.

Definition 9.17. For a measure $Q \ll P$, the *upper variation process* for $\mathscr{S}$ is the increasing process A^Q defined by

$$A_0^Q := 0 \quad \text{and} \quad A_{t+1}^Q - A_t^Q := \operatorname*{ess\,sup}_{\xi \in \mathscr{S}} [\xi_{t+1} \cdot (E_Q[\,X_{t+1} \mid \mathscr{F}_t\,] - X_t)]$$

for $t = 0, \dots, T-1$. By $\mathscr{Q}_{\mathscr{S}}$, we denote the set of all $Q \approx P$ such that

$$E_Q[\,A_T^Q\,] < \infty \tag{9.16}$$

and such that

$$E_Q[\,|X_{t+1} - X_t| \mid \mathscr{F}_t\,] < \infty \quad P\text{-a.s. for all } t.$$ △

Clearly, the upper variation process of any measure $Q \approx P$ satisfies

$$A_{t+1}^Q - A_t^Q = \operatorname*{ess\,sup}_{\xi \in \mathscr{S}^\infty} [\xi_{t+1} \cdot (E_Q[\,X_{t+1} \mid \mathscr{F}_t\,] - X_t)],$$

where $\mathscr{S}^\infty$ are the bounded processes in $\mathscr{S}$. Hence for $Q \in \mathscr{Q}_{\mathscr{S}}$ and $\xi \in \mathscr{S}^\infty$, the condition $E_Q[\,|X_{t+1} - X_t| \mid \mathscr{F}_t\,] < \infty$ guarantees that

$$\xi_{t+1} \cdot (E_Q[\,X_{t+1} \mid \mathscr{F}_t\,] - X_t) = E_Q[\,\xi_{t+1} \cdot (X_{t+1} - X_t) \mid \mathscr{F}_t\,],$$

and it follows that

$$A^Q_{t+1} - A^Q_t = \operatorname*{ess\,sup}_{\xi \in \mathcal{S}^\infty} E_Q[\, \xi_{t+1} \cdot (X_{t+1} - X_t) \mid \mathcal{F}_t \,] \quad \text{for } Q \in \mathcal{Q}_{\mathcal{S}}. \tag{9.17}$$

Moreover, we have

$$A^{\widetilde{P}}_T = 0 \quad \text{for } \widetilde{P} \in \mathcal{P}_{\mathcal{S}}, \tag{9.18}$$

which implies the inclusion

$$\mathcal{P} \subseteq \mathcal{P}_{\mathcal{S}} \subseteq \mathcal{Q}_{\mathcal{S}}. \tag{9.19}$$

Proposition 9.18. *If $Q \in \mathcal{Q}_{\mathcal{S}}$, and V is the value process of a trading strategy in $\overline{\mathcal{S}}$, then $V - A^Q$ is a local Q-supermartingale. If, moreover, $E_Q[\, V_T^- \,] < \infty$, then $V - A^Q$ is a Q-supermartingale.*

Proof. Let V be the value process of $\overline{\xi} = (\xi^0, \xi) \in \overline{\mathcal{S}}$. Denote by $\tau_n(\omega)$ the first time t at which

$$|\xi_{t+1}(\omega)| > n \quad \text{or} \quad E_Q[\, |X_{t+1} - X_t| \mid \mathcal{F}_t \,](\omega) > n.$$

If such a t does not exist, let $\tau_n(\omega) := \infty$. Then τ_n is a stopping time, and $\tau_n \nearrow \infty$ as $n \uparrow \infty$. Since

$$\left|V^{\tau_n}_{t+1} - V^{\tau_n}_t\right| \le \mathbb{1}_{\{\tau_n \ge t+1\}} |\xi_{t+1}| \cdot |X_{t+1} - X_t| \le n \cdot |X_{t+1} - X_t| \in L^1(Q),$$

$V^{\tau_n}_t$ belongs to $L^1(Q)$, and

$$\begin{aligned} E_Q[\, V^{\tau_n}_{t+1} - V^{\tau_n}_t \mid \mathcal{F}_t \,] &= \mathbb{1}_{\{\tau_n \ge t+1\}} \xi_{t+1} \cdot (E_Q[\, X_{t+1} \mid \mathcal{F}_t \,] - X_t) \\ &\le (A^Q)^{\tau_n}_{t+1} - (A^Q)^{\tau_n}_t. \end{aligned}$$

Since $A^Q_T \in L^1(Q)$ by (9.16), this proves that $V^{\tau_n} - (A^Q)^{\tau_n}$ is a Q-supermartingale, and so $V - A^Q$ is a local Q-supermartingale. Finally, if $V_T^- \in L^1(Q)$, then $(V_T - A^Q_T)^- \le V_T^- + A^Q \in L^1(Q)$, and Proposition 9.6 yields that $V - A^Q$ is a Q-supermartingale □

Let us identify the class $\mathcal{Q}_{\mathcal{S}}$ in some special cases.

Remark 9.19. If $\mathcal{S}^\infty$ consists of all bounded predictable processes ξ with nonnegative components, then $\mathcal{Q}_{\mathcal{S}} = \mathcal{P}_{\mathcal{S}}$. To prove this, note first that $\mathcal{P}_{\mathcal{S}} \subseteq \mathcal{Q}_{\mathcal{S}}$ by (9.19). To establish the converse inclusion, take $Q \in \mathcal{Q}_{\mathcal{S}}$, and note first that $A^Q \equiv 0$. Indeed, if $Q[A^Q_t > 0] > 0$ for some t, then the definition of A^Q and the fact that $\mathcal{S}^\infty$ is a cone would imply that $Q[A^Q_T = \infty] > 0$, contradicting (9.16). Thus, value processes of strategies in $\overline{\mathcal{S}}$ are local Q-supermartingales by Proposition 9.18. By taking $\xi \in \mathcal{S}$ such that $\xi^i_t \equiv 1$ and $\xi^j_t \equiv 0$ for $j \neq i$, we get that X^i is a local Q-supermartingale, and Proposition 9.6 implies that X^i is a Q-supermartingale. In particular, X^i_t is Q-integrable, and we conclude $Q \in \mathcal{P}_{\mathcal{S}}$. ◇

Remark 9.20. If $\mathscr{S}^\infty$ consists of all bounded predictable processes ξ, then $\mathcal{Q}_{\mathscr{S}} = \mathscr{P}$. This follows by combining Remarks 9.8 and 9.19. ◇

Example 9.21. Suppose our market model contains just one risky asset, and $\mathscr{S}$ consists of all predictable processes ξ such that

$$a_t \le \xi_t \le b_t \quad P\text{-a. s. for all } t,$$

where a and b are two given predictable processes with

$$-\infty < a_t \le 0 \le b_t < \infty \quad P\text{-a. s.}$$

If we assume in addition that $E[\,|X_{t+1}-X_t| \mid \mathscr{F}_t\,] > 0$ P-a. s., then the nonredundance condition (9.1) holds, and $\mathscr{S}$ satisfies the assumptions (a) through (d) stated at the beginning of this chapter. If $Q \approx P$ is any probability measure such that

$$E_Q[\,|X_{t+1} - X_t| \mid \mathscr{F}_t\,] < \infty \quad P\text{-a. s. for all } t, \tag{9.20}$$

then

$$\begin{aligned} &\operatorname*{ess\,sup}_{\xi \in \mathscr{S}^\infty} E_Q[\,\xi_{t+1} \cdot (X_{t+1} - X_t) \mid \mathscr{F}_t\,] \\ &\qquad = b_{t+1}\,(E_Q[\,X_{t+1} - X_t \mid \mathscr{F}_t\,])^+ - a_{t+1}\,(E_Q[\,X_{t+1} - X_t \mid \mathscr{F}_t\,])^- \\ &\qquad < \infty \quad P\text{-a. s.} \end{aligned}$$

Hence, $\mathcal{Q}_{\mathscr{S}}$ consists of all measures $Q \approx P$ with (9.20), and satisfying both $E_Q[\,b_{t+1}\,(E_Q[\,X_{t+1} - X_t \mid \mathscr{F}_t\,])^+\,] < \infty$ and $E_Q[\,a_{t+1}\,(E_Q[\,X_{t+1} - X_t \mid \mathscr{F}_t\,])^-\,] > -\infty$ for all t. ◇

We now state the uniform Doob decomposition under constraints, which is the main result of this section.

Theorem 9.22. *Suppose that $\mathscr{P}_{\mathscr{S}}$ is nonempty. Then for any adapted process U with $U_T \ge 0$ P-a. s., the following conditions are equivalent.*
(a) *$U - A^Q$ is a Q-supermartingale for every $Q \in \mathcal{Q}_{\mathscr{S}}$.*
(b) *There exists $\xi \in \mathscr{S}$ and an adapted increasing process B such that $B_0 = 0$ and*

$$U_t = U_0 + \sum_{k=1}^{t} \xi_k \cdot (X_k - X_{k-1}) - B_t \quad P\text{-a. s. for all } t.$$

Proof. (b)⇒(a): Fix $Q \in \mathcal{Q}_{\mathscr{S}}$ and let

$$M_t^Q := U_0 + \sum_{k=1}^{t} \xi_k \cdot (X_k - X_{k-1}) - A_t^Q.$$

Since $M_T^Q + A_T^Q \geq U_T \geq 0$ P-a. s. and $A_T^Q \in L^1(Q)$, Proposition 9.18 yields that M^Q is a Q-supermartingale. Moreover, since

$$0 = B_0 \leq B_T = M_T^Q + A_T^Q - U_T \leq M_T^Q + A_T^Q \in L^1(Q),$$

we have $B_T \in L^1(Q)$. The fact that $0 \leq B_t \leq B_T$ furthermore yields $B_t \in L^1(Q)$ for all t. Therefore, $M - B = U - A^Q$ is a Q-supermartingale.

(a)⇒(b): We must show that for any given $t \in \{1, \dots, T\}$ there exist some $\xi \in \mathcal{S}$ and a nonnegative random variable R_t playing the role of $B_t - B_{t-1}$ such that $U_t - U_{t-1} = \xi_t \cdot (X_t - X_{t-1}) - R_t$, that is,

$$U_t - U_{t-1} \in \mathcal{K}_t^{\mathcal{S}} - L_+^0(\Omega, \mathcal{F}_t, P),$$

where

$$\mathcal{K}_t^{\mathcal{S}} = \{\xi_t \cdot (X_t - X_{t-1}) \mid \xi \in \mathcal{S}\}.$$

The formulation of this problem does not change if we switch from P to any equivalent probability measure, so we can assume without loss of generality that $P \in \mathcal{P}_{\mathcal{S}}$. In this case, $A^P \equiv 0$, and U is a P-supermartingale. In particular, $U_s \in L^1(P)$ for all s.

We assume by way of contradiction that

$$U_t - U_{t-1} \notin \mathcal{C}_t^{\mathcal{S}} = \big(\mathcal{K}_t^{\mathcal{S}} - L_+^0(\Omega, \mathcal{F}_t, P)\big) \cap L^1(P). \tag{9.21}$$

Recall that we have proved in Lemma 9.14 that $\mathcal{C}_t^{\mathcal{S}}$ is a closed convex subset of $L^1(\Omega, \mathcal{F}_t, P)$. The Hahn–Banach separation theorem, Theorem H.4, now implies the existence of a random variable $Z \in L^\infty(\Omega, \mathcal{F}_t, P)$ such that

$$\alpha := \sup_{W \in \mathcal{C}_t^{\mathcal{S}}} E[\,Z\,W\,] < E[\,Z\,(U_t - U_{t-1})\,] =: \beta < \infty. \tag{9.22}$$

Note that the function $-\lambda \mathbb{1}_{\{Z<0\}}$ belongs to $\mathcal{C}_t^{\mathcal{S}}$ for all $\lambda \geq 0$. Thus,

$$0 \leq (-\lambda)\, E[\,Z\, \mathbb{1}_{\{Z<0\}}\,] \leq \alpha$$

for every $\lambda \geq 0$, and it follows that $Z \geq 0$ P-almost surely.

In fact, we can always modify Z such that it is bounded from below by some $\varepsilon > 0$ and still satisfies (9.22). To see this, note first that every $W \in \mathcal{C}_t^{\mathcal{S}}$ is dominated by some $\xi_t \cdot (X_t - X_{t-1}) \in \mathcal{K}_t^{\mathcal{S}}$ with integrable negative part. Therefore,

$$E[\,W\,] \leq E[\,\xi_t \cdot (X_t - X_{t-1})\,] \leq \liminf_{c \uparrow \infty} E[\,\xi_t \cdot (X_t - X_{t-1}) \mathbb{1}_{\{|\xi_t| \leq c\}}\,] \leq 0,$$

where we have used Fatou's lemma and our assumption that $P \in \mathcal{P}_{\mathcal{S}}$. If we let $Z^\varepsilon := \varepsilon 1 + (1 - \varepsilon) Z$, then Z^ε still satisfies $E[\,Z^\varepsilon\, W\,] \leq 0$ for all $W \in \mathcal{C}_t^{\mathcal{S}}$ and, for ε small enough, the expectation $E[\,Z^{\varepsilon}\,(U_t - U_{t-1})\,]$ is still larger than α. So, Z^{ε} also satisfies (9.22). Therefore, we may assume from now on that our Z with (9.22) is bounded from below by some constant $\varepsilon > 0$.

For the next step, let $Z_{t-1} := E[\,Z \mid \mathscr{F}_{t-1}\,]$ and

$$\frac{dQ}{dP} := \frac{Z}{Z_{t-1}}.$$

For $s = 1, \dots, T$, suppose that $Y_s \geq 0$ is $\mathscr{F}_s$-measurable. Using Proposition B.8, one shows as in the proof of Theorem 7.5 that

$$E_Q[Y_s \mid \mathscr{F}_{s-1}] = \begin{cases} E[Y_s \mid \mathscr{F}_{s-1}] & \text{if } s \neq t, \\ \dfrac{1}{Z_{t-1}} E[Y_t Z \mid \mathscr{F}_{t-1}] & \text{if } s = t. \end{cases}$$

Since the density Z/Z_{t-1} is bounded and since $P \in \mathscr{P}_{\mathscr{S}}$, we get in particular that

$$E_Q[\,|X_s - X_{s-1}| \mid \mathscr{F}_{s-1}\,] < \infty \quad P\text{-a.s. for all } s. \tag{9.23}$$

It follows moreover that

$$E_Q[\,\xi_s \cdot (X_s - X_{s-1}) \mid \mathscr{F}_{s-1}\,] = E[\,\xi_s \cdot (X_s - X_{s-1}) \mid \mathscr{F}_{s-1}\,] \leq 0 \quad \text{for } s \neq t. \tag{9.24}$$

We now consider the case $s = t$. As we have seen in the proof of Theorem 9.9, the family

$$\{E_Q[\,\xi_t \cdot (X_t - X_{t-1}) \mid \mathscr{F}_{t-1}\,] \mid \xi \in \mathscr{S}\}$$

is directed upwards. Therefore, we may conclude as in (9.13) that

$$\begin{aligned} E_Q\Big[\,Z_{t-1} \operatorname*{ess\,sup}_{\xi \in \mathscr{S}} E_Q[\,\xi_t \cdot (X_t - X_{t-1}) \mid \mathscr{F}_{t-1}\,]\Big] &\\ \leq \sup_{\xi \in \mathscr{S}} E[\,\xi_t \cdot (X_t - X_{t-1})\, Z\,] &\\ \leq \alpha. & \end{aligned} \tag{9.25}$$

Since $Z_{t-1} \geq \varepsilon$, (9.25) implies that

$$E_Q[\,A_t^Q - A_{t-1}^Q\,] = E_Q\Big[\operatorname*{ess\,sup}_{\xi \in \mathscr{S}} E_Q[\,\xi_t \cdot (X_t - X_{t-1}) \mid \mathscr{F}_{t-1}\,]\Big] \leq \frac{\alpha}{\varepsilon}.$$

By (9.24), we may conclude that $E_Q[\,A_T^Q\,] \leq \alpha/\varepsilon$, and so (9.23) yields $Q \in \mathscr{Q}_{\mathscr{S}}$.

As a final step, we show that $U - A^Q$ cannot be a Q-supermartingale, thus leading our assumption (9.21) to a contradiction with our hypothesis (a). To this end, we use again (9.25):

$$\begin{aligned}E_Q[\,Z_{t-1}E_Q[\,U_t - U_{t-1} \mid \mathscr{F}_{t-1}\,]\,]&\\ &= E_Q[\,Z_{t-1}\,(U_t - U_{t-1})\,]\\ &= E[\,Z\,(U_t - U_{t-1})\,] = \beta\\ &> \alpha\\ &\geq E_Q[\,Z_{t-1}\operatorname*{ess\,sup}_{\xi\in\mathscr{S}} E_Q[\,\xi_t\cdot(X_t - X_{t-1}) \mid \mathscr{F}_{t-1}\,]\,]\\ &= E_Q[Z_{t-1}\,(A_t^Q - A_{t-1}^Q)\,].\end{aligned}$$

Thus, we cannot have

$$E_Q[\,U_t - U_{t-1} \mid \mathscr{F}_{t-1}\,] \leq A_t^Q - A_{t-1}^Q \quad P\text{-a.s.,}$$

so $U - A^Q$ cannot be a Q-supermartingale, in contradiction to our hypothesis (a). □

9.3 Upper Snell envelopes

From now on, we assume the no-arbitrage condition $\mathscr{P}_{\mathscr{S}} \neq \emptyset$. Let H be a discounted American claim. Our goal is to construct a superhedging strategy for H that belongs to our class $\overline{\mathscr{S}}$ of admissible strategies. The uniform Doob decomposition suggests that we should find an adapted process $U \geq H$ such that $U - A^Q$ is a Q-supermartingale for each $Q \in \mathscr{Q}_{\mathscr{S}}$. If we consider only one such Q, then the minimal process U that satisfies these requirements, is given by $\widetilde{U}^Q + A^Q$, where

$$\widetilde{U}_t^Q := \operatorname*{ess\,sup}_{\tau\in\mathcal{T}_t} E_Q[\,H_\tau - A_\tau^Q \mid \mathscr{F}_t\,], \quad t = 0,\dots,T, \tag{9.26}$$

is the Snell envelope of $H - A^Q$ with respect to Q. Thus, one may guess that

$$\operatorname*{ess\,sup}_{Q\in\mathscr{Q}_{\mathscr{S}}}(\widetilde{U}_t^Q + A_t^Q), \quad t = 0,\dots,T,$$

is the minimal process U that dominates H and for which $U - A^Q$ is a Q-supermartingale for each $Q \in \mathscr{Q}_{\mathscr{S}}$. Let us assume that

$$\sup_{Q\in\mathscr{Q}_{\mathscr{S}}} \widetilde{U}_0^Q = \sup_{Q\in\mathscr{Q}_{\mathscr{S}}}\sup_{\tau\in\mathcal{T}} E_Q[\,H_\tau - A_\tau^Q\,] < \infty.$$

Note that this condition holds if H is bounded.

Definition 9.23. The process

$$\widetilde{U}_t^{\uparrow} := \operatorname*{ess\,sup}_{Q \in \mathcal{Q}_{\mathcal{S}}} (A_t^Q + \widetilde{U}_t^Q)$$

$$= \operatorname*{ess\,sup}_{Q \in \mathcal{Q}_{\mathcal{S}}} \Big(A_t^Q + \operatorname*{ess\,sup}_{\tau \in \mathcal{T}_t} E_Q[\, H_\tau - A_\tau^Q \mid \mathcal{F}_t \,] \Big), \quad t = 0, \dots, T,$$

will be called the *upper $\mathcal{Q}_{\mathcal{S}}$-Snell envelope* of H. △

The main result of this section confirms our guess that $\widetilde{U}^{\uparrow}$ is the process we are looking for.

Theorem 9.24. *The upper $\mathcal{Q}_{\mathcal{S}}$-Snell envelope of H is the smallest process $U \geq H$ such that $U - A^Q$ is a Q-supermartingale for each $Q \in \mathcal{Q}_{\mathcal{S}}$.*

For a European claim, we have the following additional result.

Proposition 9.25. *For a discounted European claim H^E with*

$$\sup_{Q \in \mathcal{Q}_{\mathcal{S}}} E_Q[\, H^E - A_T^Q \,] < \infty,$$

the upper $\mathcal{Q}_{\mathcal{S}}$-Snell envelope takes the form

$$\widetilde{U}_t^{\uparrow} = \operatorname*{ess\,sup}_{Q \in \mathcal{Q}_{\mathcal{S}}} (E_Q[\, H^E - A_T^Q \mid \mathcal{F}_t \,] + A_t^Q), \quad t = 0, \dots, T.$$

Proposition 9.25 will follow from Lemma 9.30 below. The next result provides a scheme for the recursive calculation of $\widetilde{U}^{\uparrow}$. It will be used in the proof of Theorem 9.24.

Proposition 9.26. *For fixed $Q_0 \in \mathcal{Q}_{\mathcal{S}}$ let $\mathcal{Q}_t(Q_0)$ denote the set of all $Q \in \mathcal{Q}_{\mathcal{S}}$ that coincide with Q_0 on $\mathcal{F}_t$. Then $\widetilde{U}^{\uparrow}$ satisfies the following recursion formula:*

$$\widetilde{U}_t^{\uparrow} - A_t^{Q_0} = (H_t - A_t^{Q_0}) \vee \operatorname*{ess\,sup}_{Q \in \mathcal{Q}_t(Q_0)} E_Q[\, \widetilde{U}_{t+1}^{\uparrow} - A_{t+1}^Q \mid \mathcal{F}_t \,], \quad t = 0, \dots, T-1.$$

The proofs of this proposition and of Theorem 9.24 will be given at the end of this section. Let us recall the following concepts from Section 6.4. The *pasting* of two probability measures $Q_1 \approx Q_2$ in a stopping time $\tau \in \mathcal{T} = \{\sigma \mid \sigma \text{ is a stopping time } \leq T\}$ is the probability measure

$$\widetilde{Q}[\, A \,] = E_{Q_1}[\, Q_2[\, A \mid \mathcal{F}_\tau \,]\,], \quad A \in \mathcal{F}.$$

It was shown in Lemma 6.41 that, for all stopping times σ and $\mathcal{F}_T$-measurable $Y \geq 0$,

$$E_{\widetilde{Q}}[\, Y \mid \mathcal{F}_\sigma \,] = E_{Q_1}[\, E_{Q_2}[\, Y \mid \mathcal{F}_{\tau \vee \sigma} \,] \mid \mathcal{F}_\sigma \,]. \tag{9.27}$$

Recall also that a set $\mathcal{Q}$ of equivalent probability measures on $(\Omega, \mathcal{F})$ is called *stable* if for any pair $Q_1, Q_2 \in \mathcal{Q}$ and all $\tau \in \mathcal{T}$ the corresponding pasting also belongs to $\mathcal{Q}$. A technical inconvenience arises from the fact that our set $\mathcal{Q}_{\mathcal{S}}$ may *not* be stable. We must introduce a further condition on τ, which guarantees that the pasting of $Q_1, Q_2 \in \mathcal{Q}_{\mathcal{S}}$ in τ also belongs to $\mathcal{Q}_{\mathcal{S}}$.

Lemma 9.27. *For $\tau \in \mathcal{T}$, the pasting $\widetilde{Q}$ of $Q_1, Q_2 \in \mathcal{Q}_{\mathcal{S}}$ in τ satisfies $E_{\widetilde{Q}}[\,|X_{t+1} - X_t| \mid \mathcal{F}_t\,] < \infty$ P-a. s. for all t, and its upper variation process is given by*

$$A_t^{\widetilde{Q}} = A_{t\wedge\tau}^{Q_1} + (A_t^{Q_2} - A_\tau^{Q_2}) \cdot \mathbb{1}_{\{\tau<t\}}.$$

Moreover, we have $\widetilde{Q} \in \mathcal{Q}_{\mathcal{S}}$ under the condition that there exists $\varepsilon > 0$ such that

$$\left.\frac{dQ_2}{dQ_1}\right|_{\mathcal{F}_\tau} \geq \varepsilon \quad \text{a. s. on } \{\tau < T\}. \tag{9.28}$$

Proof. The identity (9.27) yields

$$E_{\widetilde{Q}}[\,|X_{t+1} - X_t| \mid \mathcal{F}_t\,] = E_{Q_1}[\,|X_{t+1} - X_t| \mid \mathcal{F}_t\,]\mathbb{1}_{\{\tau>t\}} + E_{Q_2}[\,|X_{t+1} - X_t| \mid \mathcal{F}_t\,]\mathbb{1}_{\{\tau\leq t\}},$$

and each of the two conditional expectations is finite almost surely.

Now we will compute the upper variation process $A^{\widetilde{Q}}$ of $\widetilde{Q}$. As above, (9.27) yields

$$E_{\widetilde{Q}}[\,\xi_t \cdot (X_{t+1} - X_t) \mid \mathcal{F}_t\,] = E_{Q_1}[\,\xi_t \cdot (X_{t+1} - X_t) \mid \mathcal{F}_t\,]\mathbb{1}_{\{\tau>t\}} + E_{Q_2}[\,\xi_t \cdot (X_{t+1} - X_t) \mid \mathcal{F}_t\,]\mathbb{1}_{\{\tau\leq t\}}.$$

Taking the essential supremum over $\xi_t \in \mathcal{S}_t^\infty$ and using the predictable convexity of $\mathcal{S}$ gives

$$A_{t+1}^{\widetilde{Q}} - A_t^{\widetilde{Q}} = (A_{t+1}^{Q_1} - A_t^{Q_1})\mathbb{1}_{\{\tau>t\}} + (A_{t+1}^{Q_2} - A_t^{Q_2})\mathbb{1}_{\{\tau\leq t\}}.$$

From this our formula for $A_t^{\widetilde{Q}}$ follows.

In the final step, we show that $\widetilde{Q}$ belongs to $\mathcal{Q}_{\mathcal{S}}$ under condition (9.28). We must show that $E_{\widetilde{Q}}[\,A_T^{\widetilde{Q}}\,] < \infty$. Let Z_t denote the density of Q_2 with respect to Q_1 on $\mathcal{F}_t$. Then, by our formula for $A_T^{\widetilde{Q}}$,

$$\begin{aligned}
E_{\widetilde{Q}}[\,A_T^{\widetilde{Q}}\,] &= E_{\widetilde{Q}}[\,A_\tau^{Q_1} + A_T^{Q_2} - A_\tau^{Q_2}\,] \\
&= E_{Q_1}[\,A_\tau^{Q_1} + E_{Q_2}[\,A_T^{Q_2} - A_\tau^{Q_2} \mid \mathcal{F}_\tau\,]\,] \\
&\leq E_{Q_1}[\,A_T^{Q_1}\,] + E_{Q_1}\left[\,\frac{1}{Z_\tau}E_{Q_1}[\,(A_T^{Q_2} - A_\tau^{Q_2})Z_T \mid \mathcal{F}_\tau\,]\,\right] \\
&\leq E_{Q_1}[\,A_T^{Q_1}\,] + \frac{1}{\varepsilon}E_{Q_2}[\,A_T^{Q_2}\,],
\end{aligned}$$

which is finite for $Q_1, Q_2 \in \mathcal{Q}_{\mathcal{S}}$. □

Lemma 9.28. *Suppose we are given $Q_1, Q_2 \in \mathcal{Q}_{\mathcal{S}}$, a stopping time $\tau \in \mathcal{T}$, and a set $B \in \mathcal{F}_\tau$ such that $dQ_2/dQ_1|_{\mathcal{F}_\tau} \geq \varepsilon$ a. s. on B. Let $\widetilde{Q}$ be the pasting of Q_1 and Q_2 in the stopping time*

$$\sigma := \tau \mathbb{1}_B + T \mathbb{1}_{B^c}.$$

Then $\widetilde{Q} \in \mathcal{Q}_{\mathcal{S}}$, and the Snell envelopes associated with these three measures via (9.26) *are related as follows:*

$$\widetilde{U}_\tau^{\widetilde{Q}} + A_\tau^{\widetilde{Q}} = (\widetilde{U}_\tau^{Q_1} + A_\tau^{Q_1}) \cdot \mathbb{1}_{B^c} + (\widetilde{U}_\tau^{Q_2} + A_\tau^{Q_2}) \cdot \mathbb{1}_B \quad P\text{-a. s.} \tag{9.29}$$

Proof. We have $dQ_2/dQ_1|_{\mathcal{F}_\sigma} \geq \varepsilon$, hence $\widetilde{Q} \in \mathcal{Q}_{\mathcal{S}}$ by Lemma 9.27. Let ρ be a stopping time in the set $\mathcal{T}_\tau$ of all stopping times that are larger than or equal to τ. The formula for $A^{\widetilde{Q}}$ in Lemma 9.27 yields

$$A_\rho^{\widetilde{Q}} = A_\tau^{Q_1} + (A_\rho^{Q_1} - A_\tau^{Q_1})\mathbb{1}_{B^c} + (A_\rho^{Q_2} - A_\tau^{Q_2})\mathbb{1}_B.$$

Moreover, (9.27) implies that

$$\begin{aligned} E_{\widetilde{Q}}[\, Y \mid \mathcal{F}_\tau \,] &= E_{Q_1}[E_{Q_2}[Y \mid \mathcal{F}_{\tau \vee \sigma}] \mid \mathcal{F}_\tau] \\ &= E_{Q_1}[E_{Q_2}[Y \mid \mathcal{F}_T]\mathbb{1}_{B^c} \mid \mathcal{F}_\tau] + E_{Q_1}[E_{Q_2}[Y \mid \mathcal{F}_\tau]\mathbb{1}_B \mid \mathcal{F}_\tau] \\ &= E_{Q_1}[\, Y \mid \mathcal{F}_\tau \,]\mathbb{1}_{B^c} + E_{Q_2}[\, Y \mid \mathcal{F}_\tau \,]\mathbb{1}_B \end{aligned}$$

for all random variables Y such that all conditional expectations make sense. Hence,

$$\begin{aligned} &E_{\widetilde{Q}}[\, H_\rho - A_\rho^{\widetilde{Q}} \mid \mathcal{F}_\tau \,] + A_\tau^{\widetilde{Q}} \\ &\quad = (E_{Q_1}[\, H_\rho - A_\rho^{Q_1} \mid \mathcal{F}_\tau \,] + A_\tau^{Q_1})\, \mathbb{1}_{B^c} + (E_{Q_2}[\, H_\rho - A_\rho^{Q_2} \mid \mathcal{F}_\tau \,] + A_\tau^{Q_2})\, \mathbb{1}_B. \end{aligned}$$

Whenever ρ_1, ρ_2 are stopping times in $\mathcal{T}_\tau$, then $\rho := \rho_1 \mathbb{1}_{B^c} + \rho_2 \mathbb{1}_B$ is also a stopping time in $\mathcal{T}_\tau$. Conversely, every $\rho \in \mathcal{T}_\tau$ can be written in that way for stopping times ρ_1 and ρ_2. Thus, taking the essential supremum over all ρ_1 and ρ_2 and applying Proposition 6.37 yields (9.29). □

In fact, we have $A_\tau^{\widetilde{Q}} = A_\tau^{Q_1}$ in (9.29), as we will have $A_\tau^{Q_k} = A_\tau^{Q_0}$ in the following lemma.

Lemma 9.29. *For any $Q_0 \in \mathcal{Q}_{\mathcal{S}}$, $\tau \in \mathcal{T}$, and $\delta > 0$, there exist a set $\Lambda_\delta \in \mathcal{F}_\tau$ such that $Q_0[\, \Lambda_\delta \,] \geq 1 - \delta$ and measures $Q_k \in \mathcal{Q}_{\mathcal{S}}$ such that $Q_k = Q_0$ on $\mathcal{F}_\tau$ and*

$$\widetilde{U}_\tau^{Q_k} + A_\tau^{Q_k} \nearrow \operatorname*{ess\,sup}_{Q \in \mathcal{Q}_{\mathcal{S}}} (\widetilde{U}_\tau^Q + A_\tau^Q) = \widetilde{U}_\tau^{\uparrow} \quad P\text{-a. s. on } \Lambda_\delta.$$

Proof. By Theorem F.1 and its proof, there exists a sequence (Q_n^0) in $\mathcal{Q}_{\mathcal{S}}$ such that

$$\lim_{k \uparrow \infty} \max_{n \leq k} (\widetilde{U}_\tau^{Q_n^0} + A_\tau^{Q_n^0}) = \widetilde{U}_\tau^{\uparrow} \quad P\text{-a. s.}$$

We will recursively define measures $Q_k \in \mathcal{Q}_{\mathcal{S}}$ and sets $\Lambda_\delta^k \in \mathcal{F}_\tau$ such that $\Lambda_\delta^k \subseteq \Lambda_\delta^{k-1}$, $Q_0[\, \Lambda_\delta^k \,] \geq 1 - (1 - 2^{-k})\delta$, and

$$\widetilde{U}_\tau^{Q_k} + A_\tau^{Q_k} = \max_{n\leq k}(\widetilde{U}_\tau^{Q_n^0} + A_\tau^{Q_n^0}) \quad P\text{-a. s. on } \Lambda_\delta^k.$$

By letting $\Lambda_\delta := \bigcap_k \Lambda_\delta^k$, this will imply the first part of the assertion. We start this recursion in $k = 0$ by taking Q_0 and $\Lambda_\delta^0 := \Omega$.

For Q_k given, the equivalence of Q_k and Q_{k+1}^0 implies that there exists some $\varepsilon > 0$ such that the set

$$D := \left\{ \frac{dQ_{k+1}^0}{dQ_k}\bigg|_{\mathscr{F}_\tau} \geq \varepsilon \right\} \in \mathscr{F}_\tau$$

satisfies $Q_0[D] \geq 1 - 2^{-(k+1)}\delta$. Thus, $\Lambda_\delta^{k+1} := \Lambda_\delta^k \cap D$ satisfies $Q_0[\Lambda_\delta^{k+1}] \geq 1 - (1 - 2^{-(k+1)})\delta$. We now define a set

$$B := \{\widetilde{U}_\tau^{Q_{k+1}^0} + A_\tau^{Q_{k+1}^0} > \widetilde{U}_\tau^{Q_k} + A_\tau^{Q_k}\} \cap D,$$

and consider the pasting Q_{k+1} of Q_k and Q_{k+1}^0 in the stopping time $\sigma := \tau\mathbb{1}_B + T\mathbb{1}_{B^c}$. By Lemma 9.28, $Q_{k+1} \in \mathcal{Q}_{\mathscr{S}}$ and

$$\begin{aligned}
\widetilde{U}_\tau^{Q_{k+1}} + A_\tau^{Q_{k+1}} &= (\widetilde{U}_\tau^{Q_k} + A_\tau^{Q_k}) \cdot \mathbb{1}_{B^c} + (\widetilde{U}_\tau^{Q_{k+1}^0} + A_\tau^{Q_{k+1}^0}) \cdot \mathbb{1}_B && P\text{-a. s.} \\
&= (\widetilde{U}_\tau^{Q_k} + A_\tau^{Q_k}) \vee (\widetilde{U}_\tau^{Q_{k+1}^0} + A_\tau^{Q_{k+1}^0}) && P\text{-a. s. on } D \\
&= \max_{n\leq k+1}(\widetilde{U}_\tau^{Q_n^0} + A_\tau^{Q_n^0}) && P\text{-a. s. on } \Lambda_\delta^{k+1}.
\end{aligned}$$

□

Now we can proceed to proving the main results in this section.

Proof of Proposition 9.26. For $Q_0 \in \mathcal{Q}_{\mathscr{S}}$ and $t \in \{t, \dots, T\}$, $\mathcal{Q}_t(Q_0)$ denotes the set of all $Q \in \mathcal{Q}_{\mathscr{S}}$ that coincide with Q_0 on $\mathscr{F}_t$. By Lemma 9.29 and by the definition of $\widetilde{U}^Q$ as the Snell envelope of $H - A^Q$,

$$\begin{aligned}
\widetilde{U}_t^\uparrow - A_t^{Q_0} &= \operatorname*{ess\,sup}_{Q\in\mathcal{Q}_t(Q_0)} \widetilde{U}_t^Q \\
&= \operatorname*{ess\,sup}_{Q\in\mathcal{Q}_t(Q_0)} ((H_t - A_t^Q) \vee E_Q[\widetilde{U}_{t+1}^Q \mid \mathscr{F}_t]) \\
&= (H_t - A_t^{Q_0}) \vee \operatorname*{ess\,sup}_{Q\in\mathcal{Q}_t(Q_0)} E_Q[\widetilde{U}_{t+1}^Q \mid \mathscr{F}_t].
\end{aligned}$$

Since $\widetilde{U}_{t+1}^Q \leq \widetilde{U}_{t+1}^\uparrow - A_{t+1}^Q$, we get

$$\widetilde{U}_t^\uparrow - A_t^{Q_0} \leq (H_t - A_t^{Q_0}) \vee \operatorname*{ess\,sup}_{Q\in\mathcal{Q}_t(Q_0)} E_Q[\widetilde{U}_{t+1}^\uparrow - A_{t+1}^Q \mid \mathscr{F}_t]. \tag{9.30}$$

For the proof of the converse inequality, let us fix an arbitrary $Q \in \mathcal{Q}_t(Q_0)$. For any $\delta > 0$, Lemma 9.29 yields a set $\Lambda_\delta \in \mathscr{F}_{t+1}$ with measure $Q[\Lambda_\delta] \geq 1 - \delta$ and $Q_k \in \mathcal{Q}_{t+1}(Q)$ such that $\widetilde{U}^{Q_k}_{t+1} \nearrow \widetilde{U}^{\uparrow}_{t+1} - A^Q_{t+1}$ P-a. s. on Λ_δ. Since Q_k coincides with Q on $\mathscr{F}_{t+1}$, we have P-a. s. on Λ_δ,

$$\begin{aligned}
E_Q[\,\widetilde{U}^{\uparrow}_{t+1} - A^Q_{t+1} \mid \mathscr{F}_t\,] &= \lim_{k\uparrow\infty} E_Q[\,\widetilde{U}^{Q_k}_{t+1} \mid \mathscr{F}_t\,] = \lim_{k\uparrow\infty} E_{Q_k}[\,\widetilde{U}^{Q_k}_{t+1} \mid \mathscr{F}_t\,] \\
&\le \limsup_{k\uparrow\infty} \widetilde{U}^{Q_k}_t \le \operatorname*{ess\,sup}_{\widetilde{Q}\in\mathcal{Q}_{t+1}(Q)} \widetilde{U}^{\widetilde{Q}}_t \\
&\le \operatorname*{ess\,sup}_{\widetilde{Q}\in\mathcal{Q}_t(Q)} \widetilde{U}^{\widetilde{Q}}_t = \widetilde{U}^{\uparrow}_t - A^Q_t \\
&= \widetilde{U}^{\uparrow}_t - A^{Q_0}_t
\end{aligned}$$

By taking $\delta \downarrow 0$ and by recalling $\widetilde{U}^{\uparrow}_t \ge H_t$, we arrive at the converse of the inequality (9.30). □

Proof of Theorem 9.24. Since $Q_0 \in \mathcal{Q}_{\mathscr{S}}$ is obviously contained in $\mathcal{Q}_t(Q_0)$, the recursion formula of Proposition 9.26 yields

$$\widetilde{U}^{\uparrow}_t - A^{Q_0}_t \ge (H_t - A^{Q_0}_t) \vee E_{Q_0}[\,\widetilde{U}^{\uparrow}_{t+1} - A^{Q_0}_{t+1} \mid \mathscr{F}_t\,] \ge E_{Q_0}[\,\widetilde{U}^{\uparrow}_{t+1} - A^{Q_0}_{t+1} \mid \mathscr{F}_t\,],$$

that is, $\widetilde{U}^{\uparrow}_t - A^{Q_0}$ is indeed a Q_0-supermartingale for each $Q_0 \in \mathcal{Q}_{\mathscr{S}}$. We also know that $\widetilde{U}^{\uparrow}$ dominates H.

Let U be any process that dominates H and for which $U - A^Q$, is a Q-supermartingale for each $Q \in \mathcal{Q}_{\mathscr{S}}$. For fixed Q, the Q-supermartingale $U - A^Q$ dominates $H - A^Q$, and hence also $\widetilde{U}^Q$, since $\widetilde{U}^Q$ is the smallest Q-supermartingale dominating $H - A^Q$ by Proposition 6.10. It follows that

$$U_t \ge \operatorname*{ess\,sup}_{Q\in\mathcal{Q}_{\mathscr{S}}} (\widetilde{U}^Q_t + A^Q_t) = \widetilde{U}^{\uparrow}_t \quad P\text{-a. s. for all } t.$$ □

Proposition 9.25 we will be implied by taking $\tau^* \equiv T$ in the following lemma.

Lemma 9.30. *Let H be a discounted American claim whose payoff is zero if it is not exercised at a given stopping time $\tau^* \in \mathcal{T}$, that is, $H_t(\omega) = 0$ if $t \ne \tau^*(\omega)$. Then its upper $\mathcal{Q}_{\mathscr{S}}$-Snell envelope is given by*

$$\widetilde{U}^{\uparrow}_t = \mathbb{1}_{\{\tau^*\ge t\}} \operatorname*{ess\,sup}_{Q\in\mathcal{Q}_{\mathscr{S}}} (E_Q[\,H_{\tau^*} - A^Q_{\tau^*} \mid \mathscr{F}_t\,] + A^Q_t), \quad t = 0,\dots,T.$$

Proof. By definition,

$$\widetilde{U}^{\uparrow}_t = \operatorname*{ess\,sup}_{\tau\in\mathcal{T}_t} \operatorname*{ess\,sup}_{Q\in\mathcal{Q}_{\mathscr{S}}} (E_Q[\,H_\tau - A^Q_\tau \mid \mathscr{F}_t\,] + A^Q_t).$$

Since each process A^Q is increasing, it is clearly optimal to take $\tau = t$ on $\{\tau^* \le t\}$. Hence,

$$\widetilde{U}_t^{\uparrow} = \begin{cases} 0 & \text{on } \{\tau^* < t\}, \\ H_t & \text{on } \{\tau^* = t\}. \end{cases}$$

So, we have to show that choosing $\tau \equiv \tau^*$ is optimal on $\{\tau^* > t\}$. If $\sigma \in \mathcal{T}_t$ is a stopping time with $P[\sigma > \tau^*] > 0$, then $\tau := \sigma \wedge \tau^*$ is as least as good as σ, since each process A^Q is increasing. So, it remains to rule out the case that there exists a stopping time $\sigma \in \mathcal{T}_t$ with $\sigma \le \tau^*$ on $\{\tau^* > t\}$ and $P[\sigma < \tau^*] > 0$, such that σ yields a strictly better result than τ^*. If there is such a σ, there exists some $Q_1 \in \mathcal{Q}_{\mathcal{S}}$ such that

$$E_{Q_1}[\,H_\sigma - A_\sigma^{Q_1} \mid \mathcal{F}_t\,] + A_t^{Q_1} > \operatorname*{ess\,sup}_{Q \in \mathcal{Q}_{\mathcal{S}}} (E_Q[\,H_{\tau^*} - A_{\tau^*}^Q \mid \mathcal{F}_t\,] + A_t^Q) \tag{9.31}$$

with strictly positive probability on $\{\tau^* > t\}$. Take any $\widetilde{P} \in \mathcal{P}_{\mathcal{S}}$ and $\varepsilon > 0$, and define

$$B_\varepsilon := \left\{ \frac{d\widetilde{P}}{dQ_1}\bigg|_{\mathcal{F}_\sigma} \ge \varepsilon \right\}.$$

Now let Q^ε be the pasting Q_1 and $\widetilde{P}$ in the stopping time $\sigma \mathbb{1}_{B_\varepsilon} + T\mathbb{1}_{B_\varepsilon^c}$. According to Lemmas 9.27 and 9.28, $Q^\varepsilon \in \mathcal{Q}_{\mathcal{S}}$, and its upper variation process satisfies $A_t^{Q^\varepsilon} = A_t^{Q_1}$ and $A_\sigma^{Q^\varepsilon} = A_\sigma^{Q_1}$ as well as

$$A_{\tau^*}^{Q^\varepsilon} = A_{\tau^*}^{Q_1} \mathbb{1}_{B_\varepsilon^c} + A_\sigma^{Q_1} \mathbb{1}_{B_\varepsilon} \quad P\text{-a. s. on } \{\tau^* > t\}.$$

By using our assumption that $H_\sigma \le H_{\tau^*}$, we get

$$E_{Q_1}[\,H_\sigma - A_\sigma^{Q_1} \mid \mathcal{F}_t\,] + A_t^{Q_1} \le E_{Q^\varepsilon}[\,H_{\tau^*} - A_{\tau^*}^{Q^\varepsilon} \mid \mathcal{F}_t\,] + A_t^{Q^\varepsilon} \quad P\text{-a. s. on } B_\varepsilon.$$

By letting $\varepsilon \downarrow 0$, the P-measure of B_ε becomes arbitrarily close to 1, and we arrive at a contradiction to (9.31). □

9.4 Superhedging and risk measures

Let H be a discounted American claim such that

$$\widetilde{U}_0^{\uparrow} = \sup_{Q \in \mathcal{Q}_{\mathcal{S}}} \widetilde{U}_0^Q = \sup_{Q \in \mathcal{Q}_{\mathcal{S}}} \sup_{\tau \in \mathcal{T}} E_Q[\,H_\tau - A_\tau^Q\,] < \infty.$$

Our aim in this section is to construct superhedging strategies for claims H that belong to our set $\overline{\mathcal{S}}$ of admissible trading strategies. Recall that a superhedging strategy for H is any self-financing trading strategy whose value process dominates H. If applied with $t = 0$, the following theorem shows that $\widetilde{U}_0^{\uparrow}$ is the minimal amount for which a superhedging strategy is available.

Denote by $\widetilde{\mathcal{U}}_t^{\uparrow}(H)$ the set of all $\mathcal{F}_t$-measurable random variables $U_t \ge 0$ for which there exists some $\eta \in \mathcal{S}$ such that

$$U_t + \sum_{k=t+1}^{u} \eta_k \cdot (X_k - X_{k-1}) \geq H_u \quad \text{for all } u \geq t \text{ } P\text{-a.s.} \tag{9.32}$$

Theorem 9.31. *The upper $\mathcal{Q}_{\mathcal{S}}$-Snell envelope $\widetilde{U}_t^{\uparrow}$ of H is the minimal element of $\widetilde{\mathcal{U}}_t^{\uparrow}(H)$. More precisely,*
(a) $\widetilde{U}_t^{\uparrow} \in \widetilde{\mathcal{U}}_t^{\uparrow}(H)$,
(b) $\widetilde{U}_t^{\uparrow} = \operatorname{ess\,inf} \widetilde{\mathcal{U}}_t^{\uparrow}(H)$.

Proof. The uniform Doob decomposition in Section 9.2 combined with Theorem 9.24 yields an increasing adapted process B and some $\xi \in \mathcal{S}$ such that

$$\widetilde{U}_u^{\uparrow} = \widetilde{U}_t^{\uparrow} + \sum_{k=t+1}^{u} \xi_k \cdot (X_k - X_{k-1}) + B_t - B_u \quad P\text{-a.s. for } u \geq t.$$

So, the fact that $\widetilde{U}^{\uparrow}$ dominates H proves (a).

As to part (b), we first get $\widetilde{U}_t^{\uparrow} \geq \operatorname{ess\,inf} \widetilde{\mathcal{U}}_t^{\uparrow}(H)$ from (a). For the proof of the converse inequality, take $U_t \in \widetilde{\mathcal{U}}_t^{\uparrow}(H)$ and choose a predictable process $\eta \in \mathcal{S}$ for which (9.32) holds. We must show that the set $B := \{\widetilde{U}_t^{\uparrow} \leq U_t\}$ satisfies $P[\,B\,] = 1$. Let

$$\hat{U}_t := \widetilde{U}_t^{\uparrow} \wedge U_t = \widetilde{U}_t^{\uparrow} \cdot \mathbb{1}_B + U_t \cdot \mathbb{1}_{B^c}.$$

Then $\hat{U}_t \leq \widetilde{U}_t^{\uparrow}$, and our claim will follow if we can show that $U_t^{\uparrow} \leq \hat{U}_t$. Let ξ denote the predictable process obtained from the uniform Doob decomposition of the $\mathcal{P}$-supermartingale $\widetilde{U}^{\uparrow}$, and define

$$\hat{\xi}_s := \begin{cases} \xi_s & \text{if } s \leq t, \\ \xi_s \cdot \mathbb{1}_B + \eta_s \cdot \mathbb{1}_{B^c} & \text{if } s > t. \end{cases}$$

With this choice, $\hat{\xi} \in \mathcal{S}$ by predictable convexity, and $\hat{U}_t$ satisfies (9.32), that is, $\hat{U}_t \in \widetilde{\mathcal{U}}_t^{\uparrow}(H)$. Let

$$\hat{V}_s := \widetilde{U}_0^{\uparrow} + \sum_{k=1}^{s} \hat{\xi}_k \cdot (X_k - X_{k-1}).$$

Then $\hat{V}_s \geq H_s \geq 0$ for all s, and so $\hat{V} - A^Q$ is a Q-supermartingale for each $Q \in \mathcal{Q}_{\mathcal{S}}$ by Propositions 9.18 and 9.6. Hence,

$$\begin{aligned} \widetilde{U}_t^{\uparrow} &= \operatorname*{ess\,sup}_{Q \in \mathcal{Q}_{\mathcal{S}}} \operatorname*{ess\,sup}_{\tau \in \mathcal{T}_t} E_Q[\, H_\tau - A_\tau^Q + A_t^Q \mid \mathcal{F}_t \,] \\ &\leq \operatorname*{ess\,sup}_{Q,\tau} E_Q\left[\, \hat{U}_t + \sum_{k=t+1}^{\tau} \hat{\xi}_k \cdot (X_k - X_{k-1}) - A_\tau^Q + A_t^Q \,\middle|\, \mathcal{F}_t \right] \\ &\leq \hat{U}_t. \end{aligned}$$

This proves $\widetilde{U}_t^{\uparrow} \leq \operatorname{ess\,inf} \widetilde{\mathcal{U}}_t^{\uparrow}(H)$. □

For European claims, the upper $\mathcal{Q}_{\mathcal{S}}$-Snell envelope takes the form

$$\widetilde{U}_t^{\uparrow} = \operatorname*{ess\,sup}_{Q\in\mathcal{Q}_{\mathcal{S}}}(E_Q[\,H^E - A_T^Q \mid \mathcal{F}_t\,] + A_t^Q), \quad t = 0,\dots,T.$$

By taking $t = 0$, it follows that

$$\widetilde{U}_0^{\uparrow} = \sup_{Q\in\mathcal{Q}_{\mathcal{S}}} (E_Q[\,H^E\,] - E_Q[\,A_T^Q\,]) \tag{9.33}$$

is the smallest initial investment that suffices for superhedging the claim H^E. In fact, the formula above can be regarded as a special case of the representation theorem for *convex risk measures* in our financial market model. This will be explained next. Let us take $L^\infty := L^\infty(\Omega, \mathcal{F}, P)$ as the space of all financial positions. A position $Y \in L^\infty$ will be regarded as *acceptable* if it can be hedged with a strategy in $\overline{\mathcal{S}}$ at no additional cost. Thus, we introduce the *acceptance set*

$$\mathcal{A}^{\mathcal{S}} := \left\{ Y \in L^\infty \;\middle|\; \exists\, \xi \in \mathcal{S} : \; Y + \sum_{t=1}^{T} \xi_t \cdot (X_t - X_{t-1}) \ge 0 \text{ } P\text{-a. s.} \right\}.$$

Due to the convexity of $\mathcal{S}$, this set $\mathcal{A}^{\mathcal{S}}$ is convex, and under the mild condition

$$\inf\{m \in \mathbb{R} \mid m \in \mathcal{A}^{\mathcal{S}}\} > -\infty, \tag{9.34}$$

$\mathcal{A}^{\mathcal{S}}$ induces a convex risk measure $\rho^{\mathcal{S}} := \rho_{\mathcal{A}^{\mathcal{S}}}$ via

$$\rho^{\mathcal{S}}(Y) := \inf\{m \in \mathbb{R} \mid m + Y \in \mathcal{A}^{\mathcal{S}}\};$$

see Section 4.1. Note that condition (9.34) holds in particular if $\mathcal{S}$ does not contain arbitrage opportunities. In this case, we have in fact

$$\rho^{\mathcal{S}}(0) = \inf\{m \in \mathbb{R} \mid m \in \mathcal{A}^{\mathcal{S}}\} = 0,$$

that is, $\rho^{\mathcal{S}}$ is *normalized*. The main results of this chapter can now be restated in terms of $\rho^{\mathcal{S}}$:

Corollary 9.32. *Under condition* (9.6), *the following conditions are equivalent:*
(a) $\rho^{\mathcal{S}}$ *is sensitive.*
(b) $\overline{\mathcal{S}}$ *contains no arbitrage opportunities.*
(c) $\mathcal{P}_{\mathcal{S}} \neq \emptyset$.
If these equivalent conditions hold, then

$$\rho^{\mathcal{S}}(Y) = \sup_{Q\in\mathcal{Q}_{\mathcal{S}}} (E_Q[-Y\,] - E_Q[\,A_T^Q\,]), \quad Y \in L^\infty. \tag{9.35}$$

In other words, $\rho^{\mathcal{S}}$ can be represented in terms of the penalty function

$$\alpha(Q) = \begin{cases} E_Q[\,A_T^Q\,] & \textit{if } Q \in \mathcal{Q}_{\mathcal{S}}, \\ +\infty & \textit{otherwise.} \end{cases}$$

Proof. That (a) implies (b) is obvious. The equivalence between (b) and (c) was shown in Theorem 9.9. Since both sides of (9.35) are cash invariant, it suffices to prove (9.35) for $Y \leq 0$. But then the representation for $\rho^{\mathcal{S}}$ is just a special case of the superhedging duality (9.33). Finally, (9.35) and (c) imply that $\rho^{\mathcal{S}}(X) \geq \sup_{\widetilde{P} \in \mathcal{P}_{\mathcal{S}}} \widetilde{E}[\,-X\,]$, and the sensitivity of $\rho^{\mathcal{S}}$ follows. □

10 Minimizing the hedging error

In this chapter, we present an alternative approach to the problem of hedging in an incomplete market model. Instead of controlling the downside risk, we simply aim at minimizing the quadratic hedging error. We begin with a local version of the minimization problem, which may be viewed as a sequential regression procedure. Its solution involves an orthogonal decomposition of a given contingent claim; this extends a classical decomposition theorem for martingales known as the Kunita–Watanabe decomposition. Often, the value process generated by a locally risk-minimizing strategy can be described as the martingale of conditional expectations of the given contingent claim for a special choice of an equivalent martingale measure. Such "minimal" martingale measures will be studied in Section 10.2. In Section 10.3, we investigate the connection between local risk minimization and the problem of variance-optimal hedging where one tries to minimize the global quadratic hedging error. The local and the global versions coincide if the underlying measure is itself a martingale measure.

10.1 Local quadratic risk

In this section, we no longer restrict our discussion to strategies that are self-financing. Instead, we admit the possibility that the value of a position is readjusted at the end of each period by an additional investment in the numéraire asset. This means that, in addition to the initial investment at time $t = 0$, we allow for a cash flow throughout the trading periods up to the final time T. In particular, it will now be possible to replicate any given European claim, simply by matching the difference between the payoff of the claim and the value generated by the preceding strategy with a final transfer at time T.

Definition 10.1. A *generalized trading strategy* is a pair of two stochastic process (ξ^0, ξ) such that $\xi^0 = (\xi_t^0)_{t=0,\dots,T}$ is adapted, and such that $\xi = (\xi_t)_{t=1,\dots,T}$ is a d-dimensional predictable process. The (discounted) *value process* V of (ξ^0, ξ) is defined as

$$V_0 := \xi_0^0 \quad \text{and} \quad V_t := \xi_t^0 + \xi_t \cdot X_t \quad \text{for } t \geq 1. \qquad \triangle$$

For such a generalized trading strategy (ξ^0, ξ), the gains and losses accumulated up to time t by investing into the risky assets are given by the sum

$$\sum_{k=1}^{t} \xi_k \cdot (X_k - X_{k-1}).$$

The value process V takes the form

$$V_0 = \xi_1^0 + \xi_1 \cdot X_0 \quad \text{and} \quad V_t = V_0 + \sum_{k=1}^{t} \xi_k \cdot (X_k - X_{k-1}), \quad t = 1, \dots, T,$$

https://doi.org/10.1515/9783111045283-012

if and only if $\overline{\xi} = (\xi_t^0, \xi_t)_{t=1,\dots,T}$ is a *self-financing* trading strategy with initial investment $V_0 = \xi_0^0 = \xi_1^0 + \xi_1 \cdot X_0$. In this case, $(\xi_t^0)_{t=1,\dots,T}$ is a predictable process. In general, however, the difference

$$V_t - \sum_{k=1}^{t} \xi_k \cdot (X_k - X_{k-1})$$

is now nontrivial, and it can be interpreted as the cumulative cost up to time t. This motivates the following definition.

Definition 10.2. The *gains process* G of a generalized trading strategy (ξ^0, ξ) is given by

$$G_0 := 0 \quad \text{and} \quad G_t := \sum_{k=1}^{t} \xi_k \cdot (X_k - X_{k-1}), \qquad t = 1, \dots, T.$$

The *cost process* C of (ξ^0, ξ) is defined as the difference

$$C_t := V_t - G_t, \qquad t = 0, \dots, T,$$

of the value process V and the gains process G. △

In this and in the following sections, we will measure the risk of a strategy in terms of quadratic criteria for the hedging error, based on the "objective" measure P. Our aim will be to minimize such criteria within the class of those generalized strategies (ξ^0, ξ) that replicate a given discounted European claim H in the sense that their value process satisfies

$$V_T = H \quad P\text{-a. s.}$$

The claim H will be fixed for the remainder of this section. As usual, we assume that the σ-field $\mathscr{F}_0$ is trivial, that is, $\mathscr{F}_0 = \{\emptyset, \Omega\}$. In contrast to the previous sections of Part II, however, our approach does *not* exclude a priori the existence of arbitrage opportunities, even though the interesting cases will be those in which there exist equivalent martingale measures. Since our approach is based on L^2-techniques, another set of hypotheses is needed.

Assumption 10.3. Throughout this section, we assume that the discounted claim H and the discounted price process X of the risky assets are both square-integrable with respect to the objective measure P:
(a) $H \in \mathscr{L}^2(\Omega, \mathscr{F}_T, P) =: \mathscr{L}^2(P)$.
(b) $X_t \in \mathscr{L}^2(\Omega, \mathscr{F}_t, P; \mathbb{R}^d)$ for all t.

In addition to these assumptions, the quadratic optimality criteria we have in mind require the following integrability conditions for strategies.

Definition 10.4. An $\mathscr{L}^2$*-admissible strategy* for H is a generalized trading strategy (ξ^0, ξ) whose value process V satisfies

$$V_T = H \ P\text{-a.s.} \quad \text{and} \quad V_t \in \mathscr{L}^2(P) \text{ for each } t,$$

and whose gains process G is such that

$$G_t \in \mathscr{L}^2(P) \quad \text{for each } t. \qquad \triangle$$

We can now introduce the local version of a quadratic criterion for the hedging error of an $\mathscr{L}^2$-admissible strategy.

Definition 10.5. The *local risk process* of an $\mathscr{L}^2$-admissible strategy (ξ^0, ξ) is the process

$$R_t^{\text{loc}}(\xi^0, \xi) := E[\,(C_{t+1} - C_t)^2 \mid \mathscr{F}_t\,], \qquad t = 0, \dots, T-1.$$

An $\mathscr{L}^2$-admissible strategy $(\hat{\xi}^0, \hat{\xi})$ is called a *locally risk-minimizing strategy* if, for all t,

$$R_t^{\text{loc}}(\hat{\xi}^0, \hat{\xi}) \le R_t^{\text{loc}}(\xi^0, \xi) \quad P\text{-a.s.}$$

for each $\mathscr{L}^2$-admissible strategy (ξ^0, ξ) whose value process satisfies $V_{t+1} = \hat{\xi}^0_{t+1} + \hat{\xi}_{t+1} \cdot X_{t+1} = \hat{V}_{t+1}$. $\triangle$

Remark 10.6. The reason for fixing the value $V_{t+1} = \hat{V}_{t+1}$ in the preceding definition becomes clear if we try to construct a locally risk-minimizing strategy $(\hat{\xi}^0, \hat{\xi})$ backwards in time. At time T, we want to construct $\hat{\xi}^0_{T-1}, \hat{\xi}^0_T, \hat{\xi}_{T-1}, \hat{\xi}_T$ as a minimizer for the local risk $R_{T-1}^{\text{loc}}(\xi^0, \xi)$. Since the terminal value of every $\mathscr{L}^2$-admissible strategy must be equal to H, this minimization requires the side condition $\xi^0_T + \xi_T \cdot X_T = H = \hat{V}_T$. As we will see in the proof of Theorem 10.9 below, minimality of $R_{T-1}^{\text{loc}}(\hat{\xi}^0, \hat{\xi})$ completely determines $\hat{\xi}^0_T$ and $\hat{\xi}_T$ and $\hat{V}_{T-1}$, but one is still free to choose $\hat{\xi}^0_{T-1}$ and $\hat{\xi}_{T-1}$ among all ξ^0_{T-1}, ξ_{T-1} with $\xi^0_{T-1} + \xi_{T-1} \cdot X_{T-1} = \hat{V}_{T-1}$. In the next step, it is therefore natural to minimize $R_{T-2}^{\text{loc}}(\xi^0, \xi)$ under the condition that V_{T-1} is equal to the value $\hat{V}_{T-1}$ obtained from the preceding step. Moreover, the problem will now be of the same type as the previous one. $\diamond$

Although locally risk-minimizing strategies are generally not self-financing, it will turn out that they are “self-financing on average” in the following sense.

Definition 10.7. An $\mathscr{L}^2$-admissible strategy is called *mean self-financing* if its cost process C is a P-martingale, that is, if

$$E[\,C_{t+1} - C_t \mid \mathscr{F}_t\,] = 0 \quad P\text{-a.s. for all } t. \qquad \triangle$$

In order to formulate conditions for the existence of a locally risk-minimizing strategy, let us first introduce some notation. The *conditional covariance* of two random variables W and Z with respect to $\mathscr{F}_t$ and P is defined as

$$\operatorname{cov}(W, Z \mid \mathscr{F}_t) := E[\, W Z \mid \mathscr{F}_t\,] - E[\, W \mid \mathscr{F}_t\,] E[\, Z \mid \mathscr{F}_t\,],$$

provided that the conditional expectations and their difference make sense. Moreover, if $Z = (Z^1, \dots, Z^d)$ is a random vector and W is a scalar random variable, then $\operatorname{cov}(W, Z \mid \mathscr{F}_t)$ will be understood in a component-wise manner. That is, $\operatorname{cov}(W, Z \mid \mathscr{F}_t)$ is the vector with components $\operatorname{cov}(W, Z^i \mid \mathscr{F}_t)$, $i = 1, \dots, d$. We also define the *conditional variance* of W as

$$\begin{aligned}\operatorname{var}(W \mid \mathscr{F}_t) &= E[\, W^2 \mid \mathscr{F}_t\,] - E[\, W \mid \mathscr{F}_t\,]^2 \\ &= \operatorname{cov}(W, W \mid \mathscr{F}_t);\end{aligned}$$

see also Exercise 5.2.6.

Definition 10.8. Two adapted processes U and Y are called *strongly orthogonal* with respect to P if the conditional covariances

$$\operatorname{cov}(U_{t+1} - U_t, Y_{t+1} - Y_t \mid \mathscr{F}_t), \quad t = 0, \dots, T-1,$$

are well-defined and vanish P-almost surely. △

When we consider the strong orthogonality of two processes U and Y in the sequel, then often one of them will be a P-martingale. In this case, their conditional covariances reduce to

$$\operatorname{cov}(U_{t+1} - U_t, Y_{t+1} - Y_t \mid \mathscr{F}_t) = E[\,(U_{t+1} - U_t)(Y_{t+1} - Y_t) \mid \mathscr{F}_t\,]. \tag{10.1}$$

After these preparations, we are now ready to state our first result, namely the following characterization of locally risk-minimizing strategies. Criteria for the existence of such strategies will be provided later on.

Theorem 10.9. *An $\mathscr{L}^2$-admissible strategy is locally risk-minimizing if and only if it is mean self-financing and its cost process is strongly orthogonal to X.*

Proof. The local risk process of any $\mathscr{L}^2$-admissible strategy (ξ^0, ξ) can be expressed as a sum of two nonnegative terms:

$$R_t^{\mathrm{loc}}(\xi^0, \xi) = \operatorname{var}(C_{t+1} - C_t \mid \mathscr{F}_t) + E[\, C_{t+1} - C_t \mid \mathscr{F}_t\,]^2.$$

Since the conditional variance does not change if we add $\mathscr{F}_t$-measurable random variables to its argument, the first term on the right-hand side takes the form

$$\operatorname{var}(C_{t+1} - C_t \mid \mathscr{F}_t) = \operatorname{var}(V_{t+1} - \xi_{t+1} \cdot (X_{t+1} - X_t) \mid \mathscr{F}_t). \tag{10.2}$$

The second term satisfies

$$E[\, C_{t+1} - C_t \mid \mathscr{F}_t\,]^2 = \big(E[\, V_{t+1} \mid \mathscr{F}_t\,] - \xi_{t+1} \cdot E[\, X_{t+1} - X_t \mid \mathscr{F}_t\,] - V_t\big)^2. \tag{10.3}$$

In a second step, we fix t and V_{t+1}, and we consider ξ_{t+1} and V_t as parameters. Our purpose is to derive necessary conditions for the minimality of $R_t^{\mathrm{loc}}(\xi^0, \xi)$ with respect to variations of ξ_{t+1} and V_t. To this end, note first that it is possible to change the parameters ξ_t^0 and ξ_t in such a way that V_t takes any given value, that the modified strategy is still an $\mathscr{L}^2$-admissible strategy for H, and that the values of ξ_{t+1} and V_{t+1} remain unchanged. In particular, the value in (10.2) is not affected by such a modification, and so it is necessary for the optimality of $R_t^{\mathrm{loc}}(\xi^0, \xi)$ that V_t minimizes (10.3). This is the case if and only if

$$V_t = E[\, V_{t+1} \mid \mathscr{F}_t \,] - \xi_{t+1} \cdot E[X_{t+1} - X_t \mid \mathscr{F}_t \,]. \tag{10.4}$$

The value of (10.2) is independent of V_t and a quadratic form in terms of the $\mathscr{F}_t$-measurable random vector ξ_{t+1}. Thus, (10.2) is minimal if and only if ξ_{t+1} solves the linear equation

$$0 = \operatorname{cov}\big(V_{t+1} - \xi_{t+1} \cdot (X_{t+1} - X_t),\, X_{t+1} - X_t \mid \mathscr{F}_t\big). \tag{10.5}$$

Note that (10.4) is equivalent to

$$E[\, C_{t+1} - C_t \mid \mathscr{F}_t \,] = E[\, V_{t+1} - \xi_{t+1} \cdot (X_{t+1} - X_t) \mid \mathscr{F}_t \,] - V_t = 0.$$

Moreover, given (10.4), the condition (10.5) holds if and only if

$$E[\, (C_{t+1} - C_t)(X_{t+1} - X_t) \mid \mathscr{F}_t \,] = 0$$

where we have used the fact that the conditional covariance in (10.5) is not changed by subtracting the $\mathscr{F}_t$-measurable random variable V_t from the first argument. Backward induction on t concludes the proof. □

The previous proof provides a recipe for a recursive construction of a locally risk-minimizing strategy: If V_{t+1} is already given, minimize

$$E[\, (C_{t+1} - C_t)^2 \mid \mathscr{F}_t \,] = E[\, \big(V_{t+1} - (V_t + \xi_{t+1} \cdot (X_{t+1} - X_t))\big)^2 \mid \mathscr{F}_t \,]$$

with respect to V_t and ξ_{t+1}. This is just a conditional version of the standard problem of determining the *linear regression* of V_{t+1} on the increment $X_{t+1} - X_t$. Let us now consider the case

$$d = 1,$$

where our market model contains just one risky asset. Then a heuristic computation yields the following recursive scheme for a solution to our minimization problem:

$$
\begin{aligned}
\hat{V}_T &:= H, \\
\hat{\xi}_{t+1} &:= \frac{\operatorname{cov}(\hat{V}_{t+1}, X_{t+1} - X_t \mid \mathscr{F}_t)}{\sigma_{t+1}^2} \cdot \mathbb{1}_{\{\sigma_{t+1} \neq 0\}}, \\
\hat{V}_t &:= E[\, \hat{V}_{t+1} \mid \mathscr{F}_t\,] - \hat{\xi}_{t+1} \cdot E[\, X_{t+1} - X_t \mid \mathscr{F}_t\,].
\end{aligned}
\tag{10.6}
$$

Here, σ_{t+1}^2 is a shorthand notation for the conditional variance

$$\sigma_{t+1}^2 := \operatorname{var}(X_{t+1} - X_t \mid \mathscr{F}_t).$$

Defining $\hat{\xi}_t^0 := \hat{V}_t - \hat{\xi}_t \cdot X_t$, we obtain a generalized trading strategy $(\hat{\xi}^0, \hat{\xi})$ whose terminal portfolio value $\hat{V}_T$ coincides with H. However, an extra condition is needed to conclude that this strategy is indeed $\mathscr{L}^2$-admissible.

Proposition 10.10. *Consider a market model with a single risky asset and assume that there exists a constant κ such that*

$$\left(E[\, X_t - X_{t-1} \mid \mathscr{F}_{t-1}\,]\right)^2 \leq \kappa \cdot \sigma_t^2 \quad P\text{-a. s. for all } t. \tag{10.7}$$

Then the recursion (10.6) *defines a locally risk-minimizing strategy $(\hat{\xi}^0, \hat{\xi})$. Moreover, any other locally risk-minimizing strategy coincides with $(\hat{\xi}^0, \hat{\xi})$ up to modifications of $\hat{\xi}_t$ on the set $\{\sigma_t^2 = 0\}$.*

Proof. We have to show that $(\hat{\xi}^0, \hat{\xi})$ is $\mathscr{L}^2$-admissible. To this end, observe that the recursion (10.6) and the condition (10.7) imply that

$$
\begin{aligned}
&E[\, (\hat{\xi}_t \cdot (X_t - X_{t-1}))^2\,] \\
&\quad = E\left[\frac{\operatorname{cov}(\hat{V}_t, X_t - X_{t-1} \mid \mathscr{F}_{t-1})^2}{\sigma_t^4} \cdot E[\, (X_t - X_{t-1})^2 \mid \mathscr{F}_{t-1}\,]\, \mathbb{1}_{\{\sigma_t^2 \neq 0\}} \right] \\
&\quad \leq (1 + \kappa) \cdot E\left[\frac{\operatorname{cov}(\hat{V}_t, X_t - X_{t-1} \mid \mathscr{F}_{t-1})^2}{\sigma_t^2} \right] \\
&\quad \leq (1 + \kappa) \cdot E\big[\operatorname{var}(\hat{V}_t \mid \mathscr{F}_{t-1})\big].
\end{aligned}
$$

The last expectation is finite if $\hat{V}_t$ is square-integrable. In this case, $\hat{\xi}_t \cdot (X_t - X_{t-1}) \in \mathscr{L}^2(P)$ and in turn $\hat{V}_{t-1} \in \mathscr{L}^2(P)$. Hence, $\mathscr{L}^2$-admissibility of $(\hat{\xi}^0, \hat{\xi})$ follows by backward induction. The claim that $(\hat{\xi}^0, \hat{\xi})$ is locally risk-minimizing as well as the uniqueness assertion follow immediately from the construction. □

Remark 10.11. The predictable process

$$\sum_{s=1}^{t} \frac{(E[\, X_s - X_{s-1} \mid \mathscr{F}_{s-1}\,])^2}{\operatorname{var}(X_s - X_{s-1} \mid \mathscr{F}_{s-1})}, \qquad t = 1, \dots, T,$$

is called the *mean-variance trade-off process* of X, and condition (10.7) is known as the assumption of *bounded mean-variance trade-off.* ◇

Remark 10.12. The assumption of bounded mean-variance trade-off is equivalent to the existence of some $\delta < 1$ such that

$$\left(E[X_t - X_{t-1} \mid \mathcal{F}_{t-1}]\right)^2 \leq \delta \cdot E[(X_t - X_{t-1})^2 \mid \mathcal{F}_{t-1}] \quad P\text{-a.s. for all } t. \tag{10.8}$$

Indeed, with $\alpha_t = E[X_t - X_{t-1} \mid \mathcal{F}_{t-1}]$, the assumption of bounded mean-variance trade-off is equivalent to

$$\alpha_t^2 \leq \kappa(E[(X_t - X_{t-1})^2 \mid \mathcal{F}_{t-1}] - \alpha_t^2),$$

which is seen to be equivalent to (10.8) by choosing $\delta = \kappa/(1+\kappa)$. ◇

Example 10.13. Let us consider a market model consisting of a single risky asset S^1 and a riskless bond

$$S_t^0 = (1+r)^t, \quad t = 0, \ldots, T,$$

with constant return $r > -1$. We assume that $S_0^1 = 1$, and that the returns

$$R_t := \frac{S_t^1 - S_{t-1}^1}{S_{t-1}^1}, \quad t = 1, \ldots, T,$$

of the risky asset are independent and identically distributed random variables in $\mathcal{L}^2(P)$. Under these assumptions, the discounted price process X, defined by

$$X_t = \prod_{s=1}^{t} \frac{1+R_s}{1+r}, \quad t = 0, \ldots, T,$$

is square-integrable. Denoting by $\widetilde{m}$ the mean of R_t and by $\widetilde{\sigma}^2$ its variance, we get

$$E[X_t - X_{t-1} \mid \mathcal{F}_{t-1}] = X_{t-1} \cdot \frac{\widetilde{m} - r}{1+r},$$
$$\operatorname{var}(X_t - X_{t-1} \mid \mathcal{F}_{t-1}) = X_{t-1}^2 \cdot \frac{\widetilde{\sigma}^2}{(1+r)^2}.$$

Thus, the condition of bounded mean-variance trade-off holds without any further assumptions, and a locally risk-minimizing strategy exists. Moreover, P is a martingale measure if and only if $\widetilde{m} = r$. ◇

Let us return to our general market model with an arbitrary number of risky assets

$$X = (X^1, \ldots, X^d).$$

The following result characterizes the existence of locally risk-minimizing strategies in terms of a decomposition of the claim H.

Corollary 10.14. *There exists a locally risk-minimizing strategy if and only if H admits a decomposition*

$$H = c + \sum_{t=1}^{T} \xi_t \cdot (X_t - X_{t-1}) + L_T \quad P\text{-a. s.}, \tag{10.9}$$

where c is a constant, ξ is a d-dimensional predictable process such that

$$\xi_t \cdot (X_t - X_{t-1}) \in \mathcal{L}^2(P) \quad \text{for all } t,$$

and where L is a square-integrable P-martingale that is strongly orthogonal to X and satisfies $L_0 = 0$. In this case, a locally risk-minimizing strategy $(\hat{\xi}^0, \hat{\xi})$ is given by $\hat{\xi} = \xi$ and by the adapted process $\hat{\xi}^0$ defined via $\hat{\xi}^0_0 = c$ and

$$\hat{\xi}^0_t = c + \sum_{s=1}^{t} \xi_s \cdot (X_s - X_{s-1}) + L_t - \xi_t \cdot X_t, \quad t = 1, \ldots, T.$$

Proof. If $(\hat{\xi}^0, \hat{\xi})$ is a given locally risk-minimizing strategy with cost process $\hat{C}$, then $L_t := \hat{C}_t - \hat{C}_0$ is a square-integrable P-martingale that is strongly orthogonal to X by Theorem 10.9. Hence, we obtain a decomposition (10.9). Conversely, if such a decomposition exists, then the strategy $(\hat{\xi}^0, \hat{\xi})$ has the cost process $\hat{C} = c + L$, and Theorem 10.9 implies that $(\hat{\xi}^0, \hat{\xi})$ is locally risk-minimizing. □

A decomposition of the form (10.9) will be called an *orthogonal decomposition* of the contingent claim H with respect to the process X. If X is itself a P-martingale, then the orthogonal decomposition reduces to the *Kunita–Watanabe decomposition*, which we will explain next. To this end, we will need some preparation.

Lemma 10.15. *For two square-integrable martingales M and N, the following two conditions are equivalent:*

(a) *M and N are strongly orthogonal.*

(b) *The product MN is a martingale.*

Proof. The martingale property of M and N gives

$$E[\,(M_{t+1} - M_t)(N_{t+1} - N_t) \mid \mathcal{F}_t\,] = E[\,M_{t+1}N_{t+1} \mid \mathcal{F}_t\,] - M_t N_t,$$

and this expression vanishes if and only if MN is a martingale. □

Let $\mathcal{H}^2$ denote the space of all square-integrable P-martingales. Via the identity $M_t = E[\,M_T \mid \mathcal{F}_t\,]$, each $M \in \mathcal{H}^2$ can be identified with its terminal value $M_T \in \mathcal{L}^2(P)$.

With the standard identification of random variables that coincide P-a. s. (see (H.2)), $\mathcal{H}^2$ becomes a Hilbert space isomorphic to $L^2(P)$, if endowed with the inner product

$$(M,N)_{\mathcal{H}^2} := E[\,M_T N_T\,], \quad M,N \in \mathcal{H}^2.$$

Recall from Definition 6.14 that, for a stopping time τ, the stopped process M^τ is defined as

$$M_t^\tau := M_{\tau\wedge t}, \quad t = 0,\dots,T.$$

Definition 10.16. A subspace $\mathcal{S}$ of $\mathcal{H}^2$ is called *stable* if $M^\tau \in \mathcal{S}$ for each $M \in \mathcal{S}$ and every stopping time τ. △

Proposition 10.17. *For a stable subspace $\mathcal{S}$ of $\mathcal{H}^2$ and for $L \in \mathcal{H}^2$ with $L_0 = 0$, the following conditions are equivalent:*
(a) *L is orthogonal to $\mathcal{S}$, that is,*

$$(L,M)_{\mathcal{H}^2} = 0 \quad \textit{for all } M \in \mathcal{S}.$$

(b) *L is strongly orthogonal to $\mathcal{S}$, that is, for each $M \in \mathcal{S}$*

$$E[\,(L_{t+1} - L_t)(M_{t+1} - M_t) \mid \mathcal{F}_t\,] = 0 \quad P\text{-a. s. for all } t.$$

(c) *The product LM is a martingale for each $M \in \mathcal{S}$.*

Proof. The equivalence of (b) and (c) follows from Lemma 10.15. To prove (a)⇔(c), we will show that LM is a martingale for fixed $M \in \mathcal{S}$ if and only if $(L,M^\tau)_{\mathcal{H}^2} = 0$ for all stopping times $\tau \le T$. By the stopping theorem in the form of Proposition 6.37,

$$(L,M^\tau)_{\mathcal{H}^2} = E[\,L_T M_\tau\,] = E[L_\tau M_\tau].$$

Using the fact that $L_0 M_0 = 0$ and applying the stopping theorem in the form of Theorem 6.15, we conclude that $(L,M^\tau)_{\mathcal{H}^2} = 0$ for all stopping times $\tau \le T$ if and only if LM is a martingale. □

If P is the unique martingale measure in a complete market model, then the following martingale representation holds according to Theorem 5.38: For every P-martingale M, there exists a d-dimensional predictable process ξ such that M can be represented as the sum of M_0 and the gains process of ξ, that is,

$$M_t = M_0 + \sum_{s=1}^{t} \xi_s \cdot (X_s - X_{s-1}), \quad t = 0,\dots,T.$$

Theorem 5.38 even states that this martingale representation property is equivalent to the completeness of the market model. The following discrete-time version of the

Kunita–Watanabe decomposition theorem shows that, in an incomplete market, every square-integrable martingale M can be decomposed into the sum of M_0, the gains process of a d-dimensional predictable process ξ, and a square-integrable martingale L that is strongly orthogonal to X.

Theorem 10.18. *If the process X is a square-integrable martingale under P, then every martingale $M \in \mathscr{H}^2$ is of the form*

$$M_t = M_0 + \sum_{s=1}^{t} \xi_s \cdot (X_s - X_{s-1}) + L_t,$$

where ξ is a d-dimensional predictable process such that $\xi_t \cdot (X_t - X_{t-1}) \in \mathscr{L}^2(P)$ for each t, and where L is a square-integrable P-martingale that is strongly orthogonal to X and satisfies $L_0 = 0$. Moreover, this decomposition is unique in the sense that L is uniquely determined.

Proof. Denote by $\mathscr{S}$ the set of all d-dimensional predictable processes ξ such that $\xi_t \cdot (X_t - X_{t-1}) \in \mathscr{L}^2(P)$ for each t, and denote by

$$G_t(\xi) := \sum_{s=1}^{t} \xi_s \cdot (X_s - X_{s-1}), \quad t = 0, \dots, T,$$

the "stochastic integral" of $\xi \in \mathscr{S}$ with respect to X. Since for $\xi \in \mathscr{S}$ the process $G(\xi)$ is a square-integrable P-martingale, the set $\mathscr{G}$ of all those martingales can be regarded as a linear subspace of the Hilbert space $\mathscr{H}^2$. In fact, $\mathscr{G}$ is a closed subspace of $\mathscr{H}^2$. To prove this claim, note that the martingale property of $G(\xi)$ implies that

$$(G(\xi), G(\xi))_{\mathscr{H}^2} = E[\,(G_T(\xi))^2\,] = \sum_{t=1}^{T} E[\,(\xi_t \cdot (X_t - X_{t-1}))^2\,].$$

Thus, if $\xi^{(n)}$ is such that $G(\xi^{(n)})$ is a Cauchy sequence in $\mathscr{H}^2$, then $\xi_t^{(n)} \cdot (X_t - X_{t-1})$ is a Cauchy sequence in $L^2(P)$ for each t. Since P is a martingale measure, we may apply Lemma 1.71 to conclude that any limit point of $\xi_t^{(n)} \cdot (X_t - X_{t-1})$ is of the form $\xi_t \cdot (X_t - X_{t-1})$ for some $\xi_t \in L^0(\Omega, \mathscr{F}_{t-1}, P; \mathbb{R}^d)$. Hence, $\mathscr{G}$ is closed in $\mathscr{H}^2$. Moreover, $\mathscr{G}$ is stable. Indeed, if $\xi \in \mathscr{S}$ and τ is a stopping time, then $G_{t\wedge\tau}(\xi) = G_t(\tilde{\xi})$ where

$$\tilde{\xi}_s := \xi_s \mathbb{1}_{\{\tau \geq s\}}, \quad s = 1, \dots, T.$$

Furthermore, we have $\tilde{\xi} \in \mathscr{S}$ since

$$E[(\tilde{\xi}_t \cdot (X_t - X_{t-1}))^2] \leq E[(\xi_t \cdot (X_t - X_{t-1}))^2] < \infty.$$

Since $\mathscr{G}$ is closed, the orthogonal projection N of $M - M_0$ onto $\mathscr{G}$ is well-defined by standard Hilbert space techniques. The martingale N belongs to $\mathscr{G}$, and the difference

$L := M - M_0 - N$ is orthogonal to $\mathscr{G}$. By Proposition 10.17, L is strongly orthogonal to $\mathscr{G}$ and hence strongly orthogonal to X. Therefore, $M = M_0 + N + L$ is the desired decomposition of M.

To show that L is uniquely determined, suppose that there exists another decomposition of M in terms of $\widetilde{\xi}$ and $\widetilde{L}$. Then

$$N_t := L_t - \widetilde{L}_t = \sum_{s=1}^{t} (\widetilde{\xi}_s - \xi_s) \cdot (X_s - X_{s-1})$$

is a square-integrable P-martingale that is strongly orthogonal to X. By (10.1), strong orthogonality implies

$$E[\,(\widetilde{\xi}_t - \xi_t) \cdot (X_t - X_{t-1})(X_t - X_{t-1}) \mid \mathscr{F}_{t-1}\,] = 0.$$

Multiplying this identity with $\widetilde{\xi}_t - \xi_t$ gives

$$E[\,((\widetilde{\xi}_t - \xi_t) \cdot (X_t - X_{t-1}))^2 \mid \mathscr{F}_{t-1}\,] = 0,$$

and so

$$N_t - N_{t-1} = (\widetilde{\xi}_t - \xi_t) \cdot (X_t - X_{t-1}) = 0 \quad P\text{-almost surely.}$$

In view of $\widetilde{L}_0 = L_0 = 0$, we thus get $\widetilde{L} = L$. □

Remark 10.19. In dimension $d = 1$, the assumption of bounded mean-variance tradeoff (10.7) is clearly satisfied if X is a square-integrable P-martingale. Combining Proposition 10.10 with Corollary 10.14 then yields an alternative proof of Theorem 10.18. Moreover, the recursion (10.6) identifies the predictable process ξ appearing in the Kunita–Watanabe decomposition of a martingale M:

$$\xi_t = \frac{E[\,(M_t - M_{t-1})(X_t - X_{t-1}) \mid \mathscr{F}_{t-1}\,]}{E[\,(X_t - X_{t-1})^2 \mid \mathscr{F}_{t-1}\,]} \mathbb{1}_{\{E[\,(X_t - X_{t-1})^2 \mid \mathscr{F}_{t-1}\,] \neq 0\}}. \qquad \diamond$$

10.2 Minimal martingale measures

If P is itself a martingale measure, Theorem 10.18 combined with Corollary 10.14 yields immediately a solution to our original problem of constructing locally risk-minimizing strategies.

Corollary 10.20. *If P is a martingale measure, then there exists a locally risk-minimizing strategy. Moreover, this strategy is unique in the sense that its value process $\hat{V}$ is uniquely determined as*

$$\hat{V}_t = E[\,H \mid \mathscr{F}_t\,], \quad t = 0, \ldots, T, \tag{10.10}$$

and that its cost process is given by

$$\hat{C}_t = \hat{V}_0 + L_t, \quad t = 0, \dots, T,$$

where L is the strongly orthogonal P-martingale arising in the Kunita–Watanabe decomposition of $\hat{V}$.

The identity (10.10) allows for a time-consistent interpretation of $\hat{V}_t$ as an arbitrage-free price for H at time t. In the general case in which X is not a martingale under P, one may ask whether there exists an equivalent martingale measure $\hat{P}$ such that the value process $\hat{V}$ of a locally risk-minimizing strategy can be obtained in a similar manner as the martingale

$$\hat{E}[H \mid \mathcal{F}_t], \quad t = 0, \dots, T. \tag{10.11}$$

Definition 10.21. An equivalent martingale measure $\hat{P} \in \mathcal{P}$ is called a *minimal martingale measure* if

$$E\left[\left(\frac{d\hat{P}}{dP}\right)^2\right] < \infty,$$

and if every P-martingale $M \in \mathcal{H}^2$ that is strongly orthogonal to X is also a $\hat{P}$-martingale. △

The following result shows that a minimal martingale measure provides the desired representation (10.11)—*if* such a minimal martingale measure exists.

Theorem 10.22. *If* $\hat{P}$ *is a minimal martingale measure, and if* $\hat{V}$ *is the value process of a locally risk-minimizing strategy, then*

$$\hat{V}_t = \hat{E}[H \mid \mathcal{F}_t], \quad t = 0, \dots, T.$$

Proof. Denote by

$$H = c + \sum_{t=1}^{T} \xi_t \cdot (X_t - X_{t-1}) + L_T$$

an orthogonal decomposition of H as in Corollary 10.14. Then $\hat{V}$ is given by

$$\hat{V}_t = c + \sum_{s=1}^{t} \xi_s \cdot (X_s - X_{s-1}) + L_t.$$

The process L is a $\hat{P}$-martingale, because it is a square-integrable P-martingale strongly orthogonal to X. Moreover, $\xi_s \cdot (X_s - X_{s-1}) \in \mathcal{L}^1(\hat{P})$, because both $\xi_s \cdot (X_s - X_{s-1})$ and $d\hat{P}/dP$ are square-integrable with respect to P. It follows that $\hat{V}$ is a $\hat{P}$-martingale. In view of $\hat{V}_T = H$, the assertion follows. □

Our next goal is to derive a characterization of a minimal martingale measure and to use it in order to obtain criteria for its existence. To this end, we have to analyze the effect of an equivalent change of measure on the structure of martingales. The results we will obtain in this direction are of independent interest, and their continuous-time analogues have a wide range of applications in stochastic analysis.

Lemma 10.23. *Let $\widetilde{P}$ be a probability measure equivalent to P. An adapted process $\widetilde{M}$ is a $\widetilde{P}$-martingale if and only if the process*

$$\widetilde{M}_t \cdot E\left[\frac{d\widetilde{P}}{dP} \,\Big|\, \mathcal{F}_t \right], \qquad t = 0, \dots, T,$$

is a P-martingale.

Proof. Let us denote

$$Z_t := E\left[\frac{d\widetilde{P}}{dP} \,\Big|\, \mathcal{F}_t \right].$$

Observe that $\widetilde{M}_t \in \mathcal{L}^1(\widetilde{P})$ if and only if $\widetilde{M}_t Z_t \in \mathcal{L}^1(P)$. Moreover, the process Z is P-a. s. strictly positive by the equivalence of $\widetilde{P}$ and P. Hence, Proposition B.8 yields that

$$Z_t \cdot \widetilde{E}[\,\widetilde{M}_{t+1} \mid \mathcal{F}_t\,] = E[\,\widetilde{M}_{t+1} Z_{t+1} \mid \mathcal{F}_t\,],$$

and it follows that $\widetilde{E}[\,\widetilde{M}_{t+1} \mid \mathcal{F}_t\,] = \widetilde{M}_t$ if and only if $E[\,\widetilde{M}_{t+1} Z_{t+1} \mid \mathcal{F}_t\,] = \widetilde{M}_t Z_t$. □

The following representation (10.13) of the density process may be viewed as the discrete-time version of the Doléans–Dade stochastic exponential from continuous-time stochastic calculus.

Proposition 10.24. *If $\widetilde{P}$ is a probability measure equivalent to P, then there exists a P-martingale Λ such that*

$$\Lambda_0 = 1 \quad \text{and} \quad \Lambda_{t+1} - \Lambda_t > -1 \quad P\text{-a. s. for all } t, \tag{10.12}$$

and such that the martingale

$$Z_t := E\left[\frac{d\widetilde{P}}{dP} \,\Big|\, \mathcal{F}_t \right], \qquad t = 0, \dots, T,$$

can be represented as

$$Z_t = \prod_{s=1}^{t} (1 + \Lambda_s - \Lambda_{s-1}), \qquad t = 0, \dots, T. \tag{10.13}$$

Conversely, if Λ is a P-martingale with (10.12) *and such that* (10.13) *defines a P-martingale Z, then*

$$d\widetilde{P} := Z_T \, dP$$

defines a probability measure $\widetilde{P} \approx P$.

Proof. For $\widetilde{P} \approx P$ given, define Λ by $\Lambda_0 = 1$ and

$$\Lambda_{t+1} := \Lambda_t + \frac{Z_{t+1} - Z_t}{Z_t} = \Lambda_t - 1 + \frac{Z_{t+1}}{Z_t}, \qquad t = 0, \dots, T-1.$$

Clearly, (10.13) holds with this choice of Λ. In particular, Λ satisfies (10.12), because the equivalence of P and $\widetilde{P}$ implies that Z_t is P-a. s. strictly positive for all t.

In the next step, we show by induction on t that $\Lambda_t \in \mathscr{L}^1(P)$. For $t = 0$, this holds by definition. Suppose that $\Lambda_t \in \mathscr{L}^1(P)$. Since Z is nonnegative, the conditional expectation of Z_{t+1}/Z_t is well-defined and satisfies P-a. s.

$$E\left[\frac{Z_{t+1}}{Z_t} \,\middle|\, \mathscr{F}_t \right] = \frac{1}{Z_t} \cdot E[\, Z_{t+1} \mid \mathscr{F}_t \,] = 1.$$

It follows that $Z_{t+1}/Z_t \in \mathscr{L}^1(P)$, and in turn that

$$\Lambda_{t+1} = \Lambda_t - 1 + \frac{Z_{t+1}}{Z_t} \in \mathscr{L}^1(P).$$

Now it is easy to derive the martingale property of Λ: Since Z_t is strictly positive, we may divide both sides of the equation $E[\, Z_{t+1} \mid \mathscr{F}_t \,] = Z_t$ by Z_t, and we arrive at $E[\, \Lambda_{t+1} - \Lambda_t \mid \mathscr{F}_t \,] = 0$.

As for the second part of the assertion, it is clear that $E[\, Z_t \,] = Z_0 = 1$ for all t, provided that Λ is a P-martingale such that (10.12) holds and such that (10.13) defines a strictly positive P-martingale Z. □

The following theorem shows how a martingale M is affected by an equivalent change of the underlying probability measure P. Typically, M will no longer be a martingale under the new measure $\widetilde{P}$, and so a nontrivial predictable process $(A_t)_{t=1,\dots,T}$ will appear in the *Doob decomposition*

$$M = \widetilde{M} + A$$

of M under $\widetilde{P}$. Alternatively, $-A$ may be viewed as the predictable process arising in the Doob decomposition of the $\widetilde{P}$-martingale $\widetilde{M}$ under the measure P. The following result, a discrete-time version of the *Girsanov formula*, describes A in terms of the martingale Λ arising in the representation (10.13) of the successive densities.

Theorem 10.25. *Let P and $\widetilde{P}$ be two equivalent probability measures, and let Λ denote the P-martingale arising in the representation* (10.13) *of the successive densities $Z_t := E[\, d\widetilde{P}/dP \mid \mathscr{F}_t \,]$. If $\widetilde{M}$ is a $\widetilde{P}$-martingale such that $\widetilde{M}_t \in \mathscr{L}^1(P)$ for all t, then*

$$M_t := \widetilde{M}_t + \sum_{s=1}^{t} E[\,(\Lambda_s - \Lambda_{s-1})(\widetilde{M}_s - \widetilde{M}_{s-1}) \mid \mathscr{F}_{s-1}\,]$$

is a P-martingale.

Proof. Note first that

$$(\Lambda_t - \Lambda_{t-1})(\widetilde{M}_t - \widetilde{M}_{t-1}) = \frac{1}{Z_{t-1}}(Z_t(\widetilde{M}_t - \widetilde{M}_{t-1})) - (\widetilde{M}_t - \widetilde{M}_{t-1})\,. \tag{10.14}$$

According to Lemma 10.23, $Z_t(\widetilde{M}_t - \widetilde{M}_{t-1})$ is a martingale increment, and hence belongs to $\mathscr{L}^1(P)$. If we let

$$\tau_n := \inf\{t \mid Z_t < 1/n\}, \quad n = 2, 3, \dots,$$

it follows that

$$(\Lambda_t - \Lambda_{t-1})(\widetilde{M}_t - \widetilde{M}_{t-1})\,\mathbb{1}_{\{\tau_n \geq t\}} \in \mathscr{L}^1(P)\,.$$

In particular, the conditional expectations appearing in the statement of the theorem are P-a. s. well-defined. Moreover, the identity (10.14) implies that P-a. s. on $\{\tau_n \geq t\}$,

$$\begin{aligned}
E[\,\widetilde{M}_t - \widetilde{M}_{t-1} \mid \mathscr{F}_{t-1}\,] \\
&= \frac{1}{Z_{t-1}} E[\,Z_t(\widetilde{M}_t - \widetilde{M}_{t-1}) \mid \mathscr{F}_{t-1}\,] - E[\,(\Lambda_t - \Lambda_{t-1})(\widetilde{M}_t - \widetilde{M}_{t-1}) \mid \mathscr{F}_{t-1}\,] \\
&= -E[\,(\Lambda_t - \Lambda_{t-1})(\widetilde{M}_t - \widetilde{M}_{t-1}) \mid \mathscr{F}_{t-1}\,].
\end{aligned}$$

By sending n to infinity, we obtain the Doob decomposition of $\widetilde{M}$ under P. □

The preceding theorem allows us to characterize those equivalent measures $P^* \approx P$ that are martingale measures. Let

$$X = Y + B \tag{10.15}$$

denote the Doob decomposition of X under P, where Y is a d-dimensional P-martingale, and $(B_t)_{t=1,\dots,T}$ is a d-dimensional predictable process given by $B_0 = 0$ and

$$B^i_{t+1} - B^i_t = E[\,X^i_{t+1} - X^i_t \mid \mathscr{F}_t\,], \quad t = 0, \dots, T-1,\ i = 1, \dots, d.$$

Corollary 10.26. *Let $P^* \approx P$ be such that $E^*[\,|X_t|\,] < \infty$ for each t, and denote by Λ the P-martingale arising in the representation* (10.13) *of the successive densities $Z_t := E[\,dP^*/dP \mid \mathscr{F}_t\,]$. Then P^* is an equivalent martingale measure if and only if the predictable process B in the Doob decomposition* (10.15) *satisfies*

$$B_t = -\sum_{s=1}^{t} E[\,(\Lambda_s - \Lambda_{s-1})(Y_s - Y_{s-1}) \mid \mathscr{F}_{s-1}\,]$$
$$= -\sum_{s=1}^{t} E[\,(\Lambda_s - \Lambda_{s-1})(X_s - X_{s-1}) \mid \mathscr{F}_{s-1}\,]$$

P-a. s. for $t = 1, \dots, T$.

Proof. If P^* is an equivalent martingale measure, then our formula for B is an immediate consequence of Theorem 10.25. For the proof of the converse implication, we denote by

$$X = Y^* + B^* \tag{10.16}$$

the Doob decomposition of X under P^*. Then Y^* is a P^*-martingale. Using Theorem 10.25 with $\widetilde{P} := P^*$ and $\widetilde{M} := Y^*$, we see that $M := Y^* + \hat{B}$ is a P-martingale where

$$\hat{B}_t = \sum_{s=1}^{t} E[\,(\Lambda_s - \Lambda_{s-1})(Y_s^* - Y_{s-1}^*) \mid \mathscr{F}_s\,] = -B_t.$$

That is, $M = Y^* - B$ is a P-martingale. Plugging this identity into (10.16) gives $X = M + B^* + B$. Since M is a P-martingale and $B^* + B$ is predictable with $B_0^* + B_0 = 0$, the identity $X = M + B^* + B$ must be the Doob decomposition of X with respect to P. By the uniqueness of the Doob decomposition, it must coincide with (10.15). In particular, we must have $B^* + B = B$, and hence $B^* = 0$. Plugging this result into (10.16) yields that P^* is a martingale measure. □

We can now return to our initial task of characterizing a minimal martingale measure.

Theorem 10.27. *Let $\hat{P} \in \mathscr{P}$ be an equivalent martingale measure whose density $d\hat{P}/dP$ is square-integrable. Then $\hat{P}$ is a minimal martingale measure if and only if the P-martingale Λ of* (10.13) *admits a representation as a "stochastic integral" with respect to the P-martingale Y arising in the Doob decomposition of X:*

$$\Lambda_t = 1 + \sum_{s=1}^{t} \lambda_s \cdot (Y_s - Y_{s-1}), \quad t = 0, \dots, T, \tag{10.17}$$

for some d-dimensional predictable process λ.

Proof. To prove sufficiency of (10.17), we have to show that if $M \in \mathscr{H}^2$ is strongly orthogonal to X, then M is a $\hat{P}$-martingale. By Lemma 10.23, this follows if we can show that MZ is a P-martingale where

$$Z_t := E\Big[\, \frac{d\hat{P}}{dP} \,\Big|\, \mathscr{F}_t \,\Big].$$

Clearly, $M_t Z_t \in \mathscr{L}^1(P)$ since M and Z are both square-integrable.

We now consider again the Doob decomposition $X = Y + B$ under P and show that $E[\,|Y_t|^2\,] < \infty$ for all t. Indeed, Jensen's inequality for conditional expectations yields that for $t = 0, \dots, T-1$ and $i = 1, \dots, d$,

$$E[\,(B^i_{t+1} - B^i_t)^2\,] = E[\,(E[\,X^i_{t+1} - X^i_t \mid \mathscr{F}_t\,])^2\,] \le E[\,(X^i_{t+1} - X^i_t)^2\,] < \infty.$$

Thus, $B^i_t \in \mathscr{L}^2(P)$ for all t and i, and hence $Y^i_t = X^i_t - B^i_t \in \mathscr{L}^2(P)$.

For the next step, we introduce the stopping times

$$\tau_n := \inf\{t \ge 0 \mid |\lambda_{t+1}| > n\}.$$

By stopping the martingale Λ at τ_n, we obtain the P-martingale Λ^{τ_n}, which is square-integrable since λ^{τ_n} is bounded and Y is square-integrable. It follows that $M\Lambda^{\tau_n}$ is integrable. Furthermore, M is strongly orthogonal to Y, because M is strongly orthogonal to X, and X and Y differ only by a predictable process. Thus, MY is a d-dimensional P-martingale by Lemma 10.15. Hence,

$$\begin{aligned} &E[\,M_{t+1}(\Lambda^{\tau_n}_{t+1} - \Lambda^{\tau_n}_t) \mid \mathscr{F}_t\,] \\ &\qquad = \mathbb{1}_{\{t+1 \le \tau_n\}} \lambda_{t+1} \cdot (E[\,M_{t+1} Y_{t+1} \mid \mathscr{F}_t\,] - E[\,M_{t+1} \mid \mathscr{F}_t\,] Y_t) = 0. \end{aligned} \tag{10.18}$$

Noting that

$$Z^{\tau_n}_t = \prod_{s=1}^{t} (1 + \Lambda^{\tau_n}_s - \Lambda^{\tau_n}_{s-1}),$$

and that Z^{τ_n} is square-integrable, we conclude from (10.18) that

$$\begin{aligned} E[\,M_{t+1} Z^{\tau_n}_{t+1} \mid \mathscr{F}_t\,] &= Z^{\tau_n}_t E[\,M_{t+1}(1 + \Lambda^{\tau_n}_{t+1} - \Lambda^{\tau_n}_t) \mid \mathscr{F}_t\,] \\ &= Z^{\tau_n}_t E[\,M_{t+1} \mid \mathscr{F}_t\,] \\ &= Z^{\tau_n}_t M_t. \end{aligned}$$

Thus, $Z^{\tau_n} M$ is a P-martingale for each n. By Doob's stopping theorem, the process

$$(Z^{\tau_n} M)^{\tau_n} = (ZM)^{\tau_n}$$

is also a P-martingale. Since $\tau_n \nearrow \infty$ P-a. s. and

$$|M^{\tau_n}_{t+1} Z^{\tau_n}_{t+1}| \le \sum_{s=0}^{T} |M_s Z_s| \in \mathscr{L}^1(P),$$

we may apply the dominated convergence theorem for conditional expectations to obtain the desired martingale property of MZ:

$$E[\, M_{t+1} Z_{t+1} \mid \mathscr{F}_t \,] = \lim_{n \uparrow \infty} E[\, M_{t+1}^{\tau_n} Z_{t+1}^{\tau_n} \mid \mathscr{F}_t \,] = \lim_{n \uparrow \infty} M_t^{\tau_n} Z_t^{\tau_n} = M_t Z_t.$$

Thus, $\hat{P}$ is a minimal martingale measure.

For the proof of the converse implication of the theorem, denote by

$$Z_t = 1 + \sum_{s=1}^{t} \eta_s \cdot (Y_s - Y_{s-1}) + L_t$$

the Kunita–Watanabe decomposition of the density process Z with respect to the measure P and the square-integrable martingale Y, as explained in Theorem 10.18. The process L is a square-integrable P-martingale strongly orthogonal to Y, and hence to X. Thus, the assumption that $\hat{P}$ is a minimal martingale measure implies that L is also a $\hat{P}$-martingale. Applying Lemma 10.23, it follows that

$$L_t Z_t = L_t + L_t \sum_{s=1}^{t} \eta_s \cdot (Y_s - Y_{s-1}) + L_t^2$$

is a P-martingale. According to Lemma 10.15, the strong orthogonality of L and Y yields that

$$L_t \sum_{s=1}^{t} \eta_s \cdot (Y_s - Y_{s-1})$$

is a P-martingale; recall that $\eta_s \cdot (Y_s - Y_{s-1}) \in \mathscr{L}^2(P)$ for all s. But then L_t^2 must also be a martingale. In particular, the expectation of L_t^2 is independent of t and so

$$E[\, L_t^2 \,] = L_0^2 = 0$$

from which we get that L vanishes P-almost surely. Hence, $Z - 1$ is equal to the "stochastic integral" of η with respect to Y, and we conclude that

$$\Lambda_{t+1} - \Lambda_t = \frac{Z_{t+1} - Z_t}{Z_t} = \frac{1}{Z_t} \eta_{t+1} \cdot (Y_{t+1} - Y_t).$$

This proves that (10.17) holds with $\lambda_t := \eta_t / Z_{t-1}$. □

Corollary 10.28. *There exists at most one minimal martingale measure.*

Proof. Let $\hat{P}$ and $\hat{P}'$ be two minimal martingale measures, and denote the martingales in the representation (10.13) by Λ and Λ', respectively. On the one hand, it follows from Corollary 10.26 that the martingale $N := \Lambda - \Lambda'$ is strongly orthogonal to Y. On the other hand, Theorem 10.27 implies that N admits a representation as a "stochastic integral" with respect to the P-martingale Y:

$$N_t = \sum_{s=1}^{t} (\lambda_s - \lambda_s') \cdot (Y_s - Y_{s-1}), \quad t = 0, \dots, T.$$

Let $\tau_n := \inf\{t \mid |\lambda_{t+1} - \lambda_{t+1}'| > n\}$, so that N^{τ_n} is in $\mathscr{L}^2(P)$. Then it follows as in the proof of Theorem 10.18 that N^{τ_n} vanishes P-almost surely. Hence, the densities of $\hat{P}$ and $\hat{P}'$ coincide. □

Recall that, for $d = 1$, we denote by

$$\sigma_t^2 = \operatorname{var}(X_t - X_{t-1} \mid \mathscr{F}_{t-1})$$

the conditional variance of the increments of X.

Corollary 10.29. *In dimension $d = 1$, the following two conditions are implied by the existence of a minimal martingale measure $\hat{P}$:*

(a) *The predictable process λ arising in the representation formula* (10.17) *is of the form*

$$\lambda_t = \frac{-E[\, X_t - X_{t-1} \mid \mathscr{F}_{t-1} \,]}{\sigma_t^2} \quad P\text{-a. s. on } \{\sigma_t^2 \neq 0\}. \tag{10.19}$$

(b) *For each t, P-a. s. on $\{\sigma_t^2 \neq 0\}$,*

$$(X_t - X_{t-1}) \cdot E[\, X_t - X_{t-1} \mid \mathscr{F}_{t-1} \,] < E[\, (X_t - X_{t-1})^2 \mid \mathscr{F}_{t-1} \,].$$

Proof. (a): Denote by $X = Y + B$ the Doob decomposition of X with respect to P. According to Corollary 10.26, the P-martingale Λ arising in the representation (10.13) of the density $d\hat{P}/dP$ must satisfy

$$\begin{aligned} B_t - B_{t-1} &= -E[\, (\Lambda_t - \Lambda_{t-1})(Y_t - Y_{t-1}) \mid \mathscr{F}_{t-1} \,] \\ &= -\lambda_t E[\, (Y_t - Y_{t-1})^2 \mid \mathscr{F}_{t-1} \,]. \end{aligned}$$

On the one hand, using the properties of the conditional variance gives

$$\sigma_t^2 = \operatorname{var}(X_t - X_{t-1} \mid \mathscr{F}_{t-1}) = \operatorname{var}(Y_t - Y_{t-1} \mid \mathscr{F}_{t-1}) = E[\, (Y_t - Y_{t-1})^2 \mid \mathscr{F}_{t-1} \,].$$

On the other hand, $B_t - B_{t-1} = E[\, X_t - X_{t-1} \mid \mathscr{F}_{t-1} \,]$. Combining these facts yields our formula for λ_t.

(b): By Proposition 10.24, the P-martingale Λ must be such that P-a. s. on $\{\sigma_t \neq 0\}$,

$$\begin{aligned} -1 < \Lambda_t - \Lambda_{t-1} &= \lambda_t \cdot (Y_t - Y_{t-1}) \\ &= \frac{-E[\, X_t - X_{t-1} \mid \mathscr{F}_{t-1} \,]}{\sigma_t^2} \cdot (X_t - X_{t-1} - E[\, X_t - X_{t-1} \mid \mathscr{F}_{t-1} \,]). \end{aligned} \tag{10.20}$$

Using the formulas for the conditional variance obtained in Exercise 5.2.6 gives

$$\sigma_t = E[\,(X_t - X_{t-1})^2 \mid \mathscr{F}_{t-1}\,] - E[\,X_t - X_{t-1} \mid \mathscr{F}_{t-1}\,]^2.$$

Plugging this fact into (10.20) yields the assertion. □

Note that condition (b) of Corollary 10.29 is rather restrictive as it imposes an almost-sure bound on the $\mathscr{F}_t$-measurable increment $X_t - X_{t-1}$ in terms of $\mathscr{F}_{t-1}$-measurable quantities.

Theorem 10.30. *Consider a market model with a single risky asset satisfying condition* (b) *of Corollary* 10.29 *and the assumption*

$$\big(E[\,X_t - X_{t-1} \mid \mathscr{F}_{t-1}\,]\big)^2 \le C \cdot \sigma_t^2, \quad P\text{-a. s. for all } t \ge 1,$$

of bounded mean-variance trade-off. Then there exists a unique minimal martingale measure $\hat{P}$ whose density $d\hat{P}/dP = Z_T$ is given by

$$Z_t = \prod_{s=1}^{t} (1 + \lambda_s \cdot (Y_s - Y_{s-1})) \tag{10.21}$$

for λ as in (10.19).

Proof. Denote by $X = Y + B$ the Doob decomposition of X under P. For λ_t defined via (10.19), the assumption of bounded mean-variance trade-off yields that

$$E[\,(\lambda_t \cdot (Y_t - Y_{t-1}))^2 \mid \mathscr{F}_{t-1}\,] \le C \quad P\text{-a. s.} \tag{10.22}$$

Hence, Λ_t defined according to (10.17) is a square-integrable P-martingale. As observed in the second part of the proof of Corollary 10.29, its condition (b) holds if and only if $\lambda_s \cdot (Y_s - Y_{s-1}) > -1$ for all t, so that Z defined by (10.21) is P-a. s. strictly positive. Moreover, the bound (10.22) guarantees that Z is a square-integrable P-martingale. We may thus conclude from Proposition 10.24 that Z is the density process of a probability measure $\hat{P} \approx P$ with a square-integrable density $d\hat{P}/dP$. In particular, X_t is $\hat{P}$-integrable for all t. Our choice of λ implies that

$$\begin{aligned} E[\,(\Lambda_t - \Lambda_{t-1})(Y_t - Y_{t-1}) \mid \mathscr{F}_{t-1}\,] &= -E[\,X_t - X_{t-1} \mid \mathscr{F}_{t-1}\,] \\ &= -(B_t - B_{t-1}), \end{aligned}$$

and so $\hat{P}$ is an equivalent martingale measure by Corollary 10.26. Finally, Theorem 10.27 states that $\hat{P}$ is a minimal martingale measure, while uniqueness was already established in Corollary 10.28. □

Example 10.31. Let us consider again the market model of Example 10.13 with independent and identically distributed returns $R_t \in \mathscr{L}^2(P)$. We have seen there that the condition of bounded mean-variance trade-off is satisfied without further assumptions. Let $\widetilde{m} := E[\,R_1\,]$ and $\widetilde{\sigma}^2 := \operatorname{var}(R_1)$. A short calculation using the formulas for

$E[X_t - X_{t-1} \mid \mathscr{F}_{t-1}]$ and $\operatorname{var}(X_t - X_{t-1} \mid \mathscr{F}_{t-1})$ obtained in Example 10.13 as well as an application of Exercise 5.2.6 show that the crucial condition (b) of Corollary 10.29 is equivalent to

$$(\widetilde{m} - r)(R_1 - \widetilde{m}) < \widetilde{\sigma}^2 \quad P\text{-a. s.} \tag{10.23}$$

By Corollary 10.29 and Theorem 10.30, the condition (10.23) is hence equivalent to the existence of the minimal martingale measure. For $\widetilde{m} > r$, the condition (10.23) is an upper bound on R_1, while we obtain a lower bound for $\widetilde{m} < r$. In the case $\widetilde{m} = r$, the measure P is itself the minimal martingale measure, and the condition (10.23) is automatically satisfied. If the distribution of R_1 is given, and if

$$a \leq R_1 \leq b \quad P\text{-a. s.}$$

for certain constants $a > -1$ and $b < \infty$, then (10.23) is satisfied for all r in a certain neighborhood of $\widetilde{m}$. ◇

Exercise 10.2.1. In the context of Example 10.31, derive formulas for λ_t and for the density process of the minimal martingale measure, provided that condition (10.23) holds. ◇

Remark 10.32. The purpose of condition (b) of Corollary 10.29 is to ensure that the density Z defined via

$$Z_t = \prod_{s=1}^{t} (1 + \lambda_s \cdot (Y_s - Y_{s-1}))$$

is strictly positive. In cases where this condition is violated, Z may still be a square-integrable P-martingale and can be regarded as the density of a *signed* measure

$$d\hat{P} = Z_T \, dP,$$

which shares some properties with the minimal martingale measure of Definition 10.21; see, for example, [299]. ◇

In the remainder of this section, we consider briefly another quadratic criterion for the risk of an $\mathscr{L}^2$-admissible strategy.

Definition 10.33. The *remaining conditional risk* of an $\mathscr{L}^2$-admissible strategy (ξ^0, ξ) with cost process C is given by the process

$$R_t^{\mathrm{rem}}(\xi^0, \xi) := E[\,(C_T - C_t)^2 \mid \mathscr{F}_t\,], \quad t = 0, \ldots, T.$$

We say that an $\mathscr{L}^2$-admissible strategy (ξ^0, ξ) *minimizes the remaining conditional risk* if

$$R_t^{\mathrm{rem}}(\xi^0, \xi) \leq R_t^{\mathrm{rem}}(\eta^0, \eta) \quad P\text{-a. s.}$$

for all t and for each $\mathscr{L}^2$-admissible strategy (η^0, η) that coincides with (ξ^0, ξ) up to time t.
$\triangle$

The next result shows that minimizing the remaining conditional risk for a martingale measure is essentially the same as minimizing the local risk. In this case, Corollary 10.20 yields formulas for the value process and the cost process of a minimizing strategy.

Proposition 10.34. *For $P \in \mathscr{P}$, an $\mathscr{L}^2$-admissible strategy minimizes the remaining conditional risk if and only if it is locally risk-minimizing.*

Proof. Let $(\hat{\xi}^0, \hat{\xi})$ be a locally risk-minimizing strategy, which exists by Corollary 10.20, and write $\hat{V}$ and $\hat{C}$ for its value and cost processes. Take another $\mathscr{L}^2$-admissible strategy (η^0, η) whose value and cost processes are denoted by V and C. Since $\hat{V}_T = H = V_T$, the cost process C satisfies

$$\begin{aligned} C_T - C_t &= V_T - V_t - \sum_{k=t+1}^{T} \eta_k \cdot (X_k - X_{k-1}) \\ &= \hat{V}_t - V_t + \sum_{k=t+1}^{T} (\hat{\xi}_k - \eta_k) \cdot (X_k - X_{k-1}) + \hat{C}_T - \hat{C}_t. \end{aligned}$$

Since X and $\hat{C}$ are strongly orthogonal martingales, the remaining conditional risk of (η^0, η) satisfies

$$\begin{aligned} & R_t^{\text{rem}}(\eta^0, \eta) \\ &= (\hat{V}_t - V_t)^2 + E\left[\sum_{k=t+1}^{T} ((\hat{\xi}_k - \eta_k) \cdot (X_k - X_{k-1}))^2 | \mathscr{F}_t \right] + E[\, (\hat{C}_T - \hat{C}_t)^2 \mid \mathscr{F}_t\,], \end{aligned}$$

and this expression is minimal if and only if $V_t = \hat{V}_t$ and $\hat{\xi}_k \cdot (X_k - X_{k-1}) = \eta_k \cdot (X_k - X_{k-1})$ for all $k \geq t+1$ P-almost surely. The result now follows by applying the preceding argument to all $t \in \{0, \dots, T\}$. $\square$

In general, however, an $\mathscr{L}^2$-admissible strategy minimizing the remaining conditional risk does not exist; see Remark 10.42 at the end of the following Section 10.3.

10.3 Variance-optimal hedging

Let $H \in \mathscr{L}^2(P)$ be a square-integrable discounted claim. Throughout this section, we assume that the discounted price process X of the risky asset is square-integrable with respect to P:

$$E[\, |X_t|^2] < \infty \quad \text{for all } t.$$

As in the previous section, there is no need to exclude the existence of arbitrage opportunities, even though the cases of interest will of course be arbitrage-free.

Informally, the problem of variance-minimal hedging is to minimize the quadratic hedging error defined as the squared $L^2(P)$-distance

$$\| H - V_T \|_2^2 = E[\,(H - V_T)^2\,]$$

between H and the terminal value of the value process V of a self-financing trading strategy.

Remark 10.35. Mean-variance hedging is closely related to the discussion in the preceding sections, where we considered the problems of minimizing the local conditional risk or the remaining conditional risk within the class of $\mathscr{L}^2$-admissible strategies for H. To see this, let (ξ^0, ξ) be an $\mathscr{L}^2$-admissible strategy for H in the sense of Definition 10.4, and denote by V, G, and C the resulting value, gains, and cost processes. The quantity

$$R(\xi^0, \xi) := E[\,(C_T - C_0)^2\,]$$

may be called the "global quadratic risk" of (ξ^0, ξ). It coincides with the initial value of the process $R_0^{\mathrm{rem}}(\xi^0, \xi)$ of the remaining conditional risk introduced in Definition 10.33. Since

$$C_T - C_0 = V_T - G_T - (V_0 - G_0) = H - G_T - V_0,$$

the global quadratic risk is independent of the values of the numéraire component ξ^0 at the times $t = 1, \dots, T$. Thus, the global quadratic risk of the generalized trading strategy (ξ^0, ξ) coincides with the quadratic hedging error

$$E[\,(H - \widetilde{V}_T)^2\,]$$

where $\widetilde{V}$ is the value process of the *self-financing* trading strategy arising from the d-dimensional predictable process ξ and the initial investment $\widetilde{V}_0 = V_0 = \xi_0^0$. ◇

Let us rephrase the problem of mean-variance hedging in a form that can be interpreted both within the class of self-financing trading strategies and within the context of Section 10.1. For a d-dimensional predictable process ξ, we denote by $G(\xi)$ the gains process

$$G_t(\xi) = \sum_{s=1}^{t} \xi_s \cdot (X_s - X_{s-1}), \quad t = 0, \dots, T,$$

associated with ξ. Let us introduce the class

$$\mathscr{S} := \{\xi \mid \xi \text{ is predictable and } G_t(\xi) \in \mathscr{L}^2(P) \text{ for all } t\}.$$

Definition 10.36. A pair (V_0^*, ξ^*) where $V_0^* \in \mathbb{R}$ and $\xi^* \in \mathscr{S}$ is called a *variance-optimal strategy* for the discounted claim H if

$$E[\,(H - V_0^* - G_T(\xi^*))^2\,] \le E[\,(H - V_0 - G_T(\xi))^2\,]$$

for all $V_0 \in \mathbb{R}$ and all $\xi \in \mathscr{S}$. △

Our first result identifies a variance-optimal strategy if P is itself a martingale measure.

Proposition 10.37. *Assume that $P \in \mathscr{P}$, and let $(\hat{\xi}^0, \hat{\xi})$ be a locally risk-minimizing $\mathscr{L}^2$-admissible strategy as constructed in Corollary 10.20. Then $(V_0^*, \xi^*) := (\hat{\xi}_0^0, \hat{\xi})$ is a variance-optimal strategy.*

Proof. Recall from Remark 10.35 that if (ξ^0, ξ) is an $\mathscr{L}^2$-admissible strategy for H with value process V, then the expression

$$E[\,(H - V_0 - G_T(\xi))^2\,]$$

is equal to the initial value $R_0^{\mathrm{rem}}(\xi^0, \xi)$ of the remaining conditional risk process of (ξ^0, ξ). But according to Proposition 10.34, R_0^{rem} is minimized by $(\hat{\xi}^0, \hat{\xi})$. □

The general case where X is not a P-martingale will be studied under the simplifying assumption that the market model contains only one risky asset. We will first derive a general existence result, and then determine an explicit solution in a special setting. The key idea for showing the existence of a variance-optimal strategy is to minimize the functional

$$\mathscr{S} \ni \xi \longmapsto E[\,(H - V_0 - G_T(\xi))^2\,],$$

first for fixed V_0, and then to vary the parameter V_0. The first step will be accomplished by projecting $H - V_0$ onto the space of "stochastic integrals"

$$\mathscr{G}_T := \{G_T(\xi) \mid \xi \in \mathscr{S}\}.$$

Clearly, $\mathscr{G}_T$ is a linear subspace of $L^2(P)$. Thus, we can obtain the optimal $\xi = \xi(V_0)$ by using the orthogonal projection of $H - V_0$ on $\mathscr{G}_T$ as soon as we know that $\mathscr{G}_T$ is *closed* in $L^2(P)$. In order to formulate a criterion for the closedness on $\mathscr{G}_T$, we denote by

$$\sigma_t^2 := \mathrm{var}(X_t - X_{t-1} \mid \mathscr{F}_{t-1}), \quad t = 1, \dots, T,$$

the conditional variance, and by

$$\alpha_t := E[\,X_t - X_{t-1} \mid \mathscr{F}_{t-1}\,], \quad t = 1, \dots, T,$$

the conditional mean of the increments of X.

Proposition 10.38. *Suppose that $d = 1$, and assume the condition of bounded mean-variance trade-off*

$$\alpha_t^2 \leq \kappa\sigma_t^2 \quad P\text{-a. s. for all } t. \tag{10.24}$$

Then $\mathcal{G}_T$ is a closed linear subspace of $L^2(P)$.

Proof. Let $X = Y + B$ be the Doob decomposition of X into a P-martingale Y and a process B such that $B_0 = 0$ and $(B_t)_{t=1,\dots,T}$ is predictable. Since

$$\sigma_T^2 = \operatorname{var}(X_T - X_{T-1} \mid \mathcal{F}_{T-1}) = E[\,(Y_T - Y_{T-1})^2 \mid \mathcal{F}_{T-1}\,],$$

we get for $\xi \in \mathcal{S}$,

$$\begin{aligned} E[G_T(\xi)^2\,] &= E[(G_{T-1}(\xi) + \xi_T \cdot (X_T - X_{T-1}))^2\,] \\ &= E[(G_{T-1}(\xi) + \xi_T \cdot (B_T - B_{T-1}))^2\,] + E[(\xi_T \cdot (Y_T - Y_{T-1}))^2\,] \\ &\geq E[\xi_T^2 \cdot \sigma_T^2\,]. \end{aligned} \tag{10.25}$$

Suppose now that ξ^n is a sequence in $\mathcal{S}$ such that $G_T(\xi^n)$ converges in $L^2(P)$ to some random variable Ψ. Applying the inequality (10.25) to $G_T(\xi^n) - G_T(\xi^m) = G_T(\xi^n - \xi^m)$, we find that $\phi^n := \xi_T^n \cdot \sigma_T$ is a Cauchy sequence in $L^2(P)$. Denote $\phi := \lim_n \phi^n$, and let

$$\xi_T := \mathbb{1}_{\{\sigma_T > 0\}} \frac{\phi}{\sigma_T}.$$

By using our assumption (10.24), we obtain

$$\begin{aligned} &E[\,(\xi_T^n \cdot (X_T - X_{T-1}) - \xi_T \cdot (X_T - X_{T-1}))^2\,] \\ &\qquad = E[\,(\xi_T^n - \xi_T)^2 \cdot E[\,(X_T - X_{T-1})^2 \mid \mathcal{F}_{T-1}\,]\,] \\ &\qquad = E[\,(\xi_T^n - \xi_T)^2 \cdot (\sigma_T^2 + \alpha_T^2)\,] \\ &\qquad \leq (1 + \kappa)\,E[\,(\xi_T^n \sigma_T - \xi_T \sigma_T)^2\,] \\ &\qquad = (1 + \kappa)\,E[\,(\phi^n - \phi)^2\,]. \end{aligned}$$

Since the latter term converges to 0, it follows that

$$G_{T-1}(\xi^n) = G_T(\xi^n) - \xi_T^n \cdot (X_T - X_{T-1}) \longrightarrow \Psi - \xi_T \cdot (X_T - X_{T-1})$$

in $L^2(P)$. A backward iteration of this argument yields a predictable process $\xi \in \mathcal{S}$ such that $\Psi = G_T(\xi)$. Hence, $\mathcal{G}_T$ is closed in $L^2(P)$. □

Theorem 10.39. *In dimension $d = 1$, the condition (10.24) of bounded mean-variance trade-off guarantees the existence of a variance-optimal strategy (V_0^*, ξ^*). Such a strategy is P-a. s. unique up to modifications of ξ_t^* on $\{\sigma_t = 0\}$.*

Proof. Let $p : L^2(P) \to \mathscr{G}_T$ denote the orthogonal projection onto the closed subspace $\mathscr{G}_T$ of the Hilbert space $L^2(P)$, that is, $p : L^2(P) \to \mathscr{G}_T$ is a linear operator such that

$$E[\,(\Psi - p(\Psi))^2\,] = \min_{Z \in \mathscr{G}_T} E[\,(\Psi - Z)^2\,] \tag{10.26}$$

for all $\Psi \in L^2(P)$.

For any $V_0 \in \mathbb{R}$, there exists some $\xi(V_0) \in \mathscr{S}$ such that $G_T(\xi(V_0)) = p(H - V_0)$. The identity (10.26) shows that $\xi(V_0)$ minimizes the functional

$$E[\,(H - V_0 - G_T(\xi))^2\,]$$

over all $\xi \in \mathscr{S}$. Note that

$$V_0 \longmapsto G_T(\xi(V_0)) = p(H - V_0) = p(H) - V_0 \cdot p(1)$$

is an affine mapping. Hence,

$$E[\,(H - V_0 - G_T(\xi(V_0)))^2\,]$$

is a quadratic function of V_0 and there exists a minimizer V_0^*. For any $V_0 \in \mathbb{R}$ and $\xi \in \mathscr{S}$, we clearly have

$$\begin{aligned} E[\,(H - V_0 - G_T(\xi))^2\,] &\geq E[\,(H - V_0 - G_T(\xi(V_0)))^2\,] \\ &\geq E[\,(H - V_0^* - G_T(\xi(V_0^*)))^2\,]. \end{aligned}$$

Hence, $(V_0^*, \xi^*) := (V_0^*, \xi(V_0^*))$ is a variance-optimal strategy. Uniqueness follows from (10.25) and an induction argument. □

Under the additional assumption that

$$\frac{\alpha_t^2}{\sigma_t^2} \text{ is deterministic for each } t \tag{10.27}$$

(here we use the convention $\frac{0}{0} := 0$), the variance-optimal strategy (ξ^*, V_0^*) can be determined explicitly. It turns out that (ξ^*, V_0^*) is closely related to the locally risk-minimizing strategy $(\hat{\xi}^0, \hat{\xi})$ for the discounted claim H. Recall from Proposition 10.10 that $(\hat{\xi}^0, \hat{\xi})$ and its value process $\hat{V}$ are determined by the following recursion:

$$\begin{aligned} \hat{V}_T &= H, \\ \hat{\xi}_{t+1} &= \frac{\operatorname{cov}(\hat{V}_{t+1}, X_{t+1} - X_t \mid \mathscr{F}_t)}{\sigma_{t+1}^2} \cdot \mathbb{1}_{\{\sigma_{t+1} \neq 0\}}, \\ \hat{V}_t &= E[\,\hat{V}_{t+1} \mid \mathscr{F}_t\,] - \hat{\xi}_{t+1} \cdot E[\,X_{t+1} - X_t \mid \mathscr{F}_t\,]; \end{aligned}$$

the numéraire component $\hat{\xi}^0$ is given by $\hat{\xi}_t^0 := \hat{V}_t - \hat{\xi}_t \cdot X_t$.

Theorem 10.40. *Under condition* (10.27), $V_0^* := \hat{V}_0$ *and*

$$\xi_t^* := \hat{\xi}_t + \frac{\alpha_t}{\alpha_t^2 + \sigma_t^2}\,(\hat{V}_{t-1} - \hat{V}_0 - G_{t-1}(\xi^*)), \quad t = 1, \dots, T,$$

defines a variance-optimal strategy (ξ^*, V_0^*). *Moreover,*

$$E\big[\,(H - V_0^* - G_T(\xi^*))^2\big] = \sum_{t=1}^{T} \gamma_t \cdot E[\,(\hat{C}_t - \hat{C}_{t-1})^2\,],$$

where $\hat{C}$ *denotes the cost process of* $(\hat{\xi}^0, \hat{\xi})$, *and* γ_t *is given by*

$$\gamma_t := \prod_{k=t+1}^{T} \frac{\sigma_k^2}{\sigma_k^2 + \alpha_k^2}.$$

Proof. We proceed by induction on T. For $T = 1$, the problem is just a particular case of Proposition 10.10, which yields $\xi_1^* = \hat{\xi}_1$ and $V_0^* = \hat{V}_0$.

For $T > 1$, we use the orthogonal decomposition

$$H = \hat{V}_T = \hat{V}_0 + \sum_{t=1}^{T} \hat{\xi}_t \cdot (X_t - X_{t-1}) + \hat{C}_T - \hat{C}_0, \tag{10.28}$$

of the discounted claim H as constructed in Corollary 10.14. Suppose that the assertion is proved for $T - 1$. Let us consider the minimization of

$$\xi_T \mapsto E\big[\,(\hat{V}_T - V_{T-1} - \xi_T \cdot (X_T - X_{T-1}))^2\big], \tag{10.29}$$

where V_{T-1} is any random variable in $\mathscr{L}^2(\Omega, \mathscr{F}_{T-1}, P)$. By (10.28) and Theorem 10.9, we may write $\hat{V}_T$ as

$$\hat{V}_T = \hat{V}_{T-1} + \hat{\xi}_T \cdot (X_T - X_{T-1}) + \hat{C}_T - \hat{C}_{T-1},$$

where $\hat{C}$ is a P-martingale strongly orthogonal to X. Thus,

$$\begin{aligned} &E\big[\,(\hat{V}_T - V_{T-1} - \xi_T \cdot (X_T - X_{T-1}))^2\big] \\ &\quad = E\big[\,(\hat{V}_{T-1} - V_{T-1} + (\hat{\xi}_T - \xi_T) \cdot (X_T - X_{T-1}) + \hat{C}_T - \hat{C}_{T-1})^2\big]. \end{aligned}$$

The expectation conditional on $\mathscr{F}_{T-1}$ of the integrand on the right-hand side is equal to

$$\begin{aligned} &(\hat{V}_{T-1} - V_{T-1})^2 + 2(\hat{V}_{T-1} - V_{T-1})(\hat{\xi}_T - \xi_T) \cdot \alpha_T \\ &\quad + (\hat{\xi}_T - \xi_T)^2(\sigma_T^2 + \alpha_T^2) + E[\,(\hat{C}_T - \hat{C}_{T-1})^2 \mid \mathscr{F}_{T-1}]. \end{aligned} \tag{10.30}$$

This expression is minimized by

$$\xi_T(V_{T-1}) := \hat{\xi}_T + \frac{\alpha_T}{\alpha_T^2 + \sigma_T^2}(\hat{V}_{T-1} - V_{T-1}), \tag{10.31}$$

which must also be the minimizer in (10.29). The minimal value in (10.30) is given by

$$(\hat{V}_{T-1} - V_{T-1})^2 \frac{1}{1 + \alpha_T^2/\sigma_T^2} + E[\,(\hat{C}_T - \hat{C}_{T-1})^2 \mid \mathscr{F}_{T-1}].$$

Using our assumption (10.27) that α_T^2/σ_T^2 is constant, we can compute the expectation of the latter expression, and we arrive at the following identity:

$$\begin{aligned} &E[\,(\hat{V}_T - V_{T-1} - \xi_T(V_{T-1}) \cdot (X_T - X_{T-1}))^2] \\ &\quad = E[\,(\hat{V}_{T-1} - V_{T-1})^2\,] \cdot \frac{\sigma_T^2}{\sigma_T^2 + \alpha_T^2} + E[\,(\hat{C}_T - \hat{C}_{T-1})^2]. \end{aligned} \tag{10.32}$$

So far, we have not specified V_{T-1}. Let us now consider the minimization of

$$E[\,(\hat{V}_T - V_{T-1} - \xi_T(V_{T-1}) \cdot (X_T - X_{T-1}))^2]$$

with respect to V_{T-1}, if V_{T-1} is of the form $V_{T-1} = V_0 + G_{T-1}(\xi)$ for $\xi \in \mathscr{S}$ and $V_0 \in \mathbb{R}$. According to our identity (10.32), this problem is equivalent to the minimization of

$$(V_0, \xi) \longmapsto E[\,(H_{T-1} - V_0 - G_{T-1}(\xi))^2],$$

where $H_{T-1} := \hat{V}_{T-1}$. By the induction hypotheses, this problem is solved by V_0^* and $(\xi_t^*)_{t=1,\dots,T-1}$ as defined in the assertion. Inserting our formula (10.31) for $\xi_T(V_0^* + G_{T-1}(\xi^*))$ completes the induction argument. □

Remark 10.41. The martingale property of $\hat{C}$ implies

$$E[\,(\hat{C}_T - \hat{C}_0)^2\,] = \sum_{t=1}^{T} E[\,(\hat{C}_t - \hat{C}_{t-1})^2\,].$$

If $\hat{C} \not\equiv \hat{C}_0$ and if $\gamma_1 < 1$, then $E[\,(\hat{C}_T - \hat{C}_0)^2\,]$ must be strictly larger than the minimal global risk

$$E[\,(H - (V_0^* + G_T(\xi^*)))^2] = \sum_{t=1}^{T} \gamma_t \cdot E[\,(\hat{C}_t - \hat{C}_{t-1})^2\,].$$ ◇

Remark 10.42. It follows from Theorem 10.40 as well as from the preceding remark that the component ξ^* of a variance-optimal strategy (ξ^*, V_0^*) will differ from the corresponding component $\hat{\xi}$ of a locally risk-minimizing strategy if α_t does not vanish for all t, that is, if P is not a martingale measure. This explains why there may be no strategy that minimizes the remaining conditional risk in the sense of Definition 10.33: For the minimality of

$$R_0^{\text{rem}}(\xi^0, \xi) = E[\,(C_T - C_0)^2\,] = E[\,(H - \xi_0^0 - G_T(\xi))^2\,],$$

we need that $\xi = \xi^*$, while the minimality of

$$R_{T-1}^{\text{rem}}(\xi^0, \xi) = E[\,(C_T - C_{T-1})^2 \mid \mathscr{F}_{T-1}\,] = R_{T-1}^{\text{loc}}(\xi^0, \xi)$$

requires $\xi_T = \hat{\xi}_T$. Hence, the two minimality requirements are in general incompatible. ◇

11 Dynamic risk measures

In this chapter, we return to the quantification of financial risk in terms of monetary risk measures. As we saw in Chapter 4, a monetary risk measure specifies the capital that is needed in order to make a given financial position acceptable. In our present dynamic setting, the capital requirement at a given time will depend on the available information and will thus become a random variable. In Section 11.1, we introduce the notion of a conditional monetary risk measure. In the convex case, we prove a conditional version of the robust representation theorem, which involves the conditional expected losses under various models Q and a conditional penalization of these models.

As time varies, both the capital requirement and the penalization of a given probabilistic model become stochastic processes. The structure of these processes depends on how the risk assessments at different times are connected with each other. In Section 11.2, we focus on a strong notion of time consistency, which amounts to a recursive property of the successive capital requirements. For a typical model Q, time consistency implies that the corresponding penalization process is a supermartingale under Q, and that its Doob decomposition takes a special form. Time consistency is also characterized by a combined supermartingale property of the capital requirements and the penalization process.

11.1 Conditional risk measures and their robust representation

As in the preceding chapters, we fix a filtration $(\mathscr{F}_t)_{t=0,\dots,T}$ on our probability space $(\Omega, \mathscr{F}, P)$ such that $\mathscr{F}_0 = \{\emptyset, \Omega\}$ and $\mathscr{F}_T = \mathscr{F}$. As the set $\mathscr{X}$ of all financial positions, we take the space $L^\infty := L^\infty(\Omega, \mathscr{F}, P)$. The subspace $L_t^\infty := L^\infty(\Omega, \mathscr{F}_t, P)$ consists of those positions whose outcome only depends on the history up to time t. All inequalities and equalities applied to random variables are meant to hold P-a. s. if not stated otherwise.

Definition 11.1. A map $\rho_t : L^\infty \to L_t^\infty$ will be called a *monetary conditional risk measure* if it satisfies the following properties for all $X, Y \in L^\infty$:

- *Conditional cash invariance*: $\rho_t(X + X_t) = \rho_t(X) - X_t$ for any $X_t \in L_t^\infty$.
- *Monotonicity*: $X \le Y \Rightarrow \rho_t(X) \ge \rho_t(Y)$.
- *Normalization:* $\rho_t(0) = 0$.

A monetary conditional risk measure will be called *convex* if it satisfies

- *Conditional convexity*: $\rho_t(\lambda X + (1-\lambda)Y) \le \lambda\rho_t(X) + (1-\lambda)\rho_t(Y)$ for $\lambda \in L_t^\infty$ with $0 \le \lambda \le 1$.

A convex conditional risk measure will be called *coherent* if it satisfies in addition

- *Conditional positive homogeneity*: $\rho_t(\lambda X) = \lambda\rho_t(X)$ for $\lambda \in L_t^\infty$ such that $\lambda \ge 0$. △

For $t = 0$, we recover our previous definition of (unconditional and normalized) monetary, convex, and coherent risk measures given in Section 4.1, since the space L_0^∞ reduces to the real line.

https://doi.org/10.1515/9783111045283-013

To any conditional monetary risk measure ρ_t, we associate its *acceptance set*

$$\mathscr{A}_t := \{X \in L^\infty \mid \rho_t(X) \le 0\}. \tag{11.1}$$

One easily checks that $\mathscr{A}_t$ has the following properties:
- $X \in \mathscr{A}_t, Y \ge X \ \Rightarrow \ Y \in \mathscr{A}_t$,
- $\operatorname{ess\,inf}\{X \in L_t^\infty \mid X \in \mathscr{A}_t\} = 0$, and $0 \in \mathscr{A}_t$.

Note that ρ_t is uniquely determined by its acceptance set since

$$\rho_t(X) = \operatorname{ess\,inf}\{Y \in L_t^\infty \mid X + Y \in \mathscr{A}_t\}. \tag{11.2}$$

A conditional monetary risk measure can thus be viewed as the conditional capital requirement needed at time t to make a financial position X acceptable at that time. Moreover, ρ_t is a convex conditional risk measure if and only if $\mathscr{A}_t$ satisfies the additional property of
- *Conditional convexity:* $X, Y \in \mathscr{A}_t$ and $\lambda \in L_t^\infty$ with $0 \le \lambda \le 1$ implies $\lambda X + (1-\lambda)Y \in \mathscr{A}_t$.

Conversely, one can use acceptance sets to define conditional monetary risk measures: If a given acceptance set $\mathscr{A}_t \subset L^\infty$ satisfies the above conditions then the functional $\rho_t : L^\infty \to L_t^\infty$ defined via (11.2) is a conditional monetary risk measure.

Exercise 11.1.1. Let $\rho_t : L^\infty \to L_t^\infty$ be a conditional monetary risk measure. For $X \in L^\infty$, we define

$$\|X\|_{\mathscr{F}_t} := \operatorname{ess\,inf}\{m_t \in L_t^\infty \mid |X| \le m_t \ P\text{-a. s.}\}.$$

Show that for $X, Y \in L^\infty$,

$$\big|\rho_t(X) - \rho_t(Y)\big| \le \|X - Y\|_{\mathscr{F}_t}.$$

◇

Exercise 11.1.2. Show that every conditional monetary risk measure $\rho_t : L^\infty \to L_t^\infty$ has the following property:

$$\rho_t(\mathbb{1}_A X) = \mathbb{1}_A \rho_t(X) \quad \text{for } X \in L^\infty \text{ and } A \in \mathscr{F}_t. \tag{11.3}$$

Then show that (11.3) is equivalent to the following *local property:* if $X, Y \in L^\infty$ and $A \in \mathscr{F}_t$, then

$$\rho_t(\mathbb{1}_A X + \mathbb{1}_{A^c} Y) = \mathbb{1}_A \rho_t(X) + \mathbb{1}_{A^c} \rho_t(Y).$$

Hint: It can be helpful to use Exercise 11.1.1. ◇

By $\mathscr{M}_1(P)$, we denote the set of all probability measures on $(\Omega, \mathscr{F})$ that are absolutely continuous with respect to P. As we have shown in Theorem 4.33, an unconditional convex risk measure that has the Fatou property admits a *robust representation* of the form

$$\rho(X) = \sup_{Q \in \mathcal{M}_1(P)} (E_Q[-X] - \alpha(Q)) \tag{11.4}$$

with some penalty function $\alpha : \mathcal{M}_1(P) \to \mathbb{R} \cup \{+\infty\}$. In fact, we can take the *minimal penalty function*

$$\alpha^{\min}(Q) = \sup_{X \in L^\infty, \rho(X) \le 0} E_Q[-X]. \tag{11.5}$$

In this section, we are going to prove a robust representation for conditional convex risk measures, which is analogous to (11.4). Here, the penalty function will depend on the history up to time t, and so it will be given by a map α_t from (a subset of) $\mathcal{M}_1(P)$ to the set of $\mathcal{F}_t$-measurable random variables with values in $\mathbb{R} \cup \{+\infty\}$. In analogy to (11.5), we are going to take the minimal penalty function, defined as the worst conditional loss over all positions that are acceptable at time t:

$$\alpha_t^{\min}(Q) = \operatorname*{ess\,sup}_{X \in \mathcal{A}_t} E_Q[-X \mid \mathcal{F}_t]. \tag{11.6}$$

The conditional expectations on the right are a priori defined only Q-a. s., but they also need to be well-defined under the reference measure P. This is the case if Q belongs to the set

$$\mathcal{Q}_t := \{Q \in \mathcal{M}_1(P) \mid Q \approx P \text{ on } \mathcal{F}_t\}.$$

Theorem 11.2. *For a convex conditional risk measure ρ_t, the following properties are equivalent:*

(a) *ρ_t has the representation*

$$\rho_t(X) = \operatorname*{ess\,sup}_{Q \in \mathcal{Q}_t} (E_Q[-X \mid \mathcal{F}_t] - \alpha_t^{\min}(Q)), \tag{11.7}$$

where for $Q \in \mathcal{Q}_t$ the penalty function $\alpha_t^{\min}(Q)$ is given by (11.6)*. In fact, the representation* (11.7) *also holds if we replace $\mathcal{Q}_t$ in* (11.7) *by the smaller set*

$$\mathcal{P}_t := \{Q \in \mathcal{Q}_t \mid Q = P \text{ on } \mathcal{F}_t\}.$$

(b) *ρ_t has the following* Fatou property*:*

$$\rho_t(X) \le \liminf_{n \to \infty} \rho_t(X_n)$$

for any bounded sequence (X_n) in L^∞ that converges P-a. s. to $X \in L^\infty$.

(c) *ρ_t* is continuous from above, *that is,*

$$X_n \searrow X \quad \Longrightarrow \quad \rho_t(X_n) \nearrow \rho_t(X)$$

for any sequence (X_n) in L^∞ and $X \in L^\infty$.

Proof. (a)⇒(b): Lebesgue's dominated convergence theorem for conditional expectations yields

$$E_Q[X_n \mid \mathscr{F}_t] \longrightarrow E_Q[X \mid \mathscr{F}_t] \quad Q\text{-a. s.,}$$

and hence P-a. s., for each $Q \in \mathcal{Q}_t$. Thus, P-a. s.,

$$\begin{aligned} E_Q[-X \mid \mathscr{F}_t] - \alpha_t^{\min}(Q) &\le \liminf_{n\uparrow\infty} \operatorname*{ess\,sup}_{Q\in\mathcal{Q}_t}(E_Q[-X_n \mid \mathscr{F}_t] - \alpha_t^{\min}(Q)) \\ &= \liminf_{n\uparrow\infty} \rho_t(X_n), \end{aligned}$$

and so the representation (11.7) implies $\rho_t(X) \le \liminf_n \rho_t(X_n)$.

(b)⇒(c): The Fatou property yields $\liminf_n \rho_t(X_n) \ge \rho_t(X)$. On the other hand, we have $\limsup_n \rho_t(X_n) \le \rho_t(X)$ by monotonicity, and this implies $\rho_t(X_n) \to \rho_t(X)$.

(c)⇒(a): Since $X + \rho_t(X) \in \mathscr{A}_t$ for any $X \in L^\infty$ by conditional cash invariance, the definition (11.6) of $\alpha_t^{\min}$ implies

$$\alpha_t^{\min}(Q) \ge E_Q[-X \mid \mathscr{F}_t] - \rho_t(X),$$

hence

$$\rho_t(X) \ge E_Q[-X \mid \mathscr{F}_t] - \alpha_t^{\min}(Q)$$

for any $Q \in \mathcal{Q}_t$. This yields the inequality

$$\rho_t(X) \ge \operatorname*{ess\,sup}_{Q\in\mathcal{Q}_t}(E_Q[-X \mid \mathscr{F}_t] - \alpha_t^{\min}(Q)). \tag{11.8}$$

In order to prove the equality in (11.7), both for $\mathcal{Q}_t$ and for the smaller set $\mathscr{P}_t$, it is therefore enough to show that

$$E[\rho_t(X)] \le E\Big[\operatorname*{ess\,sup}_{Q\in\mathscr{P}_t^\rho}(E_Q[-X \mid \mathscr{F}_t] - \alpha_t^{\min}(Q))\Big], \tag{11.9}$$

where

$$\mathscr{P}_t^\rho := \{Q \in \mathscr{P}_t \mid E[\alpha_t^{\min}(Q)] < \infty\} \subseteq \mathscr{P}_t \subseteq \mathcal{Q}_t. \tag{11.10}$$

To this end, note first that

$$E[\alpha_t^{\min}(Q)] = E_Q[\alpha_t^{\min}(Q)] \quad \text{for } Q \in \mathscr{P}_t. \tag{11.11}$$

Next, we consider the map $\rho : L^\infty \to \mathbb{R}$ defined by

$$\rho(X) := E[\rho_t(X)].$$

It is easy to check that ρ is a convex risk measure. Property (c) implies that ρ is continuous from above; equivalently, it has the Fatou property. Thus, Theorem 4.33 states that ρ has the robust representation

$$\rho(X) = \sup_{Q\in\mathcal{M}_1(P)} (E_Q[-X] - \alpha(Q)),$$

where the penalty function $\alpha(Q)$ is given by

$$\alpha(Q) = \sup_{\rho(X)\leq 0} E_Q[-X]. \tag{11.12}$$

Next, we prove that $Q = P$ on $\mathcal{F}_t$, and hence $Q \in \mathcal{P}_t$, if $\alpha(Q) < \infty$. Indeed, take $A \in \mathcal{F}_t$ and $\lambda > 0$. Then

$$-\lambda P[A] = E[\rho_t(\lambda \mathbb{1}_A)] = \rho(\lambda \mathbb{1}_A) \geq E_Q[-\lambda \mathbb{1}_A] - \alpha(Q),$$

hence

$$P[A] \leq Q[A] + \frac{1}{\lambda}\alpha(Q)$$

for any $\lambda > 0$. Thus, $\alpha(Q) < \infty$ implies $P[A] \leq Q[A]$ for any $A \in \mathcal{F}_t$, and hence $P = Q$ on $\mathcal{F}_t$.

Furthermore,

$$E[\alpha_t^{\min}(Q)] \leq \alpha(Q) \tag{11.13}$$

holds for every $Q \in \mathcal{P}_t$. Indeed,

$$E_Q[\alpha_t^{\min}(Q)] = \sup_{Y\in\mathcal{A}_t} E_Q[-Y]$$

holds by equation (11.17) in Lemma 11.3 below, and so (11.12) and (11.11) imply (11.13) since $\rho(X) \leq 0$ for all $X \in \mathcal{A}_t$.

Thus, we have shown

$$\alpha(Q) < \infty \quad \Longrightarrow \quad Q \in \mathcal{P}_t^{\rho}. \tag{11.14}$$

We can now conclude the proof of inequality (11.9) as follows, using first (11.14) and then (11.13):

$$\begin{aligned} E[\rho_t(X)] &= \rho(X) \\ &= \sup_{Q\in\mathcal{M}_1(P)} (E_Q[-X] - \alpha(Q)) \\ &= \sup_{Q\in\mathcal{P}_t^{\rho}} (E_Q[-X] - \alpha(Q)) \end{aligned} \tag{11.15}$$

$$\leq \sup_{Q \in \mathscr{P}_t^\rho} E[\, E_Q[\, -X \mid \mathscr{F}_t\,] - \alpha_t^{\min}(Q)\,]$$
$$\leq E\Big[\, \operatorname*{ess\,sup}_{Q \in \mathscr{P}_t^\rho} (E_Q[\, -X \mid \mathscr{F}_t\,] - \alpha_t^{\min}(Q))\,\Big]. \qquad \square$$

The following lemma was used in the preceding proof.

Lemma 11.3. *For $Q \in \mathcal{Q}_t$ and $0 \leq s \leq t$,*

$$E_Q[\, \alpha_t^{\min}(Q) \mid \mathscr{F}_s\,] = \operatorname*{ess\,sup}_{Y \in \mathscr{A}_t} E_Q[\, -Y \mid \mathscr{F}_s\,], \tag{11.16}$$

and in particular

$$E_Q[\, \alpha_t^{\min}(Q)\,] = \sup_{Y \in \mathscr{A}_t} E_Q[\, -Y\,]. \tag{11.17}$$

Proof. First, we show that the family

$$\{E_Q[\, -X \mid \mathscr{F}_t\,] \mid X \in \mathscr{A}_t\}$$

is directed upward for any $Q \in \mathcal{Q}_t$ (see Appendix F). Indeed, for $X, Y \in \mathscr{A}_t$ we can take $Z := X\mathbb{1}_A + Y\mathbb{1}_{A^c}$, where $A := \{E_Q[\, -X \mid \mathscr{F}_t\,] \geq E_Q[\, -Y \mid \mathscr{F}_t\,]\} \in \mathscr{F}_t$. Conditional convexity of ρ_t implies $Z \in \mathscr{A}_t$, and clearly we have

$$E_Q[\, -Z \mid \mathscr{F}_t\,] = E_Q[\, -X \mid \mathscr{F}_t\,] \vee E_Q[\, -Y \mid \mathscr{F}_t\,].$$

Thus the family is directed upward. Hence, Theorem F.1 implies that its essential supremum can be computed as the supremum of an increasing sequence within this family, that is, there exists a sequence (X_n^Q) in $\mathscr{A}_t$ such that

$$\alpha_t^{\min}(Q) = \lim_{n \uparrow \infty} E_Q[\, -X_n^Q \mid \mathscr{F}_t\,].$$

By monotone convergence, we get

$$\begin{aligned} E_Q[\, \alpha_t^{\min}(Q) \mid \mathscr{F}_s\,] &= \lim_{n \uparrow \infty} E_Q[\, E_Q[\, -X_n^Q \mid \mathscr{F}_t\,] \mid \mathscr{F}_s\,] \\ &= \lim_{n \uparrow \infty} E_Q[\, -X_n^Q \mid \mathscr{F}_s\,] \leq \operatorname*{ess\,sup}_{Y \in \mathscr{A}_t} E_Q[\, -Y \mid \mathscr{F}_s\,]. \end{aligned}$$

The converse inequality follows directly from the definition of $\alpha_t^{\min}(Q)$. This shows (11.16), and for $s = 0$ we get (11.17). $\square$

Remark 11.4. The inequalities (11.8) and (11.9) in the proof of the implication (c)⇒(a) in Theorem 11.2 show that the representation (11.7) also holds if we replace $\mathcal{Q}_t$ in (11.7) not only by $\mathscr{P}_t$ but by the even smaller set

$$\mathscr{P}_t^\rho = \{Q \in \mathscr{P}_t \mid E[\,\alpha_t^{\min}(Q)\,] < \infty\}. \tag{11.18}$$

In view of (11.8) and (11.9), we can conclude that the representation (11.7) holds for any set $\mathcal{Q}$ such that $\mathscr{P}_t^\rho \subseteq \mathcal{Q} \subseteq \mathcal{Q}_t$. ◇

Example 11.5. Suppose that preferences are characterized by an exponential utility function $u(x) = 1 - \exp(-\beta x)$ with $\beta > 0$. At time t, the conditional expected utility of a financial position $X \in L^\infty$ is then given by the $\mathscr{F}_t$-measurable random variable

$$U_t(X) = E[\,1 - e^{-\beta X} \mid \mathscr{F}_t\,].$$

The set

$$\mathscr{A}_t := \{X \in L^\infty \mid U_t(X) \geq U_t(0)\} = \{X \in L^\infty \mid E[\,e^{-\beta X} \mid \mathscr{F}_t\,] \leq 1\}$$

satisfies the conditions required from an acceptance set. In analogy to Example 4.34, the induced convex conditional risk measure ρ_t is given by

$$\begin{aligned}\rho_t(X) &= \operatorname{ess\,inf}\{\,Y \in L_t^\infty \mid X + Y \in \mathscr{A}_t\,\}\\ &= \operatorname{ess\,inf}\{\,Y \in L_t^\infty \mid E[\,e^{-\beta X} \mid \mathscr{F}_t\,] \leq e^{\beta Y}\,\},\end{aligned}$$

that is,

$$\rho_t(X) = \frac{1}{\beta}\log E[\,e^{-\beta X} \mid \mathscr{F}_t\,].$$

Clearly, ρ_t is a monetary conditional risk measure. It will be shown in Exercise 11.1.3 that it is also convex and admits a robust representation with minimal penalty function

$$\alpha_t^{\min}(Q) = \frac{1}{\beta}\widehat{H}_t(Q|P), \quad Q \in \mathscr{P}_t, \tag{11.19}$$

where

$$\widehat{H}_t(Q|P) := E_Q\Big[\,\log\frac{dQ}{dP}\,\Big|\,\mathscr{F}_t\,\Big] = E\Big[\,\frac{dQ}{dP}\log\frac{dQ}{dP}\,\Big|\,\mathscr{F}_t\,\Big]. \tag{11.20}$$

denotes the *conditional relative entropy* of $Q \in \mathscr{P}_t$ with respect to P, given $\mathscr{F}_t$. For this reason, ρ_t is called the *conditional entropic risk measure.* ◇

Exercise 11.1.3. Argue as in Theorem C.5 to show that for any $Q \in \mathscr{P}_t$, the conditional relative entropy (11.20) satisfies the following variational identity:

$$\begin{aligned}\widehat{H}_t(Q|P) &= \sup_{Z\in L^\infty}(E_Q[\,Z \mid \mathscr{F}_t\,] - \log E[\,e^Z \mid \mathscr{F}_t\,])\\ &= \sup\{E_Q[\,Z \mid \mathscr{F}_t\,] - \log E[\,e^Z \mid \mathscr{F}_t\,] \mid e^Z \in L^1(P)\},\end{aligned}$$

where the second supremum is attained by $Z := \log \frac{dQ}{dP}$. Then proceed as in Example 4.34 to conclude that the conditional entropic risk measure is convex and that (11.19) describes indeed its minimal penalty function for $Q \in \mathcal{P}_t$. ◇

In the *coherent* case, we obtain the following representation result.

Corollary 11.6. *Let ρ_t be a coherent conditional risk measure that has the Fatou property. Then ρ_t has the representation*

$$\rho_t(X) = \operatorname*{ess\,sup}_{Q \in \mathcal{P}_t^\rho} E_Q[\,-X \mid \mathcal{F}_t\,], \tag{11.21}$$

where the set $\mathcal{P}_t^\rho$ is given by

$$\mathcal{P}_t^\rho = \{Q \in \mathcal{P}_t \mid \alpha_t^{\min}(Q) = 0 \ P\text{-a. s.}\}$$

and coincides with the set in (11.18).

Proof. In the coherent case, the penalty function $\alpha_t^{\min}(Q)$ can only take the values 0 or ∞ for $Q \in \mathcal{Q}_t$. Indeed, for $A := \{\alpha_t^{\min}(Q) > 0\}$, $X \in \mathcal{A}_t$, and any $\lambda > 0$ we have $\lambda \mathbb{1}_A X \in \mathcal{A}_t$ due to conditional positive homogeneity of ρ_t, hence

$$\begin{aligned}
\alpha_t^{\min}(Q) &= \operatorname*{ess\,sup}_{X \in \mathcal{A}_t} E_Q[\,-X \mid \mathcal{F}_t\,] \\
&\geq \operatorname*{ess\,sup}_{X \in \mathcal{A}_t} E_Q[\,-\lambda \mathbb{1}_A X \mid \mathcal{F}_t\,] \\
&= \lambda \mathbb{1}_A \alpha_t^{\min}(Q),
\end{aligned}$$

and the lower bound converges to infinity on A as $\lambda \uparrow \infty$. Thus, $\alpha_t^{\min}(Q) = \infty$ on A. Next, let $\mathcal{P}_t^\rho$ be as in (11.18). Then $E_P[\,\alpha_t^{\min}(Q)\,] < \infty$ for $Q \in \mathcal{P}_t^\rho$, and this implies $P[\,A\,] = 0$. Thus $\mathcal{P}_t^\rho$ coincides with the set of all $Q \in \mathcal{Q}_t$ for which $\alpha_t^{\min}(Q) = 0$ P-a. s., and so the representation (11.7), where we are free to take $\mathcal{P}_t^\rho$ instead of $\mathcal{Q}_t$ by Remark 11.4, reduces to (11.21). □

Example 11.7. For any $\lambda \in (0, 1)$, the acceptance set

$$\mathcal{A}_t = \{X \in L^\infty \mid P[\,X < 0 \mid \mathcal{F}_t\,] \geq \lambda\}$$

defines a monetary conditional risk measure, which we call *conditional Value at Risk at level λ*:

$$V@R_\lambda(X|\mathcal{F}_t) = \operatorname{ess\,inf}\{m_t \in L_t^\infty \mid P[\,X + m_t < 0 \mid \mathcal{F}_t\,] \leq \lambda\}. \tag{11.22}$$

The conditional risk measure $\mathrm{V@R}_\lambda(\,\cdot \mid \mathcal{F}_t)$ is conditionally positively homogenous, but it is not conditionally convex. While the terminology “conditional Value at Risk” seems quite natural in our present dynamic setting, one should be aware of a possible confusion because the *unconditional* risk measure AV@R, which was introduced in Definition 4.48, is sometimes also called Conditional Value at Risk and denoted by *CV@R*. ◇

Definition 11.8. For $\lambda \in (0,1]$ let $\mathcal{Q}_t^\lambda$ denote the set of all measures $Q \in \mathcal{P}_t$ whose density dQ/dP is P-a. s. bounded by $1/\lambda$. The resulting coherent conditional risk measure

$$\mathrm{AV@R}_\lambda(X|\mathcal{F}_t) = \operatorname*{ess\,sup}_{Q\in\mathcal{Q}_t^\lambda} E_Q[\,-X \mid \mathcal{F}_t\,] \tag{11.23}$$

is called *conditional Average Value at Risk at level* λ. △

We have the following representation result for the coherent conditional risk measure $\mathrm{AV@R}_\lambda(\,\cdot \mid \mathcal{F}_t)$.

Proposition 11.9. *For $X \in L^\infty$, the essential supremum in* (11.23) *is attained for any $Q^X \in \mathcal{Q}_t^\lambda$ whose density satisfies*

$$\frac{dQ^X}{dP} = \begin{cases} \dfrac{1}{\lambda} & \text{on } \{-X > \mathrm{V@R}_\lambda(X \mid \mathcal{F}_t)\}, \\ 0 & \text{on } \{-X < \mathrm{V@R}_\lambda(X \mid \mathcal{F}_t)\}, \end{cases} \tag{11.24}$$

and the set of such maximizers is nonempty. In particular,

$$\mathrm{AV@R}_\lambda(X|\mathcal{F}_t) = \frac{1}{\lambda}E[\,(X + \mathrm{V@R}_\lambda(X|\mathcal{F}_t))^-\,|\mathcal{F}_t] + \mathrm{V@R}_\lambda(X|\mathcal{F}_t). \tag{11.25}$$

Proof. Fix $X \in L^\infty$ and note that

$$\begin{aligned} P[\,-X > \mathrm{V@R}_\lambda(X|\mathcal{F}_t) \mid \mathcal{F}_t\,] &= P[\,X + \mathrm{V@R}_\lambda(X|\mathcal{F}_t) < 0 \mid \mathcal{F}_t\,] \\ &= \lim_{n\uparrow\infty} P\Big[\,X + \mathrm{V@R}_\lambda(X|\mathcal{F}_t) + \frac{1}{n} < 0 \,\Big|\, \mathcal{F}_t\Big] \\ &\le \lambda, \end{aligned} \tag{11.26}$$

where we have used (11.22) in the last step. On the other hand, (11.22) also implies that

$$P\Big[\,X + \mathrm{V@R}_\lambda(X|\mathcal{F}_t) - \frac{1}{n} < 0 \,\Big|\, \mathcal{F}_t\Big] > \lambda$$

for each n, and hence that

$$\begin{aligned} P[\,-X \ge \mathrm{V@R}_\lambda(X|\mathcal{F}_t) \mid \mathcal{F}_t\,] &= P[\,X + \mathrm{V@R}_\lambda(X|\mathcal{F}_t) \le 0 \mid \mathcal{F}_t\,] \\ &= \lim_{n\uparrow\infty} P\Big[\,X + \mathrm{V@R}_\lambda(X|\mathcal{F}_t) - \frac{1}{n} < 0 \,\Big|\, \mathcal{F}_t\Big] \\ &\ge \lambda. \end{aligned} \tag{11.27}$$

The two inequalities (11.26) and (11.27) imply that the set of measures $Q^X \in \mathcal{Q}_t^\lambda$ satisfying (11.24) is nonempty. Now let $Z^X = dQ^X/dP$ be the density of such a measure, and let Z be the density of any other $Q \in \mathcal{Q}_t^\lambda$. We furthermore define $\widetilde{X} := X + \mathrm{V@R}_\lambda(X|\mathcal{F}_t)$. Then $\mathrm{V@R}_\lambda(\widetilde{X}|\mathcal{F}_t) = 0$, and arguing as in the proof of the general Neyman–Pearson lemma in the form of Theorem E.3,

$$(-\widetilde{X})Z^X - (-\widetilde{X})Z = (-\widetilde{X})\left(\frac{1}{\lambda} - Z\right)\mathbb{1}_{\{\widetilde{X}<0\}} + (-\widetilde{X})(-Z)\mathbb{1}_{\{\widetilde{X}>0\}} \geq 0,$$

with P-a. s. equality if and only if Z is also of the form (11.24). The conditional cash invariance of $\mathrm{AV@R}_\lambda(\cdot|\mathcal{F}_t)$ thus yields the first part of the assertion. The identity (11.25) now follows from the following chain of identities:

$$\begin{aligned}\mathrm{AV@R}_\lambda(X|\mathcal{F}_t) - \mathrm{V@R}_\lambda(X|\mathcal{F}_t) &= \mathrm{AV@R}_\lambda(\widetilde{X}|\mathcal{F}_t)\\ &= E[\,(-\widetilde{X})Z^X|\mathcal{F}_t\,]\\ &= \frac{1}{\lambda}E[\,(\widetilde{X})^-|\mathcal{F}_t\,].\end{aligned}$$

□

11.2 Time consistency

In this section, we consider a sequence of convex conditional risk measures

$$\rho_t : L^\infty \to L_t^\infty, \quad t = 0, 1, \ldots, T.$$

In such a dynamic setting, the key question is how the risk assessments of a financial position at different times are connected to each other.

Definition 11.10. A sequence of conditional risk measures $(\rho_t)_{t=0,1,\ldots,T}$ is called *(strongly) time-consistent* if for any $X, Y \in L^\infty$ and for all $t \geq 0$ the following condition holds:

$$\rho_{t+1}(X) \leq \rho_{t+1}(Y) \quad \Longrightarrow \quad \rho_t(X) \leq \rho_t(Y).$$

△

Lemma 11.11. *Time consistency is equivalent to each of the following two properties:*
(a) $\rho_{t+1}(X) = \rho_{t+1}(Y) \quad \Longrightarrow \quad \rho_t(X) = \rho_t(Y)$ *for* $t = 0, 1, \ldots, T-1$.
(b) Recursiveness*:* $\rho_t = \rho_t(-\rho_{t+1})$ *for* $t = 0, 1, \ldots, T-1$.

Proof. Time consistency clearly implies (a).

(a)⇒(b): Note that $\rho_{t+1}(-\rho_{t+1}(X)) = \rho_{t+1}(X)$, due to conditional cash invariance and normalization. Applying (a) with $Y := -\rho_{t+1}(X)$, we thus obtain $\rho_t(X) = \rho_t(-\rho_{t+1}(X))$.

(b) implies time consistency: If $\rho_{t+1}(X) \leq \rho_{t+1}(Y)$, then $\rho_t(-\rho_{t+1}(X)) \leq \rho_t(-\rho_{t+1}(Y))$ by monotonicity, and so $\rho_t(X) \leq \rho_t(Y)$ follows from (b). □

Example 11.12. For fixed $\beta > 0$, the sequence of *conditional entropic risk measures*

$$\rho_t(X) = \frac{1}{\beta} \log E[\, e^{-\beta X} \mid \mathscr{F}_t\,], \quad t = 0, \dots, T,$$

is time-consistent. Let us check recursiveness:

$$\begin{aligned}\rho_t(-\rho_{t+1}(X)) &= \frac{1}{\beta} \log E\Big[\, \exp\Big(-\beta\Big(-\frac{1}{\beta} \log E[\, \exp(-\beta X) \mid \mathscr{F}_{t+1}\,]\Big)\Big) \,\Big|\, \mathscr{F}_t\,\Big] \\ &= \frac{1}{\beta} \log E[\, E[\, e^{-\beta X} \mid \mathscr{F}_{t+1}\,] \mid \mathscr{F}_t\,] \\ &= \rho_t(X).\end{aligned}$$

Note, however, that time consistency will be lost if the constant risk aversion parameter β is replaced by an adapted process (β_t). Moreover, under a suitable condition of dynamic law invariance, the conditional entropic risk measure is in fact the only time-consistent dynamic convex risk measure; see [229], where this result is reduced to an application of Proposition 2.46. Here, we include the ordinary conditional expectation with respect to P as the limiting case $\beta = 0$. ◇

Example 11.13. The sequence of coherent conditional risk measures

$$\rho_t(X) = \mathrm{AV@R}_\lambda(X|\mathscr{F}_t), \quad t = 0, \dots, T,$$

given by *conditional Average Value at Risk* at some level $\lambda \in (0, 1)$ is *not* time-consistent, and the same is true for *conditional Value at Risk* and for conditional mean-standard deviation risk measures defined in terms of *Sharpe ratios*. To see this, note first that all these risk measures are well-defined on $L^2(P)$, and that they are of the form

$$\rho_t(X) = -E[\,-X \mid \mathscr{F}_t\,] + \gamma \sqrt{E[\,(X - E[\,X \mid \mathscr{F}_t\,])^2\,]}$$

if X has a conditional Gaussian distribution with respect to $\mathscr{F}_t$ (compare Examples 4.11, 4.12, and Exercise 4.5.3(b)). Now consider a position of the form $X = X_1 + X_2$ for two independent Gaussian random variables X_i with distribution $N(0, \sigma_i^2)$ and assume that $\mathscr{F}_1$ is the σ-field generated by X_1. Then

$$\rho_1(X) = -X_1 + \rho_1(X_2) = -X_1 + \gamma\sigma_2,$$

hence

$$\rho_0(-\rho_1(X)) = \rho_0(X_1) + \gamma\sigma_2 = \gamma(\sigma_1 + \sigma_2).$$

On the other hand,

$$\rho_0(X) = \gamma(\sigma_1^2 + \sigma_2^2)^{\frac{1}{2}},$$

and so we have $\rho_0(-\rho_1(X)) > \rho_0(X)$ unless X_1 or X_2 are constant and, therefore, $\sigma_1 = 0$ or $\sigma_2 = 0$. Note that the preceding argument works also if it is possible to represent ρ_t in the form

$$\rho_t(X) = \operatorname*{ess\,sup}_{\mu \in \mathscr{M}} \int_{(0,1]} \mathrm{AV@R}_\lambda(X|\mathscr{F}_t)\,\mu(d\lambda)$$

for some set $\mathscr{M}$ of Borel probability measures on $(0,1]$. Such a representation would extend the representation of law-invariant coherent risk measures given in Corollary 4.69 to the conditional setting. ◇

Exercise 11.2.1. Show that recursiveness as defined in part (b) of Lemma 11.11 is equivalent to the following condition: for $0 \le s < t \le T$ and $X \in L^\infty$,

$$\rho_s(X) = \rho_s(-\rho_t(X)).$$ ◇

The next exercise relies on and generalizes the preceding one.

Exercise 11.2.2. For a stopping time $\tau : \Omega \to \{0, \dots, T\}$, we define

$$\rho_\tau(X) := \sum_{t=0}^{T} \mathbb{1}_{\{\tau=t\}}\rho_t(X).$$

Show that recursiveness is equivalent to the following property: if σ is another stopping time such that $\sigma \le \tau$, then

$$\rho_\sigma(X) = \rho_\sigma(-\rho_\tau(X)).$$

Hint: Use Exercise 11.2.1 and (11.3). ◇

Exercise 11.2.3. Let ρ_t, $t = 0, \dots, T$, be any sequence of convex conditional risk measures. Show that the recursive definition

$$\widetilde{\rho}_T := \rho_T \quad \text{and} \quad \widetilde{\rho}_t := \rho_t(-\widetilde{\rho}_{t+1}) \quad \text{for } t = 0, \dots, T-1,$$

yields a time-consistent sequence of convex conditional risk measures. ◇

We are now going to characterize time consistency of the sequence $(\rho_t)_{t=0,1,\dots,T}$ both in terms of the corresponding acceptance sets $(\mathscr{A}_t)_{t=0,1,\dots,T}$ and in terms of the penalty processes $(\alpha_t^{\min}(Q))_{t=0,1,\dots,T}$. Our main goal is to prove a supermartingale criterion, which characterizes time consistency in terms of the joint behavior of the stochastic processes $(\rho_t(X))_{t=0,1,\dots,T}$ and $(\alpha_t(Q))_{t=0,1,\dots,T}$.

To this end, we introduce some notation. Suppose that we look just one step ahead and assess the risk only for those positions whose outcome will be known by the end of the next period. This means that we restrict the conditional convex risk measure ρ_t to the space L^∞_{t+1}. The corresponding one-step acceptance set is given by

$$\mathscr{A}_{t,t+1} := \{ X \in L^\infty_{t+1} \mid \rho_t(X) \le 0 \},$$

and

$$\alpha^{\min}_{t,t+1}(Q) := \operatorname*{ess\,sup}_{X \in \mathscr{A}_{t,t+1}} E_Q[\, -X \mid \mathscr{F}_t \,]$$

is the resulting one-step penalty function.

The following lemma holds for any sequence of monetary conditional risk measures. The equivalences between set inclusions for the acceptance sets and inequalities for the risk measures can be used as starting points for various departures from the strong notion of time consistency we are considering here; see, for example, [275, 315], and [320]. For two sets $\mathscr{A}, \mathscr{B} \subseteq L^\infty$, we will use the following notation:

$$\mathscr{A} + \mathscr{B} := \{X + Y \mid X \in \mathscr{A},\, Y \in \mathscr{B}\}.$$

Lemma 11.14. *Let $(\rho_t)_{t=0,1,\dots,T}$ be a sequence of monetary conditional risk measures. Then the following equivalences hold for all $t = 0, \dots, T-1$ and all $X \in L^\infty$:*

(a) $X \in \mathscr{A}_{t,t+1} + \mathscr{A}_{t+1} \iff -\rho_{t+1}(X) \in \mathscr{A}_{t,t+1}$.

(b) $\mathscr{A}_t \subseteq \mathscr{A}_{t,t+1} + \mathscr{A}_{t+1} \iff \rho_t(-\rho_{t+1}) \le \rho_t$.

(c) $\mathscr{A}_t \supseteq \mathscr{A}_{t,t+1} + \mathscr{A}_{t+1} \iff \rho_t(-\rho_{t+1}) \ge \rho_t$.

Proof. (a) To prove "⇒," take $X = X_{t,t+1} + X_{t+1}$ with $X_{t,t+1} \in \mathscr{A}_{t,t+1}$ and $X_{t+1} \in \mathscr{A}_{t+1}$. Then

$$\rho_{t+1}(X) = \rho_{t+1}(X_{t+1}) - X_{t,t+1} \le -X_{t,t+1}$$

by cash invariance, and monotonicity implies

$$\rho_t(-\rho_{t+1}(X)) \le \rho_t(X_{t,t+1}) \le 0,$$

hence $-\rho_{t+1}(X) \in \mathscr{A}_{t+1}$. The converse implication follows immediately from the decomposition $X = X + \rho_{t+1}(X) - \rho_{t+1}(X)$, since $X + \rho_{t+1}(X) \in \mathscr{A}_{t+1}$ for all $X \in L^\infty$ and $-\rho_{t+1}(X) \in \mathscr{A}_{t,t+1}$ by assumption.

(b) In order to show "⇒," take $X \in L^\infty$. Since $X + \rho_t(X) \in \mathscr{A}_t \subseteq \mathscr{A}_{t,t+1} + \mathscr{A}_{t+1}$, we obtain

$$-\rho_{t+1}(X) + \rho_t(X) = -\rho_{t+1}(X + \rho_t(X)) \in \mathscr{A}_{t,t+1},$$

by (a) and by cash invariance. This implies

$$\rho_t(-\rho_{t+1}(X)) - \rho_t(X) = \rho_t(-(\rho_{t+1}(X) - \rho_t(X))) \le 0.$$

To prove "⇐," take $X \in \mathscr{A}_t$. Then $-\rho_{t+1}(X) \in \mathscr{A}_{t,t+1}$ by the right-hand side of (b), and hence $X \in \mathscr{A}_{t,t+1} + \mathscr{A}_{t+1}$ by (a).

(c) Take $X \in L^\infty$ and assume $\mathscr{A}_t \supseteq \mathscr{A}_{t,t+1} + \mathscr{A}_{t+1}$. Then

$$\rho_t(-\rho_{t+1}(X)) + X = \rho_t(-\rho_{t+1}(X)) - \rho_{t+1}(X) + \rho_{t+1}(X) + X \in \mathscr{A}_{t,t+1} + \mathscr{A}_{t+1}$$

belongs to $\mathscr{A}_t$. This implies

$$\rho_t(X) - \rho_t(-\rho_{t+1}(X)) = \rho_t(X + \rho_t(-\rho_{t+1}(X))) \leq 0$$

by cash invariance, and so we have shown "⇒." For the converse implication, take $X \in \mathscr{A}_{t,t+1} + \mathscr{A}_{t+1}$. Since $-\rho_{t+1}(X) \in \mathscr{A}_{t,t+1}$ by (a), we obtain

$$\rho_t(X) \leq \rho_t(-\rho_{t+1}(X)) \leq 0,$$

hence $X \in \mathscr{A}_t$. □

The preceding lemma implies immediately the following result.

Proposition 11.15. *Let $(\rho_t)_{t=0,1,\dots,T}$ be a sequence of convex conditional risk measures such that each ρ_t has the Fatou property. Then the following conditions are equivalent:*
(a) *$(\rho_t)_{t=0,1,\dots,T}$ is time-consistent.*
(b) *$\mathscr{A}_t = \mathscr{A}_{t,t+1} + \mathscr{A}_{t+1}$ for $t = 0, \dots, T-1$.*

In the sequel, we will investigate the time consistency of a sequence $(\rho_t)_{t=0,1,\dots,T}$ of convex conditional risk measures in terms of the dynamics of their penalty functions. In doing this, we need to assume that every element of the sequence can be represented in terms of the *same* set $\mathcal{Q}$ of probability measures:

$$\rho_t(X) = \operatorname*{ess\,sup}_{Q\in\mathcal{Q}}(E_Q[\,-X \mid \mathscr{F}_t\,] - \alpha_t^{\min}(Q)), \quad \text{for } t = 0, \dots, T. \tag{11.28}$$

These representations are only well-defined if $\mathcal{Q} \subseteq \mathcal{Q}_t$ for all t. Since $T < \infty$, every $Q \in \mathcal{Q}$ must be equivalent to P on $\mathscr{F}_T = \mathscr{F}$, and so we may as well assume

$$\mathcal{Q} = \{Q \in \mathscr{M}_1(P) \mid Q \approx P\}. \tag{11.29}$$

Definition 11.16. A sequence $(\rho_t)_{t=0,1,\dots,T}$ of convex conditional risk measures is called *sensitive* if the representation (11.28) holds in terms of the set $\mathcal{Q}$ in (11.29). △

Sensitivity is also called *relevance*. It generalizes the concept of sensitivity introduced in Section 4.4 to the case of conditional risk measures, and it can be characterized along the lines of Theorem 4.43. In fact, it can be shown that a time-consistent sequence $(\rho_t)_{t=0,1,\dots,T}$ of conditional convex risk measures is sensitive as soon as ρ_0 is sensitive in the sense of Definition 4.42; see [145].

The following theorem, and in particular the equivalence of its conditions (a) and (c), is the main result of this section.

Theorem 11.17. *Let $(\rho_t)_{t=0,1,\dots,T}$ be a sensitive sequence of convex conditional risk measures. Then the following conditions are equivalent:*

(a) *$(\rho_t)_{t=0,1,\dots,T}$ is time-consistent.*

(b) *For any $Q \in \mathcal{Q}$,*

$$\alpha_t^{\min}(Q) = \alpha_{t,t+1}^{\min}(Q) + E_Q[\,\alpha_{t+1}^{\min}(Q) \mid \mathcal{F}_t\,] \quad \text{for } t = 0,1,\dots,T-1.$$

(c) *For any $Q \in \mathcal{Q}$ and any $X \in L^\infty$, the process*

$$U_t^{Q,X} := \rho_t(X) + \alpha_t^{\min}(Q), \quad t = 0,1,\dots,T,$$

satisfies

$$E_Q[\,U_{t+1}^{Q,X} \mid \mathcal{F}_t\,] \leq U_t^{Q,X} \quad Q\text{-a. s.},$$

and is a Q-supermartingale for any Q with $\alpha_0^{\min}(Q) < \infty$.

We prepare the proof with the following two lemmas. The first lemma deals with the pasting of two probability measures Q_1 and Q_2 in the deterministic time t. It is a special case of the results in Section 6.4, but we state and prove it here to keep the discussion self-contained.

Lemma 11.18. *For $Q_1, Q_2 \in \mathcal{Q}$, and $t \in \{0,\dots,T\}$ there exists $Q \in \mathcal{Q}$ such that $Q = Q_1$ on $\mathcal{F}_t$ and*

$$E_Q[\,Y \mid \mathcal{F}_t\,] = E_{Q_2}[\,Y \mid \mathcal{F}_t\,] \quad P\text{-a. s. for all } Y \in L^\infty. \tag{11.30}$$

In particular, we have

$$\alpha_t^{\min}(Q) = \alpha_t^{\min}(Q_2). \tag{11.31}$$

Proof. When setting

$$Z := \left(\frac{dQ_2}{dQ_1}\bigg|_{\mathcal{F}_t} \right)^{-1} \frac{dQ_2}{dQ_1}$$

then $E_{Q_1}[\,Z \mid \mathcal{F}_t\,] = 1$, and so $dQ = Z\,dQ_1$ defines a measure $Q \in \mathcal{Q}$ that coincides with Q_1 on $\mathcal{F}_t$. Moreover, Proposition B.8 yields (11.30). The identity (11.31) now follows from the definition of $\alpha_t^{\min}$. □

Lemma 11.19. *The family*

$$\{E_Q[\,-X \mid \mathcal{F}_t\,] - \alpha_t^{\min}(Q) \mid Q \in \mathcal{Q}\}$$

is directed upward.

Proof. By applying Lemma 11.18 with $Q_1 := P$, we see that it is enough to show that the family

$$\{E_Q[-X \mid \mathscr{F}_t] - \alpha_t^{\min}(Q) \mid Q \in \mathcal{Q} \cap \mathscr{P}_t\}$$

is directed upward, where $\mathscr{P}_t$ denotes again the set of measures Q that coincide with P on $\mathscr{F}_t$. To this end, fix $Q_1, Q_2 \in \mathcal{Q} \cap \mathscr{P}_t$ and let $Y_i := E_{Q_i}[-X \mid \mathscr{F}_t] - \alpha_t^{\min}(Q_i)$ and $A := \{Y_2 \geq Y_1\}$. We then define

$$Z := \frac{dQ_2}{dP} \cdot \mathbb{1}_A + \frac{dQ_1}{dP} \cdot \mathbb{1}_{A^c}.$$

This random variable is P-a. s. strictly positive and satisfies

$$E[Z \mid \mathscr{F}_t] = \mathbb{1}_A E\left[\frac{dQ_2}{dP} \,\middle|\, \mathscr{F}_t\right] + \mathbb{1}_{A^c} E\left[\frac{dQ_1}{dP} \,\middle|\, \mathscr{F}_t\right] = \mathbb{1}_A + \mathbb{1}_{A^c} = 1.$$

It follows that Z is the density of a probability measure $Q \in \mathcal{Q} \cap \mathscr{P}_t$. Under Q, we have

$$E_Q[-X \mid \mathscr{F}_t] = \mathbb{1}_A E_{Q_2}[-X \mid \mathscr{F}_t] + \mathbb{1}_{A^c} E_{Q_1}[-X \mid \mathscr{F}_t].$$

Similarly,

$$\alpha_t^{\min}(Q) = \mathbb{1}_A \alpha_t^{\min}(Q_2) + \mathbb{1}_{A^c} \alpha_t^{\min}(Q_1).$$

Therefore,

$$E_Q[-X \mid \mathscr{F}_t] - \alpha_t^{\min}(Q) = \mathbb{1}_A Y_2 + \mathbb{1}_{A^c} Y_1 = Y_0 \vee Y_1,$$

which concludes the proof. □

Proof of Theorem 11.17. (a)⇒(b): By Proposition 11.15, any $X \in \mathscr{A}_t$ can be written as the sum of some $X_{t,t+1} \in \mathscr{A}_{t,t+1}$ and some $X_{t+1} \in \mathscr{A}_{t+1}$. Hence, we obtain

$$\begin{aligned}
\alpha_t^{\min}(Q) &= \operatorname*{ess\,sup}_{X \in \mathscr{A}_t} E_Q[-X \mid \mathscr{F}_t] \\
&= \operatorname*{ess\,sup}_{X_{t,t+1} \in \mathscr{A}_{t,t+1}} E_Q[-X_{t,t+1} \mid \mathscr{F}_t] + \operatorname*{ess\,sup}_{X_{t+1} \in \mathscr{A}_{t+1}} E_Q[-X_{t+1} \mid \mathscr{F}_t] \\
&= \alpha_{t,t+1}^{\min}(Q) + E_Q[\,\alpha_{t+1}^{\min}(Q) \mid \mathscr{F}_t\,]
\end{aligned}$$

for any $Q \in \mathcal{Q}$, using Lemma 11.3 in the last step.

(b)⇒(c): Fix $X \in L^\infty$. By Lemma 11.19 and Theorem F.1, we can find a sequence $Q_n \in \mathcal{Q}$ such that $\rho_{t+1}(X)$ can be identified as the limit of an increasing sequence:

$$(E_{Q_n}[-X \mid \mathscr{F}_{t+1}] - \alpha_{t+1}^{\min}(Q_n)) \nearrow \rho_{t+1}(X) \quad P\text{-a. s.} \tag{11.32}$$

Now take $Q \in \mathcal{Q}$ and write $U_t := U_t^{Q,X}$. We have to show that $U_t \geq E_Q[\, U_{t+1} \mid \mathscr{F}_t\,]$. Using property (b), we see that

$$\begin{aligned} E_Q[\, U_{t+1} \mid \mathscr{F}_t\,] &= E_Q[\,\rho_{t+1}(X) + \alpha_{t+1}^{\min}(Q) \mid \mathscr{F}_t\,] \\ &= E_Q[\,\rho_{t+1}(X) \mid \mathscr{F}_t\,] - \alpha_{t,t+1}^{\min}(Q) + \alpha_t^{\min}(Q). \end{aligned} \tag{11.33}$$

Now we use the sequence of measures Q_n appearing in (11.32). By Lemma 11.18, we are free to assume that $Q_n = Q$ on $\mathscr{F}_{t+1}$, and this implies

$$\alpha_{t,t+1}^{\min}(Q_n) = \alpha_{t,t+1}^{\min}(Q). \tag{11.34}$$

Using the approximation (11.32) and monotone convergence for conditional expectations, and applying property (b) together with equation (11.34) to each Q_n, we obtain

$$\begin{aligned} E_Q[\,\rho_{t+1}(X) \mid \mathscr{F}_t\,] &= \lim_{n\uparrow\infty}(E_{Q_n}[\,-X \mid \mathscr{F}_t\,] + \alpha_{t,t+1}^{\min}(Q_n) - \alpha_t^{\min}(Q_n)) \\ &= \lim_{n\uparrow\infty}(E_{Q_n}[\,-X \mid \mathscr{F}_t\,] - \alpha_t^{\min}(Q_n)) + \alpha_{t,t+1}^{\min}(Q) \\ &\leq \rho_t(X) + \alpha_{t,t+1}^{\min}(Q). \end{aligned}$$

Together with (11.33), this yields

$$E_Q[\, U_{t+1} \mid \mathscr{F}_t\,] \leq \rho_t(X) + \alpha_t^{\min}(Q) = U_t.$$

If in addition $\alpha_0(Q) < \infty$, then (b) implies $E_Q[\,\alpha_t^{\min}(Q)\,] \leq \alpha_0(Q) < \infty$. Moreover, $\rho_t(X)$ is bounded by $\|X\|_\infty$, and so U_t is integrable, and hence a supermartingale.

(c)⇒(a): Take $X, Y \in L^\infty$ such that $\rho_{t+1}(X) \leq \rho_{t+1}(Y)$; we have to show $\rho_t(X) \leq \rho_t(Y)$. For each $Q \in \mathcal{Q}$,

$$\begin{aligned} \rho_t(Y) + \alpha_t^{\min}(Q) &\geq E_Q[\,\rho_{t+1}(Y) + \alpha_{t+1}^{\min}(Q) \mid \mathscr{F}_t\,] \\ &\geq E_Q[\,\rho_{t+1}(X) + \alpha_{t+1}^{\min}(Q) \mid \mathscr{F}_t\,] \\ &\geq E_Q[\, E_Q[\,-X \mid \mathscr{F}_{t+1}\,] \mid \mathscr{F}_t\,] \\ &= E_Q[\,-X \mid \mathscr{F}_t\,]. \end{aligned}$$

The first inequality follows from the supermartingale property in (c), and in the third step, we have used the inequality

$$\rho_{t+1}(X) + \alpha_{t+1}^{\min}(Q) \geq E_Q[\,-X \mid \mathscr{F}_{t+1}\,],$$

which is valid for any $Q \in \mathcal{Q}$; see (11.8). Thus,

$$\rho_t(Y) \geq E_Q[\,-X \mid \mathscr{F}_t\,] - \alpha_t^{\min}(Q)$$

for all $Q \in \mathcal{Q}$, and this implies the desired inequality $\rho_t(Y) \geq \rho_t(X)$ due to our assumption that ρ_t admits the representation (11.28). □

Remark 11.20. It follows from property (b) of Theorem 11.17 that

$$E_Q[\,\alpha_{t+1}^{\min}(Q) \mid \mathscr{F}_t\,] \leq \alpha_t^{\min}(Q) \quad \text{for any } Q \in \mathcal{Q}. \tag{11.35}$$

This in turn implies $E_Q[\,\alpha_t^{\min}(Q)\,] < \infty$ for all $t \geq 0$ if $\alpha_0^{\min}(Q) < \infty$. Thus, the process $(\alpha_t^{\min}(Q))_{t=0,1,\dots,T}$ is a Q-supermartingale for any such Q. This can be seen as a built-in learning effect: If the dynamics are in fact driven by the probability measure Q, then the penalization of these measures decreases "on average" in the sense that it is a supermartingale under Q. Note that property (b) provides more information than just the supermartingale property: It actually yields an explicit description of the predictable increasing process in the Doob decomposition of the supermartingale $(\alpha_t^{\min}(Q))_{t=0,1,\dots,T}$ in terms of the "one-step" penalty functions $\alpha_{t,t+1}^{\min}(Q)$. More precisely,

$$\alpha_t^{\min}(Q) = M_t^Q - \sum_{k=0}^{t-1} \alpha_{k,k+1}^{\min}(Q),$$

where M^Q is a martingale under Q. ◇

Exercise 11.2.4. Show that the cost of superhedging defines a time-consistent sequence of conditional convex risk measures. More precisely, in the situation and with the notation of Chapter 9, let

$$\rho_t^{\mathscr{S}}(Y) := \inf\left\{ m_t \in L_t^\infty \;\middle|\; \exists \xi \in \mathscr{S} :\; m_t + Y + \sum_{k=t}^{T} \xi_k \cdot (X_k - X_{k-1}) \geq 0 \; P\text{-a.s.} \right\}$$

for $Y \in L^\infty$. Show that $(\rho_t^{\mathscr{S}})_{t=0,\dots,T}$ is a time-consistent sequence of conditional convex risk measures if the conditions of Corollary 9.32 are satisfied. ◇

Remark 11.21. The weaker version (11.35) of property (b) is in fact equivalent to a weaker notion of time consistency, called *weak time consistency*, which is defined by the implication

$$\rho_{t+1}(X) \leq 0 \quad \Longrightarrow \quad \rho_t(X) \leq 0,$$

and that is equivalent to the following weaker version:

$$\mathscr{A}_{t+1} \subseteq \mathscr{A}_t, \quad t = 0, 1, \dots, T-1,$$

of property (b) in Proposition 11.15; see Exercise 11.2.5 ◇

Exercise 11.2.5. Let $(\rho_t)_{t=0,1,\dots,T}$ be a sensitive sequence of convex conditional risk measures. Show that the following conditions are equivalent:

(a) $(\rho_t)_{t=0,1,\dots,T}$ is weakly time-consistent in the sense that $\rho_{t+1}(X) \le 0$ implies $\rho_t(X) \le 0$.
(b) The acceptance sets satisfy the inclusion $\mathscr{A}_{t+1} \subseteq \mathscr{A}_t$ for $t = 0, \dots, T-1$.
(c) The penalty functions satisfy $E_Q[\,\alpha_{t+1}^{\min}(Q) \mid \mathscr{F}_t\,] \le \alpha_t^{\min}(Q)$ for $Q \in \mathcal{Q}$. ◇

We will now consider the coherent case in Theorem 11.17. An initial discussion of this case was already given at the end of Section 6.5 in connection with our analysis of stability under pasting. Recall from Definition 6.39 that the *pasting* of two equivalent probability measures Q_1 and Q_2 in a stopping time $\sigma \le T$ is the probability measure

$$\widetilde{Q}[\,A\,] := E_{Q_1}[\,Q_2[\,A \mid \mathscr{F}_\sigma\,]\,], \quad A \in \mathscr{F}_T.$$

According to Definition 6.42, a set $\mathcal{Q}^\rho$ of equivalent probability measures is called *stable* if, for any $Q_1, Q_2 \in \mathcal{Q}$ and $\sigma \in \mathcal{T}$, also their pasting in σ is belongs to $\mathcal{Q}$. We can now state the following characterization of time consistency for coherent dynamic risk measures. It extends and provides a converse of Theorem 6.52.

Theorem 11.22. *Let $(\rho_t)_{t=0,1,\dots,T}$ be a sensitive sequence of convex conditional risk measures, and assume that ρ_0 is coherent. Then the following conditions are equivalent:*
(a) *$(\rho_t)_{t=0,1,\dots,T}$ is time-consistent.*
(b) *There exists a stable set $\mathcal{Q}^\rho \subseteq \mathcal{Q}$ such that*

$$\rho_t(X) = \operatorname*{ess\,sup}_{Q \in \mathcal{Q}^\rho} E_Q[\,-X \mid \mathscr{F}_t\,] \quad \text{for } t = 0, \dots, T \text{ and } X \in L^\infty. \tag{11.36}$$

In particular, each ρ_t is coherent if these equivalent conditions hold.

Proof. It was shown in Theorem 6.52 that (b) implies recursiveness and in turn (a).

To show the implication (a)⇒(b), let

$$\mathcal{Q}^\rho := \mathcal{Q} \cap \mathscr{P}_0^\rho = \mathcal{Q} \cap \{Q \in \mathscr{M}_1(P) \mid \alpha_0^{\min}(Q) < \infty\}.$$

Then $\alpha_t^{\min}(Q)$ is a nonnegative Q-supermartingale for any $Q \in \mathcal{Q}^\rho$ by Theorem 11.17. Since ρ_0 is coherent, we have $\alpha_0^{\min}(Q) = 0$, and hence $\alpha_t^{\min}(Q) = 0$ Q-a. s. (and hence P-a. s.) for each t, due to the supermartingale property (compare Exercise 6.1.2). Therefore, the representation (11.36) follows with Remark 11.4. In particular, each ρ_t is coherent.

To show the stability of $\mathcal{Q}^\rho$, we recall from Proposition 6.44 that the stability of $\mathcal{Q}^\rho$ is equivalent to the following fact: For $Q_1, Q_2 \in \mathcal{Q}^\rho$, $t \in \{0, \dots, T\}$, and $B \in \mathscr{F}_t$, also the probability measure

$$Q[\,A\,] := E_{Q_1}[\,Q_2[\,A \mid \mathscr{F}_t\,] \cdot \mathbb{1}_B + \mathbb{1}_{A \cap B^c}\,], \quad A \in \mathscr{F}_T,$$

belongs to $\mathcal{Q}^\rho$. Clearly, this measure Q is equivalent to P, and hence belongs to $\mathcal{Q}$. To prove that $\alpha_0^{\min}(Q) < \infty$, and in turn $Q \in \mathcal{Q}^\rho$, we show first that $\alpha_t^{\min}(Q) = 0$. Indeed, for $X \in \mathscr{A}_t$,

$$E_Q[-X \mid \mathscr{F}_t] = \mathbb{1}_B E_{Q_2}[-X \mid \mathscr{F}_t] + \mathbb{1}_{B^c} E_{Q_1}[-X \mid \mathscr{F}_t] \le 0,$$

because for $i = 1, 2$,

$$0 = \alpha_t^{\min}(Q_i) = \operatorname*{ess\,sup}_{X \in \mathscr{A}_t} E_{Q_i}[-X \mid \mathscr{F}_t].$$

This shows our claim $\alpha_t^{\min}(Q) = 0$. Next, part (b) of Theorem 11.17 yields that

$$\begin{aligned}
\alpha_{t-1}^{\min}(Q) &= \alpha_{t-1,t}^{\min}(Q) + E_Q[\,\alpha_t^{\min}(Q) \mid \mathscr{F}_{t-1}\,] \\
&= \alpha_{t-1,t}^{\min}(Q) \\
&= \operatorname*{ess\,sup}_{X \in \mathscr{A}_{t-1} \cap L_t^\infty} E_Q[-X \mid \mathscr{F}_t] \\
&= \operatorname*{ess\,sup}_{X \in \mathscr{A}_{t-1} \cap L_t^\infty} E_{Q_1}[-X \mid \mathscr{F}_t] \\
&\le \alpha_{t-1}^{\min}(Q_1) \\
&= 0,
\end{aligned}$$

where we have used the fact that $Q = Q_1$ on $\mathscr{F}_t$ in the fourth identity. An iteration of this argument shows $\alpha_0^{\min}(Q) = 0$. □

Remark 11.23. As in the other chapters of Part II, we have limited the discussion to a finite horizon $T < \infty$. If we pass to an infinite horizon, the characterization of time consistency in terms of supermartingale properties in Theorem 11.17 yields convergence results for risk measures, which can be seen as nonlinear extensions of martingale convergence; cf. [145]. ◇

12 Long-run performance of rebalanced portfolios

In this final chapter, we provide an introduction to T. Cover's theory of universal portfolios [68], which develops a nonparametric approach to portfolio selection via online statistical learning. In contrast to the preceding chapters, the focus will now be on the performance of portfolio strategies in the long run, that is, on the asymptotic growth rate of their value process as time tends to infinity. In Section 12.1, we look at constant rebalanced portfolio strategies and study the maximal growth rate that could be achieved in hindsight, using at each time the value of the constant rebalanced strategy that would have been best so far. In Section 12.2, it is shown that this maximal growth rate in retrospect is actually achieved by the universal portfolio, defined as a weighted average of constant rebalanced portfolio strategies, where the weights are updated depending on past performance. While this result does not rely on specific probabilistic assumptions, it does assume some qualitative stability of the financial market, namely convergence of the empirical distributions of the performance vectors as time goes to infinity. Numerical aspects of implementing universal portfolios are discussed in Section 12.3.

12.1 Constant rebalanced portfolio strategies

Let us consider a discrete-time financial market model $S_0, S_1, \dots$ with infinite time horizon, where

$$S_t = (S_t^1, \dots, S_t^d)$$

is the price vector of d assets at time t. These assets may or may not include a riskless asset. Since discounting will not be used throughout this section, we do not need to specify a numéraire as asset number 0. The asset prices S_t^i can be modeled as random variables on a given probability space, just as in the previous chapters of this book. However, such an assumption will not be needed for large parts of this chapter. Alternatively, each S_t^i can be thought of as the price of the ith asset observed at time t, and hence simply as a real number. For the moment, our only assumption is that

$$S_t^i > 0 \quad \text{for all } i \text{ and } t. \tag{12.1}$$

Note that the issue of model uncertainty does not arise in such a setting.

Let $\xi = (\xi^1, \dots, \xi^d)$ be a trading strategy. That is, ξ_t^i shares of the ith asset are held from time $t-1$ to time t. Again, for the time being, ξ_t^i may simply be a real number. We will work under the no-short-sales constraint

$$\xi_t^i \geq 0 \quad \text{for all } i \text{ and } t. \tag{12.2}$$

Moreover, the strategy ξ must satisfy the self-financing condition,

https://doi.org/10.1515/9783111045283-014

$$V_t := \xi_t \cdot S_t = \sum_{i=1}^{d} \xi_t^i S_t^i = \sum_{i=1}^{d} \xi_{t+1}^i S_t^i = \xi_{t+1} \cdot S_t \quad \text{for all } t \geq 1. \tag{12.3}$$

We also set

$$V_0 := \xi_1 \cdot S_0.$$

Then (12.3) is equivalent to the following condition:

$$V_t - V_{t-1} = \xi_t \cdot (S_t - S_{t-1}) \quad \text{for } t = 1, 2, \ldots; \tag{12.4}$$

see Proposition 5.7. If $V_{t-1} > 0$, then

$$\pi_t^i := \frac{\xi_t^i S_{t-1}^i}{V_{t-1}} \tag{12.5}$$

is the proportion of the total portfolio value that is invested into the ith asset from time $t-1$ to time t. Note that by (12.3)

$$\pi_t^i \geq 0 \quad \text{and} \quad \sum_{i=1}^{d} \pi_t^i = \frac{1}{V_{t-1}} \xi_t \cdot S_{t-1} = \frac{1}{V_{t-1}} \xi_{t-1} \cdot S_{t-1} = 1. \tag{12.6}$$

That is, $\pi_t = (\pi_t^1, \ldots, \pi_t^d)$ is an element of the $(d-1)$-dimensional standard simplex

$$\Delta := \left\{ \pi = (\pi^1, \ldots, \pi^d) \;\middle|\; \pi^i \geq 0, \sum_{i=1}^{d} \pi^i = 1 \right\}.$$

Definition 12.1. A *portfolio strategy* is a sequence $\pi_1, \pi_2, \ldots$ such that $\pi_t \in \Delta$ for all t. △

Let us introduce the *performance vectors* $Y_1, Y_2, \ldots$ with components

$$Y_t^i := \frac{S_t^i}{S_{t-1}^i}, \quad i = 1, \ldots, d, \; t = 1, 2, \ldots. \tag{12.7}$$

Proposition 12.2. *For every $V_0 > 0$, there exists a one-to-one correspondence between all portfolio strategies and all self-financing strategies satisfying the no-short-sales constraint* (12.2) *and whose value process satisfies $V_t > 0$ for all t. It is given by the identity*

$$V_{t-1}\pi_t^i = \xi_t^i S_{t-1}^i \quad i = 1, \ldots, d, \; t = 1, 2, \ldots. \tag{12.8}$$

Moreover, the value process of ξ can be computed as follows:

$$V_t = V_0 \prod_{k=1}^{t} \pi_k \cdot Y_k, \quad t = 1, 2, \ldots. \tag{12.9}$$

Proof. We have already seen in (12.5) and (12.6) how a self-financing strategy gives rise to a portfolio strategy. From (12.5), we get moreover,

$$\pi_t \cdot Y_t = \frac{1}{V_{t-1}} \xi_t \cdot S_t = \frac{V_t}{V_{t-1}},$$

and this implies (12.9).

Conversely, let $\pi_1, \pi_2, \ldots$ be a given portfolio strategy. We define V_t via (12.9) for $t > 0$. Then $V_t > 0$ for all t, and so we can define $\xi_1, \xi_2, \ldots$ through the identity (12.8). Then the no-short-sales constraint (12.2) is satisfied. Furthermore, for $t = 1, 2, \ldots,$

$$\xi_t \cdot S_{t-1} = V_{t-1} \sum_{i=1}^{d} \pi_t^i = V_{t-1}.$$

Moreover,

$$\xi_t \cdot S_t = V_{t-1} \sum_{i=1}^{d} \frac{\pi_t^i}{S_{t-1}^i} S_t^i = V_{t-1} \pi_t \cdot Y_t = V_t.$$

Hence, V is the value process of ξ, condition (12.4) is satisfied, and so ξ is a self-financing strategy. □

Definition 12.3. A portfolio strategy $(\pi_t)_{t=1,2,\ldots}$ that is constant in t is called a *constant rebalanced portfolio strategy.* △

If $(\pi_t)_{t=1,2,\ldots}$ is a constant rebalanced portfolio strategy, then there is $\pi \in \Delta$ such that $\pi_t = \pi$ for all t. In the sequel, we will often identify the strategy $(\pi_t)_{t=1,2,\ldots}$ with π. Note that, although the portfolio weights $\pi_t^1, \ldots, \pi_t^d$ are constant in t, the number of shares held in each asset is typically not constant. Indeed, by (12.8), the number of shares held in the ith asset is given by

$$\xi_t^i = \frac{\pi^i V_{t-1}}{S_{t-1}^i},$$

and this number fluctuates as asset prices and portfolio values rise and fall in t.

Example 12.4. The *equal-weighted portfolio* is the constant rebalanced portfolio strategy corresponding to $\pi^i = 1/d$ for $i = 1, \ldots, d$. That is, each asset is allocated the same amount of capital; see Figure 12.1. ◇

For $\pi \in \Delta$, we denote the value process arising from the constant rebalanced portfolio strategy $\pi \in \Delta$ by

$$V_0(\pi) := 1 \quad \text{and} \quad V_t(\pi) := \prod_{k=1}^{t} \pi \cdot Y_k \,. \tag{12.10}$$

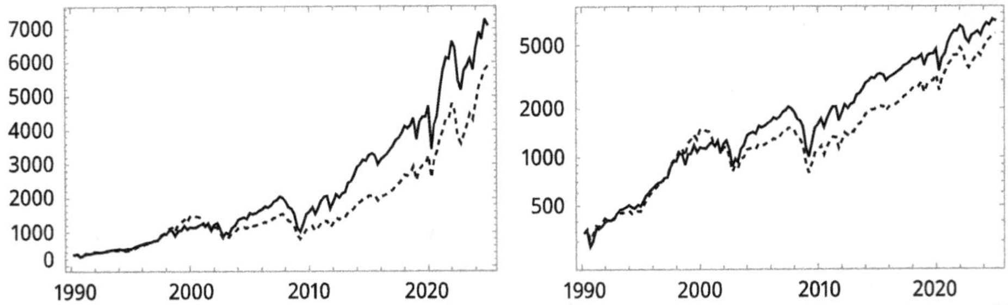

Figure 12.1: The S&P 500 Equal Weight Index (solid) is the equal-weight version of the market-cap-weighted S&P 500 (dashed, see Example 1.21). The S&P 500 Equal Weight Index is rebalanced quarterly. The left-hand panel shows quarterly data of the two indices on a linear scale, the right-hand panel shows the same data on a logarithmic scale.

Exercise 12.1.1. For given $\pi \in \Delta$, show that $V_t(\pi)$ exceeds the π-weighted geometric mean of the individual asset performances. More precisely,

$$V_t(\pi) \geq \prod_{i=1}^{d} \left(\frac{S_t^i}{S_0^i} \right)^{\pi^i} . \qquad \diamond$$

In the sequel, we are interested in the asymptotic growth rate of $V_t(\pi)$ as $t \uparrow \infty$. To this end, we introduce the empirical distribution of the performance vectors Y_k up to time t, that is, the probability measure on $(0, \infty)^d$ defined by

$$\rho_t := \frac{1}{t} \sum_{k=1}^{t} \delta_{Y_k} = \frac{1}{t} \sum_{k=1}^{t} \delta_{(Y_k^1, \dots, Y_k^d)}, \qquad t = 1, 2, \dots .$$

Then

$$\begin{aligned} V_t(\pi) &= \prod_{k=1}^{t} \pi \cdot Y_t = \exp\left(t \cdot \frac{1}{t} \sum_{k=1}^{t} \log(\pi \cdot Y_k) \right) \\ &= \exp\left(t \int \log(\pi \cdot y) \, \rho_t(dy) \right) = \exp(t \cdot F(\pi, \rho_t)) , \end{aligned} \tag{12.11}$$

where we define

$$F(\pi, \mu) := \int \log(\pi \cdot y) \, \mu(dy) ,$$

for any Borel probability measure μ on $(0, \infty)^d$ for which the integral is well-defined. Due to (12.11),

$$F(\pi, \rho_t) = \frac{1}{t} \log V_t(\pi)$$

describes the growth rate of the portfolio value of the constant rebalanced strategy π until time t. The functional F will therefore play a key role in the sequel. As a first step toward analyzing the properties of F, we need to find a suitable space of measures μ such that $F(\pi, \mu)$ is well-defined for all $\pi \in \Delta$. To this end, we introduce the continuous gauge function $\psi : (0, \infty)^d \to [d, \infty)$ via

$$\psi(x) := d + \sum_{i=1}^{d} |\log x^i|, \quad x = (x^1, \ldots, x^d) \in (0, \infty)^d. \tag{12.12}$$

Then we let

$$\mathcal{M}_1^\psi := \mathcal{M}_1^\psi((0, \infty)^d) = \left\{ \mu \in \mathcal{M}_1((0, \infty)^d) \,\middle|\, \int \psi \, d\mu < \infty \right\}$$

and denote by $C_\psi := C_\psi((0, \infty)^d)$ the set of all continuous functions f on $(0, \infty)^d$ such that $|f(x)| \leq C\psi(x)$ for some constant C and all x. The measure space $\mathcal{M}_1^\psi$ will be endowed with the ψ-weak topology, which is defined as the coarsest topology for which all the maps $\mu \mapsto \int f \, d\mu$ with $f \in C_\psi$ are continuous; see Appendix G.2. Corollary G.12 states that $\mathcal{M}_1^\psi$ is a separable and metrizable space for the ψ-weak topology. Note that $\mathcal{M}_1^\psi$ contains, for instance, all multivariate log-normal distributions, that is, the laws of all random vectors $(e^{Z^1}, \ldots, e^{Z^d})$, where $Z = (Z^1, \ldots, Z^d)$ has a multivariate normal distribution.

Lemma 12.5. *For all $\pi \in \Delta$ and $x \in (0, \infty)^d$, we have*

$$|\log(\pi \cdot x)| \leq \psi(x). \tag{12.13}$$

In particular, the function $x \mapsto \log(\pi \cdot x)$ belongs to C_ψ and $F(\pi, \mu) = \int \log(\pi \cdot x) \, \mu(dx)$ is well-defined and finite for every $\pi \in \Delta$ and any $\mu \in \mathcal{M}_1^\psi$.

Proof. We claim first that, if f is a concave function with $f(0) = 0$, then f is subadditive in the sense that $f(u + v) \leq f(u) + f(v)$ for all $u, v > 0$. To prove this auxiliary claim, we note that for all $z \geq 0$ and $\alpha \in [0, 1]$,

$$\alpha f(z) = \alpha f(z) + (1 - \alpha) f(0) \leq f\big(\alpha z + (1 - \alpha)0\big) = f(\alpha z).$$

Thus, for $\alpha = u/(u + v) \in (0, 1)$,

$$\begin{aligned} f(u) &= f\big(\alpha(u + v)\big) \geq \alpha f(u + v), \\ f(v) &= f\big((1 - \alpha)(u + v)\big) \geq (1 - \alpha) f(u + v). \end{aligned}$$

Adding the preceding two inequalities yields the desired subadditivity.

Now we apply the subadditivity property to the concave function $f(u) := \log(1 + u)$,

$$\log(\pi \cdot x) \leq \log(1 + \pi \cdot x) \leq \log\left(1 + \sum_{i=1}^{d} x^i\right) \leq \sum_{i=1}^{d} \log(1 + x^i).$$

Since $\log(1+u) \le 1 + |\log u|$ for all $u > 0$, we get

$$\sum_{i=1}^{d} \log(1+x^i) \le \sum_{i=1}^{d} (1 + |\log x^i|) \le \psi(x),$$

and so we arrive at $\log(\pi \cdot x) \le \psi(x)$. On the other hand, the concavity of the logarithm yields that

$$\log(\pi \cdot x) = \log \sum_{i=1}^{d} \pi^i x^i \ge \sum_{i=1}^{d} \pi^i \log x^i \ge -\sum_{i=1}^{d} \pi^i |\log x^i| \ge -\sum_{i=1}^{d} |\log x^i| > -\psi(x).$$

This proves (12.13). The remaining assertions are immediate consequences of (12.13). □

Lemma 12.6. *The functional* $(\pi, \mu) \mapsto F(\pi, \mu)$ *is continuous on* $\Delta \times \mathscr{M}_1^{\psi}$, *affine in* μ, *and concave in* π. *In particular,* $F(\cdot, \mu)$ *attains its maximum on* Δ.

Proof. Clearly, F is affine in μ and concave in π. To show continuity, we take a sequence (π_n, μ_n) that converges to (π_∞, μ_∞) in $\Delta \times \mathscr{M}_1^{\psi}$ (here, the metrizability of $\mathscr{M}_1^{\psi}$ allows us to focus on sequences when considering convergence). By Corollary G.15, there exists a probability space $(\Omega, \mathscr{F}, P)$ and random variables X_n such that each X_n has law μ_n (including $n = \infty$), $X_n \to X_\infty$ P-a. s., and $\psi(X_n) \to \psi(X_\infty)$ in L^1. The Vitali convergence theorem, Theorem H.17, thus yields that the sequence $(\psi(X_n))_{n\in\mathbb{N}}$ is uniformly integrable. Letting $Z_n := \log(\pi_n \cdot X_n)$, the inequality (12.13) states that $|Z_n| \le \psi(X_n)$, and so the sequence $(Z_n)_{n=1,2,\ldots}$ is also uniformly integrable. Since $Z_n \to Z_\infty$ P-a. s., another application of Theorem H.17 gives that $Z_n \to Z_\infty$ in L^1, which in turn implies that

$$F(\pi_n, \mu_n) = E[\log(\pi_n \cdot X_n)] = E[Z_n] \longrightarrow E[Z_\infty] = E[\log(\pi_\infty \cdot X_\infty]) = F(\pi_\infty, \mu_\infty).$$

This proves the continuity of F. Finally, since Δ is a compact set, it is now clear that the function $\pi \mapsto F(\pi, \mu)$ must attain its maximum for each given $\mu \in \mathscr{M}_1^{\psi}$. □

Let us now introduce

$$V_t^* := \max_{\pi\in\Delta} V_t(\pi),$$

which is the maximal possible value we could have achieved in hindsight when trading with any constant rebalanced portfolio strategy. Of course, neither the value V_t^* nor any maximizer $\pi_t^* \in \operatorname{argmax}_{\pi\in\Delta} V_t(\pi)$ will typically be known before time t.

Remark 12.7. Let $\varepsilon_i \in \Delta$ denote the constant rebalanced portfolio strategy that invests only into the ith asset, that is,

$$\varepsilon_i = (0, \ldots, 0, 1, 0, \ldots, 0).$$

Then it is clear that V_t^* outperforms the best asset in the sense that

$$V_t^* \geq \max_{i=1,\dots,d} V_t(\varepsilon_i) = \max_{i=1,\dots,d} \frac{S_t^i}{S_0^i}. \tag{12.14}$$

The preceding inequality implies that

$$V_t^* \geq \sum_{i=1}^d \beta^i V_t(\varepsilon_i) = \sum_{i=1}^d \beta^i \frac{S_t^i}{S_0^i},$$

for any $\beta \in \Delta$. In particular, V^* outperforms any stock market index that is weighted by price or market capitalization, such as the Dow Jones Industrial Average or the S&P 500, respectively. The inequality between the (weighted) arithmetic and geometric means now yields that V_t^* exceeds any geometric average of the individual assets. That is,

$$V_t^* \geq \prod_{i=1}^d V_t(\varepsilon_i)^{\beta^i} = \prod_{i=1}^d \left(\frac{S_t^i}{S_0^i}\right)^{\beta^i}, \tag{12.15}$$

where $\beta \in \Delta$ is arbitrary; see also Exercise 12.1.1. This applies in particular to geometrically averaged stock market indices such as the Value Line Geometric Index. ◇

By (12.11), the time-t growth rate of the hindsight-optimal portfolio value V_t^* is given by

$$\frac{1}{t}\log V_t^* = \max_{\pi\in\Delta} \frac{1}{t}\log V_t(\pi) = \max_{\pi\in\Delta} F(\pi, \rho_t).$$

We therefore introduce the functional

$$F^*(\mu) = \max_{\pi\in\Delta} F(\pi,\mu), \quad \mu \in \mathscr{M}_1^\pi,$$

so that

$$\frac{1}{t}\log V_t^* = F^*(\rho_t).$$

Proposition 12.8. *The functional $F^* : \mathscr{M}_1^\psi \to \mathbb{R}$ is continuous and convex. In particular, if $Y_1, Y_2, \dots$ is a sequence of performance vectors in $(0,\infty)^d$ whose empirical distributions ρ_t converge ψ-weakly to some $\mu \in \mathscr{M}_1^\psi$, then*

$$\frac{1}{t}\log V_t^* \longrightarrow F^*(\mu). \tag{12.16}$$

Proof. F^* is convex and lower semicontinuous as a supremum of continuous affine functions. Thus, it suffices to show its upper semicontinuity. To this end, we take a sequence $(\mu_n)_{n\in\mathbb{N}}$ in $\mathscr{M}_1^\psi$ that converges ψ-weakly to some $\mu \in \mathscr{M}_1^\psi$. Lemma 12.6 states that to each μ_n there corresponds some $\pi_n \in \Delta$ with $F^*(\mu_n) = F(\pi_n, \mu_n)$. Since Δ is compact, we may assume without loss of generality that π_n converges to some $\pi \in \Delta$; otherwise, we extract a convergent subsequence. Hence, by Lemma 12.6,

$$\limsup_{n\uparrow\infty} F^*(\mu_n) = \limsup_{n\uparrow\infty} F(\pi_n, \mu_n) = F(\pi, \mu) \le F^*(\mu) .$$

Therefore, F^* is upper semicontinuous, and in turn, continuous. Finally, the continuity of F^* yields (12.16). □

Remark 12.9. There are many situations in which the convergence of empirical distributions assumed in the preceding proposition is satisfied. Consider a probabilistic setting where the performance vectors Y_k are interpreted as random variables on some probability space $(\Omega, \mathscr{F}, P)$. If $Y_1, Y_2, \dots$ is an i. i. d. sequence with common law $\mu \in \mathscr{M}_1^\psi$, then the empirical distributions

$$\rho_t = \frac{1}{t}\sum_{k=1}^{t} \delta_{Y_k}, \quad t = 1, 2, \dots,$$

converge P-a. s. in the ψ-weak topology to μ; see Corollary G.16. Convergence also holds if the sequence $Y_1, Y_2, \dots$ is stationary in the sense that $(Y_1, Y_2, \dots)$ has the same law as $(Y_k, Y_{k+1}, \dots)$ for any $k \in \mathbb{N}$. This includes the special case in which the $Y_1, Y_2, \dots$ form a stationary Markov chain; see Remark G.17. ◇

By (12.16), $F^*(\mu)$ is the optimal growth rate that could possibly be achieved in hindsight with a constant rebalanced strategy. But are there perhaps more general adaptive portfolio strategies that can yield an even better growth rate? The following result gives a negative answer for the case of i. i. d. performance vectors.

Theorem 12.10. *Suppose that the performance vectors $Y_1, Y_2, \dots$ are i. i. d. random vectors on some probability space $(\Omega, \mathscr{F}, P)$ with common distribution $\mu \in \mathscr{M}_1^\psi$. Let $(\pi_t)_{t=0,1,2,\dots}$ be any portfolio strategy that is predictable with respect to the filtration $\mathscr{F}_t := \sigma(Y_1, \dots, Y_t)$ $(t = 0, 1, \dots)$ and denote by V_t^π its value process with initial capital $V_0^\pi = 1$. Then the asymptotic growth rate of V_t^π is at most $F^*(\mu)$. That is,*

$$\limsup_{t\uparrow\infty} \frac{1}{t}\log V_t^\pi \le F^*(\mu) \quad P\text{-a. s.}$$

The proof relies on two auxiliary lemmas.

Lemma 12.11. *Suppose $\mu \in \mathscr{M}_1^\psi$ is given. Then $\pi^* \in \Delta$ maximizes $\pi \mapsto F(\pi, \mu)$ if and only if*

$$\int \frac{\pi \cdot y}{\pi^* \cdot y}\,\mu(dy) \le 1 \quad \text{for all } \pi \in \Delta. \tag{12.17}$$

Proof. For given $\pi \in \Delta$ and $\alpha \in [0, 1]$, let $\pi_\alpha := \alpha\pi + (1-\alpha)\pi^*$ and $h(\alpha) := F(\pi_\alpha, \mu)$. We claim first that π^* is a maximizer if and only if $h'_+(0) \le 0$ for all $\pi \in \Delta$. Here, h'_+

denotes the right-hand derivative of the concave function h, which exists due to Proposition A.7(b) and the fact that $-h$ is convex. Indeed, if π^* is a maximizer, then $h(0) \geq h(\alpha)$ for all $\alpha \in [0,1]$, and so

$$h'_+(0) = \lim_{\alpha \downarrow 0} \frac{h(\alpha) - h(0)}{\alpha} \leq 0.$$

Conversely, suppose that $h'_+(0) \leq 0$ for all π. Then Proposition A.7(b) yields that $h'_+(\alpha) \leq h'_+(0) \leq 0$ for all α. Moreover, part (d) of Proposition A.7 states that for each $\alpha \in (0,1)$ and all $\delta \in (0,\alpha)$,

$$h(\alpha) = h(\delta) + \int_\delta^\alpha h'_+(s)\,ds \leq h(\delta).$$

In the limit $\delta \downarrow 0$, the continuity of h yields that $h(\alpha) \leq h(0)$.

Now we compute $h'_+(0)$. Since $\alpha \mapsto \log(\pi_\alpha \cdot y)$ is a concave function on $[0,1]$ for each $y \in (0,\infty)^d$,

$$g(y,\alpha) := \frac{\log(\pi_\alpha \cdot y) - \log(\pi_0 \cdot y)}{\alpha}$$

increases to

$$g'(y,0) = \frac{(\pi - \pi^*) \cdot y}{\pi^* \cdot y} = \frac{\pi \cdot y}{\pi^* \cdot y} - 1$$

as $\alpha \downarrow 0$; see (A.5) for details. Moreover, by Lemma 12.5,

$$g(y,\alpha) \geq g(y,1) \geq -2\psi(y) \in L^1(\mu).$$

Hence, the monotone convergence theorem yields that

$$h'_+(0) = \lim_{\alpha \downarrow 0} \frac{h(\alpha) - h(0)}{\alpha} = \lim_{\alpha \downarrow 0} \int g(\alpha, y)\,\mu(dy) = \int \frac{\pi \cdot y}{\pi^* \cdot y}\,\mu(dy) - 1.$$

This completes the proof. □

Lemma 12.12. *In the setting of Theorem 12.10, let $\pi^* \in \Delta$ denote a maximizer of $\pi \mapsto F(\pi,\mu)$. Then the process*

$$\frac{V_t^\pi}{V_t(\pi^*)}, \quad t = 0,1,\ldots$$

is a supermartingale with respect to the filtration $\mathscr{F}_t := \sigma(Y_1,\ldots,Y_t)$ $(t = 0,1,\ldots)$.

Proof. Recall from (12.9) that $V_t^\pi = V_{t-1}^\pi \pi_{t-1} \cdot Y_t$. Therefore, and by Lemma 12.11, for P-a. e. $\omega \in \Omega$,

$$
\begin{aligned}
E\Big[\frac{V_t^\pi}{V_t(\pi^*)}\,\Big|\,\mathscr{F}_{t-1}\Big](\omega) &= E\Big[\frac{V_{t-1}^\pi}{V_{t-1}(\pi^*)}\cdot\frac{\pi_{t-1}\cdot Y_t}{\pi^*\cdot Y_t}\,\Big|\,\mathscr{F}_{t-1}\Big](\omega)\\
&= \frac{V_{t-1}^\pi(\omega)}{V_{t-1}(\pi^*)(\omega)}E\Big[\frac{\pi_{t-1}(\omega)\cdot Y_t}{\pi^*\cdot Y_t}\Big]\\
&= \frac{V_{t-1}^\pi(\omega)}{V_{t-1}(\pi^*)(\omega)}\int\frac{\pi_{t-1}(\omega)\cdot y}{\pi^*\cdot y}\,\mu(dy)\\
&\le \frac{V_{t-1}^\pi(\omega)}{V_{t-1}(\pi^*)(\omega)}.
\end{aligned}
$$

Here, we have used in the second step that Y_t is independent of $\mathscr{F}_{t-1}$. Also, note that all conditional expectations are well-defined, because all quantities are nonnegative. Induction on t now gives

$$
0 \le E\Big[\frac{V_t^\pi}{V_t(\pi^*)}\Big] = E\Big[\frac{V_t^\pi}{V_t(\pi^*)}\,\Big|\,\mathscr{F}_0\Big] \le \frac{V_0^\pi}{V_0(\pi^*)} = 1.
$$

In particular, $V_t^\pi/V_t(\pi^*) \in L^1$, and the assertion follows. □

Proof of Theorem 12.10. Let π^* denote a maximizer of $\pi \mapsto F(\pi,\mu)$. By Lemma 12.12 and Doob's martingale convergence theorem (see, e. g., Theorem 19.1 in [21]), the nonnegative supermartingale $V_t^\pi/V_t(\pi^*)$ converges P-a. s. to a finite limit. In particular,

$$
Z := \sup_{t=1,2,\dots}\frac{V_t^\pi}{V_t(\pi^*)} < \infty \quad P\text{-a. s.}
$$

Consequently,

$$
\limsup_{t\uparrow\infty}\frac{1}{t}\log\frac{V_t^\pi}{V_t(\pi^*)} \le \limsup_{t\uparrow\infty}\frac{1}{t}\log Z = 0 \quad P\text{-a. s.},
$$

and thus

$$
\limsup_{t\uparrow\infty}\frac{1}{t}\log V_t^\pi \le \liminf_{t\uparrow\infty}\frac{1}{t}\log V_t(\pi^*) \quad P\text{-a. s.} \tag{12.18}
$$

On the other hand, let

$$
\rho_t = \frac{1}{t}\sum_{k=1}^t \delta_{Y_k}, \quad t = 1,2,\dots
$$

denote again the empirical distribution of the performance vectors Y_k. Then $\rho_t \to \mu$ P-a. s. in the ψ-weak topology; see Remark 12.9 and Corollary G.16. Thus, the continuity of F, as established in Lemma 12.6, implies that

$$
\frac{1}{t}\log V_t(\pi^*) = F(\pi^*,\rho_t) \longrightarrow F(\pi^*,\mu) = F^*(\mu) \quad \text{as } t\uparrow\infty
$$

Using this result in (12.18) implies the assertion. □

12.2 The universal portfolio

In the preceding section, we have analyzed the asymptotic growth rate of constant rebalanced portfolio strategies under the assumption that the empirical distributions,

$$\rho_t = \frac{1}{t}\sum_{k=1}^{t}\delta_{Y_k}, \quad t = 1, 2, \ldots,$$

of the performance vectors Y_k converge ψ-weakly to some $\mu \in \mathcal{M}_1^\psi$. See Remark 12.9 for a discussion of situations in which this assumption is satisfied. Our main results identified the functional

$$F^*(\mu) = \max_{\pi\in\Delta} F(\pi,\mu) = \max_{\pi\in\Delta}\int \log(\pi\cdot y)\,\mu(dy)$$

as the maximal asymptotic growth rate that could be obtained in hindsight by any $\pi \in \Delta$. We have even seen in Theorem 12.10 that the asymptotic growth rate $F^*(\mu)$ cannot be beaten by any adaptive portfolio strategy if the performance vectors Y_k are i. i. d. with common distribution μ. In this section, we now present an adaptive portfolio strategy, which can be constructed in a straightforward manner and which asymptotically achieves the target growth rate $F^*(\mu)$. Note that the construction is model-free in the sense that it does not involve any probabilistic assumptions on the performance vectors Y_k.

Definition 12.13. Let ν be any probability distribution on the $(d-1)$-dimensional standard simplex Δ. Then

$$\hat{\pi}_t := \frac{1}{\int V_{t-1}(\pi)\,\nu(d\pi)}\int \pi\, V_{t-1}(\pi)\,\nu(d\pi), \quad t = 1, 2, \ldots$$

is called the *universal portfolio strategy* corresponding to ν. △

The universal portfolio $\hat{\pi}_t$ is a weighted average over the set of all possible constant rebalancing portfolio strategies, and the weights are taken in correspondence to the past performance of the corresponding portfolio values. While the definition of the universal portfolio depends on the choice of ν, it will turn out later that, at least in many cases, ν is irrelevant for the asymptotic growth of the corresponding value process, which, according to (12.9), is given by

$$\hat{V}_0 := 1 \quad\text{and}\quad \hat{V}_t := \prod_{k=1}^{t}\hat{\pi}_k\cdot Y_k, \quad t = 1, 2, \ldots.$$

Remark 12.14. For any ν that is invariant under permutations of the indices of $\pi = (\pi^1, \ldots, \pi^d)$, we have

$$\hat{\pi}_1^i = \frac{\int \pi^i V_0(\pi)\,\nu(d\pi)}{\int V_0(\pi)\,\nu(d\pi)} = \int \pi^i\,\nu(d\pi) = \frac{1}{d}, \quad i = 1, \ldots, d.$$

That is, the initial portfolio is the equal-weighted portfolio. ◇

We start analyzing the universal portfolio strategy by deriving a formula for its value process.

Proposition 12.15. *The value process $\hat{V}_t$ of the universal portfolio strategy $\hat{\pi}_t$ is equal to the ν-average of the values $V_t(\pi)$, that is,*

$$\hat{V}_t = \int V_t(\pi)\,\nu(d\pi).$$

Proof. Since $V_t(\pi) = V_{t-1}(\pi)\pi \cdot Y_t$,

$$\hat{\pi}_t \cdot Y_t = \left(\frac{\int \pi\, V_{t-1}(\pi)\,\nu(d\pi)}{\int V_{t-1}(\pi)\,\nu(d\pi)} \right) \cdot Y_t = \frac{\int (\pi \cdot Y_t)\, V_{t-1}(\pi)\,\nu(d\pi)}{\int V_{t-1}(\pi)\,\nu(d\pi)} = \frac{\int V_t(\pi)\,\nu(d\pi)}{\int V_{t-1}(\pi)\,\nu(d\pi)}.$$

Thus, together with our assumptions $\hat{V}_0 = 1$, $V_0(\pi) = 1$, and $\nu(\Delta) = 1$, we get that

$$\hat{V}_t = \prod_{k=1}^{t} \hat{\pi}_k \cdot Y_k = \frac{\int V_t(\pi)\,\nu(d\pi)}{\int V_0(\pi)\,\nu(d\pi)} = \int V_t(\pi)\,\nu(d\pi).$$

This proves the proposition. □

The following exercise can be compared with Remark 12.7 and in particular with Equation (12.15).

Exercise 12.2.1. For $i = 1, \ldots, d$, let $\beta_i := \int \pi^i\,\nu(d\pi)$, so that $\beta = (\beta^1, \ldots, \beta^d) \in \Delta$. Show that $\hat{V}_t$ exceeds the geometric average of the underlying assets $S^1, \ldots, S^d$ in the sense that

$$\hat{V}_t \geq \prod_{i=1}^{d} \left(\frac{S_t^i}{S_0^i} \right)^{\beta_i} \quad \text{for } t = 0, 1, 2, \ldots.$$ ◇

The following theorem is the main result of this chapter. It states that the value process of the universal portfolio grows asymptotically with the a posteriori rate $F^*(\mu)$, where μ is the limit of the empirical distribution.

Theorem 12.16. *Suppose ν has full support and $Y_1, Y_2, \ldots$ is a sequence of performance vectors in $(0, \infty)^d$ whose empirical distributions ρ_t converge ψ-weakly to some $\mu \in \mathcal{M}_1^\psi$ as $t \uparrow \infty$. Then*

$$\frac{1}{t} \log \hat{V}_t \longrightarrow F^*(\mu) = \lim_{t \uparrow \infty} \frac{1}{t} \log V_t^*.$$

Proof. By Proposition 12.15,

$$\hat{V}_t = \int V_t(\pi)\, dv(\pi) \leq V_t^*,$$

and hence $\limsup_t \frac{1}{t} \log \hat{V}_t \leq F^*(\mu)$ by Proposition 12.8.

For the proof of the converse inequality, let $\pi^* \in \Delta$ denote a maximizer of the function $\pi \mapsto F(\pi,\mu)$ and let $\varepsilon > 0$ be given. By the continuity of F, there exists an open neighborhood U of (π^*,μ) in $\Delta \times \mathscr{M}_1^\psi$ such that

$$F(\pi,\nu) \geq F(\pi^*,\mu) - \varepsilon = F^*(\mu) - \varepsilon \quad \text{for all } (\pi,\nu) \in U.$$

We can assume without loss of generality that U is of the form $U = U_1 \times U_2$ for open sets $U_1 \subset \Delta$ and $U_2 \subset \mathscr{M}_1^\psi$. Then there will be some $t_0 \in \mathbb{N}$ such that $\rho_t \in U_2$ for all $t \geq t_0$. Hence, by Proposition 12.15 and (12.11), for $t \geq t_0$,

$$\begin{aligned}\frac{1}{t}\log \hat{V}_t &= \int V_t(\pi)\,\nu(d\pi)\\ &= \frac{1}{t}\log \int e^{tF(\pi,\rho_t)}\,\nu(d\pi)\\ &\geq \frac{1}{t}\log \int_{U_1} e^{tF(\pi,\rho_t)}\,\nu(d\pi)\\ &\geq F^*(\mu) - \varepsilon + \frac{1}{t}\log \nu(U_1).\end{aligned}$$

Since ε was arbitrary and $\nu(U_1) > 0$ by our assumption that ν has full support, we conclude

$$\liminf_{t\uparrow\infty} \frac{1}{t}\log \hat{V}_t \geq F^*(\mu).$$

This completes the proof. □

12.3 Computing the universal portfolio

Recall from (12.10) that the value process of the constant rebalanced portfolio strategy $\pi \in \Delta$ is given by

$$V_t(\pi) = \prod_{k=1}^{t} \pi \cdot Y_k = \prod_{k=1}^{t}\sum_{i=1}^{d} \pi^i Y_k^i.$$

Hence, $V_t(\pi)$ is a polynomial of degree t of the components $\pi^1,\dots,\pi^d$ of $\pi \in \Delta$. According to Definition 12.13, computing the universal portfolio involves the integration of the polynomials $V_t(\pi)$ and $\pi^i V_t(\pi)$ over the simplex Δ and with respect to the measure ν. Let us first consider the case in which ν is equal to the Lebesgue measure λ on Δ. Since Δ is a $(d-1)$-dimensional surface embedded into $\mathbb{R}^d$, it is a null set for the d-dimensional Lebesgue measure on $\mathbb{R}^d$. But λ can be constructed as follows by embedding Δ into $\mathbb{R}^{d-1}$ via dropping the dth coordinate of vectors. More precisely, consider

$$\widetilde{\Delta} := \left\{ x = (x^1, \dots, x^{d-1}) \;\middle|\; x^i \geq 0,\ \sum_{i=1}^{d-1} x^i \leq 1 \right\} \subset \mathbb{R}^{d-1} \tag{12.19}$$

and the bijective map $\varphi : \widetilde{\Delta} \to \Delta$ defined by

$$\varphi(x^1, \dots, x^{d-1}) = \left(x^1, \dots, x^{d-1}, 1 - \sum_{i=1}^{d-1} x^i \right). \tag{12.20}$$

Then λ is the image measure under φ of the $(d-1)$-dimensional Lebesgue measure restricted to $\widetilde{\Delta}$. Thus, if $\widetilde{\lambda}$ denotes the $(d-1)$-dimensional Lebesgue measure restricted to $\widetilde{\Delta}$, then $\lambda = \widetilde{\lambda} \circ \varphi^{-1}$.

Exercise 12.3.1. Prove that $\lambda(\Delta) = \widetilde{\lambda}(\widetilde{\Delta}) = 1/(d-1)!$. ◇

Since $V_t(\pi)$ is a polynomial of $\pi^1, \dots, \pi^d$, an exact computation of the integrals $\int V_t(\pi)\,\lambda(d\pi)$ and $\int \pi V_t(\pi)\,\lambda(d\pi)$, and hence of the universal portfolio, can be achieved via the following proposition.

Proposition 12.17. *Let $k_1, \dots, k_d$ be nonnegative integers. Then*

$$\int_{\Delta} x_1^{k_1} \cdots x_d^{k_d}\, \lambda(dx_1, \dots, dx_d) = \frac{k_1! \cdots k_d!}{(k_1 + \cdots + k_d + d - 1)!}. \tag{12.21}$$

The formula (12.21) will be explored in greater generality in Theorem 12.20. In particular, Proposition 12.17 will follow as a corollary of that more general result; see Remark 12.22. Note that taking $k_1 = \cdots = k_d = 0$ in (12.21) yields the formula $\lambda(\Delta) = 1/(d-1)!$ we already encountered in Exercise 12.3.1.

To compute the universal portfolio via Proposition 12.17, one first needs to expand the polynomials $V_t(\pi)$ and $\pi V_t(\pi)$ into their standard monomial basis representations. The leading terms of this expansion will have exponents $k_1, \dots, k_d$ with $k_1 + \cdots + k_d = t$, so that the denominator on the right-hand side of (12.21) is $(t + d - 1)!$. The computation of such a factorial number becomes practically impossible already for relatively moderate values of $t + d$. Moreover, expanding the polynomial $V_t(\pi)$ into its monomial basis representation is computationally costly. For these reasons, Proposition 12.17 is not a suitable method for computing the universal portfolio in applications. In the sequel, we will therefore explore the numerical computation of the integrals $\int V_t(\pi)\,\nu(d\pi)$ and $\int \pi V_t(\pi)\,\nu(d\pi)$ via a Monte Carlo approach. This method will allow us to efficiently evaluate those integrals for large t (in some of our simulations we will take $t = 10{,}000$), and it also works for a large class of probability measures ν (see Remark 12.24).

The general idea of the Monte Carlo method is to exploit the law of large numbers. Specifically, suppose that $(\Omega, \mathscr{F}, P)$ is a probability space supporting a sequence $\Pi_1, \Pi_2, \dots$ of independent Δ-valued random vectors with common distribution ν. When defining

$$\hat{V}_t^{(n)} := \frac{1}{n}\sum_{k=1}^{n} V_t(\Pi_k) \quad \text{and} \quad \hat{\pi}_t^{(n)} := \frac{\sum_{k=1}^{n} \Pi_k V_{t-1}(\Pi_k)}{\sum_{k=1}^{n} V_{t-1}(\Pi_k)}, \tag{12.22}$$

then Kolmogorov's law of large numbers yields that P-a. s. for $n \uparrow \infty$,

$$\begin{aligned} \hat{V}_t^{(n)} &\longrightarrow \int V_t(\pi)\,\nu(d\pi) = \hat{V}_t, \\ \hat{\pi}_t^{(n)} &\longrightarrow \frac{\int \pi\, V_{t-1}(\pi)\,\nu(d\pi)}{\int V_{t-1}(\pi)\,\nu(d\pi)} = \hat{\pi}_t. \end{aligned} \tag{12.23}$$

Thus, for sufficiently large n, the quantities $\hat{V}_t^{(n)}$ and $\hat{\pi}_t^{(n)}$ will provide reasonable estimates for $\hat{V}_t$ and $\hat{\pi}_t$. Note also that $\hat{\pi}^{(n)}$ can be regarded as the universal portfolio strategy with respect to the empirical distribution

$$\nu^{(n)} := \frac{1}{n}\sum_{k=1}^{n} \delta_{\Pi_k}.$$

The following lemma states that V_t^* can be approximated by the successive maxima of $V_t(\Pi_1), \dots, V_t(\Pi_n)$.

Lemma 12.18. *Suppose that ν has full support on Δ. Then P-a. s. for all t,*

$$\lim_{n\uparrow\infty} V_t(\Pi_1) \vee \dots \vee V_t(\Pi_n) = V_t^*.$$

Proof. Let $\varepsilon > 0$ be given. Then

$$\begin{aligned} P\big[V_1(\Pi_1) \vee \dots \vee V_t(\Pi_n) \le V_t^* - \varepsilon\big] &= P\Bigg[\bigcap_{k=1}^{n}\{V_t(\Pi_k) \le V_t^* - \varepsilon\}\Bigg] \\ &= P\big[V_t(\Pi_1) \le V_t^* - \varepsilon\big]^n. \end{aligned}$$

Since $\pi \mapsto V_t(\pi)$ is continuous by Lemma 12.6, the set $U := \{\pi \in \Delta \mid V_t(\pi) > V_t^* - \varepsilon\}$ is a nonempty open subset of Δ. Since ν has full support on Δ, we must have $\nu(U) > 0$, and hence

$$P\big[V_t(\Pi_1) \le V_t^* - \varepsilon\big] = 1 - \nu(U) < 1.$$

Therefore,

$$\sum_{n=1}^{\infty} P\big[V_1(\Pi_1) \vee \dots \vee V_t(\Pi_n) \le V_t^* - \varepsilon\big] < \infty,$$

and the Borel–Cantelli lemma yields that

$$\lim_{n\uparrow\infty} V_1(\Pi_1) \vee \cdots \vee V_t(\Pi_n) \geq V_t^* - \varepsilon \quad P\text{-a. s.}$$

Taking countable many ε_k that tend to zero as $k \uparrow \infty$ now gives the assertion. □

The approximations (12.23) of $\hat{V}_t$ and $\hat{\pi}_t$ rely on the generation of independent Δ-valued random vectors $\Pi_1, \ldots, \Pi_n$ with common distribution ν. Let us now focus on the task of creating such random vectors from simple univariate random variables, which can easily be generated via standard software. We start by considering the following class of Borel measures ν on Δ.

Definition 12.19. The *Dirichlet distribution* with parameters $\alpha_1, \ldots, \alpha_d > 0$ is the Borel measure ν on Δ with density

$$\frac{d\nu}{d\lambda}(x) = \frac{\Gamma(\alpha_1 + \cdots + \alpha_d)}{\Gamma(\alpha_1)\cdots\Gamma(\alpha_d)} \prod_{i=1}^{d} x_i^{\alpha_i - 1} \quad \text{for } x = (x_1, \ldots, x_d) \in \Delta, \tag{12.24}$$

where Γ is the standard gamma function. △

For $d = 2$, the Dirichlet distribution is simply a beta distribution, and hence a probability distribution. Another special case occurs if all parameters α_i are equal to 1, that is, $\alpha_1 = \cdots = \alpha_d = 1$. Then

$$\frac{d\nu}{d\lambda} = \frac{\Gamma(d)}{\Gamma(1)^d} = (d-1)!,$$

and so ν is simply the normalized Lebesgue measure on Δ; see Exercise 12.3.1. Hence, also in this case ν is a probability measure. In fact, it will follow from Theorem 12.20 that the Dirichlet distribution is always a probability measure on Δ. To prepare for the statement of that theorem, let $(\Omega, \mathscr{F}, P)$ be a probability space. A random variable X is called *gamma distributed* with parameters $\alpha, \beta > 0$, if its probability density function is given by

$$f(x) = \frac{\beta^\alpha x^{\alpha-1} e^{-\beta x}}{\Gamma(\alpha)} \mathbb{1}_{(0,\infty)}(x), \quad x \in \mathbb{R}.$$

For $\alpha = 1$, we recover the case of an exponentially distributed random variable with parameter β.

Theorem 12.20. *Let $X_1, \ldots, X_d$ be independent random variables such that each X_i is gamma distributed with parameters $\alpha_i > 0$ and $\beta = 1$. We also define*

$$S := X_1 + \cdots + X_d \quad \text{and} \quad \Pi^i := \frac{X_i}{S}, \quad i = 1, \ldots, d.$$

Then the Δ-valued random vector $(\Pi^1, \dots, \Pi^d)$ has a Dirichlet distribution with parameters $\alpha_1, \dots, \alpha_d$ and is independent of S. In particular, the Dirichlet distribution defined in (12.24) *is a probability measure on Δ.*

Proof. With $\widetilde{\Delta}$ as defined in (12.19), we introduce the diffeomorphism

$$\begin{aligned}\Phi : \widetilde{\Delta} \times (0,\infty) &\longrightarrow (0,\infty)^d \\ (x_1, \dots, x_{d-1}, s) &\longmapsto \Phi(x_1, \dots, x_{d-1}, s) = \big(sx_1, \dots, sx_{d-1}, s(1 - x_1 - \dots - x_{d-1})\big).\end{aligned}$$

Then $(X_1, \dots, X_d) = \Phi(\Pi^1, \dots, \Pi^{d-1}, S)$ and $(\Pi^1, \dots, \Pi^{d-1}, S) = \Phi^{-1}(X_1, \dots, X_d)$. If f is a measurable function from $\widetilde{\Delta} \times (0,\infty)$ to $[0,\infty)$, our goal is to compute

$$\begin{aligned}E[f(\Pi^1, \dots, \Pi^{d-1}, S)] &= E[f(\Phi^{-1}(X_1, \dots, X_d))] \\ &= \int\limits_{(0,\infty)^d} f(\Phi^{-1}(z_1, \dots, z_d))\bigg(\prod_{i=1}^d \frac{z_i^{\alpha_i - 1} e^{-z_i}}{\Gamma(\alpha_i)}\bigg)\, dz_1 \cdots dz_d. \end{aligned} \tag{12.25}$$

Denoting the integrand of the latter integral by $g(z_1, \dots, z_d)$ and noting that $(0,\infty)^d = \Phi(\widetilde{\Delta} \times (0,\infty))$, the transformation rule for multidimensional integrals states that

$$\begin{aligned}&E[f(\Pi^1, \dots, \Pi^{d-1}, S)] \\ &= \int\limits_{(0,\infty)^d} g(z_1, \dots, z_d)\, dz_1 \cdots dz_d \\ &= \int\limits_{\widetilde{\Delta} \times (0,\infty)} g(\Phi(x_1, \dots, x_{d-1}, s)) \big|\det(D\Phi(x_1, \dots, x_{d-1}, s))\big|\, dx_1, \dots dx_{d-1}\, ds,\end{aligned} \tag{12.26}$$

where $\det(D\Phi)$ is the determinant of the Jacobi matrix $D\Phi$ of Φ. Denoting $x_d := 1 - \sum_{i=1}^{d-1} x_i$, we have

$$\begin{aligned}g(\Phi(x_1, \dots, x_{d-1}, s)) &= f(x_1, \dots, x_{d-1}, s) \prod_{i=1}^d \frac{(sx_i)^{\alpha_i - 1} e^{-sx_i}}{\Gamma(\alpha_i)} \\ &= f(x_1, \dots, x_{d-1}, s) \bigg(\prod_{i=1}^d \frac{x_i^{\alpha_i - 1}}{\Gamma(\alpha_i)}\bigg) s^{\alpha_1 + \dots + \alpha_d - d} e^{-s}.\end{aligned} \tag{12.27}$$

Next, we need to compute the determinant of

$$D\Phi(x_1, \dots, x_{d-1}, s) = \begin{pmatrix} s & 0 & \cdots & \cdots & x_1 \\ 0 & s & 0 & \cdots & x_2 \\ \vdots & \ddots & \ddots & \ddots & \vdots \\ 0 & \cdots & 0 & s & x_{d-1} \\ -s & \cdots & \cdots & -s & 1 - \sum_i x_i \end{pmatrix}.$$

To this end, recall that the determinate of a matrix remains unchanged if one row is added to another. When adding rows 1 through $d-1$ of $D\Phi$ to row d, we get

$$\det(D\Phi(x_1,\dots,x_{d-1},s)) = \det\begin{pmatrix} s & 0 & \cdots & \cdots & x_1 \\ 0 & s & 0 & \cdots & x_2 \\ \vdots & \ddots & \ddots & \ddots & \vdots \\ 0 & \cdots & 0 & s & x_{d-1} \\ 0 & \cdots & \cdots & 0 & 1 \end{pmatrix}.$$

Using Laplace expansion along the dth row, we arrive at

$$\det(D\Phi(x_1,\dots,x_{d-1},s)) = s^{d-1}.$$

Plugging this result and (12.27) into (12.26) yields

$$E[f(\Pi^1,\dots,\Pi^{d-1},S)] = \int\limits_{\tilde{\Delta}\times(0,\infty)} f(x_1,\dots,x_{d-1},s)\left(\prod_{i=1}^{d}\frac{x_i^{\alpha_i-1}}{\Gamma(\alpha_i)}\right)s^{\alpha_1+\dots+\alpha_d-1}e^{-s}\,dx_1\cdots dx_{d-1}\,ds,$$

where we use again the shorthand notation $x_d = 1-\sum_{i=1}^{d-1}x_i$. If $f(x_1,\dots,x_{d-1},s) = f_1(x_1,\dots,x_{d-1})f_2(s)$ for two nonnegative measurable functions f_1 and f_2, then the integral on the right-hand side factorizes into two integrals, and we obtain the asserted independence of S and $(\Pi^1,\dots,\Pi^d)$. If f does not depend on s, then the definition of the gamma function implies

$$\begin{aligned} E[f(\Pi^1,\dots,\Pi^{d-1})] &= \int\limits_{\tilde{\Delta}\times(0,\infty)} f(x_1,\dots,x_{d-1})\left(\prod_{i=1}^{d}\frac{x_i^{\alpha_i-1}}{\Gamma(\alpha_i)}\right)s^{\alpha_1+\dots+\alpha_d-1}e^{-s}\,dx_1\cdots dx_{d-1}\,ds \\ &= \int\limits_{\tilde{\Delta}} f(x_1,\dots,x_{d-1})\left(\prod_{i=1}^{d}\frac{x_i^{\alpha_i-1}}{\Gamma(\alpha_i)}\right)dx_1\cdots dx_{d-1}\int\limits_0^\infty s^{\alpha_1+\dots+\alpha_d-1}e^{-s}\,ds \\ &= \frac{\Gamma(\alpha_1+\dots+\alpha_d)}{\Gamma(\alpha_1)\cdots\Gamma(\alpha_d)}\int\limits_{\tilde{\Delta}} f(x_1,\dots,x_{d-1})\left(\prod_{i=1}^{d}x_i^{\alpha_i-1}\right)dx_1\cdots dx_{d-1}. \end{aligned}$$

By using the map $\varphi : \tilde{\Delta} \to \Delta$ introduced in (12.20), we can transform the latter integral into an integral over Δ and thus complete the proof. □

Remark 12.21. The proof of Theorem 12.20 shows that $S = X_1 + \dots + X_d$ is gamma distributed with parameters $\alpha_1 + \dots + \alpha_d$ and $\beta = 1$. Of course, this fact also follows from the well-known convolution property of the gamma distribution. This observation also yields the following converse of Theorem 12.20: If $(\Pi^1,\dots,\Pi^d)$ has a Dirichlet distribution with parameters $\alpha_1,\dots,\alpha_d$ and S is an independent gamma distributed random variable with parameters $\alpha_1 + \cdots + \alpha_d$ and $\beta = 1$, then $X_i := S\Pi^i$ is gamma distributed with parameters α_i and 1, and the random variables $X_1, \ldots, X_d$ are independent. ◇

Remark 12.22. The fact that the Dirichlet distribution ν is a probability measure yields as a corollary that for $\alpha_1, \ldots, \alpha_d > 0$,

$$\int_\Delta x_1^{\alpha_1-1} \cdots x_d^{\alpha_d-1} \lambda(dx_1, \ldots, dx_d) = \frac{\Gamma(\alpha_1) \cdots \Gamma(\alpha_d)}{\Gamma(\alpha_1 + \cdots + \alpha_d)}.$$

Using the fact that $\Gamma(k+1) = k!$ for every nonnegative integer k, we especially obtain the formula stated in Proposition 12.17. ◇

There are several methods for generating gamma-distributed random variables with parameters $\alpha > 0$ and $\beta = 1$. When combining such a method with Theorem 12.20, one obtains a technique for generating Δ-valued random vectors with a general Dirichlet distribution. In this way, Theorem 12.20 provides a tool for numerically generating random vectors from a very versatile family of probability distributions on Δ. A particularly simple situation occurs if all parameters α_i in Theorem 12.20 are equal to one. In this case, we have the following immediate corollary on the construction of random variables with a uniform distribution on Δ.

Corollary 12.23. *Let $X_1, \ldots, X_d$ be independent random variables such that each X_i is exponentially distributed with parameter $\alpha = 1$. Then the Δ-valued random vector $(\Pi^1, \ldots, \Pi^d)$ with components*

$$\Pi^i = \frac{X_i}{X_1 + \cdots + X_d}, \qquad i = 1, \ldots, d, \tag{12.28}$$

has law $(d-1)!\lambda$.

The preceding corollary provides the following straightforward method for constructing random vectors with a uniform distribution on Δ from d independent random variables with a uniform distribution on $(0,1)$. The generation of such random variables $U_1, \ldots, U_d$ is readily available in most standard software packages. First, the exponential distribution with parameter 1 has the cumulative distribution function

$$F(x) = (1 - e^{-x})\mathbb{1}_{[0,\infty)}(x),$$

and hence the quantile function

$$q(t) = \log \frac{1}{1-t}.$$

Next, the random variables $X_i := q(U_i)$ are independent and have the desired exponential distribution; see Lemma D.7. From here, we now get the uniformly distributed random vector $\Pi = (\Pi^1, \ldots, \Pi^d)$ via (12.28).

Remark 12.24. There are several approaches for modifying the above Monte Carlo approach so that it applies to the computation of integrals with respect to a more general probability distribution ν on Δ. Here, we briefly sketch the method of *importance sampling*. To this end, suppose that ν admits the density φ with respect to the normalized Lebesgue measure $(d-1)!\lambda$ on Δ. Let $\Pi_1, \Pi_2, \ldots$ be a sequence of independent random vectors with a uniform distribution on Δ. We now replace the estimation formulas (12.22) by

$$\hat{V}_t^{(n)} := \frac{1}{n}\sum_{k=1}^{n} V_t(\Pi_k)\varphi(\Pi_k) \quad \text{and} \quad \hat{\pi}_t^{(n)} := \frac{\sum_{k=1}^{n} \Pi_k V_{t-1}(\Pi_k)\varphi(\Pi_k)}{\sum_{k=1}^{n} V_{t-1}(\Pi_k)\varphi(\Pi_k)}.$$

Then Kolmogorov's law of large numbers yields that P-a. s. for each t,

$$\lim_{n\uparrow\infty} \hat{V}_t^{(n)} = E\big[V_t(\Pi_1)\varphi(\Pi_1)\big] = (d-1)! \int V_t(\pi)\varphi(\pi)\,\lambda(d\pi) = \int V_t(\pi)\,\nu(d\pi),$$

which by Proposition 12.15 is equal to the value process $\hat{V}_t$ of the universal portfolio strategy $\hat{\pi}$ with respect to ν. In the same way, one gets $\lim_n \hat{\pi}_t^{(n)} = \hat{\pi}_t$. ◇

In the remainder of this section, we provide two numerical case studies for the performance of the universal portfolio strategy. One case study will be based on historical stock market prices and the other one on synthetically generated data. For the case study with historic stock market prices, we considered the 11 sectors of the S&P 500 Index. These sectors can be traded in the form of the 11 exchange-traded funds (ETFs) listed in Table 12.1. We used the closing prices, adjusted for dividends and stock splits, for the 351 weeks between June 22, 2018, and March 7, 2025. From these data, the universal portfolio was computed using the method derived in Corollary 12.23. To this end, we generated 1,100,000 independent standard exponential random variables. These were grouped into 100,000 independent random vectors of dimension $d = 11$, from which we derived 100,000 independent random vectors with a uniform distribution on the

Table 12.1: S&P 500 Sector ETFs.

Ticker	Sector name
XLB	Materials
XLC	Communication Services
XLE	Energy
XLF	Financials
XLI	Industrials
XLK	Information Technology
XLP	Consumer Staples
XLRE	Real Estate
XLU	Utilities
XLV	Health Care
XLY	Consumer Discretionary

d-dimensional standard simplex. Then we used (12.22) to compute approximations of order $n = 100,000$ for the universal portfolio strategy and its value process; see Figures 12.2 and 12.3. Figure 12.4 compares the growth rates $\frac{1}{t}\log \hat{V}_t$ and $\frac{1}{t}\log V_t^*$ over time.

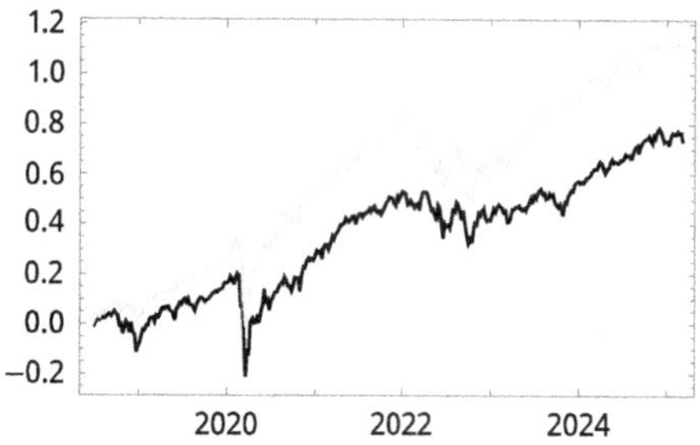

Figure 12.2: Logarithmic plot of the universal portfolio $\hat{V}_t$ (black) and the hindsight-optimal strategy V_t^* (gray) for the 11 sector ETFs.

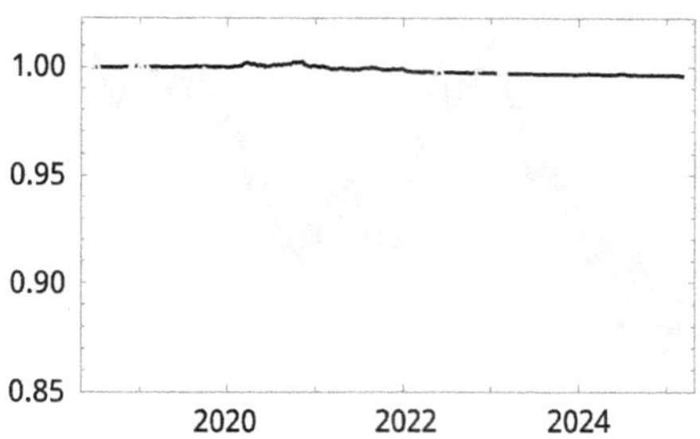

Figure 12.3: Ratios $\hat{V}_t/V_t(\pi_{\mathrm{eq}})$ (black) and $\hat{V}_t/V_t^{\mathrm{SPY}}$ (gray), where π_{eq} is the equal-weighted portfolio of the 11 sector ETFs and V_t^{SPY} denotes time-t value of a unit investment at $t = 0$ into the ETF with ticker symbol SPY, which tracks the performance of the S&P 500. While SPY appears to outperform both the universal and equal-weighted portfolios, Figure 12.1 suggests that this outperformance may be episodic.

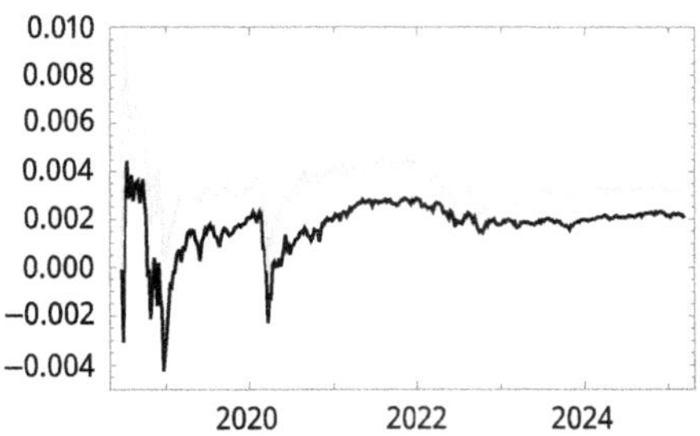

Figure 12.4: Growth rates $\frac{1}{t}\log \hat{V}_t$ (black) and $\frac{1}{t}\log V_t^*$ (gray) for the 11 sector ETFs.

For our case study with synthetic data, we took $d = 10$ and generated 10,000 independent performance vectors Y_k with a multivariate log-normal distribution. That is, $Y_k^i = \exp(X_k^i)$, where the X_k are i. i. d. random vectors with a multivariate normal distribution with random covariance matrix and random mean vector. Then we used again Corollary 12.23 to generate 100,000 independent random vectors with a uniform distribution on the d-dimensional standard simplex and used those to compute approximations of order $n = 100,000$ for the universal portfolio strategy and its value process. In this simulation, the universal portfolio outperformed the equal-weighted portfolio; see Figures 12.5 and 12.6. However, Figures 12.5 and 12.7 show that, despite the large data

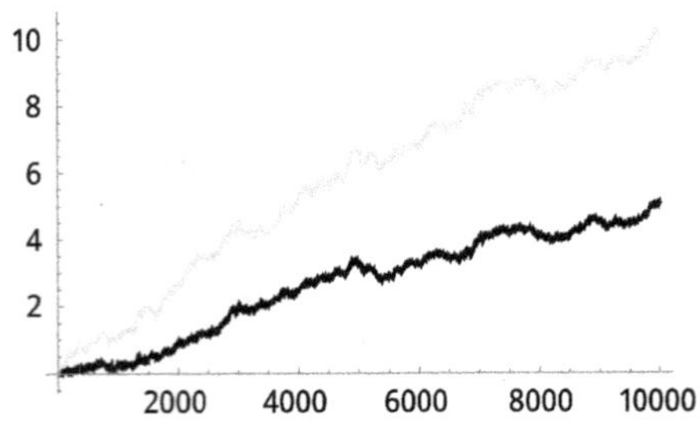

Figure 12.5: Logarithmic plot of the universal portfolio $\hat{V}_t$ (black) and the hindsight-optimal strategy V_t^* (gray) for synthetic data.

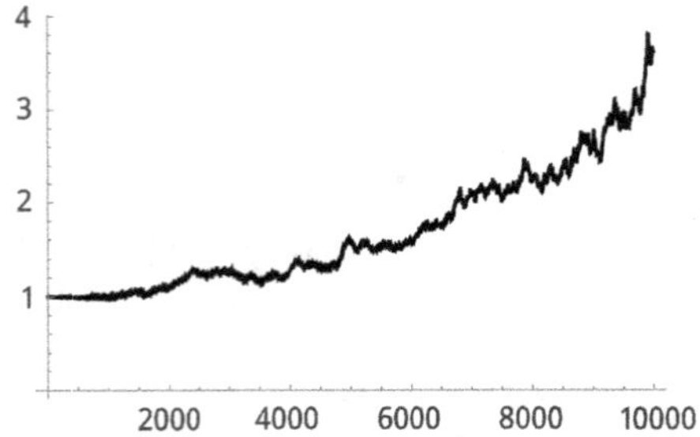

Figure 12.6: Relative performance $\hat{V}_t/V_t(\pi_{\text{eq}})$, where π_{eq} is the equal-weighted portfolio, for synthetic data.

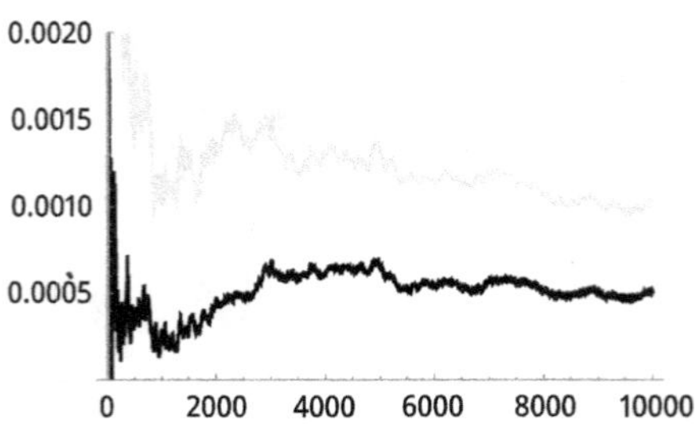

Figure 12.7: Growth rates $\frac{1}{t}\log\hat{V}_t$ (black) and $\frac{1}{t}\log V_t^*$ (gray) for synthetic data.

set, the growth rate of the universal portfolio still lags behind the one of the hindsight-optimal portfolio V^*. This suggests that the convergence stated in Theorem 12.16 may be slow.

A Convexity

A.1 Convex sets

This section contains a few basic facts on convex functions and on convex sets. For further facts, we refer to [273], from where we also took some of the material in this section. In the sequel, E will denote a general real vector space.

Definition A.1. A subset C of E is called *convex* if $\lambda x + (1-\lambda)y \in C$ whenever $x, y \in C$ and $0 \le \lambda \le 1$. △

A *convex combination* of $x_1, \dots, x_n \in E$ is of the form

$$\lambda_1 x_1 + \cdots + \lambda_n x_n, \tag{A.1}$$

with real numbers $\lambda_i \ge 0$ satisfying $\lambda_1 + \cdots + \lambda_n = 1$.

Proposition A.2. *A subset C of E is convex if and only if it contains all convex combinations of its elements.*

Proof. If C contains all convex combinations of its elements, then it is convex, as $\lambda x + (1-\lambda)y$ for $0 \le \lambda \le 1$ is clearly a convex combination of x and y, and hence belongs to C. Conversely, we use induction on n in (A.1). The base case $n = 2$ follows from Definition A.1. Next, suppose that C contains all convex combinations for some $n \ge 2$, and consider $x = \lambda_1 x_1 + \cdots + \lambda_{n+1} x_{n+1}$, where $x_i \in C$, $\lambda_i \ge 0$, and $\lambda_1 + \cdots + \lambda_{n+1} = 1$. If $\lambda_{n+1} = 0$, we are done. Otherwise,

$$x = \lambda_1 x_1 + \cdots + \lambda_{n-1} x_{n-1} + \tilde{\lambda}_n \tilde{x}_n,$$

where $\tilde{\lambda}_n = \lambda_n + \lambda_{n+1} > 0$ and $\tilde{x}_n = \alpha x_n + (1-\alpha) x_{n+1}$ for $\alpha := \lambda_n / \tilde{\lambda}_n \in [0, 1)$. The convexity of C implies $\tilde{x}_n \in C$. Since moreover $\lambda_1 + \cdots + \lambda_{n-1} + \tilde{\lambda}_n = 1$, the induction hypothesis now implies $x \in C$. □

Definition A.3. Let A be a nonempty subset of E. The *convex hull* of $A \subseteq E$ is a set denoted by $\operatorname{conv} A$ and defined as the following intersection:

$$\operatorname{conv} A = \bigcap_{\substack{C \subseteq E \text{ convex} \\ A \subseteq C}} C.$$ △

Since the vector space E is a convex set containing A, it is clear that $\operatorname{conv} A$ is well-defined. It follows in particular that $A \subseteq \operatorname{conv} A$.

Proposition A.4. *The convex hull of a nonempty set $A \subseteq E$ admits the following alternative characterizations:*

(a) $\operatorname{conv} A$ *is the smallest convex set containing* A.

https://doi.org/10.1515/9783111045283-015

(b) conv A *is equal to the set of all convex combinations of elements of A, that is,*

$$\operatorname{conv} A = \left\{ \sum_{i=1}^{n} \lambda_i x_i \,\middle|\, \lambda_i \geq 0,\ \sum_{i=1}^{n} \lambda_i = 1,\ x_i \in A,\ n \in \mathbb{N} \right\}. \tag{A.2}$$

Proof. (a) It is easy to see that the intersection of convex sets is again convex. Hence, conv A is indeed a convex set. It is moreover clear from the definition that there can be no smaller convex set C containing A.

(b) Let C_0 denote the right-hand side of (A.2). It is easy to see that C_0 is convex and contains A. Therefore, $\operatorname{conv} A \subseteq C_0$. Conversely, let C be any convex set containing A. By Proposition A.2, C must contain every convex combination of elements of A. Hence, $C_0 \subseteq C$. This implies $C_0 \subseteq \operatorname{conv} A$ and concludes the proof. □

Now we consider specifically $E = \mathbb{R}^n$ and denote by

$$x \cdot y = \sum_{i=1}^{n} x^i y^i$$

the Euclidean inner product of $x = (x^1, \dots, x^n) \in \mathbb{R}^n$ and $y = (y^1, \dots, y^n) \in \mathbb{R}^n$. By

$$|x| := \sqrt{x \cdot x}$$

we denote the Euclidean norm of $x \in \mathbb{R}^n$.

Exercise A.1.1. Suppose that $\mathscr{C}$ is a nonempty convex subset of $\mathbb{R}^n$. Show that its closure $\overline{\mathscr{C}}$ is also convex. ◇

The following proposition is a version of Minkowski's separating hyperplane theorem.

Proposition A.5. *Suppose that $\mathscr{C} \subseteq \mathbb{R}^n$ is a nonempty convex set with $0 \notin \mathscr{C}$. Then there exists $\eta \in \mathbb{R}^n$ with $\eta \cdot x \geq 0$ for all $x \in \mathscr{C}$, and with $\eta \cdot x_0 > 0$ for at least one $x_0 \in \mathscr{C}$. If, moreover, $\inf_{x \in \mathscr{C}} |x| > 0$, then one can find $\eta \in \mathbb{R}^n$ with $\inf_{x \in \mathscr{C}} \eta \cdot x > 0$.*

Proof. First, we consider the case in which $y := \inf_{x \in \mathscr{C}} |x| > 0$. This infimum is attained by some nonzero y in the closure $\overline{\mathscr{C}}$ of $\mathscr{C}$. Indeed, $x \mapsto |x|$ is a continuous function, and hence attains its minimum on the compact set that is obtained by intersecting $\overline{\mathscr{C}}$ with the closed ball of radius $y + 1$ centered in 0. Moreover,

$$0 < y = |y| = \inf_{x \in \overline{\mathscr{C}}} |x|. \tag{A.3}$$

Since $\overline{\mathscr{C}}$ is a convex set by Exercise A.1.1, we have $\alpha x + (1 - \alpha) y \in \overline{\mathscr{C}}$ for each $x \in \mathscr{C}$ and all $\alpha \in [0, 1]$. Thus, by (A.3),

$$0 \leq \left|\alpha x + (1 - \alpha) y\right|^2 - |y|^2 = 2\alpha y \cdot (x - y) + \alpha^2 |x - y|^2.$$

Dividing by α and sending α to 0 yields $y \cdot x \geq y \cdot y > 0$, and so we can take $\eta := y$.

The case in which $\inf_{x\in\mathscr{C}}|x| = 0$ is more subtle. For the first step, let $\{e_1,\dots,e_k\}$ be a maximal collection of linearly independent vectors in $\mathscr{C}$. Then each $x \in \mathscr{C}$ can be expressed as a linear combination of $e_1,\dots,e_k$. Let

$$z := -\sum_{i=1}^{k} e_i.$$

We claim that βz does not belong to $\overline{\mathscr{C}}$ for each $\beta > 0$. To prove this claim, we assume by way of contradiction that $\beta z \in \overline{\mathscr{C}}$ for some $\beta > 0$. Then there are $z_n \in \mathscr{C}$ converging to βz. If we write $z_n = \sum_{i=1}^k \lambda_n^i e_i$, then $z_n \to \beta z$ is equivalent to the convergence $\lambda_n^i \to -\beta$ for all i. It follows that for some $n_0 \in \mathbb{N}$ all coefficients $\lambda_{n_0}^i$ are strictly negative. Let

$$\alpha_0 := \frac{1}{1-\sum_{i=1}^k \lambda_{n_0}^i} \quad \text{and} \quad \alpha_j := \frac{-\lambda_{n_0}^j}{1-\sum_{i=1}^k \lambda_{n_0}^i} \quad \text{for } j = 1,\dots,k.$$

Then $\alpha_j > 0$ for each j and $\alpha_0 + \cdots + \alpha_k = 1$. Thus, the convexity of $\mathscr{C}$ implies that

$$0 = \alpha_0 z_{n_0} + \alpha_1 e_1 + \cdots + \alpha_k e_k \in \mathscr{C},$$

which is a contradiction to our assumption $0 \notin \mathscr{C}$. This proves that βz is not contained in $\overline{\mathscr{C}}$.

In the second step, we can now prove the existence of a separating η for the case in which 0 is a boundary point of $\mathscr{C}$, and thus $\inf_{x\in\mathscr{C}}|x| = 0$. We may assume without loss of generality that $\mathscr{C}$ is not contained in a proper affine subspace of $\mathbb{R}^n$; otherwise, we replace $\mathbb{R}^n$ with the Euclidean space spanned by the vectors $e_1,\dots,e_k$ constructed in the preceding step. Take a sequence $\beta_m \downarrow 0$ and let $\mathscr{C}_m := \mathscr{C} - \beta_m z$, where z is as in the preceding step. Then $0 \notin \overline{\mathscr{C}} - \beta_m z = \overline{\mathscr{C}}_m$, and consequently, $\inf_{x\in\mathscr{C}_m}|x| > 0$. Hence, the first part of the proof yields $\eta_m \in \mathbb{R}^n$ such that

$$\inf_{x\in\mathscr{C}_m} \eta_m \cdot x > 0.$$

We may assume without loss of generality that $|\eta_m| = 1$ for all m. By compactness of the $(n-1)$-dimensional unit sphere, there exists a convergent subsequence (η_{m_k}) with limit η. Thus,

$$\eta \cdot x = \lim_{k\uparrow\infty} \eta_{m_k} \cdot x = \lim_{k\uparrow\infty} \eta_{m_k} \cdot (x - \beta_{m_k} z) \geq \liminf_{k\uparrow\infty} \inf_{y\in\mathscr{C}_{m_k}} \eta_{m_k} \cdot y \geq 0$$

for all $x \in \mathscr{C}$. Since η is nonzero and $\mathscr{C}$ is not contained in a proper affine subspace of $\mathbb{R}^n$, the case $\eta \cdot x = 0$ for all $x \in \mathscr{C}$ cannot occur, and so there must be some $x_0 \in \mathscr{C}$ with $\eta \cdot x_0 > 0$. □

A.2 Convex functions

Now we consider again the situation of a general real vector space E.

Definition A.6. Let C be a convex subset of E. A function $f : C \to [-\infty, +\infty]$ is called *convex* if its *epigraph*,

$$\operatorname{epi} f := \{(x, \alpha) \in E \times \mathbb{R} \mid f(x) \le \alpha\},$$

is a convex subset of $E \times \mathbb{R}$. The *effective domain* of f is defined as

$$\operatorname{dom} f := \{x \in C \mid f(x) < \infty\}.$$

A convex function f is called *proper* if $\operatorname{dom} f \neq \emptyset$ and if f does not take the value $-\infty$. A function $u : C \to [-\infty, +\infty]$ is called *concave* if $f := -u$ is convex. A concave function is called *proper* if $f := -u$ is a proper convex function. △

Exercise A.2.1. Let C be a convex subset of E and $f : C \to \mathbb{R}$ a function. Show that the following conditions are equivalent:
(a) f is convex.
(b) For all $x, y \in C$ and $\alpha \in [0, 1]$, we have

$$f(\alpha x + (1-\alpha)y) \le \alpha f(x) + (1-\alpha) f(y). \tag{A.4}$$

(c) For $x_1, \dots, x_n \in E$ and $\lambda_1, \dots, \lambda_n \ge 0$ with $\lambda_1 + \cdots + \lambda_n = 1$,

$$f(\lambda_1 x_1 + \cdots + \lambda_n x_n) \le \lambda_1 f(x_1) + \cdots + \lambda_n f(x_n).$$ ◇

Often the convexity of a function is defined by requiring the condition (A.4) instead of requiring the convexity of its epigraph. Note, however, that the right-hand side of (A.4) may not be well-defined if f takes both $-\infty$ and $+\infty$ as values.

A proper convex function $f : C \to \mathbb{R} \cup \{+\infty\}$, defined on the convex set $C \subseteq E$, can be extended to the entire space E by putting $f(x) := +\infty$ if $x \in E \setminus C$. It is easy to verify that this extension is again a proper convex function. It is therefore sufficient to study proper convex functions that are defined on the full space E.

Now we consider the special situation in which f is a proper convex function on $E = \mathbb{R}$. In this case, $\operatorname{dom} f$ is a real interval. The following proposition summarizes continuity and differentiability properties of a proper convex function on its effective domain.

Proposition A.7. *Let f be a proper convex function on $\mathbb{R}$, and denote by D the interior of* $\operatorname{dom} f$.
(a) *f is upper semicontinuous on* $\operatorname{dom} f$ *and locally Lipschitz continuous on D.*

(b) *f admits left- and right-hand derivatives,*

$$f'_-(y) := \lim_{x \uparrow y} \frac{f(x) - f(y)}{x - y} \quad \text{and} \quad f'_+(y) := \lim_{z \downarrow y} \frac{f(z) - f(y)}{z - y},$$

at each $y \in D$. Both f'_+ and f'_- are increasing functions and satisfy $f'_- \le f'_+$.

(c) *The right-hand derivative f'_+ is right-continuous, the left-hand derivative f'_- is left-continuous.*

(d) *f is differentiable a. e. in D, and for any $x_0 \in D$,*

$$f(x) = f(x_0) + \int_{x_0}^{x} f'_+(y)\, dy = f(x_0) + \int_{x_0}^{x} f'_-(y)\, dy, \quad x \in D.$$

Proof. We first prove part (b). For $x, y, z \in D$ with $x < y < z$, we take $\alpha \in (0,1)$ such that $y = \alpha z + (1-\alpha)x$. Using the convexity of f, one gets

$$\frac{f(y) - f(x)}{y - x} \le \frac{f(z) - f(x)}{z - x} \le \frac{f(z) - f(y)}{z - y}. \tag{A.5}$$

Thus, the difference quotient

$$\frac{f(x) - f(y)}{x - y}$$

is an increasing function of x, which shows the existence of the left- and right-hand derivatives. Moreover, we get $f'_-(y) \le f'_+(y) \le f'_-(z)$ for $y < z$.

(a): Let $z \in \operatorname{dom} f$, and take a sequence (x_n) in $\operatorname{dom} f$ such that $x_n \to z$. Without loss of generality, we may assume that $x_n \downarrow z$ or $x_n \uparrow z$. In either case, $x_n = \delta_n x_1 + (1 - \delta_n)z$, where $\delta_n \downarrow 0$. Convexity of f yields

$$\limsup_{n\uparrow\infty} f(x_n) \le \limsup_{n\uparrow\infty} (\delta_n f(x_1) + (1 - \delta_n) f(z)) = f(z),$$

and so f is upper semicontinuous. To prove local Lipschitz continuity, take $a \le x < y \le b$ such that $[a,b] \subset D$. We get from part (b) that

$$f'_+(a) \le f'_+(x) \le \frac{f(x) - f(y)}{x - y} \le f'_-(y) \le f'_+(b).$$

Hence, f is a Lipschitz continuous function on $[a,b]$ with Lipschitz constant $L := |f'_+(a)| \vee |f'_-(b)|$.

(c): The continuity of f implies that for $x < z$,

$$\frac{f(z) - f(x)}{z - x} = \lim_{y\downarrow x} \frac{f(z) - f(y)}{z - y} \ge \limsup_{y\downarrow x} f'_+(y).$$

Taking $z \downarrow x$ yields $f'_+(x) \geq \limsup_{y\downarrow x} f'_+(y)$. Since f'_+ is increasing, we must in fact have $f'_+(y) \to f'_+(x)$ as $y \downarrow x$. In the same way, one shows left-continuity of f'_-.

(d): Since the function f is locally Lipschitz continuous, it is absolutely continuous on every compact interval in D. By Lebesgue's differentiation theorem, f is hence a. e. differentiable and equal to the integral of its derivative, which is equal to $f'_-(x) = f'_+(x)$ for a. e. $x \in D$. □

Exercise A.2.2. Let $f : (0,1) \to \mathbb{R}$ be a function that is bounded from above and satisfies

$$f\left(\frac{x+y}{2}\right) \leq \frac{1}{2}f(x) + \frac{1}{2}f(y) \quad \text{for } x,y \in (0,1). \tag{A.6}$$

Prove that f is continuous and conclude then that f is convex.
Note: See [172, p. 96] for the construction of a function $f : \mathbb{R} \to \mathbb{R}$ that satisfies $f(x+y) = f(x) + f(y)$ for $x,y \in \mathbb{R}$, and hence (A.6), but that is neither convex nor continuous. ◇

Definition A.8. The *Fenchel–Legendre transform* of a function $f : \mathbb{R} \to \mathbb{R} \cup \{+\infty\}$ is defined as

$$f^*(y) := \sup_{x\in\mathbb{R}}(yx - f(x)), \quad y \in \mathbb{R}.$$ △

If $f \not\equiv +\infty$, then f^* is convex and lower semicontinuous as the supremum of the affine functions $y \mapsto yx - f(x)$. If f is a proper convex function, then f^* is also called the *conjugate function* of f.

Proposition A.9. *Let f be a proper convex function on $\mathbb{R}$.*
(a) *For all $x,y \in \mathbb{R}$,*

$$xy \leq f(x) + f^*(y) \tag{A.7}$$

with equality if x belongs to the interior of $\operatorname{dom} f$ and if $y \in [f'_-(x), f'_+(x)]$.
(b) *If f is lower semicontinuous, then $f^{**} = f$, that is,*

$$f(x) = \sup_{y\in\mathbb{R}}(xy - f^*(y)), \quad x \in \mathbb{R}.$$

Proof. (a): The inequality (A.7) is obvious. Now suppose that x_0 belongs to the interior of $\operatorname{dom} f$. Proposition A.7 yields $f(x) \geq f(x_0) + f'_\pm(x_0)(x - x_0)$ for all x in the interior of $\operatorname{dom} f$ and, by upper semi-continuity, for all $x \in \mathbb{R}$. Hence, $f(x) \geq f(x_0) + y_0\,(x - x_0)$ whenever $y_0 \in [f'_-(x_0), f'_+(x_0)]$. This shows that $xy_0 - f(x) \leq x_0y_0 - f(x_0)$ for all $x \in \mathbb{R}$, that is,

$$x_0y_0 - f(x_0) = \sup_{x\in\mathbb{R}}(x\,y_0 - f(x)) = f^*(y_0).$$

(b): We first show the following auxiliary claim: If $\beta < f(x_0)$, then there exists an affine function h such that $h(x_0) = \beta$ and $h(x) < f(x)$ for all x. For the proof of this claim, consider the epigraph,

$$\operatorname{epi} f = \{(x, a) \in \mathbb{R}^2 \mid f(x) \le a\}.$$

It is a nonempty convex set since f is convex and proper, and it is closed due to the lower semicontinuity of f. The point (x_0, β) does not belong to $\operatorname{epi} f$, and Proposition A.5, applied to $\mathscr{C} := \operatorname{epi} f - (x_0, \beta)$, thus yields some $\eta = (\eta_1, \eta_2) \in \mathbb{R}^2$ such that

$$\delta := \inf_{(x,a)\in \operatorname{epi} f} (\eta_1 x + \eta_2 a) > \eta_1 x_0 + \eta_2 \beta.$$

Since $(x, f(x)) \in \operatorname{epi} f$ for all $x \in \operatorname{dom} f$, it follows that

$$\inf_{x\in \operatorname{dom} f} (\eta_1 x + \eta_2 f(x)) \ge \delta > \eta_1 x_0 + \eta_2 \beta.$$

If $f(x_0) < \infty$, we get $\eta_1 x_0 + \eta_2 f(x_0) > \eta_1 x_0 + \eta_2 \beta$. Hence, $\eta_2 > 0$, and one checks that

$$h(x) := -\frac{\eta_1}{\eta_2}(x - x_0) + \beta$$

is as desired. If $f(x_0) = \infty$ and $\eta_2 > 0$, then the same definition works. Now assume that $f(x_0) = \infty$ and $\eta_2 = 0$. Since f is proper, the first step of the proof of our claim, applied with some $\widetilde{x}_0 \in \operatorname{dom} f$, allows us to construct an affine function g with $g < f$. If $g(x_0) \ge \beta$, then $h := g + \beta - g(x_0)$ is as desired. Otherwise, we let $\widetilde{h}(x) := \delta - \eta_1 x$. Then $\widetilde{h}(x_0) > 0$ and $\widetilde{h}(x) \le 0$ for $x \in \operatorname{dom} f$. It follows that $h(x) := g(x) + \lambda \widetilde{h}(x)$ for $\lambda := (\beta - g(x_0))/\widetilde{h}(x_0) > 0$ is as desired. This concludes the proof of our auxiliary claim.

Now we can complete the proof of part (b). By (A.7), we have $f(x) \ge xy - f^*(y)$ for all y, and thus $f \ge f^{**}$. Suppose there exists a point x_0 such that $f(x_0) > f^{**}(x_0)$. Take β strictly between $f^{**}(x_0)$ and $f(x_0)$. By the auxiliary claim, there exists an affine function $h < f$ such that $h(x_0) = \beta$. Let us write $h(x) = y_0 x + \alpha$. Then it follows that

$$f^*(y_0) = \sup_{x\in\mathbb{R}} (xy_0 - f(x)) \le \sup_{x\in\mathbb{R}} (xy_0 - h(x)) = -\alpha,$$

and hence

$$f^{**}(x_0) \ge y_0 x_0 - f^*(y_0) \ge h(x_0) = \beta > f(x_0),$$

which is a contradiction. □

B Absolutely continuous probability measures

B.1 Radon–Nikodym derivatives

Suppose that P and Q are two probability measures on a measurable space $(\Omega, \mathscr{F})$.

Definition B.1. Q is said to be *absolutely continuous* with respect to P on the σ-algebra $\mathscr{F}$, and we write $Q \ll P$, if for all $A \in \mathscr{F}$,

$$P[\,A\,] = 0 \quad \Longrightarrow \quad Q[\,A\,] = 0.$$

If both $Q \ll P$ and $P \ll Q$ hold, we will say that Q and P are *equivalent*, and we will write $Q \approx P$. △

The following characterization of absolute continuity is known as the Radon–Nikodym theorem.

Theorem B.2 (Radon–Nikodym). *Q is absolutely continuous with respect to P on $\mathscr{F}$ if and only if there exists an $\mathscr{F}$-measurable function $\varphi \geq 0$ such that*

$$E_Q[\,F\,] = \int F\,dQ = \int F\varphi\,dP = E[\,F\varphi\,] \tag{B.1}$$

for all $\mathscr{F}$-measurable functions $F \geq 0$.

Proof. See, for example, § 17 of [22]. □

The function φ is called the *density* or *Radon–Nikodym derivative* of Q with respect to P, and we will write

$$\frac{dQ}{dP} := \varphi.$$

Clearly, the Radon–Nikodym derivative is uniquely determined through (B.1) up to a P-null set and it satisfies

$$E\left[\,\frac{dQ}{dP}\,\right] = \int \frac{dQ}{dP}\,dP = Q[\,\Omega\,] = 1.$$

Remark B.3. If $Q \ll P$, then

$$Q\left[\,\frac{dQ}{dP} = 0\,\right] = \int_{\{\frac{dQ}{dP}=0\}} \frac{dQ}{dP}\,dP = 0.$$ ◇

Proposition B.4. *Let P, Q, and R be three probability measures on $(\Omega, \mathscr{F})$ and suppose that $Q \ll P$.*

https://doi.org/10.1515/9783111045283-016

(a) *If* $R \ll Q$*, then* $R \ll P$ *and the following* chain rule for Radon–Nikodym derivatives *holds:*

$$\frac{dR}{dP} = \frac{dR}{dQ} \cdot \frac{dQ}{dP} \quad P\text{-a. s.} \tag{B.2}$$

(b) *If* $R \ll P$*, then* $R \ll Q$ *holds if and only if*

$$\frac{dQ}{dP} > 0 \quad R\text{-a. s.}, \tag{B.3}$$

and in this case

$$\frac{dR}{dQ} = \frac{dR}{dP} \cdot \left(\frac{dQ}{dP}\right)^{-1} \quad Q\text{-a. s.} \tag{B.4}$$

Proof. (a) If $A \in \mathscr{F}$ is a P-null set, then $Q[A] = 0$ and in turn $R[A] = 0$. Hence, $R \ll P$. For proving (B.2), we need to make sure that the right-hand side of (B.2) is P-a. s. well-defined, because dR/dQ is a priori only Q-a. s. uniquely defined. To this end, let φ be a nonnegative and $\mathscr{F}$-measurable random variable such that $\int F\varphi\, dQ = \int F\, dR$ for all $\mathscr{F}$-measurable $F \geq 0$. That is, φ is a version of the Radon–Nikodym derivative dR/dQ. Then

$$\int F\, dR = \int F\varphi\, dQ = \int F\varphi \cdot \frac{dQ}{dP}\, dP \tag{B.5}$$

for all $\mathscr{F}$-measurable $F \geq 0$. It follows that

$$\frac{dR}{dP} = \varphi \cdot \frac{dQ}{dP} \quad P\text{-a. s.}$$

Since the left-hand side of (B.5) does not depend on the particular version φ of dR/dQ, the same must be true of its right-hand side. That is,

$$\varphi \cdot \frac{dQ}{dP} = \widetilde{\varphi} \cdot \frac{dQ}{dP} \quad P\text{-a. s.},$$

whenever $\widetilde{\varphi}$ is another version of dR/dQ. This establishes (B.2) and shows that the right-hand side of (B.2) is P-a. s. well-defined.

(b) If $R \ll Q$, then (B.3) follows from Remark B.3. Conversely, let us assume that (B.3) holds. We denote $\varphi := dQ/dP$. Then, for any $\mathscr{F}$-measurable $F \geq 0$,

$$\begin{aligned}\int F\, dR &= \int F\mathbb{1}_{\{\varphi>0\}}\, dR = \int F\mathbb{1}_{\{\varphi>0\}} \frac{dR}{dP}\, dP = \int F\mathbb{1}_{\{\varphi>0\}} \frac{dR}{dP}\frac{1}{\varphi} \cdot \varphi\, dP \\ &= \int F\mathbb{1}_{\{\varphi>0\}} \frac{dR}{dP}\frac{1}{\varphi}\, dQ = \int F \frac{dR}{dP}\frac{1}{\varphi}\, dQ,\end{aligned}$$

where we have used (B.3) in the first step and Remark B.3 in the last. This proves $R \ll Q$ and (B.4). □

By taking $R := P$ in Proposition B.4, we obtain the following corollary.

Corollary B.5. *If $Q \ll P$ on $\mathscr{F}$, then*

$$Q \approx P \quad \Longleftrightarrow \quad \frac{dQ}{dP} > 0 \text{ } P\text{-a. s.}$$

In this case, the density of P with respect to Q is given by

$$\frac{dP}{dQ} = \left(\frac{dQ}{dP}\right)^{-1}.$$

Remark B.6. Let us stress that absolute continuity depends on the underlying σ-field $\mathscr{F}$. For example, let P be the Lebesgue measure on $\Omega := [0, 1)$. Then every probability measure Q is absolutely continuous with respect to P on any σ-algebra $\mathscr{F}_0$ that is generated by finitely many intervals $[a_{i-1}, a_i)$ with $0 = a_0 < a_1 < \cdots < a_n = 1$. However, if we take for $\mathscr{F}$ the Borel σ-algebra on Ω, then for instance a Dirac point mass $Q = \delta_x$ is clearly not absolutely continuous with respect to P on $\mathscr{F}$. ◇

While the preceding remark shows that, in general, absolute continuity is not preserved under an enlargement of the underlying σ-algebra, the next proposition states that it is safe to take smaller σ-algebras. This proposition involves the notion of a *conditional expectation*

$$E[\, F \mid \mathscr{F}_0 \,]$$

of an $\mathscr{F}$-measurable function $F \geq 0$ with respect to a probability measure P and a σ-algebra $\mathscr{F}_0 \subseteq \mathscr{F}$. Recall that $E[\, F \mid \mathscr{F}_0 \,]$ may be defined as the P-a. s. unique $\mathscr{F}_0$-measurable random variable F_0 such that

$$E[\, F\, ; A_0 \,] = E[\, F_0\, ; A_0 \,] \quad \text{for all } A_0 \in \mathscr{F}_0; \tag{B.6}$$

see, for example, § 15 of [21]. Note also our convention of writing

$$E[\, F\, ; A_0 \,] := E[\, F\, \mathbb{1}_{A_0} \,].$$

Clearly, we can replace in (B.6) the class of all indicator functions of sets in $\mathscr{F}_0$ by the class of all bounded $\mathscr{F}_0$-measurable functions or by the class of all nonnegative $\mathscr{F}_0$-measurable functions.

Proposition B.7. *Suppose that Q and P are two probability measures on the measurable space $(\Omega, \mathscr{F})$ and that $Q \ll P$ on $\mathscr{F}$ with density φ. If $\mathscr{F}_0$ is a σ-algebra contained in $\mathscr{F}$, then $Q \ll P$ on $\mathscr{F}_0$, and the corresponding density is given by*

$$\left.\frac{dQ}{dP}\right|_{\mathscr{F}_0} = E[\, \varphi \mid \mathscr{F}_0 \,] \quad P\text{-a. s.}$$

Proof. $Q \ll P$ on $\mathscr{F}_0$ follows immediately from the definition of absolute continuity. Since φ is the density on $\mathscr{F} \supseteq \mathscr{F}_0$, it follows for $A \in \mathscr{F}_0$ that

$$Q[A] = E[\varphi; A] = E[E[\varphi \mid \mathscr{F}_0]; A].$$

Therefore, the $\mathscr{F}_0$-measurable random variable $E[\varphi \mid \mathscr{F}_0]$ must coincide with the density on $\mathscr{F}_0$. □

Now we prove a formula for computing a conditional expectation $E_Q[F \mid \mathscr{F}_0]$ under a measure Q in terms of a conditional expectation with respect to another measure P with $Q \ll P$.

Proposition B.8. *Suppose that $Q \ll P$ on $\mathscr{F}$ with density φ, and that $\mathscr{F}_0 \subseteq \mathscr{F}$ is another σ-algebra. Then, for any $\mathscr{F}$-measurable $F \geq 0$,*

$$E_Q[F \mid \mathscr{F}_0] = \frac{1}{E[\varphi \mid \mathscr{F}_0]} E[F\varphi \mid \mathscr{F}_0] \quad Q\text{-a. s.}$$

Proof. We must show that

$$E_Q[G_0 F] = E_Q\left[G_0 \frac{1}{\varphi_0} E[F\varphi \mid \mathscr{F}_0]\right] \tag{B.7}$$

for all $\mathscr{F}_0$-measurable $G_0 \geq 0$, where $\varphi_0 := E[\varphi \mid \mathscr{F}_0]$ is the density of Q with respect to P on $\mathscr{F}_0$ by Proposition B.7. Since $\varphi_0 > 0$ Q-a. s. by Remark B.3, we may assume without loss of generality that $G_0 = 0$ on $\{\varphi_0 = 0\}$. Then, on the one hand,

$$E_Q[G_0 F] = E[G_0 F\varphi] = E[G_0 E[F\varphi \mid \mathscr{F}_0]]. \tag{B.8}$$

On the other hand,

$$E_Q\left[G_0 \frac{1}{\varphi_0} E[F\varphi \mid \mathscr{F}_0]\right] = E\left[\varphi_0 G_0 \frac{1}{\varphi_0} E[F\varphi \mid \mathscr{F}_0]\right] = E[G_0 E[F\varphi \mid \mathscr{F}_0]]. \tag{B.9}$$

Comparing (B.8) with (B.9) yields (B.7). □

B.2 The Lebesgue decomposition

For two arbitrary probability measures P and Q on a measurable space $(\Omega, \mathscr{F})$, the following *Lebesgue decomposition* leads to a decomposition of P into a part that is absolutely continuous with respect to Q and another part that is singular with respect to Q.

Theorem B.9. *For any two probability measures Q and P on $(\Omega, \mathscr{F})$, there exists an $\mathscr{F}$-measurable function φ with values in $[0, \infty]$ such that $Q[\varphi < \infty] = 1$ and*

$$P[A] = P[A \cap \{\varphi = \infty\}] + E_Q[\varphi \mathbb{1}_A] \quad \textit{for all } A \in \mathscr{F}. \tag{B.10}$$

The restriction of P to the Q-null set

$$N := \{\varphi = \infty\}$$

is called the singular part *of P with respect to Q, and the restriction of P to N^c is called the* absolutely continuous part. *The generalized density φ will also be denoted as follows:*

$$\frac{dP}{dQ} := \varphi. \tag{B.11}$$

Proof. Let $R := \frac{1}{2}(Q + P)$. Then both Q and P are absolutely continuous with respect to R with respective densities dQ/dR and dP/dR. Let

$$N := \left\{\frac{dQ}{dR} = 0\right\}.$$

Then $Q[\,N\,] = 0$. We define

$$\varphi := \begin{cases} \dfrac{dP}{dR} \cdot \left(\dfrac{dQ}{dR}\right)^{-1} & \text{on } N^c, \\ +\infty & \text{on } N. \end{cases}$$

Then, for $A \in \mathscr{F}$,

$$\begin{aligned} P[\,A\,] &= P[\,A \cap N\,] + \int_{N^c} \mathbb{1}_A \frac{dP}{dR}\,dR \\ &= P[\,A \cap N\,] + \int_{N^c} \mathbb{1}_A \frac{dP}{dR} \cdot \left(\frac{dQ}{dR}\right)^{-1} dQ \\ &= P[\,A \cap N\,] + \int \mathbb{1}_A\, \varphi\, dQ, \end{aligned}$$

where we have used the fact that $Q[\,N\,] = 0$ in the final step. Moreover, $\varphi < \infty$ R-a. s. on N^c, and so we have $N = \{\varphi = \infty\}$ R-a. s., hence Q-almost surely. □

Exercise B.2.1. In addition to the generalized density φ in the Lebesgue decomposition (B.10) of P with respect to Q, we consider the generalized density φ_0 on a smaller σ-algebra $\mathscr{F}_0 \subseteq \mathscr{F}$. Show that

$$\{\varphi_0 = \infty\} \subseteq \{\varphi = \infty\}\ P\text{-a. s.} \quad \text{and} \quad \varphi_0 \geq E_Q[\,\varphi \mid \mathscr{F}_0]\ Q\text{-a. s.}$$

◇

C Relative entropy

In this section, we collect some basic facts about the relative entropy between two probability measures P and Q on a measurable space $(\Omega, \mathscr{F})$. Consider the convex function $h(x) := x \log x$ for $x > 0$, and note that $h(x) \geq -1/e$, with equality if and only if $x = 1/e$. Note moreover that $h(x) \to 0$ as $x \downarrow 0$, and so we can extend h to a continuous and strictly convex function $h : [0, \infty) \to [-1/e, \infty)$ by putting $h(0) = 0 \log 0 := 0$.

Definition C.1. The *relative entropy* of a probability measure Q with respect to P is defined as

$$H(Q|P) := \begin{cases} E_Q\Big[\log \dfrac{dQ}{dP}\Big] = E\Big[\dfrac{dQ}{dP}\log \dfrac{dQ}{dP}\Big] & \text{if } Q \ll P, \\ +\infty & \text{otherwise.} \end{cases} \qquad \triangle$$

The relative entropy is also called the *Kullback–Leibler divergence*. The following inequality is sometimes called the *Gibbs inequality*.

Lemma C.2. *For any probability measure Q,*

$$H(Q|P) \geq 0,$$

with equality if and only if $Q = P$.

Proof. For $Q \ll P$, Jensen's inequality applied with the convex function $h(x) = x \log x$ yields that

$$H(Q|P) = E\Big[h\Big(\frac{dQ}{dP}\Big)\Big] \geq h(1) = 0. \tag{C.1}$$

Since h is strictly convex, equality holds if and only if $dQ/dP = 1$ P-a. s., that is, $Q = P$. □

Example C.3. Let Ω be a finite set and $\mathscr{F}$ be its power set. Every probability Q on $(\Omega, \mathscr{F})$ is absolutely continuous with respect to the uniform distribution P. Let us denote $Q(\omega) := Q[\,\{\omega\}\,]$. Then

$$H(Q|P) = \sum_{\omega \in \Omega} Q(\omega) \log \frac{Q(\omega)}{P(\omega)} = \sum_{\omega \in \Omega} Q(\omega) \log Q(\omega) + \log |\Omega|.$$

The quantity

$$S(Q) := -\sum_{\omega \in \Omega} Q(\omega) \log Q(\omega)$$

is usually called the *entropy* of Q. Observe that $S(P) = \log |\Omega|$, so that

$$H(Q|P) = S(P) - S(Q).$$

https://doi.org/10.1515/9783111045283-017

Since the left-hand side is nonnegative by (C.1), the uniform distribution P has maximal entropy among all probability distributions on $(\Omega, \mathscr{F})$. ◇

Example C.4. Let $\mu = N(m, \sigma^2)$ denote the normal distribution with mean m and variance σ^2 on $\mathbb{R}$. Then, for $\widetilde{\mu} = N(\widetilde{m}, \widetilde{\sigma}^2)$,

$$\frac{d\widetilde{\mu}}{d\mu}(x) = \frac{\sigma}{\widetilde{\sigma}} \exp\left[-\frac{(x-\widetilde{m})^2}{2\widetilde{\sigma}^2} + \frac{(x-m)^2}{2\sigma^2}\right],$$

and hence

$$H(\widetilde{\mu}|\mu) = \frac{1}{2}\left(\frac{\widetilde{\sigma}^2}{\sigma^2} - \log\frac{\widetilde{\sigma}^2}{\sigma^2} - 1\right) + \frac{1}{2}\left(\frac{m-\widetilde{m}}{\sigma}\right)^2.$$ ◇

The following theorem provides a dual representation of the relative entropy.

Theorem C.5. *For any probability measure* Q,

$$H(Q|P) = \max_{Z \in \mathscr{Z}}(E_Q[Z] - \log E[e^Z]) = \sup_{Z \in L^\infty(\Omega,\mathscr{F},P)}(E_Q[Z] - \log E[e^Z]), \tag{C.2}$$

where $\mathscr{Z}$ *denotes the set of all* $[-\infty, \infty]$*-valued* $\mathscr{F}$*-measurable random variables for which* $E[e^Z] < \infty$ *and* $E_Q[Z^-] < \infty$. *The maximum is attained by* $Z = \log dQ/dP$ *if* $Q \ll P$ *and otherwise by* $Z = \infty \cdot \mathbb{1}_A$ *for any* $A \in \mathscr{F}$ *such that* $Q[A] > 0$ *and* $P[A] = 0$.

Proof. We first show that

$$H(Q|P) \geq E_Q[Z] - \log E[e^Z] \quad \text{for any given } Z \in \mathscr{Z}. \tag{C.3}$$

This inequality is obvious if $H(Q|P) = \infty$, so let us assume that $H(Q|P) < \infty$. Our condition $E_Q[Z^-] < \infty$ implies that $Q[Z > -\infty] = 1$. Since $Q \ll P$, we hence must have $P[Z > -\infty] > 0$ and in turn $E[e^Z] > 0$. Therefore, we may define a probability measure $P^Z \ll P$ via its Radon–Nikodym density,

$$\frac{dP^Z}{dP} = \frac{e^Z}{E[e^Z]}. \tag{C.4}$$

Since $Q[e^Z > 0] = Q[Z > -\infty] = 1$, Proposition B.4(b) implies that $Q \ll P^Z$. Hence, using (B.2) and taking logarithms yields that

$$\log\frac{dQ}{dP} = \log\frac{dQ}{dP^Z} + \log\frac{dP^Z}{dP} \quad Q\text{-a.s.}$$

Taking expectations with respect to Q gives

$$H(Q|P) = H(Q|P^Z) + E_Q[Z] - \log E[e^Z] \geq E_Q[Z] - \log E[e^Z], \tag{C.5}$$

where we have used the fact that $H(Q|P^Z) \geq 0$ by (C.1). This establishes (C.3).

Now we show that there exists $Z \in \mathscr{X}$ for which $H(Q|P) = E_Q[Z] - \log E[e^Z]$. If $Q \ll P$, then $Z := \log dQ/dP$ satisfies $E[e^Z] = 1$. Moreover,

$$E_Q[Z^-] = E\left[\frac{dQ}{dP}\left(\log\frac{dQ}{dP}\right)^-\right]$$

is finite, because $x \log x \geq -1/e$. Therefore, $Z \in \mathscr{X}$. Observe that the corresponding measure P^Z as defined in (C.4) coincides with Q, and so there is equality in (C.5). Thus, Z attains the maximum in (C.2). If $Q \not\ll P$, then there exists $A \in \mathscr{F}$ such that $Q[A] > 0$ and $P[A] = 0$. It follows that $Z = \infty \cdot \mathbb{1}_A$ attains the maximum in (C.2) since $E_Q[Z] = \infty$ and $E[e^Z] = 1$.

Next, since $L^\infty(\Omega, \mathscr{F}, P) \subseteq \mathscr{X}$, the maximum in (C.2) is clearly larger than or equal to the supremum. Thus, the proof will be complete if we can show that the supremum in (C.2) is larger than or equal to $H(Q|P)$. To this end, we first consider the case $Q \not\ll P$. Take $Z_n := n\mathbb{1}_A$ where A is such that $Q[A] > 0$ and $P[A] = 0$. Then, as $n \uparrow \infty$,

$$E_Q[Z_n] - \log E[e^{Z_n}] = n \cdot Q[A] \longrightarrow \infty = H(Q|P).$$

Now suppose that $Q \ll P$ with density $\varphi = dQ/dP$. We let $Z_n = (-n) \vee (\log\varphi) \wedge n \in L^\infty(P)$. We split the expectation $E[e^{Z_n}]$ according to the two sets $\{\varphi \geq 1\}$ and $\{\varphi < 1\}$,

$$E[e^{Z_n}] = E[e^{Z_n}; \varphi \geq 1] + E[e^{Z_n}; \varphi < 1] = E[\varphi \wedge e^n; \varphi \geq 1] + E[e^{-n} \vee \varphi; \varphi < 1].$$

Monotone convergence implies that $\lim_n E[\varphi \wedge e^n; \varphi \geq 1] = E[\varphi; \varphi \geq 1]$ and dominated convergence yields $\lim_n E[e^{-n} \vee \varphi; \varphi < 1] = E[\varphi; \varphi < 1]$. Hence, as $n \uparrow \infty$,

$$E[e^{Z_n}] \longrightarrow E[\varphi; \varphi \geq 1] + E[\varphi; \varphi < 1] = E[\varphi] = 1. \tag{C.6}$$

Since $x \log x \geq -1/e$, we have $\varphi Z_n \geq -1/e$ uniformly in n, and Fatou's lemma yields

$$\liminf_{n\uparrow\infty} E_Q[Z_n] = \liminf_{n\uparrow\infty} E[\varphi Z_n] \geq E\left[\varphi \liminf_{n\uparrow\infty} Z_n\right] = E[\varphi \log\varphi] = H(Q|P). \tag{C.7}$$

Putting both (C.6) and (C.7) together shows that

$$\liminf_{n\uparrow\infty}(E_Q[Z_n] - \log E[e^{Z_n}]) \geq H(Q|P),$$

and the proof is complete. □

Remark C.6. Theorem C.5 shows that the relative entropy is monotone with respect to an increase of the underlying σ-algebra: Let P and Q be two probability measures on a measurable space $(\Omega, \mathscr{F})$, and denote by $H(Q|P)$ their relative entropy. Suppose that $\mathscr{F}_0$ is a σ-field such that $\mathscr{F}_0 \subseteq \mathscr{F}$ and denote by $H_0(Q|P)$ the relative entropy of Q with respect to P considered as probability measures on the smaller space $(\Omega, \mathscr{F}_0)$. Then the relation $L^\infty(\Omega, \mathscr{F}_0, P) \subseteq L^\infty(\Omega, \mathscr{F}, P)$ implies that the first supremum on the right-hand side of (C.2) increases if we pass from $\mathscr{F}_0$ to $\mathscr{F}$. Hence, we must also have

$$H_0(Q|P) \le H(Q|P). \qquad \diamond$$

Exercise C.0.1. In the context of Remark C.6, show that the inequality $H_0(Q|P) \le H(Q|P)$ is strict if and only if $H_0(Q|P) < \infty$ and the density φ_0 of Q with respect to P on $(\Omega, \mathscr{F}_0)$ is not P-a. s. equal to the density φ of Q with respect to P on $(\Omega, \mathscr{F})$. ◇

Corollary C.7. *For any Z such that $E[e^Z] \in (0, \infty)$ and $E[Ze^Z] < \infty$,*

$$\max_Q (E_Q[Z] - H(Q|P)) = \log E[e^Z], \tag{C.8}$$

where the maximum is taken over all probability measures Q such that $H(Q|P) < \infty$. The maximum is attained by Q if and only if Q coincides with the probability measure P^Z defined by

$$\frac{dP^Z}{dP} = \frac{e^Z}{E[e^Z]}.$$

In particular, P^Z has minimal relative entropy $H(\cdot|P)$ among all probability measures Q such that $E_Q[Z] = E_{P^Z}[Z]$.

Proof. Let us extend the function $h(x) = x \log x$ to all of $\mathbb{R}$ by letting $h(x) := +\infty$ for $x < 0$. Then the Fenchel–Legendre transform h^* of the strictly convex function h is given by $h^*(y) = e^{y-1}$; see Definition A.8. If Q is a probability measure with $H(Q|P) < \infty$ and density $\varphi = dQ/dP$, then Proposition A.9(a) yields that

$$\varphi Z \le h(\varphi) + h^*(Z) = \varphi \log \varphi + e^{Z-1} \in L^1(P).$$

Thus, the expectation $E_Q[Z] = E[\varphi Z]$ is well-defined in $[-\infty, \infty)$. If $E_Q[Z] = -\infty$, then clearly

$$E_Q[Z] - H(Q|P) \le \log E[e^Z]. \tag{C.9}$$

If $E_Q[Z] > -\infty$, then (C.3) implies

$$H(Q|P) \ge E_Q[Z] - \log E[e^Z].$$

Hence, (C.9) also holds in this case.

Now suppose that equality holds in (C.9). Since, by (C.5),

$$H(Q|P) = H(Q|P^Z) + E_Q[Z] - \log E[e^Z],$$

equality in (C.9) implies that $H(Q|P^Z) = 0$, hence $Q = P^Z$.

Conversely, our assumption $E[Ze^Z] < \infty$ and the fact that $ze^z \geq -1/e$ for all $z \in \mathbb{R}$ imply $Ze^Z \in L^1(P)$. Hence, $Z \in L^1(P^Z)$,

$$H(P^Z|P) = E_{P^Z}[Z] - \log E[e^Z] < \infty,$$

and

$$\log E[e^Z] = E_{P^Z}[Z] - H(P^Z|P).$$

Thus, the maximum in (C.8) is attained by $Q = P^Z$. □

Remark C.8. If $Z \in L^\infty(P)$, then $E_Q[Z]$ is finite for every probability measure $Q \ll P$, and so we may replace (C.8) with the following identity:

$$\max_{Q \ll P}(E_Q[Z] - H(Q|P)) = \log E[e^Z]. \qquad \diamond$$

Remark C.9. In Corollary C.7, the probability measure P^Z is characterized by the property that it maximizes the functional $F(Q) := E_Q[Z] - H(Q|P)$. In the context of statistical mechanics, this is often referred to as the *Gibbs variational principle*, and P^Z is called the *Gibbs measure* with respect to Z. For $Z = -\beta X$, it takes the form

$$\frac{1}{\beta}\log E[e^{-\beta X}] = \max_Q\left(E_Q[-X] - \frac{1}{\beta}H(Q|P)\right). \tag{C.10}$$

In the context of Chapter 4, the left-hand side of (C.10) will be viewed as a convex risk measure, and (C.10) yields its dual representation in terms of relative entropy; cf. Examples 4.13, 4.34, and 4.125. ◇

Remark C.10. Note that neither the proof of (C.3) nor the proof of Corollary C.7 make use of the fact that the measure P is a probability measure. Hence, (C.8) remains valid if P is replaced by an arbitrary reference measure $R \geq 0$, for example, by the counting measure on some countable set Ω, or by the Lebesgue measure on $\Omega = \mathbb{R}^d$. In the two latter cases,

$$S(Q) := -H(Q|R) = -\int \log\frac{dQ}{dR}\,dQ$$

is often called the *Shannon–Boltzmann entropy* of a probability measure $Q \ll R$. Thus, (C.8) shows that the Gibbs measure R^Z, defined as in (C.4) with R instead of P and for suitable Z, has maximal Shannon–Boltzmann entropy among all probability measures Q with the same expectation $E_Q[Z]$. More precisely, assume that $E_R[e^Z] \in (0,\infty)$ and $E_R[|Z|e^Z] < \infty$, and note that

$$S(Q) = -H(Q|R^Z) + \log E_R[e^Z] - E_Q[Z]$$

is well-defined in $[-\infty, \infty)$ for any probability measure $Q \ll R$ such that $E_Q[Z^-] < \infty$. Since $H(Q|R^Z) \geq 0$, we obtain $S(Q) \leq S(R^Z)$ for any Q that has the same expectation $E_Q[Z]$ as the Gibbs measure R^Z. Note that Example C.3 can be seen as a special case, with R = counting measure, $Z = 0$ and $R^Z = P$. As another example, consider the normal distribution $\mu = N(m, \sigma^2)$ on $\Omega = \mathbb{R}$. Its Shannon–Boltzmann entropy is given by

$$S(\mu) = \frac{1}{2}(\log(2\pi\sigma^2) + 1),$$

and this is maximal among all probability measures ν on $\mathbb{R}$ that have the same variance σ^2. ◇

D Quantile functions and their applications

D.1 Inverse functions

Let us fix a nonempty open interval $(a, b) \subseteq \mathbb{R}$ and an increasing function $F : (a, b) \to \mathbb{R}$. We define

$$c := \lim_{x \downarrow a} F(x) \in [-\infty, \infty) \quad \text{and} \quad d := \lim_{x \uparrow b} F(x) \in (-\infty, +\infty]. \tag{D.1}$$

For instance, F can be the *cumulative distribution function* of a real-valued random variable X on some probability space $(\Omega, \mathscr{F}, P)$, that is, $F(x) = P[X \le x]$ for $x \in \mathbb{R}$. In this case, $a = -\infty$, $b = +\infty$, $c = 0$, and $d = 1$.

Our goal is to define a function $q : (c, d) \to (a, b)$ that plays the role of an inverse function of F. This is standard if F is continuous and strictly increasing. In general, however, F may contain discontinuities and horizontal stretches. This more general situation will be addressed by the following definition.

Definition D.1. A function $q : (c, d) \to (a, b)$ is called an *inverse function* for F if

$$F(q(s)-) \le s \le F(q(s)+) \quad \text{for all } s \in (c, d). \tag{D.2}$$

The functions

$$q^-(s) := \sup\{x \in (a, b) \mid F(x) < s\} \quad \text{and} \quad q^+(s) := \inf\{x \in (a, b) \mid F(x) > s\} \tag{D.3}$$

are called the *left-* and *right-continuous inverse functions*. △

The following lemma characterizes all inverse functions in terms of q^- and q^+ and explains the reason for calling q^- and q^+ the left- and right-continuous inverse functions of F. See Figure D.1 for an illustration.

Lemma D.2. *A function $q : (c, d) \to (a, b)$ is an inverse function for F if and only if*

$$q^-(s) \le q(s) \le q^+(s) \quad \textit{for all } s \in (c, d). \tag{D.4}$$

In particular, q^- and q^+ are inverse functions. Moreover, q^- is left-continuous, q^+ is right-continuous, and every inverse function q is increasing and satisfies $q(s-) = q^-(s)$ and $q(s+) = q^+(s)$ for all $s \in (c, d)$. In particular, all inverse functions coincide in all but countably many points in (c, d).

Proof. First, we prove the left- and right-continuity of q^- and q^+. It is clear from (D.3) that both q^- and q^+ are increasing. Let us now show that q^+ is right-continuous, that is, $q^+(s+) = q^+(s)$ for any given $s \in (c, d)$. Since q^+ is increasing, we have $q^+(s+) \ge q^+(s)$. To prove the converse inequality, we take a sequence (x_n) in $\{x \in (a, b) \mid F(x) > s\}$ that decreases toward $q^+(s) = \inf\{x \in (a, b) \mid F(x) > s\}$. If we define

https://doi.org/10.1515/9783111045283-018

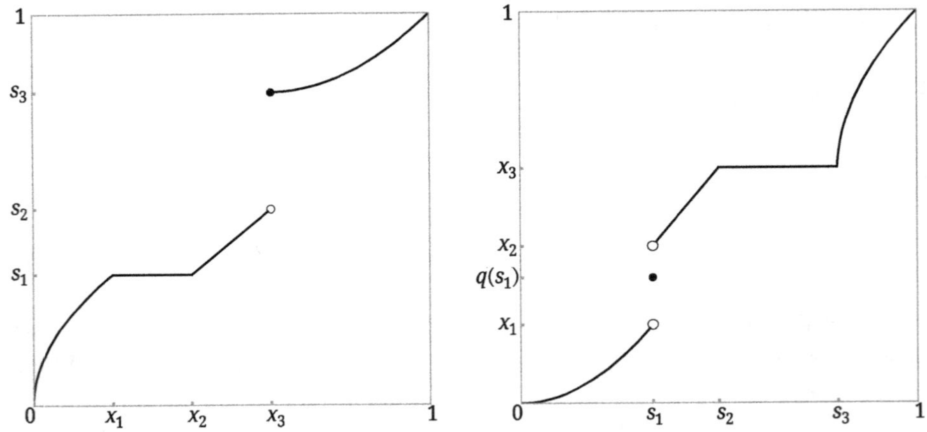

Figure D.1: The left-hand panel shows an increasing right-continuous function $F : (0,1) \to (0,1)$ and the right-hand panel displays a corresponding inverse function q. Note how vertical jumps of F translate into horizontal stretches for q and vice versa. Every inverse function must coincide with q everywhere except at the point s_1, where it can take any value between x_1 and x_2. In particular, q^- and q^+ are characterized through their respective values in s_1, namely $q^-(s_1) = x_1$ and $q^+(s_1) = x_2$.

$$\varepsilon_n := \frac{F(x_n) - s}{2},$$

then $F(x_n) > s + \varepsilon_n$, and in turn, $x_n \geq q^+(s + \varepsilon_n)$. It follows that

$$q^+(s) = \lim_{n\uparrow\infty} x_n \geq \limsup_{n\uparrow\infty} q^+(s + \varepsilon_n) \geq q^+(s+).$$

This proves the right-continuity of q^+. An analogous argument shows the left-continuity of q^-.

Now we prove that $q^-(s+) = q^+(s)$ for any given $s \in (c,d)$. The inequality $q^-(s+) \leq q^+(s)$ follows from the fact that $q^- \leq q^+$ and the right-continuity of q^+. To prove the converse inequality, let (s_n) be a sequence in (c,d) that strictly decreases to s. Then, for $n \geq 2$ and any $\varepsilon > 0$ such that $q^-(s_1) + \varepsilon \in (a,b)$,

$$q^-(s_n) + \varepsilon \in \{x \in (a,b) \mid F(x) < s_n\}^c = \{x \in (a,b) \mid F(x) \geq s_n\} \supseteq \{x \in (a,b) \mid F(x) > s_{n-1}\}.$$

Hence,

$$q^-(s_n) + \varepsilon \geq \inf\{x \in (a,b) \mid F(x) > s_{n-1}\} = q^+(s_{n-1}).$$

By sending $n \uparrow \infty$ and $\varepsilon \downarrow 0$, we get $q^-(s+) \geq q^+(s+) = q^+(s)$. This proves $q^-(s+) = q^+(s)$. In the same way, we can prove $q^+(s-) = q^-(s)$.

We now prove that every inverse function q must satisfy (D.4). We already know from (D.3) that $q^-(s) \le q^+(s)$ for each given $s \in (c, d)$. It follows from (D.2) that for every $\varepsilon > 0$ such that $s - \varepsilon > c$ and $q(s) + \varepsilon < b$ we have $F(q(s) + \varepsilon) \ge s > s - \varepsilon$, and hence

$$q(s) + \varepsilon \ge \sup\{x \in (a, b) \mid F(x) < s - \varepsilon\} = q^-(s - \varepsilon).$$

Sending $\varepsilon \downarrow 0$ yields $q(s) \ge q^-(s-) = q^-(s)$. In the same way, one shows that $q(s) \le q^+(s)$. Thus, (D.4) necessarily holds for every inverse function.

Next, let q be any function satisfying (D.4). Then $F(q(s)-) \le F(q^+(s)-)$ and $F(q(s)+) \ge F(q^-(s)+)$. Hence, q will satisfy (D.2) if

$$F(q^+(s)-) \le s \le F(q^-(s)+) \quad \text{for all } s. \tag{D.5}$$

But this inequality follows immediately from the fact that $x < q^+(s)$ implies $F(x) \le s$ and $y > q^-(s)$ implies $F(y) \ge s$.

The claim that every inverse function q is increasing and satisfies $q(s-) = q^-(s)$ and $q(s+) = q^+(s)$ for all $s \in (c, d)$ is now obvious. Since every increasing function can have at most countably many discontinuity points, it now follows from the fact that q^+ is the right-hand limit of q^- that $q^-(s) = q^+(s)$ for all but countably many s. Thus, our final claim follows from (D.4). □

The following lemma allows us to extend the domain of an inverse function q by putting

$$q(c) := a \quad \text{and} \quad q(d) := b. \tag{D.6}$$

Lemma D.3. *Every inverse function satisfies $q(s) \searrow a$ as $s \downarrow c$ and $q(s) \nearrow b$ as $s \uparrow d$.*

Proof. The definitions (D.1) and (D.3) imply that $q^-(s) \searrow a$ as $s \downarrow c$ and $q^+(s) \nearrow b$ as $s \uparrow d$. The assertion now follows from the fact that any other inverse function satisfies (D.4) and coincides with the increasing functions q^- and q^+ everywhere except in at most countably many discontinuity points. □

Remark D.4. The left- and right-continuous inverse functions can also be represented as

$$q^-(s) = \inf\{x \in (a, b) \mid F(x) \ge s\} \quad \text{and} \quad q^+(s) = \sup\{x \in (a, b) \mid F(x) \le s\}.$$

To see this, note first that $q^-(s)$ is clearly dominated by the infimum on the right. On the other hand, we have $F(q^-(s) + \varepsilon) \ge s$ for every $\varepsilon > 0$ such that $q^-(s) + \varepsilon < b$. Hence, $q^-(s) + \varepsilon \ge \inf\{x \in (a, b) \mid F(x) \ge s\}$, and sending $\varepsilon \downarrow 0$ yields the first identity. The proof for q^+ is analogous. ◇

Lemma D.5. *Let q be an inverse function for F. Then F is an inverse function for q. In particular,*

$$F(x+) = \inf\{s \in (c, d) \mid q(s) > x\} \quad \textit{for } x \textit{ with } F(x) < d. \tag{D.7}$$

Proof. If $s > F(x)$, then $q(s) \geq q^-(s) \geq x$, and hence $q(F(x)+) \geq x$. Conversely, $s < F(x)$ implies $q(s) \leq q^+(s) \leq x$, and thus $q(F(x)-) \leq x$. This proves that F is an inverse function for q. □

Lemma D.6. *The function q is an inverse function for F if and only if it is an inverse function for any other function $\widetilde{F} : (a,b) \to (c,d)$ with $F(x-) \leq \widetilde{F}(x) \leq F(x+)$ for all $x \in (a,b)$. In particular,*

$$\begin{aligned} q^-(t) &= \sup\{x \in (a,b) \mid F(x-) < t\} = \sup\{x \in (a,b) \mid F(x+) < t\} \\ &= \inf\{x \in (a,b) \mid F(x-) \geq t\} = \inf\{x \in (a,b) \mid F(x+) \geq t\}. \end{aligned} \tag{D.8}$$

Proof. If q is an inverse function of F, then the Lemmas D.2 and D.5 imply that $F^-(x) := F(x-)$ and $F^+(x) := F(x+)$ are the smallest and largest inverse functions of q, respectively, and that every function $\widetilde{F}$ with $F^- \leq \widetilde{F} \leq F^+$ is also an inverse function of q. Applying once again Lemma D.5 shows that q is an inverse function of $\widetilde{F}$. The converse implication follows by taking $\widetilde{F} := F$. The identities in (D.8) are now a consequence of (D.3) and Remark D.4. □

Exercise D.1.1. Give a direct proof of the identities

$$\sup\{x \in (a,b) \mid F(x-) < t\} = \sup\{x \in (a,b) \mid F(x+) < t\}$$

and

$$\inf\{x \in (a,b) \mid F(x-) \geq t\} = \inf\{x \in (a,b) \mid F(x+) \geq t\}.$$ ◇

D.2 Quantile functions

From now on, we will assume that

$$F : \mathbb{R} \longrightarrow [0,1] \text{ is increasing and right-continuous,}$$

and that F is normalized in the sense that $c = 0$ and $d = 1$. This assumption always holds if $F(x) = P[\, X \leq x\,]$ is the cumulative distribution function of a random variable X on probability space $(\Omega, \mathscr{F}, P)$, which will be fixed throughout this section. The following lemma shows in particular that also the converse is true: any normalized increasing right-continuous functions $F : \mathbb{R} \to [0,1]$ is the cumulative distribution function of some random variable.

Lemma D.7. *Let U be a random variable with a uniform distribution on $(0,1)$, that is, $P[\, U \leq s\,] = s$ for all $s \in (0,1)$. If q is an inverse function of a normalized increasing right-continuous function $F : \mathbb{R} \to [0,1]$, then*

$$X(\omega) := q(U(\omega))$$

has the cumulative distribution function F.

Proof. First, note that any inverse function for F is measurable because it is increasing. Since $q(F(x)-) \le x$ by Lemma D.5, we have $q(s) \le x$ for $s < F(x)$. Moreover, the monotonicity of F and (D.2) yield that $q(s) \le x$ implies $F(x) \ge F(q(s)) = F(q(s)+) \ge s$. In other words,

$$(0, F(x)) \subseteq \{s \in (0,1) \mid q(s) \le x\} \subseteq (0, F(x)].$$

Hence,

$$\begin{aligned} F(x) = P[\, U \in (0, F(x)) \,] &\le P[\, U \in \{s \mid q(s) \le x\} \,] \\ &\le P[\, U \in (0, F(x)] \,] = F(x). \end{aligned}$$

The assertion now follows from the identity $P[\, U \in \{s \mid q(s) \le x\} \,] = P[\, X \le x \,]$. □

Remark D.8. By considering the laws of real-valued random variables, Lemma D.7 also yields a one-to-one correspondence $F(x) = \mu((-\infty, x])$ between all Borel probability measures μ on $\mathbb{R}$ and all normalized increasing right-continuous functions $F : \mathbb{R} \to [0,1]$. This correspondence can be extended beyond the case of probability measures to the case of positive Radon measures. Recall that a positive Radon measure on a real interval I is a σ-additive measure μ that assigns to each Borel set $A \subseteq I$ a value in $\mu(A) \in [0, \infty]$ that is finite if A is compact. Consider, for instance, the Lebesgue measure λ on $I = \mathbb{R}$. Via the identity $\lambda([a,b]) = b - a$ for $-\infty < a \le b < \infty$, it corresponds to the continuous and increasing function $F(x) = x$ on $\mathbb{R}$. Conversely, every increasing right-continuous function F on a nonempty open interval I gives rise to a corresponding positive Radon measure μ on I. This follows by applying Lemma D.7 on compact subintervals of I and to the corresponding probability measure $\frac{1}{\mu(I)}\mu(\,\cdot \cap I)$ on I. ◇

Definition D.9. An inverse function $q : (0,1) \to \mathbb{R}$ of a cumulative distribution function F is called a *quantile function*. That is, q is a function with

$$F(q(s)-) \le s \le F(q(s)) \quad \text{for all } s \in (0,1).$$

The left- and right-continuous inverses,

$$q^-(s) = \sup\{x \in \mathbb{R} \mid F(x) < s\} \quad \text{and} \quad q^+(s) = \inf\{x \in \mathbb{R} \mid F(x) > s\},$$

are called the *lower* and *upper quantile functions*. △

We will often use the generic notation F_X for the distribution function of a random variable X. If the emphasis is on the law μ of X, we will also write F_μ. In the same manner, we will write q_X or q_μ for the corresponding quantile functions. The value $q_X(\lambda)$ of a quantile function at a given level $\lambda \in (0,1)$ is often called a *λ-quantile* of X.

Exercise D.2.1. Compute the upper and lower quantile functions for the following cumulative distribution functions:

(a) The cumulative distribution function of a Dirac point mass: $F(x) = \mathbb{1}_{\{x \geq x_0\}}$ for some $x_0 \in \mathbb{R}$.

(b) The exponential distribution function: $F(x) = 0$ for $x < 0$ and $F(x) = \int_0^x \theta e^{-\theta y}\,dy$ for $x \geq 0$, where $\theta > 0$.

(c) The cumulative distribution function of a Gumbel distribution: $F(x) = e^{-e^{-(x-m)/\beta}}$, where $m \in \mathbb{R}$ and $\beta > 0$.

(d) The cumulative distribution function of a Pareto distribution with scale parameter $x_m > 0$ and shape parameter $\alpha > 0$,

$$F(x) = \begin{cases} 1 - (x_m/x)^\alpha & \text{if } x \geq x_m, \\ 0 & \text{if } x < x_m. \end{cases}$$ ◇

The following result complements Lemma D.7. It implies that a probability space supports a random variable with uniform distribution on $(0,1)$ if and only if it supports a random variable X with a continuous distribution.

Lemma D.10. *Let X be a random variable with a continuous cumulative distribution function F_X and with quantile function q_X. Then $U := F_X(X)$ is uniformly distributed on* $(0,1)$, *and $X = q_X(U)$ P-almost surely.*

Proof. Let $(\widetilde{\Omega}, \widetilde{\mathcal{F}}, \widetilde{P})$ be any probability space that supports a random variable $\widetilde{U}$ with a uniform distribution on $(0,1)$. Then $\widetilde{X} := q_X(\widetilde{U})$ has the same distribution as X due to Lemma D.7. Hence, $F_X(X)$ and $F_X(\widetilde{X})$ also have the same distribution. But if F_X is continuous, then $F_X(q_X(s)) = s$ by (D.2), and thus $F_X(\widetilde{X}) = \widetilde{U}$. This shows that $F_X(X)$ is uniformly distributed.

To show that $X = q_X(U)$ P-a. s., note first that $q_X^+(F_X(t)) \geq t$ by Lemma D.5 and that $q_X(U) = q_X^+(U)$ P-a. s. by the final claim in Lemma D.2. Hence, $q_X(U) = q_X^+(U) \geq X$ P-almost surely. Now let $f : \mathbb{R} \to (0,1)$ be any strictly increasing function. Since, by Lemma D.7, $q_X(U)$ and X have the same law, we have $E[f(q_X(U))] = E[f(X)]$ and get

$$P[f(q_X(U)) > f(X)] = P[q_X(U) > X] = 0.$$

This proves the assertion. □

There are several possibilities how the preceding lemma can be generalized to the case of discontinuous cumulative distribution functions F_X. A first possibility is provided in the following exercise. It requires the existence of an independent random variable with a uniform distribution on $(0,1)$. A second possibility will be given in Lemma D.17, which will only assume the existence of some uniformly distributed random variable, not its independence of X.

Exercise D.2.2. Let X be a random variable with cumulative distribution function F_X. The *modified cumulative distribution function* of X is defined by

$$\widetilde{F}_X(x,\lambda) := P[\,X < x\,] + \lambda P[\,X = x\,], \quad x \in \mathbb{R},\ \lambda \in [0,1].$$

Suppose that $\widetilde{U}$ is a random variable that is independent of X and uniformly distributed on $(0,1)$. Show that

$$U := \widetilde{F}_X(X, \widetilde{U})$$

is uniformly distributed on $(0,1)$ and that

$$X = q_X(U) \quad P\text{-a.s.}$$

◇

The following lemma uses the concept of the Fenchel–Legendre transform of a convex function as introduced in Definition A.8.

Lemma D.11. *Let X be a random variable with cumulative distribution function F_X and quantile function q_X and suppose $E[\,|X|\,] < \infty$. Then the Fenchel–Legendre transform of the convex function*

$$\Psi(x) := \int_{-\infty}^{x} F_X(z)\,dz = E[\,(x-X)^+\,]$$

is given by

$$\Psi^*(y) = \sup_{x\in\mathbb{R}}(xy - \Psi(x)) = \begin{cases} \int_0^y q_X(t)\,dt & \text{if } 0 \le y \le 1,\\ +\infty & \text{otherwise.} \end{cases}$$

Moreover, for $0 < y < 1$, the supremum is attained in x if and only if x is a y-quantile of X.

Proof. Note first that, by Fubini's theorem and Lemma D.7,

$$\Psi(x) = E\left[\int_{-\infty}^{x} \mathbb{1}_{\{X\le z\}}\,dz\right] = E[\,(x-X)^+\,] = \int_0^1 (x - q_X(t))^+\,dt. \tag{D.9}$$

It follows that $\Psi^*(y) = +\infty$ for $y < 0$, $\Psi^*(0) = -\inf_x \Psi(x) = 0$,

$$\Psi^*(1) = \sup_{x\in\mathbb{R}}(x - \Psi(x)) = \lim_{x\uparrow\infty}\int_0^1 (x - (x - q_X(t))^+)\,dt = \int_0^1 q_X(t)\,dt,$$

and $\Psi^*(y) = \infty$ for $y > 1$. To prove our formula for $0 < y < 1$, note that the right-hand and left-hand derivatives of the concave function $f(x) = xy - \Psi(x)$ are given by $f'_+(x) = y - F_X(x)$ and $f'_-(x) = y - F_X(x-)$. A point x_y is a maximizer of f if $f'_+(x_y) \le 0$ and $f'_-(x_y) \ge 0$, which is equivalent to x_y being a y-quantile. Writing $x_y = q_X(y)$ and using (D.9) gives

$$\Psi^*(y) = x_y y - \Psi(x_y) = x_y y - \int_0^y (x_y - q_X(t))\,dt = \int_0^y q_X(t)\,dt,$$

and our formula follows. □

Lemma D.12. *If $X = f(Y)$ for an* increasing *function f and q_Y is a quantile function for Y, then $f(q_Y(t))$ is a quantile function for X. In particular, for any quantile function q_X of X,*

$$q_X(t) = q_{f(Y)}(t) = f(q_Y(t)) \quad \textit{for a. e. } t \in (0,1).$$

If f is decreasing, *then $f(q_Y(1-t))$ is a quantile function for X. In particular,*

$$q_X(t) = q_{f(Y)}(t) = f(q_Y(1-t)) \quad \textit{for a. e. } t \in (0,1).$$

Proof. If f is increasing, then

$$\begin{aligned} F_X(f(q_Y(t))) &= P[\,f(Y) \le f(q_Y(t))\,] \ge P[\,Y \le q_Y(t)\,] \\ &= F_Y(q_Y(t)) \ge t \ge F_Y(q_Y(t)-) \\ &= P[\,Y < q_Y(t)\,] \ge P[\,f(Y) < f(q_Y(t))\,] \\ &= F_X(f(q_Y(t)-)). \end{aligned}$$

Hence, $f(q_Y(t))$ is a quantile function for $X = f(Y)$. If f is deceasing, then $-f$ is increasing and the preceding argument yields that $-f(q_Y(t))$ is a quantile function for $-X$. Thus, the assertion will follow once we have shown that $-q_Y(1-t)$ is a quantile function for $-Y$. Indeed, we have

$$F_{-Y}(-q_Y(1-1)) = P[\,-Y \le -q_Y(1-t)\,] = 1 - F_Y(q_Y(1-t)-) \ge t,$$

and a similar argument yields $F_{-Y}(-q_Y(1-1)) \le t$. □

Exercise D.2.3. Let X, Y, and Z by random variables such that $X = f(Z)$ and $Y = g(Z)$ for two increasing functions f and g. Show that if q_X, q_Y, and q_{X+Y} are quantile functions for X, Y, and $X+Y$, then

$$q_{X+Y}(t) = q_X(t) + q_Y(t) \quad \text{for a. e. } t.$$

Note: Two random variables X and Y in the form considered in this exercise are called *comonotone*. See Section 4.9 for more background on the notion of comonotonicity. ◇

D.3 The Hardy–Littlewood inequalities

The following theorem is a version of the *Hardy–Littlewood inequalities*. They provide upper and lower bounds on $E[XY]$ in terms of quantile functions q_X and q_Y. The probability space $(\Omega, \mathscr{F}, P)$ will be fixed throughout this section.

Theorem D.13. *Let X and Y be two random variables with quantile functions q_X and q_Y. Then*

$$\int_0^1 q_X(1-s)q_Y(s)\,ds \le E[XY] \le \int_0^1 q_X(s)q_Y(s)\,ds,$$

provided that all integrals are well-defined. If $X = f(Y)$ and the lower (upper) bound is finite, then the lower (upper) bound is attained if and only if f can be chosen as a decreasing (increasing) function.

Proof. We first prove the result for $X, Y \ge 0$. By Fubini's theorem,

$$\begin{aligned} E[XY] &= E\left[\int_0^\infty \mathbb{1}_{\{X>x\}}\,dx \int_0^\infty \mathbb{1}_{\{Y>y\}}\,dy\right] \\ &= \int_0^\infty\int_0^\infty P[X>x,\, Y>y]\,dx\,dy. \end{aligned}$$

Since

$$\begin{aligned} P[X>x,\, Y>y] &\ge (P[X>x] - P[Y\le y])^+ \\ &= \int_0^1 \mathbb{1}_{\{F_Y(y)\le s\}}\,\mathbb{1}_{\{s\le 1-F_X(x)\}}\,ds, \end{aligned}$$

and since

$$q_Z^+(s) = \sup\{x\ge 0 \mid F_Z(x)\le s\} = \int_0^\infty \mathbb{1}_{\{F_Z(x)\le s\}}\,dx$$

for any random variable $Z \ge 0$, another application of Fubini's theorem yields

$$E[XY] \ge \int_0^1 q_X^+(1-s)\,q_Y^+(s)\,ds = \int_0^1 q_X(1-s)\,q_Y(s)\,ds.$$

In the same way, the upper estimate follows from the inequality

$$P[X > x, Y > y] \leq P[X > x] \wedge P[Y > y]$$
$$= \int_0^1 \mathbb{1}_{\{F_X(x) \leq s\}} \mathbb{1}_{\{F_Y(y) \leq s\}} \, ds.$$

For $X = f(Y)$,

$$E[XY] = E[f(Y)Y] = \int_0^1 f(q_Y(t))q_Y(t) \, dt, \tag{D.10}$$

due to Lemma D.7, and so Lemma D.12 implies that the upper and lower bounds are attained for increasing and decreasing functions, respectively.

Conversely, assume that $X = f(Y)$, and that the upper bound is attained and finite:

$$E[f(Y)Y] = \int_0^1 q_X(t)q_Y(t) \, dt < \infty. \tag{D.11}$$

Our aim is to show that

$$X = f(Y) = \tilde{f}(Y) \quad P\text{-a. s.}, \tag{D.12}$$

where $\tilde{f}$ is an increasing function on $[0, \infty)$ such that

$$\tilde{f}(x) = q_X(F_Y(x))$$

if $0 < F_Y(x) < 1$ and x is a continuity point of F_Y, and

$$\tilde{f}(x) = \frac{1}{F_Y(x) - F_Y(x-)} \int_{F_Y(x-)}^{F_Y(x)} q_X(t) \, dt$$

if $F_Y(x-) < F_Y(x)$. It is shown in Exercise 3.4.2 that

$$\tilde{f}(q_Y) = E_\lambda[\, q_X \mid q_Y \,], \tag{D.13}$$

where $E_\lambda[\,\cdot \mid q_Y\,]$ denotes the conditional expectation with respect to q_Y under the Lebesgue measure λ on $(0, 1)$. Therefore, (D.11) implies that

$$\infty > \int_0^1 f(q_Y(t))q_Y(t) \, dt = E[f(Y)Y] = \int_0^1 q_X(t)q_Y(t) \, dt = \int_0^1 \tilde{f}(q_Y(t))q_Y(t) \, dt, \tag{D.14}$$

where we have used Lemma D.7 in the first identity. After these preparations, we can proceed to proving (D.12). Let ν denote the distribution of Y. By introducing the positive measures $d\mu = f \, d\nu$ and $d\tilde{\mu} = \tilde{f} \, d\nu$, (D.14) can be written as

$$\int_0^\infty \mu([y,\infty))\,dy = \int x\,\mu(dx) = \int x\,\widetilde{\mu}(dx) = \int_0^\infty \widetilde{\mu}([y,\infty))\,dy. \tag{D.15}$$

On the other hand, with g denoting the increasing function $\mathbb{1}_{[y,\infty)}$, the upper Hardy–Littlewood inequality, Lemma D.12, and (D.13) yield

$$\begin{aligned}\mu([y,\infty)) &= E[\,g(Y)f(Y)\,]\\ &\le \int_0^1 q_{g(Y)}(t)q_X(t)\,dt\\ &= \int_0^1 g(q_Y(t))\widetilde{f}(q_Y(t))\,dt\\ &= \widetilde{\mu}([y,\infty)).\end{aligned}$$

In view of (D.15), we obtain $\mu = \widetilde{\mu}$, and hence $f = \widetilde{f}$ ν-a. s. and $X = \widetilde{f}(Y)$ P-almost surely. An analogous argument applies to the lower bound, and the proof for $X, Y \ge 0$ is concluded.

The result for general X and Y is reduced to the case of nonnegative random variables by separately considering the positive and negative parts of X and Y:

$$\begin{aligned}E[\,XY\,] &= E[\,X^+Y^+\,] - E[\,X^+Y^-\,] - E[\,X^-Y^+\,] + E[\,X^-Y^-\,]\\ &\le \int_0^1 q_{X^+}(t)q_{Y^+}(t)\,dt - \int_0^1 q_{X^+}(t)q_{Y^-}(1-t)\,dt\\ &\quad - \int_0^1 q_{X^-}(1-t)q_{Y^+}(t)\,dt + \int_0^1 q_{X^-}(t)q_{Y^-}(t)\,dt,\end{aligned} \tag{D.16}$$

where we have used the upper Hardy–Littlewood inequality on the positive terms and the lower one on the negative terms. Since $q_{Z^+}(t) = (q_Z(t))^+$ and $q_{Z^-}(t) = (q_Z(1-t))^-$ for all random variables due to Lemma D.12, one checks that the right-hand side of (D.16) is equal to $\int_0^1 q_X(t)q_Y(t)\,dt$, and we obtain the general form of the upper Hardy–Littlewood inequality. The same argument also works for the lower one.

Now suppose that $X = f(Y)$. We first note that (D.10) still holds, and so Lemma D.12 implies that the upper and lower bounds are attained for increasing and decreasing functions, respectively. Conversely, let us assume that the upper Hardy–Littlewood inequality is an identity. Then all four inequalities used in (D.16) must also be equalities. Using the fact that $XY^+ = f(Y^+)Y^+$ and $XY^- = f(-Y^-)Y^-$, the assertion is reduced to the case of nonnegative random variables, and one checks that f can be chosen as an increasing function. The same argument applies if the lower Hardy–Littlewood inequality is attained. □

Remark D.14. For indicator functions of two sets A and B in $\mathscr{F}$, the Hardy–Littlewood inequalities reduce to the elementary inequalities

$$(P[A] + P[B] - 1)^+ \leq P[A \cap B] \leq P[A] \wedge P[B]; \tag{D.17}$$

note that these estimates were used in the preceding proof. Applied to the sets $\{X \leq x\}$ and $\{Y \leq y\}$, where X and Y are random variables with cumulative distribution functions F_X and F_Y and joint cumulative distribution function $F_{X,Y}$ defined by $F_{X,Y}(x,y) = P[X \leq x,\ Y \leq y]$, they take the form

$$(F_X(x) + F_Y(y) - 1)^+ \leq F_{X,Y}(x,y) \leq F_X(x) \wedge F_Y(y). \tag{D.18}$$

The estimates (D.17) and (D.18) are often called *Fréchet bounds*. The Hardy–Littlewood inequalities, which are also called *Hoeffding–Fréchet bounds*, provide their natural extension from sets to random variables. ◇

Exercise D.3.1. Let X and Y be two random variables for which all terms appearing in the Hardy–Littlewood inequalities make sense. Suppose moreover that there exists a third random variable Z and functions f and g such that $X = f(Z)$ and $Y = g(Z)$.

(a) Show that the upper Hardy–Littlewood inequality reduces to an equality if X and Y are *comonotone* in the sense that f and g are both increasing or both decreasing.

(b) Show that the lower Hardy–Littlewood inequality reduces to an equality if X and Y are *anticomonotone* in the sense that one of the functions f and g is increasing and the other one is decreasing.

Note: See Section 4.9 for more background on the notion of comonotonicity. ◇

Exercise D.3.2. This exercise complements the Fréchet bounds (D.17) and (D.18).

(a) Derive similar bounds as in (D.17) for the probability $P[A \cup B]$ of a union of two events $A, B \in \mathscr{F}$.

(b) Show that the bounds in (D.17) admit the following extension to the case of n events:

$$(P[A_1] + \cdots + P[A_n] - (n-1))^+ \leq P[A_1 \cap \cdots \cap A_n] \leq P[A_1] \wedge \cdots \wedge P[A_n]$$

for $A_1, \ldots, A_n \in \mathscr{F}$. Then derive an extension of (D.18) to the case of n random variables $X_1, \ldots, X_n$. ◇

D.4 Atomless probability spaces

Definition D.15. Let $(\Omega, \mathscr{F}, P)$ be a probability space.

(a) A set $A \in \mathscr{F}$ is called an *atom* of $(\Omega, \mathscr{F}, P)$, if $P[A] > 0$ and if each $B \in \mathscr{F}$ with $B \subseteq A$ satisfies either $P[B] = 0$ or $P[B] = P[A]$.

(b) The probability space $(\Omega, \mathscr{F}, P)$ is called *atomless* if it contains no atoms. △

Proposition D.16. *For any probability space, the following conditions are equivalent:*
(a) *$(\Omega, \mathscr{F}, P)$ is atomless.*
(b) *There exists an i. i. d. sequence $X_1, X_2, \dots$ of random variables with symmetric Bernoulli distribution,*

$$P[X_1 = 1] = P[X_1 = 0] = \frac{1}{2}.$$

(c) *For any $\mu \in \mathscr{M}_1(\mathbb{R})$, there exist i. i. d. random variables $Y_1, Y_2, \dots$ with common distribution μ.*
(d) *$(\Omega, \mathscr{F}, P)$ supports a random variable with a continuous distribution.*

Proof. (a)⇒(b): We need the following intuitive fact from measure theory: If $(\Omega, \mathscr{F}, P)$ is atomless, then for every $A \in \mathscr{F}$ and all δ with $0 \le \delta \le P[A]$ there exists a measurable set $B \subseteq A$ such that $P[B] = \delta$; see Theorem 9.51 of [3] for a proof. Thus, we may take a set $A \in \mathscr{F}$ such that $P[A] = 1/2$ and define $X_1 := 1$ on A and $X_1 := 0$ on A^c. The random variables $X_2, X_3, \dots$ will now be constructed recursively as follows. If $X_1, \dots, X_n$ have already been constructed, then

$$P[X_1 = x_1, \dots, X_n = x_n] = 2^{-n}$$

for all $x_1, \dots, x_n \in \{0, 1\}$, and this property is equivalent to $X_1, \dots, X_n$ being independent with the desired symmetric Bernoulli distribution. For all $x_1, \dots, x_n \in \{0, 1\}$, we may choose a set

$$B \subseteq \{X_1 = x_1, \dots, X_n = x_n\}$$

such that $P[B] = 2^{-(n+1)}$ and define $X_{n+1} := 1$ on B and $X_{n+1} := 0$ on $B^c \cap \{X_1 = x_1, \dots, X_n = x_n\}$. Clearly, the collection $X_1, \dots, X_{n+1}$ is again i. i. d. with a symmetric Bernoulli distribution.

(b)⇒(c): By relabeling the sequence $X_1, X_2, \dots$, we may obtain a double-indexed sequence $(X_{i,j})_{i,j \in \mathbb{N}}$ of independent Bernoulli-distributed random variables. If we let

$$U_i := \sum_{n=1}^{\infty} 2^{-n} X_{i,n},$$

then it is straightforward to check that U_i has a uniform distribution. Let q be a quantile function for μ. Lemma D.7 shows that the i. i. d. sequence $Y_i := q(U_i)$, $i = 1, 2, \dots$, has common distribution μ.

The proofs of the implications (c)⇒(d) and (d)⇒(a) are straightforward. □

Lemma D.17. *If X is a random variable on an atomless probability space, then there exists a random variable U with uniform distribution on $(0, 1)$ such that $X = q_X(U)$ P-a. s. for any quantile function q_X of X.*

Proof. Let us fix a quantile function $q := q_X$ and denote by λ the Lebesgue measure on $(0,1)$. For each $x \in \mathbb{R}$, $I_x := \{t \in (0,1) \mid q(t) = x\}$ is a (possibly empty or degenerate) real interval, which has Lebesgue measure $\lambda(I_x) = P[X = x]$ by Lemma D.7. Consider the set $D := \{x \in \mathbb{R} \mid P[X = x] > 0\}$, which is at most countable. For each $x \in D$, the probability space $(\Omega, \mathscr{F}, P[\,\cdot \mid X = x\,])$ is again atomless, and hence supports a random variable $U_x : \Omega \to I_x$ with a uniform law on I_x. That is, $P[\,U_x \in A \mid X = x\,] = \lambda(A \cap I_x)/\lambda(I_x)$ or equivalently,

$$P[\,U_x \in A,\, X = x\,] = \lambda(A \cap I_x) \quad \text{for all measurable } A \subseteq (0,1). \tag{D.19}$$

Let $F := F_X$ be the cumulative distribution function of X and define the random variable

$$U := F(X)\mathbb{1}_{\{X \notin D\}} + \sum_{x \in D} U_x \mathbb{1}_{\{X = x\}}.$$

Since $q(U_x) = x$ for any $x \in D$, we have $q(U) = X$ on $\{X \in D\}$. Let us show next that $q(F(X)) = X$ P-a. s. on $\{X \notin D\}$, which will then yield that $q(U) = X$ P-a. s. as desired. To this end, let $\Delta := \{t \in (0,1) \mid q(t+) > q(t-)\}$ denote the set of discontinuities of q. If $F(x) \notin \Delta$, then $q(F(x)+) = q(F(x)-)$, while Lemma D.5 implies that $q(F(x)-) \le x \le q(F(x)+)$. Combining these two facts gives

$$q(F(x)) = x \quad \text{if } F(x) \notin \Delta. \tag{D.20}$$

Thus, we will have $q(F(X)) = X$ P-a. s. on $\{X \notin D\}$ if we can prove that

$$P[\,F(X) \in \Delta,\, X \notin D\,] = 0. \tag{D.21}$$

To prove (D.21), fix $t \in \Delta$. Then Remark D.4 and Lemma D.2 yield that the set $J_t := \{x \in \mathbb{R} \mid F(x) = t\}$ is a real interval with endpoints $q(t-)$ and $q(t+)$. On the one hand, by the right-continuity of F,

$$P[\,q(t-) < X < q(t+)\,] = F(q(t+)-) - F(q(t-)) = t - t = 0.$$

On the other hand, $P[\,X \in \{q(t-), q(t+)\},\, X \notin D\,] = 0$. Thus, $P[\,X \in J_t,\, X \notin D\,] = 0$ for each $t \in \Delta$. Since Δ is countable, our claim (D.21) follows.

It remains to show that U has a uniform law. To this end, take a measurable subset A of $(0,1)$. Using (D.19), we get

$$\begin{aligned} P[\,U \in A\,] &= P[\,U \in A,\, X \notin D\,] + \sum_{x \in D} P[\,U \in A,\, X = x\,] \\ &= P[\,F(X) \in A,\, X \notin D\,] + \sum_{x \in D} \lambda(A \cap I_x). \end{aligned}$$

Thus, the proof will be completed by showing that $P[\,F(X) \in A, X \notin D\,] = \lambda(A \cap I^c)$, where $I := \{t \mid q(t) \in D\} = \bigcup_{x \in D} I_x$. To this end, recall first that $\lambda \circ q^{-1} = P \circ X^{-1}$ by Lemma D.7. Now, by (D.21),

$$\begin{aligned} P[\,F(X) \in A, X \notin D\,] &= P[\,F(X) \in A, F(X) \notin \Delta, X \notin D\,] \\ &= \lambda(F(q) \in A, F(q) \notin \Delta, q \notin D) \\ &= \lambda(\{F(q) \in A\} \cap \{F(q) \notin \Delta\} \cap I^c), \end{aligned} \tag{D.22}$$

as $\{t \mid q(t) \notin D\} = I^c$. We now claim that

$$\{F(q) \in A\} \cap \{F(q) \notin \Delta\} \cap I^c = A \cap \{F(q) \notin \Delta\} \cap I^c. \tag{D.23}$$

To prove this claim, note first that $\{F(q) \in A\} \subseteq \{q(F(q)) \in q(A)\}$ and $q(F(q)) = q$ on $\{F(q) \notin \Delta\}$ by (D.20). Therefore, the intersection on the left-hand side of (D.23) is contained in $\{q \in q(A)\} \cap \{F(q) \notin \Delta\} \cap I^c$. Next, suppose $q(t) \in q(A)$ for some $t \in \{F(q) \notin \Delta\} \cap I^c$. Then $q(t) = q(a)$ for some $a \in A$. But since $t \in I^c$, we have $q(t) \notin D$ and, therefore, $\lambda(I_{q(t)}) = 0$. Hence, the interval $I_{q(t)}$ is degenerate, which implies $t = a$, and in turn, $t \in A$. This gives "$\subseteq$" in (D.23). Conversely, suppose that t belongs to the intersection on the right-hand side of (D.23). Since $t \in I^c$, we have $q(t) \notin D$, and hence $F(q(t)-) = F(q(t))$. Combining the latter fact with $F(q(t)-) \le t \le F(q(t))$ gives $F(q(t)) = t \in A$, and in turn, "$\supseteq$." This completes the proof of (D.23).

Plugging (D.23) into (D.22) gives

$$\lambda(\{F(q) \in A\} \cap \{F(q) \notin \Delta\} \cap I^c) = \lambda(A \cap \{F(q) \notin \Delta\} \cap I^c).$$

Applying Lemma D.7 to (D.21) yields

$$\lambda(\{F(q) \in \Delta\} \cap I^c) = P[\,F(X) \in \Delta, X \notin D\,] = 0.$$

Therefore, $\lambda(A \cap \{F(q) \notin \Delta\} \cap I^c) = \lambda(A \cap I^c)$, and the proof is complete. □

E The Neyman–Pearson lemma

Suppose that P and Q are two probability measures on $(\Omega, \mathscr{F})$, and consider the Lebesgue decomposition

$$P[\,A\,] = P[\,A \cap \{\varphi = \infty\}\,] + E_Q[\,\varphi \mathbb{1}_A\,], \quad A \in \mathscr{F},$$

of P with respect to Q as in (B.10). Here, $\varphi = dP/dQ$ denotes the generalized density of P with respect to Q. Recall that we always have $Q[\varphi < \infty] = 1$, and that $P[\varphi < \infty] = 1$ holds if and only if $P \ll Q$. For fixed $c \in [0, \infty)$, we let

$$A^0 := \{\varphi > c\}.$$

Proposition E.1 (Neyman–Pearson lemma). *If $A \in \mathscr{F}$ is such that $Q[\,A\,] \le Q[\,A^0\,]$, then $P[\,A\,] \le P[\,A^0\,]$.*

Proof. Let $F := \mathbb{1}_{A^0} - \mathbb{1}_A$. Then $F \ge 0$ on $\{\varphi = \infty\}$, and $(\varphi - c)F \ge 0$. Hence,

$$\begin{aligned} P[\,A^0\,] - P[\,A\,] &= E[\,F\,] \\ &= E[\,\mathbb{1}_{\{\varphi=\infty\}}F\,] + E_Q[\,\varphi F\,] \\ &\ge cE_Q[\,F\,] \\ &= c(Q[\,A^0\,] - Q[\,A\,]). \end{aligned}$$

This proves the proposition. □

Remark E.2. In statistical test theory, A^0 is interpreted as the *likelihood ratio test* of the null hypothesis Q against the alternative hypothesis P: If the outcome ω of a statistical experiment is in A^0, then the null hypothesis is rejected. There are two possible kinds of error that can occur in such a test. A *type* 1 *error*, or *false positive*, occurs if the null hypotheses is rejected despite the fact that Q is the “true” probability. Similarly, a *type* 2 *error*, or *false negative*, occurs if the null hypothesis is not rejected, although Q is not the “true” probability. The probability of a type 1 error is given by $Q[\,A^0\,]$. This quantity is usually called the *size* or the *significance level* of the statistical test A^0. A type 2 error occurs with probability $P[\,(A^0)^c\,]$. The complementary probability $P[\,A^0\,] = 1 - P[\,(A^0)^c\,]$ is called the *power* of the test A^0. In this setting, the set A of Proposition E.1 can be regarded as another statistical test to which our likelihood ratio test is compared. The proposition can thus be restated as follows: *A likelihood ratio test has maximal power on its significance level.* ◇

Indicator functions of sets take only the values 0 and 1. We now generalize Proposition E.1 by considering $\mathscr{F}$-measurable functions $\psi : \Omega \to [0, 1]$; let $\mathscr{R}$ denote the set of all such functions.

Theorem E.3. *Let $R := \frac{1}{2}(P + Q)$.*

https://doi.org/10.1515/9783111045283-019

(a) *Take $c \in [0, \infty)$, and suppose that $\psi^0 \in \mathscr{R}$ satisfies R-a. s.*

$$\psi^0 = \begin{cases} 1 & \text{on } \{\varphi > c\}, \\ 0 & \text{on } \{\varphi < c\}. \end{cases} \tag{E.1}$$

Then, for any $\psi \in \mathscr{R}$,

$$E_Q[\,\psi\,] \le E_Q[\,\psi^0\,] \quad \Longrightarrow \quad E[\,\psi\,] \le E[\,\psi^0\,]. \tag{E.2}$$

(b) *For any $\alpha_0 \in (0,1)$ there is some $\psi^0 \in \mathscr{R}$ of the form* (E.1) *such that $\int \psi^0 \, dQ = \alpha_0$. More precisely, if c is an $(1-\alpha_0)$-quantile of φ under Q, we can define ψ^0 by*

$$\psi^0 = \mathbb{1}_{\{\varphi > c\}} + \kappa \mathbb{1}_{\{\varphi = c\}},$$

where κ is defined as

$$\kappa := \begin{cases} 0 & \text{if } Q[\,\varphi = c\,] = 0, \\ \dfrac{\alpha_0 - Q[\,\varphi > c\,]}{Q[\,\varphi = c\,]} & \text{otherwise.} \end{cases}$$

(c) *Suppose conversely that $\psi^0 \in \mathscr{R}$ satisfies* (E.2) *and $E_Q[\,\psi^0\,] > 0$. Then ψ^0 is R-a. s. of the form* (E.1) *for some $c \ge 0$.*

Proof. (a): Take $F := \psi^0 - \psi$ and repeat the argument in the proof of Proposition E.1.

(b): Let F denote the distribution function of φ under Q. Then $Q[\,\varphi > c\,] = 1 - F(c) \le \alpha_0$ and

$$\begin{aligned} Q[\,\varphi = c\,] &= F(c) - F(c-) \\ &\ge F(c) - 1 + \alpha_0 \\ &= \alpha_0 - Q[\,\varphi > c\,]. \end{aligned}$$

Hence, $0 \le \kappa \le 1$ and ψ^0 belongs to $\mathscr{R}$. The fact that $E_Q[\,\psi^0\,] = \alpha_0$ is obvious.

(c): Suppose that $\psi^0 \in \mathscr{R}$ satisfies (E.2) and $\alpha_0 := E_Q[\,\psi^0\,] > 0$. The case $\alpha_0 = 1$ is trivial. For $0 < \alpha_0 < 1$, we let

$$\widetilde{\psi} := \mathbb{1}_{\{\varphi > c\}} + \kappa \mathbb{1}_{\{\varphi = c\}},$$

where c and κ are as in part (b). Then $\alpha_0 = E_Q[\,\psi^0\,] = E_Q[\,\widetilde{\psi}\,]$. One also has that $E[\,\psi^0\,] = E[\,\widetilde{\psi}\,]$, as can be seen by applying (E.2) to both ψ^0 and $\widetilde{\psi}$ with reversed roles. Hence, for $G = \widetilde{\psi} - \psi^0$,

$$0 = E[\,G\,] - cE_Q[\,G\,] = E[\,G\mathbb{1}_{\{\varphi = \infty\}}\,] + E_Q[\,(\varphi - c)G\,].$$

But the fact that $\widetilde{\psi}$ satisfies (E.1) implies that both $G \geq 0$ P-a. s. on $\{\varphi = \infty\}$, and $(\varphi - c)G \geq 0$ Q-a. s. Hence, G vanishes R-a. s. on $\{\varphi \neq c\}$, and so ψ^0 must also be R-a. s. of the form (E.1). □

Remark E.4. In the context of Remark E.2, an element ψ of $\mathscr{R}$ is interpreted as a *randomized statistical test*: If ω is the outcome of a statistical experiment and $p := \psi(\omega)$, then the null hypothesis is rejected with probability p, that is, after performing an independent random coin toss with success probability p. *Significance level* and *power* of a randomized test are defined as above, and a test of the form (E.1) is called a *generalized likelihood ratio test*. Thus, the general Neyman–Pearson lemma in the form of Theorem E.3 can be stated as follows: *A randomized test has maximal power on its significance level, if and only if it is a generalized likelihood ratio test.* ◇

F The essential supremum of a family of random variables

In this section, we discuss the essential supremum of an arbitrary family Φ of random variables on a given probability space $(\Omega, \mathscr{F}, P)$. Consider first the case in which the set Φ is countable. Then $\varphi^*(\omega) := \sup_{\varphi\in\Phi} \varphi(\omega)$ will also be a random variable, that is, φ^* is measurable. Measurability of the pointwise supremum, however, is not guaranteed if Φ is uncountable. Even if the pointwise supremum is measurable, it may not be the right concept when we focus on almost sure properties. This can be illustrated by taking P as the Lebesgue measure on $\Omega := [0,1]$ and $\Phi := \{\mathbb{1}_{\{x\}} \mid 0 \le x \le 1\}$. Then $\sup_{\varphi\in\Phi} \varphi(x) \equiv 1$ whereas $\varphi = 0$ P-a. s. for each single $\varphi \in \Phi$. This suggests the following notion of an *essential supremum* defined in terms of almost sure inequalities.

Theorem F.1. *Let Φ be any set of random variables on $(\Omega, \mathscr{F}, P)$.*

(a) *There exists a $[-\infty, +\infty]$-valued random variable φ^* with the following two properties:*
   (i) *$\varphi^* \ge \varphi$ P-a. s. for all $\varphi \in \Phi$.*
   (ii) *$\varphi^* \le \psi$ P-a. s. for every random variable ψ satisfying $\psi \ge \varphi$ P-a. s. for all $\varphi \in \Phi$.*

(b) *Suppose in addition that Φ is directed upward, that is, for $\varphi, \widetilde{\varphi} \in \Phi$ there exists $\psi \in \Phi$ with $\psi \ge \varphi \vee \widetilde{\varphi}$ P-almost surely. Then there exists an increasing sequence $\varphi_1 \le \varphi_2 \le \cdots$ in Φ such that $\varphi^* = \lim_n \varphi_n$ P-almost surely.*

Definition F.2. The random variable φ^* in Theorem F.1 is called the *essential supremum* of Φ with respect to P, and we write

$$\operatorname{ess\,sup} \Phi = \operatorname*{ess\,sup}_{\varphi\in\Phi} \varphi := \varphi^*.$$

The *essential infimum* of Φ with respect to P is defined as

$$\operatorname{ess\,inf} \Phi = \operatorname*{ess\,inf}_{\varphi\in\Phi} \varphi := -\operatorname*{ess\,sup}_{\varphi\in\Phi}(-\varphi). \qquad \triangle$$

Proof of Theorem F.1. If Φ is empty, then $\varphi^* := -\infty$ is as desired. So, let us suppose that Φ is nonempty. We may assume without loss of generality that each $\varphi \in \Phi$ takes values in $[0,1]$; otherwise we may consider $\widetilde{\Phi} := \{f \circ \varphi \mid \varphi \in \Phi\}$ with $f : [-\infty, +\infty] \to [0,1]$ strictly increasing and take $\varphi^* := f^{-1}(\operatorname{ess\,sup} \widetilde{\Phi})$.

If $\Psi \subseteq \Phi$ is countable, let $\varphi_\Psi(\omega) := \sup_{\varphi\in\Psi} \varphi(\omega)$. Then φ_Ψ is measurable. We claim that the upper bound

$$c := \sup\{E[\,\varphi_\Psi\,] \mid \Psi \subseteq \Phi \text{ countable}\}$$

is attained by some countable $\Psi^* \subseteq \Phi$. To see this, take Ψ_n with $E[\,\varphi_{\Psi_n}\,] \to c$ and let $\Psi^* := \bigcup_n \Psi_n$. Then Ψ^* is countable and $E[\,\varphi_{\Psi^*}\,] = c$.

https://doi.org/10.1515/9783111045283-020

We now show that $\varphi^* := \varphi_{\Psi^*}$ satisfies (i). Suppose that (i) does not hold. Then there exists $\varphi \in \Phi$ such that $P[\varphi > \varphi^*] > 0$. Hence, $\Psi' := \Psi^* \cup \{\varphi\}$ satisfies

$$E[\varphi_{\Psi'}] > E[\varphi_{\Psi^*}] = c,$$

in contradiction to the definition of c. Furthermore, if ψ is any other random variable satisfying $\psi \geq \varphi$ P-a. s. for all $\varphi \in \Phi$, then obviously $\psi \geq \varphi^*$.

Finally, the construction shows that φ_{Ψ^*} can be approximated by an increasing sequence if Ψ is directed upwards. □

Remark F.3. For a given random variable X, let Φ be the set of all constants c such that $P[X > c] > 0$. The number

$$\operatorname{ess\,sup} X := \sup \Phi$$

is the smallest constant $c \leq +\infty$ such that $X \leq c$ P-a. s. and called the *essential supremum of X* with respect to P. The *essential infimum* of X is defined as

$$\operatorname{ess\,inf} X := \operatorname{ess\,sup}(-X).$$

◇

G Spaces of measures

G.1 Weak convergence

A topological space S is called *metrizable* if there exists a metric d on S that generates the topology of S. That is, the open d-balls

$$B_\varepsilon(x) := \{y \in S \mid d(x,y) < \varepsilon\}, \quad x \in S, \varepsilon > 0,$$

form a *base* for the topology of S in the sense that a set $U \subseteq S$ is open if and only if it can be written as a union of such d-balls. A convenient feature of metrizable spaces is that their topological properties can be characterized via convergent sequences. For instance, a subset A of the metrizable space S is closed if and only if the limit of every convergent sequence in A also belongs to A. Moreover, a function $f : S \to \mathbb{R}$ is continuous at $y \in S$ if and only if $f(y_n)$ converges to $f(y)$ for every sequence (y_n) converging to y. We write $C_b(S)$ for the set of all bounded and continuous functions on S.

The metrizable space S is called *separable* if there exists a countable dense subset $\{x_1, x_2, \dots\}$ of S. In this case, the Borel σ-algebra $\mathscr{S}$ of S is generated by the open d-balls $B_\varepsilon(x)$ with radii $\varepsilon > 0$, $\varepsilon \in \mathbb{Q}$, and centered in $x \in \{x_1, x_2, \dots\}$. In what follows, we will always assume that S is separable and metrizable. If, moreover, the metric d can be chosen to be complete, that is, if every Cauchy sequence with respect to d converges to some point in S, then S is called a *Polish space*. Clearly, $\mathbb{R}^d$ with the Euclidean distance is a complete and separable metric space, hence a Polish space.

Let us denote by

$$\mathscr{M}(S) := \mathscr{M}(S, \mathscr{S})$$

the set of all nonnegative finite measures on $(S, \mathscr{S})$. Every $\mu \in \mathscr{M}(S)$ is of the form $\mu = \alpha\nu$ for some factor $\alpha \in [0, \infty)$ and some probability measure ν on the measurable space $(S, \mathscr{S})$. The space of all probability measures on $(S, \mathscr{S})$ is denoted by

$$\mathscr{M}_1(S) = \mathscr{M}_1(S, \mathscr{S}).$$

Definition G.1. The *weak topology* on $\mathscr{M}(S)$ is the coarsest topology for which all mappings

$$\mathscr{M}(S) \ni \mu \mapsto \int f \, d\mu, \quad f \in C_b(S),$$

are continuous. △

It follows from this definition that the sets

$$U_\varepsilon(\mu; f_1, \dots, f_n) := \bigcap_{i=1}^{n} \left\{ \nu \in \mathscr{M}(S) \,\middle|\, \left| \int f_i \, d\nu - \int f_i \, d\mu \right| < \varepsilon \right\} \tag{G.1}$$

https://doi.org/10.1515/9783111045283-021

for $\mu \in \mathcal{M}(S)$, $\varepsilon > 0$, $n \in \mathbb{N}$, and $f_1, \dots, f_n \in C_b(S)$ form a base for the weak topology on $\mathcal{M}(S)$; for details see, for example, Section 2.13 of [3]. Since the constant function 1 is continuous,

$$\mathcal{M}_1(S) = \left\{ \mu \in \mathcal{M}(S) \;\middle|\; \mu(S) = \int 1\,d\mu = 1 \right\}$$

is a closed subset of $\mathcal{M}(S)$. A well-known example for weak convergence of probability measures is the classical *central limit theorem*; the following version is needed in Section 5.7.

Theorem G.2. *Suppose that for each $N \in \mathbb{N}$ we are given a probability space $(\Omega_N, \mathcal{F}_N, P_N)$ with N independent random variables $Y_1^{(N)}, \dots, Y_N^{(N)}$ that satisfy the following conditions:*
- *There are constants γ_N such that $\gamma_N \to 0$ and $|Y_k^{(N)}| \le \gamma_N$ P_N-a. s.*
- $\sum_{k=1}^N E_N[\, Y_k^{(N)} \,] \longrightarrow m.$
- $\sum_{k=1}^N \mathrm{var}_N(Y_k^{(N)}) \longrightarrow \sigma^2$, *where var_N denotes the variance with respect to P_N.*

Then the distributions of

$$Z_N := \sum_{k=1}^N Y_k^{(N)}, \quad N = 1, 2, \dots,$$

converge weakly to the normal distribution with mean m and variance σ^2.

Proof. See, for instance, the corollary to Theorem 7.1.2 of [65]. □

The following theorem allows us to examine the weak topology in terms of weakly converging *sequences* of measures.

Theorem G.3. *The space $\mathcal{M}(S)$ is separable and metrizable for the weak topology. If S is Polish, then so is $\mathcal{M}(S)$. Moreover, if S_0 is a dense subset of S, then the set*

$$\left\{ \sum_{i=1}^n \alpha_i \delta_{x_i} \;\middle|\; \alpha_i \in \mathbb{Q}^+,\ x_i \in S_0,\ n \in \mathbb{N} \right\}$$

of simple measures on S_0 with rational weights is dense in $\mathcal{M}(S)$ for the weak topology.

Proof. In most textbooks on measure theory, the previous result is proved for $\mathcal{M}_1(S)$ instead of $\mathcal{M}(S)$; see, for example, Theorem 14.12 of [3]. The general case requires only minor modifications. It is treated in full generality in Chapter IX, § 5, of [39]. □

The following characterization of weak convergence in $\mathcal{M}(S)$ is known as the "portmanteau theorem."

Theorem G.4. *For any sequence $\mu, \mu_1, \mu_2, \dots$ of measures in $\mathcal{M}(S)$, the following conditions are equivalent:*

(a) *The sequence $(\mu_n)_{n \in \mathbb{N}}$ converges weakly to μ.*

(b) $\mu_n(S) \to \mu(S)$ *and*

$$\limsup_{n\uparrow\infty} \mu_n(A) \le \mu(A) \quad \text{for every closed set } A \subseteq S.$$

(c) $\mu_n(S) \to \mu(S)$ *and*

$$\liminf_{n\uparrow\infty} \mu_n(U) \ge \mu(U) \quad \text{for every open set } U \subseteq S.$$

(d) $\mu_n(B) \to \mu(B)$ *for every Borel set* B *whose boundary* ∂B *is not charged by* μ *in the sense that* $\mu(\partial B) = 0$.

(e) $\int f\, d\mu_n \to \int f\, d\mu$ *for every bounded measurable function* f *that is* μ*-a. e. continuous.*

(f) $\int f\, d\mu_n \to \int f\, d\mu$ *for every bounded and uniformly continuous function* f.

Proof. The result is proved for $\mathscr{M}_1(S)$ in [3], Theorem 14.3. The general case requires only minor modifications; see Chapter IX of [39]. □

As an application of the portmanteau theorem, we now state a law of large numbers for the weak convergence of the empirical distributions of a sequence of independent and identically distributed random variables. It uses the notion of a *Dirac measure* or *Dirac point mass* in $y \in S$, which is defined as

$$\delta_y(A) = \mathbb{1}_A(y) \quad \text{for all } A \subseteq S.$$

Clearly, δ_y is a measure, which can be defined on the entire power set of S.

Theorem G.5. *Let* S *be a separable and metrizable space and suppose that* $X_1, X_2, \dots$ *are independent random variables with common distribution* μ. *Then, with probability one, the empirical distributions,*

$$\varrho_n := \frac{1}{n}\sum_{k=1}^{n} \delta_{X_k}, \quad n = 1, 2, \dots,$$

converge weakly to μ *as* $n \uparrow \infty$.

Proof. Let $(\Omega, \mathscr{F}, P)$ be the probability space supporting the random variables X_n. For any measurable set $A \subseteq S$, Kolmogorov's law of large numbers states that

$$\varrho_n(\omega)(A) = \frac{1}{n}\sum_{k=1}^{n} \mathbb{1}_A(X_k(\omega)) \longrightarrow E[\mathbb{1}_A(X_1)] = \mu(A), \tag{G.2}$$

for P-a. e. $\omega \in \Omega$. However, the null set of all $\omega \in \Omega$ for which (G.2) does not hold depends on A. Unless S is a finite space, there are uncountably many measurable subsets A of S, and so the convergence (G.2) by itself is not sufficient to establish the assertion.

Since S is separable and metrizable, there exists a countable topological base, i. e., a collection $V_1, V_2, \dots$ of open sets such that every open subset of S can be represented as a certain union of sets V_i. For instance, if d is a metric for the topology of S, we can take the collection of open d-balls that have rational radii and are centered in the members of a countable dense subset of S. We let

$$\mathscr{U} := \left\{ \bigcup_{i \in I} V_i \;\middle|\; I \subset \mathbb{N} \text{ is finite} \right\}.$$

Then $\mathscr{U}$ is countable and, in view of (G.2), there exists a measurable set $\Omega_0 \in \mathscr{F}$ such that $P[\Omega_0] = 1$ and $\rho_1(\omega)(V) \to \mu(V)$ for each $V \in \mathscr{U}$ and all $\omega \in \Omega_0$. Moreover, for every open set $U \subseteq S$ there are $U_1, U_2, \dots \in \mathscr{U}$ such $U_1 \subseteq U_2 \subseteq \dots \subseteq U$ and $U = \bigcup_k U_k$. It follows that for each k and all $\omega \in \Omega_0$,

$$\liminf_{n \uparrow \infty} \rho_n(\omega_0)(U) \geq \liminf_{n \uparrow \infty} \rho_n(\omega_0)(U_k) = \mu(U_k).$$

When passing to the limit $k \uparrow \infty$, we have $\mu(U_k) \to \mu(U)$, and so we arrive at

$$\liminf_{n \uparrow \infty} \rho_n(\omega_0)(U) \geq \mu(U).$$

Since clearly $\rho_n(\omega)(S) = 1 = \mu(S)$ for all n and ω, condition (c) of Theorem G.4 is satisfied, and we obtain that $\rho_n(\omega) \to \mu$ weakly for each $\omega \in \Omega_0$. □

Remark G.6. It is possible to establish the convergence $\rho_n \to \mu$ in stronger topologies than the weak topology. This is the content of the Glivenko–Cantelli theorem and its generalizations; see, for example, [317]. A slightly stronger statement than Theorem G.5 will be provided in Corollary G.16. On the other hand, it is easy to relax the assumption of independence between the random variables (X_k). To this end, one can simply replace Kolmogorov's law of large numbers in (G.2) by some other almost-sure limit theorem for the sample mean, while retaining all other arguments in the proof. For instance, Birkhoff's ergodic theorem (e. g., Theorem 1 in § 3 of Chapter V in [302]) yields the P-a. s. convergence of $\rho_n(A)$ to a (possibly random) limit, if the sequence $X_1, X_2, \dots$ is stationary in the sense that $(X_1, X_2, \dots)$ has the same law as $(X_k, X_{k+1}, \dots)$ for all $k \in \mathbb{N}$. This includes the special case in which the $X_1, X_2, \dots$ form a stationary Markov chain. ◇

The following exercise explores the implications of the portmanteau theorem for weak convergence of Borel probability measures on $\mathbb{R}$.

Exercise G.1.1. Use the portmanteau theorem to show that the following conditions are equivalent for any sequence $(\mu_n)_{n=0,1,\dots}$ in $\mathscr{M}_1(\mathbb{R})$:

(a) μ_n converges weakly to μ_0.

(b) The corresponding cumulative distribution functions satisfy

$$F_{\mu_0}(x-) \leq \liminf_{n\uparrow\infty} F_{\mu_n}(x) \leq \limsup_{n\uparrow\infty} F_{\mu_n}(x) \leq F_{\mu_0}(x)$$

for all $x \in \mathbb{R}$.

(c) $F_{\mu_n}(x) \to F_{\mu_0}(x)$ for any continuity point x of F_{μ_0}.

(d) For any choice of quantile functions q_{μ_n} and all $t \in (0,1)$,

$$q_{\mu_0}(t-) \leq \liminf_{n\uparrow\infty} q_{\mu_n}(t) \leq \limsup_{n\uparrow\infty} q_{\mu_n}(t) \leq q_{\mu_0}(t+).$$

(e) We have $q_{\mu_n}(t) \to q_{\mu_0}(t)$ for any continuity point t of q_{μ_0}. ◇

The following result is often called the *Skorokhod representation theorem*. While a full proof of this result is more involved, Exercise G.1.2 will explore a special case in which an easy proof can be obtained by way of Exercise G.1.1 and Lemma D.7.

Theorem G.7 (Skorokhod). *If (μ_n) is a sequence of Borel probability measures on the separable metric space S that converge weakly to the probability measure μ_∞, then there exists a probability space $(\Omega, \mathscr{F}, P)$ and S-valued random variables $X_1, X_2, \ldots, X_\infty$ such that each X_n has law μ_n (including $n = \infty$) and $X_n \to X_\infty$ P-almost surely.*

Proof. See, for example, Theorem 4.30 in [199]. □

Exercise G.1.2. Use Exercise G.1.1 and Lemma D.7 to prove Theorem G.7 for $S = \mathbb{R}$. ◇

The next theorem can be regarded as a stability result for weak convergence.

Theorem G.8 (Slutsky). *Suppose that, for $n \in \mathbb{N}$, we are given a probability space $(\Omega_n, \mathscr{F}_n, P_n)$ with real-valued random variables X_n and Y_n such that the laws of X_n converge weakly to the law of X, and the laws of Y_n converge weakly to the Dirac measure δ_y for some $y \in \mathbb{R}$. Then:*

(a) *The laws of $X_n + Y_n$ converge weakly to the law of $X + y$.*

(b) *The laws of $X_n \cdot Y_n$ converge weakly to the law of $X \cdot y$.*

Proof. See, for instance, Section 8.1 of [64]. □

We turn now to the fundamental characterization of the relative compact subsets of $\mathscr{M}(S)$ known as *Prohorov's theorem*.

Theorem G.9 (Prohorov). *Let S be a Polish space. A nonempty subset M of $\mathscr{M}(S)$ is relatively compact for the weak topology if and only if*

$$\sup_{\mu\in M} \mu(S) < \infty$$

and if M is tight, *that is, if for every $\varepsilon > 0$ there exists a compact subset K of S such that*

$$\sup_{\mu \in M} \mu(K^c) \le \varepsilon.$$

In particular, $\mathcal{M}_1(S)$ is weakly compact if S is a compact metric space.

Proof. For a proof in the context of probability measures see, for instance, Theorem 1 in § III.2 of [302]. The general case requires only minor modifications; see Chapter IX of [39]. □

Example G.10. Take for S the positive half axis $[0, \infty)$ and define

$$\mu_n := \frac{n-1}{n}\,\delta_0 + \frac{1}{n}\,\delta_n \quad \text{and} \quad \mu := \delta_0.$$

Clearly,

$$\int f\, d\mu_n \longrightarrow \int f\, d\mu \quad \text{for all } f \in C_b(S)$$

so that μ_n converges weakly to μ. However, if we take the continuous but unbounded function $f(x) = x$, then $\int f\, d\mu_n = 1$ for all n so that

$$\lim_{n \uparrow \infty} \int f\, d\mu_n = 1 \neq 0 = \int f\, d\mu. \qquad \diamond$$

The preceding example shows that the weak topology is not an appropriate topology for ensuring the convergence of integrals against unbounded test functions. A simple criterion that guarantees the convergence for certain unbounded test functions on $S = \mathbb{R}$ is stated in Lemma 5.61. A more general approach will be developed in the following section.

G.2 The ψ-weak topology

In the sequel, we introduce a general transformation of the weak topology that will allow us to deal with certain classes of unbounded functions. We let S be a separable and metrizable space and fix a continuous function

$$\psi : S \mapsto [1, \infty),$$

which will serve as a *gauge function*, and we denote by

$$C_\psi(S)$$

the linear space of all continuous functions f on S for which there exists a constant c such that

$$|f(x)| \leq c \cdot \psi(x) \quad \text{for all } x \in S.$$

Furthermore, we denote by

$$\mathscr{M}^{\psi}(S)$$

the set of all measures $\mu \in \mathscr{M}(S)$ such that $\int \psi \, d\mu < \infty$.

Definition G.11. The *ψ-weak topology* on $\mathscr{M}^{\psi}(S)$ is the coarsest topology for which all mappings

$$\mathscr{M}^{\psi}(S) \ni \mu \longmapsto \int f \, d\mu, \quad f \in C_{\psi}(S),$$

are continuous. △

Since the gauge function ψ takes values in $[1, \infty)$, every bounded continuous function f belongs to $C_{\psi}(S)$. It follows that all mappings

$$\mathscr{M}^{\psi}(S) \ni \mu \longmapsto \int f \, d\mu, \quad f \in C_b(S),$$

are continuous. In particular, the set

$$\mathscr{M}_1^{\psi}(S) := \{\mu \in \mathscr{M}^{\psi}(S) \mid \mu(S) = 1\}$$

of all Borel probability measures in $\mathscr{M}^{\psi}(S)$ is closed for the ψ-weak topology.

As in the case of the weak topology, it follows that the sets

$$U_{\varepsilon}^{\psi}(\mu; f_1, \dots, f_n) := \bigcap_{i=1}^{n} \left\{ \nu \in \mathscr{M}^{\psi}(S) \;\middle|\; \left| \int f_i \, d\nu - \int f_i \, d\mu \right| < \varepsilon \right\}$$

for $\mu \in \mathscr{M}^{\psi}(S)$, $\varepsilon > 0$, $n \in \mathbb{N}$, and $f_1, \dots, f_n \in C_{\psi}(S)$ form a base for the ψ-weak topology on $\mathscr{M}^{\psi}(S)$.

Consider the map

$$\Psi : \mathscr{M}(S) \longrightarrow \mathscr{M}^{\psi}(S)$$

defined through

$$\Psi(\mu)(A) := \int \mathbb{1}_A \psi^{-1} \, d\mu, \quad A \in \mathscr{S}.$$

Clearly, Ψ is a bijective mapping between the two sets $\mathscr{M}(S)$ and $\mathscr{M}^{\psi}(S)$. Moreover, if we apply Ψ to an open neighborhood for the weak topology as in (G.1), we get

$$\Psi(U_{\varepsilon}(\mu; f_1, \dots, f_n)) = U_{\varepsilon}^{\psi}(\Psi(\mu); f_1\psi, \dots, f_n\psi).$$

Since $f\psi \in C_\psi(S)$ for each bounded and continuous function f, and since every function in $C_\psi(S)$ arises in this way, we conclude that a subset U of $\mathcal{M}(S)$ is weakly open if and only if $\Psi(U)$ is open for the ψ-weak topology. Hence, Ψ is a homeomorphism. This observation allows us to translate statements for the weak topology into results for the ψ-weak topology. For instance, the following result is immediately implied by Theorem G.3.

Corollary G.12. *For separable and metrizable S, the space $\mathcal{M}^\psi(S)$ is separable and metrizable for the ψ-weak topology. If S is Polish, then so is $\mathcal{M}^\psi(S)$. Moreover, if S_0 is a dense subset of S, then the set*

$$\left\{ \sum_{i=1}^{n} \alpha_i \delta_{x_i} \;\middle|\; \alpha_i \in \mathbb{Q}^+,\, x_i \in S_0,\, n \in \mathbb{N} \right\}$$

of simple measures on S_0 with rational weights is dense in $\mathcal{M}^\psi(S)$ for the ψ-weak topology.

The preceding corollary implies in particular that it suffices to consider ψ-weakly converging sequences when studying the ψ-weak topology. The following corollary is implied by the portmanteau theorem.

Corollary G.13. *A sequence $(\mu_n)_{n\in\mathbb{N}}$ in $\mathcal{M}^\psi(S)$ converges ψ-weakly to μ if and only if*

$$\int f\, d\mu_n \longrightarrow \int f\, d\mu$$

for every measurable function f that is μ-a. e. continuous and for which exists a constant c such that $|f| \le c \cdot \psi$ μ-almost everywhere.

Prohorov's theorem translates as follows to our present setting.

Corollary G.14. *Let S be a Polish space and M be a subset of $\mathcal{M}^\psi(S)$. The following conditions are equivalent:*

(a) *M is relatively compact for the ψ-weak topology.*

(b) *We have*

$$\sup_{\mu\in M} \int \psi\, d\mu < \infty\,,$$

and for every $\varepsilon > 0$ there exists a compact subset K of S such that

$$\sup_{\mu\in M} \int_{K^c} \psi\, d\mu \le \varepsilon\,.$$

(c) *There exists a measurable function $\phi : S \to [1,\infty]$ such that each set*

$$\{x \in S \mid \phi(x) \le n\,\psi(x)\}\,, \quad n \in \mathbb{N}\,,$$

is relatively compact in S, and such that

$$\sup_{\mu \in M} \int \phi \, d\mu < \infty .$$

Proof. (a)⇔(b): This follows immediately from Theorem G.9 and the fact that Ψ is a homeomorphism.

(b)⇒(c): Take an increasing sequence $K_1 \subseteq K_2 \subseteq \cdots$ of compact sets in S such that

$$\sup_{\mu \in M} \int_{K_n^c} \psi \, d\mu \leq 2^{-n},$$

and define ϕ by

$$\phi(x) := \psi(x) + \sum_{n=1}^{\infty} \mathbb{1}_{K_n^c}(x)\psi(x).$$

Then $\{\phi \leq n\psi\} \subseteq K_n$. Moreover,

$$\sup_{\mu \in M} \int \phi \, d\mu \leq \sup_{\mu \in M} \int \psi \, d\mu + 1 < \infty.$$

(c)⇒(b): Since $\{\phi \leq \psi\}$ is relatively compact, we have that

$$c := \sup\{\psi(x) \mid x \in S, \, \phi(x) \leq \psi(x)\} < \infty.$$

Hence, $\psi \leq c + \phi \leq (1 + c)\phi$ and, therefore,

$$\sup_{\mu \in M} \int \psi \, d\mu \leq (1 + c) \sup_{\mu \in M} \int \phi \, d\mu < \infty.$$

Moreover, for $n \geq \varepsilon^{-1} \sup_{\mu \in M} \int \phi \, d\mu$, the relatively compact set $K := \{\phi \leq n\psi\}$ satisfies

$$\sup_{\mu \in M} \int_{K^c} \psi \, d\mu \leq \frac{1}{n} \sup_{\mu \in M} \int_{K_n^c} \phi \, d\mu \leq \varepsilon,$$

and so condition (b) is satisfied. □

Now we state a version of the Skorokhod representation theorem for the ψ-weak topology.

Corollary G.15. *Suppose that S is separable and metrizable and μ_n are Borel probability measures on S for $n = 1, 2, \ldots$ and $n = \infty$. Then the following conditions are equivalent:*

(a) *$\mu_n \to \mu_\infty$ in the ψ-weak topology.*

(b) *$\mu_n \to \mu_\infty$ weakly and $\int \psi \, d\mu_n \to \int \psi \, d\mu_\infty$.*

(c) *There exists a probability space $(\Omega, \mathscr{F}, P)$ and random variables X_n such that each X_n has law μ_n (including $n = \infty$), $X_n \to X_\infty$ P-a. s., and $E[\psi(X_n)] \to E[\psi(X_\infty)]$.*

(d) *There exists a probability space* $(\Omega, \mathscr{F}, P)$ *and random variables* X_n *such that each* X_n *has law* μ_n *(including* $n = \infty$*),* $X_n \to X_\infty$ *P-a. s., and* $E[|f(X_n) - f(X_\infty)|] \to 0$ *for all* $f \in C_\psi$.

Proof. The implication (a)⇒(b) follows from the fact that C_ψ contains all bounded continuous functions and the definition of ψ-weak convergence.

To see that (b)⇒(c), we apply the Skorokhod representation theorem in the form of Theorem G.7 to obtain random variables X_n with respective laws μ_n such that $X_n \to X_\infty$ P-almost surely. Moreover,

$$E[\psi(X_n)] = \int \psi \, d\mu_n \longrightarrow \int \psi \, d\mu_\infty = E[\psi(X_\infty)].$$

(c)⇒(d): Theorem H.18 states that the sequence $(\psi(X_n))_{n\in\mathbb{N}}$ is uniformly integrable. Now we take $f \in C_\psi(S)$. Then $f(X_n) \to f(X_\infty)$ P-almost surely. Moreover, there exists a constant c such that $|f(X_n)| \le c\psi(X_n)$ for all n, which implies that the sequence $(f(X_n))_{n\in\mathbb{N}}$ is uniformly integrable. Another application of Theorem H.18 now yields $f(X_n) \to f(X_\infty)$ in L^1.

Finally, (d) implies

$$\int f \, d\mu_n = E[f(X_n)] \longrightarrow E[f(X_\infty)] = \int f \, d\mu_\infty$$

and in turn (a). □

We conclude this section with a law of large numbers for the ψ-weak convergence of the empirical distributions of a sequence of independent and identically distributed random variables. It strengthens Theorem G.5.

Corollary G.16. *Let* S *be a separable and metrizable space and suppose that* $X_1, X_2, \dots$ *are independent random variables with common distribution* $\mu \in \mathscr{M}_1^\psi$. *Then, with probability one, the empirical distributions,*

$$\rho_n := \frac{1}{n} \sum_{k=1}^{n} \delta_{X_k}, \quad n = 1, 2, \dots,$$

converge ψ*-weakly to* μ *as* $n \uparrow \infty$.

Proof. By Kolmogorov's law of large numbers,

$$\int \psi \, d\rho_n = \frac{1}{n} \sum_{k=1}^{n} \psi(X_k) \longrightarrow E[\psi(X_1)] = \int \psi \, d\mu,$$

P-almost surely. Hence, the assertion follows from Theorem G.5 and the characterization of ψ-weak convergence in Corollary G.13(b). □

Remark G.17. As discussed in Remark G.6, the weak convergence of the empirical distribution to a (possibly random) limit measure μ can be extended to the case in which the sequence $X_1, X_2, \dots$ is stationary in the sense that $(X_1, X_2, \dots)$ has the same law as $(X_k, X_{k+1}, \dots)$ for all $k \in \mathbb{N}$. This includes the special case in which the $X_1, X_2, \dots$ form a stationary Markov chain. If in addition $E[\psi(X_1)] < \infty$, then Birkhoff's ergodic theorem (e. g., Theorem 1 in § 3 of Chapter V in [302]) yields the convergence of

$$\int \psi \, d\rho_n = \frac{1}{n} \sum_{k=1}^{n} \psi(X_k).$$

The same argument as in the preceding proof can thus be used to establish the ψ-weak convergence of ρ_n. ◇

G.3 The theorems of Riesz and Daniell–Stone

We turn now to the task of identifying a linear functional on a space of functions as the integral with respect to a suitable measure.

Theorem G.18 (Riesz). *Let S be a compact metric space and suppose that I is a linear functional on $C(S)$ that is nonnegative in the sense that $f \geq 0$ everywhere on S implies $I(f) \geq 0$. Then there exists a unique positive Borel measure μ on S such that*

$$I(f) = \int f \, d\mu \quad \text{for all } f \in C(S).$$

The preceding theorem can be deduced from the next result; see Exercise G.3.1 below. To state this result, we need the notion of a *vector lattice* of real-valued functions on an arbitrary set S. This is a linear space $\mathscr{L}$ that is stable under the operation of taking the pointwise maximum: for $f, g \in \mathscr{L}$ also $f \vee g \in \mathscr{L}$. The vector lattice $\mathscr{L}$ is called a *Stone vector lattice* if $f \wedge 1 \in \mathscr{L}$ for all $f \in \mathscr{L}$. One example is the space of all bounded measurable functions on $(S, \mathscr{S})$. Another one is the space $C_b(S)$ of all bounded continuous functions on a separable metric space S. In this case, the σ-algebra $\sigma(\mathscr{L})$ generated by $\mathscr{L} = C_b(S)$ coincides with the Borel σ-algebra of S. We now state the fundamental Daniell–Stone theorem for the special case in which a linear functional can be represented as the integral with respect to a probability measure.

Theorem G.19 (Daniell–Stone). *Let I be a linear functional on a Stone vector lattice $\mathscr{L}$ of functions on S such that the following conditions hold:*

(a) $1 \in \mathscr{L}$ *and* $I(1) = 1$.

(b) *I is nonnegative in the sense that $f \geq 0$ everywhere on S implies $I(f) \geq 0$.*

(c) *If (f_n) is a sequence in $\mathscr{L}$ such that $f_n \searrow 0$ pointwise on S, then $I(f_n) \searrow 0$.*

Then there exists a unique probability measure μ on the measurable space $(S, \sigma(\mathscr{L}))$ such that

$$I(f) = \int f \, d\mu \quad \textit{for all } f \in \mathscr{L}.$$

Proof. See, for example, Theorem 7.8.1 of [36]. □

Exercise G.3.1. Use Dini's lemma, as recalled in Lemma 4.24, to deduce Theorem G.18 from Theorem G.19. ◇

G.4 Finitely-additive set functions

Without the continuity assumption (c) in Theorem G.19, the representation of positive linear functionals on the space of bounded measurable functions takes a different form, as we will discuss now.

Definition G.20. Let $(\Omega, \mathscr{F})$ be a measurable space. A mapping $\mu : \mathscr{F} \to \mathbb{R}$ is called a *finitely additive set function* if $\mu(\emptyset) = 0$, and if for any finite collection $A_1, \dots, A_n \in \mathscr{F}$ of mutually disjoint sets

$$\mu\left(\bigcup_{i=1}^{n} A_i\right) = \sum_{i=1}^{n} \mu(A_i).$$

We denote by $\mathscr{M}_{1,f} := \mathscr{M}_{1,f}(\Omega, \mathscr{F})$ the set of all those finitely additive set functions $\mu : \mathscr{F} \to [0,1]$ that are normalized to $\mu(\Omega) = 1$. The *total variation* of a finitely additive set function μ is defined as

$$\|\mu\|_{\text{var}} := \sup\left\{\sum_{i=1}^{n} |\mu(A_i)| \;\middle|\; A_1, \dots, A_n \text{ disjoint sets in } \mathscr{F}, n \in \mathbb{N}\right\}.$$

The space of all finitely additive measures μ whose total variation is finite is denoted by $ba(\Omega, \mathscr{F})$. △

We will now give a brief outline of the integration theory with respect to a measure $\mu \in ba := ba(\Omega, \mathscr{F})$; for details, we refer to Chapter III in [117]. The space $\mathscr{X}$ of all bounded measurable functions on $(\Omega, \mathscr{F})$ is a Banach space if endowed with the *supremum norm*,

$$\|F\| := \sup_{\omega \in \Omega} |F(\omega)|, \quad F \in \mathscr{X}.$$

Let $\mathscr{X}_0$ denote the linear subspace of all finitely valued step functions that can be represented in the form

$$F = \sum_{i=1}^{n} \alpha_i \mathbb{1}_{A_i},$$

for some $n \in \mathbb{N}$, $\alpha_i \in \mathbb{R}$, and sets $A_1, \dots, A_n \in \mathscr{F}$. For this F, we define

$$\int F\,d\mu := \sum_{i=1}^{n} \alpha_i\,\mu(A_i),$$

and one can check that this definition is independent of the particular representation of F. Moreover,

$$\left|\int F\,d\mu\right| \le \|F\| \cdot \|\mu\|_{\mathrm{var}}. \tag{G.3}$$

Since $\mathscr{X}_0$ is dense in $\mathscr{X}$ with respect to $\|\cdot\|$, this inequality allows us to define the integral on the full space $\mathscr{X}$ as the extension of the continuous linear functional $\mathscr{X}_0 \ni F \mapsto \int F\,d\mu$. Clearly, $\mathscr{M}_{1,f}$ is contained in *ba*, and we will denote the integral of a function $F \in \mathscr{X}$ with respect to $Q \in \mathscr{M}_{1,f}$ by

$$E_Q[\,F\,] := \int F\,dQ. \tag{G.4}$$

Theorem G.21. *The integral*

$$\ell(F) := \int F\,d\mu, \quad F \in \mathscr{X},$$

defines a one-to-one correspondence between continuous linear functionals ℓ on $\mathscr{X}$ and finitely additive set functions $\mu \in ba$.

Proof. By definition of the integral and by (G.3), it is clear that any $\mu \in ba$ defines a continuous linear functional on $\mathscr{X}$. Conversely, if a continuous linear functional ℓ is given, then we can define a finitely additive set function μ on $(\Omega, \mathscr{F})$ by

$$\mu(A) := \ell(\mathbb{1}_A), \quad A \in \mathscr{F}.$$

If $L \ge 0$ is such that $\ell(F) \le L$ for all $F \in \mathscr{X}$ with $\|F\| \le 1$, then $\|\mu\|_{\mathrm{var}} \le L$, and so $\mu \in ba$. One then checks that the integral with respect to μ coincides with ℓ on $\mathscr{X}_0$. Since $\mathscr{X}_0$ is dense in $\mathscr{X}$, we see that $\int F d\mu$ and $\ell(F)$ coincide for all $F \in \mathscr{X}$. □

Remark G.22. Theorem G.21 yields in particular a one-to-one correspondence between set functions $Q \in \mathscr{M}_{1,f}$ and continuous linear functionals ℓ on $\mathscr{X}$ such that $\ell(1) = 1$ and $\ell(X) \ge 0$ for $X \ge 0$. ◇

Example G.23. Clearly, the set $\mathscr{M}_{1,f}$ coincides with the set $\mathscr{M}_1 := \mathscr{M}_1(\Omega, \mathscr{F})$ of all σ-additive probability measures if $(\Omega, \mathscr{F})$ can be reduced to a finite set in the sense that $\mathscr{F}$ is generated by a finite partition of Ω. Otherwise, $\mathscr{M}_{1,f}$ is strictly larger than $\mathscr{M}_1$. Suppose in fact that there are infinitely many disjoint sets $A_1, A_2, \ldots \in \mathscr{F}$, take $\omega_n \in A_n$, and define

$$\ell_n(X) := \frac{1}{n}\sum_{i=1}^{n} X(\omega_i), \quad n = 1, 2, \ldots.$$

The continuous linear functionals ℓ_n on $\mathscr{X}$ belong to the unit ball B_1 in the dual Banach space $\mathscr{X}'$, which according to Theorem G.21 can be identified with *ba*. By Theorem H.10, there exists a cluster point ℓ of (ℓ_n). That is, for every open neighborhood U of ℓ, there exists $n \in \mathbb{N}$ such that $\ell_n \in U$. For each $A \in \mathscr{F}$, the map $\mu \mapsto \mu(A)$ defines a continuous linear functional on *ba*, and so

$$U_k := \left\{\mu \in ba \,\middle|\, |\mu(A) - \ell(\mathbb{1}_A)| < \frac{1}{k}\right\}$$

is an open set for each $k \in \mathbb{N}$. Hence, there exists a subsequence (ℓ_{n_k}) such that $\ell_{n_k} \in U_k$ for each k. It follows that $\ell_{n_k}(\mathbb{1}_A) \to \ell(\mathbb{1}_A)$. This implies that $\ell(\mathbb{1}_A) \geq 0$ for $A \in \mathscr{F}$ and $\ell(1) = 1$. Hence, $\ell(\mathbb{1}_A) = Q[A]$ for some $Q \in \mathscr{M}_{1,f}$. But Q is not σ-additive, since $Q[A_n] = \ell(\mathbb{1}_{A_n}) = 0$ and $Q[\bigcup_n A_n] = 1$. ◇

H Some functional analysis

H.1 Topological vector spaces

Numerous arguments in this book involve *infinite-dimensional* vector spaces. Typical examples arising in connection with a probability space $(\Omega, \mathscr{F}, P)$ are the spaces $L^p := L^p(\Omega, \mathscr{F}, P)$ for $0 \leq p \leq \infty$, which we will introduce below. To this end, we first take $p \in (0, \infty]$ and denote by $\mathscr{L}^p(\Omega, \mathscr{F}, P)$ the set of all $\mathscr{F}$-measurable functions Z on $(\Omega, \mathscr{F}, P)$ such that $\|Z\|_p < \infty$, where

$$\|Z\|_p := \begin{cases} E[\,|Z|^p\,]^{1/p}, & \text{if } 0 < p < \infty, \\ \inf\{c \geq 0 \mid P[\,|Z| > c\,] = 0\,\}, & \text{if } p = \infty. \end{cases} \tag{H.1}$$

Let us also introduce the space $\mathscr{L}^0(\Omega, \mathscr{F}, P)$, defined as the set of all P-a. s. finite random variables. If no ambiguity with respect to σ-algebra and measure can arise, we may sometimes write $\mathscr{L}^p(P)$ or just $\mathscr{L}^p$ instead of $\mathscr{L}^p(\Omega, \mathscr{F}, P)$. For $p \in [0, \infty]$, the space $L^p(\Omega, \mathscr{F}, P)$, or just L^p, is obtained from $\mathscr{L}^p$ by identifying random variables which coincide up to a P-null set. Thus, L^p consists of all equivalence classes with respect to the equivalence relation

$$Z \sim \widetilde{Z} \quad :\Longleftrightarrow \quad Z = \widetilde{Z}\ P\text{-a. s.} \tag{H.2}$$

If $p \in [1, \infty]$, then the vector space L^p is a *Banach space* with respect to the norm $\|\cdot\|_p$ defined in (H.1), that is, every Cauchy sequence with respect to $\|\cdot\|_p$ converges to some element in L^p. In principle, one should distinguish between a random variable $Z \in \mathscr{L}^p$ and its associated equivalence class $[Z] \in L^p$, of which Z is a representative element. In order to keep things simple, we will follow the usual convention of identifying Z with its equivalence class, that is, we will just write $Z \in L^p$.

On the space L^0, we use the topology of convergence in P-measure. This topology is generated by the metric

$$d(X, Y) := E[\,|X - Y| \wedge 1\,], \quad X, Y \in L^0. \tag{H.3}$$

Note, however, that d is not a norm.

Definition H.1. A linear space E that carries a topology is called a *topological vector space* if every singleton $\{x\}$ for $x \in E$ is a closed set, and if the vector space operations are continuous in the following sense:

$$(x, y) \longmapsto x + y$$

is a continuous mapping from $E \times E$ into E, and

https://doi.org/10.1515/9783111045283-022

$$(\alpha, x) \longmapsto \alpha x$$

is a continuous mapping from $\mathbb{R} \times E$ into E. △

Clearly, every Banach space is a topological vector space. The following result is a generalization of the separation argument in Proposition A.5 to an infinite-dimensional setting.

Theorem H.2. *In a topological vector space E, any two disjoint convex sets $\mathscr{B}$ and $\mathscr{C}$, one of which has an interior point, can be separated by a nonzero continuous linear functional ℓ on E, that is,*

$$\ell(x) \leq \ell(y) \quad \textit{for all } x \in \mathscr{C} \textit{ and all } y \in \mathscr{B}. \tag{H.4}$$

Proof. See [117], Theorem V.2.8. □

If one wishes to *strictly* separate two convex sets by a linear functional in the sense that one has a strict inequality in (H.4), then one needs additional conditions both on the convex sets and on the underlying space E.

Definition H.3. A topological vector space E is called a *locally convex space* if its topology has a base consisting of convex sets. △

If E is a Banach space with norm $\|\cdot\|$, then the open balls

$$\{y \in E \mid \|y - x\| < r\}, \quad x \in E, \ r > 0,$$

form by definition a base for the topology of E. Since such balls are convex sets, any Banach space is locally convex. The space $L^0(\Omega, \mathscr{F}, P)$ with the topology of convergence in P-measure, however, is *not* locally convex if $(\Omega, \mathscr{F}, P)$ has no atoms; see, for example, Theorem 12.41 of [3].

The following theorem is one variant of the classical Hahn–Banach theorem on the existence of “separating hyperplanes.”

Theorem H.4 (Hahn–Banach). *Suppose that $\mathscr{B}$ and $\mathscr{C}$ are two nonempty, disjoint, and convex subsets of a locally convex space E. Then, if $\mathscr{B}$ is compact and $\mathscr{C}$ is closed, there exists a continuous linear functional ℓ on E such that*

$$\sup_{x \in \mathscr{C}} \ell(x) < \inf_{y \in \mathscr{B}} \ell(y).$$

Proof. See, for instance, [288], page 65, or [117], Theorem V.2.10. □

One corollary of the preceding result is that on a locally convex space E the collection

$$E' := \{\ell : E \to \mathbb{R} \mid \ell \text{ is continuous and linear}\}$$

separates the points of E. That is, for any two distinct points $x, y \in E$ there exists some $\ell \in E'$ such that $\ell(x) \neq \ell(y)$. The space E' is called the *dual* or the *dual space* of E. For instance, if $p \in [1, \infty)$ it is well known that the dual of $L^p := L^p(\Omega, \mathscr{F}, P)$ is given by L^q, where $\frac{1}{p} + \frac{1}{q} = 1$. More precisely, for every $\ell \in (L^p)'$ there exists $Y \in L^q$ such that

$$\ell(X) = E[XY] \quad \text{for all } X \in L^p; \tag{H.5}$$

see, for example, Theorems IV.8.1 and IV.8.5 in [117]. The following definition describes a natural way in which locally convex topologies often arise.

Definition H.5. Let E be linear space, and suppose that F is a linear class of linear functionals on E that separates the points of E. The *F-topology* on E, denoted by $\sigma(E, F)$, is the topology on E that is obtained by taking as a base all sets of the form

$$\{y \in E \mid |\ell_i(y) - \ell_i(x)| < r,\ i = 1, \ldots, n\},$$

where $n \in \mathbb{N}$, $x \in E$, $\ell_i \in F$, and $r > 0$. If E already carries a locally convex topology, then the E'-topology $\sigma(E, E')$ is called the *weak topology* on E. △

If E is infinite-dimensional, then E is typically *not* metrizable in the F-topology. In this case, it may not suffice to consider converging sequences when making topological assertions; see, however, Theorem H.13 below. The following proposition summarizes a few elementary properties of the F-topology.

Proposition H.6. *Consider the situation of Definition* H.5. *Then:*
(a) *E is a locally convex space for the F-topology.*
(b) *The F-topology is the coarsest topology on E for which every $\ell \in F$ is continuous.*
(c) *The dual of E for the F-topology is equal to F.*

Proof. See, for example, Section V.3 of [117]. □

Theorem H.7. *Suppose that E is a locally convex space and that $\mathscr{C}$ is a convex subset of E. Then $\mathscr{C}$ is weakly closed if and only if $\mathscr{C}$ is closed in the original topology of E.*

Proof. If the convex set $\mathscr{C}$ is closed in the original topology then, by Theorem H.4, it is equal to the intersection of the half-spaces $H = \{\ell \leq c\}$ such that $H \supseteq \mathscr{C}$, and thus closed in the weak topology $\sigma(E, E')$. Conversely, the original topology is finer than the weak topology, and so $\mathscr{C}$ is closed in the original topology if it is weakly closed. □

For a given locally convex space E, we can turn things around and consider E as a set of linear functionals on the dual space E' by letting $x(\ell) := \ell(x)$ for $\ell \in E'$ and $x \in E$. The E-topology $\sigma(E', E)$ obtained in this way is called the *weak* topology* on E'. According to part (c) of Proposition H.6, E is then the topological dual of $(E', \sigma(E', E))$. For example, according to (H.5), the Banach space $L^\infty := L^\infty(\Omega, \mathscr{F}, P)$ is the dual of L^1, but the converse is generally not true; see our discussion on finitely additive set functions in Section G and, for a precise statement, Theorem IV.8.16 in [117]. However, L^1 becomes the dual of L^∞ if we endow L^∞ with the weak* topology $\sigma(L^\infty, L^1)$.

The mutual duality between E and E' allows us to state a general version of part (b) of Proposition A.9.

Definition H.8. Let E be a locally convex space with dual E'. The *Fenchel–Legendre transform* of a function $f : E \to \mathbb{R} \cup \{+\infty\}$ is the function f^* on E' defined by

$$f^*(\ell) := \sup_{x \in E}(\ell(x) - f(x)). \qquad \triangle$$

If $f \not\equiv +\infty$, then f^* is a proper convex and lower semicontinuous function as the supremum of affine functions. If f is itself a proper convex function, then f^* is also called the *conjugate function* of f.

Theorem H.9. *Let f be a proper convex function on a locally convex space E. If f is lower semicontinuous with respect to $\sigma(E, E')$, then $f = f^{**}$.*

It is straightforward to adapt the proof we gave in the one-dimensional case of Proposition A.9 to the infinite-dimensional situation of Theorem H.9; all one has to do is to replace the separating hyperplane lemma by the Hahn–Banach separation theorem in the form of Theorem H.4.

One of the reasons for considering the weak topology on a Banach space or, more generally, on a locally convex space is that typically more sets are compact for the weak topology than for the original topology. The following result shows that the unit ball in the dual of a Banach space is weak* compact. Here, we use the fact that a Banach space $(E, \|\cdot\|_E)$ defines the following norm on its dual E':

$$\|\ell\|_{E'} := \sup_{\|x\|_E \le 1} \ell(x), \qquad \ell \in E'.$$

Theorem H.10 (Banach–Alaoglu). *Let E be a Banach space with dual E'. Then $\{\ell \in E' \mid \|\ell\|_{E'} \le r\}$ is weak* compact for every $r \ge 0$.*

Proof. See, for example, Theorem IV.21 in [271]. □

Theorem H.11 (Krein–Šmulian). *Let E be a Banach space and suppose that $\mathscr{C}$ is a convex subset of the dual space E'. Then $\mathscr{C}$ is weak* closed if and only if*

$$\mathscr{C} \cap \{\ell \in E' \mid \|\ell\|_{E'} \le r\}$$

is weak closed for each $r > 0$.*

Proof. See Theorem V.5.7 in [117]. □

The preceding theorem implies the following characterization of weak* closed sets in $L^\infty = L^\infty(\Omega, \mathscr{F}, P)$ for a given probability space $(\Omega, \mathscr{F}, P)$.

Lemma H.12. *A convex subset $\mathscr{C}$ of L^∞ is weak* closed if for every $r > 0$,*

$$\mathscr{C}_r := \mathscr{C} \cap \{X \in L^\infty \mid \|X\|_\infty \le r\}$$

is closed in L^1.

Proof. Since $\mathscr{C}_r$ is convex and closed in L^1, it is weakly closed in L^1 by Theorem H.7. Since the natural injection

$$(L^\infty, \sigma(L^\infty, L^1)) \longrightarrow (L^1, \sigma(L^1, L^\infty))$$

is continuous, $\mathscr{C}_r$ is $\sigma(L^\infty, L^1)$-closed in L^∞. Thus, $\mathscr{C}$ is weak* closed due to the Krein–Šmulian theorem, applied with $E = L^1$ and $E' = L^\infty$. □

We now state a fundamental theorem on *weakly compact sets.*

Theorem H.13 (Eberlein–Šmulian). *For any subset A of a Banach space E, the following conditions are equivalent:*

(a) *A is* weakly sequentially compact, *that is, any sequence in A has a subsequence that converges weakly in E.*
(b) *A is* weakly relatively compact, *that is, the weak closure of A is weakly compact.*

Proof. See [117], Theorem V.6.1. □

H.2 Uniform integrability

In this section of the Appendix, we study the notion of uniform integrability, which is very useful for understanding both the strong and weak topologies of the Banach space $L^1 = L^1(\Omega, \mathscr{F}, P)$.

Definition H.14. A set $\mathscr{X}$ of random variables on $(\Omega, \mathscr{F}, P)$ is called *uniformly integrable* if

$$\lim_{c\uparrow\infty} \sup_{X\in\mathscr{X}} E[|X|\mathbb{1}_{\{|X|>c\}}] = 0. \qquad \triangle$$

Remark H.15. Suppose that $\mathscr{X}$ is a uniformly integrable set of random variables. Then there exists $c > 0$ such that $E[|X|\mathbb{1}_{\{|X|>c\}}] \le 1$ for all $X \in \mathscr{X}$. It follows that

$$\|X\|_1 = E[|X|] = E[|X|\mathbb{1}_{\{|X|\le c\}}] + E[|X|\mathbb{1}_{\{|X|>c\}}] \le c + 1$$

for all $X \in \mathscr{X}$. That is, $\mathscr{X}$ is a bounded subset of L^1. ◇

Exercise H.2.1. Show that a set $\mathscr{X}$ of random variables is uniformly integrable if there exists a nonnegative random $Y \in L^1$ such that $|X| \le Y$ P-a. s. for all $X \in \mathscr{X}$. ◇

Exercise H.2.2. Let $\mathscr{X}$ be a uniformly integrable subset of L^1 and $\mathscr{Y} \subset L^1$ be finite. Show that $\mathscr{X} \cup \mathscr{Y}$ is uniformly integrable. In particular, every finite subset of L^1 is uniformly integrable. ◇

Exercise H.2.3. Sow that a set $\mathscr{X}$ of random variables is uniformly integrable if and only if the following two conditions hold:

(a) $\mathscr{X}$ is a bounded subset of L^1.

(b) For every $\varepsilon > 0$, there exists $\delta > 0$ such that for every $X \in \mathscr{X}$, we have $E[|X|\mathbb{1}_A] < \varepsilon$ whenever $A \in \mathscr{F}$ satisfies $P[A] < \delta$.

Condition (b) can be regarded as a requirement of *uniform absolute continuity* of the measures $\mu_X(A) := E[|X|\mathbb{1}_A]$. ◇

The following criterion for uniform integrability is often very useful.

Proposition H.16 (de la Vallée Poussin). *A set $\mathscr{X}$ of random variables is uniformly integrable if there exists a function $h : [0,\infty) \to \mathbb{R}$ that is bounded from below and has superlinear growth, i. e., $h(x)/x \to \infty$ as $x \uparrow \infty$, such that*

$$\sup_{X\in\mathscr{X}} E[h(|X|)] < \infty.$$

Proof. Let $M := \sup_{X\in\mathscr{X}} E[h(|X|)]$. For $\varepsilon > 0$, we let $a := M/\varepsilon$ and take $c > 0$ such that $h(x)/x > a$ for all $x \geq c$. Then

$$E[|X|\mathbb{1}_{\{|X|>c\}}] \leq \frac{1}{a}E[h(|X|)\mathbb{1}_{\{|X|>c\}}] \leq \frac{M}{a} = \varepsilon$$

for all $X \in \mathscr{X}$. Thus, $\mathscr{X}$ is uniformly integrable. □

Exercise H.2.4. Show that equivalence holds in Proposition H.16. That is, if $\mathscr{X}$ is uniformly integrable, then there exists a function $h : [0,\infty) \to \mathbb{R}$ that is bounded from below and has superlinear growth such that

$$\sup_{X\in\mathscr{X}} E[h(|X|)] < \infty.$$ ◇

Suppose that the random variables $X_n \in L^1$ converge in L^1 to the random variable X. Then Markov's inequality yields that for each $\varepsilon > 0$,

$$P[|X_n - X| > \varepsilon] \leq \frac{1}{\varepsilon}E[|X_n - X|] \longrightarrow 0 \quad \text{as } n \uparrow \infty.$$

Thus, $X_n \to X$ in probability. The following *Vitali convergence theorem* shows that convergence in probability implies L^1-convergence, if and only if the sequence $(X_n)_{n\in\mathbb{N}}$ is uniformly integrable. The property of uniform integrability is hence fundamental for strong convergence in L^1.

Theorem H.17 (Vitali). *Suppose that the random variables $X_n \in L^1$ converge in probability to the random variable $X \in L^1$. Then the following conditions are equivalent:*
(a) *The sequence $(X_n)_{n\in\mathbb{N}}$ is uniformly integrable.*
(b) *$X_n \to X$ in L^1.*

Proof. To prove that (a) implies (b), let $f_c(x) := (-c) \vee x \wedge c$ for $c > 0$. The Lipschitz continuity of f_c yields that, for each $\varepsilon > 0$,

$$P[|f_c(X_n) - f_c(X)| \geq \varepsilon] \leq P[|X_n - X| \geq \varepsilon] \longrightarrow 0 \quad \text{as } n \uparrow \infty.$$

That is, $f_c(X_n) \to f_k(X)$ in probability as $n \uparrow \infty$. By dominated convergence, we therefore get $f_c(X_n) \to f_c(X)$ in L^1.

Now let $\varepsilon > 0$ be given. Our goal is to show that there exists $n_0 \in \mathbb{N}$ such that

$$E[|X_n - X|] < \varepsilon \quad \text{for all } n \geq n_0. \tag{H.6}$$

To this end, we first pick $c > 0$ such that

$$E[|Y|\mathbb{1}_{\{|Y|>c\}}] < \frac{\varepsilon}{3}$$

for all $Y \in \{X_n \mid n \in \mathbb{N}\} \cup \{X\}$. Such a c exists due to the uniform integrability of the sequence $(X_n)_{n\in\mathbb{N}}$ and Exercise H.2.2. Since $|Y - f_c(Y)| \leq |Y|\mathbb{1}_{\{|Y|>c\}}$, we then get

$$E[|Y - f_c(Y)|] \leq E[|Y|\mathbb{1}_{\{|Y|>c\}}] < \frac{\varepsilon}{3}$$

for all $Y \in \{X_n \mid n \in \mathbb{N}\} \cup \{X\}$. Next, we take n_0 such that

$$E[|f_c(X_n) - f_c(X)|] < \frac{\varepsilon}{3} \quad \text{for all } n \geq n_0,$$

which is possible due to the L^1 convergence $f_c(X_n) \to f_c(X)$. Consequently,

$$E[|X_n - X|] \leq E[|X_n - f_c(X_n)|] + E[|f_c(X_n) - f_c(X)|] + E[|X - f_c(X)|] < \varepsilon,$$

which establishes (H.6).

For the proof that (b) implies (a), let $\varepsilon > 0$ be given. We first take $n_0 \in \mathbb{N}$ such that $E[|X_n - X|] < \varepsilon/2$ for all $n \geq n_0$. Then we choose $\delta > 0$ such that

$$E[|X|\mathbb{1}_A] < \frac{\varepsilon}{2} \quad \text{whenever } A \in \mathscr{F} \text{ satisfies } P[A] < \delta.$$

This is possible, for example, by Exercise H.2.3 applied to the uniformly integrable set $\mathscr{X} := \{X\}$. Now we choose $c > 0$ such that

$$\sup_{n\in\mathbb{N}} E[|X_n|] < \delta c.$$

Markov's inequality then yields that for each n,

$$P[|X_n| > c] \le \frac{1}{c}E[|X_n|] < \delta.$$

It follows that for $n \ge n_0$,

$$E[|X_n|\mathbb{1}_{\{|X_n|>c\}}] \le E[|X|\mathbb{1}_{\{|X_n|>c\}}] + E[|X_n - X|] < \varepsilon.$$

Hence, the family $(X_n)_{n\ge n_0}$ is uniformly integrable. The uniform integrability of the entire sequence now follows from Exercise H.2.2. □

The following theorem shows in particular that the convergence $E[|X_n|] \to E[|X|]$ is necessary and sufficient for L^1-convergence, if the random variables X_n converge P-a. s. to $X \in L^1$. An L^p-version of this theorem will be explored in Exercise H.2.5.

Theorem H.18 (Riesz). *Suppose that a sequence of random variables $X_n \in L^1$ converges P-a. s. to the random variable $X \in L^1$. Then the following conditions are equivalent:*
(a) $E[|X_n|] \to E[|X|]$.
(b) $X_n \to X$ *in* L^1.
(c) *The sequence* $(X_n)_{n\in\mathbb{N}}$ *is uniformly integrable.*

Proof. To show that (a) implies (b), let

$$Y_n := |X_n| + |X| - |X_n - X|.$$

Then $Y_n \ge 0$ and $Y_n \to 2|X|$ P-almost surely. Hence, Fatou's lemma and assumption (a) imply that

$$\begin{aligned} 2E[|X|] &\le \liminf_{n\uparrow\infty} E[Y_n] = \liminf_{n\uparrow\infty}(E[|X_n|] + E[|X|] - E[|X_n - X|] \\ &= 2E[|X|] - \limsup_{n\uparrow\infty} E[|X_n - X|]. \end{aligned}$$

That is,

$$\limsup_{n\uparrow\infty} E[|X_n - X|] \le 0,$$

which yields (b).

The converse implication, (b)⇒(a), follows by taking the expectation of all terms in the following inequality:

$$-|X_n - X| \le |X_n| - |X| \le |X_n - X|$$

and then passing to the limit $n \uparrow \infty$.

The equivalence of (c) and (b) is an immediate consequence of Theorem H.17. □

Exercise H.2.5. Extend the equivalence of (a) and (b) in Theorem H.18 to the case of L^p-convergence for $1 < p < \infty$. That is, show that the following conditions are equivalent if the random variables $X_n \in L^p$ converge P-a. s. to the random variable $X \in L^p$:

(a) $E[|X_n|^p] \to E[|X|^p]$.

(b) $X_n \to X$ in L^p. ◇

The property of uniform integrability is not only fundamental for strong convergence in L^1. The following *Dunford–Pettis theorem* shows that uniform integrability also characterizes those subsets of the Banach space $L^1 = L^1(\Omega, \mathcal{F}, P)$ that are weakly relatively compact. Recall from (H.5) that L^∞ is the dual of L^1, so that the weak topology on L^1 is given by $\sigma(L^1, L^\infty)$.

Theorem H.19 (Dunford–Pettis). *A subset of $L^1(\Omega, \mathcal{F}, P)$ is weakly relatively compact if and only if it is uniformly integrable.*

Proof. See, for example, Theorem IV.8.9 or Corollary IV.8.11 in [117] and recall from Remark H.15 that every uniformly integrable subset of random variables on a probability space $(\Omega, \mathcal{F}, P)$ is bounded in L^1. □

Bibliographical notes

In these notes, we do not make any attempt to give a systematic account of all the sources that have been relevant for the development of the field. We simply mention a number of references that had a direct influence on our decisions how to present the topics discussed in this book. More comprehensive lists of references can be found, for example, by Delbaen and Schachermayer [99], Jeanblanc, Yor, and Chesney [190], Karatzas and Shreve [205], and McNeil, Frey, and Embrechts [245].

Chapter 1: The proof of Theorem 1.7 is based on Dalang, Morton, and Willinger [78]. Remark 1.18 and Example 1.19 are taken from Schachermayer [285]. The Taylor formulas in Example 1.27 and Example 1.3.3 go back at least to Lacroix [231]. Their financial interpretation was emphasized by Bick [32]. Exercise 1.5.3 was communicated to us by Martin Barbie. Section 1.6 is mainly based on [285], with the exception of Lemma 1.66, which is taken from Kabanov and Stricker [196]. Our proof of Lemma 1.70 combines ideas from [196] with the original argument in [285], as suggested to us by Irina Penner. For a historical overview of the development of arbitrage pricing and for an outlook to continuous-time developments, we refer to [99]. For some mathematical connections between superhedging of call options as discussed in Section 1.3 and bounds on stop-loss premiums in insurance, see Chapter 5 of Goovaerts et al. [169].

Chapter 2: The results on the structure of preferences developed in this chapter are, to a large extent, standard topics in mathematical economics. We refer to textbooks on expected utility theory such as Fishburn [135, 136], Kreps [224], or Savage [284], and to the survey articles in [13, 18]. The ideas and results of Section 2.1 go back to classical references such as Debreu [88], Eilenberg [119], Milgram [249], and Rader [270]. The theory of affine numerical representations in Section 2.2 was initiated by von Neumann and Morgenstern [256] and further developed by Herstein and Milnor [180]. The characterization of certain equivalents with the translation in Proposition 2.46 goes back to de Finetti [134]; see also Kolmogorv [212] and Nagumo [254]. The drastic consequences of the assumption that a favorable bet is rejected at any level of wealth, as explained in Proposition 2.49, were stressed by Rabin [269]. The discussion of the partial orders $\succcurlyeq_{icv}$, $\succcurlyeq_{i}$, and $\succcurlyeq_{cx}$ in Sections 2.4 and 2.6 has a long history. A first version of Theorem 2.57 is already contained in Hardy, Littlewood, and Polya [172]. A complete treatment was given by Strassen [309]; this paper is also the source for Section 2.6. The economic interpretation of Theorem 2.57 was developed by Rothschild and Stiglitz [278, 279]. The analysis of robust preferences in Section 2.5 is mainly based on the ideas of Savage [284], Anscombe and Aumann [6], and Gilboa and Schmeidler [166]. See Gilboa [161] for an alternative axiomatic approach to the characterization in part (b) of Theorem 2.78, and Karni and Schmeidler [207] and Gilboa [162] for surveys of related developments. In the context of robust statistics, a special case of Proposition 2.84 appears in Huber [183]. The relaxed axiom of weak certainty independence and the corresponding extension of Theorem 2.87 is due to Maccheroni, Marinacci, and Rustichini [241]; see also [148].

https://doi.org/10.1515/9783111045283-023

Chapter 3: Given a preference relation of von Neumann–Morgenstern type, the analysis of optimal portfolios in Section 3.1 or, more generally, of optimal asset profiles in Section 3.3 is a standard exercise, both in microeconomic theory and in convex optimization. Section 3.1 shows that the existence of a solution is equivalent to the absence to arbitrage; here we follow Rogers [276]. In the special case of exponential utility, the construction of the optimal portfolio is equivalent to the minimization of relative entropy as discussed in Section 3.2. This may be viewed as the financial interpretation of general results on entropy minimization in Csiszar [70, 71]. The methods for characterizing optimal asset profiles in Section 3.3 in terms of "first-order conditions" are well known; see, for example, [120] or [205], where they are developed in greater generality. The optimization problem in Theorem 3.38, which is formulated in terms of the partial order $\succcurlyeq_{\text{icv}}$ and involves the Hardy–Littlewood inequalities of Theorem D.13, is less standard. Our discussion is based on Dybvig [118], Jouini and Kallal [192], and on a proof of Dana and Meilijson, and we are obliged to Rose-Anne Dana for introducing us to this topic; see Dana [80, 82] for further developments. Exercise 3.4.3 is taken from Bernard, Boyle, and Vanduffel [29]. The characterization of a least-favorable measure in Proposition 3.50 is due to Huber and Strassen [185]. The application to robust portfolio choice is taken from Schied [291]. The concept of weak information in Example 3.46 was suggested by Baudoin [20]. As to the existence of Arrow–Debreu equilibria discussed in Section 3.7, we refer to the classical version in Debreu [87] and to the survey articles in [13]. In our financial context, equilibrium allocations do no longer involve commodity bundles in Euclidean space as in [87] but asset profiles described by random variables on a probability space. This formulation of the equilibrium problem goes back to Borch [38], where it was motivated by the problem of risk exchange in a reinsurance market. The systematic analysis of the equilibrium problem in an infinite-dimensional setting was developed by Bewley, Mas-Colell, and others; see [31] and, for example, [4, 243, 244]. In our introductory approach, the existence proof is reduced to an application of Brouwer's fixed-point theorem. Here, we benefited from discussions with Peter Bank; see also Dana [79] and Carlier and Dana [49]. Examples 3.58 and Example 3.59 are based on Bühlmann [43]; see also [44] and Deprez and Gerber [107]. As mentioned in Remark 3.67, the equilibrium discussion of interest rates requires an intertemporal setting; for a systematic discussion see, for example, Duffie [114] and, in a different conceptual framework, Bank and Riedel [17]. For a utility-maximizing agent, Theorem 3.42 is due to Arrow [12]. Its extension to the optimality with respect to the increasing concave order was obtained by Gollier and Schlesinger [168]; a variant of their original proof is sketched in Exercise 3.5.1. Corollary 3.44 was obtained by Kahn [197].

Chapter 4: The axiomatic approach to coherent measures of risk and their acceptance sets was initiated by Artzner, Delbaen, Eber, and Heath [14], and most results of Section 4.1 are based on this seminal paper. The extension to convex risk measures was given independently by Heath [176], Heath and Ku [178], Föllmer and Schied [146], and

Frittelli and Rosazza Gianin [155]. In the actuarial literature, risk measures ρ have appeared in the form of premium principles $H = -\rho$; see Deprez and Gerber [107] and, for example, Denneberg [105] and Wang and Dhaene [318]. In the coherent case, they have also appeared in the context of robust statistics; see Huber [183]. The robust representation theorems in Section 4.2 are taken from [147]. Exercise 4.1.5 was suggested by E. Delage. The discussion in Section 4.3 of convex risk measures on a space of continuous functions corrects an error in [147] and in the first edition of this book; see also Krätschmer [215] for a further analysis. Exercise 4.2.2 is taken from El Karoui and Ravanelli [122]. The representation theory on L^∞ as presented in Section 4.4 was developed by Delbaen [89, 90]; for the connection to the general duality theory as explained in Remarks 4.18 and 4.44, see [89, 90, 155, 156]. Among the results on Value at Risk and its various modifications in Section 4.5, Proposition 4.47 and Theorem 4.73 are taken from [14] and [89]. Average Value at Risk is discussed, for example, by Acerbi and Tasche [2], Delbaen [89], and Rockafellar and Uryasev [274]. Remark 4.49 was pointed out to us by Ruszczynski; see [259]. The notation V@R is taken from Pflug and Ruszczynski [264]. Theorem 4.58 is due to Jouini, Schachermayer, and Touzi [193], where it was established under an additional condition. The proof of Theorem 4.58 in this book follows the argument given in Delbaen [93], where Proposition 4.59 was introduced to eliminate the need for that condition. While our proof of Theorem 4.58 is essentially the same as in [93], we provide a new proof of Proposition 4.59 based on the classical theory of sequences that are uniformly distributed modulo 1. For an alternative proof of Theorem 4.58 that uses an extension of the Hardy-Littlewood bound in Proposition 4.66 to a finitely additive setting we refer to Liebrich and Munari [239]. The representations of law-invariant risk measures given in Section 4.7 were first obtained in the coherent case by Kusuoka [230]; see also Kunze [228] and Frittelli and Rosazza Gianin [156, 157] for the extension to the general convex case. Theorem 4.76 in Section 4.8 was first proved in [230]. The representations of the core of a concave distortion in Theorem 4.85 and Corollary 4.86 are due to Carlier and Dana [50]. See also [51] for further applications. The alternative representation in Exercise 4.8.8 is due to Cherny [59]. The definition of the risk measures MINVAR, MAXVAR, MINMAXVAR, and MAXMINVAR and the representations given in Exercises 4.1.9, 4.8.1, 4.8.5, 4.8.6, and 4.8.7 are due to Cherny and Madan [61]. The statement of Exercise 4.8.2 is due to Castagnoli, Maccheroni, and Marinacci [53]. Exercises 4.1.8 and 4.8.4 are inspired by Example A.1 in Pflug and Kovacevic [262]. The study of Choquet integrals with respect to general set functions as used in Section 4.9 was started by Choquet [63]. The connections with coherent risk measures were observed by Delbaen [89, 90]. The two implications in Theorem 4.94 are due to Dellacherie [100] and Schmeidler [295], respectively. The proof via Lemma 4.95, which we give here, is taken from Denneberg [106]. Theorem 4.99 is due to Kusuoka [230]. The equivalence between (b) and (d) in Theorem 4.100 was first proved in [63], item (c) was added in [295]. The proof given here is based on [106]. The first part of Section 4.10 is based on [146], the second on Carr, Geman, and Madan [52]. Exercise 4.10.3 is taken from Barrieu and El Karoui [19].

The representation of utility-based shortfall risk in Section 4.11 is taken from [146]. Theorem 4.126 is an extension of a classical result for Orlicz spaces; see Theorem 10.5 in Chapter II of Krasnoselskii and Rutickii [222]. Exercise 4.11.2 is taken from Bellini, Klar, and Müller [24], Exercise 4.11.1 from Frittelli and Scandolo [160]. The dual representation of the expectile risk measure in Example 4.128 was first obtained in Example 5 of Chapter 4 in [93]. The distorted Bayesian preferences in Exercise 4.11.4 were introduced in Klibanoff, Marinacci, and Mukerji [209]; see also Bäuerle and Mahayni [23]. Divergence risk measures were considered by Ben-Tal and Teboulle [27, 28] under the name of *optimized certainty equivalents*. The proof of Theorem 4.137 is based on Schied [293].

Chapter 5: Martingales in Finance have a long history; see, for example, Samuelson [282]. In the context of dynamic arbitrage theory, martingales and martingale measures are playing a central role, both in discrete and continuous time; for a historical overview, we refer again to Delbaen and Schachermayer [99]. The first four sections of this chapter are based on Harrison and Kreps [173], Kreps [223], Harrison and Pliska [174], Dalang, Morton, and Willinger [78], Stricker [311], Schachermayer [285], Jacka [187], Rogers [276], Ansel and Stricker [10], and Kabanov and Kramkov [195]. Exercises 5.2.10 and 5.2.11 are taken from Gerhold and Krühner [159] and were suggested to us by S. Gerhold. The binomial model of Section 5.5 was introduced by Cox, Ross, and Rubinstein in [69]. Geometric Brownian motion, which appears in Section 5.7 as the diffusion limit of binomial models, was proposed since the late 1950s by Samuelson and others as a model for price fluctuations in continuous time, following the rediscovery of the linear Brownian motion model of Bachelier [16]; see Samuelson [283] and Cootner [67]. The corresponding dynamic theory of arbitrage pricing in continuous time goes back to Black and Scholes [35] and Merton [247]. A Black–Scholes-type formula for option pricing appears in Sprenkle [308] in an ad hoc manner, without the arbitrage argument introduced by Black and Scholes. The approximation of Black–Scholes prices for various options by arbitrage-free prices in binomial models goes back to Cox, Ross, and Rubinstein [69]. A functional version of Theorem 5.54, based on Donsker's invariance principle, can be found in [114].

Chapter 6: The dynamic arbitrage theory for American options begins with Bensoussan [26] and Karatzas [200]. A survey is given in Myeni [253]. The theory of optimal stopping problems as presented in Sections 6.1 and 6.2 was initiated by Snell [307]; see [257] for a systematic introduction. Stability under pasting as discussed in Section 6.4 has appeared under several names in various contexts; see Delbaen [91] for a number of references and for an extension to continuous time. Our discussion of upper and lower Snell envelopes in Section 6.5 uses ideas from Karatzas and Kou [203] and standard techniques from dynamic programming. The application to the time consistency of dynamic coherent risk measures in Remark 6.54 recovers results by Artzner, Delbaen, Eber, Heath, and Ku [15], Delbaen [91], and Riedel [272].

Chapter 7: Optional decompositions, or uniform Doob decompositions as we call them, and the resulting construction of superhedging strategies were first obtained by El Karoui and Quenez [121] in a jump-diffusion model. In a general semimartingale setting, the theory was developed by Kramkov [219], Föllmer and Kabanov [141], and Delbaen and Schachermayer [99]. From a mathematical point of view, the existence of martingale measures with marginals determined by given option prices in Theorem 7.25 is a corollary of Strassen [309]; continuous-time analoga were proved by Doob [111] and Kellerer [208]. For the economic interpretation see, for example, Breeden and Litzenberger [40]. The results on superhedging of exotic derivatives by means of plain vanilla options stated in Theorems 7.27, 7.31, 7.33, and Corollary 7.34 are due to Hobson [181] and Brown, Hobson, and Rogers [41]; they are related to martingale inequalities of Dubins and Gilat [112].

Chapter 8: The analysis of quantile hedging was triggered by a talk of D. Heath in March 1995 at the Isaac Newton Institute on the results in Kulldorf [226], where an optimization problem for Brownian motion with drift is reduced to the Neyman–Pearson lemma. Section 8.1 is based on Föllmer and Leukert [143]; see also Karatzas [201], Cvitanic and Spivak [77], Cvitanic and Karatzas [76], and Browne [42]. The results in Section 8.2 on minimizing the shortfall risk are taken from Föllmer and Leukert [144]; see also Leukert [237], Cvitanic and Karatzas [75], Cvitanic [72], and Pham [265]. Theorem 8.26 is due to Sekine [300]. The proof given here is taken from Schied [292] and Theorem 8.27 from [290].

Chapter 9: In continuous-time models, dynamic arbitrage pricing with portfolio constraints was considered by Cvitanic and Karatzas [73, 74]. Proposition 9.6 and Exercise 9.1.1 are based on Jacod and Shiryaev [188]. In a discrete-time model with convex constraints, absence of arbitrage was characterized by Carrassus, Pham, and Touzi [46]. In a general semimartingale setting, Föllmer and Kramkov [142] proved a uniform Doob decomposition and superhedging duality theorems for a predictably convex set of admissible trading strategies and for American contingent claims; see also Karatzas and Kou [203].

Chapter 10: The idea of quadratic risk minimization for hedging strategies goes back to Föllmer and Sondermann [151], where the optimality criterion was formulated with respect to a martingale measure. Extensions to the general case and the construction of minimal martingale measures were developed by Föllmer and Schweizer [150] and Schweizer; see, for example, [296, 297]. Our exposition also uses arguments from Föllmer and Schweizer [149], Schäl [289], and Li and Xia [238]. Variance-optimal hedging was introduced by Duffie and Richardson [116] and further developed by Schweizer and others; the discrete-time theory as presented in Section 10.3 is based on Schweizer [298]. Melnikov and Nechaev [246] give an explicit formula for a variance-optimal strategy without condition (10.27); in fact, they show that their formula always defines a variance-optimal strategy if one does not insist on the square-integrability of the gains process at intermediate times. For a survey on related results, we refer to Schweizer [299].

Chapter 11: Representations of conditional risk measures as in Section 11.1 were discussed in Artzner, Delbaen, Eber, Heath, and Ku [15], Bion-Nadal [33, 34], Burgert [45], Cheridito, Delbaen, and Kupper [54–56], Delbaen [91], Detlefsen and Scandolo [108], Föllmer and Penner [145], Penner [260], and Riedel [272]. Exercise 11.1.2 is taken from [56]. The characterization of time consistency in Section 11.2 is due to [15] and [91] in the coherent case and to [33, 34, 56, 145, 260] in the general convex case; for a detailed survey, we refer to Acciao and Penner [1].

Chapter 12: The theory of universal portfolio strategies as presented in this chapter was developed by Cover [68].

Appendix. Our exposition of convex sets and functions in Appendix A is mainly based on Rockafellar [273]. Additional background on absolutely continuous measures as considered in Appendix B can be found in the books [21, 22] by Bauer. The representation of relative entropy in Appendix C is sometimes called the Donsker–Varadhan representation; see [110]. It can be seen as a reformulation, in terms of general measure theory, of a classical result of J. W. Gibbs; see Theorem III in Chapter XI of Gibbs [165], and also the proof of Theorem I, where the Gibbs inequality appears. For further connections to Statistical Mechanics see, for example, Simon [306]. In analysis, a quantile function as discussed in Appendix D is also known as an increasing rearrangement of a measurable function. We refer to Hardy, Littlewood, and Pólya [172] and to Chong and Rice [62] for further results and an historical overview. For Lemma D.11, see Pflug [261], Rockafellar and Uryasev [274], and [2]. Exercise D.2.2 is due to Rüschendorf [280]; see also Ferguson [130]. The statement of Lemma D.17 is due to Ryff [281]. The results in Section E go back to Neyman and Pearson [258]. Comprehensive treatments of the material from Appendix G and Appendix H can be found in Bourbaki [39] and Dunford and Schwartz [117].

Bibliography

[1] Acciao, B., Penner, I. Dynamic risk measures. In: Advanced Mathematical Methods for Finance, Springer, Berlin, 2011.

[2] Acerbi, C., Tasche, D. On the coherence of expected shortfall. J. Bank. Finance 26 (2002), 1487–1503.

[3] Aliprantis, C. D., Border, K. C. Infinite Dimensional Analysis. A Hitchhiker's Guide, 2nd edition. Springer-Verlag, Berlin, 1999.

[4] Aliprantis, C., Brown, D., Burkinshaw, O. Existence and Optimality of Competitive Equilibria. Springer-Verlag, Berlin, 1989.

[5] Allais, M. Le comportement de l'homme rationell devant de risque: critique des postulats et axiomes de l'école Américaine. Econometrica 21 (1953), 503–546.

[6] Anscombe, F. J., Aumann, R. J. A definition of subjective probability. Ann. Math. Stat. 34 (1963), 199–205.

[7] Ansel, J.-P., Stricker, C. Quelques remarques sur un théorème de Yan. In: Séminaire de Probabilités XXIV, 1988/89 (J. Azéma, P. André and M. Yor, eds.). Lecture Notes in Math. 1426, Springer-Verlag, Berlin, 1990, 266–274.

[8] Ansel, J.-P., Stricker, C. Lois de martingale, densités et décomposition de Föllmer-Schweizer. Ann. Inst. Henri Poincaré Probab. Stat. 28 (1992), 375–392.

[9] Ansel, J.-P., Stricker, C. Unicité et existence de la loi minimale. In: Séminaire de Probabilités XXVII (J. Azéma et al., eds.). Lecture Notes in Math. 1557, Springer-Verlag, Berlin, 1993, 22–29.

[10] Ansel, J.-P., Stricker, C. Couverture des actifs contingents et prix maximum. Ann. Inst. Henri Poincaré Probab. Stat. 30 (1994), 303–315.

[11] Ararat, C., Hamel, A., Rudloff, B. Set-valued shortfall and divergence risk measures. Int. J. Theor. Appl. Finance 20 (2017), 1750026.

[12] Arrow, K. J. Essays in the Theory of Risk Bearing. Markham, Chicago, 1971.

[13] Arrow, K., Hildebrandt, W., Intriligator, D., Sonnenschein, H. (eds.), Handbook of Mathematical Economics, Vol. I–IV. Elsevier Science Publishers, 1991.

[14] Artzner, P., Delbaen, F., Eber, J.-M., Heath, D. Coherent measures of risk. Math. Finance 9 (1999), 203–228.

[15] Artzner, P., Delbaen, F., Eber, J.-M., Heath, D., Ku, H. Coherent multiperiod risk adjusted values and Bellman's priciple. Ann. Oper. Res. 152 (2007), 5–22.

[16] Bachelier, L. Théorie de la spéculation. Ann. Sci. Éc. Norm. Supér. 17 (1900), 21–86.

[17] Bank, P., Riedel, F. Existence and Structure of Stochastic Equilibria with Intertemporal Substitution. Finance Stoch. 5 (2001), 487–509.

[18] Barbera, S., Hammond, P. J., Seidl, C. (eds.), Handbook of Utility Theory, Vol. 1: Principles. Kluwer Academic Publishers, 1998.

[19] Barrieu, P., El Karoui, N. Inf-convolution of risk measures and optimal risk transfer. Finance Stoch. 9 (2005), no. 2, 269–298.

[20] Baudoin, F. Conditioned stochastic differential equations: theory, examples and application to finance. Stoch. Process. Appl. 100 (2002), 109–145.

[21] Bauer, H. Probability Theory. De Gruyter Stud. Math. 23, Walter de Gruyter, Berlin–New York, 1996.

[22] Bauer, H. Measure and Integration Theory. De Gruyter Stud. Math. 26, Walter de Gruyter, Berlin–New York, 2001.

[23] Bäuerle, N., Mahayni, A. Optimal investment in ambiguous financial markets with learning. Eur. J. Oper. Res. 315 (2024), 393–410.

[24] Bellini, F., Klar, B., Müller, A., Rosazza Gianin, E. Generalized quantiles as risk measures. Insur. Math. Econ. 54 (2014), 41–48.

[25] Belomestny, D., Krätschmer, V. Central limit theorems for law-invariant coherent risk measures. J. Appl. Probab. 49 (2012), no. 1, 1–21.

[26] Bensoussan, A. On the theory of option pricing. Acta Appl. Math. 2 (1984), 139–158.

https://doi.org/10.1515/9783111045283-024

[27] Ben-Tal, A., Teboulle, M. Penalty functions and duality in stochastic programming via ϕ-divergence functionals. Math. Oper. Res. 12 (1987), 224–240.
[28] Ben-Tal, A., Teboulle, M. An old-new concept of convex risk measures: the optimized certainty equivalent. Math. Finance 17 (2007), no. 3, 449–476.
[29] Bernard, C., Boyle, P., Vanduffel, S. Explicit representation of cost-efficient strategies. Finance 35 (2014), 5–55.
[30] Bernoulli, D. Specimen theoriae novae de mensura sortis. Commentarii Academiae Scientiarum Imperialis Petropolitanae 5 (1738), 175–192. Translated by L. Sommer: Econometrica 22 (1954), 23–36.
[31] Bewley, T. Existence of equilibria in economies with infinitely many commodities. J. Econ. Theory 43 (1972), 514–540.
[32] Bick, A. Comments on the valuation of derivative assets. J. Financ. Econ. 10 (1982), 331–345.
[33] Bion-Nadal, J. Dynamic risk measures: time consistency and risk measures from BMO martingales. Finance Stoch. 12 (2008), 219–244.
[34] Bion-Nadal, J. Time consistent dynamic risk processes. Stoch. Process. Appl. 119 (2008), 633–654.
[35] Black, F., Scholes, M. The pricing of options and corporate liabilities. J. Polit. Econ. 81 (1973), 637–654.
[36] Bogachev, V. I. Measure Theory, Vol. I, II. Springer-Verlag, Berlin, 2007.
[37] Blackwell, D., Dubins, L. A converse to the dominated convergence theorem. Ill. J. Math. 7 (1963), 508–514.
[38] Borch, K. Equilibrium in a reinsurance market. Econometrica 30 (1962), 424–444.
[39] Bourbaki, N. Intégration. Hermann, Paris, 1969.
[40] Breeden, D., Litzenberger, R. Prices of state-contingent claims implicit in option prices. J. Bus. 51 (1978), 621–651.
[41] Brown, H., Hobson, D., Rogers, L. C. G. Robust hedging of barrier options. Math. Finance 11 (2001), 285–314.
[42] Browne, S. The risks and rewards of minimizing shortfall probability. J. Portf. Manag. 25 (1999), 76–85.
[43] Bühlmann, H. The general economic premium priciple. ASTIN Bull. 14 (1984), 13–21.
[44] Bühlmann, H., Jewell, S. Optimal risk exchanges. ASTIN Bull. 10 (1979), 243–262.
[45] Burgert, C. Darstellungssätze für statische und dynamische Risikomaße mit Anwendungen. Dissertation, Universität Freiburg, 2005.
[46] Carassus, L., Pham, H., Touzi, N. No-arbitrage in discrete-time under portfolio constraints. Math. Finance 11 (2001), 315–329.
[47] Carlier, G., Dana, R.-A. Law invariant concave utility functions and optimization problems with monotonicity and comonotonicity constraints. Stat. Decis. 24 (2006), no. 1, 127–152.
[48] Carlier, G., Dana, R.-A. Rearrangement inequalities in non-convex insurance models. J. Math. Econ. 41 (2005), no. 4–5, 483–503.
[49] Carlier, G., Dana, R. A. Two-persons efficient risk-sharing and equilibria for concave law-invariant utilities. Econom. Theory 36 (2008), no. 2, 189–223.
[50] Carlier, G., Dana, R.-A. Core of convex distortions of a probability. J. Econ. Theory 113 (2003), no. 2, 199–222.
[51] Carlier, G., Dana, R.-A. Insurance contracts with deductibles and upper limits. Preprint, Ceremade, Université Paris Dauphine, 2002.
[52] Carr, P., Geman, H., Madan, D. Pricing and hedging in incomplete markets. J. Financ. Econ. 62 (2001), 131–167.
[53] Castagnoli, E., Maccheroni, F., Marinacci, M. Choquet insurance pricing: a caveat. Math. Finance 14 (2004), 481–485.
[54] Cheridito, P., Delbaen, F., Kupper, M. Coherent and convex monetary risk measures for bounded càdlàg processes. Stoch. Process. Appl. 112 (2004), no. 1, 1–22.
[55] Cheridito, P., Delbaen, F., Kupper, M. Coherent and convex monetary risk measures for unbounded càdlàg processes. Finance Stoch. 9 (2005), 1713–1732.

[56] Cheridito, P., Delbaen, F., Kupper, M. Dynamic monetary risk measures for bounded discrete-time processes. Electron. J. Probab. 11 (2006), 57–106.
[57] Cheridito, P., Kupper, M. Recursiveness of indifference prices and translation-invariant preferences. Math. Financ. Econ. 2 (2009), no. 3, 173–188.
[58] Cheridito, P., Li, T. Risk measures on Orlicz hearts. Math. Finance 19 (2009), no. 2, 189–214.
[59] Cherny, A. S. Weighted V@R and its properties. Finance Stoch. 10 (2006), no. 3, 367–393.
[60] Cherny, A. S., Grigoriev, P. Dilatation monotone risk measures are law invariant. Finance Stoch. 11 (2007), 291–298.
[61] Cherny, A., Madan, D. New measures for performance evaluation. Rev. Financ. Stud. 22 (2009), 2571–2606.
[62] Chong, K. M., Rice, N. M. Equimeasurable Rearrangements of Functions. Queen's Papers in Pure and Applied Mathematics 28, Queen's University, Kingston, Ont., 1971.
[63] Choquet, G. Theory of capacities. Ann. Inst. Fourier 5 (1953/54), 131–295.
[64] Chow, Y. S., Teicher, H. Probability Theory: Independence, Interchangeability, Martingales. Springer-Verlag, New York–Heidelberg–Berlin, 1978.
[65] Chung, K.-L. A Course in Probability Theory. 2nd edition. Academic Press, New York–London, 1974.
[66] Cont, R. Model uncertainty and its impact on the pricing of derivative instruments. Math. Finance 16 (2006), 519–542.
[67] Cootner, P. H. (ed.), The Random Character of Stock Market Prices. MIT Press, Cambridge, MA, 1964.
[68] Cover, T. Universal portfolios. Math. Finance 1 (1991), 1–29.
[69] Cox, J., Ross, S., Rubinstein, M. Option pricing: a simplified approach. J. Financ. Econ. 7 (1979), 229–263.
[70] Csisźar, I. I-divergence geometry of probability distributions and minimization problems. Ann. Probab. 3 (1975), 146–158.
[71] Csisźar, I. Sanov property, generalized I-projection and a conditional limit theorem. Ann. Probab. 12 (1984), 768–793.
[72] Cvitanic, J. Minimizing expected loss of hedging in incomplete and constrained markets. SIAM J. Control Optim. 38 (2000), 1050–1066.
[73] Cvitanic, J., Karatzas, I. Convex duality in constrained portfolio optimization. Ann. Appl. Probab. 2 (1992), 767–818.
[74] Cvitanic, J., Karatzas, I. Hedging contingent claims with constrained portfolios. Ann. Appl. Probab. 3 (1993), 652–681.
[75] Cvitanic, J., Karatzas, I. On dynamic measures of risk. Finance Stoch. 3 (1999), 451–482.
[76] Cvitanic, J., Karatzas, I. Generalized Neyman-Pearson lemma via convex duality. Bernoulli 7 (2001), 79–97.
[77] Cvitanic, J., Spivak, G. Maximizing the probability of perfect hedge. Ann. Appl. Probab. 9 (1999), 1303–1328.
[78] Dalang, R., Morton, A., Willinger, W. Equivalent martingale measures and no-arbitrage in stochastic securities market models. Stoch. Stoch. Rep. 29 (1990), 185–201.
[79] Dana, R.-A. Existence and uniqueness of equilibria when preferences are additively separable and gross substitute. Econometrica 61 (1993), 953–957.
[80] Dana, R.-A. Market behavior when preferences are generated by second-order stochastic dominance. J. Math. Econ. 40 (2004), no. 6, 619–639.
[81] Dana, R.-A. Ambiguity uncertainty aversion and equilibrium welfare. Econom. Theory 23 (2004), no. 3, 569–587.
[82] Dana, R.-A. Stochastic dominance, individual decision making, and equilibrium asset pricing. Preprint, Ceremade, Université Paris Dauphine, 2002.
[83] Dana, R.-A. A representation result for concave Schur concave functions. Math. Finance 15 (2005), 613–634.

[84] Dana, R.-A., Jeanblanc, M. Financial Markets in Continuous Time. Translated from the 1998 French original by Anna Kennedy. Springer Finance. Springer-Verlag, Berlin, 2003.

[85] Dana, R.-A., Scarsini, M. Optimal risk sharing with background risk. J. Econ. Theory 133 (2007), no. 1, 152–176.

[86] Davis, M. H. A. Option pricing in incomplete markets. In: Mathematics of Derivative Securities (M. A. H. Dempster and S. R. Pliska, eds.). Publ. Newton Inst., Cambridge University Press, Cambridge, 1997, 216–226.

[87] Debreu, G. Theory of Value: An Axiomatic Analysis of Economic Equilibrium. Cowles Foundation for Research in Economics at Yale University, Monograph 17, John Wiley & Sons, Inc., New York; Chapman & Hall, Ltd., London, 1959.

[88] Debreu, G. Continuity properties of Paretian utility. Int. Econ. Rev. 5 (1964), 285–293.

[89] Delbaen, F. Coherent measures of risk on general probability spaces. In: Advances in Finance and Stochastics. Essays in Honour of Dieter Sondermann, Springer-Verlag, 2002, 1–37.

[90] Delbaen, F. Coherent Risk Measures. Cattedra Galileiana. Scuola Normale Superiore, Classe di Scienze, Pisa, 2000.

[91] Delbaen, F. The structure of m-stable sets and in particular of the set of risk neutral measures. In: In Memoriam Paul-André Meyer: Séminaire de Probabilités XXXIX. Lecture Notes in Math. 1874, Springer, Berlin (2006), 215–258.

[92] Delbaen, F. Risk measures for non-integrable random variables. Math. Finance 19 (2009), 329–333.

[93] Delbaen, F. Monetary Utility Functions. Osaka University CSFI Lecture Notes Series 3, Osaka University Press, 2012.

[94] Delbaen, F., Bellini, F., Bignozzi, V., Ziegel, J. Risk measures with the CxLS property. Finance Stoch. 20 (2016), 433–453.

[95] Delbaen, F., Grandits, P., Rheinländer, T., Samperi, D., Schweizer, M. Stricker, C. Exponential hedging and entropic penalties. Math. Finance 12 (2002), 99–123.

[96] Delbaen, F., Owari, K. Convex functions on dual Orlicz spaces (English summary). Positivity 23 (2019), no. 5, 1051–1064.

[97] Delbaen, F., Schachermayer, W. A general version of the fundamental theorem of asset pricing. Math. Ann. 300 (1994), 463–520.

[98] Delbaen, F., Schachermayer, W. The fundamental theorem of asset pricing for unbounded stochastic processes. Math. Ann. 312 (1998), 215–250.

[99] Delbaen, F., Schachermayer, W. The Mathematics of Arbitrage. Springer Finance. Springer-Verlag, Berlin, 2006.

[100] Dellacherie, C. Quelques commentaires sur les prolongements de capacités. In: Séminaire Probabilités V, Strasbourg. Lecture Notes in Math. 191, Springer-Verlag, Berlin, 1970.

[101] Dellacherie, C., Meyer, P.-A. Probabilités et potentiel. Chapitres I à IV. Édition entièrement refondue. Publications de l'Institut de Mathématique de l'Université de Strasbourg, No. XV Actualités Sci. Indust. 1372, Hermann, Paris, 1975.

[102] Dellacherie, C., Meyer, P.-A. Probabilités et potentiel. Chapitres IX à XI. Théorie discrète du potential. Édition entièrement refondue. Publications de l'Institut de Mathématiques de l'Université de Strasbourg, XVIII. Actualités Sci. Indust. 1410, Hermann, Paris, 1983.

[103] Dembo, A., Zeitouni, O. Large Deviations Techniques and Applications, 2nd edition. Appl. Math. 38, Springer-Verlag, New York, 1998.

[104] Denis, L., Martini, C. A theoretical framework for the pricing of contingent claims in the presence of model uncertainty. Ann. Appl. Probab. 16 (2006), no. 2, 827–852.

[105] Denneberg, D. Distorted probabilities and insurance premium. Methods Oper. Res. 52 (1990), 21–42.

[106] Denneberg, D. Non-Additive Measure and Integral. Theory Decis. Lib. Ser. B Math. Statist. Meth. 27, Kluwer Academic Publishers, Dordrecht, 1994.

[107] Deprez, O., Gerber, H. U. On convex principles of premium calculation. Insur. Math. Econ. 4 (1985), 179–189.

[108] Detlefsen, K., Scandolo, G. Conditional and dynamic convex risk measures. Finance Stoch. 9 (2005), no. 4, 539–561.
[109] Deuschel, J.-D., Stroock, D. Large Deviations. Pure Appl. Math. 137, Academic Press, Inc., Boston, MA, 1989.
[110] Donsker, M. D., Varadhan, S. R. S. Asymptotic evaluation of certain Markov process expectations for large time IV. Commun. Pure Appl. Math. 36 (1983), 183–212.
[111] Doob, J. Generalized sweeping-out and probability. J. Funct. Anal. 2 (1968), 207–225.
[112] Dubins, L., Gilat, D. On the distribution of maxima of martingales. Proc. Am. Math. Soc. 68 (1978), 337–338.
[113] Dudley, R. Real Analysis and Probability. Wadsworth & Brooks/Cole, Pacific Grove, CA, 1989.
[114] Duffie, D. Security Markets. Econom. Theory Econometrics Math. Econom. Academic Press, Boston, MA, 1988.
[115] Duffie, D. Dynamic Asset Pricing Theory, 2nd edition. Princeton University Press, 1996.
[116] Duffie, D., Richardson, H. R. Mean-variance hedging in continuous time. Ann. Appl. Probab. 1 (1991), 1–15.
[117] Dunford, N., Schwartz, J.. Linear Operators. Part I: General Theory. Interscience Publishers, New York, 1958.
[118] Dybvig, P. Distributional analysis of portfolio choice. J. Bus. 61 (1988), 369–393.
[119] Eilenberg, S. Ordered topological spaces. Am. J. Math. LXIII (1941), 39–45.
[120] Ekeland, I., Temam, R. Convex Analysis and Variational Problems. Stud. Math. Appl. 1, North-Holland Publishing Co., Amsterdam–Oxford; American Elsevier Publishing Co., Inc., New York, 1976.
[121] El Karoui, N., Quenez, M.-C. Dynamic programming and pricing of contingent claims in an incomplete market. SIAM J. Control Optim. 33 (1995), 29–66.
[122] El Karoui, N., Ravanelli, C. Cash subadditive risk measures and interest rate ambiguity. Math. Finance 19 (2009), 561–590.
[123] Elliott, R., Föllmer, H. Orthogonal martingale representation. In: Stochastic Analysis. Liber Amicorum for Moshe Zakai, Academic Press, Boston, MA, 1991, 139–152.
[124] Elliott, R., Kopp, P. Mathematics of Financial Markets. Springer Finance, Springer-Verlag, New York, 1999.
[125] Ellsberg, D. Risk, ambiguity, and the savage axioms. Q. J. Econ. 75 (1961), 643–669.
[126] Embrechts, P., Klüppelberg, C., Mikosch, T. Modelling Extremal Events. Springer, Berlin, 1997.
[127] Embrechts, P., Schied, A., Wang, R. Robustness in the optimization of risk measures. Oper. Res. 70 (2022), 95–110.
[128] Embrechts, P., Wang, R. Seven proofs for the subadditivity of expected shortfall. Depend. Model. 3 (2015), no. 1, 126–140.
[129] Feller, W. An Introduction to Probability Theory and Its Applications, 3rd edition. John Wiley & Sons, New York, 1968.
[130] Ferguson, T. S. Mathematical Statistics: A Decision Theoretic Approach. Probability and Mathematical Statistics 1, Academic Press, New York, 1967.
[131] Fertis, A., Baes, M., Lüthi, H.-J. Robust risk management. Eur. J. Oper. Res. 222 (2012), 663–672.
[132] Filipović, D., Kupper, M. Monotone and cash-invariant convex functions and hulls. Insur. Math. Econ. 41 (2007), no. 1, 1–16.
[133] Filipović, D., Svindland, G. The canonical model space for law-invariant convex risk measures is L^1. Math. Finance 22 (2012), no. 3, 585–589.
[134] de Finetti, B. Sul concetto di media. G. Ist. Ital. Attuari 2 (1931), 369–396.
[135] Fishburn, P. Utility Theory for Decision Making. Publ. Operations Res. 18, John Wiley, New York, 1970.
[136] Fishburn, P. Nonlinear Preference and Utility Theory. The John Hopkins University Press, Baltimore, MD, 1988.
[137] Fishburn, P. Stochastic dominance and moments of distributions. Math. Oper. Res. 5 (1980), no. 1, 94–100.

[138] Föllmer, H. Calcul d'Ito sans probabilités. In: Seminaire de probabilites XV, Univ. Strasbourg 1979/80. Lect. Notes Math. 850, 1981, 143–150.
[139] Föllmer, H. Probabilistic aspects of financial risk. In: Plenary Lecture at the Third European Congress of Mathematics. Proceedings of the European Congress of Mathematics, Barcelona 2000, Birkhäuser, Basel, 2001.
[140] Föllmer, H., Gundel, A. Robust projections in the class of martingale measures, Ill. J. Math. 50 (2006), no. 2, 439–472.
[141] Föllmer, H., Kabanov, Yu. M. Optional decomposition and Lagrange multipliers. Finance Stoch. 2 (1998), 69–81.
[142] Föllmer, H., Kramkov, D. Optional decompositions under constraints. Probab. Theory Relat. Fields 109 (1997), 1–25.
[143] Föllmer, H., Leukert, P. Quantile hedging. Finance Stoch. 3 (1999), 251–273.
[144] Föllmer, H., Leukert, P. Efficient hedging: cost versus shortfall risk. Finance Stoch. 4 (2000), 117–146.
[145] Föllmer, H., Penner, I. Convex risk measures and the dynamics of their penalty functions. Stat. Decis. 24 (2006), no. 1, 61–96.
[146] Föllmer, H., Schied, A. Convex measures of risk and trading constraints. Finance Stoch. 6 (2002), no. 4, 429–447.
[147] Föllmer, H., Schied, A. Robust representation of convex measures of risk. In: Advances in Finance and Stochastics. Essays in Honour of Dieter Sondermann, Springer-Verlag, 2002, 39–56.
[148] Föllmer, H., Schied, A., Weber, S. Robust preferences and robust portfolio choice. In: Mathematical Modelling and Numerical Methods in Finance (P. Ciarlet, A. Bensoussan and Q. Zhang, eds.). Handbook of Numerical Analysis 15, 2009, 29–88.
[149] Föllmer, H., Schweizer, M. Hedging by sequential regression: an introduction to the mathematics of option trading. ASTIN Bull. 18 (1989), 147–160.
[150] Föllmer, H., Schweizer, M. Hedging of contingent claims under incomplete information. In: Applied Stochastic Analysis, London. Stochastic Monogr. 5, 1990, 389–414.
[151] Föllmer, H., Sondermann, D. Hedging of non-redundant contingent claims. In: Contributions to Mathematical Economics, Hon. G. Debreu, 1986, 206–223.
[152] Fouque, J.-P., Papanicolaou, G., Sircar, R. Derivatives in Financial Markets with Stochastic Volatilities. Cambridge University Press, Cambridge, 2000.
[153] Frittelli, M. Introduction to a theory of value coherent with the no arbitrage principle. Finance Stoch. 4 (2000), no. 3, 275–297.
[154] Frittelli, M. The minimal entropy martingale measure and the valuation problem in incomplete markets. Math. Finance 10 (2000), 39–52.
[155] Frittelli, M., Rosazza Gianin, E. Putting order in risk measures. J. Bank. Finance 26 (2002), 1473–1486.
[156] Frittelli, M., Rosazza Gianin, E. Dynamic convex risk measures. In: New Risk Measures in Investment and Regulation (G. Szegö, ed.). John Wiley & Sons, New York, 2003.
[157] Frittelli, M., Rosazza Gianin, E. Law-invariant convex risk measures. Adv. Math. Econ. 7 (2005), 33–46.
[158] Gao, N., Leung, D., Munari, C., Xanthos, F. Fatou property, representations, and extensions of law-invariant risk measures on general Orlicz spaces. Finance Stoch. 22 (2018), no. 2, 395–415.
[159] Gerhold, S., Krühner, P. Dynamic trading under integer constraints. Finance Stoch. 22 (2018), 919–957.
[160] Frittelli, M., Scandolo, G. Risk measures and capital requirements for processes. Math. Finance 16 (2006), 589–612.
[161] Gilboa, I. Expected utility with purely subjective non-additive probabilities. J. Math. Econ. 16 (1987), 65–88.
[162] Gilboa, I. Theory of Decision under Uncertainty. Cambridge University Press, 2009.
[163] Ghossoub, M. Arrow's theorem of the deductible with heterogeneous beliefs. N. Am. Actuar. J. 21 (2017), no. 1, 15–35.

[164] Ghossoub, M. A Neyman-Pearson problem with ambiguity and nonlinear pricing. Math. Financ. Econ. 12 (2018), no. 3, 365–385.
[165] Gibbs, J. Elementary Principles in Statistical Mechanics: Developed with Especial Reference to the Rational Foundations of Thermodynamics. C. Scribner's Sons, 1902.
[166] Gilboa, I., Schmeidler, D. Maxmin expected utility with non-unique prior. J. Math. Econ. 18 (1989), 141–153.
[167] Gneiting, T. Making and evaluating point forecasts. J. Am. Stat. Assoc. 106 (2011), no. 494, 746–762.
[168] Gollier, C., Schlesinger, H. Arrow's theorem on the optimality of deductibles: a stochastic dominance approach. Econ. Theory 7 (1996), 359–363.
[169] Goovaerts, M. J., De Vylder, F., Haezendonck, J. Insurance Premiums. Theory and Applications. North-Holland Publishing Co., Amsterdam, 1984.
[170] Gundel, A. Robust utility maximization in complete and incomplete market models. Finance Stoch. 9 (2005), no. 2, 151–176.
[171] Halmos, P., Savage, L. Application of the Radon–Nikodym theorem to the theory of sufficient statistics. Ann. Math. Stat. 20 (1949), 225–241.
[172] Hardy, G. H., Littlewood, J. E., Pólya, G. Inequalities. Cambridge University Press, Cambridge, 1934.
[173] Harrison, J., Kreps, D. Martingales and arbitrage in multiperiod securities markets. J. Econ. Theory 20 (1979), 381–408.
[174] Harrison, J., Pliska, S. Martingales and stochastic integrals in the theory of continuous trading. Stoch. Process. Appl. 11 (1981), 215–260.
[175] He, X.-D., Zhou, X.-Y. Portfolio choice via quantiles. Math. Finance 21 (2011), no. 2, 203–231.
[176] Heath, D. Back to the future. Plenary lecture, First World Congress of the Bachelier Finance Society, Paris, 2000.
[177] Heath, D., Platen, E. Introduction to Quantitative Finance: A Benchmark Approach. Springer Finance. Springer, Berlin, 2006.
[178] Heath, D., Ku, H. Pareto equilibria with coherent measures of risk. Math. Finance 14 (2004), no. 2, 163–172.
[179] Herdegen, M., Munari, C. An elementary proof of the dual representation of expected shortfall. Math. Financ. Econ. 17 (2023), no. 4, 655–662.
[180] Herstein, I. N., Milnor, J. An axiomatic approach to measurable utility. Econometrica 21 (1953), 291–297.
[181] Hobson, D. Robust hedging of the loockback option. Finance Stoch. 2 (1998), 329–347.
[182] Hoeffding, W. Maszstabinvariante Korrelationstheorie. Schr. Math. Inst. Inst. Angew. Math. Univ. Berlin 5 (1940), 181–233.
[183] Huber, P. Robust Statistics. Wiley Ser. Probab. Math. Statist. John Wiley & Sons, Inc., New York, 1981.
[184] Huber, P., Ronchetti, E. Robust Statistics, 2nd edition. Wiley Ser. Probab. Stat. John Wiley & Sons, Inc., Hoboken, NJ, 2009.
[185] Huber, P., Strassen, V. Minimax tests and the Neyman-Pearson lemma for capacities. Ann. Stat. 1 (1973), 251–263.
[186] Hull, J. Options, Futures, and Other Derivatives, 4th edition. Prentice Hall, London, 2000.
[187] Jacka, S. A martingale representation result and an application to incomplete financial markets. Math. Finance 2 (1992), 239–250.
[188] Jacod, J. and Shiryaev, A. N. Local martingales and the fundamental asset pricing theorems in the discrete-time case. Finance Stoch. 2 (1998), 259–273.
[189] Jaschke, S., Küchler, U. Coherent risk measures and good-deal bounds. Finance Stoch. 5 (2001), no. 2, 181–200.
[190] Jeanblanc, M., Yor, M., Chesney, M. Mathematical Methods for Financial Markets. Springer-Verlag, 2009.
[191] Jouini, E., Kallal, H. Arbitrage in securities markets with short-sales constraints. Math. Finance 3 (1995), 197–232.

[192] Jouini, E., Kallal, H. Efficient trading strategies in the presence of market frictions. Rev. Financ. Stud. 14 (2001), 343–369.
[193] Jouini, E., Schachermayer, W., Touzi, N. Law invariant risk measures have the Fatou property. In: Advances in Mathematical Economics. Adv. Math. Econ. 9, Springer, Tokyo, 2006, 49–71.
[194] Jouini, E., Schachermayer, W., Touzi, N. Optimal risk sharing for law invariant monetary utility functions. Math. Finance 18 (2008), no. 2, 269–292.
[195] Kabanov, Yu. M., Kramkov, D. O. No-arbitrage and equivalent martingale measures: an elementary proof of the Harrison-Pliska theorem. Theory Probab. Appl. 39 (1994), 523–527.
[196] Kabanov, Yu. M., Stricker, C. A teacher's note on no-arbitrage criteria. In: Sém. Probab. 35. Lecture Notes in Math. 1755, Springer-Verlag, 2001, 149–152.
[197] Kahn, P. M. Some remarks on a recent paper by Borch. ASTIN Bull. 1 (1961), 265–272.
[198] Kahneman, D., Tversky, A. Prospect theory: an analysis of decision under risk. Econometrica 47 (1979), 263–291.
[199] Kallenberg, O. Foundations of Modern Probability, 2nd edition. Springer Verlag, 2002.
[200] Karatzas, I. On the pricing of American options. Appl. Math. Optim. 17 (1988), 37–60.
[201] Karatzas, I. Lectures on the Mathematics of Finance. CRM Monogr. Ser. 8, American Mathematical Society, Providence, RI, 1996.
[202] Karatzas, I., Kou, S. G. On the pricing of contingent claims under constraints. Ann. Appl. Probab. 6 (1996), 321–369.
[203] Karatzas, I., Kou, S. G. Hedging American contingent claims with constraint portfolios. Finance Stoch. 2 (1998), 215–258.
[204] Karatzas, I., Shreve, S. Brownian Motion and Stochastic Calculus. Grad. Texts in Math. 113, Springer-Verlag, New York, 1991.
[205] Karatzas, I., Shreve, S. Methods of Mathematical Finance. Appl. Math. Springer-Verlag, Berlin, 1998.
[206] Karni, E., Schmeidler, D., Vind, K. On state-dependent preferences and subjective probabilities. Econometrica 51 (1983), 1021–1031.
[207] Karni, E., Schmeidler, D. Utility theory with uncertainty. In: Handbook of Mathematical Economics, Vol. IV (W. Hildebrandt and H. Sonnenschein, eds.). Elsevier Science Publishers, 1991, 1763–1831.
[208] Kellerer, H. Markov-Komposition und eine Anwendung auf Martingale. Math. Ann. 198 (1972), 99–122.
[209] Klibanoff, P., Marinacci, M., Mukerji, S. A smooth model of decision making under uncertainty. Econometrica 73 (2005), 1849–1892.
[210] Knight, F. Risk, Uncertainty, and Profit. Houghton Mifflin, Boston, 1921.
[211] Koch-Medina, P., Munari, C. Market-Consistent Prices—An Introduction to Arbitrage Theory. Birkhäuser/Springer, 2020.
[212] Kolmogorov, A. Sur la notion de moyenne. Rend. Lincei, 2° Sem. Fasc. 9 (1930).
[213] Komlós, J. A generalization of a problem of Steinhaus. Acta Math. Acad. Sci. Hung. 18 (1967), 217–229.
[214] Korn, R. Optimal Portfolio. World Scientific, Singapore, 1997.
[215] Krätschmer, V. Robust representation of convex risk measures by probability. Finance Stoch. 9 (2005), 597–608.
[216] Krätschmer, V. Compactness in spaces of inner regular measures and a general portmanteau lemma. J. Math. Anal. Appl. 351 (2009), no. 2, 792–803.
[217] Krätschmer, V., Schied, A., Zähle, H. Comparative and qualitative robustness for law-invariant risk measures. Finance Stoch. 18 (2014), no. 2, 271–295.
[218] Krätschmer, V., Schied, A., Zähle, H. Domains of weak continuity of statistical functionals with a view toward robust statistics. J. Multivar. Anal. 158 (2017), 1–19.
[219] Kramkov, D. O. Optional decomposition of supermartingales and hedging contingent claims in incomplete security markets. Probab. Theory Related Fields 105 (1996), 459–479.
[220] Kramkov, D., Schachermayer, W. The asymptotic elasticity of utility functions and optimal investment in incomplete markets. Ann. Appl. Probab. 9 (1999), no. 3, 904–950.

[221] Kramkov, D., Schachermayer, W. Necessary and sufficient conditions in the problem of optimal investment in incomplete markets. Ann. Appl. Probab. 13 (2003), no. 4, 1504–1516.
[222] Krasnosel'skii, M., Rutickii, Ya. Convex Functions and Orlicz Spaces. Gordon and Breach Science Publishers, 1961.
[223] Kreps, D. Arbitrage and equilibrium in economies with infinitely many commodities. J. Math. Econ. 8 (1981), 15–35.
[224] Kreps, D. Notes on the Theory of Choice. Westview Press, Boulder, CO, 1988.
[225] Kuipers, L., Niederreiter, H. Uniform Distribution of Sequences. John Wiley & Sons, New York–London–Sydney, 1974.
[226] Kulldorff, M. Optimal control of favorable games with a time limit. SIAM J. Control Optim. 31 (1993), 52–69.
[227] Kullback, S., Leibler, R. A. On information and sufficiency. Ann. Math. Stat. 22 (1951), 79–86.
[228] Kunze, M. Verteilungsinvariante konvexe Risikomaße. Diplomarbeit, Humboldt-Universität zu Berlin, 2003.
[229] Kupper, M., Schachermayer, W. Representation results for law invariant time consistent functions. Math. Financ. Econ. 2 (2009), no. 3, 189–210.
[230] Kusuoka, S. On law invariant coherent risk measures. Adv. Math. Econ. 3 (2001), 83–95.
[231] Lacroix, S. F. Traité du calcul différentiel et du calcul intégral. Courcier, 1810.
[232] Lamberton, D., Lapeyre, B. Introduction to Stochastic Calculus Applied to Finance, 2nd edition. English edition, translated by N. Rabeau and F. Mantion, Chapman & Hall, London, 2007.
[233] Lamberton, D., Lapeyre, B. Introduction au calcul stochastique appliqué à la finance, 2ème éd. Ellipses, Paris, 1997.
[234] Leitner, J. A short note on second-order stochastic dominance preserving coherent risk measures. Math. Finance 15 (2005), 649–651.
[235] Leland, H. Who should buy portolio insurance? J. Finance 35 (1980), 581–594.
[236] Lembcke, J. The necessity of strongly subadditive capacities for Neyman-Pearson minimax tests. Monatshefte Math. 105 (1988), 113–126.
[237] Leukert, P. Absicherungsstrategien zur Minimierung des Verlustrisikos. Dissertation, Humboldt-Universität zu Berlin, 1999.
[238] Li, P., Xia, J. Minimal martingale measures in discrete time. Acta Math. Appl. Sin. (2002).
[239] Liebrich, F.-B., Munari, C. Short communication: revisiting the automatic Fatou property of law-invariant functionals. SIAM J. Financ. Math. 16 (2025), no. 1, SC1–SC11.
[240] Liu, P., Schied, A., Wang, R. Distributional transforms, probability distortions, and their applications. Math. Oper. Res. 46 (2021), no. 4, 1490–1512.
[241] Maccheroni, F., Marinacci, M., Rustichini, A. Ambiguity aversion, robustness, and the variational representation of preferences. Econometrica 74 (2006), 1447–1498.
[242] Mao, T., Wang, R. Risk aversion in regulatory capital principles. SIAM J. Financ. Math. 11 (2020), 169–200.
[243] Mas-Colell, A. The Theory of General Equilibrium: A Differentiable Approach. Cambridge University Press, Cambridge, 1985.
[244] Mas-Colell, A., Zame, W. Equilibrium Theory in Infinite Dimensional Spaces. Handbook of Mathematical Economics 4, North-Holland, 1991.
[245] McNeil, A. J., Frey, R., Embrechts, P. Risk Management, 2nd edition. Cambridge University Press, Cambridge, 2015.
[246] Melnikov, A. V., Nechaev, M. L. On the mean-variance hedging problem. Theory Probab. Appl. 43 (1998), 588–603; translation from Teor. Veroyatnost. i Primenen. 43 (1998), 672–691.
[247] Merton, R. Theory of rational option pricing. Bell J. Econ. Manag. Sci. 4 (1973), 141–183.
[248] Meyer, P. A. Probability and Potentials. Blaisdell Publishing Co, N. Y., 1966.
[249] Milgram, A. N. Partially ordered sets, separating systems and inductiveness. Rep. Math. Colloq. II Ser. 1, University of Notre Dame, 1939, 18–30.

[250] Monat, P., Stricker, C. Föllmer-Schweizer decomposition and mean-variance hedging for general claims. Ann. Probab. 23 (1995), 605–628.
[251] Mosler, K., Scarsini, M. Some theory of stochastic dominance. In: Stochastic Orders and Decision Under Risk (Hamburg, 1989). IMS Lecture Notes Monogr. Ser. 19, Inst. Math. Statist., Hayward, CA, 1991, 261–284.
[252] Musiela, M., Rutkowski, M. Martingale Methods in Financial Modelling. Appl. Math., Springer-Verlag, Berlin, 1997.
[253] Myeni, R. The pricing of the American option. Ann. Appl. Probab. 2 (1992), 1–23.
[254] Nagumo, M. Über eine Klasse der Mittelwerte. Jpn. J. Math. 7 (1930), 71–79.
[255] Nendel, M. Lower semicontinuity of monotone functionals in the mixed topology on C_b. Finance Stoch. 29 (2025), no. 1, 261–287.
[256] von Neumann, J., Morgenstern, O. Theory of Games and Economic Behavior. 2nd edition. Princeton University Press, Princeton, NJ, 1947.
[257] Neveu, J. Discrete-Parameter Martingales. North-Holland, Amsterdam–Oxford, 1975.
[258] Neyman, J., Pearson, E. S. On the problem of the most efficient tests of statistical hypotheses. Philos. Trans. R. Soc. Lond. Ser. A 231 (1933), 289–337.
[259] Ogryczak, W., Ruszczynski, A. Dual stochastic dominance and related mean-risk models. SIAM J. Optim. 13 (2002), no. 1, 60–78.
[260] Penner, I. Dynamic convex risk measures: time consistency, prudence, and sustainability. Dissertation, Humboldt-Universität zu Berlin, 2007.
[261] Pflug, G. Some remarks on the value-at-risk and the conditional value-at-risk. In: Probabilistic Constrained Optimization (S. Uryasev, ed.). Springer, 2000, 272–281.
[262] Kovacevic, R. M., Pflug, G. C. Are time consistent valuations information monotone? Int. J. Theor. Appl. Finance 17 (2014), 1450003.
[263] Pflug, G., Römisch, W. Modeling, Measuring and Managing Risk. World Scientific Publishing Co. Pte. Ltd., Hackensack, NJ, 2007.
[264] Pflug, G., Ruszczynski, A. Measuring risk for income streams. Comput. Optim. Appl. 32 (2005), no. 1–2, 161–178.
[265] Pham, H. Dynamic L^p-hedging in discrete time under cone constraints. SIAM J. Control Optim. 38 (2000), 665–682.
[266] Pliska, S. Introduction to Mathematical Finance. Blackwell Publishers, 1997.
[267] Protter, P. Stochastic Integration and Differential Equations. Version 2.1, 2nd edition. Applications of Mathematics. Stochastic Modelling and Applied Probability 21, Springer, Berlin, 2005.
[268] Quiggin, J. A theory of anticipated utility. J. Econ. Behav. Organ. 3 (1982), 225–243.
[269] Rabin, M. Risk aversion and expected-utility theory: a calibration theorem. Econometrica 68 (2000), no. 5. 1281–1292.
[270] Rader, J. The existence of a utility function to represent preferences. Rev. Econ. Stud. 30 (1963), 229–232.
[271] Reed, M., Simon, B. Methods of Modern Mathematical Physics. Vol. 1: Functional Analysis. Academic Press, New York–London, 1972.
[272] Riedel, F. Dynamic convex risk measures. Stoch. Process. Appl. 112 (2004), 185–200.
[273] Rockafellar, R. T. Convex Analysis. Princeton University Press, Princeton, NJ, 1970.
[274] Rockafellar, R. T., Uryasev, S. Conditional value-at-risk for general loss distributions. J. Bank. Finance 26 (2002), 1443–1471.
[275] Roorda, B., Schumacher, J. M. Time consistency conditions for acceptability measures, with an application to Tail Value at Risk. Insur. Math. Econ. 40 (2007), 209–230.
[276] Rogers, L. C. G. Equivalent martingale measures and no-arbitrage. Stoch. Stoch. Rep. 51 (1994), 41–49.
[277] Ross, S. A. The arbitrage theory of capital asset pricing. J. Econ. Theory 13 (1976), 341–360.
[278] Rothschild, M., Stiglitz, J. Increasing risk: I. A definition. J. Econ. Theory 2 (1970), 225–243.

[279] Rothschild, M., Stiglitz, J. Increasing risk: II. Its economic consequences. J. Econ. Theory 3 (1971), 66–84.

[280] Rüschendorf, L. On the distributional transform, Sklar's theorem, and the empirical copula process. J. Stat. Plan. Inference 139 (2009), 3921–3927.

[281] Ryff, J. Orbits of L^1-functions under doubly stochastic transformations. Trans. Am. Math. Soc. 117 (1965), 92–100.

[282] Samuelson, P. A. Proof that properly anticipated prices fluctuate randomly. Ind. Manage. Rev. 4 (1965), 41–50.

[283] Samuelson, P. A. Rational theory of warrant pricing. Ind. Manage. Rev. 6 (1965), 13–31.

[284] Savage, L. J. The Foundations of Statistics. Wiley Publ. Stat., John Wiley and Sons, New York, 1954.

[285] Schachermayer, W. A Hilbert space proof of the fundamental theorem of asset pricing in finite discrete time. Insur. Math. Econ. 11 (1992), 249–257.

[286] Schachermayer, W. Optimal investment in incomplete financial markets. In: Proceedings of the First World Congress of the Bachelier Society, Paris 2000, Springer-Verlag, Berlin, 2001.

[287] Schachermayer, W. Introduction to the Mathematics of Financial Markets. Lectures on Probability Theory and Statistics (Saint-Flour, 2000). Lecture Notes in Math. 1816, Springer, Berlin, 2003, 107–179.

[288] Schaefer, H. Topological Vector Spaces. Grad. Texts in Math. 3, Springer-Verlag, New York, 1971.

[289] Schäl, M. On quadratic cost criteria for option hedging. Math. Oper. Res. 19 (1994), 121–131.

[290] Schied, A. On the Neyman-Pearson problem for law-invariant risk measures and robust utility functionals. Ann. Appl. Probab. 14 (2004), 1398–1423.

[291] Schied, A. Optimal investments for robust utility functionals in complete market models. Math. Oper. Res. 30 (2005), no. 3, 750–764.

[292] Schied, A. Risk measures and robust optimization problems. Stoch. Models 22 (2006), 753–831.

[293] Schied, A. Optimal investments for risk- and ambiguity-averse preferences: a duality approach. Finance Stoch. 11 (2007), 107–129.

[294] Schied, A., Wu, C.-T. Duality theory for optimal investments under model uncertainty. Stat. Decis. 23 (2005), no. 3, 199–217.

[295] Schmeidler, D. Integral representation without additivity. Proc. Am. Math. Soc. 97 (1986), no. 2, 255–261.

[296] Schweizer, M. Hedging of options in a general semimartingale model. Diss ETH Zürich no. 8615, 1988.

[297] Schweizer, M. Risk-minimality and orthogonality of martingales. Stoch. Stoch. Rep. 30 (1990), 123–131.

[298] Schweizer, M. Variance-optimal hedging in discrete time. Math. Oper. Res. 20 (1995), 1–32.

[299] Schweizer, M. A guided tour through quadratic hedging approaches. In: Option Pricing, Interest Rates, and Risk Management (E. Jouini, M. Musiela and J. Cvitanic, eds.). Cambridge University Press, Cambridge, 2001, 538–574.

[300] Sekine, J. Dynamic minimization of worst conditional expectation of shortfall. Math. Finance 14 (2004), 605–618.

[301] Shaked, M., Shanthikumar, J. Stochastic Orders. Springer Series in Statistics. Springer, New York, 2007.

[302] Shiryaev, A. N. Probability, 2nd edition. Grad. Texts in Math. 95, Springer-Verlag, New York, 1995.

[303] Shiryaev, A. N. Essentials of Stochastic Finance. Adv. Ser. Stat. Sci. Appl. Probab. 3, World Scientific, Singapore, 1999.

[304] Shreve, S. Stochastic Calculus Models for Finance: Discrete Time. Springer, 2004.

[305] Shreve, S. Stochastic Calculus Models for Finance, II: Continuous Time Models. Springer, 2004.

[306] Simon, B. The Statistical Mechanics of Lattice Gases, Vol. I. Princeton University Press, Princeton, 1993.

[307] Snell, L. Applications of the martingale systems theorem. Trans. Am. Math. Soc. 73 (1952), 293–312.

[308] Sprenkle, C. Warrant prices as indicators of expectations and preferences. Yale Econ. Essays 1 (1964), 178–231.

[309] Strassen, V. The existence of probability measures with given marginals. Ann. Math. Stat. 36 (1965), 423–439.
[310] Staum, J. Fundamental theorems of asset pricing for good deal bounds. Math. Finance 14 (2004), no. 2, 141–161.
[311] Stricker, C. Arbitrage et lois de martingale. Ann. Inst. Henri Poincaré Probab. Stat. 26 (1990), 451–460.
[312] Talay, D., Zheng, Z. Worst case model risk management. Finance Stoch. 6 (2002), 517–537.
[313] Tao, T. Analysis I, 4th edition. Springer-Verlag, 2022.
[314] Turner, A. The Turner Review: A Regulartory Response to the Global Banking Crisis. Financial Services Authority, London, 2009.
[315] Tutsch, S. Update rules for convex risk measures. Quant. Finance 8 (2008), 833–843.
[316] Tversky, A., Kahneman, D. Advances in prospect theory: cumulative representation of uncertainty. J. Risk Uncertain. 5 (1992), 297–323.
[317] van der Vaart, A. W., Wellner, J. A. Weak Convergence and Empirical Processes. With Applications to Statistics, 2nd edition. Springer-Verlag, 2023.
[318] Wang, S., Dhaene, J. Comonotonicity, correlation order and premium principles. Insur. Math. Econ. 22 (1998), 235–242.
[319] Wang, S., Young, V. Ordering risks: expected utility theory versus Yaari's dual theory of risk. Insur. Math. Econ. 22 (1998), no. 2, 145–161.
[320] Weber, S. Distribution-invariant dynamic risk measures. Math. Finance 16 (2006), 419–442.
[321] von Weizsäcker, H. Can one drop L^1-boundedness in Komlós' subsequence theorem? Am. Math. Mon. 111 (2004), 900–903.
[322] Yaari, M. The dual theory of choice under risk. Econometrica 55 (1987), 95–116.
[323] Yan, J.-A. Caractérisation d'une classe d'ensembles convexes de L^1 ou H^1. In: Seminaire de Probabilités XIV, 1978/79 (J. Azéma and M. Yor, eds.). Lecture Notes in Math. 784, Springer-Verlag, Berlin, 1980, 220–222.
[324] Ziegel, J. Coherence and elicitability. Math. Finance 26 (2016), no. 4, 901–918.

List of symbols

$\vee$ 21
$\wedge$ 21
$\sim$ 55
$\succ$ 55
$\succeq$ 55
$\approx$ 557
$\succcurlyeq_{i}$ 98
$\succcurlyeq_{icv}$ 87, 123
$\succcurlyeq_{cv}$ 95, 125
$\succcurlyeq_{cx}$ 95
$\ll$ 557
$\|\cdot\|$ 602
$\|\cdot\|_p$ 605
$\|\cdot\|_\infty$ 605
$\|\cdot\|_{\mathrm{var}}$ 602
$|x|$ 550
A^Q 461
AV@R 226
AV@R$_\lambda(X|\mathscr{F}_t)$ 513
$ba(\Omega,\mathscr{F})$ 602
cov 478
$C_\psi(S)$ 596
$C_b(S)$ 591
$C_t(K)$ 412
CV@R 226
ess inf 589, 590
ess sup 589, 590
ES_λ 226
$E[\,F\,;A_0\,]$ 559
$E[\,F\,|\,\mathscr{F}_0\,]$ 559
$E_Q[\,F\,]$ 603
F_X 573
F_μ 573
$\mathscr{F}_\tau$ 385
$\mathscr{F}_t$ 301
$H(Q|P)$ 140, 563
$\widehat{H}_t(Q|P)$ 511
$\mathscr{H}^2$ 483
I^+ 150
$I_g(Q|P)$ 218
L_t^∞ 505
$\mathscr{L}^p(\Omega,\mathscr{F},P)$ 605
$M(\mu)$ 35
$M_b(\mu)$ 35
$\mathscr{M}(S,\mathscr{S})$ 591
$\mathscr{M}^\psi(S)$ 597
$\mathscr{M}_1^\psi(S)$ 119, 597
$\mathscr{M}_1(P)$ 214
$\mathscr{M}_1(S)$ 61
$\mathscr{M}_1(S,\mathscr{S})$ 61, 591
$\mathscr{M}_1(\Omega,\mathscr{F})$ 107
$\mathscr{M}_{1,f}(\Omega,\mathscr{F})$ 602
$\mathscr{M}_{1,f}$ 107, 602
$\mathscr{M}_b(S)$ 68
$N(m,\sigma^2)$ 91
$N^\perp$ 49, 452
$\mathscr{P}$ 7, 308
$\mathscr{P}_{\mathscr{S}}$ 455
$\mathscr{P}_\Phi$ 415
$\mathscr{P}_{\Phi_T}$ 416
$\mathscr{Q}_{\mathscr{S}}$ 461
$R(V)$ 13
supp μ 36
$\mathscr{S}$ 264, 451, 498
$\mathscr{S}^\infty$ 456
$\mathscr{S}_t^\infty$ 456
$\mathscr{S}_t$ 451
$\mathcal{T}$ 371
$\mathcal{T}_\tau$ 387
$\mathcal{T}_t$ 372
U^Q 367
$U^\uparrow$ 466
$U^\downarrow$ 391
U^{P^*} 366
var 479
var(μ) 95
V@R 223
V@R$_\lambda(X|\mathscr{F}_t)$ 512
$x\cdot y$ 550
$\Gamma(\mu)$ 37
δ_x 593
μQ 88
$\pi(V)$ 12
$\pi_{\inf}(C)$ 23
$\pi_{\inf}(H)$ 320, 381
$\pi_{\sup}(C)$ 23
$\pi_{\sup}(H)$ 320, 381
$\overline{\pi}$ 3
$\Pi(C)$ 23
$\Pi(H)$ 320, 381
$\sigma(E,F)$ 607
$\sigma(L^\infty,L^1)$ 608

https://doi.org/10.1515/9783111045283-025

Index

absolute continuity 557
- uniform 610
absolute risk aversion 78
absolutely continuous measure 557
acceptance set 191, 215, 263, 279, 409, 433, 474
- for conditional risk measure 506
- one-step 516
act 104
adapted process 302
admissible strategy 264, 426
affine hull of a convex set 37
affine numerical representation 61
Allais paradox 69, 102
American claim
- arbitrage-free price of 375, 381
- attainable 384
- hedging strategy 384
American option 363
- call 365
- put 365, 376
arbitrage bounds 24
arbitrage opportunity 5, 41, 266, 306, 412
arbitrage-free price
- of a contingent claim 23
- of a European claim 320
- of American claim 381
- of an American claim 375
- universal bounds 26, 416
arbitrage-free prices
- jointly 29
Archimedean axiom 61, 107
Arrow–Debreu equilibrium 172
- and interest rate 185
- w. r. t. general utility functionals 186
Arrow–Pratt coefficient 78
Arrow's theorem of the deductible 160
Asian option 315, 336
asymmetric relation 55
at-the-money option 27
atom of a probability space 30, 231, 328, 580
atomless probability space 231, 580
attainable
- American claim 384
- European claim 318
- payoff 12
average
- price option 316
- strike option 316
Average Value at Risk 226
- as divergence risk measure 293
- comonotonicity 256, 258
- sensitivity 222
- time-inconsistency 515

backwards martingale 86
balayage order 95, 410, 414
Banach space 605
Banach–Alaoglu theorem 608
barrier option 316, 335
- Black–Scholes price of 360
- down-and-in put 343
- up-and-in call 341, 421
- up-and-out call 342, 421, 423
barycenter of a measure 35
basket option 152
bear put spread 18, 29
Bermuda option 365
Bernoulli, Daniel 54, 74
Bernoulli, Nicholas 74
beta distribution 540
binomial model 333, 347, 377
Birkhoff's ergodic theorem 594, 601
Black–Scholes formula 98, 349, 351
Black–Scholes p. d. e. 353, 358
Black–Scholes price 348
- of a European call option 349, 351
Bolzano–Weierstraß theorem
- randomized version 48
boundedly supported measures 68
Brownian motion 356
bull call spread 18
butterfly spread 19, 29

call option 17, 152, 411
- American 365, 376
- Asian 316
- Bermuda 365
- Black–Scholes price 349, 351
- European 315, 326, 376, 438
- universal bounds on price 26
- with barrier 316, 421
call overwriting 18
call spread 18, 177
call-put parity 18

https://doi.org/10.1515/9783111045283-026

cap 21
CARA utility 78, 80, 84, 130, 136, 137, 147, 152, 197, 274, 280, 511
cash additivity 190
cash invariance 110, 189
- conditional 505
catastrophe bond 22
central limit theorem 346, 592
- functional 356
- multiplicative 345
certainty equivalent 73, 80, 108, 148, 195
- as reservation price 159
- as risk measure 196
- cash-invariant 197
- quasi-convex 197
- relative 84
- translation property 79
certainty independence 105
- weak 106
chain rule for Radon–Nikodym derivatives 557
Charm of a call option 355
Choquet boundary 410
Choquet integral 198, 199, 248, 254
coherent conditional risk measure 512
coherent measure of risk 191
- and superhedging 409
- representation theorem 204, 218
coherent monetary utility functional 112
coherent risk measure 112, 191
- conditional 396, 505
- dynamic 396
- dynamically consistent 396
- representation theorem 204
- time-consistent 396
comonotone random variables 106, 116, 254, 576, 580
comonotonic
- functional 116
- independence 106, 116
- risk measure 254
complete
- market 29, 328
- relation 55
compound lottery 61
concave distortion 199, 244
concave function 552
- proper 552
concave monetary utility functional 118
concave order 95, 125
conditional Average Value at Risk
- time-inconsistency 515
conditional cash invariance 505
conditional coherent risk measure 396
conditional convexity 505
conditional covariance 478
conditional entropic risk measure 511
conditional expectation 42, 559
conditional mean-standard deviation risk measure 515
conditional relative entropy 511
conditional risk measure
- coherent 512
- relevant 518
- representation 507
- sensitive 518
- time-consistent 514
conditional Value at Risk 226, 512
- time-inconsistency 515
conditional variance 313, 479
conjugate function 277, 554, 608
- of a convex risk measure 203, 222
connected topological space 59
constant absolute risk aversion 78, 80
constant rebalanced portfolio strategy 527
constant relative risk aversion 78
contingent claim 22
- American 364
- arbitrage-free price of 23
- attainable 25
- European 315
- redundant 25
- replicable 25
contingent utility function 154
continuation region 379
continuity axiom 62, 107
continuity from above 115, 206, 215, 507
continuity from below 107, 115, 207, 276
convex combination 549
convex function 552
- duality theory 608
- proper 552
convex hull 549
convex measure of risk 190
- continuous from above 206, 215
- continuous from below 207, 276
- in financial market 262, 433, 474
- law-invariant 231
- penalty function for 201, 205, 214, 277, 279

- relevant 220
- representation theorem 201, 214
- sensitive 220, 474
- tight 211
convex order 95, 410, 414
convex risk measure 118, 190
- conditional 505
- on L^p 222
- representation theorem 201
convex set 549
convexity
- conditional 505
core of a set function 260
cost of superhedging 403
cost process 477
cost-efficient payoff profile 158
countably convex set 46, 47
covariance
- conditional 478
covered call 18, 21
Cox–Ross–Rubinstein model 333, 347, 377
credit spread 18
criterion of de la Vallée Poussin 610
CRR model 333, 347, 377
CRRA utility 78
cumulative distribution function 569
- and stochastic dominance 99
- and stochastic order 88
- modified 575
CxLS 276

de la Vallée Poussin
- criterion 610
debit spread 29
debit spreads 18
declining a favorable bet 81
deductible 160
Delta hedging 337, 351, 357
Delta of a call option 351
density function 557
density process 307, 386
derivative 16
derivative securities 16, 22, 315
digital option 316, 417
dilatation monotonicity 242
Dini's lemma 208
Dirac measure 593
Dirac point mass 593
Dirichlet distribution 540
discount certificate 21
discounted American claim 366
discounted claim 317
discounting 11, 186
distortion function 247
distortion of a probability measure 247
distribution
- beta 540
- Dirichlet 540
- gamma 540
- Gumbel 574
- log-normal 96, 159, 345
- normal 91
- Pareto 198, 574
- Poisson 27
distribution function 569
divergence 218, 293
- Kullback–Leibler 563
divergence risk measure 218
- sensitivity 222
divergence risk measures 293
diversification 190
Doléans–Dade exponential 488
Donsker's invariance principle 356
Doob decomposition 363, 399, 489
- uniform 399, 463
Doob's stopping theorem 369, 386
Doob's systems theorem 308
down-and-in option 316, 336, 343
down-and-out option 317
dual representation
- on L^p 222
dual representation of a risk measure 204
dual space 607
duality theory of convex functions 608
Dunford–Pettis theorem 613
dynamic coherent risk measure 396
dynamic consistency of a risk measure 396
dynamic risk measure
- relevant 518
- sensitive 518
- time-consistent 514
- weak time consistency 522
dynamically consistent 396

early exercise premium 376
Eberlein–Šmulian theorem 609
effective domain of a convex function 552
efficient market hypothesis 310

Ellsberg paradox 103
enlagement of filtration 327
entropic risk measure 197, 216, 218, 277, 278
- as divergence risk measure 294
- conditional 511
- sensitivity 222
entropy
- relative 136, 173, 216
- Shannon–Boltzmann 567
entropy of a measure 563
entropy-minimizing risk-neutral measure 141, 152
equal-weighted portfolio 527
equivalent martingale measure 7, 308
equivalent measure 557
Esscher transform 141
essential infimum 589
- of a random variable 590
essential supremum 589
- of a random variable 590
ETF 17, 544
Euclidean inner product 550
Euclidean norm 550
European call option 326, 376, 411
- Black–Scholes price 349, 351
- efficient hedging of 438
- intrinsic value of 322
- time value of 322
European contingent claim 315
- arbitrage-free price 320
- attainable 318, 323
- redundant 318
- replicable 318
European put option 348
exchange-traded fund 17, 544
exercise strategy for American claim 365
Expected Shortfall 226
expected utility representation 72
expectile 276
expectile risk measure 277, 279
expiration date of a contingent claim 315
exponential distribution 574
exponential family 138
exponential utility function 78, 84, 152
extreme point 328
- of comonotonic convex risk measures 259

fair premium 71
fair price 71
Fatou property 206, 215, 507
- of law-invariant risk measures 231
- on L^p 222
- on L^p 222
favorable bet 81
feasible allocation 171
Fenchel–Legendre transform 142, 143, 277, 554, 575, 608
- of a convex risk measure 203, 222
filtered probability space 302
filtration 302
- enlargement of 327
financial position 189
finitely additive set function 107, 200, 602
first welfare theorem 180
first-order stochastic dominance 99, 241
- for general ordered space 126
fork convex sets 389
forward contract 16, 412
forward starting option 337
fractional part of a real number 232
Fréchet bounds 580
FTAP 7, 310
- second 29, 328
- under portfolio constraints 455
- with contingent initial data 43
- without a priori measure 415, 416
function
- concave 552
- convex 552
fundamental theorem of asset pricing
- for multi-period market 310
- geometric form 38
- in one period 7
- second 29, 328
- under portfolio constraints 455
- with contingent initial data 43
- without a priori measure 415, 416
futures contract 17

g-divergence 166, 218, 293
gains process 304, 477
gamma distribution 540
Gamma of a call option 352
gauge function 68, 119, 596
Gaussian error function 198
generalized likelihood ratio test 587
generalized trading strategy 476
- mean self-financing 478

geometric Brownian motion 356
Gibbs inequality 563
Gibbs measure 567
Gibbs variational principle 567
Gini coefficient 197
Gini's mean difference 197
Girsanov formula 489
Glivenko–Cantelli theorem 594
global quadratic risk 498
Greeks 355
- and Black–Scholes equation 353
- Charm 355
- Delta 351
- Gamma 352
- Rho 354
- Theta 353
- Vanna 355
- Vega 355
- Volga 355
- Vomma 355
Gumbel distribution 574

Hahn–Banach theorem 606
Halmos–Savage theorem 47
HARA utility 78, 85, 130, 153
Hardy–Littlewood inequality 157, 577
Hausdorff space 58
hedging strategy for American claim 384
Hoeffding–Fréchet bounds 580
hyperbolic absolute risk aversion 78

importance sampling 544
in-the-money option 27
Inada conditions 150
increasing concave order 123, 155
- and Average Value at Risk 226
- and risk measures 242
increasing order 98, 241
- for general ordered space 126
indemnity 160
independence axiom 61, 102, 105, 108
- on $\widetilde{\mathscr{X}}$ 115
index option 17
indicator function 9
indifference relation 55
inequality
- Gibbs 563
- Hardy–Littlewood 157, 577
inf-convolution of two risk measures 268
inner product
- Euclidean 550
insider information 327
insurance 22, 76, 83
integrability
- uniform 609
interest rate 186
intrinsic value 27
inverse function 569
- left-continuous 569
- right-continuous 569
iron condor 19
Itô integral 359
Itô's formula 358

jointly arbitrage-free prices 29

Knightian uncertainty 5, 105, 200
knock-in option 316, 421
knock-out option 316, 427
Kolmogorov's law of large numbers 53, 84, 86, 539, 544, 593
Krein–Šmulian theorem 608
Kreps–Yan theorem 47, 458
Kullback–Leibler divergence 216, 563
Kunita–Watanabe decomposition 485

λ-efficient allocation 179
λ-quantile 222, 574
$\mathscr{L}^2$-admissible strategy 478
law of large numbers 53, 84, 86, 539, 544, 593
law of one price 12, 26
law-based risk measure 231
law-determined risk measure 231
law-equivariant risk measure 231
least-favorable measure 164
- and hypothesis testing 167
Lebesgue decomposition 560
Lebesgue property 210, 217, 219
left-continuous inverse 569
Legendre transform 575
leverage effect 32, 353
lexicographical order 57, 58
likelihood ratio test 585
liquid option 411
local martingale 452
local risk process 478
locally convex space 606
locally risk-minimizing strategy 478
locally riskless bond 303

log-normal distribution 96, 159, 345
– change of measure 313
lookback call 336
lookback option 317, 343
lookback put 335
loss function 273, 432
lottery 54, 61
– variance of 95
lower quantile function 573
lower Snell envelope 391, 406

m-stable sets 389
market clearing condition 171
market portfolio 171
Markov kernel 87
Markov property 415
married put 18, 32
martingale 43, 307
– backwards 86
– convergence theorem 86
– local 452
– of successive densities 307, 386
– reversed 86
– strongly orthogonal 479
martingale measure 7, 42, 307
– as extension of a pricing rule 415
– minimal 487
martingale representation 328, 484
maturity of a contingent claim 315
MAXMINVAR 250
MAXVAR 249
mean self-financing strategy 478
mean-preserving spread 94, 125, 126
mean-standard deviation risk measure 196, 198
– conditional 515
mean-variance trade-off process 482, 500
measure
– absolutely continuous 557
– Dirac 593
– entropy of 563
– equivalent 557
– simple 592, 598
– support of 36
measure of risk
– acceptance set of 191
– cash invariant 189
– coherent 191, 409
– comonotonic 254
– continuous from above 206, 215
– continuous from below 207
– convex 190
– in financial market 262, 433, 474
– monetary 189
– monotone 189
– normalized 190
– on L^p 222
– penalty function for 201
– positive homogeneous 191
– relevant 220
– representation theorem 201
– sensitive 220, 474
– subadditive 191
– translation invariant 189
metric space 591
minimal martingale measure 487
Minkowski's separating hyperplane theorem 550
MINMAXVAR 250
MINVAR 199, 246
mixture space 66
model uncertainty 108
modified cumulative distribution function 575
moment generating function 136
monetary conditional risk measure 505
monetary measure of risk 189
monetary risk measure 189
monetary utility functional
– coherent 112
– concave 118
monotone preference relation
– on assets 101
– on distributions 71
monotone set function 247

n^{th}-order stochastic dominance 100
negative part 45
negative transitive relation 55
Neyman–Pearson lemma 428, 585
– generalized 585
nonredundance condition 13, 452
norm
– Euclidean 550
normal distribution 91, 96, 138, 564
normalized set function 247
numéraire 11, 41, 303
– change of 305, 311
numerical representation 56, 59
– affine 61
– of robust Savage form 393

- of Savage form 101, 371
- of von Neumann–Morgenstern form 61, 67, 69, 71, 108

optimal stopping problem 371
- continuation region 379
- stopping region 379
option 16
- Black–Scholes price 348
optional decomposition 402
- theorem 399
- under constraints 463
optional sampling theorem 369, 386
order dense set 56
orthogonal decomposition
- of a contingent claim 483
out-of-the-money option 27

ψ-weak topology 69, 119, 597
P&L 189
paradox
- Allais 69, 102
- Ellsberg 103
- St. Petersburg 73
Pareto distribution 198, 574
Pareto optimality 180
partial order 87
pasting of two probability measures 387, 467, 523
penalty function 118, 201, 507
- one-step 517
performance vector 526
PnL 189
point mass
- Dirac 593
Poisson distribution 27, 138
Polish space 591
portfolio 4
portfolio insurance 19
portfolio strategy
- constant rebalanced 527
- universal 535
portmanteau theorem 592
positive affine transformations 62
power of a statistical test 585, 587
predictable process 302
predictable representation property 328, 484
predictably convex set 451
preference order 55
preference relation 55
- asymmetry of 55
- continuous 57
- continuous from above 115
- continuous from below 115
- lexicographical 57, 58
- monotone 71, 101
- negative transitivity of 55
- numerical representation of 56, 59, 61, 67, 69, 71
- random 154
- risk averse 71
- state dependent 154
- weak 55
price density 172
price system 3
pricing measure 7
probability kernel 87
probability measure
- absolutely continuous 557
- equivalent 557
probability space
- atomless 231, 580
profits and losses 189
Prohorov's theorem 595
- for ψ-weak topology 598
proper concave function 552
proper convex function 552
put option 17
- American 376
- Asian 316
- European 315, 348
- universal bounds on price 27
- with barrier 316
put spread 18
put-call parity 18, 322

$\mathcal{Q}$-martingale 398
$\mathcal{Q}$-submartingale 398
$\mathcal{Q}$-supermartingale 398
$\mathcal{Q}_{\mathcal{S}}$-Snell envelope
- upper 473
quadratic risk
- global 498
- local 478
- remaining 498
quantile function 222, 573
- and stochastic order 88
- for nonadditive set functions 254
quantile hedging 441
quasi-convexity 197

Radon measure 20, 573
Radon–Nikodym
- chain rule 557
- density 557
- derivative 557
- theorem 557
random preference relation 154
random utility function 154
random variable
- gamma distributed 540
random walk 338
randomized statistical test 429, 434, 449, 587
ratio spread 18
recursiveness 514
reference portfolio 49
reflection principle 338
regular conditional distribution 49, 90, 99
relation
- antisymmetric 87
- asymmetric 55
- complete 55
- equivalence 55
- indifference 55
- negative transitive 55
- reflexive 87
- transitive 55, 87
relative entropy 136, 140, 173, 216, 277, 294, 563
- conditional 511
relative interior of a convex set 37
relative risk aversion 78
relevant risk measure 220, 518
remaining conditional risk 496
replicating
- portfolio 25
- strategy 318
representation theorem
- Skorokhod 595
representation theorem for risk measures 201
- on L^∞ 214
reservation price 159
residue modulo 1 232
return 13
reverse convertible bond 21
reversed martingale 86
Rho of a call option 354
right-continuous inverse 569
risk 5
risk aversion 71, 87, 104, 129, 432
- absolute 78
- relative 78
risk measure 409
- acceptance set of 191
- cash invariant 189
- coherent 112, 191
- comonotonic 254
- conditional 396, 505
- continuous from above 206, 215
- continuous from below 207, 276
- convex 118, 190
- dilatation-monotone 242
- dual representation 204
- dynamic 396
- dynamically consistent 396
- entropic 197, 216, 218, 277, 278, 294
- in financial market 262, 433, 474
- law-based 231
- law-determined 231
- law-equivariant 231
- monetary 189
- monotone 189
- normalized 190
- on L^p 222
- penalty function for 201
- positive homogeneous 191
- relevant 220, 518
- representation theorem 201
- sensitive 220, 474, 518
- subadditive 191
- tight 211
- time-consistent 396
- translation invariant 189
- worst-case 219, 290
risk neutrality 78
risk premium 73
risk transfer 22
risk-neutral measure 7, 42
- entropy-minimizing 141, 152
robust representation
- of conditional risk measure 507
- on L^p 222
robust Savage representation 107, 114, 393
robust shortfall risk 279
running maximum 335, 338

S&P 500 17, 544
Savage representation 101, 371
- robust 107, 114, 393

second fundamental theorem of asset pricing 29, 328
second-order stochastic dominance 87, 123, 155
self-financing strategy 302
– in continuous time 359
sensitive risk measure 220, 474, 518
separable metric space 591
separating hyperplane theorem 550
separation theorems 606
sequence
– uniformly distributed modulo 1 232
set
– convex 549
set function
– core of 260
– finitely additive 107, 200, 602
– monotone 247, 254
– normalized 247, 254
– strongly additive 247
– strongly subadditive 165, 199, 260
– submodular 165, 199, 247, 260
Shannon–Boltzmann entropy 567
Sharpe ratio 196
short sales 4
– constraints 264, 452
shortfall 273
shortfall risk 273, 433
– robust 279
significance level of a statistical test 585, 587
simple measure 592, 598
simple probability distribution 62
simple random walk 338
simplex
– standard 526
single-crossing property 93, 162
size of a statistical test 585
Skorokhod representation theorem 595
Snell envelope 366, 372, 466
– lower 391
– upper 391, 398, 403, 405
– upper $\mathcal{Q}_{\mathcal{S}}$- 467
space of reference portfolios 49
SPX 17
St. Petersburg paradox 73
stability under pasting 389
stable set of measures 389, 467
stable subspace of $\mathcal{H}^2$ 484
Standard and Poor's 500 17, 544
standard normal distribution 91
standard simplex 526
state-dependent preference relation 154
state-dependent utility function 154
statistical test 585
– for composite hypothesis 167, 449
– randomized 429, 434, 587
stochastic differential equation 358
stochastic dominance 99
– first-order 99, 126, 241
– n^{th}-order 100
– of the first kind 99
– of the second kind 87
– second-order 87, 123, 155
stochastic kernel 87, 104
stochastic order
– concave 95, 125
– convex 95, 410, 414
– increasing 98, 126, 241
– increasing concave 87, 123, 155, 226
– usual 126
stochastic process 302
– adapted 302
– predictable 302
Stone vector lattice 601
stop-loss contract 77, 160
stopping region 379
stopping theorem 369, 386
stopping time 369
straddle 19
strategy
– $\mathcal{L}^2$-admissible 478
– locally risk-minimizing 478
– mean self-financing 478
– self-financing 302
– self-financing in continuous time 359
stress test measures 270
strike price 17, 315
strong law of large numbers 53, 84, 86, 593
strong orthogonality 479
strongly additive set function 247
strongly subadditive set function 165, 199, 260
submartingale 364
– local 453
submodular 165
submodular set function 199, 247, 260
substitution axiom 61
success ratio 429
success set 427

superhedging 25, 368
- duality 404, 408, 473
- strategy 402, 473
supermartingale 364
- local 453
superreplication strategy 402, 403
support of a measure 36
supremum norm 602
systems theorem 308

theorem
- Birkhoff 594, 601
- Doob's systems theorem 308
- Dunford–Pettis 613
- Eberlein–Šmulian 609
- ergodic 594, 601
- Glivenko–Cantelli 594
- Hahn–Banach 606
- Halmos–Savage 47
- Krein–Šmulian 608
- Kreps–Yan 47, 458
- of the deductible 160
- optional decomposition 399
- optional sampling 369
- portmanteau 592
- Radon–Nikodym 557
- randomized Bolzano–Weierstraß 48
- separating hyperplane 550
- separation 606
- Skorokhod representation 595
- stopping 369, 386
- superhedging duality 404
- Vitali 610
- welfare 180
Theta of a call option 353
tightness
- of a set of measures 595
- of risk measure 211
time consistency 514
- weak 522
time consistency of a risk measure 396
time value 27, 322
- of a European call option 322
topological space
- connected 59
- Hausdorff 58
topological vector space 605
- locally convex 606
topology
- weak* 607
- weak 607
total variation 203, 602
trading strategy 302
- $\mathscr{L}^2$-admissible 478
- generalized 476
- replicating 318
- self-financing 302
transitive relation 55
translation invariance 189
translation property 189
- of certainty equivalent 79
type 1 error 585
type 2 error 585

uncertainty 5
uncertainty aversion 105
uniform absolute continuity 610
uniform Doob decomposition 399
- under constraints 463
uniform integrability 609
universal arbitrage bounds 27, 411
universal portfolio strategy 535
up-and-in call 421
up-and-in option 335, 341
- Black–Scholes price of 360
up-and-out call 421, 423
up-and-out option 316, 335, 342
upper $\mathcal{Q}_{\mathscr{S}}$-Snell envelope 467
- characterization of 473
upper hedging price 403
upper quantile function 573
upper Snell envelope 391, 398
- characterization of 405
- optional decomposition of 403
USDEUR 305
utility function 72, 87, 129
- CARA 78, 80, 84, 130, 136, 137, 147, 152, 197, 274, 280, 511
- contingent 154
- CRRA 78
- decreasingly risk averse 84
- exponential 78, 80, 84, 130, 136, 137, 147, 152, 197, 274, 280, 511
- HARA 78, 130, 153
- random 154, 437
- state dependent 154
utility-based shortfall risk measure 195, 275

Value at Risk 188, 195, 223, 225, 243, 274, 426
- comonotonicity 258
- conditional 512
- time-inconsistency 515
value process 304
Vanna of a call option 355
variance 95
- conditional 313, 479
variance-optimal strategy 499
variational principle
- Gibbs 567
vector lattice 601
Vega of a call option 355
Vitali convergence theorem 610
volatility 355, 358
Volga of a call option 355
Vomma of a call option 355
von Neumann–Morgenstern representation 61, 67, 69, 71, 108
weak certainty independence 106
weak information 164
weak preference order 55
weak time consistency 522
weak topology
- on a Banach space 607
- on a locally convex space 607
weak topology for measures 591
weak* topology 607
weakly compact set 609
welfare theorem 180
Wiener measure 357
Wiener process 356
worst conditional expectation 220, 244
worst-case risk measure 194
- on L^∞ 219, 227, 290

www.ingramcontent.com/pod-product-compliance
Lightning Source LLC
Chambersburg PA
CBHW081248060825
30705CB00007B/189
9783111044811